Atlantis Studies in Differential Equations

Volume 1

Series editor

Michel Chipot, Zürich, Switzerland

For further volumes:
www.atlantis-press.com

Aims and Scope of the Series

The ''Atlantis Studies in Differential Equations'' publishes monographs in the area of differential equations, written by leading experts in the field and useful for both students and researchers. Books with a theoretical nature will be published alongside books emphasizing applications.

For more information on this series and our other book series, please visit our website www.atlantis-press.com/publications/books

AMSTERDAM – PARIS – BEIJING
ATLANTIS PRESS
Atlantis Press
29, avenue Laumière
75019 Paris, France

Anvarbek Meirmanov

Mathematical Models for Poroelastic Flows

ATLANTIS PRESS

Anvarbek Meirmanov
Mathematics and Cybernetics
Kazakh-British Technical University
Almaty
Kazakhstan

ISSN 2214-6253 ISSN 2214-6261 (electronic)
ISBN 978-94-6239-014-0 ISBN 978-94-6239-015-7 (eBook)
DOI 10.2991/978-94-6239-015-7

Library of Congress Control Number: 2013949487

Printed on acid-free paper

To my wife Liudmila and my daughters Anastasia and Aliya

Preface

This book is devoted to the rigorous mathematical modeling of physical processes in underground continuous media, namely, the correct description of porous elastic solids with fluid-filled pores. Recently this subject has been attracting increased attention for many reasons: the recovery of oil and gas, liquid waste disposal into the ground, seismic phenomena, acoustic wave propagation in the water-saturated porous bottom of the ocean, diffusion-convection in porous media, etc.

Such continuous media are called heterogeneous continuous media. That is, those continuous media which consist of two or more different components (phases) and in any sufficiently small amount of a continuum there are different phases. The minimum size of this volume is different for different heterogeneous media, but usually it is in the range from several microns to several tens of microns.

There are two different approaches to the description of heterogeneous media. The first approach, which we call the *phenomenological approach*, is based upon the notion of a continuous medium as a kind of conglomerate, where at each point all phases of such a medium are present. In this approach, the main difficulty is the physical modeling: the choice of axioms that define the dependence of the stress tensor on the basic characteristics of motion and thermodynamic relations. The second approach is based on precise physical modeling with further simplification of the mathematical model using the methods of mathematical analysis. As a rule, the differential equations of the exact mathematical model contain a small parameter. Therefore, the main methods of simplifying exact mathematical models are the methods of linearization and homogenization. Roughly speaking, these are methods of constructing approximate mathematical models from the original one, when the small parameter tends to zero. In this approach we must keep in mind the limits of applicability of the physical models and methods of mathematical analysis. The more precise and more rigorous methods provide the more trustable mathematical models.

The phenomenological poroelastic equations derived by K. von Terzaghi and M. Biot have long been regarded as standard and have formed the basis for solving particular problems in poroelasticity. Terzaghi's and Biot's poroelastic equations take into account the displacement of both the fluid in the pores and the solid skeleton and the coupling between them. These works have been rather heuristic. Hence, several authors (R. Burridge and J. Keller [1], E. Sanchez-Palencia [2], T. Levy [3–6]) have attempted to apply the second approach and derive the macroscopic poroelastic equations on the basis of the fundamental laws of continuum mechanics and rigorous homogenization methods. The idea is quite natural: one first must describe the joint motion of the elastic skeleton and the fluid in pores at the microscopic level by means of classical continuum mechanics, and then use homogenization to find appropriate approximation models (homogenized equations). The Navier-Stokes equations still hold at this scale of the pore size in the order of 5–15 microns [7, 8]. Thus, as we have mentioned above, the macroscopic mathematical models obtained are still within the limits of physical applicability.

In this book we follow the method suggested by R. Burridge and J. Keller and E. Sanchez-Palencia and systematically study filtration and acoustic processes in poroelastic media.

Some parts of the book have been written in cooperation with my Ph.D. students. Chapter 4 has been written in cooperation with I. Nekrasova, the first part of the Chap. 5 has been written in cooperation with N. Erygina, the first part of the Chap. 8 has been written in cooperation with A. Guerus, and Chap. 10 has been written in cooperation with R. Zimin.

I am grateful to J. R. Ockendon for his always useful advices and my special thanks for the book title.

I am also grateful to R. Perkins who carefully read the manuscript and essentially improved my English. Researches which form the basis of the present book have been partially supported by the Federal Program "Research and scientific—pedagogical brainpower of Innovative Russia" for 2009–2013 (State Contract 14.A18.21.0357) and Grants N^0 0691/GF and N^0 0751/GF of the Ministry of Education and Science of Kazakhstan.

Almaty, April 2013 Anvarbek Meirmanov

References

1. Burridge, R., & Keller, J. B. (1981). Poroelasticity equations derived from microstructure. *Journal of the Acoustical Society of America*, *70*, N4, 1140–1146.
2. Sanchez-Palencia, E. (1980) Non-Homogeneous Media and Vibration Theory. *Lecture Notes in Physics*, *129*, Berlin: Springer.
3. Levy, T. (1977). Acoustic phenomena in elastic porous media. *Mechanics Research Communications*, *4*, 4, 253–257.

4. Levy, T. (1987). Fluids in porous media and suspensions. In Homogenization Techniques for Composite Media, Lecture Notes In Physics, 272, Berlin: Springer.
5. Levy, T. (1979). Propagation waves in a fluid-saturated porous elastic solid. *International Journal of Engineering Science*, *17*, 1005–1014.
6. Levy, T., & Sanchez–Palencia, E. (1977). Equations and interface conditions for acoustic phenomena in porous media. *Journal of Mathematical Analysis and Applications*, *61*, 813–834.
7. Lissandrello, C., Yakhot, V., Ekinci, K. L. (2012). Crossover from Hydrodynamics to the Kinetic Regime in Confined Nanoflows. *Physical review letters*, *108*, 084501.
8. Rustom, B., & Jane Wang, Z. (2004). Effect of fluids on the Q factor and resonance frequency of oscillating micrometer and nanometer scale beams. *Physical Review E*, *69*, 036307.

Contents

Introduction

In the present book we derive different mathematical models describing flows in poroelastic media. To explain the method we consider a bounded domain $\Omega \subset \mathbb{R}^3$ perforated by pores. A pore space (a liquid domain) Ω_f is filled with a viscous liquid and there is a solid skeleton $\Omega_s = \Omega \backslash \overline{\Omega_f}$ which is supposed to be an elastic body. Then the joint motion in Ω is described by the system [1]

$$\frac{\partial}{\partial t}(\rho\,\mathbf{v}) + \nabla \cdot \big(\rho\,\mathbf{v}\otimes\mathbf{v} - \tilde{\chi}\,\mathbb{P}_f + (1-\tilde{\chi})\mathbb{P}_s\big) = \rho\,\mathbf{F}, \tag{0.0.1}$$

$$\frac{\partial \rho}{\partial t} + \nabla \cdot (\rho\,\mathbf{v}) = 0, \tag{0.0.2}$$

where $\nabla \cdot \mathbf{u}$ is the divergence of $\mathbf{u}$:

$$\nabla \cdot \mathbf{u} = \mathrm{tr}(\nabla \mathbf{u}),$$

a matrix $\mathbf{a}\otimes\mathbf{b}$ is defined as

$$(\mathbf{a}\otimes\mathbf{b})\cdot\mathbf{c} = \mathbf{a}(\mathbf{b}\cdot\mathbf{c}),$$

for any vectors $\mathbf{a}$, $\mathbf{b}$, and $\mathbf{c}$, and a divergence $\nabla \cdot \mathbb{P}$ for any smooth tensor $\mathbb{P}$ is defined as

$$\mathbf{c}\cdot(\nabla \cdot \mathbb{P}) = \nabla \cdot (\mathbb{P}^*\cdot\mathbf{c})$$

for any constant vector $\mathbf{c}$.

The function $\tilde{\chi}$ is the characteristic function of the pore space Ω_f, $\mathbb{P}_f$ and $\mathbb{P}_s$ are stress tensors in the liquid domain and in the solid skeleton respectively, $\mathbf{v}$ is the velocity, ρ is the density of the medium, and $\mathbf{F}$ is a given vector of distributed mass forces.

Equations (0.0.1, 0.0.2) are understood in the sense of distributions (as corresponding integral identities) and contain dynamic equations for the liquid

$$\rho\,\frac{d\mathbf{v}}{dt} = \nabla \cdot \mathbb{P}_f + \rho\,\mathbf{F}, \quad \frac{d\rho}{dt} + \rho\,\nabla \cdot \mathbf{v} = 0,$$

in the pore space Ω_f for $t > 0$, dynamic equations for the solid component

$$\rho\,\frac{d\mathbf{v}}{dt}=\nabla\cdot\mathbb{P}_s+\rho\,\mathbf{F},\quad \frac{d\rho}{dt}+\rho\,\nabla\cdot\mathbf{v}=0,$$

in Ω_s for $t>0$, and the continuity condition for normal tensions

$$(\mathbb{P}_s-\mathbb{P}_f)\cdot\mathbf{n}=0$$

on the common boundary $\Gamma(t)$ "pore space-solid skeleton". Here $\mathbf{n}$ is a unit normal to $\Gamma(t)$.

We do not intend to specify state equations for stress tensors and for the density. Let us just outline the problem, because it is completely nonlinear and contains one more unknown subject: the common boundary "pore space-solid skeleton". The main postulate here is that the solid and the liquid components are immiscible. Therefore, the unknown (free) boundary $\Gamma(t)$ is a surface of a **contact discontinuity** [1], which is defined by the Cauchy problem for the characteristic function $\tilde{\chi}$:

$$\frac{d\tilde{\chi}}{dt}\equiv\frac{\partial\tilde{\chi}}{\partial t}+\nabla\tilde{\chi}\cdot\mathbf{v}=0,\quad \tilde{\chi}(\mathbf{x},0)=\chi_0(\mathbf{x}), \tag{0.0.3}$$

in the whole domain Ω for $t>0$.

It is clear that, even if one knows how to solve the problem (0.0.1–0.0.3), this mathematical model would not be useful for practical needs, since the function $\tilde{\chi}$ changes its value from 0 to 1 on the scale of a few microns. Thus, the most suitable way to get a practically significant mathematical model is a homogenization. But for this case the problem (0.0.1–0.0.3) becomes absolutely unsolvable. To get something solvable and still reasonable, we use the scheme suggested in [2, 3] and linearize the basic system.

That is, we approximate the characteristic function $\tilde{\chi}$ of the liquid domain Ω_f by its value at the initial time moment

$$\tilde{\chi}\simeq\chi_0(\mathbf{x}),$$

and the free boundary $\Gamma(t)$ by its initial position Γ_0.

Next we suppose that

$$\mathbf{v}\simeq\frac{\partial\mathbf{w}}{\partial t},$$

where $\mathbf{w}$ is a displacement vector of the medium,

$$\frac{\partial}{\partial t}(\rho\,\mathbf{v})\simeq\big(\rho_f\chi_0+\rho_s(1-\chi_0)\big)\frac{\partial^2\mathbf{w}}{\partial t^2},$$

$$\frac{d\rho}{dt}+\rho\,\nabla\cdot\mathbf{v}\simeq\frac{1}{c_f^2}\frac{\partial p}{\partial t}+\rho_f\,\nabla\cdot\frac{\partial\mathbf{w}}{\partial t}$$

in the liquid part,

$$\frac{d\rho}{dt} + \rho \nabla \cdot \mathbf{v} \simeq \frac{1}{c_s^2} \frac{\partial p}{\partial t} + \rho_s \nabla \cdot \frac{\partial \mathbf{w}}{\partial t}$$

in the solid part, where ρ_f and ρ_s are constant average densities of the liquid in pores and of the solid skeleton respectively, c_f and c_s are the speed of compression sound waves in the pore liquid and in the solid skeleton respectively, and that

$$\mathbb{P}_f = \mu \mathbb{D}(x, \mathbf{v}) + \left(\nu (\nabla \cdot \mathbf{v}) - p\right) \mathbb{I}, \tag{0.0.4}$$

$$\mathbb{P}_s = \lambda \mathbb{D}(x, \mathbf{w}) - p \mathbb{I}. \tag{0.0.5}$$

Here $\mathbb{D}(x, \mathbf{u}) = \frac{1}{2}(\nabla \mathbf{u} + \nabla \mathbf{u}^*)$ is the symmetric part of $\nabla \mathbf{u}$, $\mathbb{I}$ is a unit tensor, μ is the dynamic viscosity, ν is the bulk viscosity, and λ is the elastic constant.

To apply the well-known homogenization results [4], we must consider special liquid domains Ω_f and impose the following constraints.

Assumption 0.1

(1) Let $\chi(\mathbf{y})$ be some 1-periodic function, $Y_s = \{\mathbf{y} \in Y : \chi(\mathbf{y}) = 0\}$ be the "solid part" of the unit cube $Y = (0,1)^3 \subset \mathbb{R}^3$, and let the "liquid part" $Y_f = \{\mathbf{y} \in Y : \chi(\mathbf{y}) = 1\}$ of Y be its open complement. We write $\gamma = \partial Y_f \cap \partial Y_s$ and assume that γ is a Lipschitz continuous surface.

(2) The domain E_f^ε is a periodic repetition in $\mathbb{R}^3$ of the elementary cell $Y_f^\varepsilon = \varepsilon Y_f$ and the domain E_s^ε is a periodic repetition in $\mathbb{R}^3$ of the elementary cell $Y_s^\varepsilon = \varepsilon Y_s$.

(3) The pore space $\Omega_f^\varepsilon \subset \Omega = \Omega \cap E_f^\varepsilon$ is a periodic repetition in Ω of the elementary cell εY_f, and the solid skeleton $\Omega_s^\varepsilon \subset \Omega = \Omega \cap E_s^\varepsilon$ is a periodic repetition in Ω of the elementary cell εY_s. The Lipschitz continuous boundary $\Gamma^\varepsilon = \partial\Omega_s^\varepsilon \cap \partial\Omega_f^\varepsilon$ is a periodic repetition in Ω of the boundary $\varepsilon\gamma$.

(4) Y_s and Y_f are connected sets.

Under this assumption

$$\chi_0(\mathbf{x}) = \chi^\varepsilon(\mathbf{x}) = \varsigma(\mathbf{x}) \chi(\frac{\mathbf{x}}{\varepsilon}),$$

where $\varsigma(\mathbf{x})$ is the characteristic function of the domain Ω.

In dimensionless variables

$$\mathbf{x} \to \frac{\mathbf{x}}{L}, \ \mathbf{w} \to \frac{\mathbf{w}}{L}, \ t \to \frac{t}{\tau}, \ \mathbf{F} \to \frac{\mathbf{F}}{g}, \ \rho \to \frac{\rho}{\rho^0},$$

where L is the characteristic size of the physical domain in consideration, τ is the characteristic time of the physical process, ρ^0 is the mean density of water, and g is acceleration due gravity, the dynamic system takes the form

$$\alpha_\tau \rho^\varepsilon \frac{\partial^2 \mathbf{w}}{\partial t^2} = \nabla \cdot \mathbb{P} + \rho^\varepsilon \mathbf{F}, \tag{0.0.6}$$

$$\mathbb{P} = \chi^\varepsilon \alpha_\mu \mathbb{D}(x, \frac{\partial \mathbf{w}}{\partial t}) + (1 - \chi^\varepsilon)\alpha_\lambda \mathbb{D}(x, \mathbf{w}) + \big(\chi^\varepsilon \alpha_\nu (\nabla \cdot \frac{\partial \mathbf{w}}{\partial t}) - p\big)\, \mathbb{I}, \tag{0.0.7}$$

$$p + \alpha_p^\varepsilon \nabla \cdot \mathbf{w} = 0. \tag{0.0.8}$$

In (0.0.6–0.0.8) $\varepsilon = \dfrac{l}{L}$ is the dimensionless pore size, l is the average size of pores,

$$\alpha_p^\varepsilon = \alpha_{p,f}\chi^\varepsilon + \alpha_{p,s}(1 - \chi^\varepsilon), \quad \rho^\varepsilon = \rho_f\, \chi^\varepsilon + \rho_s\, (1 - \chi^\varepsilon),$$

$$\alpha_\tau = \frac{L}{g\tau^2}, \quad \alpha_\mu = \frac{2\mu}{\tau L g\, \rho^0}, \quad \alpha_\lambda = \frac{2\lambda}{L g\, \rho^0},$$

$$\alpha_\nu = \frac{2\nu}{\tau L g\, \rho^0}, \quad \alpha_{p,f} = \frac{\rho_f\, c_f^2}{L g}, \quad \alpha_{p,s} = \frac{\rho_s\, c_s^2}{L g},$$

ρ_f and ρ_s are the respective mean dimensionless densities of the liquid in pores and the solid skeleton, correlated with the mean density of water ρ^0.

Various particular cases of the linearization of (0.0.1–0.0.3) have been intensively studied by many authors: Buchanan–Gilbert-Lin [5, 6], Buckingham [7], Burridge–Keller [2], Clopeau–Ferrin–Gilbert–Mikelić–Paoli [8, 9, 10], Levy [11], Nguetseng [12], Sanchez-Hubert [13], Sanchez-Palencia [3].

The present book is based on the author's ideas [14–20]. We systematically investigate the special form (0.0.6–0.0.8) of the linearization of (0.0.1–0.0.3), containing the dimensionless parameter α_τ, which is responsible for the type of the physical process. For very slow and long-term processes, such as the filtration of liquids, $\alpha_\tau \sim 0$. For fast (short-term) processes, such as in acoustics or hydraulic shock, $\alpha_\tau \sim 1$, or $\alpha_\tau \sim \infty$.

Theoretically the system (0.0.6–0.0.8), with corresponding initial and boundary conditions, is one of the most adequate mathematical models, describing the motion of the viscous liquid in the pore space of the elastic solid skeleton. But, as we have mentioned above, such a model has no practical significance, since it is necessary to solve the problem in the physical domain of a few hundred meters, while the coefficients oscillate on the scale of a few tens of microns. The practical significance of the model appears only after homogenization. So, we have to let all dimensionless criteria α_τ, α_μ, $\alpha_\lambda, \ldots$ be variable functions depending on the small parameter ε, and find all the limiting regimes of (0.0.6–0.0.8) as $\varepsilon \to 0$. It is clear, that these limiting regimes depend on criteria α_τ, α_μ, $\alpha_\lambda \ldots$, or more precisely, on their limiting values at $\varepsilon = 0$.

In this book we analyze all possible limiting regimes and all possible physical processes described by the system (0.0.6–0.0.8). To separate physical processes and the possible different types of continuous media, we introduce the following criteria

$$\tau_0 = \lim_{\varepsilon \searrow 0} \alpha_\tau(\varepsilon), \quad \mu_0 = \lim_{\varepsilon \searrow 0} \alpha_\mu(\varepsilon), \quad \lambda_0 = \lim_{\varepsilon \searrow 0} \alpha_\lambda(\varepsilon),$$

$$c_{f,0}^2 = \lim_{\varepsilon \searrow 0} \alpha_{p,f}(\varepsilon), \quad c_{s,0}^2 = \lim_{\varepsilon \searrow 0} \alpha_{p,s}(\varepsilon),$$

$$\mu_1 = \lim_{\varepsilon \searrow 0} \frac{\alpha_\mu}{\varepsilon^2}, \quad \lambda_1 = \lim_{\varepsilon \searrow 0} \frac{\alpha_\lambda}{\varepsilon^2}.$$

For filtration processes $\tau_0 = 0$ and instead of (0.0.6) we may consider the equation

$$\nabla \cdot \mathbb{P} + \rho^\varepsilon \mathbf{F} = 0. \tag{0.0.9}$$

The system (0.0.7–0.0.9) describes the slow motion of compressible viscous liquids in pores. Usually, for such motion the velocity of the liquid is about 6–12 m per year.

As a rule in classical mechanics we are trying to use the simplest equations that take into account all the possible simplifying assumptions. Here, the next simplifying assumption is the incompressibility of the medium. As is well known, **the measure of incompressibility of any given medium is its speed of sound**. Incompressible media have an infinite speed of sound. In particular, for long-term physical processes the behavior of acoustic waves is not so important, and for many real liquids we may accept the assumption that the given liquid is incompressible. On the other hand, the speed of sound of compression waves in the solid skeleton is two or three times more than the speed of sound of compression waves in the liquid. So the first assumption implies that the given solid skeleton is also incompressible. Thus

$$c_{f,0} = \infty, \quad c_{s,0} = \infty,$$

and the filtration of an incompressible liquid in an elastic solid skeleton is described by the system

$$\nabla \cdot \big(\chi^\varepsilon \alpha_\mu \mathbb{D}(x, \frac{\partial \mathbf{w}}{\partial t}) + (1 - \chi^\varepsilon)\alpha_\lambda \mathbb{D}(x, \mathbf{w}) - p\,\mathbb{I}\big) + \rho^\varepsilon \mathbf{F} = 0, \tag{0.0.10}$$

$$\nabla \cdot \mathbf{w} = 0. \tag{0.0.11}$$

Of course, the simplest case, is the motion of a viscous liquid in an absolutely rigid solid body. This case is described by the criterion

$$\lambda_0 = \infty.$$

The corresponding simplification of (0.0.7–0.0.9) for a compressible liquid is the system

$$\frac{\partial p}{\partial t} + \alpha_{p,f} \nabla \cdot \mathbf{v} = 0, \tag{0.0.12}$$

$$\nabla \cdot \big(\alpha_\mu \mathbb{D}(x, \mathbf{v}) - p\,\mathbb{I}\big) + \rho_f \mathbf{F} = 0 \tag{0.0.13}$$

for the liquid velocity $\mathbf{v}$ and liquid pressure p in the domain Ω_f for $t > 0$.

An incompressible viscous liquid in an absolutely rigid solid skeleton is described by the Eq. (0.0.13) and the continuity equation

$$\nabla \cdot \mathbf{v} = 0. \tag{0.0.14}$$

Note that all these simplified models can be derived rigorously as appropriate asymptotic limits for the basic model ($\alpha_\lambda \to \infty$, or $\alpha_{p,f} \to \infty$).

For $\tau_0 > 0$ we may rescale variables by setting

$$\alpha_\tau \mathbf{w} \to \mathbf{w},$$

and get the system

$$\rho^\varepsilon \frac{\partial^2 \mathbf{w}}{\partial t^2} = \nabla \cdot \mathbb{P} + \rho^\varepsilon \mathbf{F}, \tag{0.0.15}$$

$$\mathbb{P} = \chi^\varepsilon \,\bar{\alpha}_\mu \,\mathbb{D}(x, \frac{\partial \mathbf{w}}{\partial t}) + (1 - \chi^\varepsilon)\bar{\alpha}_\lambda \,\mathbb{D}(x, \mathbf{w}) + \big(\bar{\alpha}_\nu \,(\nabla \cdot \mathbf{v}) - p\big)\,\mathbb{I}, \tag{0.0.16}$$

$$p + \bar{\alpha}_p^\varepsilon \nabla \cdot \mathbf{w} = 0, \tag{0.0.17}$$

which describes short-term processes like acoustics or hydraulic shock.

In (0.0.15–0.0.17)

$$\bar{\alpha}_\mu = \frac{\alpha_\mu}{\alpha_\tau}, \quad \bar{\alpha}_\nu = \frac{\alpha_\nu}{\alpha_\tau}, \quad \bar{\alpha}_\lambda = \frac{\alpha_\lambda}{\alpha_\tau}, \quad \bar{\alpha}_p^\varepsilon = \frac{\alpha_p^\varepsilon}{\alpha_\tau}.$$

Now, after the first simplification, we may pass to the limit as $\varepsilon \to 0$ and get the desired homogenized models. But first and foremost we have to decide what kind of model do we want to get? Everything depends on dimensionless criteria τ_0, μ_0, λ_0, etc. For example, for the system (0.0.10, 0.0.11) we have two variable quantities α_μ and α_λ and four criteria μ_0, λ_0, μ_1 and λ_1. We emphasize again that the system (0.0.10, 0.0.11) is the basic one and all its homogenized systems are just approximations of different degrees of exactness. If we are going to get the simplest system, then we look for the limit as $\varepsilon \to 0$ with

$$\lambda_0 = \infty, \quad \mu_0 = 0, \quad 0 < \mu_1 < \infty.$$

This case corresponds to the usual *Darcy system of filtration*

$$\mathbf{v} = \frac{1}{\mu_1} \mathbb{B}\Big(- \nabla p + \rho_f \mathbf{F}\Big), \tag{0.0.18}$$

$$\nabla \cdot \mathbf{v} = 0 \tag{0.0.19}$$

in the domain Ω for $t \in (0, T)$ (see Sect. 1.1). We may refer to this system as the **first level of approximation** of the system (0.0.10, 0.0.11).

For

$$0<\lambda_0<\infty, \quad \mu_0=0, \quad 0<\mu_1<\infty$$

one has the **second level of approximation** of the basic system, which is the *Terzaghi–Biot system of poroelasticity*

$$\nabla\cdot\mathbf{v}+(1-m)\,\nabla\cdot\frac{\partial\mathbf{w}_s}{\partial t}=0, \tag{0.0.20}$$

$$\nabla\cdot\big(\lambda_0\,\mathfrak{N}_1^s:\mathbb{D}(x,\mathbf{w}_s)-p\,\mathbb{I}\big)+\big((m\,\rho_f+(1-m)\rho_s\big)\mathbf{F}=0, \tag{0.0.21}$$

$$\mathbf{v}=m\,\frac{\partial\mathbf{w}_s}{\partial t}+\frac{1}{\mu_1}\mathbb{B}\cdot\big(-\nabla p+\rho_f\mathbf{F}\big) \tag{0.0.22}$$

in the domain Ω for $t\in(0,T)$ for the velocity $\mathbf{v}$ of the liquid component, the displacement vector $\mathbf{w}_s$ of the solid component, and the pressure p of the mixture (see Sect. 1.2).

Finally, for

$$0<\lambda_0<\infty, \quad 0<\mu_0<\infty$$

we arrive at the **third level of approximation** of (0.0.10, 0.0.11), which is a system of poroelastic filtration

$$\nabla\cdot\mathbf{w}=0, \tag{0.0.23}$$

$$\nabla\cdot\widehat{\mathbb{P}}+\big((m\,\rho_f+(1-m)\rho_s\big)\mathbf{F}=0, \tag{0.0.24}$$

$$\begin{aligned}\widehat{\mathbb{P}}&=-p\,\mathbb{I}+\mathfrak{N}_1:\mathbb{D}(x,\frac{\partial\mathbf{w}}{\partial t})+\mathfrak{N}_2:\mathbb{D}(x,\mathbf{w})\\&\quad+\int_0^t\mathfrak{N}_3(t-\tau):\mathbb{D}(x,\mathbf{w}(\mathbf{x},\tau))d\tau\end{aligned} \tag{0.0.25}$$

in the domain Ω for $t\in(0,T)$ for the displacements $\mathbf{w}$ and the pressure p of the mixture (see Sect. 1.4).

Thus we have a set of approximate models, from a simple one to quite complex ones. The choice of model depends on the needs of the researcher. Of course, in practical applications all physical parameters are fixed and we cannot determine theoretical limits of the dimensionless complexes, which we have defined above. For this reason we just put

$$\mu_0=\alpha_\mu=\frac{2\mu}{\tau Lg\,\rho^0}, \quad \mu_1=\frac{\mu_0}{\varepsilon^2}=\frac{2\mu}{\tau Lg\,\rho^0}\frac{L^2}{l^2}, \quad \varepsilon=\frac{l}{L},$$

$$\lambda_0=\alpha_\lambda=\frac{2\lambda}{Lg\,\rho^0}.$$

We apply the same scheme to the model (0.0.15–0.0.17) of short-term processes, and to all other mathematical models at the microscopic level considered here.

So, the main aims of the book are the following:

(1) To find the most adequate and correct mathematical models at the microscopic level for each of the physical process under consideration, based on the basic principles of continuum mechanics.
(2) To fulfill rigorously the limiting procedures from the microscopic level to the macroscopic ones.

The choice of the model at the microscopic level must be based upon the definition of theoretical small parameters and dimensionless criteria, describing the process (long-term or short-term process, compressible or incompressible medium and so on).

Under correct mathematical model we will understand the initial boundary-value problem for the system of differential equations, which has a unique solution in some appropriate sense (classical, weak, or very weak).

Once again we emphasize that all our results are based upon the Nguetseng method of the two-scale convergence. All details of Nguetseng's theory and corresponding references may be found in [21, 4]. Appendix B lists all basic definitions and statements of this theory. Here we only note the principal advantage of the method. As is well known, very often the main difficulty in PDE problems is the limit as $\varepsilon \to 0$ in integrals I^{ε} of the form

$$I^{\varepsilon} = \int_{\Omega} u^{\varepsilon}(\mathbf{x})\, v^{\varepsilon}(\mathbf{x})\varphi(\mathbf{x})dx, \tag{0.0.26}$$

where sequences $\{u^{\varepsilon}\}$ and $\{v^{\varepsilon}\}$ converge only weakly in $L_2(\Omega)$. The current theory cannot give the answer what will be the exact limit. As a rule, in homogenization theory $u^{\varepsilon}(\mathbf{x}) = u(\frac{\mathbf{x}}{\varepsilon})$. For this case Nguetseng has suggested a new notion of convergence in $L_2(\Omega)$, the so called **two-scale convergence**, where all functions of the type $u^{\varepsilon}(\mathbf{x}) = u(\mathbf{x}, \frac{\mathbf{x}}{\varepsilon})$ form the class of test functions. Thus, Nguetseng has transformed a very difficult problem into a not less difficult one, but has solved the latter in a brilliant manner.

Proper homogenization theory has attracted the attention of a very large number of researchers, and we refer to the books of Bensoussan, Lions, and Papanicolau [22], Jikov, Kozlov, and Oleinik [23], Hornung [24], Bakhvalov and Panasenko [25].

Note that there are reasonable objections to mathematical modeling via the homogenization of periodic structures: the method does not possess the necessary commonality, since the global periodicity is not inherent in physical reality. Fortunately, from the very beginning we understood the importance of such objections. In Chaps. 5 and 9 it will be shown that homogenized mathematical models of liquid filtration and acoustics possess the property of locality. That is, the physical medium being modeling may have different physical parameters of the solid skeleton, such as elasticity, density, geometry, etc., in different parts of the domain under consideration.

A brief description of the organization of the book is as follows. The material consists of eleven chapters and two Appendices.

In Chap. 1 we deal with different models for the isothermal filtration of compressible and incompressible liquids in a solid skeleton, and derive homogenized models for different types of continuous media. We have roughly divided the continuous media under consideration into the following groups.

(1) The liquid is **slightly viscous**, if $\mu_0 = 0$.
(2) The liquid is **viscous**, if $0 < \mu_0 < \infty$.
(3) The solid body is **extremely elastic** if $\lambda_0 = 0$.
(4) The solid body is **elastic** if $0 < \lambda_0 < \infty$.
(5) The solid body is **absolutely rigid** if $\lambda_0 = \infty$.

This Chap. 1 is the main base for the book and for different applications, which will be discussed in later chapters.

The Chap. 2 deals with homogenized models for non-isothermal filtration of a compressible liquid in a solid skeleton.

In the Chap. 3 we consider hydraulic shock (short-term processes) in incompressible media. Hydraulic shock is a sharp rise of pressure in some fluid-filled system such as pipes, fractures and pores. This process is used in oil well fracturing. There are some engineering models (formulae) to calculate the pressure in the pipe system during a hydraulic shock. But these models do not work for more complex systems, such as an oil well. Existing mathematical models of hydraulic shock in porous media [26–28] are nothing more than the same engineering models of pipe systems. For the basic model at the microscopic level we systematically derive all possible homogenized models.

The Chap. 4 deals with double-porosity models of liquid filtration describing a liquid filtration in a solid body, perforated by pores and fractures. Already there are many different mathematical models describing this physical process. They take into account the geometry of the space occupied by the liquid (liquid domain), and the physical properties of the liquid and solid components (see, for example, [29–34].

Note that pores differ from fractures only by their characteristic size: if l_p is the characteristic size of pores and l_c is the characteristic size of fractures, then $l_p \ll l_c$. The well-known phenomenological double-porosity model, suggested by G. I. Barenblatt, Iu. P. Zheltov and I. N. Kochina [29], describes a two-velocity liquid continuum in an absolutely rigid body, where the macroscopic velocity $\mathbf{v}_p$ and the pressure q_p in pores and the macroscopic velocity $\mathbf{v}_c$ and the pressure q_c in fractures satisfy two different Darcy's laws

$$\mathbf{v}_p = \frac{k_p}{\mu}(-\nabla q_p + \rho_f \mathbf{F}), \quad \mathbf{v}_c = \frac{k_c}{\mu}(-\nabla q_c + \rho_f \mathbf{F}), \tag{0.0.27}$$

and two continuity equations

$$\nabla \cdot \mathbf{v}_p = J, \quad \nabla \cdot \mathbf{v}_c = -J. \tag{0.0.28}$$

The model is completed by the postulate

$$J = \beta(q_c - q_p), \quad \beta = \text{ const }.$$

The scientific and practical value of mathematical models describing such complicated processes is obvious. But their physical reliability is also very important. Pioneering work of L. Tartar (see Appendix in [3]), where Darcy's law of filtration has been rigorously derived, was an example that stimulated many authors to repeat the same result for double-porosity models (T. Arbogast et al. [35], A. Bourgeat et al. [36] and Z. Chen [37]). That is, firstly find an adequate mathematical model at the microscopic level with corresponding small parameters, and, secondly, rigorously fulfill the homogenization procedure.

Because the last two papers repeat the ideas of the first one, let us briefly discuss the main points in [35]. As an initial model at the microscopic level the authors have considered a periodic structure, consisting of "solid" blocks of a size δ surrounded by fluid. The solid component is assumed to be already homogenized: there is no pore space and the motion of the fluid in blocks is governed by usual Darcy equations of filtration. So the authors have used the method of *reiterated homogenization*; when the first homogenization procedure is applied to the solid matrix and liquid in pores, and then to the mixture "solid" blocks—liquid in fractures. The motion of the fluid in the fracture space (the space between "solid" blocks) is described by some artificial system, similar to the Darcy equations of filtration. There is no physical basis, but from a mathematical point of view such a choice of equations of fluid dynamics in fractures has a very solid basis. It is impossible to find reasonable boundary conditions on the common boundary of the "solid" block—fracture space if the fluid dynamics is described by the Stokes equations, but there are reasonable boundary conditions if the liquid motion is described by the Darcy equations of filtration. Therefore, the final homogenized models in [35, 36], and in [37] have no connection with the fundamental laws of continuum mechanics. But there is sense in the idea of the method of reiterated homogenization for such problems. Unfortunately, the authors have used the wrong models, describing the motion of "solid" blocks and the liquid in the fracture space. We will use this method for acoustics in Chap. 9.

In Chap. 4 we follow the chosen method and first formulate the mathematical model at the microscopic level. The difference from the first chapter is only in the geometry of the liquid domain, because differential equations for the motion of the liquid and the solid components must be the same. To model the geometry we postulate that there are two small parameters: the dimensionless size of pores ε and the dimensionless size of fractures δ and $\varepsilon \ll \delta$. As usual, we suppose the

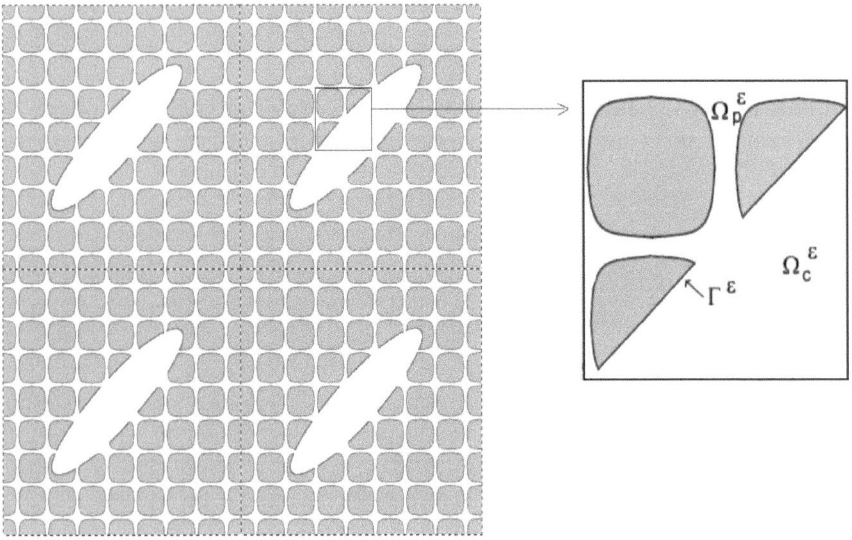

Fig. 0.1 Double porosity geometry: isolated fractures

periodicity of the pore and fracture spaces, so that the characteristic function χ_0 of the liquid domain has a form

$$\chi_0(\mathbf{x}) = \tilde{\chi}(\mathbf{x}, \frac{\mathbf{x}}{\varepsilon}, \frac{\mathbf{x}}{\delta}),$$

where $\tilde{\chi}(\mathbf{x}, \mathbf{y}, \mathbf{z})$ is a 1-periodic function with respect to variables $\mathbf{y}$ and $\mathbf{z}$ (see Figs. 0.1 and 0.2).

Thus, for this mathematical model at the microscopic level one of the main problems is the limit in integrals of the form

$$I^{\varepsilon,\delta} = \int_\Omega u(\frac{\mathbf{x}}{\varepsilon}, \frac{\mathbf{x}}{\delta})\, v^{\varepsilon,\delta}(\mathbf{x})\varphi(\mathbf{x})dx, \tag{0.0.29}$$

when ε and δ tend to zero. Compared with (0.0.26), in (0.0.29) there appears a new scale of the fast variable, associated with the parameter δ. This problem has been solved in [21, 38] by introducing the method of reiterated homogenization. For the slightly viscous incompressible liquid in an elastic incompressible skeleton the homogenized system consists of the differential equations

$$\mathbf{v} = \mathbf{v}_c + (1 - m_c)\frac{\partial \mathbf{w}_s}{\partial t}, \quad \mathbf{v}_p = (1 - m_c)\, m_p\, \frac{\partial \mathbf{w}_s}{\partial t}, \tag{0.0.30}$$

$$\nabla \cdot \mathbf{v} = 0, \tag{0.0.31}$$

$$\lambda_0 \nabla \cdot \left(\mathfrak{B}_0^{(s)} : \mathbb{D}(\mathbf{w}_s)\right) - \frac{1}{m}\nabla q_f + \hat{\rho}\mathbf{F} = 0, \tag{0.0.32}$$

$$\mathbf{v}_c = m_c \frac{\partial \mathbf{w}_s}{\partial t} + \frac{1}{\mu_2} \mathbb{B}^{(c)} \left(\rho_f \mathbf{F} - \frac{1}{m} \nabla q_f \right), \qquad (0.0.33)$$

for the velocity $\mathbf{v}_p$ in pores, velocity $\mathbf{v}_c$ and pressure q_f in fractures, and for the displacements $\mathbf{w}_s$ of the solid component (for details see Chap. 4). For an absolutely rigid skeleton

$$\lambda_0 \to \infty, \quad \mathbf{w}_s \to 0,$$

and we arrive at the system

$$\mathbf{v} = \mathbf{v}_c, \quad \mathbf{v}_p = 0, \quad \mathbf{w}_s = 0, \qquad (0.0.34)$$

$$\nabla \cdot \mathbf{v}_c = 0, \qquad (0.0.35)$$

$$\mathbf{v}_c = \frac{1}{\mu_2} \mathbb{B}^{(c)} \left(\rho_f \mathbf{F} - \frac{1}{m} \nabla q_f \right), \qquad (0.0.36)$$

for the velocity $\mathbf{v}_c$ and pressure q_f in fractures. The last system describes the motion of the slightly viscous incompressible liquid in the pore–fracture space of an absolutely rigid skeleton, and obviously asymptotically closed to the basic equations of the continuum mechanics.

But the equations obtained contradict to the model (0.0.27, 0.0.28) from [29]. In (0.0.34–0.0.36) the liquid in pores is blocked and unmoved,and there is no second pressure in fractures.

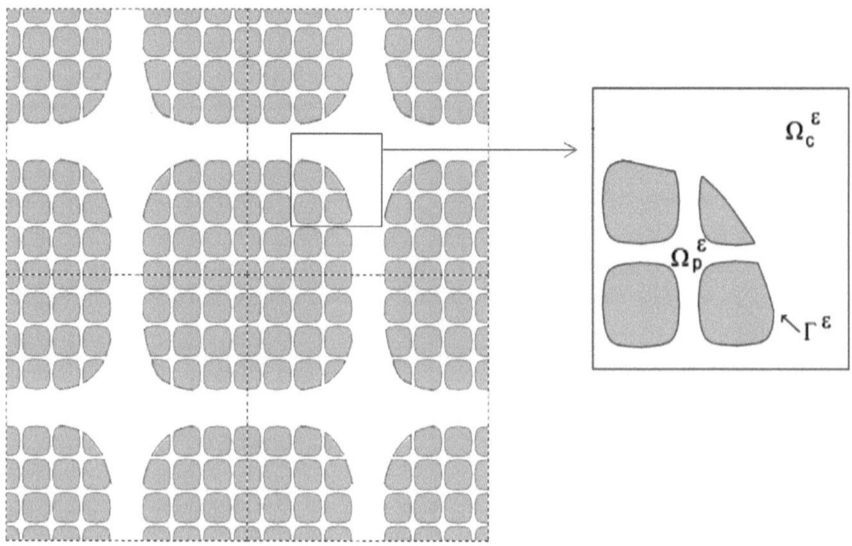

Fig. 0.2 Double porosity geometry: connected fracture space

Later, one of the authors [29] G. I. Barenblatt in one of his books on the filtration theory ([39], p. 187), noted that "... the liquid motion in such a medium (fractured–porous) is realized **mainly along fractures**, while the volume of fractures is small and **main reserves of the liquid are in the porous blocks**". This observation, based on the deep physical intuition of the author, is not confirmed by when the correct mechanism for extracting liquid from the pores is specified. Our result shows that for a more accurate description of real physical processes one must take into account the elastic properties of the solid skeleton, namely, the elastic stresses in the solid skeleton (but not the second pressure in the pores) are the main factor that allow the liquid to flow from the pores into the fractures.

In Chap. 5 we investigate liquid filtration in composite domains. That is, liquid filtration in at least two domains with a common boundary, and with different properties.

For example, filtration in a poroelastic medium, which has a common boundary with some elastic body, or with a water reservoir. The main problem here is the boundary conditions on the common boundary for the solutions of homogenized equations. P. Polubarinova-Kochina [40] uses the Darcy system of filtration in the porous medium and simply postulates the hydrostatics in the reservoir and the continuity of the pressure on the common boundary "reservoir-porous medium". There are some particular results obtained by W. Jäger and A. Mikelić [41–43] for special geometry of pore space (disconnected solid skeleton) and only for domains in $\mathbb{R}^2$. We study the complete problem in $\mathbb{R}^3$ for the arbitrary geometry of corresponding pore spaces.

Next, if we consider filtration from a reservoir into the poroelastic medium (see Fig. 0.3), then for the basic model at the microscopic level there is a flow from a reservoir Ω^0 into the porous medium Ω and maybe backwards, and its can be calculated. The same property remains valid for the homogenized model of poroelastic filtration ($0 < \mu_0, \lambda_0 < \infty$). But for Darcy's system of filtration in Ω (first level of approximation with $\mu_0 = 0$, $\lambda_0 = \infty$), or for the Terzaghi–Biot

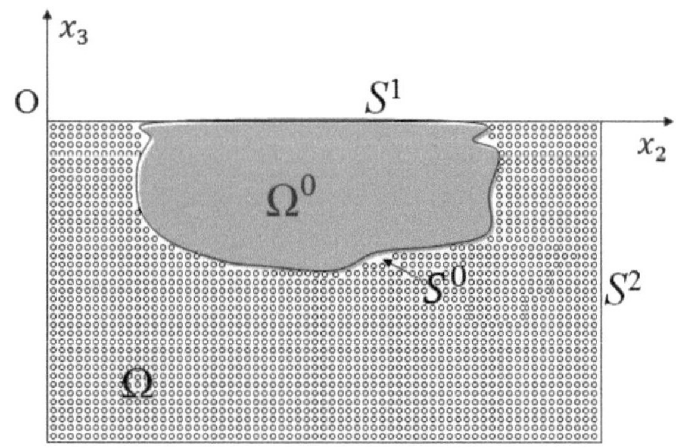

Fig. 0.3 Filtration from reservoir

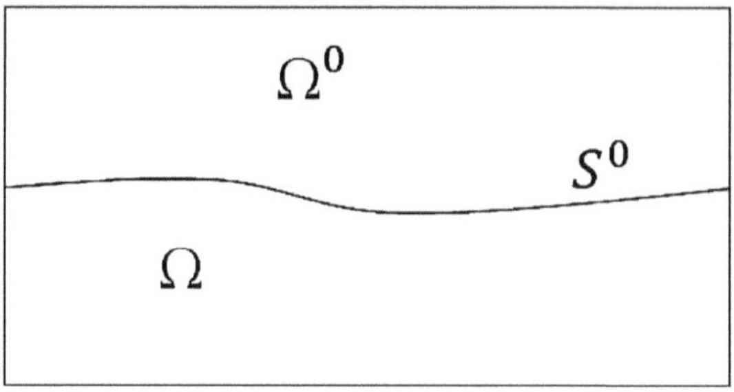

Fig. 0.4 Two different poroelastic media

system of poroelasticity in Ω (second level of approximation with $\mu_0 = 0,\ 0 < \lambda_0 < \infty$), the motion in the reservoir Ω^0 is automatically approximated by hydraulics, the limiting pressure on the common boundary S^0 takes the value of the hydraulic pressure at S^0 (the pressure is continuous!), and there is no information about the flow from Ω^0 into Ω and back.

Another example involves two poroelastic media with the same properties of the liquid. Here everything depends on the structure of the pore spaces in Ω and Ω^0. Let $\mu_0 = 0$. If Y_f^0 defines the pore space in Ω^0, and Y_f defines the pore space in Ω, then for $Y_f \cap Y_f^0 \neq \emptyset$ the pressure is continuous on the common boundary S^0 and there is a flow from Ω^0 into Ω or vice-versa. For $Y_f \cap Y_f^0 = \emptyset$ instead of continuity of the pressure, one has another condition, which shows that there is neither flow from Ω^0 into Ω nor back (Fig. 0.4).

As is known, the most rigorous results of homogenization theory are obtained for very special physical media, when the local heterogeneity has a periodic structure. Therefore, the objections from opponents of such a method of mathematical modeling—that the models have a low practical value—are quite reasonable. The situation is similar to the situation with differential equations with constant coefficients, and differential equations with variable coefficients. The practical value of the first is not comparable to the practical value of the latter. But the history of mathematics shows that the theory of equations with variable coefficients cannot be constructed without a complete theory of equations with constant coefficients. The above analogy suggests a way to solve the problem of mathematical modeling of physical processes in macroscopic inhomogeneous media. The detailed analysis of the homogenized problems in composite domains permits the derivation of homogenized models allowing for the variable geometry and elasticity of the solid component.

Expressly, let Ω be a domain in consideration and $\Pi^{(\delta)} = \{K_1^{(\delta)}, \ldots, K_{N_\delta}^{(\delta)}\}$ be a partition of Ω into nonintersecting subdomains with a diameter δ. All physical and geometrical characteristics of the medium are assumed to be constant in the given subdomain $K_n^{(\delta)}$. The problem as formulated for a fixed δ is defined by the characteristic function of the pore space $\chi^{(\delta)}(\mathbf{x},\mathbf{y})$. This function is 1-periodic in $\mathbf{y}$ and piecewise-constant in $\mathbf{x}$. The homogenized model obtained, depending on the parameter δ, has been already studied above and admits a subsequent limit as $\delta \to 0$, which leads to the final homogenized model, taking into account the macroscopic inhomogeneity of the continuum.

Note that the formal justification of the symmetry of the diagram (the limit as $\delta \to 0$ for fixed ε and then the limit as $\varepsilon \to 0$) is not physically rigorous, because the diameter δ of the subdomain $K_n^{(\delta)}$ cannot be less than the characteristic size ε of pores in the solid skeleton. Nevertheless, for sufficiently reasonable agreements the limit as $n \to \infty$ leads to the homogenization problem with a characteristic function of the pore space $\chi(\mathbf{x},\mathbf{y}) = \lim_{\delta\to 0}\chi^{(\delta)}(\mathbf{x},\mathbf{y})$ (in each $\mathbf{x}_0 \in \Omega$ there is a proper pore space, defined by the characteristic function $\chi(\mathbf{x}_0,\mathbf{y})$). The homogenization of this problem coincides with the final homogenized model, obtained before. This proves the correctness of our approach.

Chapters 6–9 are devoted to homogenized models in acoustics. In Chap. 6 we consider isothermal acoustics and in Chap. 7 nonisothermal acoustics. Next, Chap. 8 repeats the results of Chap. 5 for acoustics, and in Chap. 9 we derive double-porosity models for acoustics.

For example, in Chap. 6 we derive mathematical models $(\mathbb{IA})_1$–$(\mathbb{IA})_{16}$. Obviously, all these models describe the same physical process, but with varying degrees of approximation. We may also say that due to an incomplete description we allocate in each model the various components of the process. For example, in the first models $(\mathbb{IA})_1$–$(\mathbb{IA})_4$ we describe only compression sound waves in the fluid in pores. In models $(\mathbb{IA})_5$–$(\mathbb{IA})_{12}$ we study the interaction of compression sound waves in the solid skeleton and in the fluid in pores. Vice-versa, in the model $(\mathbb{IA})_{13}$ we describe compression sound waves and shear waves in the solid skeleton. Finally, in models $(\mathbb{IA})_{15}$ and $(\mathbb{IA})_{16}$ we study the interaction of compression sound waves in the solid skeleton and in the fluid in pores, and shear waves in the solid skeleton.

In the last two Chaps. 10 and 11 we consider some applications of our theory: diffusion-convection in porous media, and the free boundary problem of a joint motion of two immiscible incompressible liquids (the Muskat problem).

Diffusion-convection processes in porous media $\Omega \subset \mathbb{R}^3$ is described by a diffusion-convection equation in the liquid domain Ω_f (pore space)

$$\frac{\partial c}{\partial t} + \mathbf{v}\cdot\nabla c = D\,\triangle\, c, \tag{0.0.37}$$

for the concentration c of an admixture. Here D is the given diffusion coefficient, and $\mathbf{v}$ is the velocity of the liquid.

If we consider the most general case of the motion of continuous media, which is a generalized motion with strong discontinuity, then the boundary condition on the surface of strong discontinuity $\Gamma_t = \partial\Omega_s \cap \partial\Omega_f$ (the common boundary "pore space–solid skeleton") at the time moment $t > 0$ has a form

$$c\left((\mathbf{v}\cdot\mathbf{n})-V_n\right)=D\,(\nabla c\cdot\mathbf{n}). \tag{0.0.38}$$

In (0.0.38) $\mathbf{n}$ is a unit normal vector to Γ_t, and V_n is a velocity of replacement of surface Γ_t in the direction of normal $\mathbf{n}$.

In the general case the velocity field is defined by the mathematical model (0.0.1–0.0.3), which is a free boundary problem. In particular, one of the boundary condition on the free surface of a contact discontinuity has the form

$$\mathbf{v}\cdot\mathbf{n}=V_n, \tag{0.0.39}$$

and for this case (0.0.38) transforms to

$$\nabla c\cdot n=0 \tag{0.0.40}$$

As we have mentioned above, the mathematical model (0.0.1–0.0.3) obviously would not be suitable for practical use and in dimensionless variables the most appropriate dynamic system, coupled with a convection-diffusion equation has the form

$$\nabla\cdot\mathbf{w}=0, \tag{0.0.41}$$

$$\nabla\cdot\mathbb{P}+\tilde{\rho}\mathbf{F}=0, \tag{0.0.42}$$

$$\mathbb{P}=\chi_0\alpha_\mu\mathbb{D}\Big(x,\frac{\partial\mathbf{w}}{\partial t}\Big)+(1-\chi_0)\alpha_\lambda\mathbb{D}(x,\mathbf{w})-p\,\mathbb{I}, \tag{0.0.43}$$

$$\frac{\partial c}{\partial t}+\frac{\partial\mathbf{w}}{\partial t}\cdot\nabla c=\alpha_D\,\Delta\,c, \tag{0.0.44}$$

where

$$\alpha_D=\frac{D\tau}{L^2},\quad \alpha_\mu=\alpha_\mu(c),\quad \tilde{\rho}=\chi_0(\rho_f+\delta\,c)+(1-\chi_0)\rho_s.$$

We must complete the model with the boundary conditions on the common (and fixed) boundary Γ. The boundary conditions for dynamic equations have already been discussed earlier. For the convection-diffusion equation one has a choice. By supposition, $V_n=0$, and this postulate and the boundary condition (0.0.38) imply

$$\Big(\alpha_D\nabla c-c\,\frac{\partial\mathbf{w}}{\partial t}\Big)\cdot\mathbf{n}=0. \tag{0.0.45}$$

This is the first choice. The second choice is a condition (0.0.40). Thus, for the same process we have two different models—the mathematical model (0.0.40–0.0.44), and the mathematical model (0.0.41–0.0.45). The difference is caused by the fact that for our approximation both the boundary conditions for the concentration are not quite exact.

Note that for an absolutely rigid solid skeleton $\mathbf{w}=\mathbf{v}=0$ in the solid part (0.0.40) coincides with (0.0.45). That is, both models coincide.

In Chap. 10 we derive homogenized models for diffusion-convection in an absolutely rigid solid skeleton, and in poroelastic media for the model (0.0.40–0.0.44) with $\alpha_\mu = \mu_0 = const > 0$.

The results in Chap. 10 are based upon the author's papers [22–24, 36, 44, 48–55].

The homogenization procedure for (0.0.37) with a given velocity field $\mathbf{v}$ has been described in [22–24, 36, 44, 48–55].

In the last Chap. 11 we study the joint motion of immiscible liquids, which are modeled as one nonhomogeneous liquid. We consider two different incompressible viscous liquids with the same viscosity and different constant densities. The exact mathematical model at the microscopic level consists of the dynamic equations

$$\nabla \cdot \big(\tilde{\chi}\,\mathbb{P}_f + (1-\tilde{\chi})\mathbb{P}_s\big) + \big(\rho_f\,\tilde{\chi} + \rho_s\,(1-\tilde{\chi})\big)\mathbf{F} = 0, \tag{0.0.46}$$

$$\nabla \cdot \mathbf{v} = 0, \tag{0.0.47}$$

$$\frac{\partial \tilde{\chi}}{\partial t} + \mathbf{v} \cdot \nabla\,\tilde{\chi} = 0, \quad \tilde{\chi}(\mathbf{x},0) = \chi_0(\mathbf{x}), \tag{0.0.48}$$

which is a sub-model of (0.0.1–0.0.3) for the case of the filtration of an incompressible liquid in an incompressible solid skeleton, completed with the Cauchy problem

$$\frac{\partial \rho_f}{\partial t} + \mathbf{v} \cdot \nabla\,\rho_f = 0, \quad \rho_f(\mathbf{x},0) = \rho_f^{(0)}(\mathbf{x}) \tag{0.0.49}$$

for the density of the nonhomogeneous liquid in the liquid domain Ω_f for $t > 0$.

The last problem is equivalent to the the Cauchy problem

$$\frac{\partial \rho}{\partial t} + \mathbf{v} \cdot \nabla\,\rho = 0, \quad \rho(\mathbf{x},0) = \rho_f^{(0)}(\mathbf{x})\,\chi_0(\mathbf{x}) + \rho_s\,\big(1-\chi_0(\mathbf{x})\big) \tag{0.0.50}$$

for the density $\rho = \rho_f\,\tilde{\chi} + \rho_s\,(1-\tilde{\chi})$ of the medium.

Let Π_0 be a smooth surface dividing Ω into two subdomains Ω^+ and Ω^- and

$$\rho_f^{(0)}(\mathbf{x}) = \rho_f^+ = const \quad \text{for} \quad \mathbf{x} \in \Omega^+, \;\; \rho_f^{(0)}(\mathbf{x}) = \rho_f^- = const \quad \text{for} \quad \mathbf{x} \in \Omega^-.$$

Then for the smooth velocity field $\mathbf{v}(\mathbf{x},t)$ there exists a smooth surface of the strong discontinuity $\Pi(t)$, $\Pi(0) = \Pi_0$, dividing Ω_f into two subdomains $\Omega_f^+(t)$ and $\Omega_f^-(t)$, such that

$$\rho(\mathbf{x},t) = \rho_f^+ \quad \text{for} \quad \mathbf{x} \in \Omega_f^+(t), \quad \text{and} \quad \rho(\mathbf{x},t) = \rho_f^- \quad \text{for} \quad \mathbf{x} \in \Omega^-.$$

That is, the problem (0.0.46–0.0.48, 0.0.50) really describes the joint motion of two immiscible incompressible liquids with the different constant densities separated by the free boundary $\Pi(t)$.

It is obvious that the resulting problem is too complicated. To simplify the model and get simpler, but still reasonable one, we replace (0.0.46–0.0.48) by the system

$$\nabla \cdot \left(\chi_0\, \alpha_\mu\, \mathbb{D}(x, \frac{\partial \mathbf{w}}{\partial t}) + (1-\chi_0)\alpha_\lambda\, \mathbb{D}(x, \mathbf{w}) - p\,\mathbb{I}\right) + \rho\, \mathbf{F} = 0, \tag{0.0.51}$$

$$\nabla \cdot \mathbf{v} = 0, \quad \mathbf{v} = \frac{\partial \mathbf{w}}{\partial t}, \tag{0.0.52}$$

completed with the Cauchy problem

$$\frac{\partial \rho}{\partial t} + \mathbf{v} \cdot \nabla\, \rho = 0, \quad \rho(\mathbf{x}, 0) = \rho_f^{(0)}(\mathbf{x})\, \chi_0 + \rho_f\, (1-\chi_0) \tag{0.0.53}$$

for the density ρ of the medium.

For $\chi_0 = \chi^\varepsilon(\mathbf{x})$ this mathematical problem at the microscopic level has at least one weak solution $\{\mathbf{v}^\varepsilon, p^\varepsilon, \rho^\varepsilon\}$ and the main problem is to get homogenized equations in the limit as $\varepsilon \to 0$ in the transport equation for the density ρ^ε. The structure of the system does not permit any uniform estimates for the density with respect to the small parameter ε, except its boundedness:

$$0 < \rho_* = const < \rho^\varepsilon(\mathbf{x}, t) < \rho_*^{-1}.$$

Thus, we may expect only the weak compactness of $\{\rho^\varepsilon\}$, and the limit in (0.0.53) is possible, if $\{\mathbf{v}^\varepsilon\}$ converges strongly in $L_2(\Omega_T)$. It is almost impossible to get such a property if $\mu_0 = 0$.

So, in the Chap. 11 under the restrictions

$$0 < \mu_0,\ \lambda_0 < \infty$$

we derive a homogenized system, which we refer as *the Muskat problem for a viscoelastic filtration*.

All results of that chapter are based on [56].

The case of an absolutely rigid solid skeleton is a sub-model of (0.0.51–0.0.53) when $\lambda_0 = \infty$ ($\alpha_\lambda \to \infty$) and is described by the problem

$$\nabla \cdot \left(\chi_0\, \alpha_\mu\, \mathbb{D}(x, \mathbf{v}) - p\,\mathbb{I}\right) + \rho_f\, \mathbf{F} = 0, \tag{0.0.54}$$

$$\nabla \cdot \mathbf{v} = 0, \tag{0.0.55}$$

$$\frac{\partial \rho_f}{\partial t} + \mathbf{v} \cdot \nabla\, \rho_f = 0, \tag{0.0.56}$$

$$\rho_f(\mathbf{x}, 0) = \rho_f^{(0)}(\mathbf{x}) \tag{0.0.57}$$

for the velocity $\mathbf{v}$, pressure p, and density ρ_f of the liquid in the domain Ω_f for $t > 0$. The existence of a smooth free boundary for this problem has been proved in [57].

The nontrivial homogenization of the dynamic system (0.0.54, 0.0.56) makes sense only for $\chi_0 = \chi^{\varepsilon}(\mathbf{x})$, $\mu_0 = 0$, and $0 < \mu_1 < \infty$ and, as was shown earlier in Chap. 1, leads to the Darcy system of filtration.

We have already mentioned above, that for this case our method does not permit the correct limit in the transport equation (0.0.57). But the formal homogenization of (0.0.54–0.0.57) under these restrictions results in the domain Ω for $t > 0$ the well-known *free boundary Muskat problem*

$$\mathbf{v} = \frac{1}{\mu_1}\mathbb{B}\Big(-\frac{1}{m}\nabla p + \rho_f \mathbf{F}\Big), \quad \nabla \cdot \mathbf{v} = 0, \tag{0.0.58}$$

$$\frac{\partial \rho_f}{\partial t} + \nabla \rho_f \cdot \mathbf{v} = 0, \quad \rho_f(\mathbf{x}, 0) = \rho_f^{(0)}(\mathbf{x}) \tag{0.0.59}$$

in its weak formulation.

This problem is easy to formulate, but almost impossible to solve. For this reason very little is known, neither in classical nor in weak solutions. There are only a few results of classical solvability locally in time or globally in time, though there are nearly explicit solutions, and there is no result for a global in time weak solvability (see, [58–62] and references there).

Numerical simulations of the problem (0.0.54–0.0.57) made in [63] for a single capillary, show the existence of a smooth free boundary (the surface of strong discontinuity) in the capillary at different times (see Fig. 0.5). This fact somewhat confirms the results of [57].

It is clear that we may consider any finite number of such disconnected capillaries and pass from the single capillary to an absolutely rigid solid skeleton, perforated by a system of disconnected capillaries.

The limiting procedure ($\varepsilon \searrow 0$) is modeled by increasing the number of capillaries. For sufficiently small ε (a sufficiently large number of capillaries) we arrive at the Muskat problem (0.0.58, 0.0.59).

We apply he same idea to an elastic body with the same geometry of the pore space, described by (0.0.51–0.0.53).

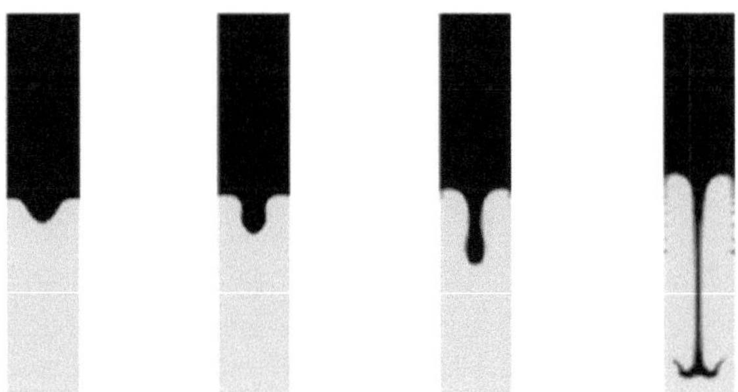

Fig. 0.5 The single capillary

Comparison of numerical simulations (see Fig. 0.6) shows the appearance of a mushy region for the free boundary Muskat problem (0.0.58, 0.0.59), and the existence of a smooth free boundary for the Muskat problem for viscoelastic filtration.

The first fact indirectly indicates a lack of classical solutions of the free boundary Muskat problem (0.0.58, 0.0.59).

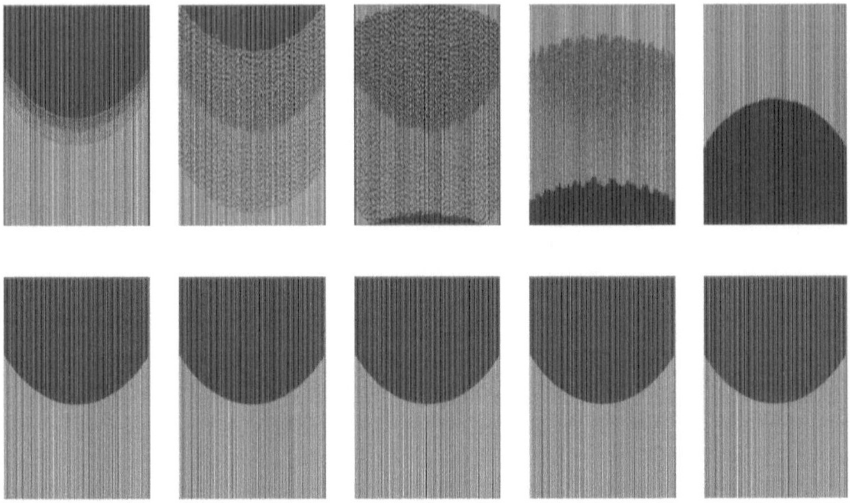

Fig. 0.6 Disconnected capillaries: absolutely rigid (above) and elastic solid skeleton

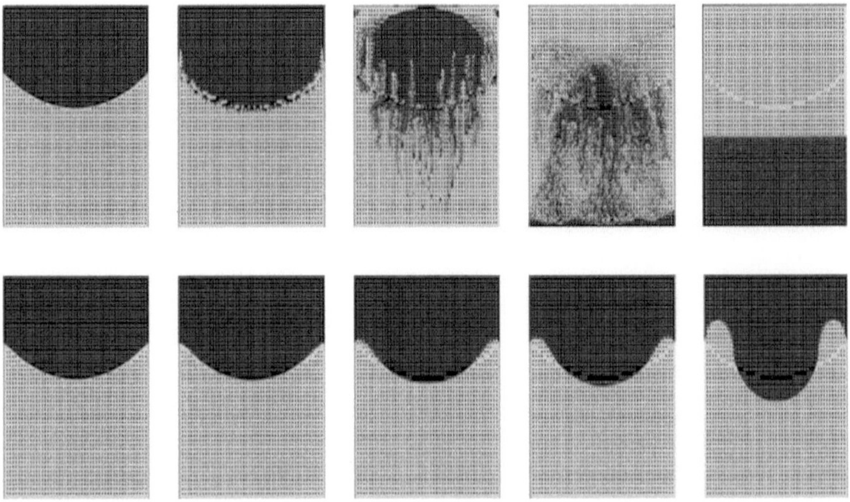

Fig. 0.7 Disconnected solid skeleton: absolutely rigid body (above) and elastic body

A similar conclusion can be reached with numerical simulation for the problems (0.0.54–0.0.57) and (0.0.51–0.0.53) for a disconnected solid skeleton, when $Y_s \subset Y$ is a cube which does not touch the boundary ∂Y (see Fig. 0.7).

In **Appendix A** we concisely list the main notions of continuum mechanics following [1], and in **Appendix B** we formulate all mathematical statements from Analysis and PDE, needed in the main text of the book.

Notations of functional spaces and norm there are the same as in [64, 65, 66]. Some of these notations are listed in Appendix B.

References

1. Ovsiannikov, L. V. (1977). Introduction to mechanics of continuous media, Parts 1, 2. Novosibirsk State University, Novosibirsk (Russian).
2. Burridge, R., & Keller, J. B. (1981). Poroelasticity equations derived from microstructure. *Journal of the Acoustical Society of America, 70*(4), 1140–1146.
3. Sanchez-Palencia, E. (1980). *Non-homogeneous media and vibration theory*. Lecture Notes in Physics, vol. 129. Springer, Berlin.
4. Nguetseng, G. (1989). A general convergence result for a functional related to the theory of homogenization. *SIAM Journal on Applied Mathematics, 20*, 608–623.
5. Buchanan, J. L., & Gilbert, R. P. (1997). Transition loss in the farfield for an ocean with a Biot sediment over an elastic substrate. *ZAMM, 77*, 121–135.
6. Gilbert, R. P., & Lin, J. Z. (1997). Acoustic waves in shallow inhomogeneous oceans with a poro-elastic seabed. *ZAMM, 79*(4), 1–12.
7. Buckingham, M. J. (1998). Seismic wave propagation in rocks and marine sediments: A new theoretical approach. In: A. Alippi, & G. B. Cannelli, (Eds.), *Underwater Acoustics*, vol. 1. CNR-IDAC, Rome.
8. Clopeau, T. H., Ferrin, J. L., Gilbert, R. P., & Mikelić, A. (2001). Homogenizing the acoustic properties of the seabed: Part II, *Mathematical and Computer Modelling, 33*, 821–841.
9. Ferrin, J. L., & Mikelić, A. (2003). Homogenizing the acoustic properties of a porous matrix containing an incompressible inviscid fluids. *Mathematical Methods in the Applied Sciences, 26*, 831–859.
10. Mikelić, A., & Paoli, L. (1999). Homogenization of the inviscid incompressible fluid flow trough a 2D porous medium. *Proceedings of the AMS, 17*, 2019–2028.
11. Levy, T. (1977). Acoustic phenomena in elastic porous media. *Mechanics Research Communications, 4*, 253–257.
12. Nguetseng, G. (1990). Asymptotic analysis for a stiff variational problem arising in mechanics. *SIAM Journal on Applied Mathematics, 21*, 1394–1414.
13. Sanchez-Hubert, J. (1980). Asymptotic study of the macroscopic behavior of a solid-liquid mixture. *Mathematical Methods in the Applied Sciences, 2*, 1–18.
14. Meirmanov, A. (2007). Nguetseng's two-scale convergence method for filtration and seismic acoustic problems in elastic porous media. *Siberian Mathematical Journal, 48*, 519–538.
15. Meirmanov, A. (2008). Acoustic and filtration properties of a thermoelastic porous medium: Biot's equations of thermo–poroelasticity. *Sbornik Mathematics, 199*(3), 1–24.
16. Meirmanov, A. (2008). Homogenized models for filtration and for acoustic wave propagation in thermo-elastic porous media. *European Journal of Applied Mathematics, 19*, 259–284.
17. Meirmanov, A. (2008). A description of seismic acoustic wave propagation in porous media via homogenization. *SIAM Journal on Applied Mathematics, 40*(3), 1272–1289.
18. Meirmanov, A. (2008). Darcy's law for a compressible thermofluid. *Asymptotic Analysis, 58*(4), 191–209.

19. Meirmanov, A. (2010). Double porosity models in incompressible poroelastic media. *Mathematical Models and Methods in Applied Sciences, 20*(4), 635–659.
20. Meirmanov, A. (2009). Derivation of equations of seismic and acoustic wave propagation and equations of filtration via homogenization of periodic structures. *Journal of Mathematical Sciences, 163*(2), 111–172.
21. Lukkassen, D., Nguetseng, G., & Wall, P. (2002). Two-scale convergence. *International Journal of Pure and Applied Mathematics, 2*(1), 35–86.
22. Bensoussan, A., Lions, J.-L., & Papanicolau, G. (1978). *Asymptotic analysis for periodic structure*. Amsterdam: North Holland.
23. Jikov, V. V., Kozlov, S. M., & Oleinik, O. A. (1969). *Homogenization of differential operators and integral functionals*. New York: Springer-Verlag.
24. Hornung, U. (1997). *Homogenization and porous media*. New York: Springer-Verlag.
25. Bakhvalov, N. S., & Panasenko, G. (1989). *Homogenization: Averaging processes in periodic media*. New York: Spinger.
26. Adachi, J. I., Detournay, E., & Peirce, A. P. (2010). Analysis of the classical pseudo-3D model for hydraulic fracture with equilibrium height growth across stress barriers. *International Journal of Rock Mechanics and Mining Sciences, 47*, 625–639.
27. Kovalyshen, Y., & Detournay, E. (2010). A reexamination of the classical PKN model of hydraulic fracture. *Transport in Porous Media, 81*, 317–339.
28. Weiguoab, L., & Yangshenga, Z. (2005). A mathematical model for solid liquid and mass transfer coupling and numerical simulation for hydraulic fracture in rock salt. *Progress in Natural Science, 15*(8), 742–748.
29. Barenblatt, G. I., Zheltov, Iu. P., & Kochina, I. N. (1960). Basic concepts in the theory of seepage of homogeneous liquids in fissures rocks.*Journal of Applied Mathematics, 24*, 1286–1303.
30. Kazemi, H. (1969). Pressure transient analysis of naturally fractured reservoirs with uniform fracture distribution. *The Journal of Petroleum Technology and Society of Petroleum, 9*, 451–462.
31. Pride, S. R., & Berryman, J. G. (2003). Linear dynamics of double-porosity dual-permeability materials I. Governing equations and acoustic attenuation. *Physical Review E, 68*, 036603.
32. Pride, S. R., & Berryman, J. G. (2003). Linear dynamics of double-porosity dual-permeability materials II. Fluid transport equations. *Physical Review E, 68*, 036604.
33. de Swaan, A. (1976). Analytic solutions for determing naturally fractured reservoir properties by well testing. *Journal of Petroleum Technology and the Society of Petroleum, 16*, 117–122.
34. Ladyzhenskaya, O. A. (1985). *The boundary-value problems of mathematical physics*. New York: Springer.
35. Arbogast, T., Douglas, J., & Hornung, U. (1990). Derivation of the double porosity model of single phase flow via homogenization theory. *SIAM Journal on Applied Mathematics, 21*(4), 823–836.
36. Bourgeat, A., Pankratov, L., & Panfilov, M. (2004). Study of the double porosity model versus the fissures thikness. *Asymptotic Analysis, 38*, 129–141.
37. Chen, Z. (2007). Homogenization and simulation for compositional flow in naturally fractured reservoirs. *Journal of Mathematical Analysis and Applications, 326*, 31–75.
38. Allaire, G., & Briane, M. (1996). Multisale convergence and reiterated homogenization. *Proceedings of Royal Society Edinburgh, 126A*, 297–342.
39. Barenblatt, G. I., Yentov, V. M., & Ryzhik, V. M. (1972). *Theory of nonstationary liquid and gas filtration*. Moscow: Nedra (Russian).
40. Polubarinova-Kochina, P. Ya. (1962). *Theory of ground water movement*. Princeton: Princeton University Press.
41. Jäger, W., & Mikelić, A. (1994). On the flow conditions at the boundary between a porous medium and an impervious solid. In *Progress in PDE: The Metz Surveys 3*. Pitman research notes in mathematics, N 314. London: Longman Scintific and Technical.

42. Jäger,W., & Mikelić, A. (1996). On the boundary conditions at the contact interface between a porous medium and a free fluid. *Annals of the Scuola Normale Superiore of Pisa—Class of Sciences, Fasc., 3*, 403–465.
43. Jäger, W., & Mikelić, A. (1999). On the boundary conditions at the contact interface between two porous media. In *PDE, Theory and Numerical Solution*. Chapman and Hall/CRC Research notes in mathematics, N 406. London: CRC Press.
44. Allaire, G., & Capdeboscq, Y. (2000). Homogenization of a spectral problem in neutronic multigroup diffusion. *Computer Methods in Applied Mechanics and Engineering, 187*, 91–117.
45. Amaziane, B., Goncharenko, M., & Pankratov, L. (2007). Homogenization of a convectiondiffusion equation in perforated domains with a weak adsorption. *The Journal of Applied Mathematics and Physics (ZAMP), 58*, 592–611.
46. Bourgeat, A., Gipouloux, O., & Marušić-Paloka, E. (2001). Mathematical modelling and numerical simulation of a non-newtonian viscous flow through a thin filter. *SIAM Journal on Applied Mathematics, 62*(2), 597–626.
47. Bourgeat, A., Gipouloux, O., & Marušić-Paloka, E. (2003). Filtration law for polymer flow through porous media. *Multiscale Modeling and Simulation, 1*(3), 132–157.
48. Bourgeat, A., Jurak, M., & Piatnitski, A. L. (2003). Averaging a transport equation with small diffusion and oscillating velocity. *Mathematical Methods in the Applied Sciences, 26*, 95–117.
49. Bourgeat, A., & Marušić-Paloka, E. (2005). A homogenized model of an underground waste repository including a disturbed zone. *Multiscale Modeling and Simulation, 1*(4), 918–939.
50. Hornung, U., & Jäger, W. (1991). Diffusion, convection, absorbtion and reaction of chemicals in porous media. *Journal on Differential Equations, 92*(2), 199–225.
51. Marušić-Paloka, E., & Piatnitski, A. (2005). Homogenization of a nonlinear convection-diffusion equation with rapidly oscilating coefficients and strong convection. *Journal of the London Mathematical Society, 72*(2), 371–409.
52. Meirmanov, A. M., & Zimin, R. (2011). Some compactness result for periodic structures and its application to the homogenization of a diffusion-convection equation. *Electronic Journal of Differential Equations, 115* 1–11.
53. Meirmanov, A. M., & Zimin, R. (2012). Mathematical models of a diffusion—convection in porous media. *Electronic Journal of Differential Equations, 105*, 1–16.
54. Meirmanov, A. (2013). A diffusion-convection in poroelastic media. Part I: the microscopic description. *Advances in Mathematical Sciences and Applications, 22*(1), 259–280.
55. Meirmanov, A. (2013). A diffusion-convection in poroelastic media. Part II: The macroscopic description. *Advances in Mathematical Sciences and Applications, 22*(1), 281–294.
56. Meirmanov, A. (2011). The Muskat problem for a viscoelastic filtration. *Interfaces and Free Boundaries, 13*(4), 463–484.
57. Antontsev, S., Meirmanov, A., & Yurinsky, V. (2000). A Free Boundary Problem for Stokes Equations: Classical Solutions. *Interfaces and Free Boundaries, 2*, 413–424.
58. Booth, R. J. S. (2011). Asymptotics for the Muskat problem. *Journal of Engineering Mathematics*. doi:10.1007/s1066501093703.
59. Escher, J., Matioc, A., & Matioc, B. (2011). odelling and analysis of the muskat problem for thin fluid layers. *Journal of Mathematical Fluid Mechanics*, Basel AG: Springer. doi:10.1007/s0002101100532.
60. Fahuai, Yi. (2003). Global classical solution of Muskat free boundary problem. *Journal of Mathematical Analysis and Applications, 288*, 442–461.
61. Radkevich, E. (1995). On the spectrum of the pencil in the Verigin-Muskat problem. *Sbornik: Mathematics, 80*(1), 33–74.
62. Siegel, M., Caflish, R. E., & Howison, S. (2004). Global existence, singular solutions, and ill-posedness for the Muskat problem. *Communications on Pure and Applied Mathematics, 57*, 1–38.

63. Galtsev, O., & Meirmanov, A. (2011). Numerical homogenization in the Rayleigh-Taylor problem of filtration two immiscible incompressible liquids. *Mathematical modelling, 23*,(10), 33–43 (Russian).
64. Warren, J. E., & Root, P. J. (1963). The behaviour of naturally fractured reservoirs. *Journal of Petroleum Technology and the Society of Petroleum, 3*, 235–255.
65. Ladyzhenskaya, O. A., Solonnikov, V. A., & Uraltseva, N. N. (1968). Linear and quasilinear equations of parabolic type. *Translations of Mathematical Monographs, 23*, American Mathematical Society Providence, Rhode Island.
66. Lions, J. L. (1969). Quelques méthodes de résolution des problémes aux limites non linéaire. Paris: Dunod, Gauthier-Villars.

Chapter 1
Isothermal Liquid Filtration

We derive all the possible homogenized equations of the model $\mathbb{M}_{14}$ of an isothermal filtration (or, simply, **a liquid filtration**)

$$\frac{1}{\tilde{\alpha}_p}\, p + \nabla \cdot \mathbf{w} = 0, \tag{1.0.1}$$

$$\nabla \cdot \mathbb{P} + \tilde{\rho}\mathbf{F} = 0, \tag{1.0.2}$$

$$\mathbb{P} = \chi_0 \alpha_\mu \mathbb{D}\Big(x, \frac{\partial \mathbf{w}}{\partial t}\Big) + (1 - \chi_0)\lambda_0 \mathbb{D}(x, \mathbf{w}) - \Big(p - \chi_0 \alpha_\nu \nabla \cdot \frac{\partial \mathbf{w}}{\partial t}\Big)\mathbb{I}, \tag{1.0.3}$$

and its submodels: the model $\mathbb{M}_{15}$ of **the filtration of an incompressible liquid**

$$\nabla \cdot \mathbf{w} = 0, \tag{1.0.4}$$

$$\nabla \cdot \mathbb{P} + \tilde{\rho}\mathbf{F} = 0, \tag{1.0.5}$$

$$\mathbb{P} = \chi_0 \alpha_\mu \mathbb{D}\Big(x, \frac{\partial \mathbf{w}}{\partial t}\Big) + (1 - \chi_0)\lambda_0 \mathbb{D}(x, \mathbf{w}) - p\,\mathbb{I}, \tag{1.0.6}$$

the model $\mathbb{M}_{17}$ of **the filtration of a compressible liquid in an absolutely rigid solid skeleton**

$$\chi_0\Big(\frac{1}{c_f^2}\, p + \nabla \cdot \mathbf{w}\Big) = 0, \tag{1.0.7}$$

$$\chi_0\big(\nabla \cdot \mathbb{P} + \rho_f \mathbf{F}\big) = 0, \tag{1.0.8}$$

$$\mathbb{P} = \alpha_\mu \mathbb{D}\Big(x, \frac{\partial \mathbf{w}}{\partial t}\Big) - \Big(p - \alpha_\nu \nabla \cdot \frac{\partial \mathbf{w}}{\partial t}\Big)\mathbb{I}, \tag{1.0.9}$$

A. Meirmanov, *Mathematical Models for Poroelastic Flows*,
Atlantis Studies in Differential Equations 1, DOI: 10.2991/978-94-6239-015-7_1,

in the bounded domain $\Omega = \Omega_f \cup \Gamma \cup \Omega_s \subset \mathbb{R}^3$, $\Gamma = \partial\Omega_f \cap \partial\Omega_s$, with a C^2 continuous boundary $S = \partial\Omega$ for $t \in (0, T)$.

In (1.0.1)–(1.0.9) $\chi_0(\mathbf{x})$ is the characteristic function of the domain Ω_f.

In this chapter we consider a homogenization procedure only for periodic structures. That is, we impose Assumption 0.1.

Under this assumption

$$\chi_0(\mathbf{x}) = \chi^\varepsilon(\mathbf{x}) = \varsigma(\mathbf{x})\chi\left(\frac{\mathbf{x}}{\varepsilon}\right), \tag{1.0.10}$$

where $\varsigma(\mathbf{x})$ is the characteristic function of the domain Ω, and

$$\tilde{\alpha}_p = \alpha_p^\varepsilon = \chi^\varepsilon c_f^2 + (1-\chi^\varepsilon)c_s^2, \ \tilde{\rho} = \rho^\varepsilon = \chi^\varepsilon \rho_f + (1-\chi^\varepsilon)\rho_s.$$

We say that the **pore space is disconnected**, if the domain E_f^1 is disconnected ($\overline{Y}_f \cap \partial Y = \emptyset$), and the **pore space is connected**, if the domain E_f^1 is connected ($\overline{Y}_f \cap \partial Y \neq \emptyset$).

Similarly, we say that the **solid skeleton is disconnected**, if the domain E_s^1 is disconnected ($\overline{Y}_s \cap \partial Y = \emptyset$), and the **solid skeleton is connected**, if the domain E_s^1 is connected ($\overline{Y}_s \cap \partial Y \neq \emptyset$).

It is assumed that all dimensionless parameters depend on the small parameter ε and the (finite or infinite) limits exist:

$$\lim_{\varepsilon \searrow 0} \alpha_\mu(\varepsilon) = \mu_0, \quad \lim_{\varepsilon \searrow 0} \frac{\alpha_\mu}{\varepsilon^2} = \mu_1, \quad \lim_{\varepsilon \searrow 0} \alpha_\nu(\varepsilon) = \nu_0.$$

We say that

(1) the liquid is **slightly viscous**, if $\mu_0 = 0$,
(2) the liquid is **viscous** , if $0 < \mu_0 < \infty$,
(3) the solid body is **extremely elastic** if $\lambda_0 = 0$,
(4) the solid body is **elastic** if $0 < \lambda_0 < \infty$,
(5) the solid body is **absolutely rigid** if $\lambda_0 = \infty$.

In what follows, we denote as C_0 any constant depending only on domains Ω, Y and Y_f.

1.1 A Compressible Slightly Viscous Liquid in an Absolutely Rigid Skeleton

In this section as a basic mathematical model at the microscopic level we consider the model $\mathbb{M}_{17}$ of the filtration of compressible liquid in an absolutely rigid solid skeleton. It is easy to show that this model is a limit of the model $\mathbb{M}_{14}$ as $\lambda_0 \to \infty$.

One of the consequences of this statement is the following:

$$\mathbf{w}(\mathbf{x},t)=0,\quad p(\mathbf{x},t)=0,\quad \mathbf{x}\in\Omega^{\varepsilon}_{s}.$$

If we put $\mathbf{v}=\dfrac{\partial\mathbf{w}}{\partial t}$, then we may rewrite the last condition and Eqs. (1.0.7)–(1.0.9) in the form

$$\frac{1}{c^2_f}\frac{\partial p}{\partial t}+\nabla\cdot\mathbf{v}=0,\ \mathbf{x}\in\Omega^{\varepsilon}_{f},\ t\in(0,T), \tag{1.1.1}$$

$$\nabla\cdot\mathbb{P}+\rho_f\mathbf{F}=0,\ \mathbf{x}\in\Omega^{\varepsilon}_{f},\ t\in(0,T), \tag{1.1.2}$$

$$\mathbb{P}=\alpha_\mu\mathbb{D}\big(x,\mathbf{v}\big)+\big(\alpha_\nu\nabla\cdot\mathbf{v}-p\big)\mathbb{I}, \tag{1.1.3}$$

$$\mathbf{v}(\mathbf{x},t)=0,\quad \mathbf{x}\in\Omega^{\varepsilon}_{s}\cup S,\ S=\partial\Omega,\ t\in(0,T), \tag{1.1.4}$$

$$p(\mathbf{x},0)=0,\quad \mathbf{x}\in\Omega. \tag{1.1.5}$$

Throughout this section we assume that conditions

$$\mu_0=0,\ 0<\mu_1\leqslant\infty,\ 0<c^2_f<\infty,\ 0\leqslant\nu_0<\infty,$$

and

$$\int_{\Omega_T}|\mathbf{F}|^2dxdt=F^2<\infty$$

hold true.

1.1.1 Statement of the Problem and Main Results

Definition 1.1 We say that the pair of functions $\{\mathbf{v}^{\varepsilon},\ p^{\varepsilon}\}$ such that $\mathbf{v}^{\varepsilon}(\mathbf{x},t)=0$ for $\mathbf{x}\in\Omega^{\varepsilon}_{s}$ and $t>0$, and

$$\mathbf{v}^{\varepsilon}\in\overset{\circ}{\mathbf{W}}{}^{1,0}_2(\Omega_T),\ p^{\varepsilon}\in L_2(\Omega_T),\ \Omega_T=\Omega\times(0,T),$$

is a weak solution of the problem (1.1.1)–(1.1.5), if it satisfies the integral identities

$$\int_{\Omega_T}\chi^{\varepsilon}\Big(\alpha_\mu\mathbb{D}(x,\mathbf{v}^{\varepsilon}):\mathbb{D}(x,\varphi)+\big(\alpha_\nu\nabla\cdot\mathbf{v}^{\varepsilon}-p^{\varepsilon}\big)\nabla\cdot\varphi-\rho_f\mathbf{F}\cdot\varphi\Big)dxdt=0, \tag{1.1.6}$$

$$\int_{\Omega_T}\Big(\nabla\xi\cdot\mathbf{v}^{\varepsilon}+\frac{1}{c^2_f}\frac{\partial\xi}{\partial t}p^{\varepsilon}\Big)dxdt=0, \tag{1.1.7}$$

for any smooth functions φ and ξ, such that φ satisfies condition (1.1.4) and ξ s atisfies condition $\xi(\mathbf{x}, T) = 0$.

In (1.1.6) the convolution $\mathbb{A} : \mathbb{B}$ of two tens ors $\mathbb{A} = (A_{ij})$ and $\mathbb{B} = (B_{ij})$ is defined as $\mathbb{A} : \mathbb{B} = \mathrm{tr}(\mathbb{A} \cdot \mathbb{B}) = \sum\limits_{i,j=1}^{3} A_{ij} B_{ij}$.

Theorem 1.1 *(1) For all $\varepsilon > 0$ and for an arbitrary time interval $[0, T]$ there exists a unique generalized solution of problem (1.1.1)–(1.1.5) and*

$$\int_{\Omega_T} \Big(\alpha_\mu |\nabla \mathbf{v}^\varepsilon|^2 + |\mathbf{v}^\varepsilon|^2 + \alpha_\nu |\nabla \cdot \mathbf{v}^\varepsilon|^2\Big) dx dt + \max_{0<t<T} \int_\Omega |p^{\,\varepsilon}|^2 dx \leqslant \frac{\varepsilon^2}{\alpha_\mu} C_0 F^2, \tag{1.1.8}$$

where the constant C_0 is independent of the small parameter ε.

(2) The nontrivial homogenization procedure for the problem (1.1.1)–(1.1.5) makes sense if and only if the pore space is connected and

$$\mu_0 = 0, \;\; 0 < \mu_1 < \infty. \tag{1.1.9}$$

Under these conditions and condition $\nu_0 > 0$, the sequences $\{\mathbf{v}^\varepsilon\}$, $\{\nabla \cdot \mathbf{v}^\varepsilon\}$, $\{p^{\,\varepsilon}\}$, and $\{q^{\,\varepsilon}\}$, where $q^{\,\varepsilon} = p^{\,\varepsilon} - \alpha_\nu \nabla \cdot \mathbf{v}^\varepsilon$, converge weakly in $\mathbf{L}_2(\Omega_T)$ and $L_2(\Omega_T)$ as $\varepsilon \to 0$ (up to some subsequences) to functions $\mathbf{v}$, $\nabla \cdot \mathbf{v}$, p and $q = p - \nu_0 \nabla \cdot \mathbf{v} \in W_2^{1,0}(\Omega_T)$ respectively and these limiting functions solve the homogenized system of equations, consisting of the continuity equation

$$\frac{m}{c_f^2} \frac{\partial p}{\partial t} + \nabla \cdot \mathbf{v} = 0, \tag{1.1.10}$$

and Darcy's law in the form

$$\mathbf{v} = \frac{1}{\mu_1} \mathbb{B}\Big(- \nabla \Big(p + \frac{\nu_0}{c_f^2} \frac{\partial p}{\partial t}\Big) + \rho_f \mathbf{F}\Big) \tag{1.1.11}$$

in the domain Ω for $t \in (0, T)$.

If $\nu_0 = 0$, then the sequences $\{\mathbf{v}^\varepsilon\}$ and $\{p^{\,\varepsilon}\}$ converge weakly in $\mathbf{L}_2(\Omega_T)$ and $L_2(\Omega_T)$ as $\varepsilon \to 0$ (up to some subsequences) to functions $\mathbf{v}$ and mp correspondingly and these limiting functions solve the homogenized system of equations, consisting of the continuity equation (1.1.10) and usual Darcy's law in the form

$$\mathbf{v} = \frac{1}{\mu_1} \mathbb{B}\big(- \nabla p + \rho_f \mathbf{F}\big) \tag{1.1.12}$$

in the domain Ω for $t \in (0, T)$.

Systems (1.1.10)–(1.1.12) are completed with boundary and initial conditions

$$\mathbf{v}(\mathbf{x},t)\cdot\mathbf{n}(\mathbf{x})=0,\ \ \mathbf{x}\in S,\ t\in(0,T),\tag{1.1.13}$$

$$p(\mathbf{x},0)=0,\ \ \mathbf{x}\in\Omega.\tag{1.1.14}$$

(3) *For a disconnected pore space, or in the case $\mu_1=\infty$, the unique limiting regime is a state of rest.*

(4) *The problems (1.1.10), (1.1.11), (1.1.13), (1.1.14) and (1.1.10), (1.1.12)–(1.1.14) have a unique solution.*
In (1.1.10)–(1.1.13) $m=\int_Y\chi(\mathbf{y})dy$ is the porosity, the symmetric strictly positive definite constant matrix $\mathbb{B}$ is given below by (1.1.27) and **n** *is the normal vector to the boundary S.*

Note, that boundary and initial conditions are understood in a weak sense as a corresponding integral identity for the continuity equation (1.1.10).

We refer to the problem (1.1.10), (1.1.11), (1.1.13), (1.1.14) as the homogenized **model** $(\mathbb{IF})_1$. Actually, this model is the one-parametric family of models, depending on the parameter $\nu_0\geqslant 0$. For $\nu_0=0$ the model $(\mathbb{IF})_1$ is described by the problem (1.1.12)–(1.1.14). So, we may just formally put this value $\nu_0=0$ in corresponding equations. In what follows we will no longer point out this fact, having in mind the above mentioned procedure.

Remark The limit as $c_f\to\infty$ results in the usual Darcy system of filtration for an incompressible liquid.

1.1.2 Proof of Theorem 1.1

1.1.2.1 Basic a Priori Estimates

The proof of existence and uniqueness results for the problem (1.1.1)–(1.1.5) is standard.

Due to the regularity of $\mathbf{v}^\varepsilon$ the function p^ε possesses a time derivative $\dfrac{\partial p^\varepsilon}{\partial t}\in L_2(\Omega_T)$ and satisfies the continuity equation (1.1.1) in the usual sense.

Let $h(\tau)=1$ for $0<\tau<t$ and $h(\tau)=0$ for $t<\tau<T$. To prove the estimate (1.1.8) we consider the integral identity (1.1.6) with the test function $\varphi(\mathbf{x},\tau)=h(\tau)\mathbf{v}^\varepsilon(\mathbf{x},\tau)$ and use the continuity equation (1.1.1):

$$\int_{\Omega_t}\chi^\varepsilon\Big(\alpha_\mu|\mathbb{D}(x,\mathbf{v}^\varepsilon)|^2+\alpha_\nu|\nabla\cdot\mathbf{v}^\varepsilon|^2\Big)dxd\tau+\frac{1}{2c_f^2}\int_\Omega|p^\varepsilon(\mathbf{x},t)|^2dx$$
$$=\int_{\Omega_t}\chi^\varepsilon\rho_f\mathbf{F}\cdot\mathbf{v}^\varepsilon dxd\tau\leqslant\frac{C_0}{\delta}F^2+\delta\int_{\Omega_t}\chi^\varepsilon|\mathbf{v}^\varepsilon|^2dxd\tau.\tag{1.1.15}$$

As a next step we estimate

$$I^{\varepsilon}=\int_{\Omega}\chi^{\varepsilon}|\mathbf{v}^{\varepsilon}|^2dx.$$

Let $G^{(\mathbf{k})}$, where $\mathbf{k}=(k_1,k_2,k_3)\in\mathbb{Z}^3$, be the intersection of Ω_f^{ε} with a set $\{\mathbf{x}:\mathbf{x}=\varepsilon(\mathbf{y}+\mathbf{k}),\ \mathbf{y}\in Y\}$. Then

$$\Omega_f^{\varepsilon}=\bigcup_{\mathbf{k}\in\mathbb{Z}^3}G^{(\mathbf{k})}\quad\text{and}\quad I^{\varepsilon}=\sum_{\mathbf{k}\in\mathbb{Z}^3}I^{\varepsilon}(\mathbf{k}),\quad I^{\varepsilon}(\mathbf{k})=\int_{G^{(\mathbf{k})}}|\mathbf{v}^{\varepsilon}|^2dx.$$

In each integral $I^{\varepsilon}(\mathbf{k})$ we change variables by $\mathbf{x}=\varepsilon\mathbf{y}$, then apply the Friedrichs–Poincaré inequality and return to the original variables:

$$\int_{G^{(\mathbf{k})}}|\mathbf{v}^{\varepsilon}|^2dx=\varepsilon^3\int_{Y^{(\mathbf{k})}}|\overline{\mathbf{v}}^{\varepsilon}|^2dy\leqslant\varepsilon^3C_0\int_{Y^{(\mathbf{k})}}|\nabla_y\overline{\mathbf{v}}^{\varepsilon}|^2dy=\varepsilon^2C_0\int_{G^{(\mathbf{k})}}|\nabla\mathbf{v}^{\varepsilon}|^2dx.$$

Finally, we arrive at the chain of inequalities

$$\begin{aligned}\int_{\Omega}\chi^{\varepsilon}|\mathbf{v}^{\varepsilon}|^2dx&\leqslant\varepsilon^2C_0\int_{\Omega}|\nabla\mathbf{v}^{\varepsilon}|^2dx\leqslant\varepsilon^2C_0\int_{\Omega}|\mathbb{D}(x,\mathbf{v}^{\varepsilon})|^2dx\\&\leqslant\frac{\varepsilon^2}{\alpha_\mu}C_0\alpha_\mu\int_{\Omega}|\mathbb{D}(x,\mathbf{v}^{\varepsilon})|^2dx=\frac{\varepsilon^2}{\alpha_\mu}C_0\alpha_\mu\int_{\Omega}\chi^{\varepsilon}|\mathbb{D}(x,\mathbf{v}^{\varepsilon})|^2dx,\end{aligned}$$

where we have used Korn's inequality (see Appendix B). The above relations and (1.1.15) for

$$\delta=\frac{\varepsilon^2}{2\alpha_\mu}C_0$$

provide (1.1.8). For the case $\mu_1=\infty$ this estimate implies

$$\mathbf{v}^{\varepsilon}\to0,\ \ p^{\varepsilon}\to0,\ \ as\ \varepsilon\to0$$

strongly in $L_2(\Omega_T)$.

1.1.3 Homogenization

Let condition (1.1.9) hold and $\nu_0>0$. Then the sequences $\{\mathbf{v}^{\varepsilon}\}$, $\{\varepsilon\mathbb{D}(x,\mathbf{v}^{\varepsilon})\}$, $\{\nabla\cdot\mathbf{v}^{\varepsilon}\}$, $\left\{\frac{\partial p^{\varepsilon}}{\partial t}\right\}$, $\{p^{\varepsilon}\}$ and $\{q^{\varepsilon}\}$, where

$$q^{\varepsilon}=-\alpha_\nu\nabla\cdot\mathbf{v}^{\varepsilon}+p^{\varepsilon}=\frac{\alpha_\nu}{c_f^2}\frac{\partial p^{\varepsilon}}{\partial t}+p^{\varepsilon},$$

are bounded in $\mathbf{L}_2(\Omega_T)$ and $L_2(\Omega_T)$. Therefore, we may extract some subsequences (for simplicity we keep the same notation) such that the sequences $\{\mathbf{v}^\varepsilon\}$, $\{\nabla\cdot\mathbf{v}^\varepsilon\}$, $\left\{\dfrac{\partial p^\varepsilon}{\partial t}\right\}$, $\{p^\varepsilon\}$ and $\{q^\varepsilon\}$ converge weakly in $\mathbf{L}_2(\Omega_T)$ and $L_2(\Omega_T)$ to functions $\mathbf{v}$, $\nabla\cdot\mathbf{v}$, $\dfrac{\partial p}{\partial t}$, p, and q respectively and

$$q = p + \frac{\nu_0}{c_f^2}\frac{\partial p}{\partial t}. \tag{1.1.16}$$

The limiting functions $\mathbf{v}$ and p obviously satisfy the continuity equation (1.1.10) and boundary and initial conditions (1.1.13) and (1.1.14).

At the same time by Nguetseng's theorem (see Appendix B) sequences $\{\mathbf{v}^\varepsilon\}$, $\{\varepsilon\mathbb{D}(x,\mathbf{v}^\varepsilon)\}$, $\{\nabla\cdot\mathbf{v}^\varepsilon\}$, $\{p^\varepsilon\}$, and $\{q^\varepsilon\}$ converge two-scale in $\mathbf{L}_2(\Omega_T)$ and $L_2(\Omega_T)$ to the functions

$$\mathbf{V}(\mathbf{x},t,\mathbf{y}),\ \mathbb{D}\big(y,\mathbf{V}(\mathbf{x},t,\mathbf{y})\big),\ \frac{1}{c_f^2}\frac{\partial P}{\partial t}(\mathbf{x},t,\mathbf{y}),\ P(\mathbf{x},t,\mathbf{y}),\ \text{and } Q(\mathbf{x},t,\mathbf{y})$$

correspondingly,

$$Q = P + \frac{\nu_0}{c_f^2}\frac{\partial P}{\partial t}, \tag{1.1.17}$$

and

$$Q,\ P \in L_2\big(\Omega_T\times Y\big),\ \mathbf{V},\ \mathbb{D}(y,\mathbf{V}) \in \mathbf{L}_2\big(\Omega_T\times Y\big). \tag{1.1.18}$$

Finally, the two-scale limit in the integral identity (1.1.7) with test functions $\xi = \xi_0(\mathbf{x},t)\xi_1\left(\dfrac{\mathbf{x}}{\varepsilon}\right)$ gives us the microscopic continuity equation

$$\nabla_y\cdot\mathbf{V} = 0,\ \ \mathbf{y}\in Y. \tag{1.1.19}$$

Lemma 1.1 *The following equalities hold true*

$$P(\mathbf{x},t,\mathbf{y}) = \chi(\mathbf{y})\,p(\mathbf{x},t),\ \ Q(\mathbf{x},t,\mathbf{y}) = \chi(\mathbf{y})\,q(\mathbf{x},t). \tag{1.1.20}$$

Proof The passage to the limit as $\varepsilon\to 0$ in (1.1.6) with test functions $\varphi = \varepsilon\,\varphi_0(\mathbf{x},t)\varphi_1\left(\dfrac{\mathbf{x}}{\varepsilon}\right)$, where $\varphi_1(\mathbf{y})$ is finite in Y_f, yields

$$\nabla_y Q(\mathbf{x},t,\mathbf{y}) = 0,\ \ \mathbf{y}\in Y_f. \tag{1.1.21}$$

Therefore

$$Q(\mathbf{x},t,\mathbf{y}) = \chi(\mathbf{y})\,q(\mathbf{x},t).$$

Equations (1.1.17) and (1.1.21) give the result

$$\nabla_y P(\mathbf{x},t,\mathbf{y}) = 0, \ \ \mathbf{y} \in Y_f,$$

and

$$P(\mathbf{x},t,\mathbf{y}) = \chi(\mathbf{y})\, p(\mathbf{x},t).$$

Now we are ready to derive equation (1.1.11). To do that we choose the test function in the integral identity (1.1.6) as $\varphi = \varphi_0(\mathbf{x},t)\varphi_1\left(\frac{\mathbf{x}}{\varepsilon}\right)$, where the 1-periodic function $\varphi_1(\mathbf{y})$ is divergence free and finite in Y_f:

$$I_1^\varepsilon + I_2^\varepsilon + I_3^\varepsilon + I_4^\varepsilon = 0,$$

where

$$I_1^\varepsilon = \frac{\alpha_\mu}{\varepsilon^2}\int_{\Omega_T} \varphi_0\left(\varepsilon\mathbb{D}(x,\mathbf{v}^\varepsilon)\right) : \mathbb{D}(y,\varphi_1)dxdt,$$

$$I_2^\varepsilon = \sqrt{\alpha_\mu}\int_{\Omega_T}\left(\sqrt{\alpha_\mu}\,\mathbb{D}(x,\mathbf{v}^\varepsilon)\right) : \left(\nabla\varphi_0\otimes\varphi_1 + \varphi_1\otimes\nabla\varphi_0\right)dxdt,$$

$$I_3^\varepsilon = -\int_{\Omega_T} q^\varepsilon\left(\varphi_1\cdot\nabla\varphi_0\right)dxdt,$$

$$I_4^\varepsilon = -\int_{\Omega_T}\rho_f\varphi_0\left(\mathbf{F}\cdot\varphi_1\right)dxdt.$$

Passage to the limit as $\varepsilon \to 0$ gives us

$$I_1^\varepsilon \to \int_{\Omega_T}\varphi_0\Big(\int_{Y_f}\mu_1\mathbb{D}(y,\mathbf{V}) : \mathbb{D}(y,\varphi_1)dy\Big)dxdt, \ \ I_2^\varepsilon \to 0,$$

$$I_3^\varepsilon \to -\int_{\Omega_T}\nabla\varphi_0\cdot\Big(\int_{Y_f} q\,\chi(\mathbf{y})\varphi_1(\mathbf{y})dy\Big)dxdt,$$

$$I_4^\varepsilon \to -\int_{\Omega_T}\varphi_0\Big(\int_{Y_f}\rho_f\left(\mathbf{F}\cdot\varphi_1(\mathbf{y})\right)dy\Big)dxdt.$$

Thus,

$$\int_{\Omega_T}\Big(\varphi_0(\mathbf{x},t)\int_{Y_f}\big(\mu_1\mathbb{D}(y,\mathbf{V}):\mathbb{D}(y,\varphi_1)-\rho_f\mathbf{F}\cdot\varphi_1(\mathbf{y})\big)dy$$

$$-\nabla\varphi_0(\mathbf{x},t)\cdot\Big(\int_{Y_f}\varphi_1(\mathbf{y})\chi(\mathbf{y})dy\Big)\,q\Big)dxdt$$

$$=\int_{\Omega_T}\big(\varphi_0(\mathbf{x},t)a(\mathbf{x},t)+q\,\nabla\varphi_0(\mathbf{x},t)\cdot\mathbf{b}\big)dxd=0, \quad (1.1.22)$$

where

$$a(\mathbf{x},t)=\int_{Y_f}\big(\mu_1\mathbb{D}(y,\mathbf{V}):\mathbb{D}(y,\varphi_1)-\rho_f\mathbf{F}\cdot\varphi_1(\mathbf{y})\big)dy,$$

$$\mathbf{b}=-\int_{Y_f}\varphi_1(\mathbf{y})dy=const.$$

Due to Lemma B. 15 (see Appendix B) we first may choose φ_1 such that $\mathbf{b}=\mathbf{e}_i,\ i=1,2,3$, where $\{\mathbf{e}_1,\mathbf{e}_2,\mathbf{e}_3\}$ is a standard Cartesian basis. Nguetseng's Theorem guarantees that $a\in L_2(\Omega_T)$. Therefore,

$$\nabla q\in\mathbf{L}_2(\Omega_T).$$

Next we reintegrate (1.1.22) with respect the variables $(\mathbf{x},t)$ and arrive at the microscopic equation

$$\frac{\mu_1}{2}\,\triangle_y\mathbf{V}-\nabla_y Q-\nabla q+\rho_f\mathbf{F}=0 \quad (1.1.23)$$

in the domain Y_f, which is understood in the sense of distributions. Here we have used the equality

$$\nabla\cdot\mathbb{D}(y,\mathbf{V})=\frac{1}{2}\triangle\mathbf{V}+\frac{1}{2}\nabla\big(\nabla\cdot\mathbf{V}\big),$$

and the continuity equation (1.1.19).

The term $\nabla_y\Pi(\mathbf{x},t,\mathbf{y})$ in (1.1.23) appears due to the orthogonality in $\mathbf{L}_2(Y_f)$ of the set of all divergence free vectors φ_1 to the set of gradients $\nabla_y\Pi$ of scalar functions Π.

The two-scale limit in the equality

$$(1-\chi^{\varepsilon})\mathbf{v}^{\varepsilon}=0$$

gives us

$$\big(1-\chi(\mathbf{y})\big)\mathbf{V}(\mathbf{x},t,\mathbf{y})=0,\ \text{ or }\mathbf{V}(\mathbf{x},t,\mathbf{y})=0,\ \ \mathbf{y}\in Y_s.$$

The last condition and the regularity condition (1.1.18) result in the boundary condition

$$\mathbf{V}(\mathbf{x}, t, \mathbf{y}) = 0, \;\; \mathbf{y} \in \gamma = \partial Y_s \cap \partial Y_s. \tag{1.1.24}$$

Let $\mathbf{e}_i$, $i = 1, 2, 3$ be the usual Cartesian basis in $\mathbb{R}^3$ and

$$\frac{2}{\mu_1}\Big(- \nabla q + \rho_f \mathbf{F}\Big) = \sum_{i=1}^{3} z_i(\mathbf{x}, t)\mathbf{e}_i.$$

Then the solution $\mathbf{V}$ of the problem (1.1.19), (1.1.23), and (1.1.24) has a form

$$\mathbf{V} = \sum_{i=1}^{3} z_i \mathbf{V}^{(i)}(\mathbf{y}) = \frac{2}{\mu_1}\Big(\sum_{i=1}^{3} \mathbf{V}^{(i)} \otimes \mathbf{e}_i\Big) \cdot \big(- \nabla q + \rho_f \mathbf{F}\big), \tag{1.1.25}$$

where the 1-periodic function $\mathbf{V}^{(i)}(\mathbf{y})$ solves the periodic boundary value problem

$$\left.\begin{array}{l} \triangle_y \mathbf{V}^{(i)} - \nabla Q^{(i)} = -\mathbf{e}_i, \quad \mathbf{y} \in Y_f, \\ \nabla \cdot \mathbf{V}^{(i)} = 0, \quad \mathbf{y} \in Y_f, \\ \mathbf{V}^{(i)} = 0, \quad \mathbf{y} \in \gamma. \end{array}\right\} \tag{1.1.26}$$

The existence and uniqueness results for the problem (1.1.26) and the properties of the matrix

$$\mathbb{B} = 2\sum_{i=1}^{3}\Big(\int_{Y_f} \mathbf{V}^{(i)}(\mathbf{y})dy\Big) \otimes \mathbf{e}_i = 2\sum_{i=1}^{3}\langle \mathbf{V}^{(i)}\rangle_{Y_f} \otimes \mathbf{e}_i, \tag{1.1.27}$$

follow from the energy equality

$$\int_{Y_f} \nabla \mathbf{V}^{(i)} : \nabla \mathbf{V}^{(j)} dy = \int_{Y_f} \mathbf{e}_i \cdot \mathbf{V}^{(j)} dy \tag{1.1.28}$$

In fact, applying in (1.1.28) for $i = j$ Hölder's and Friedrichs–Poincaré's inequalities we arrive at

$$\int_{Y_f} |\mathbf{V}^{(i)}|^2 dy \leqslant C_{Y_f}^2 \int_{Y_f} |\nabla \mathbf{V}^{(i)}|^2 dy, \; i = 1, 2, 3,$$

$$\int_{Y_f} |\nabla \mathbf{V}^{(i)}|^2 dy \leqslant m\, C_{Y_f}^2, \; i = 1, 2, 3.$$

For a disconnected pore space the unique solution of the problem (1.1.26) is

$$\mathbf{V}^{(i)} = 0, \ \ Q^{(i)} = y_i, \ \text{ and } \mathbb{B} = 0.$$

Lemma 1.2 *The matrix $\mathbb{B}$ is strictly positively definite.*

Proof Let the pore space be connected and

$$\zeta = (\zeta_1, \zeta_2, \zeta_3), \quad \eta = (\eta_1, \eta_2, \eta_3) \in \mathbb{R}^3,$$

$$\mathbf{z}_\zeta = \sum_{i=1}^{3} \zeta_i \mathbf{V}^{(i)}, \ \ \mathbf{z}_\eta = \sum_{i=1}^{3} \eta_i \mathbf{V}^{(i)}.$$

Then (1.1.27) and (1.1.28) give us

$$\frac{1}{2}(\mathbb{B} \cdot \zeta) \cdot \eta = \langle \mathbf{z}_\zeta \rangle_{Y_f} \cdot \eta, \quad \langle \mathbf{z}_\zeta \rangle_{Y_f} \cdot \eta = \langle \nabla \mathbf{z}_\zeta : \nabla \mathbf{z}_\eta \rangle_{Y_f},$$

or

$$\frac{1}{2}(\mathbb{B} \cdot \zeta) \cdot \eta = \langle \nabla \mathbf{z}_\zeta : \nabla \mathbf{z}_\eta \rangle_{Y_f}, \ \textit{and} \ \frac{1}{2}(\mathbb{B} \cdot \zeta) \cdot \zeta = \langle \nabla \mathbf{z}_\zeta : \nabla \mathbf{z}_\zeta \rangle_{Y_f} > \alpha$$

for some constant $\alpha > 0$ and any vector ζ with $|\zeta| = 1$.

In fact, otherwise there exists some vector ζ with $|\zeta| = 1$, such that

$$\nabla \mathbf{z}_\zeta = 0, \ \textit{or} \ \mathbf{z}_\zeta = \mathbb{A} \cdot \mathbf{y} + \zeta_0$$

with some constant matrix $\mathbb{A}$ and some constant vector ζ_0. But the function $\mathbf{z}_\zeta$ is a periodic solution of the problem

$$\left. \begin{array}{l} \triangle_y \mathbf{z}_\zeta - \nabla Q_\zeta = -\zeta, \quad \mathbf{y} \in Y_f, \\ \nabla \cdot \mathbf{z}_\zeta = 0, \quad \mathbf{y} \in Y_f, \\ \mathbf{z}_\zeta = 0, \quad \mathbf{y} \in \gamma. \end{array} \right\}$$

For a connected pore space any periodic linear functions can only be constant. Recalling the homogeneous boundary condition on γ for $\mathbf{z}_\zeta$ we conclude that $\mathbf{z}_\zeta = 0$. The last relation and the differential equation for $\mathbf{z}_\zeta$ result in the equality $\nabla Q_\zeta = \zeta$. Hence $Q_\zeta = \zeta \cdot \mathbf{y} + const$ and by the same arguments $Q_\zeta = const$ and $\zeta = 0$ which is impossible by supposition.

For the case $\nu_0 = 0$ the sequences $\{\mathbf{v}^\varepsilon\}$ and $\{p^\varepsilon\}$ converge weakly and two-scale in $\mathbf{L}_2(\Omega_T)$ and $L_2(\Omega_T)$ to functions $\mathbf{v}(\mathbf{x}, t)$, $\mathbf{V}(\mathbf{x}, t, \mathbf{y})$ and $p(\mathbf{x}, t)$, $\left(\frac{1}{m}\right) \chi(\mathbf{y})\, p(\mathbf{x}, t)$ correspondingly. The limiting functions satisfy (1.1.17) and (1.1.10) with boundary and initial conditions (1.1.13) and (1.1.14) in a weak sense. That is, they are solution of the integral identity

$$\int_{\Omega_T}\Big(\nabla\xi\cdot+\frac{1}{c_f^2}\frac{\partial\xi}{\partial t}\,p\Big)dxdt=0, \tag{1.1.29}$$

for any smooth functions ξ, such that $\xi(\mathbf{x},T)=0$.

Note, that the sequence $\{q^{\varepsilon}\}$ converges weakly in $L_2(\Omega_T)$ to the function $p(\mathbf{x},t)$ and two-scale in $L_2(\Omega_T)$ to the function $\chi(\mathbf{y})\,p(\mathbf{x},t)$. Repeating once more we arrive at (1.1.12).

Lemma 1.3 *The problems (1.1.10), (1.1.11), (1.1.13), (1.1.14) and (1.1.10), (1.1.12)–(1.1.14) have a unique solution.*

Proof The uniqueness of the solution to the problem (1.1.10), (1.1.11), (1.1.13), (1.1.14) follows from its linearity and the energy identity for the homogeneous problem in the form

$$\frac{d}{dt}\int_{\Omega}\Big(m\,\frac{\mu_1}{2c_f^2}\,p^2+\frac{\nu_0}{2c_f^2}\nabla p\cdot\big(\mathbb{B}\cdot\nabla p\big)\Big)dx+\int_{\Omega}\nabla p\cdot\big(\mathbb{B}\cdot\nabla p\big)dx=0. \tag{1.1.30}$$

The identity (1.1.30) is the result of a formal integration by parts over domain Ω of the equation

$$m\,\frac{\mu_1}{c_f^2}\frac{\partial p}{\partial t}=\nabla\cdot\mathbb{B}\Big(\nabla p+\frac{\nu_0}{c_f^2}\nabla\big(\frac{\partial p}{\partial t}\big)\Big)$$

after its multiplication by p. The last equation is an obvious combination of Eqs. (1.1.10) and (1.1.11).

For the problem (1.1.10), (1.1.12)–(1.1.14) we use the identity (1.1.30) with $\nu_0=0$.

In particular, the uniqueness of the limiting problems shows that any subsequences of sequences $\{\mathbf{v}^{\varepsilon}\}$, $\{q^{\varepsilon}\}$ and $\{p^{\varepsilon}\}$ converge to the same limit. Therefore each entire sequence $\{\mathbf{v}^{\varepsilon}\}$, $\{q^{\varepsilon}\}$ and $\{p^{\varepsilon}\}$ converges to a unique limit.

1.2 An Incompressible Slightly Viscous Liquid in an Incompressible Elastic Skeleton

In this section as a basic mathematical model at the microscopic level we consider the model $\mathbb{M}_{15}$ of the filtration of an incompressible liquid in an incompressible elastic solid skeleton

$$\nabla\cdot\mathbf{w}=0,\;\;\mathbf{x}\in\Omega,\;\;t\in(0,T), \tag{1.2.1}$$

$$\nabla \cdot \mathbb{P} + \rho^{\varepsilon}\mathbf{F} = 0, \ \mathbf{x} \in \Omega, \ t \in (0, T), \tag{1.2.2}$$

$$\mathbb{P} = \chi^{\varepsilon}\alpha_{\mu}\mathbb{D}\Big(x, \frac{\partial \mathbf{w}}{\partial t}\Big) + (1 - \chi^{\varepsilon})\lambda_0\mathbb{D}(x, \mathbf{w}) - p\,\mathbb{I}, \tag{1.2.3}$$

$$\mathbf{w}(\mathbf{x}, t) = 0, \ \mathbf{x} \in S, \ t \in (0, T), \tag{1.2.4}$$

$$\int_{\Omega} p(\mathbf{x}, t)dx = 0, \ t \in (0, T), \ \chi^{\varepsilon}\,\mathbf{w}(\mathbf{x}, 0) = 0, \ \mathbf{x} \in \Omega, \tag{1.2.5}$$

where

$$\rho^{\varepsilon} = \rho_f\,\chi^{\varepsilon} + \rho_s\,(1 - \chi^{\varepsilon}).$$

Throughout this section we additionally impose

Assumption 1.1 The solid skeleton Ω_s^{ε} is a connected domain.

We also assume that conditions

$$\mu_0 = 0, \ 0 < \mu_1 \leqslant \infty, \ 0 < \lambda_0 < \infty, \tag{1.2.6}$$

and

$$\int_{\Omega_T} |\mathbf{F}|^2 dxdt = F^2 < \infty \tag{1.2.7}$$

hold true.

In (1.2.3)–(1.2.6)

$$\mu_0 = \lim_{\varepsilon\searrow 0} \alpha_{\mu}(\varepsilon), \quad \mu_1 = \lim_{\varepsilon\searrow 0} \frac{\alpha_{\mu}}{\varepsilon^2}.$$

1.2.1 Statement of the Problem and Main Results

Definition 1.2 We say that the pair of functions $\{\mathbf{w}^{\varepsilon}, \ p^{\varepsilon}\}$ such that

$$\mathbf{w}^{\varepsilon} \in \overset{\circ}{\mathbf{W}}{}_2^{1,0}(\Omega_T), \ \chi^{\varepsilon}\nabla\frac{\partial \mathbf{w}^{\varepsilon}}{\partial t} \in \mathbf{L}_2(\Omega_T), \ p^{\varepsilon} \in L_2(\Omega_T),$$

is a weak solution of the problem (1.2.1)–(1.2.5), if it satisfies the continuity equation (1.2.1) almost everywhere in Ω_T, the initial and normalization conditions (1.2.5) and an integral identity

$$\int_{\Omega_T} \Big(\chi^{\varepsilon}\alpha_{\mu}\mathbb{D}\Big(x, \frac{\partial \mathbf{w}^{\varepsilon}}{\partial t}\Big) + (1 - \chi^{\varepsilon})\lambda_0\mathbb{D}(x, \mathbf{w}^{\varepsilon}) - p^{\varepsilon}\mathbb{I}\Big) : \mathbb{D}(x, \varphi)dxdt$$
$$= \int_{\Omega_T} \rho^{\varepsilon}\mathbf{F} \cdot \varphi\, dxdt, \tag{1.2.8}$$

for any functions $\varphi \in \overset{\circ}{\mathbf{W}}{}_2^{1,0}(\Omega_T)$

The solution of this problem possesses different smoothness in domains Ω_f^ε and Ω_s^ε. To preserve the best properties—which the solution possesses in the solid part—we extend the function $\mathbf{w}^\varepsilon$ from the solid part Ω_s^ε of the domain Ω onto the whole domain Ω. To do this we use the extension result (see Lemma B.4.2, Appendix B):

there exists an extension

$$\mathbf{w}_s^\varepsilon = \mathbb{E}_{\Omega_s^\varepsilon}\big(\mathbf{w}^\varepsilon\big), \quad \mathbb{E}_{\Omega_s^\varepsilon} : \mathbf{W}_2^1(\Omega_s^\varepsilon) \to \mathbf{W}_2^1(\Omega), \tag{1.2.9}$$

such that

$$\big(1 - \chi^\varepsilon(\mathbf{x})\big)\big(\mathbf{w}^\varepsilon(\mathbf{x}, t) - \mathbf{w}_s^\varepsilon(\mathbf{x}, t)\big) = 0, \quad \mathbf{x} \in \Omega, \ t \in (0, T), \tag{1.2.10}$$

and

$$\int_\Omega |\mathbf{w}_s^\varepsilon(\mathbf{x}, t)|^2 dx \leqslant C_0 \int_{\Omega_s^\varepsilon} |\mathbf{w}^\varepsilon(\mathbf{x}, t)|^2 dx,$$
$$\int_\Omega \big|\mathbb{D}\big(x, \mathbf{w}_s^\varepsilon(\mathbf{x}, t)\big)\big|^2 dx \leqslant C_0 \int_{\Omega_s^\varepsilon} \big|\mathbb{D}\big(x, \mathbf{w}^\varepsilon(\mathbf{x}, t)\big)\big|^2 dx, \quad t \in (0, T), \tag{1.2.11}$$

where C_0 is independent of ε and $t \in (0, T)$.

Theorem 1.2 *There exists a unique weak solution $\{\mathbf{w}^\varepsilon, \ p^\varepsilon\}$ of the problem (1.2.1)–(1.2.5) and*

$$\int_{\Omega_T} \Big(|\mathbf{w}_s^\varepsilon|^2 + |\mathbb{D}(x, \mathbf{w}_s^\varepsilon)|^2 + |\pi^\varepsilon|^2\Big) dx dt \leqslant C_0 F^2,$$
$$\max_{0<t<T} \varepsilon^2 \int_\Omega |\mathbb{D}(x, \mathbf{w}^\varepsilon)|^2 dx \leqslant \frac{\varepsilon^2}{\alpha_\mu} C_0 F^2, \tag{1.2.12}$$

$$\int_{\Omega_T} |\mathbf{w}^\varepsilon(\mathbf{x}, t) - \mathbf{w}_s^\varepsilon(\mathbf{x}, t)|^2 dx dt \leqslant \frac{\varepsilon^2}{\alpha_\mu} C_0 F^2, \tag{1.2.13}$$

where

$$\pi^\varepsilon(\mathbf{x}, t) = \int_0^t p^\varepsilon(\mathbf{x}, \tau) d\tau,$$

$\mathbf{w}_s^\varepsilon$ is the extension (1.2.9) and the constant C_0 is independent of the small parameter ε.

Theorem 1.3 *Let $\{\mathbf{w}^\varepsilon, \ p^\varepsilon\}$ be the weak solution of the problem (1.2.1)–(1.2.5), $\mathbf{w}_s^\varepsilon$ be an extension (1.2.9),*

$$\pi^\varepsilon(\mathbf{x}, t) = \int_0^t p^\varepsilon(\mathbf{x}, \tau) d\tau,$$

and $\mu_1 = \infty$, or $\mu_1 < \infty$, but the pore space be disconnected.

Then

(1) *up to some subsequences sequences* $\{\mathbf{w}^{\varepsilon}\}$ *and* $\{\pi^{\varepsilon}\}$ *converge weakly in* $\mathbf{L}_2(\Omega_T)$ *and* $L_2(\Omega_T)$ *to the functions* $\mathbf{w} \in \mathbf{L}_2(\Omega_T)$ *and* $\pi \in W_2^{1,0}(\Omega_T)$ *respectively, and the sequence* $\{\mathbf{w}_s^{\varepsilon}\}$ *converges weakly in* $\mathbf{W}_2^{1,0}(\Omega_T)$ *to the function* $\mathbf{w}_s = \mathbf{w} \in \overset{\circ}{\mathbf{W}}{}_2^{1,0}(\Omega_T)$;
(2) *the limiting functions solve the system of homogenized equations in the domain* Ω_T, *consisting of the continuity equation*

$$\nabla \cdot \mathbf{w}_s = 0, \tag{1.2.14}$$

and the homogenized momentum balance equation

$$\nabla \cdot \left(\lambda_0 \mathfrak{N}^s : \mathbb{D}(x, \mathbf{w}_s) - p\,\mathbb{I}\right) + \hat{\rho}\mathbf{F} = 0, \tag{1.2.15}$$

completed with homogeneous normalization and boundary conditions

$$\int_{\Omega} p(\mathbf{x}, t)dx = 0, \;\; \mathbf{w}_s(\mathbf{x}, t) = 0, \;\; \mathbf{x} \in S, \;\; t \in (0, T); \tag{1.2.16}$$

(3) *the problem (1.2.14)– (1.2.16) has a unique solution.*

In (1.2.15)

$$p = \frac{\partial \pi}{\partial t}, \quad \hat{\rho} = m\,\rho_f + (1 - m)\,\rho_s, \quad m = \int_Y \chi(\mathbf{y})dy,$$

and the symmetric strictly positively definite constant fourth-rank tensor $\mathfrak{N}^s$ *is given below by (1.2.35).*

We refer to the problem (1.2.15), (1.2.16) as the homogenized **model** $(\mathbb{IF})_2$.

Theorem 1.4 *Let* $\{\mathbf{w}^{\varepsilon},\ p^{\varepsilon}\}$ *be the weak solution of the problem (1.2.1)–(1.2.5),* $\mathbf{w}_s^{\varepsilon}$ *be an extension (1.2.9),*

$$\pi^{\varepsilon}(\mathbf{x}, t) = \int_0^t p^{\varepsilon}(\mathbf{x}, \tau)d\tau,$$

the pore space be connected, and $\mu_1 < \infty$.
Then

(1) *up to some subsequences sequences* $\{\chi^{\varepsilon}\mathbf{w}^{\varepsilon}\}$ *and* $\{\chi^{\varepsilon}\pi^{\varepsilon}\}$ *converge weakly in* $\mathbf{L}_2(\Omega_T)$ *and* $L_2(\Omega_T)$ *to functions* $\mathbf{w}^{(f)} \in \mathbf{L}_2(\Omega_T)$ *and* $m\,\pi_f \in W_2^{1,0}(\Omega_T)$ *respectively, and the sequence* $\{\mathbf{w}_s^{\varepsilon}\}$ *converges weakly in* $\mathbf{W}_2^{1,0}(\Omega_T)$ *to the function* $\mathbf{w}_s \in \overset{\circ}{\mathbf{W}}{}_2^{1,0}(\Omega_T)$;
(2) *limiting functions solve the system of homogenized equations in the domain* Ω_T, *consisting of the continuity equation*

$$\nabla \cdot \mathbf{w}^{(f)} + (1 - m)\,\nabla \cdot \mathbf{w}_s = 0, \tag{1.2.17}$$

the homogenized momentum balance equation

$$\nabla \cdot \big(\lambda_0\, \mathfrak{N}_1^s : \mathbb{D}(x, \mathbf{w}_s) - p_f\, \mathbb{I}\big) + \hat{\rho}\mathbf{F} = 0 \tag{1.2.18}$$

for the solid component, and Darcy's law in the form

$$\mathbf{w}^{(f)} = m\,\mathbf{w}_s + \frac{1}{\mu_1}\mathbb{B} \cdot \big(-\nabla \pi_f + \rho_f \int_0^t \mathbf{F}(\mathbf{x}, \tau)d\tau\big), \tag{1.2.19}$$

for the liquid component, completed with homogeneous boundary conditions (1.2.16) for the solid component and homogeneous normalization and boundary conditions

$$\int_\Omega \pi_f(\mathbf{x}, t)dx = 0,\ \ \mathbf{w}^{(f)}(\mathbf{x}, t) \cdot \mathbf{n}(\mathbf{x}) = 0,\ \ \mathbf{x} \in S,\ \ t \in (0, T) \tag{1.2.20}$$

for the liquid pressure p_f and displacements $\mathbf{w}^{(f)}$ of the fluid component;

(3) the problem (1.2.16)–(1.2.20) has a unique solution.
In (1.2.18)–(1.2.20)

$$p_f = \frac{\partial \pi_f}{\partial t} \in L_2(\Omega_T),\ \ \hat{\rho} = m\,\rho_f + (1 - m)\,\rho_s,\ \ m = \int_Y \chi(\mathbf{y})dy,$$

n *is the normal vector to the boundary S, the symmetric strictly positive definite constant matrix* $\mathbb{B}$ *is given by (1.1.27) (see Theorem 1.1), and the symmetric strictly positively definite constant fourth-rank tensor* $\mathfrak{N}_1^s$ *is given below by (1.2.38).*

We refer to the problem (1.2.16)–(1.2.20) as the homogenized **model** $(\mathbb{IF})_3$.

1.2.2 Proof of Theorem 1.2

Setting in (1.2.8) $\varphi(\mathbf{x}, \tau) = h(\tau)\mathbf{w}^\varepsilon(\mathbf{x}, \tau)$, where $h(\tau) = 1$ for $0 < \tau < t$ and $h(\tau) = 0$ for $t < \tau < T$, in the same way as in the previous section, we arrive at

$$\int_{\Omega} |\mathbf{w}^{\varepsilon}(\mathbf{x},t) - \mathbf{w}_s^{\varepsilon}(\mathbf{x},t)|^2 dx \leqslant \frac{\varepsilon^2}{\alpha_\mu} C_0 \int_{\Omega_f^{\varepsilon}} \alpha_\mu \big|\mathbb{D}\big(x, \mathbf{w}^{\varepsilon}(\mathbf{x},t)\big) - \mathbb{D}\big(x, \mathbf{w}_s^{\varepsilon})\big)\big|^2 dx,$$

$$\alpha_\mu \int_{\Omega_f^{\varepsilon}} \big|\mathbb{D}\big(x, \mathbf{w}^{\varepsilon}(\mathbf{x},t)\big)\big|^2 dx + \lambda_0 \int_0^t \int_{\Omega_s^{\varepsilon}} \big|\mathbb{D}\big(x, \mathbf{w}^{\varepsilon}(\mathbf{x},\tau)\big)\big|^2 dx d\tau$$
$$\leqslant \frac{C_0}{\delta} F^2 + \delta \int_0^t \int_{\Omega} |\mathbf{w}^{\varepsilon}(\mathbf{x},\tau)|^2 dx d\tau. \tag{1.2.21}$$

In particular,

$$\int_{\Omega} |\mathbf{w}^{\varepsilon}(\mathbf{x},t)|^2 dx \leqslant \int_{\Omega} |\mathbf{w}_s^{\varepsilon}(\mathbf{x},t)|^2 dx + C_0 \frac{\alpha_\mu}{\mu_1} \Big(\int_{\Omega_f^{\varepsilon}} |\mathbb{D}(x, \mathbf{w}^{\varepsilon})|^2 dx + \int_{\Omega} |\mathbb{D}(x, \mathbf{w}_s^{\varepsilon})|^2 dx \Big).$$

Now we use the properties of the extension operator $\mathbb{E}_{\Omega_s^{\varepsilon}}$ (see Appendix B) and Korn's inequality, which state that

$$\int_{\Omega} |\mathbf{w}_s^{\varepsilon}(\mathbf{x},t)|^2 dx \leqslant C_0 \int_{\Omega} |\nabla\, \mathbf{w}_s^{\varepsilon}(\mathbf{x},t)|^2 dx \leqslant C_0 \int_{\Omega} \big|\mathbb{D}\big(x, \mathbf{w}_s^{\varepsilon}(\mathbf{x},t)\big)\big|^2 dx,$$

and

$$\int_{\Omega} \big|\mathbb{D}\big(x, \mathbf{w}_s^{\varepsilon}(\mathbf{x},t)\big)\big|^2 dx \leqslant C_0 \int_{\Omega_s^{\varepsilon}} \big|\mathbb{D}\big(x, \mathbf{w}^{\varepsilon}(\mathbf{x},t)\big)\big|^2 dx.$$

Therefore,

$$\alpha_\mu \int_{\Omega_f^{\varepsilon}} \big|\mathbb{D}\big(x, \mathbf{w}^{\varepsilon}(\mathbf{x},t)\big)\big|^2 dx + \lambda_0 \int_0^t \int_{\Omega_s^{\varepsilon}} \big|\mathbb{D}\big(x, \mathbf{w}^{\varepsilon}(\mathbf{x},\tau)\big)\big|^2 dx d\tau$$
$$\leqslant C_0\, \delta \frac{\alpha_\mu}{\mu_1} \int_0^t \int_{\Omega_f^{\varepsilon}} \big|\mathbb{D}\big(x, \mathbf{w}^{\varepsilon}(\mathbf{x},\tau)\big)\big|^2 dx d\tau + \frac{C_0}{\delta} F^2$$
$$+ \delta\, C_0 \int_0^t \int_{\Omega_s^{\varepsilon}} \big|\mathbb{D}\big(x, \mathbf{w}^{\varepsilon}(\mathbf{x},\tau)\big)\big|^2 dx d\tau,$$

and the desired estimates (1.2.12) and (1.2.13) for the functions $\mathbf{w}^{\varepsilon}$ and $\mathbf{w}_s^{\varepsilon}$ follow now from the last inequality with $2\delta\, C_0 = \lambda_0$, Grownwall's inequality and inequality (1.2.21).

To prove the estimate (1.2.12) for the pressure $p^{\,\varepsilon}$, we consider the integral identity (1.2.8) as a relation

$$\int_{\Omega_T} \pi^{\,\varepsilon}\, \nabla \cdot \varphi\, dx dt = \int_{\Omega_T} \Big(\mathbb{F} : \mathbb{D}(x, \varphi) + \rho^{\varepsilon} \int_0^t \mathbf{F}(\mathbf{x},\tau) d\tau \cdot \varphi \Big) dx dt,$$

where

$$\frac{\partial \mathbb{F}}{\partial t} = \alpha_\mu \mathbb{D}\left(x, \frac{\partial \mathbf{w}^\varepsilon}{\partial t}\right) + (1-\chi^\varepsilon)\lambda_0 \mathbb{D}(x, \mathbf{w}^\varepsilon), \ \mathbb{F}, \ \frac{\partial \mathbb{F}}{\partial t} \in L_2(\Omega_T).$$

This relation implies

$$\left|\int_{\Omega_T} \pi^{\,\varepsilon} \nabla \cdot \varphi \, dxdt\right| \leqslant C_0 \, F \left(\int_{\Omega_T} |\nabla \varphi|^2 dxdt\right)^{\frac{1}{2}}. \tag{1.2.22}$$

Next we choose the test function φ, such that

$$\nabla \cdot \varphi = \pi^{\,\varepsilon}, \ and \ \int_{\Omega_T} |\nabla \varphi|^2 dxdt \leqslant C_0 \int_{\Omega_T} |\pi^{\,\varepsilon}|^2 dxdt.$$

Namely, we decompose the function φ to the sum of two functions φ_0 and $\nabla \psi$ such that

$$\triangle \psi = \pi^{\,\varepsilon}, \ \mathbf{x} \in \Omega, \ \psi|_S = 0, \tag{1.2.23}$$

$$\nabla \cdot \varphi_0 = 0, \ \mathbf{x} \in \Omega, \ \left(\varphi_0 + \nabla \psi\right)_S = 0. \tag{1.2.24}$$

On the strength of well-known results [56, 57] and the normalization condition

$$\int_\Omega \pi^{\,\varepsilon}(\mathbf{x}, t) dx = 0,$$

the problem (1.2.23) has a unique solution $\psi \in L_2\big((0, T); \ W_2^2(\Omega)\big)$,

$$\int_0^T \left(\|\psi\|_2^{(2)}(t)\right)^2 dt \leqslant C_0 \int_{\Omega_T} |\pi^{\,\varepsilon}|^2 dxdt,$$

and the problem (1.2.24) has at least one solution $\varphi_0 \in \mathbf{W}_2^{1,0}(\Omega_T)$,

$$\int_0^T \left(\|\varphi_0\|_2^{(1)}(t)\right)^2 dt \leqslant C_0 \int_0^T \left(\|\psi\|_2^{(2)}(t)\right)^2 dt.$$

The last two relations and (1.2.22) give us the estimate (1.2.12) for the time integral $\pi^{\,\varepsilon}$ of the pressure $p^{\,\varepsilon}$.

1.2.3 Proofs of Theorem 1.3 and 1.4

Estimate (1.2.12) guarantees the boundedness of sequences $\{\mathbf{w}^{\varepsilon}\}$, $\{\varepsilon\,\mathbb{D}(x,\mathbf{w}^{\varepsilon})\}$, $\{\chi^{\varepsilon}\,\mathbf{w}^{\varepsilon}\}$, $\{\mathbf{w}_s^{\varepsilon}\}$, $\{\mathbb{D}(x,\mathbf{w}_s^{\varepsilon})\}$, $\{\chi^{\varepsilon}\,\pi^{\varepsilon}\}$, $\{(1-\chi^{\varepsilon})\pi^{\varepsilon}\}$ and $\{\pi^{\varepsilon}\}$ in $\mathbf{L}_2(\Omega_T)$ and $L_2(\Omega_T)$. Therefore, these sequences, except $\{\varepsilon\,\chi^{\varepsilon}\,\mathbb{D}(x,\mathbf{w}^{\varepsilon})\}$, weakly converge in $\mathbf{L}_2(\Omega_T)$ and $L_2(\Omega_T)$ to functions $\mathbf{w}$, $\mathbf{w}^{(f)}$, $\mathbf{w}_s$, $\mathbb{D}(x,\mathbf{w}_s)$, $m\,\pi_f$, $(1-m)\pi_s$, and $\pi = m\,\pi_f + (1-m)\pi_s$ respectively. On the strength of the properties of the extension operator $\mathbb{E}_{\Omega_s^{\varepsilon}}$ (see Appendix B) $\mathbf{w}_s \in \overset{\circ}{\mathbf{W}}{}_2^{1,0}(\Omega_T)$.

Owing to Nguetseng's theorem, there exist 1-periodic in $\mathbf{y}$ functions

$$\mathbf{W}(\mathbf{x},t,\mathbf{y}),\ \mathbb{D}\big(y,\mathbf{W}(\mathbf{x},t,\mathbf{y})\big),\ \mathbf{W}^{(f)}(\mathbf{x},t,\mathbf{y}),\ \mathbb{D}\big(y,\mathbf{U}(\mathbf{x},t,\mathbf{y})\big),$$

$$\Pi_f(\mathbf{x},t,\mathbf{y}),\ \Pi_s(\mathbf{x},t,\mathbf{y}),\ \textit{and}\ \Pi(\mathbf{x},t,\mathbf{y}) = \Pi_f + \Pi_s$$

such that the above mentioned sequences, including $\{\varepsilon\,\chi^{\varepsilon}\,\mathbb{D}(x,\mathbf{w}^{\varepsilon})\}$, two-scale converge in $\mathbf{L}_2(\Omega_T)$ and $L_2(\Omega_T)$ as $\varepsilon \to 0$ respectively to

$$\mathbf{W},\ \chi(\mathbf{y})\,\mathbb{D}\big(y,\mathbf{W}^{(f)}\big),\ \mathbf{W}^{(f)},\ \mathbf{w}_s,\ \mathbb{D}(x,\mathbf{w}_s)+\mathbb{D}\big(y,\mathbf{U}\big),\ \Pi_f,\ \Pi_s,\ \textit{and}\ \Pi.$$

The same theorem of Nguetseng states that

$$\begin{aligned}\mathbf{W}^{(f)} &= \chi(\mathbf{y})\,\mathbf{W},\ \ \mathbf{W} = \mathbf{W}^{(f)} + \big(1-\chi(\mathbf{y})\big)\,\mathbf{w}_s(\mathbf{x},t),\\ \Pi_f &= \chi(\mathbf{y})\,\Pi,\ \ \Pi_s = \big(1-\chi(\mathbf{y})\big)\,\Pi,\end{aligned}$$

and

$$\mathbf{W},\ \mathbb{D}\big(y,\mathbf{W}\big),\ \mathbb{D}\big(y,\mathbf{U}\big),\ \Pi \in L_2(\Omega_T\times Y).$$

The two-scale limit in the continuity equation (1.2.1) in the form

$$\int_{\Omega_T} \mathbf{w}^{\varepsilon}\cdot\nabla\xi\,dxdt = 0 \tag{1.2.25}$$

with test function $\xi = \varepsilon\,\xi_0\left(\dfrac{\mathbf{x}}{\varepsilon}\right)h(\mathbf{x},t)$, where functions $\xi_0(\mathbf{y})$ are finite in Y_f, results in the microscopic continuity equation

$$\nabla_y\cdot\mathbf{W} = 0,\ \ \mathbf{y}\in Y_f. \tag{1.2.26}$$

Now, we rewrite the integral identity (1.2.8) in the form

$$\int_{\Omega_T}\Big(\alpha_{\mu}\mathbb{D}(x,\mathbf{w}^{\varepsilon}) : \mathbb{D}(x,\varphi_0^{\varepsilon}) - \pi^{\varepsilon}\,(\nabla\cdot\varphi_0^{\varepsilon}) - \rho^{\varepsilon}\int_0^t \mathbf{F}(\mathbf{x},\tau)d\tau\cdot\varphi_0^{\varepsilon}\Big)dxdt = 0,$$

with test functions

$$\varphi = \int_0^t \varphi_0\left(\mathbf{x}, \tau, \frac{\mathbf{x}}{\varepsilon}\right) d\tau, \quad \operatorname{supp} \varphi_0(\mathbf{x}, t, \mathbf{y}) \subset Y_f, \quad \text{for all } (\mathbf{x}, t) \in \Omega_T.$$

Applying the results of the previous section for

$$\varphi_0(\mathbf{x}, t, \mathbf{y}) = \varepsilon\, \tilde{\varphi}_0(\mathbf{x}, t, \mathbf{y})$$

and

$$\varphi_0(\mathbf{x}, t, \mathbf{y}) = h(\mathbf{x}, t)\, \varphi_1^{\varepsilon}(\mathbf{y}), \quad \nabla_y \cdot \varphi_1(\mathbf{y}) = 0, \quad \operatorname{supp} \varphi_1 \subset Y_f,$$

we arrive at the relations

$$\Pi_f(\mathbf{x}, t, \mathbf{y}) = \chi(\mathbf{y})\, \pi_f(\mathbf{x}, t), \;\; \nabla \pi_f \in \mathbf{L}_2(\Omega_T),$$

and at the microscopic periodic boundary value problem (1.1.19), (1.1.23), (1.1.24) in Y_f in the form

$$\mu_1 \nabla_y \cdot \mathbb{D}(y, \mathbf{W}) - \nabla_y Q - \nabla \pi_f + \rho_f \int_0^t \mathbf{F}(\mathbf{x}, \tau) d\tau = 0, \tag{1.2.27}$$

$$\mathbf{W}(\mathbf{x}, t, \mathbf{y}) = \mathbf{w}_s(\mathbf{x}, t), \;\; \mathbf{y} \in \gamma, \tag{1.2.28}$$

completed with the continuity equation (1.2.26).

For $\mu_1 < \infty$ the problem (1.2.26)–(1.2.28) for the difference $(\mathbf{W} - \mathbf{w}_s)$ results in Darcy's law (1.2.19).

For a disconnected pore space the last problem has a unique solution $\mathbf{w} = \mathbf{w}_s$.

Finally, estimate (1.2.13) and condition $\mu_1 = \infty$ imply the equality $\mathbf{w} = \mathbf{w}_s$. This fact and the passage to the limit in the continuity equation (1.2.25) for any $\xi \in W_2^{(1,0)}(\Omega_T)$, give us the limiting continuity equation (1.2.14) for the case of a disconnected pore space, or for the case of $\mu_1 = \infty$.

For the case $\mu_1 < \infty$ the two-scale limit in (1.2.25) in the form

$$\int_{\Omega_T} \left(\chi^{\varepsilon} \mathbf{w}^{\varepsilon} + (1 - \chi^{\varepsilon}) \mathbf{w}_s^{\varepsilon}\right) \cdot \nabla \xi dx dt = 0$$

with test function $\xi = \xi(\mathbf{x}, t)$ gives us the homogenized continuity equation

$$\int_{\Omega_T} \left(\mathbf{w}^{(f)} + (1 - m)\mathbf{w}_s\right) \cdot \nabla \xi dx dt = 0,$$

in the integral form, which obviously imply the differential equation (1.2.17) and the boundary condition

$$\left(\mathbf{w}^{(f)} + (1 - m)\mathbf{w}_s\right) \cdot \mathbf{n}(\mathbf{x}) = 0, \;\; \mathbf{x} \in S.$$

The properties of the extension operator $\mathbb{E}_{\Omega_s^\varepsilon}$ result in

$$\mathbf{w}_s(\mathbf{x}, t) = 0, \quad \mathbf{x} \in S.$$

Therefore, the last two conditions imply the boundary condition (1.2.20).

Next we pass to the limit as $\varepsilon \to 0$ in the integral identity (1.2.8) with two different types of test functions. First, with test functions $\varphi = \varphi(\mathbf{x}, t)$. The estimates obtained do not permit us to do it directly, because we have no compactness for the pressures. Therefore, we rewrite the identity (1.2.8) in the form

$$\int_{\Omega_T} \Big(\big(- \chi^\varepsilon \alpha_\mu \mathbb{D}(x, \mathbf{w}^\varepsilon) : \mathbb{D}\Big(x, \frac{\partial \varphi}{\partial t}\Big) + (1 - \chi^\varepsilon)\lambda_0 \mathbb{D}(x, \mathbf{w}^\varepsilon)\big) : \mathbb{D}(x, \varphi)\Big) dxdt$$
$$- \int_{\Omega_T} \pi^\varepsilon \nabla \cdot \Big(\frac{\partial \varphi}{\partial t}\Big)\Big) dxdt = \int_{\Omega_T} \rho^\varepsilon \mathbf{F} \cdot \varphi \, dxdt,$$

and only after that will pass to the limit as $\varepsilon \to 0$ and get:

$$\int_{\Omega_T} \Big(\mathbb{S} : \mathbb{D}(x, \varphi) + \pi \nabla \cdot \Big(\frac{\partial \varphi}{\partial t}\Big)\Big) dxdt = \int_{\Omega_T} \hat{\rho} \mathbf{F} \cdot \varphi \, dxdt,$$

$$\mathbb{S} = \lambda_0 \big((1 - m)\, \mathbb{D}(x, \mathbf{w}_s) + \langle \mathbb{D}(y, \mathbf{U}) \rangle_{Y_s}\big).$$

To rewrite the last identity in the usual form, we consider the mollifiers [59]

$$u_h(\mathbf{x}, t) = \frac{1}{h} \int_t^{t+h} u(\mathbf{x}, \tau) d\tau, \quad u_{\bar{h}}(\mathbf{x}, t) = \frac{1}{h} \int_{t-h}^{t} u(\mathbf{x}, \tau) d\tau,$$

and put $\varphi = (\varphi_0)_{\bar{h}}$. Carrying out smoothing and differentiation with respect to time from the test function we arrive at

$$\int_{\Omega_T} \Big(\mathbb{S}_h : \mathbb{D}(x, \varphi_0) - \frac{\partial \pi_h}{\partial t} \nabla \cdot \varphi_0 \Big) dxdt = \int_{\Omega_T} \hat{\rho} \mathbf{F} \cdot \varphi \, dxdt.$$

On the basis of this identity and the uniform boundedness in $L_2(\Omega_T)$ of $\mathbb{S}_h$ with respect to h (in the same way as in Theorem 1.2) we conclude that the sequence $\left\{ \dfrac{\partial \pi_h}{\partial t} \right\}$ is uniformly bounded in $L_2(\Omega_T)$ with respect to h.

The properties of the mollifiers imply the inclusion of

$$p = \frac{\partial \pi}{\partial t} \in L_2(\Omega_T).$$

The last relation allows us to transform the integral identity obtained above to the usual macroscopic momentum balance equation

$$\int_{\Omega_T} \Big(\mathbb{S} : \mathbb{D}(x,\varphi) - p\,\nabla\cdot\varphi\Big)dxdt = \int_{\Omega_T} \hat{\rho}\mathbf{F}\cdot\varphi\, dxdt,$$

or, to its formal expression as a differential equation

$$\nabla\cdot\Big(\lambda_0\big((1-m)\,\mathbb{D}(x,\mathbf{w}_s) + \langle\mathbb{D}(y,\mathbf{U})\rangle_{Y_s}\big) - p\mathbb{I}\Big) + \hat{\rho}\mathbf{F} = 0. \tag{1.2.29}$$

Now we repeat once more with test functions $\varphi = \varepsilon h(\mathbf{x})\varphi_0\left(\frac{\mathbf{x}}{\varepsilon}, t\right)$ in (1.2.8) and get the microscopic momentum balance equation

$$\int_0^T\int_Y \lambda_0(1-\chi)\big(\mathbb{D}(x,\mathbf{w}_s)+\mathbb{D}(y,\mathbf{U})\big) : \mathbb{D}(y,\varphi_0)dydt$$
$$= -\int_0^T\int_Y \Pi\,\nabla\cdot\left(\frac{\partial\varphi_0}{\partial t}\right)dydt, \tag{1.2.30}$$

which implies

$$P = \chi\, p_f + P_s = \frac{\partial \Pi}{\partial t} \in L_2(Y\times\Omega_T).$$

Choosing now $\varphi_0(\mathbf{y},t)$ with $supp\,\varphi_0 \subset Y_s$ we obtain

$$P_s = \frac{\partial \Pi_s}{\partial t} \in L_2(Y\times\Omega_T),$$

and all together

$$p_f = \frac{\partial \pi_f}{\partial t} \in L_2(\Omega_T), \quad P_s = \frac{\partial \Pi_s}{\partial t} \in L_2(Y\times\Omega_T).$$

The two-scale limit in the continuity equation (1.2.1) in domain Ω_s^ε in the form

$$\left(1-\chi\left(\frac{\mathbf{x}}{\varepsilon}\right)\right)\nabla\cdot\mathbf{w}_s^\varepsilon = 0$$

gives us the missing microscopic continuity equation

$$\big(1-\chi(\mathbf{y})\big)\big(\nabla\cdot\mathbf{w}_s + \nabla_y\cdot\mathbf{U}\big) = 0, \;\; \mathbf{y}\in Y. \tag{1.2.31}$$

The microscopic problem is completed with the normalization condition

$$\langle\mathbf{U}\rangle_{Y_s} = \int_{Y_s}\mathbf{U}dy = 0.$$

If $\mu_1 = \infty$, or the pore space is disconnected (the case when $\mathbf{w} = \mathbf{w}_s$) the macroscopic equation (1.2.14) holds true. Thus (1.2.31) takes the form

$$\big(1 - \chi(\mathbf{y})\big)\,\nabla_y \cdot \mathbf{U} = 0, \ \ \mathbf{y} \in Y. \tag{1.2.32}$$

Lemma 1.4 *Let $\mu_1 = \infty$, or the pore space be disconnected. Then the limiting functions $\mathbf{w}_s$ and p satisfy the homogenized momentum balance equation (1.2.15).*

Proof To derive the homogenized momentum balance equation (1.2.15) we have to express $\mathbb{D}(y, \mathbf{U})$ as an operator on $\mathbb{D}(x, \mathbf{w}_s)$ by means of Eqs. (1.2.30) and (1.2.32), and substitute it into the equation (1.2.29):

$$\mathbb{D}(y, \mathbf{U}) = \mathbb{B}_0^s(\mathbf{y}) : \mathbb{D}(x, \mathbf{w}_s),$$

and

$$\mathfrak{N}^s = (1 - m) \sum_{i,j=1}^{3} \mathbb{J}^{ij} \otimes \mathbb{J}^{ij} + \langle \mathbb{B}_0^s \rangle_{Y_s},$$

where

$$\mathbb{J}^{ij} = \frac{1}{2}\big(\mathbf{e}_i \otimes \mathbf{e}_j + \mathbf{e}_j \otimes \mathbf{e}_i\big),$$

$\{\mathbf{e}_1, \mathbf{e}_2, \mathbf{e}_3\}$ is a standard Cartesian basis, and the fourth-rank tensor $\mathbb{A} \otimes \mathbb{B}$ is the tensor (direct) product of the second-rank tensors $\mathbb{A}$ and $\mathbb{B}$: $(\mathbb{A} \otimes \mathbb{B}) : \mathbb{C} = \mathbb{A}(\mathbb{B} : \mathbb{C})$ for any second-rank tensor $\mathbb{C}$.

To this end, we rewrite (1.2.30) in the form

$$\nabla_y \cdot \Big((1 - \chi)\big(\mathbb{D}(x, \mathbf{w}_s) + \mathbb{D}(y, \mathbf{U}) - \frac{1}{\lambda_0}(P_s - p_f)\mathbb{I}\big)\Big) = 0, \tag{1.2.33}$$

and will look for a solution of (1.2.32) and (1.2.33) in the form

$$\mathbf{U} = \sum_{i,j=1}^{3} \mathbf{U}_0^{(ij)}(\mathbf{y}) D_{ij}(\mathbf{x}, t),$$

$$P_s - p_f = \lambda_0 \sum_{i,j=1}^{3} P_0^{(ij)}(\mathbf{y}) D_{ij}(\mathbf{x}, t),$$

where

$$D_{ij}(\mathbf{x},t)=\frac{1}{2}\Big(\frac{\partial u_i}{\partial x_j}(\mathbf{x},t)+\frac{\partial u_j}{\partial x_i}(\mathbf{x},t)\Big),\quad \mathbf{w}_s=(u_1,\,u_2,\,u_3),$$

$$\mathbb{D}(x,\mathbf{w}_s)=\sum_{i,j=1}^{3}D_{ij}\,\mathbb{J}^{ij},$$

and

$$\left.\begin{array}{l}\nabla_y\cdot\Big((1-\chi)\big(\mathbb{D}(y,\mathbf{U}_0^{(ij)})+\mathbb{J}^{ij}-P_0^{(ij)}\mathbb{I}\big)\Big)=0,\ \mathbf{y}\in Y,\\ (1-\chi)\nabla_y\cdot\mathbf{U}_0^{(ij)}=0,\ \mathbf{y}\in Y,\ \langle\mathbf{U}_0^{(ij)}\rangle_{Y_s}=0.\end{array}\right\}\tag{1.2.34}$$

In what follows, we understand all equations like the first one in (1.2.34) in the sense of distributions. That is, as an integral identity

$$\int_{Y_s}\big(\mathbb{D}(y,\mathbf{U}_0^{(ij)})+\mathbb{J}^{ij}-P_0^{(ij)}\mathbb{I}\big):\mathbb{D}(y,\Phi)dy=0,$$

which holds true for any smooth 1-periodic in $\mathbf{y}$ functions $\Phi(\mathbf{y})$.

The existence and uniqueness of the 1-periodic weak solution $\mathbf{U}_0^{(ij)}\in\mathbf{W}_2^1(Y_s)$ to the problem (1.2.34) follows from the a priori estimate

$$\int_{Y_s}|\nabla\mathbf{U}_0^{(ij)}(\mathbf{y})|^2dy\leqslant C_0.$$

In turn, this estimate is a consequence of an energy identity

$$\int_{Y_s}\big(|\mathbb{D}(y,\mathbf{U}_0^{(ij)})|^2+\mathbb{J}^{ij}:\mathbb{D}(y,\mathbf{U}_0^{(ij)})\big)dy=0.$$

Thus,

$$\begin{aligned}\mathbb{D}(y,\mathbf{U})&=\sum_{i,j=1}^{3}\mathbb{D}(y,\mathbf{U}_0^{(ij)})D_{ij}=\sum_{i,j=1}^{3}\mathbb{D}(y,\mathbf{U}_0^{(ij)})\big(\mathbb{J}^{ij}:\mathbb{D}(x,\mathbf{w}_s)\big)\\&=\Big(\sum_{i,j=1}^{3}\mathbb{D}(y,\mathbf{U}_0^{(ij)})\otimes\mathbb{J}^{ij}\Big):\mathbb{D}(x,\mathbf{w}_s)=\mathbb{B}_0^s(\mathbf{y}):\mathbb{D}(x,\mathbf{w}_s),\end{aligned}$$

and

$$\mathfrak{N}^s=(1-m)\sum_{i,j=1}^{3}\mathbb{J}^{ij}\otimes\mathbb{J}^{ij}+\sum_{i,j=1}^{3}\langle\mathbb{D}(y,\mathbf{U}_0^{(ij)})\rangle_{Y_s}\otimes\mathbb{J}^{ij}.\tag{1.2.35}$$

All properties of the tensor $\mathfrak{N}^s$ follow from identities

$$\int_{Y_s} \big(\mathbb{D}(y, \mathbf{U}_0^{(ij)}) : \mathbb{D}(y, \mathbf{U}_0^{(kl)}) + \mathbb{J}^{ij} : \mathbb{D}(y, \mathbf{U}_0^{(kl)})\big) dy = 0, \tag{1.2.36}$$

which are the result of multiplication of the microscopic momentum balance equation for $\mathbf{U}_0^{(ij)}$ by $\mathbf{U}_0^{(kl)}$ and integration by parts over domain Y.

Lemma 1.5 *The constant fourth-rank tensor $\mathfrak{N}^s$ is symmetric and strictly positively definite.*

Proof Let $\zeta = (\zeta_{ij})$ and $\eta = (\eta_{ij})$ be arbitrary symmetric matrices and

$$\mathbf{Y}_\zeta = \sum_{i,j=1}^{3} \mathbf{U}_0^{(ij)} \zeta_{ij}, \quad \mathbf{Y}_\eta = \sum_{i,j=1}^{3} \mathbf{U}_0^{(ij)} \eta_{ij}.$$

Then (1.2.36) transforms to

$$\langle \mathbb{D}(y, \mathbf{Y}_\zeta) : \mathbb{D}(y, \mathbf{Y}_\eta) \rangle_{Y_s} + \langle \mathbb{D}(y, \mathbf{Y}_\eta) \rangle_{Y_s} : \zeta = 0.$$

By definition

$$\big(\mathfrak{N}^s : \zeta\big) : \eta = (1 - m)\zeta : \eta + \langle \mathbb{D}(y, \mathbf{Y}_\zeta) \rangle_{Y_s} : \eta.$$

The sum of the last two equalities results in a relation

$$\big(\mathfrak{N}^s : \zeta\big) : \eta = \Big\langle \big(\mathbb{D}(y, \mathbf{Y}_\zeta) + \zeta\big) : \big(\mathbb{D}(y, \mathbf{Y}_\eta) + \eta\big) \Big\rangle_{Y_s},$$

which shows the symmetry and the positivity of the tensor $\mathfrak{N}^s$. The strict positivity of this tensor follows from the structure of the domain Y_s in the same way, as in the previous section, namely, if $\big(\mathfrak{N}^s : \eta\big) : \eta = 0$, then for some η, such that $\eta : \eta = 1$ one has $\mathbb{D}(y, \mathbf{Y}_\eta) + \eta = 0$. This equality implies the linearity of $\mathbf{Y}_\eta$. But this is impossible for a connected solid skeleton.

In fact, $\mathbf{Y}_\eta$ is a periodic solution of the problem

$$\left.\begin{array}{l} \nabla_y \cdot \Big((1 - \chi)\big(\mathbb{D}(y, \mathbf{Y}_\eta) + \eta - P_\eta \mathbb{I}\big)\Big) = 0, \ \mathbf{y} \in Y, \\ (1 - \chi)\nabla_y \cdot \mathbf{Y}_\eta = 0, \ \mathbf{y} \in Y, \ \langle \mathbf{Y}_\eta \rangle_{Y_s} = 0. \end{array}\right\}$$

As we have mentioned in the previous section, due to periodicity conditions, any linear and periodic function in the elementary cell may only be constant if the periodic repetition of this cell forms a connected domain. The normalization condition $\langle \mathbf{Y}_\eta \rangle_{Y_s} = 0$ implies $\mathbf{Y}_\eta = 0$. By construction $\mathbf{Y}_\eta + \eta = 0$. Therefore $\eta = 0$, which is impossible by the supposition $\eta : \eta = 1$.

The weak limit in the normalization condition (1.2.5) in the form

$$0 = -\int_0^T \frac{dh}{dt}(t)\Big(\int_\Omega \pi^\varepsilon dx\Big)dt \to -\int_0^T \frac{dh}{dt}(t)\Big(\int_\Omega \pi\, dx\Big)dt$$
$$= \int_0^T h(t)\Big(\int_\Omega p\, dx\Big)dt = 0$$

for arbitrary smooth functions $h(t)$ finite on $(0, T)$, results in the normalization condition (1.2.16).

Lemma 1.6 *The problem (1.2.14)–(1.2.16) has a unique solution.*

Proof The equality $\mathbf{w}_s = 0$ follows from the energy identity

$$\int_\Omega \lambda_0 \big(\mathfrak{N}^s : \mathbb{D}(x, \mathbf{w}_s)\big) : \mathbb{D}(x, \mathbf{w}_s)dx = 0$$

for $\mathbf{F} = 0$, the strict positivity of $\mathfrak{N}^s$, and the boundary condition in (1.2.16).

The differential equation (1.2.15) for $\mathbf{w}_s = 0$ defines a pressure p as some function of t. The normalization condition in (1.2.16) implies $p = 0$.

Now let $\mu_1 < \infty$ and the pore space be connected. To define $\mathbf{U}$ we have the periodic boundary value problem (1.2.31), (1.2.33) and the solution of this problem has the form

$$\mathbf{U} = \sum_{i,j=1}^{3} \mathbf{U}_0^{(ij)}(\mathbf{y})D_{ij}(\mathbf{x}, t) + \mathbf{U}_0^{(0)}(\mathbf{y})\big(\nabla\cdot \mathbf{w}_s(\mathbf{x}, t)\big),$$

$$P_s - p_f = \lambda_0 \sum_{i,j=1}^{3} P_0^{(ij)}(\mathbf{y})D_{ij}(\mathbf{x}, t) + \lambda_0 P_0^{(0)}(\mathbf{y})\big(\nabla\cdot \mathbf{w}_s(\mathbf{x}, t)\big),$$

where

$$\left.\begin{array}{l}\nabla_y\cdot\Big((1-\chi)\big(\mathbb{D}(y, \mathbf{U}_0^{(0)}) - P_0^{(0)}\mathbb{I}\big)\Big) = 0,\\ (1-\chi)(\nabla_y\cdot \mathbf{U}_0^{(0)} + 1) = 0, \quad \langle \mathbf{U}_0^{(0)}\rangle_{Y_s} = 0, \quad \mathbf{y}\in Y.\end{array}\right\} \qquad (1.2.37)$$

To solve the problem (1.2.37) we first find a 1-periodic function $\mathbf{V}_0 \in \mathbf{W}_2^1(Y_s)$ such that

$$\nabla_y\cdot \mathbf{V}_0 + 1 = 0, \quad \mathbf{y}\in Y_s.$$

There are many ways to construct the function $\mathbf{V}_0$. For example, let

$$\nabla_y\cdot \mathbf{U}_0 + 1 = 0, \quad \mathbf{y}\in Y, \quad \mathbf{U}_0(\mathbf{y}) = 0, \quad \mathbf{y}\in \partial Y.$$

The existence of $\mathbf{U}_0 \in \overset{\circ}{\mathbf{W}}{}^1_2\,(Y)$ follows from [56]. Now we extend it periodically outside of Y as

$$\mathbf{V}_0(\mathbf{y}+\mathbf{k}) = \mathbf{U}_0(\mathbf{y}),$$

where $\mathbf{y} \in Y$, $\mathbf{k} = (k_1, k_2, k_3)$ and the coordinates of the vector $\mathbf{k}$ are integers.

The correctness of the problem (1.2.37) is a consequence of the energy identity

$$\int_{Y_s} \mathbb{D}(y, \mathbf{U}_0^{(0)}) : \big(\mathbb{D}(y, \mathbf{U}_0^{(0)}) - \mathbb{D}(y, \mathbf{V}_0)\big) dy = 0.$$

The last equality is the result of the multiplication of the first equation in (1.2.37) by $(\mathbf{U}_0^{(0)} - \mathbf{V}_0)$ and integration by parts over domain Y_s.

Lemma 1.7 *The limiting functions $\mathbf{w}_s$ and p_f satisfy the homogenized momentum balance equation (1.2.18).*

Proof To find $\mathfrak{N}_1^s$ we have to calculate expressions $\langle \mathbb{D}(y, \mathbf{U})\rangle_{Y_s}$ and p as operators on $\mathbb{D}(x, \mathbf{w}_s)$ and $(\nabla \cdot \mathbf{w}_s)$:

$$\begin{aligned}(1-m)\mathbb{D}(x, \mathbf{w}_s) + \langle \mathbb{D}(y, \mathbf{U})\rangle_{Y_s} &= \mathfrak{N}^s : \mathbb{D}(x, \mathbf{w}_s) + \langle \mathbb{D}(y, \mathbf{U}_0^{(0)})\rangle_{Y_s} (\nabla \cdot \mathbf{w}_s) \\ &= \Big(\mathfrak{N}^s + \langle \mathbb{D}(y, \mathbf{U}_0^{(0)})\rangle_{Y_s} \otimes \mathbb{I}\Big) : \mathbb{D}(x, \mathbf{w}_s),\end{aligned}$$

$$\begin{aligned}p &= \langle P\rangle_Y = \langle \chi\, p_f + (1-\chi)P_s\rangle_Y = \langle p_f + (1-\chi)(P_s - p_f)\rangle_Y \\ &= p_f + \langle (P_s - p_f)\rangle_{Y_s} = p_f + \lambda_0 \left\langle \sum_{i,j=1}^{3} P_0^{(ij)} \right\rangle_{Y_s} D_{ij} + \lambda_0 \langle P_0^{(0)}\rangle_{Y_s} (\nabla \cdot \mathbf{w}_s),\end{aligned}$$

and

$$\begin{aligned}p\mathbb{I} - p_f\mathbb{I} &= \left(\lambda_0 \left\langle \sum_{i,j=1}^{3} P_0^{(ij)} \right\rangle_{Y_s} \mathbb{I} \otimes \mathbb{J}^{ij}\right) : \mathbb{D}(x, \mathbf{w}_s) + \Big(\lambda_0 \langle P_0^{(0)}\rangle_{Y_s} \mathbb{I}\Big)(\nabla \cdot \mathbf{w}_s) \\ &= \left(\lambda_0 \left\langle \sum_{i,j=1}^{3} P_0^{(ij)} \right\rangle_{Y_s} \mathbb{I} \otimes \mathbb{J}^{ij} + \lambda_0 \langle P_0^{(0)}\rangle_{Y_s} \mathbb{I} \otimes \mathbb{I}\right) : \mathbb{D}(x, \mathbf{w}_s).\end{aligned}$$

Therefore,

$$\mathfrak{N}_1^s = \mathfrak{N}^s + \langle \mathbb{D}(y, \mathbf{U}_0^{(0)})\rangle_{Y_s} \otimes \mathbb{I} - \left\langle \sum_{i,j=1}^{3} P_0^{(ij)} \right\rangle_{Y_s} \mathbb{I} \otimes \mathbb{J}^{ij} - \langle P_0^{(0)}\rangle_{Y_s} \mathbb{I} \otimes \mathbb{I}. \quad (1.2.38)$$

The basic properties of the tensor $\mathfrak{N}_1^s$ follow from the equalities

$$\langle P_0^{(0)} \rangle_{Y_s} = -\langle \mathbb{D}(y, \mathbf{U}_0^{(0)}) : \mathbb{D}(y, \mathbf{U}_0^{(0)}) \rangle_{Y_s}, \tag{1.2.39}$$

$$\langle \mathbb{D}(y, \mathbf{U}_0^{(ij)}) : \mathbb{D}(y, \mathbf{U}_0^{(0)}) \rangle_{Y_s} = 0, \tag{1.2.40}$$

$$\langle P_0^{(ij)} \rangle_{Y_s} = -\langle \mathbb{D}(y, \mathbf{U}_0^{(0)}) : \mathbb{J}^{ij} \rangle_{Y_s}, \tag{1.2.41}$$

$$\langle \mathbb{D}(y, \mathbf{U}_0^{(ij)}) : \mathbb{D}(y, \mathbf{U}_0^{(kl)}) \rangle_{Y_s} + \langle \mathbb{J}^{ij} : \mathbb{D}(y, \mathbf{U}_0^{(kl)}) \rangle_{Y_s} = 0, \tag{1.2.42}$$

for all $i,\ j,\ k,\ l\ =\ 1,\ 2,\ 3$. Equation (1.2.39) is a result of multiplication of the equation for $\mathbf{U}_0^{(0)}$ by $\mathbf{U}_0^{(0)}$ and integration by parts over domain Y_s. Equation (1.2.40) is the result of multiplication of the equation for $\mathbf{U}_0^{(0)}$ by $\mathbf{U}_0^{(ij)}$ and integration by parts over domain Y_s.

Equation (1.2.41) is the result of the multiplication of the equation for $\mathbf{U}_0^{(ij)}$ by $\mathbf{U}_0^{(0)}$ and integration by parts over domain Y_s. Here we additionally take into account the relation (1.2.40).

Finally, equation (1.2.42) is the result of the multiplication of the equation for $\mathbf{U}_0^{(ij)}$ by $\mathbf{U}_0^{(kl)}$ and integration by parts over domain Y_s.

By construction

$$\int_\Omega \langle P_s - p_f \rangle_{Y_s} dx = \lambda_0 \left\langle \sum_{i,j=1}^{3} P_0^{(ij)} \right\rangle_{Y_s} \int_\Omega D_{ij}(\mathbf{x}, t) dx$$
$$+ \lambda_0 \langle P_0^{(0)} \rangle_{Y_s} \int_\Omega \nabla \cdot \mathbf{w}_s(\mathbf{x}, t) dx = 0,$$

and

$$\int_\Omega p dx = \int_\Omega (p_f + \langle P_s - p_f \rangle_Y) dx = \int_\Omega p_f dx.$$

Thus, the normalization condition (1.2.20) follows from the last equality and the integral identity

$$\int_0^T h(t) \Big(\int_\Omega \pi^\varepsilon(\mathbf{x}, t) dx \Big) dt = 0$$

for arbitrary smooth functions $h(t)$ after taking the limit as $\varepsilon \to 0$.

Lemma 1.8 *The constant fourth-rank tensor $\mathfrak{N}_1^s$ is symmetric and strictly positively definite.*

Proof Let $\zeta = (\zeta_{ij})$ and $\eta = (\eta_{ij})$ be arbitrary symmetric matrices and

$$\mathbf{Y}_\zeta = \sum_{i,j=1}^{3} \mathbf{U}_0^{(ij)} \zeta_{ij}, \quad \mathbf{Y}_\eta = \sum_{i,j=1}^{3} \mathbf{U}_0^{(ij)} \eta_{ij}, \quad \mathbf{Y}_\zeta^0 = \mathbf{U}_0^{(0)} \operatorname{tr} \zeta, \quad \mathbf{Y}_\eta^0 = \mathbf{U}_0^{(0)} \operatorname{tr} \eta.$$

Then Eqs. (1.2.39)–(1.2.42) are transformed to equations

$$-\big(\langle P^{(0)}\rangle_{Y_s}\,\mathbb{I}\otimes\mathbb{I}:\zeta\big):\eta=\langle\mathbb{D}(y,\mathbf{Y}^0_\zeta):\mathbb{D}(y,\mathbf{Y}^0_\eta)\rangle_{Y_s},\tag{1.2.43}$$

$$\langle\mathbb{D}(y,\mathbf{Y}_\eta):\mathbb{D}(y,\mathbf{Y}^0_\zeta)\rangle_{Y_s}=0,\tag{1.2.44}$$

$$-\left(\left(\left\langle\sum_{i,j=1}^{3}P_0^{(ij)}\right\rangle_{Y_s}\mathbb{I}\otimes\mathbb{J}^{ij}\right):\zeta\right):\eta=\langle\mathbb{D}(y,\mathbf{Y}^0_\eta)\rangle_{Y_s}:\zeta,\tag{1.2.45}$$

$$\langle\mathbb{D}(y,\mathbf{Y}_\zeta):\mathbb{D}(y,\mathbf{Y}_\eta)\rangle_{Y_s}+\zeta:\langle\mathbb{D}(y,\mathbf{Y}_\eta)\rangle_{Y_s}=0.\tag{1.2.46}$$

Therefore,

$$\begin{aligned}\big(\mathfrak{N}^s_1:\zeta\big):\eta&=(1-m)\zeta:\eta+\langle\mathbb{D}(y,\mathbf{Y}_\zeta)\rangle_{Y_s}:\eta+\langle\mathbb{D}(y,\mathbf{Y}^0_\zeta)\rangle_{Y_s}:\eta\\&\quad+\zeta:\langle\mathbb{D}(y,\mathbf{Y}^0_\eta)\rangle_{Y_s}+\langle\mathbb{D}(y,\mathbf{Y}^0_\zeta):\mathbb{D}(y,\mathbf{Y}^0_\eta)\rangle_{Y_s}.\end{aligned}$$

Taking into account equalities (1.2.43)–(1.2.46) we finally get

$$\begin{aligned}\big(\mathfrak{N}^s_1:\zeta\big):\eta&=\langle\mathbb{D}(y,\mathbf{Y}_\zeta):\mathbb{D}(y,\mathbf{Y}_\eta)\rangle_{Y_s}+\langle\mathbb{D}(y,\mathbf{Y}_\eta)\rangle_{Y_s}:\zeta\\&\quad+\langle\mathbb{D}(y,\mathbf{Y}_\zeta)\rangle_{Y_s}:\eta+\langle\mathbb{D}(y,\mathbf{Y}^0_\zeta):\mathbb{D}(y,\mathbf{Y}^0_\eta)\rangle_{Y_s}+\zeta:\langle\mathbb{D}(y,\mathbf{Y}^0_\eta)\rangle_{Y_s}\\&\quad+\eta:\langle\mathbb{D}(y,\mathbf{Y}^0_\zeta)\rangle_{Y_s}+(1-m)\zeta:\eta\\&=\big\langle\big(\mathbb{D}(y,\mathbf{Y}_\zeta+\mathbf{Y}^0_\zeta)+\zeta\big):\big(\mathbb{D}(y,\mathbf{Y}_\eta+\mathbf{Y}^0_\eta)+\eta\big)\big\rangle_{Y_s}.\end{aligned}\tag{1.2.47}$$

Equation (1.2.47) shows that the tensor $\mathfrak{N}^s_1$ is symmetric:

$$\big(\mathfrak{N}^s_1:\zeta\big):\eta=\big(\mathfrak{N}^s_1:\eta\big):\zeta.$$

In particular,

$$\big(\mathfrak{N}^s_1:\zeta\big):\zeta=\big\langle\big(\mathbb{D}(y,\mathbf{Y}_\zeta+\mathbf{Y}^0_\zeta)+\zeta\big):\big(\mathbb{D}(y,\mathbf{Y}_\zeta+\mathbf{Y}^0_\zeta)+\zeta\big)\big\rangle_{Y_s}>0.$$

Therefore the tensor $\mathfrak{N}^s_1$ is strictly positively definite.

Lemma 1.9 *The problem (1.2.16)–(1.2.20) has a unique solution.*

Proof To prove the uniqueness of the problem (1.2.16)–(1.2.20) we multiply equation (1.2.18) with $\mathbf{F}=0$ by $\mathbf{w}_s$ and integrate by parts over Ω:

$$\int_\Omega\lambda_0\mathbb{D}(x,\mathbf{w}_s):\Big(\mathfrak{N}^s_1:\mathbb{D}(x,\mathbf{w}_s)\Big)dx-\int_\Omega p_f\,\nabla\cdot\mathbf{w}_s\,dx=0.$$

The combination of Darcy's law (1.2.19) with $\mathbf{F} = 0$ and the continuity equation (1.2.17) give us

$$\nabla \cdot \mathbf{w}_s = \frac{1}{\mu_1} \nabla \cdot \Big(\mathbb{B} \cdot \nabla \pi_f \Big).$$

Thus,

$$\frac{1}{2\mu_1} \frac{d}{dt} \int_\Omega \nabla \pi_f \cdot \Big(\mathbb{B} \cdot \nabla \pi_f \Big) dx + \int_\Omega p_f \, \nabla \cdot \mathbf{w}_s \, dx = 0,$$

and

$$\frac{1}{2\mu_1} \frac{d}{dt} \int_\Omega \nabla \pi_f \cdot \Big(\mathbb{B} \cdot \nabla \pi_f \Big) dx + \int_\Omega \lambda_0 \mathbb{D}(x, \mathbf{w}_s) : \Big(\mathfrak{N}_1^s : \mathbb{D}(x, \mathbf{w}_s) \Big) dx = 0.$$

The last equality and normalization condition (1.2.20) imply

$$\mathbf{w}_s = 0, \;\; \pi_f = 0.$$

As in the previous section we conclude that any subsequences of sequences $\{\mathbf{w}^\varepsilon\}$, $\{\mathbf{w}_s^\varepsilon\}$, and $\{p^\varepsilon\}$ converge to the same limits. Therefore entire sequences converge to those unique limits.

1.3 A Compressible Slightly Viscous Liquid in a Compressible Elastic Skeleton

Here, as a basic mathematical model at the microscopic level we consider the model $\mathbb{M}_{14}$ of the filtration of a compressible liquid in a compressible elastic solid skeleton

$$\frac{1}{\alpha_p^\varepsilon} p + \nabla \cdot \mathbf{w} = 0, \tag{1.3.1}$$

$$\nabla \cdot \mathbb{P} + \rho^\varepsilon \mathbf{F} = 0, \tag{1.3.2}$$

$$\mathbb{P} = \chi^\varepsilon \alpha_\mu \mathbb{D}\Big(x, \frac{\partial \mathbf{w}}{\partial t}\Big) + (1 - \chi^\varepsilon) \lambda_0 \mathbb{D}(x, \mathbf{w}) - \Big(p - \chi^\varepsilon \alpha_\nu \nabla \cdot \frac{\partial \mathbf{w}}{\partial t}\Big) \mathbb{I}, \tag{1.3.3}$$

$$\mathbf{w}(\mathbf{x}, t) = 0, \;\; \mathbf{x} \in S, \;\; t \in (0, T), \tag{1.3.4}$$

$$\alpha_\nu \chi^\varepsilon \, p(\mathbf{x}, 0) = 0, \;\; \chi^\varepsilon \, \mathbf{w}(\mathbf{x}, 0) = 0, \;\; \mathbf{x} \in \Omega. \tag{1.3.5}$$

Throughout this section assume that conditions

$$\mu_0 = 0, \;\; 0 < \mu_1 \leqslant \infty, \;\; 0 < \lambda_0, \; c_f^2, \; c_s^2 < \infty, \; 0 \leqslant \nu_0 < \infty, \tag{1.3.6}$$

and

$$\int_{\Omega_T}\left(|\mathbf{F}|^2+\left|\frac{\partial \mathbf{F}}{\partial t}\right|^2\right)dxdt = F_1^2 < \infty \tag{1.3.7}$$

hold true.

In (1.3.1)–(1.3.6)

$$\alpha_p^\varepsilon = \chi^\varepsilon c_f^2 + (1-\chi^\varepsilon)c_s^2,$$

$$\mu_0 = \lim_{\varepsilon\searrow 0}\alpha_\mu(\varepsilon),\quad \mu_1 = \lim_{\varepsilon\searrow 0}\frac{\alpha_\mu}{\varepsilon^2},\quad \nu_0 = \lim_{\varepsilon\searrow 0}\alpha_\nu(\varepsilon).$$

1.3.1 Statement of the Problem and Main Results

Definition 1.3 We say that the pair of functions $\{\mathbf{w}^\varepsilon,\ p^\varepsilon\}$ such that

$$\mathbf{w}^\varepsilon \in \overset{\circ}{\mathbf{W}}{}_2^{1,0}(\Omega_T),\ \chi^\varepsilon\nabla\left(\frac{\partial \mathbf{w}^\varepsilon}{\partial t}\right) \in \mathbf{L}_2(\Omega_T),\ p^\varepsilon \in L_2(\Omega_T),$$

is a weak solution of the problem (1.3.1)–(1.3.5), if it satisfies the continuity equation (1.3.1) almost everywhere in Ω_T, the initial condition (1.3.5) and an integral identity

$$\int_{\Omega_T}\left(\chi^\varepsilon\alpha_\mu\mathbb{D}\left(x,\frac{\partial \mathbf{w}^\varepsilon}{\partial t}\right)+(1-\chi^\varepsilon)\lambda_0\mathbb{D}(x,\mathbf{w}^\varepsilon)\right):\mathbb{D}(x,\varphi)\,dxdt$$
$$-\int_{\Omega_T}\left(p^\varepsilon-\chi^\varepsilon\alpha_\nu\nabla\cdot\frac{\partial \mathbf{w}^\varepsilon}{\partial t}\right)(\nabla\cdot\varphi)\,dxdt = \int_{\Omega_T}\rho^\varepsilon\mathbf{F}\cdot\varphi\,dxdt, \tag{1.3.8}$$

for any functions $\varphi \in \overset{\circ}{\mathbf{W}}{}_2^{1,0}(\Omega_T)$.

Theorem 1.5 *There exists a unique weak solution* $\{\mathbf{w}^\varepsilon,\ p^\varepsilon\}$ *of the problem (1.3.1)–(1.3.5) and*

$$\max_{0<t<T}\int_\Omega\left(\left|\mathbb{D}(x,\mathbf{w}_s^\varepsilon(\mathbf{x},t))\right|^2+\left|\mathbb{D}\left(x,\frac{\partial \mathbf{w}_s^\varepsilon}{\partial t}(\mathbf{x},t)\right)\right|^2+\alpha_\nu\chi^\varepsilon\left|\frac{\partial p^\varepsilon}{\partial t}(\mathbf{x},t)\right|^2\right)dx$$
$$+\int_{\Omega_T}\left(\chi^\varepsilon\alpha_\mu\left|\mathbb{D}(x,\frac{\partial^2 \mathbf{w}^\varepsilon}{\partial t^2})\right|^2+\left|\frac{\partial p^\varepsilon}{\partial t}\right|^2\right)dxdt \leqslant C_0F_1^2, \tag{1.3.9}$$

$$\int_{\Omega_T}\left|\frac{\partial \mathbf{w}^\varepsilon}{\partial t}(\mathbf{x},t)-\frac{\partial \mathbf{w}_s^\varepsilon}{\partial t}(\mathbf{x},t)\right|^2dxdt \leqslant \frac{\varepsilon^2}{\alpha_\mu}C_0F_1^2, \tag{1.3.10}$$

where $\mathbf{w}_s^\varepsilon$ *be an extension (1.2.9) and the constant* C_0 *is independent of the small parameter* ε.

Theorem 1.6 *Let* $\{\mathbf{w}^\varepsilon,\ p^\varepsilon\}$ *be the weak solution of the problem (1.3.1)–(1.3.5),* $\mathbf{w}_s^\varepsilon$ *be an extension (1.2.9), and* $\mu_1 = \infty$ *or* $\mu_1 < \infty$ *but the pore space be disconnected.*

Then

(1) *for all* $\nu_0 > 0$ *the sequences* $\{\mathbf{w}^\varepsilon\}$, $\{\nabla\cdot\mathbf{w}^\varepsilon\}$, $\{\chi^\varepsilon p^\varepsilon\}$, $\left\{\chi^\varepsilon \dfrac{\partial p^\varepsilon}{\partial t}\right\}$, *and* $\{q^\varepsilon\}$, *where* $q^\varepsilon = \chi^\varepsilon\left(p^\varepsilon + \dfrac{\alpha_\nu}{c_f^2}\dfrac{\partial p^\varepsilon}{\partial t}\right)$, *converge weakly in* $\mathbf{L}_2(\Omega_T)$ *and* $L_2(\Omega_T)$ *(up to some subsequences) to functions* $\mathbf{w}$, $\nabla\cdot\mathbf{w}$, $m\,\dfrac{\partial p_f}{\partial t}$, *and* $m\,q = m\left(p_f + \dfrac{\nu_0}{c_f^2}\dfrac{\partial p_f}{\partial t}\right)$ *respectively;*

(2) *for* $\nu_0 = 0$ *the sequences* $\{\mathbf{w}^\varepsilon\}$, $\{\nabla\cdot\mathbf{w}^\varepsilon\}$, *and* $\{\chi^\varepsilon p^\varepsilon\}$ *converge weakly in* $\mathbf{L}_2(\Omega_T)$ *and* $L_2(\Omega_T)$ *(up to some subsequences) to functions* $\mathbf{w}$, $\nabla\cdot\mathbf{w}$, *and* $m\,p_f$ *respectively;*

(3) *for all* $\nu_0 \geqslant 0$ *the sequence* $\{\mathbf{w}_s^\varepsilon\}$ *converges weakly in* $\mathbf{w}_2^{1,0}(\Omega_T)$ *(up to some subsequences) to the function* $\mathbf{w}_s \in \overset{\circ}{\mathbf{W}}{}_2^{1,0}(\Omega_T)$ *and* $\mathbf{w}_s = \mathbf{w}$;

(4) *limiting functions solve the system of homogenized equations in the domain* Ω_T, *consisting of the homogenized continuity equation*

$$\frac{m}{c_f^2}p_f + m\,\nabla\cdot\mathbf{w}_s = \mathbb{C}_0^s : \mathbb{D}(x,\mathbf{w}_s) + \frac{c_0^s}{\lambda_0}q, \tag{1.3.11}$$

the state equation

$$q = p_f + \frac{\nu_0}{c_f^2}\frac{\partial p_f}{\partial t}\quad (q = p_f \text{ for } \nu_0 = 0), \tag{1.3.12}$$

and the homogenized momentum balance equation

$$\nabla\cdot\left(\lambda_0\,\mathfrak{N}_2^s : \mathbb{D}(x,\mathbf{w}_s) - q\,\mathbb{C}_1^s\right) + \hat{\rho}\mathbf{F} = 0, \tag{1.3.13}$$

completed with homogeneous boundary and initial conditions

$$\mathbf{w}_s(\mathbf{x},t) = 0,\ \mathbf{x}\in S,\ t\in(0,T),\ \nu_0\,p_f(\mathbf{x},0) = 0,\ \mathbf{x}\in\Omega; \tag{1.3.14}$$

(5) *there exists* $\lambda_* > 0$, *such that for all* $\lambda_0 > \lambda_*$ *the problem (1.3.11)–(1.3.14) has a unique solution.*

In (1.3.11), (1.3.13)

$$\hat{\rho} = m\,\rho_f + (1-m)\,\rho_s,$$

the symmetric strictly positively definite constant fourth-rank tensor $\mathfrak{N}^s_2$, matrices $\mathbb{C}^s_0$ and $\mathbb{C}^s_1$, and the constant c^s_0 are given below by formulae (1.3.26), (1.3.27) and (1.3.31) and do not depend on λ_0.

We refer to the problem (1.3.11)–(1.3.14) as the homogenized **model** $(\mathbb{IIF})_4$.

Theorem 1.7 *Let $\{\mathbf{w}^\varepsilon,\ p^\varepsilon\}$ be the weak solution of the problem (1.3.1)–(1.3.5), $\mathbf{w}^\varepsilon_s$ be an extension (1.2.9), the pore space be connected, and $\mu_1 < \infty$.*

Then

(1) for all $\nu_0 > 0$ the sequences $\{\chi^\varepsilon\, \mathbf{w}^\varepsilon\}$, $\left\{\chi^\varepsilon \dfrac{\partial \mathbf{w}^\varepsilon}{\partial t}\right\}$, $\{\chi^\varepsilon\, p^\varepsilon\}$, $\left\{\chi^\varepsilon \dfrac{\partial p^\varepsilon}{\partial t}\right\}$, and $\{q^\varepsilon\}$, where $q^\varepsilon = \chi^\varepsilon\left(p^\varepsilon + \dfrac{\alpha_\nu}{c^2_f}\dfrac{\partial\, p^\varepsilon}{\partial t}\right)$, converge weakly in $\mathbf{L}_2(\Omega_T)$ and $L_2(\Omega_T)$ (up to some subsequences) to functions $\mathbf{w}^{(f)}$, $\dfrac{\partial \mathbf{w}^{(f)}}{\partial t}$, $m\,\dfrac{\partial p_f}{\partial t}$, and $m\, q = m\left(p_f + \dfrac{\nu_0}{c^2_f}\dfrac{\partial\, p_f}{\partial t}\right) \in W^{1,0}_2(\Omega_T)$ respectively;

(2) for $\nu_0 = 0$ the sequences $\{\chi^\varepsilon\, \mathbf{w}^\varepsilon\}$, $\left\{\chi^\varepsilon \dfrac{\partial \mathbf{w}^\varepsilon}{\partial t}\right\}$, and $\{\chi^\varepsilon\, p^\varepsilon\}$ converge weakly in $\mathbf{L}_2(\Omega_T)$ and $L_2(\Omega_T)$ (up to some subsequences) to functions $\mathbf{w}$, $\dfrac{\partial \mathbf{w}^{(f)}}{\partial t}$, and $m\, p_f \in W^{1,0}_2(\Omega_T)$ respectively;

(3) for all $\nu_0 \geqslant 0$ sequences $\{\mathbf{w}^\varepsilon_s\}$ and $\left\{\dfrac{\partial \mathbf{w}^\varepsilon_s}{\partial t}\right\}$ converge weakly in $\mathbf{W}^{1,0}_2(\Omega_T)$ (up to some subsequences) to functions $\dfrac{\partial \mathbf{w}_s}{\partial t} \in \overset{\circ}{\mathbf{W}}{}^{1,0}_2(\Omega_T)$ respectively;

(4) limiting functions solve the system of homogenized equations in the domain Ω_T, consisting of the homogenized continuity equation

$$\frac{m}{c^2_f}\, p_f + \nabla\cdot\, \mathbf{w}^{(f)} = \mathbb{C}^s_0 : \mathbb{D}(x, \mathbf{w}_s) + \frac{c^s_0}{\lambda_0} q, \tag{1.3.15}$$

the state equation (1.3.12), the homogenized momentum balance equation (1.3.13) for the solid component, and Darcy's law in the form

$$\frac{\partial \mathbf{w}^{(f)}}{\partial t} = m\,\frac{\mathbf{w}_s}{\partial t} + \mathbb{B}\cdot\big(-\nabla q + \rho_f \mathbf{F}\big), \tag{1.3.16}$$

for the liquid component, completed with the homogeneous boundary and initial conditions (1.3.14), and the homogeneous boundary condition

$$\mathbf{w}^{(f)}(\mathbf{x}, t)\cdot \mathbf{n}(\mathbf{x}) = 0,\ \ \mathbf{x}\in S,\ \ t \in (0, T) \tag{1.3.17}$$

for the pressure p^f and displacements $\mathbf{w}^{(f)}$ of the fluid component.

(5) *there exists* $\lambda_* > 0$, *such that for all* $\lambda_0 > \lambda_*$ *the problem (1.3.13)–(1.3.17) has a unique solution.*
In (1.3.16) and (1.3.17) $\mathbf{n}$ *is the normal vector to the boundary S, the symmetric strictly positively definite constant matrix* $\mathbb{B}$ *is given in Theorem 1.1 of the present chapter.*

We refer to the problem (1.3.13)–(1.3.17) as the homogenized **model** $(\mathbb{IF})_5$.

Theorem 1.8 *Let* $\nu_0 = 0$ *and* $\mu_1 = \infty$, *or* $\mu_1 < \infty$, *but the pore space be disconnected.*

Then $q = p_f$ *and the solution* $\mathbf{w}_s$ *of the problem (1.3.11)–(1.3.14) satisfies in the domain* Ω_T *the homogenized equation*

$$\nabla \cdot \big(\lambda_0\, \mathfrak{N}_3^s : \mathbb{D}(x, \mathbf{w}_s)\big) + \hat{\rho}\mathbf{F} = 0 \tag{1.3.18}$$

with the boundary condition (1.3.14).

The symmetric strictly positive definite constant fourth-rank tensor $\mathfrak{N}_3^s$ *is given below by formula (1.3.39) and does not depend on* λ_0.

We refer to the problem (1.3.14), (1.3.18) as the homogenized **model** $(\mathbb{IF})_6$.

1.3.2 Proof of Theorem 1.5

The proof of estimates (1.3.9) and (1.3.10) for the function $\left(\dfrac{\partial \mathbf{w}^\varepsilon}{\partial t} - \dfrac{\partial \mathbf{w}_s^\varepsilon}{\partial t}\right)$ repeats the proof of estimates (1.2.12) and (1.2.13) in Theorem 1.2. We just have to use the equalities

$$\begin{aligned}
&\int_\Omega \chi^\varepsilon \Big(\alpha_\mu \Big|\mathbb{D}\Big(x, \frac{\partial \mathbf{w}^\varepsilon}{\partial t}\Big)\Big|^2 + \frac{\alpha_\nu}{c_f^2}\Big|\frac{\partial p^\varepsilon}{\partial t}\Big|^2\Big)dx \\
&\qquad + \frac{1}{2}\frac{d}{dt}\int_\Omega \Big((1-\chi^\varepsilon)\lambda_0|\mathbb{D}(x, \mathbf{w}^\varepsilon)|^2 + \frac{1}{\alpha_p^\varepsilon}|p^\varepsilon|^2\Big)dx = \int_\Omega \rho^\varepsilon \mathbf{F}\cdot\frac{\partial \mathbf{w}^\varepsilon}{\partial t}dx, \\
&\frac{1}{2}\frac{d}{dt}\int_\Omega \Big((1-\chi^\varepsilon)\lambda_0\Big|\mathbb{D}\Big(x, \frac{\partial \mathbf{w}^\varepsilon}{\partial t}\Big)\Big|^2 + \frac{1}{\alpha_p^\varepsilon}\Big|\frac{\partial p^\varepsilon}{\partial t}\Big|^2\Big)dx \\
&\qquad + \int_\Omega \chi^\varepsilon\Big(\alpha_\mu\Big|\mathbb{D}\Big(x, \frac{\partial^2 \mathbf{w}^\varepsilon}{\partial t^2}\Big)\Big|^2 + \frac{\alpha_\nu}{c_f^2}\Big|\frac{\partial^2 p^\varepsilon}{\partial t^2}\Big|^2\Big)dx = \int_\Omega \rho^\varepsilon\frac{\partial \mathbf{F}}{\partial t}\cdot\frac{\partial \mathbf{w}^\varepsilon}{\partial t}dx.
\end{aligned} \tag{1.3.19}$$

1.3.3 Proof of Theorem 1.6

On the strength of Theorem 1.5 sequences $\{\mathbf{w}^\varepsilon\}$, $\{\chi^\varepsilon \mathbf{w}^\varepsilon\}$, $\{\nabla \cdot \mathbf{w}^\varepsilon\}$, $\{\mathbf{w}_s^\varepsilon\}$, $\{\mathbb{D}(x, \mathbf{w}_s^\varepsilon)\}$, $\{\chi^\varepsilon p^\varepsilon\}$, $\{(1-\chi^\varepsilon)p^\varepsilon\}$, and $\{p^\varepsilon\}$ converge weakly in $\mathbf{L}_2(\Omega_T)$ and $L_2(\Omega_T)$ (up to some subsequences) to functions $\mathbf{w}$, $\mathbf{w}^{(f)}$, $\nabla \cdot \mathbf{w}$, $\mathbf{w}_s \in \overset{\circ}{\mathbf{W}}{}_2^{1,0}(\Omega_T)$, $\mathbb{D}(x, \mathbf{w}_s)$, $m\, p_f$, $(1-m)p_s$, and $p = m\, p_f + (1-m)p_s$ respectively, and sequences $\{\mathbf{w}^\varepsilon\}$, $\{\chi^\varepsilon \mathbf{w}^\varepsilon\}$, $\{\mathbf{w}_s^\varepsilon\}$, $\{\mathbb{D}(x, \mathbf{w}_s^\varepsilon)\}$, $\{\chi^\varepsilon p^\varepsilon\}$, and $\{(1-\chi^\varepsilon)p^\varepsilon\}$ converge two-scale in $\mathbf{L}_2(\Omega_T)$ and $L_2(\Omega_T)$ respectively to functions

$$\mathbf{W}(\mathbf{x},t,\mathbf{y}) = \mathbf{W}(\mathbf{x},t,\mathbf{y})\chi(\mathbf{y}) + (1-\chi(\mathbf{y}))\mathbf{w}_s(\mathbf{x},t),\;\; \mathbf{W}(\mathbf{x},t,\mathbf{y})\chi(\mathbf{y}),$$

$$\mathbf{w}_s(\mathbf{x},t),\;\; \mathbb{D}(x,\mathbf{w}_s) + \mathbb{D}\big(y, \mathbf{U}(\mathbf{x},t,\mathbf{y})\big),$$

$$p_f(\mathbf{x},t)\chi(\mathbf{y}),\;\; \textit{and}\;\; P_s(\mathbf{x},t,\mathbf{y}).$$

For $\nu_0 = 0$ the sequence $\{q^\varepsilon\}$ converges weakly in $L_2(\Omega_T)$ to function $m\, p_f$ and two-scale in $L_2(\Omega_T)$ to function $p_f\, \chi(\mathbf{y})$.

For $\nu_0 \geqslant 0$ sequences $\left\{\chi^\varepsilon \dfrac{\partial p^\varepsilon}{\partial t}\right\}$ and $\{q^\varepsilon\}$ converge weakly in $L_2(\Omega_T)$ to functions $m\,\dfrac{\partial p_f}{\partial t}$ and mq respectively. Passing to the limit as $\varepsilon \to 0$ in the state equation in the form

$$\int_{\Omega_T} \Big((q^\varepsilon - \chi^\varepsilon p^\varepsilon)\xi(\mathbf{x},t) + \frac{\alpha_\nu}{c_f^2}\chi^\varepsilon p^\varepsilon \frac{\partial \xi}{\partial t}(\mathbf{x},t)\Big)dxdt = 0,$$

we get the homogenized state equation

$$q = p_f + \frac{\nu_0}{c_f^2}\frac{\partial p_f}{\partial t},$$

and the initial condition

$$\nu_0\, p_f(\mathbf{x},0) = 0,\;\; \mathbf{x} \in \Omega.$$

At the same time the sequence $\{q^\varepsilon\}$ converges two-scale in $L_2(\Omega_T)$ to function $q(\mathbf{x},t)\,\chi(\mathbf{y})$.

As we have shown in the proof of Theorem 1.3, weak and two-scale limits $\mathbf{W}$, p_f, and q satisfy the microscopic system (1.2.25)–(1.2.27) and the state equation (1.3.12). Therefore, if the pore space is disconnected, then $\mathbf{W}(\mathbf{x},t,\mathbf{y})\chi(\mathbf{y}) = \mathbf{w}_s(\mathbf{x},t)\chi(\mathbf{y})$ and

$$\mathbf{W}(\mathbf{x},t,\mathbf{y}) = \mathbf{w}_s(\mathbf{x},t).$$

Due to estimate (1.3.10) the same equality holds true for the case $\mu_1 = \infty$.

Lemma 1.10 *Limiting functions* $\mathbf{w}_s$ *and* p_f *satisfy in the domain* Ω_T *the macroscopic continuity equation for the liquid component*

$$\frac{m}{c_f^2}p_f + m\nabla\cdot\mathbf{w}_s = \langle\nabla_y\cdot\mathbf{U}\rangle_{Y_s}. \tag{1.3.20}$$

Proof To prove (1.3.20) we rewrite the continuity equation (1.3.1) as

$$\int_{\Omega_T}\Big(\frac{1}{c_f^2}\chi^\varepsilon p^\varepsilon\xi(\mathbf{x},t) - \mathbf{w}^\varepsilon\cdot\nabla\xi(\mathbf{x},t)\Big)dxdt \\ = \int_{\Omega_T}(1-\chi^\varepsilon)\xi(\mathbf{x},t)\nabla\cdot\mathbf{w}_s^\varepsilon dxdt.$$

Passage to the limit as $\varepsilon\to 0$ gives as

$$\int_{\Omega_T}\Big(\frac{1}{c_f^2}\chi^\varepsilon p^\varepsilon\xi(\mathbf{x},t) - \mathbf{w}^\varepsilon\cdot\nabla\xi(\mathbf{x},t)\Big)dxdt \\ \to\int_{\Omega_T}\Big(\xi\frac{m}{c_f^2}p_f - \mathbf{w}_s\cdot\nabla\xi\Big)dxdt = \int_{\Omega_T}\xi\Big(\frac{m}{c_f^2}p_f + \nabla\cdot\mathbf{w}_s\Big)dxdt,$$

$$\int_{\Omega_T}(1-\chi^\varepsilon)\nabla\cdot\mathbf{w}_s^\varepsilon\xi(\mathbf{x},t)dxdt \\ \to\int_{\Omega_T}\xi\big((1-m)\nabla\cdot\mathbf{w}_s + \langle\nabla_y\cdot\mathbf{U}\rangle_{Y_s}\big)dxdt.$$

After reintegrating we arrive at (1.3.20).

In the last relation we took into account Nguetseng's theorem (see Appendix B).

Lemma 1.11 *Limiting functions* $\mathbf{w}_s$ *and* q *satisfy in the domain* Ω_T *the homogenized momentum balance equation (1.3.13).*

Proof Initially, using the continuity equation (1.3.1) in the domain Ω_s^ε in the form

$$(1-\chi^\varepsilon)p^\varepsilon = -c_s^2(1-\chi^\varepsilon)\nabla\cdot\mathbf{w}_s^\varepsilon,$$

we rewrite the integral identity (1.3.8) as

$$I_f^\varepsilon + I_s^\varepsilon = \int_{\Omega_T}\rho^\varepsilon\mathbf{F}\cdot\varphi\,dxdt, \tag{1.3.21}$$

where

$$I_f^\varepsilon = \int_{\Omega_T} \chi^\varepsilon \alpha_\mu \mathbb{D}(x, \mathbf{v}^\varepsilon) : \mathbb{D}(x, \varphi) dx dt + \int_{\Omega_T} \chi^\varepsilon\, q^\varepsilon \nabla \cdot \varphi dx dt,$$

$$\begin{aligned} I_s^\varepsilon &= \int_{\Omega_T} (1-\chi^\varepsilon)\big(\lambda_0 \mathbb{D}(x, \mathbf{w}_s^\varepsilon) : \mathbb{D}(x, \varphi) + c_s^2 (\nabla \cdot \mathbf{w}_s^\varepsilon)(\nabla \cdot \varphi)\big) dx dt \\ &= \lambda_0 \int_{\Omega_T} (1-\chi^\varepsilon)\big(\mathfrak{N}^{(0)} : \mathbb{D}(x, \mathbf{w}_s^\varepsilon)\big) : \mathbb{D}(x, \varphi) dx dt, \end{aligned}$$

and

$$\mathfrak{N}^{(0)} = \sum_{i,j=1}^{3} \mathbb{J}^{ij} \otimes \mathbb{J}^{ij} + \frac{c_s^2}{\lambda_0} \mathbb{I} \otimes \mathbb{I}.$$

Now we pass to the limit as $\varepsilon \to 0$ in the integral identity (1.3.21) with two different types of test functions. Firstly, with test functions $\varphi = \varphi(\mathbf{x}, t)$ and secondly, with test functions $\varphi = \varepsilon h(\mathbf{x}, t) \varphi_0\Big(\dfrac{\mathbf{x}}{\varepsilon}\Big)$. After standard reintegration we obtain the macroscopic momentum balance equation

$$\nabla \cdot \Big(\lambda_0\, \mathfrak{N}^{(0)} : \big((1-m)\, \mathbb{D}(x, \mathbf{w}_s) + \langle \mathbb{D}(y, \mathbf{U}) \rangle_{Y_s}\big)\Big) - m\, \nabla q + \hat{\rho} \mathbf{F} = 0, \quad (1.3.22)$$

and the microscopic momentum balance equation

$$\nabla_y \cdot \Big((1-\chi)\big(\mathfrak{N}^{(0)} : \big(\mathbb{D}(x, \mathbf{w}_s) + \mathbb{D}(y, \mathbf{U})\big) + \frac{1}{\lambda_0}\, q\, \mathbb{I}\big)\Big) = 0. \quad (1.3.23)$$

Note, that the sequence $\{q^\varepsilon\}$ converges weakly and two-scale in $L_2(\Omega_T)$ to mq for $\nu_0 > 0$, and converges weakly and two-scale in $L_2(\Omega_T)$ to mp_f for $\nu_0 = 0$.

To calculate $\mathfrak{N}_2^s$ and $\mathbb{C}_1^s$ we have to solve (1.3.23) and find $\langle \mathbb{D}(y, \mathbf{U}) \rangle_{Y_s}$ as an operator on $\mathbb{D}(x, \mathbf{w}_s)$ and q.

Let $\mathbf{U}_2^{(ij)}(\mathbf{y})$ and $\mathbf{U}_2^{(0)}(\mathbf{y})$ be solutions of periodic problems

$$\nabla_y \cdot \Big((1-\chi)\Big(\mathfrak{N}^{(0)} : \big(\mathbb{J}^{(ij)} + \mathbb{D}(y, \mathbf{U}_2^{(ij)})\big)\Big)\Big) = 0, \quad (1.3.24)$$

and

$$\nabla_y \cdot \Big((1-\chi)\big(\mathfrak{N}^{(0)} : \mathbb{D}(y, \mathbf{U}_2^{(0)}) + \mathbb{I}\big)\Big) = 0 \quad (1.3.25)$$

in Y.

The correct solvability of problems (1.3.24) and (1.3.25) follows from the energy equalities

$$\int_{Y_s}\Big(\mathfrak{N}^{(0)}:\mathbb{D}(y,\mathbf{U}_2^{(ij)})\Big):\mathbb{D}(y,\mathbf{U}_2^{(ij)})dy=-\int_{Y_s}\Big(\mathfrak{N}^{(0)}:\mathbb{J}^{(ij)}\Big):\mathbb{D}(y,\mathbf{U}_2^{(ij)})dy,$$
$$\int_{Y_s}\Big(\mathfrak{N}^{(0)}:\mathbb{D}(y,\mathbf{U}_2^{(0)})\Big):\mathbb{D}(y,\mathbf{U}_2^{(0)})dy=-\int_{Y_s}\mathbb{I}:\mathbb{D}(y,\mathbf{U}_2^{(0)})dy(=-\langle\nabla\cdot\mathbf{U}_2^{(0)}\rangle_{Y_s}),$$

and the corresponding energy estimates.

As before, we conclude that the solution $\mathbf{U}$ to the problem (1.3.23) has a form

$$\mathbf{U}(\mathbf{x},t,\mathbf{y})=\sum_{i,j=1}^{3}\mathbf{U}_2^{(ij)}(\mathbf{y})D_{ij}(\mathbf{x},t)+\frac{1}{\lambda_0}\,q(\mathbf{x},t)\,\mathbf{U}_2^{(0)}(\mathbf{y}).$$

Then

$$\begin{aligned}\langle\mathbb{D}(y,\mathbf{U})\rangle_{Y_s}&=\sum_{i,j=1}^{3}\langle\mathbb{D}(y,\mathbf{U}_2^{(ij)})\rangle_{Y_s}D_{ij}+\frac{1}{\lambda_0}\,q\,\langle\mathbb{D}(y,\mathbf{U}_2^{(0)})\rangle_{Y_s}\\&=\Big(\sum_{i,j=1}^{3}\langle\mathbb{D}(y,\mathbf{U}_2^{(ij)})\rangle_{Y_s}\otimes\mathbb{J}^{(ij)}\Big):\mathbb{D}(x,\mathbf{w}_s)+\frac{1}{\lambda_0}\,q\,\langle\mathbb{D}(y,\mathbf{U}_2^{(0)})\rangle_{Y_s},\end{aligned}$$

and

$$\mathfrak{N}_2^s=\mathfrak{N}^{(0)}:\Big((1-m)\sum_{i,j=1}^{3}\mathbb{J}^{ij}\otimes\mathbb{J}^{ij}+\sum_{i,j=1}^{3}\langle\mathbb{D}(y,\mathbf{U}_2^{(ij)})\rangle_{Y_s}\otimes\mathbb{J}^{(ij)}\Big)\qquad(1.3.26)$$

$$\mathbb{C}_1^s=m\,\mathbb{I}-\langle\mathbb{D}(y,\mathbf{U}_2^{(0)})\rangle_{Y_s}.\qquad(1.3.27)$$

Lemma 1.12 *The constant fourth-rank tensor $\mathfrak{N}_2^s$ is symmetric and strictly positively definite.*

Proof As before (see Lemmas 1.2.2 and 1.2.5), all properties of the tensor $\mathfrak{N}_2^s$ follow from the equality

$$\begin{aligned}&\int_{Y_s}\Big(\mathfrak{N}^{(0)}:\mathbb{D}(y,\mathbf{U}_2^{(ij)})\Big):\mathbb{D}(y,\mathbf{U}_2^{(kl)})dy\\&\qquad=-\int_{Y_s}\Big(\mathfrak{N}^{(0)}:\mathbb{J}^{(ij)}\Big):\mathbb{D}(y,\mathbf{U}_2^{(kl)})dy,\end{aligned}\qquad(1.3.28)$$

which is the result of multiplying the Eq. (1.3.24) for $\mathbf{U}_2^{(ij)}$ by $\mathbf{U}_2^{(kl)}$ and integration by parts over domain Y.

Let $\zeta=(\zeta_{ij})$ and $\eta=(\eta_{ij})$ be arbitrary symmetric matrices and

$$\mathbf{Y}_\zeta = \sum_{i,j=1}^{3} \mathbf{U}_2^{(ij)} \zeta_{ij}, \quad \mathbf{Y}_\eta = \sum_{i,j=1}^{3} \mathbf{U}_2^{(ij)} \eta_{ij}.$$

Then, after multiplying the equation (1.3.28) by $\zeta_{ij}\eta_{kl}$ and summing over all indices we arrive at

$$\langle(\mathfrak{N}^{(0)} : \mathbb{D}(y, \mathbf{Y}_\zeta)) : \mathbb{D}(y, \mathbf{Y}_\eta)\rangle_{Y_s} + \langle(\mathfrak{N}^{(0)} : \zeta) : \mathbb{D}(y, \mathbf{Y}_\eta)\rangle_{Y_s} = 0. \tag{1.3.29}$$

By definition

$$(\mathfrak{N}_2^s : \zeta) : \eta = \langle(\mathfrak{N}^{(0)} : \zeta) : \eta\rangle_{Y_s} + \langle(\mathfrak{N}^{(0)} : \eta) : \mathbb{D}(y, \mathbf{Y}_\zeta)\rangle_{Y_s}. \tag{1.3.30}$$

The sum of (1.3.29) and (1.3.30) gives us the equality

$$(\mathfrak{N}_2^s : \zeta) : \eta = \int_{Y_s} \left(\mathfrak{N}^{(0)} : (\mathbb{D}(y, \mathbf{Y}_\zeta) + \eta)\right) : (\mathbb{D}(y, \mathbf{Y}_\eta) + \zeta)dy,$$

which proves the statement of the lemma, because the constant fourth-rank tensor $\mathfrak{N}^{(0)}$ is obviously symmetric and strictly positively definite.

Lemma 1.13 *Limiting functions $\mathbf{w}_s$ and q satisfy in the domain Ω_T the homogenized continuity equation (1.3.11).*

Proof To prove the lemma we just have to express the right-hand side of (1.3.20) using (1.3.26):

$$\begin{aligned}
&\langle\nabla_y \cdot \mathbf{U}\rangle_{Y_s} - \sum_{i,j=1}^{3} \langle\nabla_y \cdot \mathbf{U}_2^{(ij)}\rangle_{Y_s} D_{ij} + \frac{1}{\lambda_0} q \langle\nabla_y \cdot \mathbf{U}_2^{(0)}\rangle_{Y_s} \\
&= \left(\sum_{i,j=1}^{3} \langle\nabla_y \cdot \mathbf{U}_2^{(ij)}\rangle_{Y_s} \mathbb{J}^{ij}\right) : \mathbb{D}(x, \mathbf{w}_s) + \left(\frac{1}{\lambda_0}\langle\nabla_y \cdot \mathbf{U}_2^{(0)}\rangle_{Y_s}\right) q.
\end{aligned}$$

Therefore, (1.3.11) holds, if

$$\mathbb{C}_0^s = \sum_{i,j=1}^{3} \langle\nabla_y \cdot \mathbf{U}_2^{(ij)}\rangle_{Y_s} \mathbb{J}^{ij}, \quad c_0^s = \langle\nabla_y \cdot \mathbf{U}_2^{(0)}\rangle_{Y_s}. \tag{1.3.31}$$

Lemma 1.3.2 shows that $c_0^s < 0$.

Lemma 1.14 *The problem (1.3.11)–(1.3.14) has a unique solution.*

Proof To prove the lemma for the case $\nu_0 > 0$ let us multiply equation (1.3.13) with $\mathbf{F} = 0$ by $\mathbf{w}_s$, integrate by parts over domain Ω and estimate the term containing q using Hölder's inequality and properties of the tensor $\mathfrak{N}_2^s$:

$$\int_{\Omega} \mathbb{D}(x, \mathbf{w}_s) : \big(\mathfrak{N}_2^s : \mathbb{D}(x, \mathbf{w}_s)\big)dx = \frac{1}{\lambda_0}\int_{\Omega} q\, \mathbb{C}_1^s : \mathbb{D}(x, \mathbf{w}_s)dx$$
$$\leqslant \frac{1}{2}\int_{\Omega} \mathbb{D}(x, \mathbf{w}_s) : \big(\mathfrak{N}_2^s : \mathbb{D}(x, \mathbf{w}_s)\big)dx + \frac{C_0}{\lambda_0}\int_{\Omega} q^2 dx.$$

In a sequel, we solve (1.3.12) and find p_f as

$$p_f = \frac{\nu_0}{c_f^2}\exp\Big(-\frac{\nu_0}{c_f^2}t\Big)\int_0^t \exp\Big(\frac{\nu_0}{c_f^2}\tau\Big)\, q(\mathbf{x}, \tau)d\tau,$$

substitute it into (1.3.11), multiply the result by q, integrate over Ω, and estimate q as

$$\int_{\Omega} q^2 dx \leqslant C_0 \int_{\Omega} \mathbb{D}(x, \mathbf{w}_s) : \big(\mathfrak{N}_2^s : \mathbb{D}(x, \mathbf{w}_s)\big)dx.$$

Thus,

$$\frac{1}{2}\int_{\Omega} \mathbb{D}(x, \mathbf{w}_s) : \big(\mathfrak{N}_2^s : \mathbb{D}(x, \mathbf{w}_s)\big)dx$$
$$\leqslant \frac{C_0}{\lambda_0}\int_{\Omega} q^2 dx \leqslant \frac{C_0}{\lambda_0}\int_{\Omega} \mathbb{D}(x, \mathbf{w}_s) : \big(\mathfrak{N}_2^s : \mathbb{D}(x, \mathbf{w}_s)\big)dx,$$

which implies $\mathbf{w}_s = 0,\ q = 0$ for

$$\lambda_0 > \lambda^* = 2\, C_0.$$

The case $\nu_0 = 0$ is considered in a similar way.

1.3.4 Proof of Theorem 1.7

The sequences $\{\chi^{\varepsilon}\mathbf{w}^{\varepsilon}\}$, $\Big\{\chi^{\varepsilon}\dfrac{\partial \mathbf{w}^{\varepsilon}}{\partial t}\Big\}$, $\{\mathbf{w}_s^{\varepsilon}\}$, $\Big\{\dfrac{\partial \mathbf{w}_s^{\varepsilon}}{\partial t}\Big\}$, $\{\mathbb{D}(x, \mathbf{w}_s^{\varepsilon})\}$, $\{\chi^{\varepsilon} p^{\varepsilon}\}$, $\Big\{\chi^{\varepsilon}\dfrac{\partial p^{\varepsilon}}{\partial t}\Big\}$, and $\{q^{\varepsilon}\}$ converge weakly in $\mathbf{L}_2(\Omega_T)$ and $L_2(\Omega_T)$ up to some subsequences to functions $\mathbf{w}^{(f)}$, $\dfrac{\partial \mathbf{w}^{(f)}}{\partial t}$, $\mathbf{w}_s \in \overset{\circ}{\mathbf{W}}{}_2^{1,0}(\Omega_T)$, $\dfrac{\partial \mathbf{w}_s}{\partial t} \in \overset{\circ}{\mathbf{W}}{}_2^{1,0}(\Omega_T)$, $\mathbb{D}(x, \mathbf{w}_s)$, $m p_f$, $m\dfrac{\partial p_f}{\partial t}$, and $m\, q$ respectively, and sequences $\{\mathbf{w}^{\varepsilon}\}$, $\{\chi^{\varepsilon}\mathbf{w}^{\varepsilon}\}$, $\Big\{\chi^{\varepsilon}\dfrac{\partial \mathbf{w}^{\varepsilon}}{\partial t}\Big\}$, $\{\mathbf{w}_s^{\varepsilon}\}$, $\{\mathbb{D}(x, \mathbf{w}_s^{\varepsilon})\}$, $\{\chi^{\varepsilon} p^{\varepsilon}\}$, $\Big\{\chi^{\varepsilon}\dfrac{\partial p^{\varepsilon}}{\partial t}\Big\}$, and $\{q^{\varepsilon}\}$ converge two-scale in $\mathbf{L}_2(\Omega_T)$ and $L_2(\Omega_T)$ to functions $\mathbf{W}(\mathbf{x}, t, \mathbf{y})$, $\mathbf{W}^{(f)}(\mathbf{x}, t, \mathbf{y})$, $\dfrac{\partial \mathbf{W}^{(f)}}{\partial t}$, $\mathbf{w}_s(\mathbf{x}, t)$, $\mathbb{D}(x, \mathbf{w}_s) + \mathbb{D}\big(y, \mathbf{U}(\mathbf{x}, t, \mathbf{y})\big)$, $\chi(\mathbf{y}) p_f(\mathbf{x}, t)$, $\chi(\mathbf{y})\dfrac{\partial p_f}{\partial t}(\mathbf{x}, t)$, and $\chi(\mathbf{y}) q(\mathbf{x}, t)$ respectively.

For $\nu_0 = 0$ the sequence $\{q^\varepsilon\}$ converges weakly in $L_2(\Omega_T)$ to mp_f and two-scale in $L_2(\Omega_T)$ to $p_f(\mathbf{x},t)\,\chi(\mathbf{y})$.

As we have shown before, weak and two-scale limits $\mathbf{W}^{(f)}$, p_f, and q satisfy the microscopic system (1.2.25)–(1.2.27) and the state equation (1.3.12). Therefore, for a connected pore space the limiting functions $\mathbf{w}^{(f)}$ and $q \in W_2^{1,0}(\Omega_T)$ satisfy in the domain Ω_T Darcy's law (1.3.16) and the boundary condition (1.3.17) on the boundary S. In the same way we show that limiting functions $\mathbf{w}_s$ and q satisfy in the domain Ω_T the homogenized momentum balance equation (1.3.13). So, we only must prove the following statement.

Lemma 1.15 *Limiting functions* $\mathbf{w}_s$ *and* p_f *satisfy the homogenized continuity equation (1.3.15).*

Proof As before in Lemma 1.3.1, we rewrite the continuity equation (1.3.1) as

$$\int_{\Omega_T}\Big(\frac{1}{c_f^2}\chi^\varepsilon p^\varepsilon \xi(\mathbf{x},t) - \mathbf{w}^\varepsilon\cdot\nabla\xi(\mathbf{x},t)\Big)dxdt = \int_{\Omega_T}(1-\chi^\varepsilon)\xi(\mathbf{x},t)\nabla\cdot\mathbf{w}_s^\varepsilon dxdt. \tag{1.3.32}$$

After two-scale passage to the limit as $\varepsilon \to 0$ in (1.3.32) we arrive at the integral identity

$$\int_{\Omega_T}\Big(\frac{m}{c_f^2}p_f\xi(\mathbf{x},t) - \big(\mathbf{w}^{(f)} + (1-m)\mathbf{w}_s\big)\cdot\nabla\xi(\mathbf{x},t)\Big)dxdt = \int_{\Omega_T}\int_{\Omega_T}\xi(\mathbf{x},t)\big((1-m)\nabla\cdot\mathbf{w}_s + \langle\nabla_y\cdot\mathbf{U}\rangle_{Y_s}\big)dxdt.$$

Reintegration transforms the last identity to the macroscopic continuity equation

$$\frac{m}{c_f^2}p_f + \nabla\cdot\mathbf{w}^{(f)} = \langle\nabla_y\cdot\mathbf{U}\rangle_{Y_s}. \tag{1.3.33}$$

The rest of the proof repeats the proof of the Lemma 1.13.

Lemma 1.16 *The problem (1.3.11)–(1.3.14) has a unique solution.*

Proof Let us consider the simplest case $\nu_0 = 0$. The case $\nu_0 > 0$ is considered in a similar way (see Lemma 1.3.5 and Lemma 1.2.6).

As in Lemma 1.3.5, we multiply equation (1.3.13) with $\mathbf{F} = 0$ by $\mathbf{w}_s$, integrate by parts over domain Ω_T and estimate the term with $q = p_f$ using Hölder's inequality and properties of the tensor $\mathfrak{N}_2^s$:

$$\int_{\Omega_T} \mathbb{D}(x,\mathbf{w}_s) : \big(\mathfrak{N}_2^s : \mathbb{D}(x,\mathbf{w}_s)\big)dxdt = \frac{1}{\lambda_0}\int_{\Omega_T} q\,\mathbb{C}_1^s : \mathbb{D}(x,\mathbf{w}_s)dxdt$$
$$\leqslant \frac{1}{2}\int_{\Omega_T} \mathbb{D}(x,\mathbf{w}_s) : \big(\mathfrak{N}_2^s : \mathbb{D}(x,\mathbf{w}_s)\big)dxdt + \frac{C_0}{\lambda_0}\int_{\Omega_T} p_f^2 dxdt.$$

In a sequel, we rewrite Eqs. (1.3.15) and (1.3.16) as

$$\Big(\frac{m}{c_f^2} - \frac{c_0^s}{\lambda_0}\Big)p_f - \nabla\cdot\Big(\mathbb{B}\cdot\int_0^t \nabla p_f(\mathbf{x},\tau)d\tau\Big) = \big(\mathbb{C}_0^s - m\,\mathbb{I}\big) : \mathbb{D}(x,\mathbf{w}_s),$$

multiply by p_f, integrate over $\Omega\times(0,t)$, and estimate p_f as

$$\int_0^t\int_\Omega p_f^2(\mathbf{x},\tau)dxd\tau + \frac{1}{m}\int_\Omega |\nabla\big(\int_0^t p_f(\mathbf{x},\tau)d\tau\big)|^2dx$$
$$\leqslant C_0\int_0^t\int_\Omega \mathbb{D}\big(x,\mathbf{w}_s(\mathbf{x},\tau)\big) : \Big(\mathfrak{N}_2^s : \mathbb{D}\big(x,\mathbf{w}_s(\mathbf{x},\tau)\big)\Big)dxd\tau.$$

Thus, $\mathbf{w}_s = 0,\ \ p_f = 0$ for $\lambda_0 > \lambda^*$.

1.3.5 Proof of Theorem 1.8

If $\mu_1 = \infty$, and $\nu_0 = 0$ then

$$\mathbf{w}_s = \mathbf{w},\ \text{ and } q = p_f,$$

and the homogenized system (1.3.11)–(1.3.13) takes the form

$$p_f = \frac{1}{\beta}\big(-m\,\mathbb{I} + \mathbb{C}_0^s\big) : \mathbb{D}(x,\mathbf{w}_s) = \widetilde{\mathbb{C}} : \mathbb{D}(x,\mathbf{w}_s),$$

$$\nabla\cdot\big(\lambda_0\,\mathfrak{N}_2^s : \mathbb{D}(x,\mathbf{w}_s) - p_f\,\mathbb{C}_1^s\big) + \hat{\rho}\mathbf{F} = 0,$$

where

$$\beta = \frac{m}{c_f^2} - \frac{c_0^s}{\lambda_0} > 0.$$

The last two equations are obviously transformed to

$$\nabla\cdot\Big(\big(\lambda_0\,\mathfrak{N}_2^s + \mathbb{C}_1^s\otimes\widetilde{\mathbb{C}}\big) : \mathbb{D}(x,\mathbf{w}_s)\Big) + \hat{\rho}\mathbf{F} = 0. \tag{1.3.34}$$

It seems that the shortest way to prove the statement of the theorem is to show that the tensor $\mathfrak{N}_3^s$ in the form

$$\mathfrak{N}_3^s = \lambda_0\,\mathfrak{N}_2^s + \mathbb{C}_1^s \otimes \widetilde{\mathbb{C}}$$

is symmetric and strictly positively definite. Unfortunately, we have no idea how to do it. So, we choose a more complicated route with more technical details, but leading to the right answer.

Using the macroscopic continuity equation (1.3.20) for the liquid component we rewrite the micro-and macroscopic momentum balance equations (1.3.22) and (1.3.23) as

$$\nabla_y \cdot \Big((1-\chi)\big(\mathbb{D}(y,\mathbf{U}) + \frac{c_s^2}{\lambda_0 m}(\nabla_y \cdot \mathbf{U})\mathbb{I} + \frac{c_f^2}{\lambda_0}\langle \nabla_y \cdot \mathbf{U}\rangle_{Y_s}\mathbb{I} \\ + \mathbb{D}(x,\mathbf{w}_s) + \Big(\frac{c_s^2 - c_f^2}{\lambda_0}\Big)(\nabla \cdot \mathbf{w}_s)\,\mathbb{I}\big)\Big) = 0. \tag{1.3.35}$$

$$\nabla \cdot \Big(\big((1-m)\mathbb{D}(x,\mathbf{w}_s) + \langle \mathbb{D}(y,\mathbf{U})\rangle_{Y_s}\big) + \Big(\frac{c_s^2 - c_f^2}{\lambda_0}\Big)\langle \nabla_y \cdot \mathbf{U}\rangle_{Y_s}\mathbb{I} \\ + \Big((1-m)\frac{c_s^2}{\lambda_0} + m\,\frac{c_f^2}{\lambda_0}\Big)\Big)\Big(\nabla \cdot \mathbf{w}_s)\,\mathbb{I}\Big) + \frac{1}{\lambda_0}\hat{\rho}\mathbf{F} = 0. \tag{1.3.36}$$

Setting in (1.3.35)

$$\mathbf{U} = \sum_{i,j=1}^{3} \mathbf{U}_3^{(ij)}(\mathbf{y})D_{ij} + \mathbf{U}_3^{(0)}(\mathbf{y})(\nabla \cdot \mathbf{w}_s),$$

we arrive at the following periodic-boundary value problems in Y_s:

$$\nabla_y \cdot \Big((1-\chi)\big(\mathbb{D}(y,\mathbf{U}_3^{(ij)}) + \mathbb{J}^{ij} + \frac{c_s^2}{\lambda_0}\nabla_y \cdot \mathbf{U}_3^{(ij)}\mathbb{I} \\ + \frac{c_f^2}{\lambda_0 m}\langle \nabla_y \cdot \mathbf{U}_3^{(ij)}\rangle_{Y_s}\mathbb{I}\big)\Big) = 0, \tag{1.3.37}$$

$$\nabla_y \cdot \Big((1-\chi)\big(\mathbb{D}(y,\mathbf{U}_3^{(0)}) + (\frac{c_s^2 - c_f^2}{\lambda_0})\mathbb{I} \\ + \frac{c_s^2}{\lambda_0}\nabla_y \cdot \mathbf{U}_3^{(0)}\mathbb{I} + \frac{c_f^2}{\lambda_0 m}\langle \nabla_y \cdot \mathbf{U}_3^{(0)}\rangle_{Y_s}\mathbb{I}\big)\Big) = 0. \tag{1.3.38}$$

As before, the correctness of these problems under the normalization conditions

$$\langle \mathbf{U}_3^{(ij)} \rangle_{Y_s} = \langle \mathbf{U}_3^{(0)} \rangle_{Y_s} = 0$$

follows from the energy equalities

$$\int_{Y_s} \Big(\big(\mathbb{D}(y, \mathbf{U}_3^{(ij)}) + \mathbb{J}^{ij} \big) : \mathbb{D}(y, \mathbf{U}_3^{(ij)}) + \frac{c_s^2}{\lambda_0} (\nabla_y \cdot \mathbf{U}_3^{(ij)})^2 \Big) dy \\ + \frac{c_f^2}{\lambda_0 m} \Big(\int_{Y_s} \nabla_y \cdot \mathbf{U}_3^{(ij)} dy \Big)^2 = 0,$$

$$\int_{Y_s} \Big(\mathbb{D}(y, \mathbf{U}_3^{(0)}) : \mathbb{D}(y, \mathbf{U}_3^{(0)}) + \frac{c_s^2}{\lambda_0} (\nabla_y \cdot \mathbf{U}_3^{(0)})^2 \Big) dy \\ + \frac{c_f^2}{\lambda_0 m} \Big(\int_{Y_s} \nabla_y \cdot \mathbf{U}_3^{(0)} dy \Big)^2 + \Big(\frac{c_s^2 - c_f^2}{\lambda_0} \Big) \int_{Y_s} \nabla_y \cdot \widetilde{\mathbf{U}}^0 dy = 0.$$

Thus

$$\mathfrak{N}_3^s = (1-m) \sum_{i,j=1}^{3} \mathbb{J}^{ij} \otimes \mathbb{J}^{ij} + \Big((1-m) \frac{c_s^2}{\lambda_0} + m \frac{c_f^2}{\lambda_0} \Big) \mathbb{I} \otimes \mathbb{I} \\ + \sum_{i,j=1}^{3} \langle \mathbb{D}(y, \mathbf{U}_3^{(ij)}) \rangle_{Y_s} \otimes \mathbb{J}^{ij} + \Big(\frac{c_s^2 - c_f^2}{\lambda_0} \Big) \sum_{i,j=1}^{3} \langle \nabla \cdot \mathbf{U}_3^{(ij)} \rangle_{Y_s} \mathbb{I} \otimes \mathbb{J}^{ij} \\ + \langle \mathbb{D}(y, \mathbf{U}_3^{(0)}) \rangle_{Y_s} \otimes \mathbb{I} + \Big(\frac{c_s^2 - c_f^2}{\lambda_0} \Big) \langle \nabla \cdot \mathbf{U}_3^{(0)} \rangle_{Y_s} \mathbb{I} \otimes \mathbb{I}. \qquad (1.3.39)$$

Let $\zeta = (\zeta_{ij})$ and $\eta = (\eta_{ij})$ be arbitrary symmetric matrices and

$$\mathbf{Y}_\zeta = \sum_{i,j=1}^{3} \mathbf{U}_3^{(ij)} \zeta_{ij}, \quad \mathbf{Y}_\eta = \sum_{i,j=1}^{3} \mathbf{U}_3^{(ij)} \eta_{ij}, \quad \mathbf{Y}_\zeta^0 = \mathbf{U}_3^{(0)} \operatorname{tr} \zeta, \quad \mathbf{Y}_\eta^0 = \mathbf{U}_3^{(0)} \operatorname{tr} \eta.$$

Then

$$\big(\mathfrak{N}_3^s : \zeta \big) : \eta = (1-m) \zeta : \eta + \Big((1-m) \frac{c_s^2}{\lambda_0} + m \frac{c_f^2}{\lambda_0} \Big) \operatorname{tr} \zeta \operatorname{tr} \eta \\ + \Big(\frac{c_s^2 - c_f^2}{\lambda_0} \Big) \langle \nabla \cdot \mathbf{Y}_\zeta \rangle_{Y_s} \operatorname{tr} \eta + \langle \mathbb{D}(y, \mathbf{Y}_\zeta) \rangle_{Y_s} : \eta \\ + \langle \mathbb{D}(y, \mathbf{Y}_\zeta^0) \rangle_{Y_s} : \eta + \Big(\frac{c_s^2 - c_f^2}{\lambda_0} \Big) \langle \nabla \cdot \mathbf{Y}_\eta^0 \rangle_{Y_s} \operatorname{tr} \zeta. \qquad (1.3.40)$$

The symmetry of the tensor $\mathfrak{N}_3^s$ in the form

$$\left(\mathfrak{N}_3^s : \zeta\right) : \eta = \left(\mathfrak{N}_3^s : \eta\right) : \zeta$$

follows from the equations

$$\begin{aligned}\big(\langle\mathbb{D}(y, \mathbf{U}_3^{(ij)}) : \mathbb{D}(y, \mathbf{U}_3^{(kl)})\rangle_{Y_s} + \langle\mathbb{D}(y, \mathbf{U}_3^{(kl)})\rangle_{Y_s} : \mathbb{J}^{ij}\big) \\ + \frac{c_s^2}{\lambda_0}\langle\nabla_y \cdot \mathbf{U}_3^{(ij)}\, \nabla_y \cdot \mathbf{U}_3^{(kl)}\rangle_{Y_s} \\ + \frac{c_f^2}{\lambda_0 m}\langle\nabla_y \cdot \mathbf{U}_3^{(ij)}\rangle_{Y_s}\, \langle\nabla_y \cdot \mathbf{U}_3^{(kl)}\rangle_{Y_s} = 0,\end{aligned} \qquad (1.3.41)$$

$$\begin{aligned}\langle\mathbb{D}(y, \mathbf{U}_3^{(0)}) : \mathbb{D}(y, \mathbf{U}_3^{(0)})\rangle_{Y_s} + \frac{c_s^2}{\lambda_0}\langle(\nabla_y \cdot \mathbf{U}_3^{(0)})^2\rangle_{Y_s} \\ + \left(\frac{c_s^2 - c_f^2}{\lambda_0}\right)\langle\nabla_y \cdot \mathbf{U}_3^{(0)}\rangle_{Y_s} + \frac{c_f^2}{\lambda_0 m}\left(\langle\nabla_y \cdot \mathbf{U}_3^{(0)}\rangle_{Y_s}\right)^2 = 0,\end{aligned} \qquad (1.3.42)$$

$$\begin{aligned}\big(\langle\mathbb{D}(y, \mathbf{U}_3^{(ij)}) : \mathbb{D}(y, \mathbf{U}_3^{(0)})\rangle_{Y_s} + \langle\mathbb{D}(y, \mathbf{U}_3^{(0)})\rangle_{Y_s} : \mathbb{J}^{ij}\big) \\ + \frac{c_s^2}{\lambda_0}\langle\nabla_y \cdot \mathbf{U}_3^{(ij)}\, \nabla_y \cdot \mathbf{U}_3^{(0)}\rangle_{Y_s} + \frac{c_f^2}{\lambda_0 m}\langle\nabla_y \cdot \mathbf{U}_3^{(ij)}\rangle_{Y_s}\, \langle\nabla_y \cdot \mathbf{U}_3^{(0)}\rangle_{Y_s} = 0,\end{aligned} \qquad (1.3.43)$$

$$\begin{aligned}\langle\mathbb{D}(y, \mathbf{U}_3^{(0)}) : \mathbb{D}(y, \mathbf{U}_3^{(kl)})\rangle_{Y_s} + \left(\frac{c_s^2 - c_f^2}{\lambda_0}\right)\langle\nabla_y \cdot \mathbf{U}_3^{(kl)}\rangle_{Y_s} \\ + \frac{c_s^2}{\lambda_0}\langle\nabla_y \cdot \mathbf{U}_3^{(0)}\, \nabla_y \cdot \mathbf{U}_3^{(kl)}\rangle_{Y_s} + \frac{c_f^2}{\lambda_0 m}\langle\nabla_y \cdot \mathbf{U}_3^{(0)}\rangle_{Y_s}\, \langle\nabla_y \cdot \mathbf{U}_3^{(kl)}\rangle_{Y_s},\end{aligned} \qquad (1.3.44)$$

which appear by multiplying of Eqs. (1.3.37) and (1.3.38) by $\mathbf{U}_3^{(kl)}$ and $\mathbf{U}^{(0)}$, and integration by parts.

In fact, let us rewrite these relations in the form

$$\begin{aligned}\langle\mathbb{D}(y, \mathbf{Y}_\zeta) : \mathbb{D}(y, \mathbf{Y}_\eta)\rangle_{Y_s} + \langle\mathbb{D}(y, \mathbf{Y}_\eta)\rangle_{Y_s} : \zeta \\ + \frac{c_s^2}{\lambda_0}\langle\nabla_y \cdot \mathbf{Y}_\zeta\, \nabla_y \cdot \mathbf{Y}_\eta\rangle_{Y_s} + \frac{c_f^2}{\lambda_0 m}\langle\nabla_y \cdot \mathbf{Y}_\zeta\rangle_{Y_s}\, \langle\nabla_y \cdot \mathbf{Y}_\eta\rangle_{Y_s} = 0,\end{aligned} \qquad (1.3.45)$$

$$\langle \mathbb{D}(y, \mathbf{Y}^0_\zeta) : \mathbb{D}(y, \mathbf{Y}^0_\eta)\rangle_{Y_s} + \frac{c_s^2}{\lambda_0}\langle \nabla_y \cdot \mathbf{Y}^0_\eta \, \nabla_y \cdot \mathbf{Y}^0_\zeta\rangle_{Y_s}$$
$$+ \left(\frac{c_s^2 - c_f^2}{\lambda_0}\right)\langle \nabla_y \cdot \mathbf{Y}^0_\zeta\rangle_{Y_s} \operatorname{tr}\eta + \frac{c_f^2}{\lambda_0 m}\langle \nabla_y \cdot \mathbf{Y}^0_\zeta\rangle_{Y_s} \langle \nabla_y \cdot \mathbf{Y}^0_\eta\rangle_{Y_s} = 0, \quad (1.3.46)$$

$$\langle \mathbb{D}(y, \mathbf{Y}_\zeta) : \mathbb{D}(y, \mathbf{Y}^0_\eta)\rangle_{Y_s} + \langle\langle \mathbb{D}(y, \mathbf{Y}^0_\eta)\rangle_{Y_s} : \zeta$$
$$+ \frac{c_s^2}{\lambda_0}\langle \nabla_y \cdot \mathbf{Y}_\zeta \, \nabla_y \cdot \mathbf{Y}^0_\eta\rangle_{Y_s} + \frac{c_f^2}{\lambda_0 m}\langle \nabla_y \cdot \mathbf{Y}_\zeta\rangle_{Y_s} \langle \nabla_y \cdot \mathbf{Y}^0_\eta\rangle_{Y_s} = 0, \quad (1.3.47)$$

$$\langle \mathbb{D}(y, \mathbf{Y}^0_\zeta) : \mathbb{D}(y, \mathbf{Y}_\eta)\rangle_{Y_s} + \left(\frac{c_s^2 - c_f^2}{\lambda_0}\right)\langle \nabla_y \cdot \mathbf{Y}_\eta\rangle_{Y_s} \operatorname{tr}\zeta$$
$$+ \frac{c_s^2}{\lambda_0}\langle \nabla_y \cdot \mathbf{Y}^0_\zeta \, \nabla_y \cdot \mathbf{Y}_\eta\rangle_{Y_s} + \frac{c_f^2}{\lambda_0 m}\langle \nabla_y \cdot \mathbf{Y}^0_\zeta\rangle_{Y_s} \langle \nabla_y \cdot \mathbf{Y}_\eta\rangle_{Y_s}, \quad (1.3.48)$$

and sum Eqs. (1.3.40) and (1.3.45)–(1.3.48):

$$\begin{aligned}
(\mathfrak{N}^s_3 : \zeta) : \eta = {} & (1-m)\zeta : \eta + \langle \mathbb{D}(y, \mathbf{Y}_\zeta)\rangle_{Y_s} : \eta + \langle \mathbb{D}(y, \mathbf{Y}_\eta)\rangle_{Y_s} : \zeta \\
& + \langle \mathbb{D}(y, \mathbf{Y}_\zeta) : \mathbb{D}(y, \mathbf{Y}_\eta)\rangle_{Y_s} + \langle \mathbb{D}(y, \mathbf{Y}^0_\zeta) : \mathbb{D}(y, \mathbf{Y}^0_\eta)\rangle_{Y_s} + \langle \mathbb{D}(y, \mathbf{Y}^0_\eta)\rangle_{Y_s} : \zeta \\
& + \mathbb{D}(y, \mathbf{Y}^0_\zeta)\rangle_{Y_s} : \eta + \langle \mathbb{D}(y, \mathbf{Y}_\zeta) : \mathbb{D}(y, \mathbf{Y}^0_\eta)\rangle_{Y_s} + \langle \mathbb{D}(y, \mathbf{Y}^0_\zeta) : \mathbb{D}(y, \mathbf{Y}_\eta)\rangle_{Y_s} \\
& + \frac{c_s^2}{\lambda_0}\Big((1-m)\operatorname{tr}\zeta \operatorname{tr}\eta + \langle \nabla \cdot \mathbf{Y}_\zeta\rangle_{Y_s} \operatorname{tr}\eta + \langle \nabla_y \cdot \mathbf{Y}_\eta\rangle_{Y_s} \operatorname{tr}\zeta \\
& + \langle \nabla_y \cdot \mathbf{Y}_\zeta \, \nabla_y \cdot \mathbf{Y}_\eta\rangle_{Y_s} + \langle \nabla_y \cdot \mathbf{Y}^0_\eta \, \nabla_y \cdot \mathbf{Y}^0_\zeta\rangle_{Y_s} + \langle \nabla_y \cdot \mathbf{Y}^0_\zeta\rangle_{Y_s} \operatorname{tr}\eta \\
& + \langle \nabla \cdot \mathbf{Y}^0_\eta\rangle_{Y_s} \operatorname{tr}\zeta + \langle \nabla_y \cdot \mathbf{Y}_\zeta \, \nabla_y \cdot \mathbf{Y}^0_\eta\rangle_{Y_s} + \langle \nabla_y \cdot \mathbf{Y}^0_\zeta \, \nabla_y \cdot \mathbf{Y}_\eta\rangle_{Y_s}\Big) \\
& + \frac{c_f^2}{\lambda_0 m}\Big(m^2 \operatorname{tr}\zeta \operatorname{tr}\eta - m\,\langle \nabla_y \cdot \mathbf{Y}_\zeta\rangle_{Y_s} \operatorname{tr}\eta - m\,\langle \nabla_y \cdot \mathbf{Y}_\eta\rangle_{Y_s} \operatorname{tr}\zeta \\
& + \langle \nabla_y \cdot \mathbf{Y}_\zeta\rangle_{Y_s} \langle \nabla_y \cdot \mathbf{Y}_\eta\rangle_{Y_s} + \langle \nabla_y \cdot \mathbf{Y}^0_\zeta\rangle_{Y_s} \langle \nabla_y \cdot \mathbf{Y}^0_\eta\rangle_{Y_s} - m\,\langle \nabla_y \cdot \mathbf{Y}^0_\zeta\rangle_{Y_s} \operatorname{tr}\eta \\
& + m\,\langle \nabla_y \cdot \mathbf{Y}^0_\eta\rangle_{Y_s} \operatorname{tr}\zeta + \langle \nabla_y \cdot \mathbf{Y}_\zeta\rangle_{Y_s} \langle \nabla_y \cdot \mathbf{Y}^0_\eta\rangle_{Y_s} + \langle \nabla_y \cdot \mathbf{Y}^0_\zeta\rangle_{Y_s} \langle \nabla_y \cdot \mathbf{Y}_\eta\rangle_{Y_s}\Big).
\end{aligned}$$

Thus

$$\begin{aligned}
(\mathfrak{N}^s_3 : \zeta) : \eta = {} & \langle\big(\mathbb{D}(y, \mathbf{Z}_\zeta) + \zeta\big) : \mathbb{D}(y, \mathbf{Z}_\eta) + \eta)\rangle_{Y_s} \\
& + \frac{c_s^2}{\lambda_0}\langle(\nabla_y \cdot \mathbf{Z}_\zeta + \operatorname{tr}\zeta)(\nabla_y \cdot \mathbf{Z}_\eta + \operatorname{tr}\eta)\rangle_{Y_s} \\
& + \frac{c_f^2}{\lambda_0 m}\big(\langle \nabla_y \cdot \mathbf{Z}_\zeta\rangle_{Y_s} - m \operatorname{tr}\zeta\big)\big(\langle \nabla_y \cdot \mathbf{Z}_\eta\rangle_{Y_s} - m \operatorname{tr}\eta\big),
\end{aligned} \quad (1.3.49)$$

where $\mathbf{Z}_\zeta = \mathbf{Y}_\zeta + \mathbf{Y}^0_\zeta$.

The last relation shows the symmetry of the tensor $\mathfrak{N}^s_3$. In particular, for $\zeta = \eta$

$$(\mathfrak{N}^s_3 : \zeta) : \zeta = \langle (\mathbb{D}(y, \mathbf{Z}_\zeta) + \zeta) : \mathbb{D}(y, \mathbf{Z}_\zeta) + \zeta) \rangle_{Y_s} + \frac{c_s^2}{\lambda_0} \langle (\nabla_y \cdot \mathbf{Z}_\zeta + \operatorname{tr} \zeta)^2 \rangle_{Y_s} + \frac{c_f^2}{\lambda_0 m} \big(\langle \nabla_y \cdot \mathbf{Z}_\zeta \rangle_{Y_s} - m \operatorname{tr} \zeta \big)^2. \tag{1.3.50}$$

Therefore,

$$(\mathfrak{N}^s_3 : \mathbb{D}(x, \mathbf{w}_s)) : \mathbb{D}(x, \mathbf{w}_s) \geqslant a_0 \, \mathbb{D}(x, \mathbf{w}_s) : \mathbb{D}(x, \mathbf{w}_s), \;\; a_0 = const > 0.$$

1.4 A Viscous Liquid in an Elastic Skeleton

In the present section as basic mathematical models at the microscopic level we consider the model $\mathbb{M}_{15}$ of the filtration of an incompressible liquid in an incompressible elastic solid skeleton and the model $\mathbb{M}_{14}$ of a filtration of compressible liquid in a compressible elastic solid skeleton for the case $\alpha_\nu = 0$.

The model $\mathbb{M}_{15}$ consists of the continuity equation

$$\nabla \cdot \mathbf{w} = 0, \;\; \mathbf{x} \in \Omega, \;\; t \in (0, T), \tag{1.4.1}$$

the momentum balance equation

$$\nabla \cdot \mathbb{P} + \rho^\varepsilon \mathbf{F} = 0, \;\; \mathbf{x} \in \Omega, \;\; t \in (0, T), \tag{1.4.2}$$

the state equation

$$\mathbb{P} = \chi^\varepsilon \mu_0 \mathbb{D}\Big(x, \frac{\partial \mathbf{w}}{\partial t}\Big) + (1 - \chi^\varepsilon) \lambda_0 \mathbb{D}(x, \mathbf{w}) - p \, \mathbb{I}, \tag{1.4.3}$$

the boundary condition

$$\mathbf{w}(\mathbf{x}, t) = 0, \;\; \mathbf{x} \in S, \;\; t \in (0, T), \tag{1.4.4}$$

and normalization and initial conditions

$$\int_\Omega p(\mathbf{x}, t) dx = 0, \; t \in (0, T), \;\; \chi^\varepsilon \, \mathbf{w}(\mathbf{x}, 0) = 0, \;\; \mathbf{x} \in \Omega. \tag{1.4.5}$$

The model $\mathbb{M}_{14}$ consists of the momentum balance equation (1.4.2), the state equation (1.4.3), the boundary condition (1.4.4), the continuity equation

$$\frac{1}{\alpha_p^\varepsilon}\, p + \nabla \cdot \mathbf{w} = 0, \tag{1.4.6}$$

and the initial condition

$$\chi^\varepsilon\, \mathbf{w}(\mathbf{x}, 0) = 0, \ \ \mathbf{x} \in \Omega. \tag{1.4.7}$$

Throughout this section we assume that conditions

$$0 < \mu_0,\ \lambda_0,\ c_f^2,\ c_s^2 < \infty, \tag{1.4.8}$$

and

$$\int_{\Omega_T} \Big(|\mathbf{F}|^2 + (1-\chi^\varepsilon) \Big|\frac{\partial \mathbf{F}}{\partial t}\Big|^2 \Big) dx dt = F_1^2 < \infty \tag{1.4.9}$$

hold true.

In (1.4.6), (1.4.8)

$$\alpha_p^\varepsilon = \chi^\varepsilon c_f^2 + (1-\chi^\varepsilon) c_s^2.$$

Definition 1.4 We say that the pair of functions $\{\mathbf{w}^\varepsilon,\ p^\varepsilon\}$ such that

$$\mathbf{w}^\varepsilon \in \overset{\circ}{\mathbf{W}}{}_2^{1,0}(\Omega_T),\ \chi^\varepsilon \nabla\Big(\frac{\partial \mathbf{w}^\varepsilon}{\partial t}\Big) \in \mathbf{L}_2(\Omega_T),\ p^\varepsilon \in L_2(\Omega_T)$$

is a weak solution of the problem (1.4.1)–(1.4.5), if it satisfies the continuity equation (1.4.1) almost everywhere in Ω_T, normalization and initial conditions (1.4.5) and an integral identity

$$\int_{\Omega_T} \Big(\chi^\varepsilon \mu_0 \mathbb{D}\Big(x, \frac{\partial \mathbf{w}^\varepsilon}{\partial t}\Big) + (1-\chi^\varepsilon)\lambda_0 \mathbb{D}(x, \mathbf{w}^\varepsilon)\Big) : \mathbb{D}(x, \varphi)\, dx dt$$
$$- \int_{\Omega_T} p^\varepsilon (\nabla \cdot \varphi)\, dx dt = \int_{\Omega_T} \rho^\varepsilon \mathbf{F} \cdot \varphi\, dx dt, \tag{1.4.10}$$

for any functions $\varphi \in \overset{\circ}{\mathbf{W}}{}_2^{1,0}(\Omega_T)$.

For the problem (1.4.2)–(1.4.4), (1.4.6), (1.4.7) we change the state equation (1.4.3), excluding therefrom the pressure by means of the continuity equation (1.4.6):

$$\mathbb{P} = \chi^\varepsilon \mu_0 \mathbb{D}\Big(x, \frac{\partial \mathbf{w}}{\partial t}\Big) + \mathfrak{N}_1^{(\varepsilon)} : \mathbb{D}(x, \mathbf{w}),$$

$$\mathfrak{N}_1^{(\varepsilon)} = (1-\chi^\varepsilon)\lambda_0 \sum_{i,j=1}^{3} \mathbb{J}^{(ij)} \otimes \mathbb{J}^{(ij)} + \alpha_p^\varepsilon \mathbb{I} \otimes \mathbb{I}.$$

Definition 1.5 We say that the function $\mathbf{w}^\varepsilon$, such that

$$\mathbf{w}^{\varepsilon} \in \overset{\circ}{\mathbf{W}}{}_2^{1,0}(\Omega_T),\ \chi^{\varepsilon}\nabla\left(\frac{\partial \mathbf{w}^{\varepsilon}}{\partial t}\right) \in \mathbf{L}_2(\Omega_T),$$

is a weak solution of the problem (1.4.2)–(1.4.4), (1.4.6), (1.4.7), if it satisfies the initial condition (1.4.5) and an integral identity

$$\int_{\Omega_T} \left(\chi^{\varepsilon}\mu_0\mathbb{D}\left(x, \frac{\partial \mathbf{w}^{\varepsilon}}{\partial t}\right) + \mathfrak{N}_1^{(\varepsilon)} : \mathbb{D}(x, \mathbf{w})\right) : \mathbb{D}(x, \varphi)\, dxdt$$
$$= \int_{\Omega_T} \rho^{\varepsilon}\mathbf{F}\cdot\varphi\, dxdt, \tag{1.4.11}$$

for any functions $\varphi \in \overset{\circ}{\mathbf{W}}{}_2^{1,0}(\Omega_T)$.

Integral identities (1.4.10) and (1.4.11) show, that $\mathbf{w}^{\varepsilon}$ possesses different smoothness in domains Ω_f^{ε} and Ω_s^{ε}. As in previous sections, to preserve the best properties, which the solution now has in the liquid part, we extend the function $\dfrac{\partial \mathbf{w}^{\varepsilon}}{\partial t}$ from the liquid part of the domain Ω onto its solid part Ω_s^{ε}:

$$\mathbf{v}^{\varepsilon} = \mathbb{E}_{\Omega_f^{\varepsilon}}\left(\frac{\partial \mathbf{w}^{\varepsilon}}{\partial t}\right),$$

such that

$$\chi^{\varepsilon}(\mathbf{x})\left(\frac{\partial \mathbf{w}^{\varepsilon}}{\partial t}(\mathbf{x}, t) - \mathbf{v}^{\varepsilon}(\mathbf{x}, t)\right) = 0, \quad \mathbf{x} \in \Omega,\ t \in (0, T),$$

$$\int_{\Omega}\left|^{\varepsilon}(\mathbf{x}, t)\right|^2 dx \leqslant C_0\int_{\Omega_f^{\varepsilon}}\left|\frac{\partial \mathbf{w}^{\varepsilon}}{\partial t}(\mathbf{x}, t)\right|^2 dx,$$
$$\int_{\Omega}\left|\mathbb{D}(x, \mathbf{v}^{\varepsilon}(\mathbf{x}, t))\right|^2 dx \leqslant C_0\int_{\Omega_f^{\varepsilon}}\left|\mathbb{D}\left(x, \frac{\partial \mathbf{w}^{\varepsilon}}{\partial t}(\mathbf{x}, t)\right)\right|^2 dx, \quad t \in (0, T),$$

where $\mathbf{v}^{\varepsilon} = \mathbb{E}_{\Omega_f^{\varepsilon}}\left(\dfrac{\partial \mathbf{w}^{\varepsilon}}{\partial t}\right)$, and C_0 is independent of ε and $t \in (0, T)$.

Theorem 1.9 *There exists a unique weak solution* $\{\mathbf{w}^{\varepsilon},\ p^{\varepsilon}\}$ *of the problem (1.4.1)–(1.4.5) and*

$$\int_{\Omega_T}\left(|\mathbf{w}^{\varepsilon}|^2 + |\mathbb{D}(x,^{\varepsilon})|^2\right)dxdt$$
$$+ \max_{0<t<T}\int_{\Omega}\left(|\mathbb{D}(x, \mathbf{w}^{\varepsilon})|^2 + |p\,\mathbf{v}^{\varepsilon}|^2\right)dx \leqslant C_0F_1^2, \tag{1.4.12}$$

where $\mathbf{v}^\varepsilon = \mathbb{E}_{\Omega^\varepsilon_f}\left(\dfrac{\partial \mathbf{w}^\varepsilon}{\partial t}\right)$, *and the constant* C_0 *is independent of the small parameter* ε.

Theorem 1.10 *The statement of Theorem 1.9 holds true for a weak solution* $\mathbf{w}^\varepsilon$ *of the problem (1.4.2)–(1.4.4), (1.4.6), (1.4.7).*

Theorem 1.11 *Let the pair* $\{\mathbf{w}^\varepsilon,\ p^\varepsilon\}$ *be the weak solution of the problem (1.4.1)–(1.4.5).*

Then

(1) *the sequences* $\{\mathbf{w}^\varepsilon\}$, $\{\nabla \mathbf{w}^\varepsilon\}$, $\{\mathbf{v}^\varepsilon\}$, $\{\nabla \mathbf{v}^\varepsilon\}$, *and* $\{p^\varepsilon\}$ *converge weakly in* $\mathbf{L}_2(\Omega_T)$ *and* $L_2(\Omega_T)$ *(up to some subsequences) to functions* $\mathbf{w}$, $\nabla \mathbf{w}$, $= \dfrac{\partial \mathbf{w}^\varepsilon}{\partial t}$, $\nabla = \nabla\left(\dfrac{\partial \mathbf{w}^\varepsilon}{\partial t}\right)$, *and* p *respectively;*

(2) *limiting functions solve the system of homogenized equations in the domain* Ω_T, *consisting of the homogenized continuity equation*

$$\nabla \cdot \mathbf{w} = 0, \tag{1.4.13}$$

the homogenized momentum balance equation

$$\nabla \cdot \widehat{\mathbb{P}} + \hat{\rho}\mathbf{F} = 0, \tag{1.4.14}$$

and the state equation

$$\widehat{\mathbb{P}} = -p\,\mathbb{I} + \mathfrak{N}_1 : \mathbb{D}(x, \frac{\partial \mathbf{w}}{\partial t}) + \mathfrak{N}_2 : \mathbb{D}(x, \mathbf{w}) + \int_0^t \mathfrak{N}_3(t-\tau) : \mathbb{D}(x, \mathbf{w}(\mathbf{x}, \tau))d\tau, \tag{1.4.15}$$

completed with the homogeneous boundary condition

$$\mathbf{w}(\mathbf{x}, t) = 0,\ \ \mathbf{x} \in S,\ \ t \in (0, T), \tag{1.4.16}$$

and the homogeneous initial condition

$$\mathbf{w}(\mathbf{x}, 0) = 0,\ \ \mathbf{x} \in \Omega; \tag{1.4.17}$$

(3) *if a pore space is connected, then the symmetric tensor* $\mathfrak{N}_1$ *is strictly positively definite. For the case of a disconnected pore space (isolated pores) the symmetric positively definite tensor* $\mathfrak{N}_1$ *degenerates and the tensor* $\mathfrak{N}_2$ *becomes strictly positively definite;*

(4) *the problem (1.4.13)–(1.4.17) has a unique solution.*

In (1.4.14), (1.4.15)

$$\hat{\rho} = m\,\rho_f + (1-m)\,\rho_s,$$

and fourth-rank tensors $\mathfrak{N}_1$, $\mathfrak{N}_2$, $\mathfrak{N}_3(t)$ *are given below by formulae (1.4.30).*

We refer to the problem (1.4.13)–(1.4.17) as the homogenized **model** $(\mathbb{IF})_7$.

Theorem 1.12 *Let* $\mathbf{w}^\varepsilon$ *be the weak solution of the problem (1.4.2)–(1.4.4), (1.4.6), (1.4.7).*

Then

(1) *the sequences* $\{\mathbf{w}^\varepsilon\}$, $\{\nabla\mathbf{w}^\varepsilon\}$, $\{\mathbf{v}^\varepsilon\}$, *and* $\{\nabla\mathbf{v}^\varepsilon\}$ *converge weakly in* $\mathbf{L}_2(\Omega_T)$ *(up to some subsequences) to functions* $\mathbf{w}$, $\nabla\mathbf{w}$, $\mathbf{v} = \dfrac{\partial\mathbf{w}^\varepsilon}{\partial t}$, *and* $\nabla\mathbf{v} = \nabla\Big(\dfrac{\partial\mathbf{w}^\varepsilon}{\partial t}\Big)$ *respectively;*

(2) *limiting functions solve the system of homogenized equations in the domain* Ω_T, *consisting of the homogenized momentum balance equation*

$$\nabla\cdot\widetilde{\mathbb{P}} + \hat{\rho}\mathbf{F} = 0, \tag{1.4.18}$$

and the state equation

$$\widetilde{\mathbb{P}} = \mathfrak{N}_4 : \mathbb{D}(x, \frac{\partial\mathbf{w}}{\partial t}) + \mathfrak{N}_5 : \mathbb{D}(x,\mathbf{w}) + \int_0^t \mathfrak{N}_6(t-\tau) : \mathbb{D}(x,\mathbf{w}(\mathbf{x},\tau))d\tau, \tag{1.4.19}$$

completed with the homogeneous boundary condition

$$\mathbf{w}(\mathbf{x},t) = 0,\;\; \mathbf{x}\in S,\;\; t\in(0,T), \tag{1.4.20}$$

and the homogeneous initial condition (1.4.17).

(3) *if a pore space is connected, then the symmetric tensor* $\mathfrak{N}_4$ *is strictly positively definite. For the case of a disconnected pore space (isolated pores)* $\mathfrak{N}_4 = 0$ *and the tensor* $\mathfrak{N}_5$ *becomes strictly positively definite.*

(4) *the problem (1.4.17)–(1.4.20) has a unique solution.*
In (1.4.19) fourth-rank tensors $\mathfrak{N}_4$ *and* $\mathfrak{N}_5$, *and fourth-rank tensor* $\mathfrak{N}_6(t)$ *are given below by formulae (1.4.44).*

We refer to the problem (1.4.17)–(1.4.20) as the homogenized **model** $(\mathbb{IF})_8$.

1.4.1 Proofs of Theorem 1.9 and 1.10

The proofs of these theorems are straightforward and repeat the proof of Theorems 1.2 and 1.5. We just outline the derivation of the estimate (1.4.12) for the first case of Theorem 1.9.

To do that, we rewrite (1.4.10) as

$$\int_{\Omega}\Big(\chi^{\varepsilon}\mu_0\mathbb{D}(x,\frac{\partial \mathbf{w}^{\varepsilon}}{\partial t})+(1-\chi^{\varepsilon})\lambda_0\mathbb{D}(x,\mathbf{w}^{\varepsilon})\Big):\mathbb{D}(x,\varphi)\,dx$$
$$-\int_{\Omega}p\,\mathbf{v}^{\varepsilon}(\nabla\cdot\varphi)\,dx=\int_{\Omega}\rho^{\varepsilon}\mathbf{F}\cdot\varphi\,dx,$$

put there $\varphi = \dfrac{\partial \mathbf{w}^{\varepsilon}}{\partial t}$ and get

$$\mu_0\int_{\Omega}\chi^{\varepsilon}\Big|\mathbb{D}\Big(x,\frac{\partial \mathbf{w}^{\varepsilon}}{\partial t}\Big)\Big|^2dx+\frac{\lambda_0}{2}\frac{d}{dt}\int_{\Omega}(1-\chi^{\varepsilon})\Big|\mathbb{D}\Big(x,\mathbf{w}^{\varepsilon}\Big)\Big|^2dx$$
$$=\int_{\Omega}\rho^{\varepsilon}\mathbf{F}\cdot\frac{\partial \mathbf{w}^{\varepsilon}}{\partial t}\,dx=I.$$

Next we integrate the last relation with respect to time, rewrite the right-hand side, and using the properties of the extension $\mathbf{v}^{\varepsilon}$ estimate the result from below and from above:

$$\frac{\mu_0}{C_0}\int_0^t\int_{\Omega}|\mathbb{D}(x,\mathbf{v}^{\varepsilon})|^2(\mathbf{x},\tau)dxd\tau+\frac{\lambda_0}{2}\int_{\Omega}(1-\chi^{\varepsilon})|\mathbb{D}(x,\mathbf{w}^{\varepsilon})|^2(\mathbf{x},t)dx$$
$$\leqslant\int_0^t I d\tau=\int_0^t\int_{\Omega}\rho_f\chi^{\varepsilon}bigg(\mathbf{F}\cdot\frac{\partial \mathbf{w}^{\varepsilon}}{\partial t}\Big)(\mathbf{x},\tau)dxd\tau$$
$$+\int_0^t\int_{\Omega}\rho_s(1-\chi^{\varepsilon})\Big(\mathbf{F}\cdot\frac{\partial \mathbf{w}^{\varepsilon}}{\partial t}\Big)(\mathbf{x},\tau)dxd\tau$$
$$=\int_0^t\int_{\Omega}\rho_f\chi^{\varepsilon}(\mathbf{F}\cdot\mathbf{v}^{\varepsilon})(\mathbf{x},\tau)dxd\tau+\int_{\Omega}\rho_s(1-\chi^{\varepsilon})(\mathbf{F}\cdot\mathbf{w}^{\varepsilon})(\mathbf{x},t)dx$$
$$-\int_0^t\int_{\Omega}\rho_s(1-\chi^{\varepsilon})\Big(\frac{\partial \mathbf{F}}{\partial t}\cdot\mathbf{w}^{\varepsilon}\Big)(\mathbf{x},\tau)dxd\tau$$
$$\leqslant\frac{\mu_0}{2\,C_0}\int_0^t\int_{\Omega}|\mathbb{D}(x,\mathbf{v}^{\varepsilon})|^2(\mathbf{x},\tau)dxd\tau+\frac{\lambda_0}{4}\int_{\Omega}(1-\chi^{\varepsilon})|\mathbb{D}(x,\mathbf{w}^{\varepsilon})|^2(\mathbf{x},t)dx$$
$$+C_0\int_0^t\int_{\Omega}(1-\chi^{\varepsilon})|\mathbb{D}(x,\mathbf{w}^{\varepsilon})|^2(\mathbf{x},\tau)dxd\tau+C_0F_1^2.$$

Here for simplicity we have supposed that $\mathbf{F}(\mathbf{x},0)=0$.

The rest of the proof is standard. We just have to use the evident inequality

$$\int_{\Omega}\chi^{\varepsilon}|\mathbb{D}(x,\mathbf{w}^{\varepsilon})|^2(\mathbf{x},t)dx\leqslant C_0\int_0^t\int_{\Omega}\chi^{\varepsilon}\Big|\mathbb{D}\Big(x,\frac{\partial \mathbf{w}^{\varepsilon}}{\partial t}\Big)\Big|^2(\mathbf{x},\tau)dxd\tau,$$

and the basic property of the extension operator

$$\int_{\Omega}\chi^{\varepsilon}\Big|\mathbb{D}\Big(x,\frac{\partial \mathbf{w}^{\varepsilon}}{\partial t}\Big)\Big|^2(\mathbf{x},t)dx\leqslant C_0\int_{\Omega}|\mathbb{D}(x,^{\varepsilon})|^2(\mathbf{x},t)dx.$$

1.4.2 Proof of Theorem 1.11

First of all, we use Lemma B.11 (see Appendix B), which states that

$$\int_{\Omega} |\mathbf{v}^{\varepsilon}(\mathbf{x},t)|^2 dx \leqslant C_0 \int_{\Omega^{\varepsilon}} |\mathbb{D}(x,\mathbf{v}^{\varepsilon})|^2 dx.$$

On the strength of Theorem 1.9, Lemma B.14 ($\mathbf{v} \in \overset{\circ}{\mathbf{W}}{}_2^{1,0}(\Omega_T)$), and Nguetseng's theorem sequences $\{\mathbf{w}^{\varepsilon}\}$, $\{\mathbf{v}^{\varepsilon}\}$, $\{\mathbb{D}(x,\mathbf{w}^{\varepsilon})\}$, $\{\mathbb{D}(x,\mathbf{v}^{\varepsilon})\}$, and $\{p^{\varepsilon}\}$ converge as $\varepsilon \to 0$ weakly in $\mathbf{L}_2(\Omega_T)$ and $L_2(\Omega_T)$ (up to some subsequences) to functions $\mathbf{w} \in \overset{\circ}{\mathbf{W}}{}_2^{1,0}(\Omega_T)$, $\mathbf{v} \in \overset{\circ}{\mathbf{W}}{}_2^{1,0}(\Omega_T)$, $\mathbb{D}(x,\mathbf{w})$, $\mathbb{D}(x,\mathbf{v})$, and p respectively and converge two-scale in $\mathbf{L}_2(\Omega_T)$ and $L_2(\Omega_T)$ respectively to 1-periodic in $\mathbf{y}$ functions

$$\mathbf{w}(\mathbf{x},t),\ \mathbf{v}(\mathbf{x},t),\ \mathbb{D}(x,\mathbf{w}) + \mathbb{D}\big(y,\mathbf{W}(\mathbf{x},t,\mathbf{y})\big),$$

$$\mathbb{D}(x,\mathbf{v}) + \mathbb{D}\big(y,\mathbf{W}(\mathbf{x},t,\mathbf{y})\big),\ \textit{and}\ P(\mathbf{x},t,\mathbf{y}).$$

Lemma 1.17 *For almost all* $(\mathbf{x},t,\mathbf{y}) \in \Omega_T \times Y_f$

$$\mathbf{w}(\mathbf{x},t) = \int_0^t \mathbf{v}(\mathbf{x},\tau)d\tau,\ \ \mathbf{W}(\mathbf{x},t,\mathbf{y}) = \int_0^t \mathbf{V}(\mathbf{x},\mathbf{y},\tau)d\tau,$$

or

$$\mathbf{v}(\mathbf{x},t) = \frac{\partial \mathbf{w}}{\partial t}(\mathbf{x},t),\ \ \mathbf{V}(\mathbf{x},t,\mathbf{y}) = \frac{\partial \mathbf{W}}{\partial t}(\mathbf{x},t,\mathbf{y}),$$

and the initial condition (1.4.17) holds true.

Proof For almost all $\mathbf{x} \in \Omega$

$$\chi^{\varepsilon}\mathbf{w}^{\varepsilon}(\mathbf{x},t) = \int_0^t \chi^{\varepsilon}\mathbf{v}^{\varepsilon}(\mathbf{x},\tau)d\tau.$$

Therefore,

$$\int_{\Omega_T} \psi(t)\varphi_0(\mathbf{x}) \cdot \mathbf{w}^{\varepsilon}(\mathbf{x},t)\varphi_1\Big(\frac{\mathbf{x}}{\varepsilon}\Big)\chi\Big(\frac{\mathbf{x}}{\varepsilon}\Big)dxdt$$

$$= \int_{\Omega_T} \Big(\int_t^T \psi(\tau)d\tau\Big)\varphi_0(\mathbf{x}) \cdot \mathbf{v}^{\varepsilon}(\mathbf{x},t)\varphi_1\Big(\frac{\mathbf{x}}{\varepsilon}\Big)\chi\Big(\frac{\mathbf{x}}{\varepsilon}\Big)dxdt$$

for smooth 1-periodic in $\mathbf{y}$ arbitrary functions $\psi(t)$, $\varphi_0(\mathbf{x})$, and $\varphi_1(\mathbf{y})$.

The two-scale limit as $\varepsilon \to 0$ and reintegration results in the identity

$$0 = \int_{\Omega_T} \left(\left(\int_t^T \psi(\tau) d\tau \right) \mathbf{v}(\mathbf{x}, t) - \psi(t) \mathbf{w}(\mathbf{x}, t) \right) dx dt$$
$$= \int_{\Omega_T} \psi(t) \left(\int_0^t \mathbf{v}(\mathbf{x}, \tau) d\tau - \mathbf{w}(\mathbf{x}, t) \right) dx dt,$$

which proves the statement of the lemma for $\mathbf{w}(\mathbf{x}, t)$.

The proof of the lemma for $\mathbf{W}(\mathbf{x}, t, \mathbf{y})$ is the same.

To derive the continuity equation (1.4.13) we fulfil the usual $L_2(\Omega_T)$ limit as $\varepsilon \to 0$ in the continuity equation (1.4.1). After the two-scale limit in (1.4.1) we arrive at the microscopic continuity equation

$$\nabla_y \cdot \mathbf{W} = 0, \quad \mathbf{y} \in Y. \tag{1.4.21}$$

Next we pass to the limit as $\varepsilon \to 0$ in the integral identity (1.4.10) with two different types of test functions. First, with test functions $\varphi = \varphi(\mathbf{x}, t)$, and then with test functions $\varphi = \varepsilon h(\mathbf{x}, t) \varphi_0 \left(\frac{\mathbf{x}}{\varepsilon} \right)$. After the standard reintegrating we obtain the macroscopic momentum balance equation

$$\nabla \cdot \widehat{\mathbb{P}} + \hat{\rho} \mathbf{F} = 0, \tag{1.4.22}$$

$$\widehat{\mathbb{P}} = \mu_0 \left(m \, \mathbb{D}\left(x, \frac{\partial \mathbf{w}}{\partial t} \right) + \langle \mathbb{D}\left(y, \frac{\partial \mathbf{W}}{\partial t} \right) \rangle_{Y_f} \right)$$
$$+ \lambda_0 \big((1 - m) \, \mathbb{D}(x, \mathbf{w}) + \langle \mathbb{D}(y, \mathbf{W}) \rangle_{Y_s} \big) - p \, \mathbb{I}, \tag{1.4.23}$$

and the microscopic momentum balance equation

$$\nabla_y \cdot \left(\mu_0 \, \chi \left(\mathbb{D}(x, \frac{\partial \mathbf{w}}{\partial t}) + \mathbb{D}\left(y, \frac{\partial \mathbf{W}}{\partial t} \right) \right) \right.$$
$$\left. + \lambda_0 \, (1 - \chi) \big(\mathbb{D}(x, \mathbf{w}) + \mathbb{D}(y, \mathbf{W}) \big) - P \, \mathbb{I} \right) = 0.$$

The last one we rewrite as

$$\nabla_y \cdot \left(\chi \left(\mu_0 \, \mathbb{D}\left(y, \frac{\partial \mathbf{W}}{\partial t} \right) + \mathbb{Z} \right) + \lambda_0 \, (1 - \chi) \mathbb{D}(y, \mathbf{W}) - P \, \mathbb{I} \right) = 0, \tag{1.4.24}$$

where

$$\mathbb{Z}(\mathbf{x}, t) = \mu_0 \mathbb{D}\left(x, \frac{\partial \mathbf{w}}{\partial t} \right) - \lambda_0 \mathbb{D}(x, \mathbf{w}) = \sum_{i,j=1}^{3} Z_{ij}(\mathbf{x}, t) \mathbb{J}^{(ij)}.$$

To find tensors $\mathfrak{N}_1$, $\mathfrak{N}_2$, and $\mathfrak{N}_3(t)$ we have to solve the problem (1.4.21) and (1.4.24), find $\mathbb{D}\left(y, \dfrac{\partial \mathbf{W}}{\partial t}\right)$ and $\mathbb{D}(y, \mathbf{W})$ as operators on $\mathbb{D}\left(x, \dfrac{\partial \mathbf{w}}{\partial t}\right)$ and $\mathbb{D}(x, \mathbf{w})$, and substitute these expressions into (1.4.23).

Let $\{\mathbf{W}^{(ij)}(\mathbf{y}, t),\ P^{(ij)}(\mathbf{y}, t)\}$ and $\{\mathbf{W}_0^{(ij)}(\mathbf{y}),\ P_0^{(ij)}(\mathbf{y})\}\, i, j = 1, 2, 3$ be solutions of periodic problems

$$\left.\begin{aligned} \nabla_y \cdot \Big(\chi\, \mu_0\, \mathbb{D}\Big(y, \frac{\partial \mathbf{W}^{(ij)}}{\partial t}\Big) \\ + \lambda_0\,(1-\chi)\mathbb{D}(y, \mathbf{W}^{(ij)}) - P^{(ij)}\,\mathbb{I}\Big) &= 0, \\ \nabla_y \cdot \mathbf{w}^{(ij)} &= 0, \\ \chi(\mathbf{y})\mathbf{W}^{(ij)}(\mathbf{y}, 0) &= \mathbf{W}_0^{(ij)}(\mathbf{y}), \end{aligned}\right\} \tag{1.4.25}$$

$$\left.\begin{aligned} \nabla_y \cdot \Big(\chi\,\big(\mu_0\,\mathbb{D}(y, \mathbf{W}_0^{(ij)}) + \mathbb{J}^{(ij)} - P_0^{(ij)}\,\mathbb{I}\big)\Big) &= 0, \\ \nabla_y \cdot \mathbf{W}_0^{(ij)} = 0, \quad \int_Y \chi(\mathbf{y})\mathbf{W}_0^{(ij)}(\mathbf{y})dy &= 0, \end{aligned}\right\} \tag{1.4.26}$$

in the domain Y.

Then

$$\mathbf{W}(\mathbf{x}, t, \mathbf{y}) = \sum_{i,j=1}^{3} \int_0^t \mathbf{W}^{(ij)}(\mathbf{y}, t-\tau) Z_{ij}(\mathbf{x}, \tau)d\tau,$$

and

$$P(\mathbf{x}, t, \mathbf{y}) = \chi(\mathbf{y}) \sum_{i,j=1}^{3} P_0^{(ij)}(\mathbf{y}) Z_{ij}(\mathbf{x}, t) + \sum_{i,j=1}^{3} \int_0^t P^{(ij)}(\mathbf{y}, t-\tau) Z_{ij}(\mathbf{x}, \tau)d\tau,$$

$$\begin{aligned} \mathbb{D}(y, \mathbf{W}) &= \sum_{i,j=1}^{3} \int_0^t \mathbb{D}(y, \mathbf{W}^{(ij)}(\mathbf{y}, t-\tau)) Z_{ij}(\mathbf{x}, \tau)d\tau \\ &= \sum_{i,j=1}^{3} \int_0^t \big(\mathbb{D}(y, \mathbf{W}^{(ij)}(\mathbf{y}, t-\tau)) \otimes \mathbb{J}^{(ij)}\big) : \mathbb{Z}(\mathbf{x}, \tau)d\tau \\ &= \Big(\mu_0 \sum_{i,j=1}^{3} \mathbb{D}(y, \mathbf{W}_0^{(ij)}) \otimes \mathbb{J}^{(ij)}\Big) : \mathbb{D}(x, \mathbf{w}) \\ &\quad - \sum_{i,j=1}^{3} \int_0^t \Big(\big(\lambda_0\, \mathbb{D}(y, \mathbf{W}^{(ij)}(\mathbf{y}, t-\tau)\big) \end{aligned}$$

$$+ \mu_0 \mathbb{D}(y, \frac{\partial \mathbf{w}^{(ij)}}{\partial \tau}(\mathbf{y}, t-\tau))\big) \otimes \mathbb{J}^{(ij)}\Big) : \mathbb{D}(x, \mathbf{w}(\mathbf{x}, \tau))d\tau.$$

Using the evident relation

$$\frac{\partial \mathbf{W}^{(ij)}}{\partial \tau}(\mathbf{y}, t-\tau) = -\frac{\partial \mathbf{W}^{(ij)}}{\partial t}(\mathbf{y}, t-\tau),$$

one has

$$\mathbb{D}(y, \mathbf{W}) = \mathfrak{A}_0(\mathbf{y}) : \mathbb{D}(x, \mathbf{w}) + \int_0^t \mathfrak{A}_1(\mathbf{y}, t-\tau) : \mathbb{D}\big(x, \mathbf{ww}(\mathbf{x}, \tau)\big)d\tau, \quad (1.4.27)$$

where

$$\mathfrak{A}_0(\mathbf{y}) = \mu_0 \sum_{i,j=1}^{3} \mathbb{D}\big(y, \mathbf{W}_0^{(ij)}(\mathbf{y})\big) \otimes \mathbb{J}^{(ij)} \quad (1.4.28)$$

and

$$\mathfrak{A}_1(\mathbf{y}, t) = \sum_{i,j=1}^{3} \Big(\mu_0 \mathbb{D}\big(y, \frac{\partial \mathbf{W}^{(ij)}}{\partial t}(\mathbf{y}, t)\big) - \lambda_0 \mathbb{D}\big(y, \mathbf{W}^{(ij)}(\mathbf{y}, t)\big)\Big) \otimes \mathbb{J}^{(ij)}. \quad (1.4.29)$$

Equations (1.4.27) and (1.4.28) result in

$$\mathbb{D}\left(y, \frac{\partial \mathbf{W}}{\partial t}\right) = \mathfrak{A}_0(\mathbf{y}) : \mathbb{D}\left(x, \frac{\partial \mathbf{w}}{\partial t}(\mathbf{x}, t)\right) + \mathfrak{A}_1(\mathbf{y}, 0) : \mathbb{D}\big(x, \mathbf{w}(\mathbf{x}, t)\big)$$
$$+ \int_0^t \frac{\partial \mathfrak{A}_1}{\partial t}(\mathbf{y}, t-\tau) : \mathbb{D}\big(x, \mathbf{w}(\mathbf{x}, \tau)\big)d\tau.$$

Therefore,

$$\left.\begin{aligned} \mathfrak{N}_1 &= \mu_0 m \sum_{i,j=1}^{3} \mathbb{J}^{(ij)} \otimes \mathbb{J}^{(ij)} + \mu_0 \langle \mathfrak{A}_0 \rangle_{Y_f}, \\ \mathfrak{N}_2 &= \lambda_0 (1-m) \sum_{i,j=1}^{3} \mathbb{J}^{(ij)} \otimes \mathbb{J}^{(ij)} + \lambda_0 \langle \mathfrak{A}_0 \rangle_{Y_s} + \mu_0 \langle \mathfrak{A}_1(\mathbf{y}, 0) \rangle_{Y_f} \\ \mathfrak{N}_3(t) &= \mu_0 \left\langle \frac{\partial \mathfrak{A}_1}{\partial t}(\mathbf{y}, t) \right\rangle_{Y_f} + \lambda_0 \langle \mathfrak{A}_1(\mathbf{y}, t) \rangle_{Y_s}. \end{aligned}\right\} \quad (1.4.30)$$

Lemma 1.18 *All quantities in (1.4.30) are well-defined by virtue of the correct solvability of problems (1.4.25) and (1.4.26).*

Proof The statement of the lemma and the infinite smoothness of the solution with respect to time is a consequence of energy estimates. The latter follow from energy identities. In fact, the first chain of identities

$$\frac{1}{2}\int_Y \chi\,\mu_0\left|\mathbb{D}\big(y,\mathbf{W}^{(ij)}(\mathbf{y},t)\big)\right|^2 dy + \int_0^t\int_Y (1-\chi)\lambda_0\left|\mathbb{D}\big(y,\mathbf{W}^{(ij)}(\mathbf{y},\tau)\big)\right|^2 dy d\tau$$
$$= \frac{1}{2}\int_Y \chi\,\mu_0\left|\mathbb{D}\big(y,\mathbf{W}_0^{(ij)}(\mathbf{y})\big)\right|^2 dy,$$

$$\int_Y \chi\,\mu_0\left|\mathbb{D}\big(y,\mathbf{W}_0^{(ij)}(\mathbf{y})\big)\right|^2 dy + \int_Y \chi\mathbb{D}\big(y,\mathbf{W}_0^{(ij)}(\mathbf{y})\big) : \mathbb{J}^{(ij)} dy = 0, \qquad (1.4.31)$$

is the result of multiplying of the first equation in (1.4.25) by $\mathbf{W}^{(ij)}$ and integration by parts over $Y\times(0,t)$, and of the first equation in (1.4.26) by $\mathbf{W}_0^{(ij)}$ and integration by parts over Y. These identities provides estimates

$$\max_{0<t<T}\int_{Y_f}|\mathbb{D}\big(y,\mathbf{W}^{(ij)}(\mathbf{y},t)\big)|^2 dy + \int_0^T\int_{Y_s}|\mathbb{D}\big(y,\mathbf{W}^{(ij)}\big)|^2 dy dt \leqslant C_0.$$

Considering $(1-\chi(\mathbf{y}))\mathbf{W}^{(ij)}(\mathbf{y},0)$ as a periodic solution of the Stokes system

$$\nabla_y\cdot\left(\lambda_0\,\mathbb{D}(y,\mathbf{W}^{(ij)}) - P^{(ij)}\,\mathbb{I}\right) = 0,\ \ \nabla_y\cdot\mathbf{W}^{(ij)} = 0,$$

in Y_s, coinciding on the boundary γ with the function

$$\chi(\mathbf{y})\mathbf{W}^{(ij)}(\mathbf{y},0)\in\mathbf{W}_2^1(Y_s),$$

we obtain [56]

$$\int_{Y_s}|\mathbb{D}\big(y,\mathbf{W}^{(ij)}(\mathbf{y},0)\big)|^2 dy \leqslant C_0.$$

This estimate and (1.4.25) at $t=0$ imply

$$\int_{Y_f}\left|\mathbb{D}\Big(y,\frac{\partial\mathbf{W}^{(ij)}}{\partial t}(\mathbf{y},0)\Big)\right|^2 dy \leqslant C_0.$$

We can repeat the procedure repeatedly and finally prove the lemma.

Lemma 1.19 *If a pore space is connected, then the symmetric tensor $\mathfrak{N}_1$ is strictly positive definite. For the case of a disconnected pore space (isolated pores) the symmetric positively definite tensor $\mathfrak{N}_1$ degenerates and the symmetric tensor $\mathfrak{N}_2$ becomes strictly positive definite.*

Proof Let the pore space be connected, $\zeta = (\zeta_{ij})$ and $\eta = (\eta_{ij})$ be arbitrary symmetric matrices and

$$\mathbf{Y}_\zeta = \sum_{i,j=1}^{3} \mathbf{W}_0^{(ij)} \zeta_{ij}, \quad \mathbf{Y}_\eta = \sum_{i,j=1}^{3} \mathbf{W}_0^{(ij)} \eta_{ij}.$$

By definition

$$(\mathfrak{N}_1 : \zeta) : \eta = \mu_0 \, m \, \zeta : \eta + \mu_0^2 \langle \mathbb{D}(y, \mathbf{Y}_\zeta) \rangle_{Y_f} : \eta.$$

Next, we use equalities

$$\int_Y \chi \, \mu_0 \, \mathbb{D}(y, \mathbf{W}_0^{(ij)}) : \mathbb{D}(y, \mathbf{W}_0^{(kl)}) dy + \int_Y \chi \mathbb{D}(y, \mathbf{W}_0^{(kl)})) : \mathbb{J}^{(ij)} dy = 0$$

for $i, j = 1, 2, 3$, which are simple consequences of (1.4.31) and arrive at

$$\mu_0 \, \mathbb{D}(y, \mathbf{Y}_\eta)\rangle_{Y_f} : \zeta + \mu_0^2 \, \langle \mathbb{D}(y, \mathbf{Y}_\zeta) : \mathbb{D}(y, \mathbf{Y}_\eta)\rangle_{Y_f} = 0. \tag{1.4.32}$$

Thus,

$$(\mathfrak{N}_1 : \zeta) : \eta = \mu_0 \, \langle (\mu_0 \, \mathbb{D}(y, \mathbf{Y}_\eta) + \zeta) : (\mu_0 \, \mathbb{D}(y, \mathbf{Y}_\zeta) + \eta) \rangle_{Y_f}.$$

The first statement of the lemma follows from the last relation in the same way, as in previous sections.

Let now the pore space be disconnected. For this case the problem (1.4.26) for all $i, j = 1, 2, 3$ has a unique solution, linear in $\mathbf{y}$:

$$\begin{aligned}
\mu_0 \, \mathbf{W}_0^{(ij)} &= \frac{1}{2}(y_i \mathbf{e}_j + y_j \mathbf{e}_i), \;\; P_0^{(ij)} = 0, \;\; \text{if } i \neq j, \\
\mu_0 \, \mathbf{W}_0^{(11)} &= (-\frac{2}{3} y_1, \frac{1}{3} y_2, \frac{1}{3} y_3), \;\; P_0^{(11)} = \frac{1}{3}, \\
\mu_0 \, \mathbf{W}_0^{(22)} &= (\frac{1}{3} y_1, -\frac{2}{3} y_2, \frac{1}{3} y_3), \;\; P_0^{(22)} = \frac{1}{3}, \\
\mu_0 \, \mathbf{w}_0^{(11)} &= (\frac{1}{3} y_1, \frac{1}{3} y_2, -\frac{2}{3} y_3), \;\; P_0^{(33)} = \frac{1}{3}.
\end{aligned}$$

These equalities lead to

$$\chi(\mathbf{y}) \Big(\mu_0 \, \mathbb{D}(y, \mathbf{W}_0^{(ij)}(\mathbf{y})) + \mathbb{J}^{(ij)} - P_0^{(ij)} \mathbb{I} \Big) = 0, \tag{1.4.33}$$

$$\langle \mathfrak{A}_0 \rangle_{Y_f} = -m \sum_{i,j=1}^{3} \mathbb{J}^{(ij)} \otimes \mathbb{J}^{(ij)} + \frac{m}{3} \sum_{i=1}^{3} \mathbb{I} \otimes \mathbb{J}^{(ii)},$$

and

$$(\mathfrak{N}_1 : \zeta) : \eta = \mu_0 \frac{m}{3} \sum_{i=1}^{3} \zeta_{ii} \, \eta_{ii}.$$

Thus $\mathfrak{N}_1$ degenerates for any symmetric ζ, such that $\sum_{i=1}^{3} \zeta_{ii}^2 = 0$.

To prove the last statement, we use equations

$$\Big\langle \mu_0\, \mathbb{D}\Big(y, \frac{\partial \mathbf{W}^{(ij)}}{\partial t}(\mathbf{y}, 0)\Big) : \mathbb{D}(y, \mathbf{W}_0^{(kl)})\Big\rangle_{Y_f} + \langle \lambda_0\, \mathbb{D}(y, \mathbf{W}_0^{(ij)}) : \mathbb{D}(y, \mathbf{W}_0^{(kl)})\rangle_{Y_s} = 0,$$

for $i, j = 1, 2, 3$, which are the result of multiplying of the first equation in (1.4.25) at $t = 0$ by $\mathbf{W}_0^{(kl)}$ and integration by parts over Y.

By means of (1.4.33) we rewrite it as

$$-\Big\langle \mathbb{D}\Big(y, \frac{\partial \mathbf{Y}_\zeta}{\partial t}(\mathbf{y}, 0)\Big) : \eta\Big\rangle_{Y_f} + \langle \lambda_0 \mathbb{D}(y, \mathbf{Y}_\zeta) : \mathbb{D}(y, \mathbf{Y}_\eta)\rangle_{Y_s} = 0, \tag{1.4.34}$$

where we took into account the equality

$$\sum_{i=1}^{3} \mathbb{D}\Big(y, \frac{\partial \mathbf{W}^{(ii)}}{\partial t}\Big) : (\mathbb{I} \otimes \mathbb{J}^{(ii)}) = \sum_{i=1}^{3} \frac{\partial}{\partial t}(\nabla_y \cdot \mathbf{W}^{(ii)})\, \mathbb{J}^{(ii)} = 0.$$

Finally,

$$\begin{aligned}
(\mathfrak{N}_2 : \zeta) : \eta &= \lambda_0\, (1-m)\zeta : \eta + \lambda_0\, \langle \mathfrak{A}_0 : \zeta\rangle_{Y_s} : \eta + \mu_0\, \langle \mathfrak{A}_1(\mathbf{y}, 0) : \zeta\rangle_{Y_f} : \eta \\
&= \lambda_0\, (1-m)\zeta : \eta + \lambda_0\, \mu_0 \langle \mathbb{D}(y, \mathbf{Y}_\zeta) : \eta\rangle_{Y_s} + \mu_0^2\, \langle \mathbb{D}\big(y, \frac{\partial \mathbf{Y}_\zeta}{\partial t}(\mathbf{y}, 0)\big) : \eta\rangle_{Y_f} \\
&\quad - \lambda_0\, \mu_0 \langle \mathbb{D}(y, \mathbf{Y}_\eta) : \zeta\rangle_{Y_f} = \lambda_0\, (1-m)\zeta : \eta \\
&\quad + \lambda_0\, \mu_0 \langle \mathbb{D}(y, \mathbf{Y}_\zeta) : \eta\rangle_{Y_s} + \lambda_0\, \mu_0^2\, \langle \mathbb{D}(y, \mathbf{Y}_\zeta) : \mathbb{D}(y, \mathbf{Y}_\eta)\rangle_{Y_s} \\
&\quad - \lambda_0\, \mu_0 \langle \mathbb{D}(y, \mathbf{Y}_\eta) : \zeta\rangle_{Y_f}.
\end{aligned}$$

We recall that

$$\langle \mathbb{D}(y, \mathbf{Y}_\eta)\rangle_{Y_f} = -\langle \mathbb{D}(y, \mathbf{Y}_\eta)\rangle_{Y_s},$$

and

$$\langle \mathbb{D}(y, \mathbf{Y}_\eta)\rangle_{Y_f} : \zeta = \langle \mathbb{D}(y, \mathbf{Y}_\zeta)\rangle_{Y_f} : \eta$$

due to (1.4.32).

Therefore,

$$(\mathfrak{N}_2 : \zeta) : \eta = \lambda_0 \big\langle \big(\mu_0 \mathbb{D}(y, \mathbf{Y}_\eta) + \zeta\big) : \big(\mu_0 \mathbb{D}(y, \mathbf{Y}_\zeta) + \eta\big)\big\rangle_{Y_s},$$

which proves the lemma.

Lemma 1.20 *The problem (1.4.13)–(1.4.17) has a unique solution.*

Proof For a connected pore space the difference $\{\mathbf{w},\ p\}$ of the two possible solutions of (1.4.13)–(1.4.17) satisfies the homogeneous problem (1.4.13)–(1.4.17). Multiplication of (1.4.14) by $\mathbf{w}$ and integration by parts over Ω give us

$$\frac{1}{2}\frac{d}{dt}\int_{\Omega}\big(\mathfrak{N}_1 : \mathbb{D}(x,\mathbf{w})\big) : \mathbb{D}(x,\mathbf{w})dx$$
$$= -\int_{\Omega}\big(\mathfrak{N}_2 : \mathbb{D}(x,\mathbf{w})\big) : \mathbb{D}(x,\mathbf{w})dx$$
$$-\int_{\Omega}\Big(\int_0^t \mathfrak{N}_3(t-\tau) : \mathbb{D}(x,\mathbf{w})d\tau\Big) : \mathbb{D}(x,\mathbf{w})dx$$
$$\leqslant C_0\int_{\Omega}\mathbb{D}(x,\mathbf{w}) : \mathbb{D}(x,\mathbf{w})dx,$$

if we use Hölder's inequality and the estimate

$$\int_0^t\int_{\Omega}\mathbb{D}\big(x,\mathbf{w}(\mathbf{x},\tau)\big) : \mathbb{D}\big(x,\mathbf{w}(\mathbf{x},\tau)\big)dxd\tau$$
$$\leqslant T\int_{\Omega}\mathbb{D}\big(x,\mathbf{w}(\mathbf{x},t)\big) : \mathbb{D}\big(x,\mathbf{w}(\mathbf{x},t)\big)dx.$$

Grownwall's inequality and the properties of the tensor $\mathfrak{N}_1$ guarantee the equality

$$\int_{\Omega}\mathbb{D}\big(x,\mathbf{w}(\mathbf{x},t)\big) : \mathbb{D}\big(x,\mathbf{w}(\mathbf{x},t)\big)dx = 0$$

for all $t \in (0, T)$.

For a disconnected pore space the proof is the same:

$$C_0^{-1}\int_0^t\int_{\Omega}\mathbb{D}\big(x,\mathbf{w}(\mathbf{x},\tau)\big) : \mathbb{D}\big(x,\mathbf{w}(\mathbf{x},\tau)\big)dxd\tau$$
$$\leqslant \int_{\Omega}\Big(\mathfrak{N}_1 : \mathbb{D}\big(x,\mathbf{w}(\mathbf{x},t)\big)\Big) : \mathbb{D}\big(x,\mathbf{w}(\mathbf{x},t)\big)dx$$
$$+\int_0^t\int_{\Omega}\Big(\mathfrak{N}_2 : \mathbb{D}\big(x,\mathbf{w}(\mathbf{x},\tau)\big)\Big) : \mathbb{D}\big(x,\mathbf{w}(\mathbf{x},\tau)\big)dxd\tau$$
$$= -\int_0^t\int_{\Omega}\Big(\int_0^{\tau}\mathfrak{N}_3(\tau-\xi) : \mathbb{D}\big(x,\mathbf{w}(\mathbf{x},\xi)\big)d\xi\Big) : \mathbb{D}\big(x,\mathbf{w}(\mathbf{x},\tau)\big)dxd\tau$$
$$\leqslant C_0\Big(\int_0^t\int_0^{\tau}\int_{\Omega}\mathbb{D}\big(x,\mathbf{w}(\mathbf{x},\xi)\big) : \mathbb{D}\big(x,\mathbf{w}(\mathbf{x},\xi)\big)dxd\xi d\tau\Big)^{\frac{1}{2}}\cdot$$
$$\Big(\int_0^t\int_{\Omega}\mathbb{D}\big(x,\mathbf{w}(\mathbf{x},\tau)\big) : \mathbb{D}\big(x,\mathbf{w}(\mathbf{x},\tau)\big)dxd\tau\Big)^{\frac{1}{2}}.$$

1.4.3 Proof of Theorem 1.12

As in the proof of Theorem 1.11 there exist functions $\mathbf{w},\ \mathbf{v} \in \overset{\circ}{\mathbf{W}}{}_2^{1,0}(\Omega_T)$, $\mathbf{w}(\mathbf{x},0)=0$, $\mathbf{x}\in\Omega$, and 1-periodic in $\mathbf{y}$ functions $\mathbf{W}$, $\mathbf{V}$, such that sequences $\{\mathbf{w}^{\varepsilon}\}$ and $\{^{\varepsilon}\}$ converge as $\varepsilon\to 0$ weakly in $\mathbf{W}_2^{1,0}(\Omega_T)$ to functions $\mathbf{w}$ and $\mathbf{v}=\dfrac{\partial\mathbf{w}}{\partial t}$ respectively, and sequences $\{\mathbb{D}(x,\mathbf{w}^{\varepsilon})\}$ and $\{\mathbb{D}(x,\mathbf{v}^{\varepsilon})\}$ two-scale converge as $\varepsilon\to 0$ in $\mathbf{L}_2(\Omega_T)$ to the functions

$$\mathbb{D}(x,\mathbf{W})+\mathbb{D}\big(y,\mathbf{w}(\mathbf{x},t,\mathbf{y})\big),\ \ \textit{and}\ \mathbb{D}\Big(x,\frac{\partial\mathbf{w}}{\partial t}\Big)+\mathbb{D}\Big(y,\frac{\partial\mathbf{W}}{\partial t}\Big)$$

respectively.

To derive macro- and microscopic momentum balance equations we pass to the limit as $\varepsilon\to 0$ in the integral identity (1.4.11) with two different types of test functions. Firstly, with test functions $\varphi=\varphi(\mathbf{x},t)$, and then with test functions $\varphi=\varepsilon h(\mathbf{x},t)\varphi_0\Big(\dfrac{\mathbf{x}}{\varepsilon}\Big)$. After reintegrating we obtain

$$\begin{aligned}\nabla_y\cdot\Big(\mu_0\,\chi(\mathbf{y})\Big(\mathbb{D}\Big(x,\frac{\partial\mathbf{w}}{\partial t}\Big)+\mathbb{D}(y,\frac{\partial\mathbf{W}}{\partial t})\Big)&\\ +\mathfrak{A}_2(\mathbf{y}):\big(\mathbb{D}(x,\mathbf{w})+\mathbb{D}(y,\mathbf{W})\big)\Big)&=0,\end{aligned}\tag{1.4.35}$$

where

$$\begin{aligned}\mathfrak{A}_2(\mathbf{y})&=\big(1-\chi(\mathbf{y})\big)\Big(\lambda_0\sum_{i,j=1}^{3}\big(\mathbb{J}^{(ij)}\otimes\mathbb{J}^{(ij)}\big)+c_s^2\,(\mathbb{I}\otimes\mathbb{I})\Big)+\chi(\mathbf{y})c_f^2\,(\mathbb{I}\otimes\mathbb{I})\\&=\big(1-\chi(\mathbf{y})\big)\mathfrak{N}^{(0)}+\chi(\mathbf{y})c_f^2\,(\mathbb{I}\otimes\mathbb{I}),\\ \mathfrak{N}^{(0)}&=\lambda_0\sum_{i,j=1}^{3}\big(\mathbb{J}^{(ij)}\otimes\mathbb{J}^{(ij)}\big)+c_s^2\,(\mathbb{I}\otimes\mathbb{I}),\end{aligned}$$

and

$$\nabla\cdot\widetilde{\mathbb{P}}+\hat{\rho}\mathbf{F}=0,\tag{1.4.36}$$

where

$$\begin{aligned}\widetilde{\mathbb{P}}=\mu_0\,m\,\mathbb{D}\Big(x,\frac{\partial\mathbf{w}}{\partial t}\Big)+\mu_0\Big\langle\mathbb{D}\Big(y,\frac{\partial\mathbf{W}}{\partial t}\Big)\Big\rangle_{Y_f}&\\ +\langle\mathfrak{A}_2\rangle_Y:\mathbb{D}(x,\mathbf{w})+\langle\mathfrak{A}_2:\mathbb{D}(y,\mathbf{W})\rangle_Y.&\end{aligned}\tag{1.4.37}$$

Let

$$\mathbb{Y}(\mathbf{x},t)=\mu_0\,\mathbb{D}\Big(x,\frac{\partial\mathbf{w}}{\partial t}\Big)+\big(c_f^2\,(\mathbb{I}\otimes\mathbb{I})-\mathfrak{N}^{(0)}\big):\mathbb{D}(x,\mathbf{w})=\sum_{i,j=1}^{3}\mathbb{J}^{(ij)}\,Y_{ij}(\mathbf{x},t),$$

Then we may rewrite (1.4.35) as

$$\nabla_y\cdot\Big(\chi(\mathbf{y})\big(\mu_0\,\mathbb{D}\Big(y,\frac{\partial\mathbf{W}}{\partial t}\Big)+\mathbb{Y}(\mathbf{x},t)\big)+(1-\chi(\mathbf{y}))\mathfrak{N}_1^{(0)}:\mathbb{D}(y,\mathbf{W})\Big)=0,$$

and look for the solution of this equation in the form

$$\mathbf{W}(\mathbf{x},t,\mathbf{y})=\sum_{i,j=1}^{3}\int_0^t\widetilde{\mathbf{W}}^{(ij)}(\mathbf{y},t-\tau)Y_{ij}(\mathbf{x},\tau)d\tau,$$

where 1-periodic in $\mathbf{y}$ functions $\widetilde{\mathbf{W}}^{(ij)}(\mathbf{y},t),\ ij=1,2,3$, are solutions of the periodic initial boundary value problems

$$\left.\begin{aligned}\nabla_y\cdot\Big(\chi\,\mu_0\,\mathbb{D}\Big(y,\frac{\partial\widetilde{\mathbf{W}}^{(ij)}}{\partial t}\Big)+\mathfrak{A}_2(\mathbf{y}):\mathbb{D}(y,\widetilde{\mathbf{W}}^{(ij)})\Big)=0,\\ \chi(\mathbf{y})\widetilde{\mathbf{W}}^{(ij)}(\mathbf{y},0)=\widetilde{\mathbf{W}}_0^{(ij)}(\mathbf{y}),\end{aligned}\right\}\tag{1.4.38}$$

$$\nabla_y\cdot\Big(\chi\,\big(\mu_0\,\mathbb{D}(y,\widetilde{\mathbf{W}}_0^{(ij)})+\mathbb{J}^{(ij)}\big)\Big)=0\tag{1.4.39}$$

in the domain Y.

The proof of the existence and uniqueness of solutions to problems (1.4.38) and (1.4.39) and the infinite smoothness with respect to time of the solution of (1.4.38) is straightforward (see Lemma 1.18).

Thus,

$$\begin{aligned}\mathbb{D}(y,\mathbf{W})&=\sum_{i,j=1}^{3}\int_0^t\mathbb{D}(y,\widetilde{\mathbf{W}}^{(ij)}(\mathbf{y},t-\tau))Y_{ij}(\mathbf{x},\tau)d\tau\\&=\sum_{i,j=1}^{3}\int_0^t\mathbb{D}(y,\widetilde{\mathbf{W}}^{(ij)}(\mathbf{y},t-\tau))\otimes\mathbb{J}^{(ij)}\big):\mathbb{Y}(\mathbf{x},\tau)d\tau\\&=\mathfrak{A}_3:\mathbb{D}(x,\mathbf{w})+\int_0^t\mathfrak{A}_4(\mathbf{y},t-\tau):\mathbb{D}\big(x,\mathbf{w}(\mathbf{x},\tau)\big)d\tau,\end{aligned}\tag{1.4.40}$$

where

$$\mathfrak{A}_3(\mathbf{y})=\mu_0\sum_{i,j=1}^{3}\mathbb{D}\big(y,\widetilde{\mathbf{W}}_0^{(ij)}(\mathbf{y})\big)\otimes\mathbb{J}^{(ij)}\tag{1.4.41}$$

and

$$\mathfrak{A}_4(\mathbf{y},t) = \sum_{i,j=1}^{3}\Big(\mu_0\,\mathbb{D}\Big(y, \frac{\partial \widetilde{\mathbf{W}}^{(ij)}}{\partial t}(\mathbf{y},t)\Big)\otimes \mathbb{J}^{(ij)}\Big)$$
$$-\Big(\sum_{i,j=1}^{3}\mathbb{D}\big(y, \widetilde{\mathbf{W}}^{(ij)}(\mathbf{y},t)\big)\otimes \mathbb{J}^{(ij)}\Big) : \mathfrak{N}_1^{(0)}. \tag{1.4.42}$$

Finally,

$$\mathbb{D}(y, \frac{\partial \mathbf{W}}{\partial t}) = \mathfrak{A}_3(\mathbf{y}) : \mathbb{D}\Big(x, \frac{\partial \mathbf{w}}{\partial t}(\mathbf{x},t)\Big) + \mathfrak{A}_4(\mathbf{y},0) : \mathbb{D}\big(x, \mathbf{w}(\mathbf{x},t)\big)$$
$$+\int_0^t \frac{\partial \mathfrak{A}_4}{\partial t}(\mathbf{y}, t-\tau) : \mathbb{D}\big(x, \mathbf{w}(\mathbf{x},\tau)\big)d\tau, \tag{1.4.43}$$

and

$$\left.\begin{aligned}
\mathfrak{N}_4 &= \mu_0\, m \sum_{i,j=1}^{3} \mathbb{J}^{(ij)} \otimes \mathbb{J}^{(ij)} + \mu_0\, \langle \mathfrak{A}_3\rangle_{Y_f},\\
\mathfrak{N}_5 &= \langle \mathfrak{A}_2\rangle_Y + \langle \mathfrak{A}_2 : \mathfrak{A}_3\rangle_Y + \mu_0\, \langle \mathfrak{A}_4(\mathbf{y},0)\rangle_{Y_f}\\
\mathfrak{N}_6(t) &= \mu_0\, \langle \frac{\partial \mathfrak{A}_4}{\partial t}(\mathbf{y},t)\rangle_{Y_f} + \langle \mathfrak{A}_2 : \mathfrak{A}_4(\mathbf{y},t)\rangle_{Y_s}.
\end{aligned}\right\} \tag{1.4.44}$$

Lemma 1.21 *If a pore space is connected, then the symmetric tensor $\mathfrak{N}_4$ is strictly positive definite. For the case of a disconnected pore space (isolated pores) $\mathfrak{N}_4 = 0$ and the symmetric tensor $\mathfrak{N}_5$ becomes strictly positive definite.*

Proof Let the pore space be connected, $\zeta = (\zeta_{ij})$ and $\eta = (\eta_{ij})$ be arbitrary symmetric matrices and

$$\mathbf{Y}_\zeta = \sum_{i,j=1}^{3} \widetilde{\mathbf{W}}_0^{(ij)}\zeta_{ij}, \quad \mathbf{Y}_\eta = \sum_{i,j=1}^{3} \widetilde{\mathbf{W}}_0^{(ij)}\eta_{ij}.$$

we have

$$\big(\mathfrak{N}_4 : \zeta\big) : \eta = \mu_0\, m\, \zeta : \eta + \mu_0^2 \langle \mathbb{D}\big(y, \mathbf{Y}_\zeta)\rangle_{Y_f} : \eta,$$

and

$$\mu_0\, \mathbb{D}\big(y, \mathbf{Y}_\eta)\rangle_{Y_f} : \zeta + \mu_0^2\, \langle \mathbb{D}\big(y, \mathbf{Y}_\zeta\big) : \mathbb{D}\big(y, \mathbf{Y}_\eta)\rangle_{Y_f}.$$

which is a simple consequence of (1.4.39). Thus,

$$\big(\mathfrak{N}_4 : \zeta\big) : \eta = \mu_0\, \langle\big(\mu_0\, \mathbb{D}(y, \mathbf{Y}_\eta) + \zeta\big) : \big(\mu_0\, \mathbb{D}(y, \mathbf{Y}_\zeta) + \eta\big)\rangle_{Y_f}.$$

which proves the first statement of the lemma.

Let now the pore space be disconnected. For this case the problem (1.4.39) has a unique solution, linear in $\mathbf{y}$:

$$\mu_0\,\widetilde{\mathbf{w}}_0^{(ij)} = \frac{1}{2}(y_i\mathbf{e}_j + y_j\mathbf{e}_i),$$

$$\chi(\mathbf{y})\Big(\mu_0\,\mathbb{D}\big(y, \widetilde{\mathbf{w}}_0^{(ij)}(\mathbf{y})\big) + \mathbb{J}^{(ij)}\Big) = 0, \tag{1.4.45}$$

and

$$\langle \mathfrak{A}_3 \rangle_{Y_f} = -m \sum_{i,j=1}^{3} \mathbb{J}^{(ij)} \otimes \mathbb{J}^{(ij)}. \tag{1.4.46}$$

This equality results $\mathfrak{N}_4 = 0$.

To prove the last statement, we use equations

$$\left\langle \mu_0\,\mathbb{D}\left(y, \frac{\partial \widetilde{\mathbf{W}}^{(ij)}}{\partial t}(\mathbf{y}, 0)\right) : \mathbb{D}(y, \widetilde{\mathbf{w}}_0^{(kl)})\right\rangle_{Y_f}$$
$$+ \langle \big(\mathfrak{A}_2 : \mathbb{D}(y, \widetilde{\mathbf{W}}_0^{(ij)})\big) : \mathbb{D}(y, \widetilde{\mathbf{W}}_0^{(kl)})\rangle_Y = 0,$$

for $i, j = 1, 2, 3$, which are the result of multiplying the first equation in (1.4.38) at $t = 0$ by $\widetilde{\mathbf{W}}_0^{(kl)}$ and integration by parts over Y.

By means of (1.4.45) we rewrite it as

$$-\left\langle \mathbb{D}\left(y, \frac{\partial \mathbf{Y}_\zeta}{\partial t}(\mathbf{y}, 0)\right) : \eta\right\rangle_{Y_f} + \left\langle \big(\mathfrak{A}_2 : \mathbb{D}(y, \mathbf{Y}_\zeta)\big) : \mathbb{D}(y, \mathbf{Y}_\eta)\right\rangle_{Y_s} = 0. \tag{1.4.47}$$

We also recall that

$$\langle \mathbb{D}(y, \mathbf{Y}_\eta)\rangle_{Y_f} = -\langle \mathbb{D}(y, \mathbf{Y}_\eta)\rangle_{Y_s},$$

and

$$\langle \mathbb{D}(y, \widetilde{\mathbf{W}}_0^{(ij)}) \otimes \mathbb{J}^{(ij)}\rangle_{Y_f} = \langle \mathbb{J}^{(ij)} \otimes \mathbb{D}(y, \widetilde{\mathbf{W}}_0^{(ij)})\rangle_{Y_f}$$

due to (1.4.45).

Next, in the expression

$$\big(\mathfrak{N}_5 : \zeta\big) : \eta = \langle \big(\mathfrak{A}_2 : \zeta\big) : \eta\rangle_Y + \big(\langle \mathfrak{A}_2 : \mathfrak{A}_3\rangle_Y : \zeta\big) : \eta + \mu_0\,\langle \mathfrak{A}_4(\mathbf{y}, 0) : \zeta\rangle_{Y_f} : \eta$$

we calculate each term:

$$\langle \big(\mathfrak{A}_2 : \zeta\big) : \eta\rangle_Y = (1 - m)\big(\mathfrak{N}^{(0)} : \zeta\big) : \eta + m\,c_f^2\,(\operatorname{tr}\zeta)\,(\operatorname{tr}\eta),$$

$$
\begin{aligned}
&\big(\langle\mathfrak{A}_2 : \mathfrak{A}_3\rangle_Y : \zeta\big) : \eta \\
&= (1-m)\Big(\big(\mathfrak{N}^{(0)} : (\langle\mu_0 \sum_{i,j=1}^{3} \mathbb{D}(y, \widetilde{\mathbf{W}}_0^{(ij)})\rangle_{Y_s} \otimes \mathbb{J}^{(ij)})\big) : \zeta\Big) : \eta \\
&+ \Big(\big(c_f^2\,(\mathbb{I}\otimes\mathbb{I}) : (\langle\mu_0 \sum_{i,j=1}^{3} \mathbb{D}(y, \widetilde{\mathbf{W}}_0^{(ij)})\rangle_{Y_f} \otimes \mathbb{J}^{(ij)})\big) : \zeta\Big) : \eta \\
&= (1-m)\mu_0\langle\mathfrak{N}^{(0)} : \mathbb{D}(y, \mathbf{Y}_\eta)\rangle_{Y_s} : \zeta + c_f^2\,\mu_0\langle(\nabla\cdot\mathbf{Y}_\eta)\rangle_{Y_f}(\operatorname{tr}\zeta),
\end{aligned}
$$

$$
\begin{aligned}
\mu_0\,\langle\mathfrak{A}_4(\mathbf{y},0) : \zeta\rangle_{Y_f} : \eta &= \mu_0^2\Big\langle\mathbb{D}\Big(y, \frac{\partial \mathbf{Y}_\zeta}{\partial t}(\mathbf{y},0)\Big) : \eta\Big\rangle_{Y_f} \\
&- \Bigg(\Bigg(\Big\langle\sum_{i,j=1}^{3}\mu_0\,\mathbb{D}(y, \widetilde{\mathbf{W}}_0^{(ij)})\otimes\mathbb{J}^{(ij)}\Big\rangle_{Y_f} : \mathfrak{N}_1^{(0)}\Bigg) : \zeta\Bigg) : \eta \\
&= \mu_0^2\langle(\mathfrak{A}_2 : \mathbb{D}(y, \mathbf{Y}_\zeta)) : \mathbb{D}(y, \mathbf{Y}_\eta)\rangle_Y \\
&+ (1-m)\Bigg(\Bigg(\mathfrak{N}^{(0)} : \Big\langle\sum_{i,j=1}^{3}\mu_0\,\mathbb{D}(y, \widetilde{\mathbf{W}}_0^{(ij)})\otimes\mathbb{J}^{(ij)}\Big\rangle_{Y_s}\Bigg) : \zeta\Bigg) : \eta \\
&+ c_f^2\Bigg(\Bigg((\mathbb{I}\otimes\mathbb{I}) : \Big\langle\sum_{i,j=1}^{3}\mu_0\,\mathbb{D}(y, \widetilde{\mathbf{W}}_0^{(ij)})\otimes\mathbb{J}^{(ij)}\Big\rangle_{Y_f}\Bigg) : \zeta\Bigg) : \eta \\
&= \mu_0^2\Big\langle\big(\mathfrak{N}^{(0)} : \mathbb{D}(y, \mathbf{Y}_\zeta)\big) : \mathbb{D}(y, \mathbf{Y}_\eta)\Big\rangle_{Y_s} + \mu_0^2\,c_f^2\,\langle(\nabla\cdot\mathbf{Y}_\zeta)\cdot(\nabla\cdot\mathbf{Y}_\eta)\rangle_{Y_f} \\
&+ \mu_0\langle\mathfrak{N}^{(0)} : \mathbb{D}(y, \mathbf{Y}_\zeta)\rangle_{Y_s} : \eta + c_f^2\,\mu_0\langle(\nabla\cdot\mathbf{Y}_\zeta)\rangle_{Y_f}(\operatorname{tr}\eta).
\end{aligned}
$$

Finally we get

$$
\begin{aligned}
(\mathfrak{N}_5 : \zeta) : \eta &= \Big\langle\Big(\mathfrak{N}^{(0)} : \big(\mu_0\mathbb{D}(y, \mathbf{Y}_\zeta) + \eta\big)\Big) : \big(\mu_0\mathbb{D}(y, \mathbf{Y}_\eta) + \zeta\big)\Big\rangle_{Y_s} \\
&+ c_f^2\Big\langle\big(\mu_0(\nabla\cdot\mathbf{Y}_\zeta) + \operatorname{tr}\eta\big)\cdot\big(\mu_0(\nabla\cdot\mathbf{Y}_\eta) + \operatorname{tr}\zeta\big)\Big\rangle_{Y_f}.
\end{aligned}
$$

The uniqueness of the problem (1.4.17)–(1.4.20) is proved in the same way as in Theorem 1.11.

Chapter 2
Filtration of a Compressible Thermo-Fluid

The model $\mathbb{M}_{13}$ consists of the differential equations

$$\frac{1}{\tilde{\alpha}_p}\, p + \nabla \cdot \mathbf{w} = 0, \tag{2.0.1}$$

$$\nabla \cdot \mathbb{P} + \tilde{\rho}\mathbf{F} = 0, \tag{2.0.2}$$

$$\tilde{\eta}_0 \frac{\partial \vartheta}{\partial t} - \nabla \cdot (\tilde{\alpha}_{\varkappa} \nabla \vartheta) = \Phi - \gamma_0 \tilde{\alpha}_{\vartheta}\, \nabla \cdot \frac{\partial \mathbf{w}}{\partial t}, \tag{2.0.3}$$

$$\mathbb{P} = \chi_0 \alpha_\mu\, \mathbb{D}\left(x, \frac{\partial \mathbf{w}}{\partial t}\right) + (1 - \chi_0)\lambda_0\, \mathbb{D}(x, \mathbf{w}) - \left(p + \tilde{\alpha}_{\vartheta}\, \vartheta - \chi_0 \alpha_\nu \nabla \cdot \frac{\partial \mathbf{w}}{\partial t}\right) \mathbb{I}, \tag{2.0.4}$$

and its submodel, the mathematical model $\mathbb{M}_{16}$ of **the filtration of a compressible thermo-fluid in an non-isothermal absolutely rigid solid skeleton**. In turn, the last model consists of the differential equations

$$\chi_0 \left(\frac{1}{c_f^2}\, p + \nabla \cdot \mathbf{w}\right) = 0, \tag{2.0.5}$$

$$\chi_0 \left(\nabla \cdot \mathbb{P} + \rho_f \mathbf{F}\right) = 0, \tag{2.0.6}$$

$$\tilde{\eta}_0 \frac{\partial \vartheta}{\partial t} - \nabla \cdot (\tilde{\alpha}_{\varkappa} \nabla \vartheta) = \Phi - \chi_0 \gamma_0\, \beta_f \left(\nabla \cdot \frac{\partial \mathbf{w}}{\partial t}\right), \tag{2.0.7}$$

$$\mathbb{P} = \alpha_\mu \mathbb{D}\left(x, \frac{\partial \mathbf{w}}{\partial t}\right) - \left(p + \beta_f\, \vartheta - \alpha_\nu \nabla \cdot \left(\frac{\partial \mathbf{w}}{\partial t}\right)\right) \mathbb{I}. \tag{2.0.8}$$

These models are derived in Appendix A.

A. Meirmanov, *Mathematical Models for Poroelastic Flows*,
Atlantis Studies in Differential Equations 1, DOI: 10.2991/978-94-6239-015-7_2,

As in the previous chapter, we impose Assumption 0.1 and suppose that Ω is a domain with a C^2 continuous boundary $S = \partial\Omega$.

Under this assumption

$$\chi_0(\mathbf{x}) = \chi^{\varepsilon}(\mathbf{x}) = \varsigma(\mathbf{x})\chi\left(\frac{\mathbf{x}}{\varepsilon}\right), \quad \tilde{\rho} = \rho^{\varepsilon} = \chi^{\varepsilon}\rho_f + (1-\chi^{\varepsilon})\rho_s,$$

where $\varsigma(\mathbf{x})$ is the characteristic function of the domain Ω, and

$$\tilde{\alpha}_{\vartheta} = \alpha^{\varepsilon}_{\vartheta} = \chi^{\varepsilon}\beta_f + (1-\chi^{\varepsilon})\beta_s \quad \tilde{\eta}_0 = \eta^{\varepsilon}_0 = \tilde{\chi}\, c_{p,f} + (1-\tilde{\chi})\, c_{p,s},$$

$$\tilde{\alpha}_p = \alpha^{\varepsilon}_p = \chi^{\varepsilon}c^2_f + (1-\chi^{\varepsilon})c^2_s \quad \tilde{\alpha}_{\varkappa} = \alpha^{\varepsilon}_{\varkappa} = \chi^{\varepsilon}\varkappa_f + (1-\chi^{\varepsilon})\varkappa_s.$$

We assume that the dimensionless parameters α_μ and α_ν depend on the small parameter ε and that the (finite or infinite) limits exist:

$$\lim_{\varepsilon\searrow 0}\alpha_\mu(\varepsilon) = \mu_0, \quad \lim_{\varepsilon\searrow 0}\frac{\alpha_\mu}{\varepsilon^2} = \mu_1, \quad \lim_{\varepsilon\searrow 0}\alpha_\nu(\varepsilon) = \nu_0.$$

In what follows, we denote as C_0 any constant depending only on domains Ω, Y and Y_f.

Without lost of generality we may suppose that $\gamma_0 = 1$.

2.1 A Viscous Thermo-Fluid in a Non-isothermal Absolutely Rigid Solid Skeleton

In this section as a basic mathematical model at the microscopic level we consider the model $\mathbb{M}_{17}$ of the filtration of compressible liquid in an absolutely rigid solid skeleton. It is easy to show that this model is a limit of the model $\mathbb{M}_{14}$ as $\alpha_\lambda \to \infty$. One of the consequences of this statement is the following:

$$\mathbf{w}(\mathbf{x},t) = 0, \quad p(\mathbf{x},t) = 0, \quad \mathbf{x} \in \Omega^{\varepsilon}_s.$$

If we put $\mathbf{v} = \dfrac{\partial\mathbf{w}}{\partial t}$, then we may rewrite the last condition and Eqs. (2.0.5)–(2.0.8) in the form

$$\frac{1}{c^2_f}\frac{\partial p}{\partial t} + \nabla\cdot\mathbf{v} = 0, \ \mathbf{x}\in\Omega^{\varepsilon}_f, \ t\in(0,T), \tag{2.1.1}$$

$$\nabla\cdot\mathbb{P} + \rho_f\mathbf{F} = 0, \ \mathbf{x}\in\Omega^{\varepsilon}_f, \ t\in(0,T), \tag{2.1.2}$$

$$\eta^{\varepsilon}_0\frac{\partial\vartheta}{\partial t} - \nabla\cdot(\alpha^{\varepsilon}_{\varkappa}\nabla\vartheta) = \Phi - \chi^{\varepsilon}\beta_f\,(\nabla\cdot\mathbf{v}), \ \mathbf{x}\in\Omega, \ t\in(0,T), \tag{2.1.3}$$

$$\mathbb{P} = \alpha_\mu \mathbb{D}(x, \mathbf{v}) - q\,\mathbb{I}, \tag{2.1.4}$$

$$q = p + \beta_f\, \vartheta\, \chi^\varepsilon + \frac{\alpha_\nu}{c_f^2}\frac{\partial p}{\partial t}, \ \ \mathbf{x} \in \Omega,\ t \in (0, T), \tag{2.1.5}$$

$$\mathbf{v}(\mathbf{x}, t) = 0, \ \ \mathbf{x} \in \Omega_s^\varepsilon \cup S, \ \ t \in (0, T), \tag{2.1.6}$$

$$\vartheta(\mathbf{x}, t) = 0, \ \ \mathbf{x} \in S, \ \ t \in (0, T), \tag{2.1.7}$$

$$\vartheta(\mathbf{x}, 0) = \vartheta_0(\mathbf{x}), \ \ p(\mathbf{x}, 0) = 0, \ \ \ \mathbf{x} \in \Omega. \tag{2.1.8}$$

Throughout this section we assume that c_f^2, $\varkappa_f$, $\varkappa_s$, β_f, β_s, $c_{p,f}$, $c_{p,s}$ and λ_0 are positive constants and that conditions

$$\mu_0 = 0, \ \ 0 < \mu_1 \leqslant \infty, \ \ 0 \leqslant \nu_0 < \infty,$$

and

$$\int_{\Omega_T} |\mathbf{F}|^2 dxdt + \int_{\Omega_T} |\Phi|^2 dxdt = \mathscr{F}^2 < \infty$$

hold true.

2.1.1 Statement of the Problem and Main Results

Definition 2.1 We say that the system of four functions $\{\mathbf{v}^\varepsilon,\ p^\varepsilon,\ q^\varepsilon,\ \vartheta^\varepsilon\}$ such that

$$\mathbf{v}^\varepsilon \in \overset{\circ}{\mathbf{W}}{}_2^{1,0}(\Omega_T), \ \ p^\varepsilon,\ q^\varepsilon \in L_2(\Omega_T), \ \ \vartheta^\varepsilon \in \overset{\circ}{W}{}_2^{1,0}(\Omega_T),$$

is a weak solution of the problem (2.1.1)–(2.1.8), if it satisfies the state equation (2.1.5), the condition (2.1.6), and the integral identities

$$\int_{\Omega_T} \chi^\varepsilon \left(\alpha_\mu \mathbb{D}(x, \mathbf{v}^\varepsilon) : \mathbb{D}(x, \varphi) - q^\varepsilon\,(\nabla \cdot \varphi) - \rho_f \mathbf{F} \cdot \varphi\right) dxdt = 0, \tag{2.1.9}$$

$$\int_{\Omega_T} \left(\nabla \xi \cdot \mathbf{v}^\varepsilon + \frac{1}{c_f^2}\frac{\partial \xi}{\partial t} p^\varepsilon\right) dxdt = 0, \tag{2.1.10}$$

and

$$\int_{\Omega_T} \left(\alpha_\varkappa^\varepsilon \nabla \vartheta^\varepsilon \cdot \nabla \psi - \eta_0^\varepsilon \vartheta^\varepsilon \frac{\partial \psi}{\partial t}\right) dxdt = \int_{\Omega_T} \left(\Phi - \chi^\varepsilon \beta_f\,(\nabla \cdot \mathbf{v}^\varepsilon)\right) \psi\, dxdt, \tag{2.1.11}$$

for any functions $\varphi \in \overset{\circ}{\mathbf{W}}{}_2^{1,0}\left(\Omega_f^\varepsilon \times (0,T)\right)$ and $\xi,\ \psi \in \overset{\circ}{W}{}_2^{1,1}(\Omega_T)$, such that $\xi(\mathbf{x},T) = \psi(\mathbf{x},T) = 0$.

Theorem 2.1 *(1) For all $\varepsilon > 0$ and for an arbitrary time interval $[0,T]$ there exists a unique generalized solution of problem (2.1.1)–(2.1.8) and*

$$\int_{\Omega_T}\left(\alpha_\mu|\nabla \mathbf{v}^\varepsilon|^2 + |\nabla\vartheta^\varepsilon|^2 + |\mathbf{v}^\varepsilon)|^2 + \alpha_\mu|\nabla\cdot\mathbf{v}^\varepsilon|^2\right)dxdt \qquad (2.1.12)$$

$$+ \max_{0<t<T}\int_{\Omega}\left(|p^{\,\varepsilon}|^2 + |\vartheta^{\,\varepsilon}|^2\right)dx \leqslant \frac{\varepsilon^2}{\alpha_\mu}C_0\mathscr{F}^2,$$

where the constant C_0 is independent of the small parameter ε.

(2) The nontrivial homogenization procedure for the problem (2.1.1)–(2.1.8) makes sense if and only if the pore space is connected and

$$\mu_0 = 0,\ \ 0 < \mu_1 < \infty. \qquad (2.1.13)$$

Under these conditions the sequences $\{\mathbf{v}^\varepsilon\}$, $\{\nabla\cdot\mathbf{v}^\varepsilon\}$, $\{q^{\,\varepsilon}\}$, and $\{p^{\,\varepsilon}\}$ converge weakly in $\mathbf{L}_2(\Omega_T)$ and $L_2(\Omega_T)$ (up to some subsequences) to functions $\mathbf{v}$, $\nabla\cdot\mathbf{v}$, $q \in W_2^{1,0}(\Omega_T)$, and p respectively and the sequence $\{\vartheta^{\,\varepsilon}\}$ converges weakly in $\overset{\circ}{W}{}_2^{1,0}(\Omega_T)$ to function ϑ.

The limiting functions solve the homogenized system of equations, consisting of the continuity equation

$$\frac{m}{c_f^2}\frac{\partial p}{\partial t} + \nabla\cdot\mathbf{v} = 0, \qquad (2.1.14)$$

the state equation

$$q = p + m\,\beta_f\,\vartheta + \frac{\nu_0}{c_f^2}\frac{\partial p}{\partial t}, \qquad (2.1.15)$$

Darcy's law in the form

$$\mathbf{v} = \frac{1}{\mu_1}\mathbb{B}\left(-\nabla q + \rho_f\mathbf{F}\right), \qquad (2.1.16)$$

and the heat equation

$$\hat{c}_p\frac{\partial\vartheta}{\partial t} - \frac{\beta_f}{c_f^2}\frac{\partial p}{\partial t} = \nabla\cdot\left(\mathbb{B}^\vartheta\cdot\nabla\vartheta\right) + \Phi \qquad (2.1.17)$$

in the domain Ω for $t \in (0,T)$.

If $\nu_0 = 0$, then functions $\mathbf{v}$ and p satisfy Darcy's law in the form

$$\mathbf{v} = \frac{1}{\mu_1}\mathbb{B}\left(-\nabla(p + m\,\beta_f\,\vartheta) + \rho_f \mathbf{F}\right). \tag{2.1.18}$$

System (2.1.14)–(2.1.17) is completed with boundary and initial conditions

$$\mathbf{v}(\mathbf{x},t)\cdot\mathbf{n}(\mathbf{x}) = 0,\;\; \vartheta(\mathbf{x},t) = 0,\;\; \mathbf{x}\in S,\; t\in(0,T), \tag{2.1.19}$$

$$\vartheta(\mathbf{x},0) = 0,\;\; p(\mathbf{x},0) = 0,\;\; \mathbf{x}\in\Omega. \tag{2.1.20}$$

(3) *For a disconnected pore space, or in the case $\mu_1 = \infty$, a unique limiting regime for the liquid dynamics is a state of rest $\mathbf{v} = 0$ and $p = 0$.*
In Eqs. (2.1.14)–(2.1.20) $\hat{c}_p = m\,\hat{c}_{p,f} + (1-m)\hat{c}_{p,s}$, the symmetric strictly positive definite constant matrix $\mathbb{B}$ is the same as in Theorem 1.1, the symmetric strictly positive definite constant matrix $\mathbb{B}^{\vartheta}$ is given below by formula (2.1.27), and **n** *is the normal vector to the boundary S.*

We refer to the problem (2.1.14)–(2.1.20) as the homogenized **model** $(\mathbb{NIF})_1$.

Theorem 2.2 *For $\nu_0 > 0$ there exists $\beta^0 > 0$, such that the problem (2.1.14)–(2.1.20) has a unique solution for all $\beta_f < \beta^0$.*
If $\nu_0 = 0$, then the problem (2.1.14)–(2.1.20) has a unique solution for all $\beta_f \geqslant 0$.

2.1.2 Proof of Theorem 2.1

Existence and uniqueness results for the problem (2.1.1)–(2.1.8) are proved in the Appendix B.

The estimate (2.1.12) is proved on the basis of energy equality

$$\int_0^t\int_{\Omega}\chi^{\varepsilon}\left(\alpha_{\mu}\left|\mathbb{D}\left(x,\mathbf{v}^{\varepsilon}(\mathbf{x},\tau)\right)\right|^2 + \alpha_{\nu}\left|\nabla\cdot\mathbf{v}^{\varepsilon}(\mathbf{x},\tau)\right|^2 + \alpha^{\varepsilon}_{\varkappa}\left|\nabla\vartheta^{\varepsilon}(\mathbf{x},\tau)\right|^2\right)dxd\tau \tag{2.1.21}$$

$$+\frac{1}{2}\int_{\Omega}\left(\eta_0^{\varepsilon}\left|\vartheta^{\varepsilon}(\mathbf{x},t)\right|^2 + \frac{1}{c_f^2}\left|p^{\varepsilon}(\mathbf{x},t)\right|^2\right)dx$$

$$= \int_0^t\int_{\Omega}\left(\Phi\,\vartheta^{\varepsilon} + \chi^{\varepsilon}\rho_f\mathbf{F}\cdot\mathbf{v}^{\varepsilon}\right)dxd\tau,$$

as well as earlier in Theorem 2.1.

Therefore, sequences $\{\mathbf{v}^{\varepsilon}\}$, $\{p^{\varepsilon}\}$ and $\{q^{\varepsilon}\}$ converge weakly in $\mathbf{L}_2(\Omega_T)$ and $L_2(\Omega_T)$ (up to some subsequences) to functions **v**, mp, and mq respectively.

At the same time the sequence $\{\vartheta^{\varepsilon}\}$ converges weakly in $\overset{\circ}{W}{}^{1,0}_2(\Omega_T)$ to function $\vartheta(\mathbf{x},t)$, and sequences $\{\vartheta^{\varepsilon}\}$ and $\{\nabla\vartheta^{\varepsilon}\}$ converge two-scale in $L_2(\Omega_T)$ and $\mathbf{L}_2(\Omega_T)$ to 1-periodic in **y** functions $\vartheta(\mathbf{x},t)$ and $\nabla\,\vartheta(\mathbf{x},t) + \nabla_y\Theta(\mathbf{x},t,\mathbf{y})$ respectively.

In a similar way to the previous sections it can be shown that $q \in W_2^{1,0}(\Omega_T)$ and limiting functions $\mathbf{v}$, q, p, and ϑ obviously satisfy Eqs. (2.1.14)–(2.1.16), (2.1.18), initial condition (2.1.20) for the pressure p, and the last statement of the theorem.

Thus, we only have to prove (2.1.17). To do that we fulfill the two-scale limit as $\varepsilon \to 0$ in the integral identity (2.1.11) in the form

$$\int_{\Omega_T}\left(\alpha_{\varkappa}^{\varepsilon}\nabla\vartheta^{\varepsilon}\cdot\nabla\psi+\left(\chi^{\varepsilon}\frac{\beta_f}{c_f^2}p^{\varepsilon}-\eta_0^{\varepsilon}\vartheta^{\varepsilon}\right)\frac{\partial\psi}{\partial t}\right)dxdt=\int_{\Omega_T}\Phi\,\psi\,dxdt$$

with two different types of test functions. First, with test functions $\psi = \psi(\mathbf{x}, t)$, and then with test functions $\psi = \varepsilon h(\mathbf{x}, t)\psi_0\left(\dfrac{\mathbf{x}}{\varepsilon}\right)$.

We have

$$\int_{\Omega_T}\left(\hat{\varkappa}\nabla\vartheta+\langle\varkappa(\mathbf{y})\nabla_y\Theta\rangle_Y\right)\cdot\nabla\psi+\left(\frac{\beta_f}{c_f^2}p-\hat{c}_p\,\vartheta)\frac{\partial\psi}{\partial t}\right)dxdt \tag{2.1.22}$$

$$=\int_{\Omega_T}\Phi\,\psi\,dxdt,$$

$$\int_{\Omega_T}h(\mathbf{x},t)\left(\int_Y\varkappa(\mathbf{y})\left(\nabla\,\vartheta+\nabla_y\Theta\right)\cdot\left(\nabla_y\psi_0(\mathbf{y})\right)dy\right)dxdt=0. \tag{2.1.23}$$

After standard reintegration we obtain the macroscopic heat equation

$$\hat{c}_p\,\frac{\partial\vartheta}{\partial t}-\frac{\beta_f}{c_f^2}\frac{\partial p}{\partial t}=\nabla\cdot\left(\hat{\varkappa}\nabla\,\vartheta+\langle\varkappa(\mathbf{y})\nabla_y\Theta\rangle_Y\right)+\Phi, \tag{2.1.24}$$

and the microscopic heat equation

$$\nabla_y\cdot\left(\varkappa(\mathbf{y})\left(\nabla\,\vartheta+\nabla_y\Theta\right)\right)=0. \tag{2.1.25}$$

As usual, we look for the 1-periodic solution of the Eq. (2.1.23) in the form

$$\Theta(\mathbf{x},t,\mathbf{y})=\sum_{i=1}^{3}\Theta^{(i)}(\mathbf{y})\frac{\partial\vartheta}{\partial x_i}(\mathbf{x},t),$$

where

$$\nabla_y\cdot\left(\varkappa(\mathbf{y})\left(\nabla_y\Theta^{(i)}+\mathbf{e}_i\right)\right)=0. \tag{2.1.26}$$

Then,

$$\mathbb{B}^{\vartheta}=\hat{\varkappa}\mathbb{I}+\sum_{i=1}^{3}\left\langle\varkappa(\mathbf{y})\nabla_y\Theta^{(i)}\right\rangle_Y\otimes\mathbf{e}_i. \tag{2.1.27}$$

The homogenized heat equation (2.1.17) and initial condition (2.1.20) for the temperature follow from (2.1.24), (2.1.27), and integral identity (2.1.22).

The existence and uniqueness results for the problem (2.1.26) and properties of the matrix $\mathbb{B}^{\vartheta}$ follow from equalities

$$\int_Y \varkappa \left(\nabla\Theta^{(i)} \cdot \nabla\Theta^{(j)} + \mathbf{e}_i \cdot \Theta^{(j)}\right) dy = 0.$$

In fact, let

$$\zeta = (\zeta_1, \zeta_2, \zeta_3),\ \eta = (\eta_1, \eta_2, \eta_3) \in \mathbb{R}^3,$$

$$\mathbf{z}_\zeta = \sum_{i=1}^{3} \zeta_i \nabla\Theta^{(i)},\ \mathbf{z}_\eta = \sum_{i=1}^{3} \eta_i \nabla\Theta^{(i)}.$$

Then

$$\langle \varkappa (\mathbf{z}_\zeta \cdot \mathbf{z}_\eta)\rangle_Y + \langle \varkappa (\zeta \cdot \mathbf{z}_\eta)\rangle_Y = 0,$$

and

$$\begin{aligned}(\mathbb{B} \cdot \zeta) \cdot \eta &= \langle \varkappa (\zeta \cdot \eta)\rangle_Y + \langle \varkappa (\mathbf{z}_\zeta \cdot \eta)\rangle_Y \\ &= \langle \varkappa (\zeta \cdot \eta)\rangle_Y + \langle \varkappa (\mathbf{z}_\zeta \cdot \eta)\rangle_Y + \langle \varkappa (\mathbf{z}_\zeta \cdot \mathbf{z}_\eta)\rangle_Y + \langle \varkappa (\zeta \cdot \mathbf{z}_\eta)\rangle_Y \\ &= \left\langle \varkappa \left((\mathbf{z}_\zeta + \eta) \cdot (\mathbf{z}_\eta + \zeta)\right)\right\rangle_Y.\end{aligned}$$

2.1.3 Proof of Theorem 2.2

Let $\nu_0 > 0$. Then the uniqueness of the solution to the problem (2.1.14)–(2.1.20) follows from its linearity and corresponding energy identities.

Firstly, we rewrite Eqs. (2.1.14)–(2.1.17) as

$$\frac{m}{c_f^2}\frac{\partial p}{\partial t} = \frac{1}{\mu_1}\nabla \cdot \left(\mathbb{B} \cdot \left(\nabla p + \frac{\nu_0}{c_f^2}\nabla\left(\frac{\partial p}{\partial t}\right) + m\,\beta_f \nabla\vartheta\right)\right), \tag{2.1.28}$$

$$\hat{c}_p \frac{\partial \vartheta}{\partial t} = \nabla \cdot \left(\mathbb{B}^{\vartheta} \cdot \nabla\vartheta + \frac{\beta_f}{\mu_1}\mathbb{B} \cdot \left(\nabla p + \frac{\nu_0}{c_f^2}\nabla\left(\frac{\partial p}{\partial t}\right) + m\,\beta_f \nabla\vartheta\right)\right). \tag{2.1.29}$$

Next, we multiply (2.1.28) by $\dfrac{\partial p}{\partial t}$ and integrate by parts over Ω:

$$\frac{1}{2\,\mu_1}\frac{d}{dt}\int_\Omega \nabla p\cdot(\mathbb{B}\cdot\nabla p)\,dx+\int_\Omega\left(\frac{m}{c_f^2}\left(\frac{\partial p}{\partial t}\right)^2+\frac{\nu_0}{\mu_1\,c_f^2}\nabla\left(\frac{\partial p}{\partial t}\right)\cdot\left(\mathbb{B}\cdot\nabla\left(\frac{\partial p}{\partial t}\right)\right)\right)dx \tag{2.1.30}$$

$$=-\frac{m\,\beta_f}{\mu_1}\int_\Omega\left((\nabla\,\vartheta)\cdot\left(\mathbb{B}\cdot\nabla\left(\frac{\partial p}{\partial t}\right)\right)\right)dx$$

$$\leqslant C_0\beta_f^2\int_\Omega|\nabla\,\vartheta|^2dx+\frac{\nu_0}{4\,\mu_1\,c_f^2}\int_\Omega\left|\nabla\left(\frac{\partial p}{\partial t}\right)\right|^2dx.$$

Finally, we multiply (2.1.29) by ϑ, integrate by parts over Ω, and sum the result with (2.1.30):

$$\frac{d}{dt}\int_\Omega\left(\frac{1}{2\,\mu_1}\nabla p\cdot(\mathbb{B}\cdot\nabla p)+\frac{\hat{c}_p}{2}|\vartheta|^2\right)dx+\int_\Omega\left(\nabla\,\vartheta\cdot\left(\mathbb{B}^\vartheta+\frac{m\,\beta_f^2}{\mu_1}\mathbb{B}\right)\cdot\nabla\,\vartheta\right)dx \tag{2.1.31}$$

$$+\int_\Omega\left(\frac{m}{c_f^2}\left(\frac{\partial p}{\partial t}\right)^2+\frac{\nu_0}{\mu_1\,c_f^2}\nabla\left(\frac{\partial p}{\partial t}\right)\cdot\left(\mathbb{B}\cdot\nabla\left(\frac{\partial p}{\partial t}\right)\right)\right)dx$$

$$\leqslant-\frac{\beta_f}{\mu_1}\int_\Omega((\nabla\,\vartheta)\cdot(\mathbb{B}\cdot\nabla\,p))\,dx-\frac{\nu_0\,\beta_f}{\mu_1\,c_f^2}\int_\Omega\left((\nabla\,\vartheta)\cdot\left(\mathbb{B}\cdot\nabla\left(\frac{\partial p}{\partial t}\right)\right)\right)dx$$

$$+\,C_0\beta_f^2\int_\Omega|\nabla\,\vartheta|^2dx+\frac{\nu_0}{4\,\mu_1\,c_f^2}\int_\Omega\left|\nabla\left(\frac{\partial p}{\partial t}\right)\right|^2dx$$

$$\leqslant C_0\left(\beta_f+\beta_f^2\right)\int_\Omega|\nabla\,\vartheta|^2dx+C_0\int_\Omega|\nabla\,p|^2dx+\beta_f\int_\Omega\left|\nabla\left(\frac{\partial p}{\partial t}\right)\right|^2dx.$$

Using the properties of matrices $\mathbb{B}$ and $\mathbb{B}^\vartheta$, choosing β^0 sufficiently small, and applying Gronwall's inequality [61] we arrive at the first statement of the lemma.

Let us recall that the *Gronwall inequality* states that if a nonnegative function $y(t)$ satisfies the conditions

$$\frac{dy}{dt}(t)\leqslant c(t)\,y(t)+F(t),\quad y(0)=0$$

with nonnegative summable functions $c(t)$ and $F(t)$, then

$$y(t)\leqslant\exp\left(\int_0^t c(\tau)d\tau\right)\int_0^t F(\tau)d\tau,$$

and

$$\frac{dy}{dt}(t)\leqslant c(t)\,\exp\left(\int_0^t c(\tau)d\tau\right)\int_0^t F(\tau)d\tau+F(t).$$

For the case $\nu_0=0$, Eqs. (2.1.28) and (2.1.29) take the form

$$\frac{m}{c_f^2}\frac{\partial p}{\partial t} = \frac{1}{\mu_1}\nabla\cdot\big(\mathbb{B}\cdot(\nabla p + m\,\beta_f\nabla\vartheta)\big), \tag{2.1.32}$$

$$\hat{c}_p\,\frac{\partial\vartheta}{\partial t} = \nabla\cdot\Big(\mathbb{B}^{\vartheta}\cdot\nabla\vartheta + \frac{\beta_f}{\mu_1}\,\mathbb{B}\cdot(\nabla p + m\,\beta_f\nabla\vartheta)\Big). \tag{2.1.33}$$

Now, we multiply (2.1.32) by p, (2.1.33) by ϑ, integrate by parts over Ω, and sum results:

$$\frac{d}{dt}\int_{\Omega}\Big(\frac{m}{2\,c_f^2}\,|p|^2 + \frac{\hat{c}_p}{2}|\vartheta|^2\Big)dx + \int_{\Omega}\nabla\vartheta\cdot\big(\mathbb{B}^{\vartheta}\cdot\nabla\vartheta\big)\,dx$$
$$+\frac{1}{\mu_1}\int_{\Omega}(m\,\beta_f\nabla\vartheta + \nabla p)\big(\mathbb{B}\cdot(m\,\beta_f\nabla\vartheta + \nabla p)\big)\,dx = 0.$$

This last identity implies $\vartheta = p = 0$.

2.2 A Slightly Viscous Thermo-Fluid in a Thermo-Elastic Skeleton

Here, as a basic mathematical model at the microscopic level we consider the model $\mathbb{M}_{13}$ of a non-isothermal liquid filtration in a thermo-elastic solid skeleton, consisting of the differential equations

$$\frac{1}{\alpha_p^{\varepsilon}}\,p + \nabla\cdot\mathbf{w} = 0, \tag{2.2.1}$$

$$\nabla\cdot\mathbb{P} + \rho^{\varepsilon}\mathbf{F} = 0, \tag{2.2.2}$$

$$\eta_0^{\varepsilon}\frac{\partial\vartheta}{\partial t} - \nabla\cdot(\alpha_{\varkappa}^{\varepsilon}\nabla\vartheta) = \Phi - \alpha_{\theta}^{\varepsilon}\,\nabla\cdot\Big(\frac{\partial\mathbf{w}}{\partial t}\Big), \tag{2.2.3}$$

$$\mathbb{P} = \chi^{\varepsilon}\alpha_{\mu}\,\mathbb{D}\Big(x, \frac{\partial\mathbf{w}}{\partial t}\Big) + (1-\chi^{\varepsilon})\lambda_0\,\mathbb{D}(x,\mathbf{w}) - \tilde{q}\,\mathbb{I}, \tag{2.2.4}$$

$$\tilde{q} = p + \alpha_{\theta}^{\varepsilon}\,\vartheta + \chi^{\varepsilon}\frac{\alpha_{\nu}}{c_f^2}\frac{\partial p}{\partial t}, \tag{2.2.5}$$

$$\mathbf{w}(\mathbf{x},t) = 0,\ \ \vartheta(\mathbf{x},t) = 0,\ \ \mathbf{x}\in S,\ \ t\in(0,T), \tag{2.2.6}$$

$$\chi^{\varepsilon}\,\mathbf{w}(\mathbf{x},0) = 0,\ \ \big(\eta_0^{\varepsilon}\,\vartheta + \alpha_{\theta}^{\varepsilon}\,\nabla\cdot\mathbf{w}\big)(\mathbf{x},0) = 0,\ \ \mathbf{x}\in\Omega. \tag{2.2.7}$$

Throughout this section we additionally impose Assumption 1.1 and also assume that conditions

$$\mu_0 = 0,\ 0 < \mu_1 \leqslant \infty,\ 0 \leqslant \nu_0 < \infty, \tag{2.2.8}$$
$$0 < \lambda_0,\ c_f^2,\ c_s^2,\ c_{p,f},\ c_{p,s},\ \varkappa_f,\ \varkappa_s,\ \beta_f,\ \beta_s,\ \gamma_0 < \infty,$$

and

$$\int_{\Omega_T} \left(|\mathbf{F}|^2 dxdt + \left|\frac{\partial \mathbf{F}}{\partial t}\right|^2 \right) dxdt + \int_{\Omega_T} |\Phi|^2 dxdt = \mathscr{F}_1^2 < \infty \tag{2.2.9}$$

hold true.

In Eqs. (2.2.1)–(2.2.8)

$$\alpha_\theta^\varepsilon = \chi^\varepsilon \beta_f + (1-\chi^\varepsilon)\beta_s \quad \eta_0^\varepsilon = \chi^\varepsilon c_{p,f} + (1-\chi^\varepsilon) c_{p,s},$$
$$\alpha_p^\varepsilon = \chi^\varepsilon c_f^2 + (1-\chi^\varepsilon) c_s^2 \quad \alpha_\varkappa^\varepsilon = \chi^\varepsilon \varkappa_f + (1-\chi^\varepsilon)\varkappa_s.$$

and

$$\mu_0 = \lim_{\varepsilon \searrow 0} \alpha_\mu(\varepsilon), \quad \nu_0 = \lim_{\varepsilon \searrow 0} \alpha_\nu(\varepsilon), \quad \mu_1 = \lim_{\varepsilon \searrow 0} \frac{\alpha_\mu}{\varepsilon^2}.$$

2.2.1 Statement of the Problem and Main Results

Definition 2.2 We say that the triple of functions $\{\mathbf{w}^\varepsilon,\ p^\varepsilon,\ \vartheta^\varepsilon\}$ such that

$$\mathbf{w}^\varepsilon \in \overset{\circ}{\mathbf{W}}{}_2^{1,0}(\Omega_T),\ \vartheta^\varepsilon, \in \overset{\circ}{W}{}_2^{1,0}(\Omega_T),\ \chi^\varepsilon \nabla \left(\frac{\partial \mathbf{w}^\varepsilon}{\partial t}\right) \in \mathbf{L}_2(\Omega_T),\ p^\varepsilon,\ \frac{\partial p^\varepsilon}{\partial t} \in L_2(\Omega_T),$$

is a weak solution of the problem (2.2.1)–(2.2.7), if it satisfies the continuity and state Eqs. (2.2.1) and (2.2.5) almost everywhere in Ω_T, the initial conditions (2.2.7) and integral identities

$$\int_{\Omega_T} \left(\chi^\varepsilon \alpha_\mu \mathbb{D}\left(x, \frac{\partial \mathbf{w}^\varepsilon}{\partial t}\right) + (1-\chi^\varepsilon)\lambda_0\, \mathbb{D}(x, \mathbf{w}^\varepsilon) \right) : \mathbb{D}(x, \varphi)\, dxdt \tag{2.2.10}$$
$$- \int_{\Omega_T} \tilde{q}^{\,\varepsilon}\, (\nabla \cdot \varphi)\, dxdt = \int_{\Omega_T} \rho^\varepsilon \mathbf{F} \cdot \varphi\, dxdt,$$

and

$$\int_{\Omega_T} \left(\alpha_\varkappa^\varepsilon \nabla \vartheta^\varepsilon \cdot \nabla \psi - \left(\eta_0^\varepsilon \vartheta^\varepsilon + \alpha_\theta^\varepsilon\, \nabla \cdot \mathbf{w}^\varepsilon \right) \frac{\partial \psi}{\partial t} \right) dxdt = \int_{\Omega_T} \Phi\, \psi\, dxdt, \tag{2.2.11}$$

for any functions $\varphi \in \overset{\circ}{\mathbf{W}}{}_2^{1,0}(\Omega_T)$ and $\psi \in \overset{\circ}{W}{}_2^{1,1}(\Omega_T)$, such that $\psi(\mathbf{x}, T) = 0$.

In Eq. (2.2.10)

$$\tilde{q}^{\,\varepsilon} = p^{\,\varepsilon} + \alpha_\theta^\varepsilon \vartheta^{\,\varepsilon} + \chi^\varepsilon \frac{\alpha_\nu}{c_f^2} \frac{\partial p^{\,\varepsilon}}{\partial t},$$

Theorem 2.3 *The problem (2.2.1)–(2.2.7) has an unique weak solution* $\{\mathbf{w}^\varepsilon, p^{\,\varepsilon}, \vartheta^\varepsilon\}$, *and*

$$\max_{0<t<T} \int_\Omega \left(\alpha_\mu \chi^\varepsilon \left| \mathbb{D}\left(x, \frac{\partial \mathbf{w}^\varepsilon}{\partial t}(\mathbf{x}, t)\right)\right|^2 + |\nabla \vartheta^\varepsilon(\mathbf{x}, t)|^2 + \alpha_\nu \chi^\varepsilon \left| \frac{\partial p^{\,\varepsilon}}{\partial t}(\mathbf{x}, t)\right|^2 \right) dx \tag{2.2.12}$$

$$+ \int_{\Omega_T} \left(\left| \mathbb{D}\left(x, \frac{\partial \mathbf{w}_s^\varepsilon}{\partial t}\right)\right|^2 + \left| \frac{\partial \vartheta^\varepsilon}{\partial t}\right|^2 + \left| \frac{\partial p^{\,\varepsilon}}{\partial t}\right|^2 \right) dxdt \leqslant C_0 \mathscr{F}_1^2,$$

$$\int_{\Omega_T} \left| \frac{\partial \mathbf{w}^\varepsilon}{\partial t}(\mathbf{x}, t) - \frac{\partial \mathbf{w}_s^\varepsilon}{\partial t}(\mathbf{x}, t)\right|^2 dxdt \leqslant \frac{\varepsilon^2}{\alpha_\mu} C_0 \mathscr{F}_1^2, \tag{2.2.13}$$

where $\mathbf{w}_s^\varepsilon$ *is an extension (1.2.9), and the constant* C_0 *is independent of the small parameter* ε.

Theorem 2.4 *Let* $\{\mathbf{w}^\varepsilon,\ p^{\,\varepsilon},\ \vartheta^\varepsilon\}$ *be a weak solution of the problem (2.2.1)–(2.2.7),* $\mathbf{w}_s^\varepsilon$ *be an extension (1.2.9) and* $\mu_1 = \infty$, *or* $\mu_1 < \infty$, *but the pore space be disconnected.*

Then for all $\nu_0 \geqslant 0$

(1) up to some subsequences the sequences $\{\mathbf{w}^\varepsilon\}$, $\{\nabla \cdot \mathbf{w}^\varepsilon\}$, $\{\chi^\varepsilon p^{\,\varepsilon}\}$, *and* $\{q^\varepsilon\}$, *where* $q^\varepsilon = \chi^\varepsilon \left(p^{\,\varepsilon} + (\beta_f - \beta_s)\vartheta^\varepsilon + \left(\frac{\alpha_\nu}{c_f^2}\right) \frac{\partial p^{\,\varepsilon}}{\partial t} \right)$, *converge weakly in* $\mathbf{L}_2(\Omega_T)$ *and* $L_2(\Omega_T)$ *to functions* $\mathbf{w}$, $\nabla \cdot \mathbf{w}$, mp_f, *and* mq *respectively.*

At the same time the sequence $\{\mathbf{w}_s^\varepsilon\}$ *converges weakly in* $\mathbf{W}_2^{1,0}(\Omega_T)$ *and two-scale in* $\mathbf{L}_2(\Omega_T)$ *to function* $\mathbf{w}_s = \mathbf{w} \in \overset{\circ}{\mathbf{W}}{}_2^{1,0}(\Omega_T)$, *and the sequence* $\{\vartheta^\varepsilon\}$ *converges weakly in* $\overset{\circ}{\mathbf{W}}{}_2^{1,0}(\Omega_T)$ *and two-scale in* $L_2(\Omega_T)$ *to function* ϑ;

(2) the limiting functions solve the system of homogenized equations in the domain Ω_T, *consisting of the homogenized continuity equation*

$$\frac{m}{c_f^2} p_f + m \nabla \cdot \mathbf{w}_s = \mathbb{C}_0^s : \mathbb{D}(x, \mathbf{w}_s) + \frac{c_0^s}{\lambda_0} q, \tag{2.2.14}$$

the state equation

$$q = p_f + m\,(\beta_f - \beta_s)\,\vartheta + \frac{\nu_0}{c_f^2}\frac{\partial p_f}{\partial t} \quad (q = p_f + m\,(\beta_f - \beta_s)\,\vartheta \ \textit{for}\ \nu_0 = 0), \tag{2.2.15}$$

the homogenized momentum balance equation

$$\nabla\cdot\big(\lambda_0\,\mathfrak{N}_2^s : \mathbb{D}(x, \mathbf{w}_s) - q\,\mathbb{C}_1^s - \beta_s\,\vartheta\,\mathbb{I}\big) + \hat{\rho}\mathbf{F} = 0, \tag{2.2.16}$$

and the homogenized heat equation

$$\hat{c}_p\,\frac{\partial\vartheta}{\partial t} = \nabla\cdot\big(\mathbb{B}^\vartheta\cdot\nabla\,\vartheta\big) + \Phi + \mathbb{C}^\vartheta : \mathbb{D}\left(x, \frac{\partial\mathbf{w}_s}{\partial t}\right) + \frac{c_0^s}{\lambda_0}\frac{\partial q}{\partial t}. \tag{2.2.17}$$

The system is completed with homogeneous boundary and initial conditions

$$\mathbf{w}_s(\mathbf{x}, t) = 0,\ \ \vartheta(\mathbf{x}, t) = 0,\ \ \mathbf{x}\in S,\ \ t\in(0, T), \tag{2.2.18}$$

$$\nu_0\,p_f(\mathbf{x}, 0) = 0, \tag{2.2.19}$$

$$\hat{c}_p\,\vartheta(\mathbf{x}, 0) = \mathbb{C}^\vartheta : \mathbb{D}\,(x, \mathbf{w}_s(\mathbf{x}, 0)) + \frac{c_0^s}{\lambda_0}\,q(\mathbf{x}, 0),\ \ \mathbf{x}\in\Omega.$$

In Eqs. (2.2.14)–(2.2.17)

$$\hat{\rho} = m\,\rho_f + (1-m)\,\rho_s,$$

the symmetric strictly positively definite constant fourth-rank tensor $\mathfrak{N}_2^s$*, matrices* $\mathbb{C}_0^s$ *and* $\mathbb{C}_1^s$*, and the constant* c_0^s *are the same as in Theorem 1.6, the symmetric strictly positively definite constant matrix* $\mathbb{B}^\vartheta$ *is defined in Theorem 2.1, and the matrix* $\mathbb{C}^\vartheta$ *is given below by formula (2.2.25).*

We refer to the problem (2.2.14)–(2.2.19) as the homogenized **model** $(\mathbb{NIF})_2$.

Theorem 2.5 *Let* $\{\mathbf{w}^\varepsilon,\ p^\varepsilon,\ \vartheta^\varepsilon\}$ *be a weak solution of the problem (2.2.1)–(2.2.7),* $\mathbf{w}_s^\varepsilon$ *be an extension (1.2.9), the pore space be connected and* $\mu_1 < \infty$.

Then for all $\nu_0 \geqslant 0$

(1) up to some subsequences the sequences $\{\mathbf{w}^\varepsilon\}$, $\{\chi^\varepsilon\mathbf{w}^\varepsilon\}$, $\{\nabla\cdot\mathbf{w}^\varepsilon\}$, $\{\chi^\varepsilon p^\varepsilon\}$*, and* $\{q^\varepsilon\}$*, where* $q^\varepsilon = \chi^\varepsilon\left(p^\varepsilon + (\beta_f - \beta_s)\,\vartheta^\varepsilon + \left(\frac{\alpha_\nu}{c_f^2}\right)\frac{\partial p^\varepsilon}{\partial t}\right)$*, converge weakly in* $\mathbf{L}_2(\Omega_T)$ *and* $L_2(\Omega_T)$ *to functions* $\mathbf{w}$, $\mathbf{w}^{(f)}$, $\nabla\cdot\mathbf{w}$, mp_f*, and* $mq\in W_2^{1,0}(\Omega_T)$ *respectively. At the same time the sequence* $\{\mathbf{w}_s^\varepsilon\}$ *converges weakly in* $\mathbf{W}_2^{1,0}(\Omega_T)$ *and two-scale in* $\mathbf{L}_2(\Omega_T)$ *to function* $\mathbf{w}_s\in\overset{\circ}{\mathbf{W}}{}_2^{1,0}(\Omega_T)$*, and the sequence* $\{\vartheta^\varepsilon\}$ *converges weakly in* $\overset{\circ}{W}{}_2^{1,0}(\Omega_T)$ *and two-scale in* $L_2(\Omega_T)$ *to function* ϑ*;*

(2) the limiting functions solve the system of homogenized equations in the domain Ω_T*, consisting of the homogenized continuity equation*

$$\frac{m}{c_f^2} p_f + \nabla \cdot \mathbf{w}^{(f)} = \mathbb{C}_0^s : \mathbb{D}(x, \mathbf{w}_s) + \frac{c_0^s}{\lambda_0} q, \tag{2.2.20}$$

the state equation (2.2.15), the homogenized momentum balance equation (2.2.16) for the solid component, the homogenized heat equation

$$\hat{c}_p \frac{\partial \vartheta}{\partial t} = \nabla \cdot \left(\mathbb{B}^\vartheta \cdot \nabla \vartheta\right) + \Phi - \beta_f \left(\nabla \cdot \frac{\partial \mathbf{w}^{(f)}}{\partial t}\right) + \mathbb{C}_1^\vartheta : \mathbb{D}\left(x, \frac{\partial \mathbf{w}_s}{\partial t}\right) + \frac{c_0^s}{\lambda_0} \frac{\partial q}{\partial t}, \tag{2.2.21}$$

and Darcy's law in the form

$$\mathbf{w}^{(f)} = m\, \mathbf{w}_s + \mathbb{B} \cdot \left(\int_0^t \left(-\nabla q + \rho_f \mathbf{F}\right)(\mathbf{x}, \tau) d\tau\right), \tag{2.2.22}$$

for the liquid component.

The system is completed with homogeneous boundary conditions (2.2.18) for the solid component and for the temperature, initial conditions

$$\begin{aligned} \nu_0\, p_f(\mathbf{x}, 0) = 0,\ \hat{c}_p\, \vartheta(\mathbf{x}, 0) \\ = \mathbb{C}_1^\vartheta : \mathbb{D}\,(x, \mathbf{w}_s(\mathbf{x}, 0)) + \frac{c_0^s}{\lambda_0}\, q(\mathbf{x}, 0) \\ + \beta_f \left(\nabla \cdot \mathbf{w}^{(f)}(\mathbf{x}, 0)\right),\ \mathbf{x} \in \Omega, \end{aligned} \tag{2.2.23}$$

and homogeneous boundary condition

$$\mathbf{w}^{(f)}(\mathbf{x}, t) \cdot \mathbf{n}(\mathbf{x}) = 0,\ \mathbf{x} \in S,\ t \in (0, T) \tag{2.2.24}$$

for displacements $\mathbf{w}^{(f)}$ of the fluid component.

In Eqs. (2.2.21)–(2.2.24) $\mathbf{n}$ is the normal vector to the boundary S, the symmetric strictly positively definite constant matrix $\mathbb{B}$ is the same as in Theorem 1.1, the matrix $\mathbb{C}_0^s$ and the constant c_0^s are the same as in Theorem 1.6, the symmetric strictly positively definite constant matrix $\mathbb{B}^\vartheta$ is defined in Theorem 2.1, and the matrix $\mathbb{C}_1^\vartheta$ is given below by formula (2.2.26).

We refer to the problem (2.2.15), (2.2.16), (2.2.18), (2.2.21)–(2.2.24) as the homogenized **model** $(\mathbb{NIF})_3$.

2.2.2 Proof of Theorem 2.3

The proof of this theorem repeats the proofs of Theorem 2.1 and Theorem 1.5, and is based upon the energy equality

$$\frac{1}{2}\frac{d}{dt}\int_{\Omega}\left(\alpha_{\mu}\chi^{\varepsilon}\left|\mathbb{D}\left(x,\frac{\partial \mathbf{w}^{\varepsilon}}{\partial t}\right)\right|^{2}+\frac{\alpha_{\nu}}{c_{f}^{2}}\chi^{\varepsilon}\left|\frac{\partial p^{\varepsilon}}{\partial t}\right|^{2}+\alpha_{\varkappa}^{\varepsilon}|\nabla\vartheta^{\varepsilon}|^{2}\right)dx$$
$$+\int_{\Omega}\left((1-\chi^{\varepsilon})\lambda_{0}\left|\mathbb{D}\left(x,\frac{\partial \mathbf{w}^{\varepsilon}}{\partial t}\right)\right|^{2}+\frac{1}{\alpha_{p}^{\varepsilon}}\left|\frac{\partial p^{\varepsilon}}{\partial t}\right|^{2}+\eta_{0}^{\varepsilon}\left|\frac{\partial \vartheta^{\varepsilon}}{\partial t}\right|^{2}\right)dx$$
$$=\int_{\Omega}\left(\Phi\frac{\partial \vartheta^{\varepsilon}}{\partial t}+\rho^{\varepsilon}\frac{\partial \mathbf{F}}{\partial t}\cdot\frac{\partial \mathbf{w}^{\varepsilon}}{\partial t}\right)dx.$$

2.2.3 Proofs of Theorem 2.4 and Theorem 2.5

The proofs of these theorems are almost exact repeats of the proofs of Theorem 1.6 and Theorem 1.7. We only have to derive homogenized heat equation (2.2.17) and (2.2.21), and corresponding initial conditions (2.2.19) and (2.2.23). Here we simply repeat the proof of Theorem 2.1 with one difference for each case $\mathbf{w}=\mathbf{w}_s$ (Theorem 2.4), and $\mathbf{w}\neq\mathbf{w}_s$ (Theorem 2.5). The difference is in the macroscopic heat equations.

For both cases we start with the heat equation in the form

$$\int_{\Omega_T}\left(\alpha_{\varkappa}^{\varepsilon}\nabla\vartheta^{\varepsilon}\cdot\nabla\psi-\eta_{0}^{\varepsilon}\vartheta^{\varepsilon}\frac{\partial\psi}{\partial t}\right)dxdt=\int_{\Omega_T}\Phi\,\psi\,dxdt+I^{\varepsilon},$$

where

$$I^{\varepsilon}=\int_{\Omega_T}\alpha_{\theta}^{\varepsilon}\,(\nabla\cdot\mathbf{w}^{\varepsilon})\frac{\partial\psi}{\partial t}\,dxdt.$$

One has

$$\alpha_{\theta}^{\varepsilon}\,(\nabla\cdot\mathbf{w}^{\varepsilon})=\beta_{f}\chi^{\varepsilon}(\nabla\cdot\mathbf{w}^{\varepsilon})+\beta_{s}(1-\chi^{\varepsilon})(\nabla\cdot\mathbf{w}^{\varepsilon})$$
$$=\beta_{f}(\nabla\cdot\mathbf{w}^{\varepsilon})+(\beta_{s}-\beta_{f})(1-\chi^{\varepsilon})(\nabla\cdot\mathbf{w}^{\varepsilon}).$$

Therefore,

$$\lim_{\varepsilon\searrow 0}I^{\varepsilon}=I^{0}=\int_{\Omega_T}\Big(\beta_{f}(\nabla\cdot\mathbf{w})+(\beta_{s}-\beta_{f})\big((1-m)(\nabla\cdot\mathbf{w}_{s})$$
$$+\langle\nabla_{y}\cdot\mathbf{U}\rangle_{Y_s}\big)\Big)\frac{\partial\psi}{\partial t}\,dxdt,\ \mathbf{w}=\mathbf{w}^{(f)}+(1-m)\mathbf{w}_{s}.$$

If $\mathbf{w} = \mathbf{w}_s$, then

$$I^0 = \int_{\Omega_T} \Big(\hat{\beta}(\nabla \cdot \mathbf{w}_s) + (\beta_s - \beta_f)\langle\nabla_y \cdot \mathbf{U}\rangle_{Y_s}\Big) \frac{\partial \psi}{\partial t}\, dxdt,$$

where $\hat{\beta} = m\,\beta_f + (1-m)\beta_s$.

For $\mathbf{w} \neq \mathbf{w}_s$

$$I^0 = \int_{\Omega_T} \Big(\beta_f(\nabla \cdot \mathbf{w}^{(f)}) + (1-m)\beta_s(\nabla \cdot \mathbf{w}_s) + (\beta_s - \beta_f)\langle\nabla_y \cdot \mathbf{U}\rangle_{Y_s}\Big) \frac{\partial \psi}{\partial t}\, dxdt.$$

Therefore,

$$\hat{\beta}\left(\nabla \cdot \frac{\partial \mathbf{w}_s}{\partial t}\right) + (\beta_s - \beta_f)\left\langle \nabla_y \cdot \left(\frac{\partial \mathbf{U}}{\partial t}\right)\right\rangle_{Y_s} = \mathbb{C}^{\vartheta} : \mathbb{D}\left(x, \frac{\partial \mathbf{w}_s}{\partial t}\right) + \frac{c_0^s}{\lambda_0}\frac{\partial q}{\partial t},$$

$$(1-m)\beta_s\left(\nabla \cdot \frac{\partial \mathbf{w}_s}{\partial t}\right) + (\beta_s - \beta_f)\left\langle \nabla_y \cdot \frac{\partial \mathbf{U}}{\partial t}\right\rangle_{Y_s} = \mathbb{C}_1^{\vartheta} : \mathbb{D}\left(x, \frac{\partial \mathbf{w}_s}{\partial t}\right) + \frac{c_0^s}{\lambda_0}\frac{\partial q}{\partial t},$$

and

$$\mathbb{C}^{\vartheta} = -\hat{\beta}\mathbb{I} - (\beta_s - \beta_f)\mathbb{C}_0^s, \tag{2.2.25}$$

$$\mathbb{C}_1^{\vartheta} = -(1-m)\beta_s\mathbb{I} - \left(\beta_s - \beta_f\right)\mathbb{C}_0^s. \tag{2.2.26}$$

To derive initial conditions (2.2.19) and (2.2.23) note that the homogenized heat equations (2.2.17) and (2.2.21) are actually formal expressions (as distributions) of the corresponding integral identities

$$\int_{\Omega_T} \left(\mathbb{B}^{\vartheta} \cdot \nabla\vartheta) \cdot \nabla\psi + \left(\mathbb{C}^{\vartheta} : \mathbb{D}(x, \mathbf{w}_s) + \frac{c_0^s}{\lambda_0} q - \hat{c}_p\,\vartheta\right) \frac{\partial \psi}{\partial t}\right) dxdt = \int_{\Omega_T} \Phi\,\psi\, dxdt,$$

and

$$\int_{\Omega_T} \left(\mathbb{B}^{\vartheta} \cdot \nabla\vartheta) \cdot \nabla\psi + \left(\mathbb{C}_1^{\vartheta} : \mathbb{D}(x, \mathbf{w}_s) + \frac{c_0^s}{\lambda_0} q + \beta_f(\nabla \cdot \mathbf{w}^{(f)}) - \hat{c}_p\,\vartheta\right) \frac{\partial \psi}{\partial t}\right) dxdt$$
$$= \int_{\Omega_T} \Phi\,\psi\, dxdt.$$

The last expressions evidently contains the initial conditions (2.2.19) and (2.2.23) (in a weak sense).

2.3 A Viscous Thermo-Fluid in an Elastic Skeleton

In this section as a basic mathematical model at the microscopic level we consider the model $\mathbb{M}_{13}$ of a non-isothermal liquid filtration in a thermo-elastic solid skeleton, consisting of the differential equations

$$\frac{1}{\alpha_p^\varepsilon}\, p + \nabla \cdot \mathbf{w} = 0, \tag{2.3.1}$$

$$\nabla \cdot \mathbb{P} + \tilde{\rho}\mathbf{F} + \alpha_\theta^\varepsilon\, \nabla \vartheta = 0, \tag{2.3.2}$$

$$\eta_0^\varepsilon \frac{\partial \vartheta}{\partial t} - \nabla \cdot (\alpha_\varkappa^\varepsilon \nabla \vartheta) = \Phi - \alpha_\theta^\varepsilon\, \nabla \cdot \frac{\partial \mathbf{w}}{\partial t}, \tag{2.3.3}$$

$$\mathbb{P} = \chi^\varepsilon \mu_0\, \mathbb{D}\left(x, \frac{\partial \mathbf{w}}{\partial t}\right) + (1 - \chi^\varepsilon)\lambda_0\, \mathbb{D}(x, \mathbf{w}) - p\, \mathbb{I}, \tag{2.3.4}$$

$$\mathbf{w}(\mathbf{x}, t) = 0,\;\; \vartheta(\mathbf{x}, t) = 0,\;\; \mathbf{x} \in S,\;\; t \in (0, T), \tag{2.3.5}$$

$$\chi^\varepsilon\, \mathbf{w}(\mathbf{x}, 0) = 0,\;\; \vartheta(\mathbf{x}, 0) = 0,\;\; \mathbf{x} \in \Omega. \tag{2.3.6}$$

Throughout this section we assume that conditions

$$0 < \mu_0,\, \lambda_0,\, c_f^2,\, c_s^2,\, c_{p,f},\, c_{p,s},\, \varkappa_f,\, \varkappa_s,\, \beta_f,\, \beta_s < \infty, \tag{2.3.7}$$

and

$$\int_{\Omega_T} \left(|\mathbf{F}|^2 dxdt + \left|\frac{\partial \mathbf{F}}{\partial t}\right|^2\right) dxdt + \int_{\Omega_T} |\Phi|^2 dxdt = \mathscr{F}_1^2 < \infty \tag{2.3.8}$$

hold true.

2.3.1 Statement of the Problem and Main Results

Definition 2.3 We say that the pair of functions $\{\mathbf{w}^\varepsilon,\ \vartheta^\varepsilon\}$ such that

$$\mathbf{w}^\varepsilon \in \overset{\circ}{\mathbf{W}}{}_2^{1,0}(\Omega_T),\;\; \vartheta^\varepsilon, \in \overset{\circ}{W}{}_2^{1,0}(\Omega_T),\;\; \chi^\varepsilon \nabla\left(\frac{\partial \mathbf{w}^\varepsilon}{\partial t}\right) \in \mathbf{L}_2(\Omega_T),$$

is a weak solution of the problem (2.3.1)–(2.3.6), if it satisfies the continuity equation (2.3.1) almost everywhere in Ω_T, the initial conditions (2.3.6) and integral identities

$$\int_{\Omega_T} \left(\chi^\varepsilon \mu_0 \mathbb{D}\left(x, \frac{\partial \mathbf{w}^\varepsilon}{\partial t}\right) + \mathfrak{N}_1^{(\varepsilon)} : \mathbb{D}(x, \varphi)\right) dxdt \tag{2.3.9}$$
$$= \int_{\Omega_T} \left(\rho^\varepsilon \mathbf{F} + \alpha_\theta^\varepsilon\, \nabla \vartheta\right) \cdot \varphi\, dxdt,$$

and

$$\int_{\Omega_T}\left(\alpha_{\varkappa}^{\varepsilon}\nabla\vartheta^{\varepsilon}\cdot\nabla\psi-\eta_0^{\varepsilon}\vartheta^{\varepsilon}\frac{\partial\psi}{\partial t}\right)dxdt=\int_{\Omega_T}\left(\Phi-\alpha_{\theta}^{\varepsilon}\left(\nabla\cdot\frac{\partial\mathbf{w}^{\varepsilon}}{\partial t}\right)\right)\psi\,dxdt, \tag{2.3.10}$$

for any functions $\varphi\in\overset{\circ}{\mathbf{W}}{}_2^{1,0}(\Omega_T)$ and $\psi\in\overset{\circ}{W}{}_2^{1,1}(\Omega_T)$, such that $\psi(\mathbf{x},T)=0$.

In Eq. (2.3.9)

$$\mathfrak{N}_1^{(\varepsilon)}=(1-\chi^{\varepsilon})\lambda_0\sum_{i,j=1}^{3}\mathbb{J}^{(ij)}\otimes\mathbb{J}^{(ij)}+\alpha_p^{\varepsilon}\mathbb{I}\otimes\mathbb{I}.$$

Solution $\mathbf{w}^{\varepsilon}$ of this problem possesses different smoothness in each domain Ω_f^{ε} and Ω_s^{ε}. To preserve the best properties, which the solution now possesses in the liquid part, we use the extension lemma (see Appendix B) to extend the function $\dfrac{\partial\mathbf{w}^{\varepsilon}}{\partial t}$ from the liquid part of the domain Ω onto its solid part Ω_f^{ε}:

$$\mathbf{v}^{\varepsilon}=\mathbb{E}_{\Omega_f^{\varepsilon}}\left(\frac{\partial\mathbf{w}^{\varepsilon}}{\partial t}\right), \tag{2.3.11}$$

such that

$$\chi^{\varepsilon}(\mathbf{x})\left(\frac{\partial\mathbf{w}^{\varepsilon}}{\partial t}(\mathbf{x},t)-\mathbf{v}^{\varepsilon}(\mathbf{x},t)\right)=0,\quad \mathbf{x}\in\Omega,\ t\in(0,T),$$

and

$$\int_{\Omega}|\mathbf{v}^{\varepsilon}(\mathbf{x},t)|^2dx\leqslant C_0\int_{\Omega_f^{\varepsilon}}\left|\frac{\partial\mathbf{w}^{\varepsilon}}{\partial t}(\mathbf{x},t)\right|^2dx, \tag{2.3.12}$$
$$\int_{\Omega}\left|\mathbb{D}\left(x,\mathbf{v}^{\varepsilon}(\mathbf{x},t)\right)\right|^2dx\leqslant C_0\int_{\Omega_f^{\varepsilon}}\left|\mathbb{D}\left(x,\frac{\partial\mathbf{w}^{\varepsilon}}{\partial t}(\mathbf{x},t)\right)\right|^2dx,\quad t\in(0,T),$$

where C_0 is independent of ε and $t\in(0,T)$.

Theorem 2.6 *There exists a unique weak solution* $\{\mathbf{w}^{\varepsilon},\ \vartheta^{\varepsilon}\}$ *of the problem (2.3.1)–(2.3.6) and*

$$\int_{\Omega_T}\chi^{\varepsilon}\left(\left|\mathbb{D}(x,\mathbf{v}^{\varepsilon})\right|^2+|\nabla\vartheta^{\varepsilon}|^2\right)dxdt \tag{2.3.13}$$
$$+\max_{0<t<T}\int_{\Omega}\left(|\mathbb{D}(x,\mathbf{w}^{\varepsilon})|^2+|\vartheta^{\varepsilon}|^2\right)dx\leqslant C_0\mathscr{F}_1^2,$$

where $\mathbf{v}^{\varepsilon}$ *is an extension (2.3.12), and the constant* C_0 *is independent of the small parameter* ε.

The proof of this theorem is straightforward and repeats the proof of Theorem 1.2 and Theorem 1.5, Theorem 2.1.

Theorem 2.7 *Let* $\mathbf{w}^\varepsilon$ *be the weak solution of the problem (2.3.1)–(2.3.6) and* $\mathbf{v}^\varepsilon$ *be an extension (2.3.12).*

Then

(1) the sequences $\{\mathbf{w}^\varepsilon\}$, $\{\mathbf{v}^\varepsilon\}$, *and* $\{\vartheta^\varepsilon\}$ *converge weakly in* $\mathbf{W}_2^{1,0}(\Omega_T)$ *and* $W_2^{1,0}(\Omega_T)$ *to functions* $\mathbf{w}$, $\mathbf{v} = \dfrac{\partial \mathbf{w}}{\partial t}$, *and* ϑ *respectively;*

(2) limiting functions solve the system of homogenized equations in the domain Ω_T, *consisting of the homogenized momentum balance equation*

$$\nabla \cdot \widetilde{\mathbb{P}}^\vartheta + \hat{\rho}\mathbf{F} = 0, \tag{2.3.14}$$

the state equation

$$\widetilde{\mathbb{P}}^\vartheta = \widetilde{\mathbb{P}} - \mathbb{C}_2^\vartheta \vartheta - \int_0^t \mathbb{C}_3^\vartheta(t-\tau)\vartheta(\mathbf{x},\tau)d\tau, \tag{2.3.15}$$

$$\widetilde{\mathbb{P}} = \mathfrak{N}_4 : \mathbb{D}\Big(x, \frac{\partial \mathbf{w}}{\partial t}\Big) + \mathfrak{N}_5 : \mathbb{D}(x,\mathbf{w}) + \int_0^t \mathfrak{N}_6(t-\tau) : \mathbb{D}(x,\mathbf{w}(\mathbf{x},\tau))d\tau,$$

and the homogenized heat equation

$$\begin{aligned} &\hat{c}_p \frac{\partial \vartheta}{\partial t} - \nabla \cdot \Big(\mathbb{B}^\vartheta \cdot \nabla \vartheta - \Big(c_0^\vartheta \vartheta + \int_0^t c_1^\vartheta(t-\tau)\vartheta(\mathbf{x},\tau)d\tau\Big)\mathbb{I}\Big) \\ &= \Phi - \mathbb{C}_4^\vartheta : \mathbb{D}\Big(x, \frac{\partial \mathbf{w}}{\partial t}\Big) - \mathbb{C}_5^\vartheta : \mathbb{D}(x,\mathbf{w}) - \int_0^t \mathbb{C}_6^\vartheta(t-\tau) : \mathbb{D}(x,\mathbf{w}(\mathbf{x},\tau))\,d\tau, \end{aligned} \tag{2.3.16}$$

completed with the homogeneous boundary and initial conditions

$$\mathbf{w}(\mathbf{x},t) = 0, \ \ \vartheta(\mathbf{x},t) = 0, \ \ \mathbf{x} \in S, \ \ t \in (0,T), \tag{2.3.17}$$

$$\mathbf{w}(\mathbf{x},0) = 0, \ \ \vartheta(\mathbf{x},0) = 0, \ \ \mathbf{x} \in \Omega. \tag{2.3.18}$$

In Eqs. (2.3.14)–(2.3.16) fourth-rank tensors $\mathfrak{N}_4$, $\mathfrak{N}_5$, *and* $\mathfrak{N}_6(t)$ *are given above by formulae (1.4.44) (see Theorem 1.12), the symmetric strictly positively definite constant matrix* $\mathbb{B}^\vartheta$ *is the same as in Theorem 2.1, and matrices* $\mathbb{C}_2^\vartheta$, $\mathbb{C}_3^\vartheta(t)$, $\mathbb{C}_4^\vartheta$, $\mathbb{C}_5^\vartheta$, $\mathbb{C}_6^\vartheta(t)$, *and scalars* c_0^ϑ *and* $c_1^\vartheta(t)$ *are given below by formulae (2.3.25)–(2.3.31).*

We refer to the problem (2.3.14)–(2.3.18) as the homogenized **model** $(\mathbb{NIF})_4$.

2.3.2 *Proof of Theorem 2.7*

It is clear that the major part of the proof of this theorem repeats the proofs of Theorem 1.12, Theorem 2.1, and Theorem 2.5. The difference is in the form of the micro- and macroscopic momentum balance equations

$$\nabla_y \cdot \left(\mu_0 \chi(\mathbf{y}) \left(\mathbb{D}\left(x, \frac{\partial \mathbf{w}}{\partial t}\right) + \mathbb{D}\left(y, \frac{\partial \mathbf{W}}{\partial t}\right)\right) \right. \\ \left. - \beta(\mathbf{y})\vartheta(\mathbf{x}, t)\mathbb{I} + \mathfrak{A}_2(\mathbf{y}) : (\mathbb{D}(x, \mathbf{w}) + \mathbb{D}(y, \mathbf{W}))\right) = 0, \tag{2.3.19}$$

$$\beta(\mathbf{y}) = \chi(\mathbf{y})\beta_f + (1 - \chi(\mathbf{y}))\,\beta_s,$$

$$\mathfrak{A}_2(\mathbf{y}) = (1 - \chi(\mathbf{y})) \left(\lambda_0 \sum_{i,j=1}^{3} \left(\mathbb{J}^{(ij)} \otimes \mathbb{J}^{(ij)} \right) + c_s^2\, (\mathbb{I} \otimes \mathbb{I}) \right) + \chi(\mathbf{y}) c_f^2\, (\mathbb{I} \otimes \mathbb{I}),$$

and

$$\nabla \cdot \widetilde{\mathbb{P}}^{\vartheta} + \hat{\rho}\mathbf{F} = 0, \tag{2.3.20}$$

$$\widetilde{\mathbb{P}}^{\vartheta} = \mu_0 m \mathbb{D}\left(x, \frac{\partial \mathbf{w}}{\partial t}\right) + \mu_0 \left\langle \mathbb{D}\left(y, \frac{\partial \mathbf{W}}{\partial t}\right)\right\rangle_{Y_f} \\ + \langle \mathfrak{A}_2 \rangle_Y : \mathbb{D}(x, \mathbf{w}) + \langle \mathfrak{A}_2 : \mathbb{D}(y, \mathbf{W})\rangle_Y - \hat{\beta}\vartheta\mathbb{I}, \tag{2.3.21}$$

$$\hat{\rho} = m\rho_f + (1 - m)\rho_s, \quad \hat{\beta} = m\beta_f + (1 - m)\beta_s,$$

and in the form of the micro- and macroscopic heat equations

$$\nabla_y \cdot \left(\varkappa(\mathbf{y}) \left(\nabla \vartheta + \nabla_y \Theta \right) \right) = 0, \tag{2.3.22}$$

$$\hat{c_p} \frac{\partial \vartheta}{\partial t} - \nabla \cdot \left(\hat{\varkappa} \nabla \vartheta + \langle \varkappa(\mathbf{y}) \nabla_y \Theta \rangle_Y \right) \\ = \Phi - \hat{\beta} \nabla \cdot \left(\frac{\partial \mathbf{w}}{\partial t} \right) - \left\langle \beta(\mathbf{y}) \nabla_y \cdot \frac{\partial \mathbf{W}}{\partial t} \right\rangle_Y . \tag{2.3.23}$$

The solution of the microscopic momentum balance equation (2.3.19) is slightly different from the solution of the microscopic momentum balance equation (1.4.35) in the proof of Theorem 1.12:

$$\mathbf{W}(\mathbf{x}, t, \mathbf{y}) = \sum_{i,j=1}^{3} \int_0^t \widetilde{\mathbf{W}}^{(ij)}(\mathbf{y}, t - \tau) Y_{ij}(\mathbf{x}, \tau) d\tau - \int_0^t \widetilde{\mathbf{W}}^{(0)}(\mathbf{y}, t - \tau) \vartheta(\mathbf{x}, \tau) d\tau,$$

where $\widetilde{\mathbf{W}}^{(0)}$ is a solution of the periodic boundary value problem

$$\left.\begin{array}{r}\nabla_y\cdot\left(\chi\,\mu_0\,\mathbb{D}\left(y,\dfrac{\partial\widetilde{\mathbf{W}}^{(0)}}{\partial t}\right)+\mathfrak{A}_2(\mathbf{y}):\mathbb{D}\left(y,\widetilde{\mathbf{W}}^{(0)}\right)\right)=0,\\ \chi(\mathbf{y})\widetilde{\mathbf{W}}^{(0)}(\mathbf{y},0)=\widetilde{\mathbf{W}}_0^{(0)}(\mathbf{y}),\\ \nabla_y\cdot\left(\chi\left(\mu_0\,\mathbb{D}\left(y,\widetilde{\mathbf{W}}_0^{(0)}\right)+\beta(\mathbf{y})\mathbb{I}\right)\right)=0\end{array}\right\}\tag{2.3.24}$$

in the domain Y and $\widetilde{\mathbf{W}}^{(ij)}$ is a solution of (1.4.38), (1.4.39).

Therefore,

$$\widetilde{\mathbb{P}}^{\vartheta}=\widetilde{\mathbb{P}}-\mathbb{C}_2^{\vartheta}\vartheta-\int_0^t\mathbb{C}_3^{\vartheta}(t-\tau)\vartheta(\mathbf{x},\tau)d\tau,$$

and

$$\mathbb{C}_2^{\vartheta}=\hat{\beta}\mathbb{I}+\mu_0\left\langle\mathbb{D}\left(y,\widetilde{\mathbf{W}}_0^{(0)}\right)\right\rangle_{Y_f},\tag{2.3.25}$$

$$\mathbb{C}_3^{\vartheta}(t)=\mu_0\left\langle\mathbb{D}\left(y,\frac{\partial\widetilde{\mathbf{W}}^{(0)}}{\partial t}\right)\right\rangle_{Y_f}+\left\langle\mathfrak{A}_2:\mathbb{D}\left(y,\widetilde{\mathbf{W}}^{(0)}\right)\right\rangle_{Y}.\tag{2.3.26}$$

The solution of the microscopic heat equation (2.3.22) is the same as the solution of the microscopic heat equation (2.1.25) in the proof of Theorem 2.1.

Therefore,

$$\hat{\varkappa}\nabla\vartheta+\langle\varkappa(\mathbf{y})\nabla_y\Theta\rangle_Y=\mathbb{B}^{\vartheta}\cdot\nabla\vartheta.$$

Next,

$$\begin{aligned}\hat{\beta}\,\nabla\cdot\left(\frac{\partial\mathbf{w}}{\partial t}\right)+\left\langle\beta(\mathbf{y})\nabla_y\cdot\frac{\partial\mathbf{W}}{\partial t}\right\rangle_Y&=\hat{\beta}\,\nabla\cdot\left(\frac{\partial\mathbf{w}}{\partial t}\right)+\left\langle\sum_{i,j=1}^{3}\left(\nabla\cdot\widetilde{\mathbf{W}}_0^{(ij)}\right)\mathbb{J}^{(ij)}\right\rangle_Y:\mathbb{Y}(\mathbf{x},t)\\&+\left\langle\nabla\cdot\widetilde{\mathbf{W}}_0^{(0)}\right\rangle_Y\vartheta+\int_0^t\left(\sum_{i,j=1}^{3}\left\langle\nabla\cdot\frac{\partial\widetilde{\mathbf{W}}^{(ij)}}{\partial t}\right\rangle_Y(t-\tau)\mathbb{J}^{(ij)}:\mathbb{Y}(\mathbf{x},\tau)\right.\\&\left.+\left\langle\nabla\cdot\frac{\partial\widetilde{\mathbf{W}}^{(0)}}{\partial t}\right\rangle_Y(t-\tau)\vartheta(\mathbf{x},\tau)\right)d\tau=\mathbb{C}_4^{\vartheta}:\mathbb{D}\left(x,\frac{\partial\mathbf{w}}{\partial t}\right)+\mathbb{C}_5^{\vartheta}:\mathbb{D}(x,\mathbf{w})\\&+\int_0^t\mathbb{C}_6^{\vartheta}(t-\tau):\mathbb{D}(x,\mathbf{w}(\mathbf{x},\tau))\,d\tau+c_0^{\vartheta}\vartheta+\int_0^t c_1^{\vartheta}(t-\tau)\vartheta(\mathbf{x},\tau)d\tau,\end{aligned}$$

where

$$\mathbb{C}_4^{\vartheta}=\hat{\beta}\,\mathbb{I}+\mu_0\left\langle\sum_{i,j=1}^{3}\left(\nabla\cdot\widetilde{\mathbf{W}}_0^{(ij)}\right)\right\rangle_Y\mathbb{J}^{(ij)},\tag{2.3.27}$$

$$\mathbb{C}_5^{\vartheta} = \mu_0 \sum_{i,j=1}^{3} \left\langle \nabla \cdot \frac{\partial \widetilde{\mathbf{W}}^{(ij)}}{\partial t}(\mathbf{y}, 0) \right\rangle_Y \mathbb{J}^{(ij)} - \left(\mathfrak{N}_1^{(0)} : \mathbb{J}^{(ij)} \right) \left\langle \sum_{i,j=1}^{3} \left(\nabla \cdot \widetilde{\mathbf{W}}_0^{(ij)} \right) \right\rangle_Y , \tag{2.3.28}$$

$$\mathbb{C}_6^{\vartheta}(t) = \mu_0 \sum_{i,j=1}^{3} \mathbb{J}^{(ij)} \left\langle \nabla \cdot \frac{\partial^2 \widetilde{\mathbf{W}}^{(ij)}}{\partial t^2} \right\rangle_Y - \sum_{i,j=1}^{3} \left(\mathfrak{N}_1^{(0)} : \mathbb{J}^{(ij)} \right) \left\langle \nabla \cdot \frac{\partial \widetilde{\mathbf{W}}^{(ij)}}{\partial t} \right\rangle_Y , \tag{2.3.29}$$

$$c_0^{\vartheta} = \left\langle \nabla \cdot \widetilde{\mathbf{W}}_0^{(0)} \right\rangle_Y , \tag{2.3.30}$$

$$c_1^{\vartheta}(t) = \left\langle \nabla \cdot \frac{\partial \widetilde{\mathbf{W}}^{(0)}}{\partial t} \right\rangle_Y . \tag{2.3.31}$$

Chapter 3
Hydraulic Shock in Incompressible Poroelastic Media

3.1 The Problem Statement and Basic A Priori Estimates

As a basic mathematical model at the microscopic level here we consider the model $\mathbb{M}_{21}$ of **isothermal short-term processes in incompressible media**:

$$\nabla \cdot \mathbf{w} = 0, \tag{3.1.1}$$

$$\tilde{\rho}\frac{\partial^2 \mathbf{w}}{\partial t^2} = \nabla \cdot \mathbb{P} + \tilde{\rho}\mathbf{F}, \tag{3.1.2}$$

$$\mathbb{P} = \tilde{\chi}\bar{\alpha}_\mu \mathbb{D}\Big(x, \frac{\partial \mathbf{w}}{\partial t}\Big) + (1-\tilde{\chi})\bar{\alpha}_\lambda \mathbb{D}(x, \mathbf{w}) - p\,\mathbb{I}. \tag{3.1.3}$$

This model is derived in Appendix A.

Throughout this chapter we impose Assumption 0.1 and Assumption 1.1, completed with

Assumption 3.1 The pore space Ω_f^ε is a connected domain.

Under these assumptions

$$\tilde{\chi}(\mathbf{x}) = \chi^\varepsilon(\mathbf{x}) = \varsigma(\mathbf{x})\chi\left(\frac{\mathbf{x}}{\varepsilon}\right), \quad \tilde{\rho} = \rho^\varepsilon = \chi^\varepsilon \rho_f + (1-\chi^\varepsilon)\rho_s,$$

where $\varsigma(\mathbf{x})$ is the characteristic function of the domain Ω.

Usually, the initial impulse for the hydraulic shock is transmitted into the oil reservoir through a well filled with a liquid (Fig. 3.1).

To model this process we consider the domain Ω as a subdomain of the domain Q, such that the compliment of Ω in Q is a cylinder $\overline{\Omega}^{\,0} = \{\mathbf{x} \in \mathbb{R}^3 : x_1^2 + x_2^2 \leqslant \delta^2 < 1,\ \varphi_0(x_1, x_2) < x_3 < 0\}$. In turn, the domain Q is a subset of the half-space $\{x_3 < 0\}$ and its boundary S consists of two parts. The part S^1 is a subdomain of the plane $\{x_3 = 0\}$. The compliment $S^2 = S \backslash S^1$ is a smooth C^2 surface, and in some small

A. Meirmanov, *Mathematical Models for Poroelastic Flows*,
Atlantis Studies in Differential Equations 1, DOI: 10.2991/978-94-6239-015-7_3,

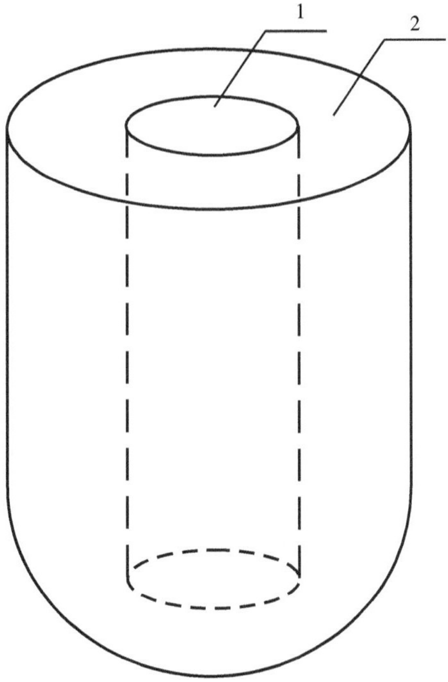

Fig. 3.1 1—domain Ω^0, 2—domain Ω

neighborhood of the plane $\{x_3 = 0\}$ it is represented by the equation $\Phi(x_1, x_2) = 0$ (that is S^2 is a cylinder near the plane $\{x_3 = 0\}$).

For the given $\varepsilon > 0$ the solid-liquid mixture in the domain Ω_T is governed by the system

$$\nabla \cdot \mathbf{w}^\varepsilon = 0, \tag{3.1.4}$$

$$\rho^\varepsilon \frac{\partial^2 \mathbf{w}^\varepsilon}{\partial t^2} = \nabla \cdot \mathbb{P}, \tag{3.1.5}$$

$$\mathbb{P} = \chi^\varepsilon \bar{\alpha}_\mu \mathbb{D}\Big(x, \frac{\partial \mathbf{w}^\varepsilon}{\partial t}\Big) + (1 - \chi^\varepsilon)\, \bar{\alpha}_\lambda \mathbb{D}(x, \mathbf{w}^\varepsilon) - p^\varepsilon\, \mathbb{I}. \tag{3.1.6}$$

In the domain Ω_T^0 the liquid motion is described by the Stokes system, consisting of the continuity equation (3.1.4) and the momentum balance equation

$$\rho_f \frac{\partial^2 \mathbf{w}^\varepsilon}{\partial t^2} = \nabla \cdot \mathbb{P}^0, \tag{3.1.7}$$

$$\mathbb{P}^0 = \bar{\alpha}_\mu \mathbb{D}\Big(x, \frac{\partial \mathbf{w}^\varepsilon}{\partial t}\Big) - p^\varepsilon\, \mathbb{I}. \tag{3.1.8}$$

On the common boundary $S^0 = \partial\Omega \cap \partial\Omega^0$ the usual continuity conditions for displacements and for the normal component of the momentum hold true

$$\lim_{\substack{\mathbf{x}\to\mathbf{x}^0\\ \mathbf{x}\in\Omega^0}} \mathbf{w}(\mathbf{x},t) = \lim_{\substack{\mathbf{x}\to\mathbf{x}^0\\ \mathbf{x}\in\Omega}} \mathbf{w}(\mathbf{x},t), \tag{3.1.9}$$

$$\lim_{\substack{\mathbf{x}\to\mathbf{x}^0\\ \mathbf{x}\in\Omega^0}} \mathbb{P}^0(\mathbf{x},t)\cdot\mathbf{n}(\mathbf{x}^0) = \lim_{\substack{\mathbf{x}\to\mathbf{x}^0\\ \mathbf{x}\in\Omega}} \mathbb{P}(\mathbf{x},t)\cdot\mathbf{n}(\mathbf{x}^0). \tag{3.1.10}$$

Here $n(\mathbf{x}^0)$ is a normal vector to the boundary S^0 at $\mathbf{x}^0 \in S^0$.

Now we impose boundary conditions on the outer boundary $S = \partial Q$. On the part S^1 we put

$$\big(\zeta\mathbb{P}^0 + (1-\zeta)\mathbb{P}\big)\cdot\mathbf{e}_3 = -p_0(\mathbf{x},t)\mathbf{e}_3, \tag{3.1.11}$$

where $p_0(\mathbf{x},t)$ is the impulse defining the hydraulic shock. We suppose that p_0 is finite in $\{\mathbf{x}\in\mathbb{R}^3 | x_1^2 + x_2^2 < \frac{\delta^2}{2} < 1, -\delta < x_3 < 0\}$.

On the rest of the outer boundary S^2

$$\mathbf{w}^\varepsilon(\mathbf{x},t) = 0 \tag{3.1.12}$$

for $t > 0$.

The problem has the homogeneous initial conditions

$$\mathbf{w}^\varepsilon(\mathbf{x},0) = 0, \quad \frac{\partial\mathbf{w}^\varepsilon}{\partial t}(\mathbf{x},0) = 0, \quad \mathbf{x}\in Q. \tag{3.1.13}$$

In the usual way we define a weak solution of the problems (3.1.4)–(3.1.13).

Definition 3.1 We say that the pair of functions $\{\mathbf{w}^\varepsilon, p^\varepsilon\}$ such that

$$\mathbf{w}^\varepsilon \in \overset{\circ}{\mathbf{W}}{}_2^{1,0}(Q_T), \quad \frac{\partial\mathbf{w}^\varepsilon}{\partial t} \in \mathbf{L}_2(Q_T), \quad p^\varepsilon \in L_2(Q_T),$$

is a weak solution of the problem (3.1.4)–(3.1.13), if it satisfies the continuity equation (3.1.4) almost everywhere in Q_T, the first initial condition in (3.1.13) for the function $\mathbf{w}^\varepsilon$, and the integral identity

$$\int_{Q_T}\Big(-\tilde{\rho}^\varepsilon\frac{\partial\mathbf{w}^\varepsilon}{\partial t}\cdot\frac{\partial\varphi}{\partial t} + \big(\zeta\mathbb{P}^0 + (1-\zeta)\mathbb{P}\big) : \mathbb{D}(x,\varphi)\Big)dxdt$$
$$= -\int_{Q_T}\nabla\cdot(\varphi\, p_0)dxdt \tag{3.1.14}$$

for all functions $\varphi \in \mathbf{W}_2^{1,0}(Q_T)$, $\dfrac{\partial \varphi}{\partial t} \in \mathbf{L}_2(Q_T)$, such that $\varphi(\mathbf{x}, t) = 0$ on the boundary S^2_T, and $\varphi(\mathbf{x}, T) = 0$ for $\mathbf{x} \in Q$.

In (3.1.12) $\tilde{\rho}^\varepsilon = \big(\zeta + (1-\zeta)\chi^\varepsilon\big)\rho_f + (1-\zeta)(1-\chi^\varepsilon)\rho_s$ and $\zeta = \zeta(\mathbf{x})$ is the characteristic function of the domain Ω^0 in Q.

The integral identity (3.1.14) contains Stokes equations in $\Omega_f \cup \Omega^0$ for $t > 0$, Lamé's equations in Ω_s for $t > 0$, the continuity condition for the normal tensions at the boundary S^0 (condition (3.1.10)), the similar condition at the common boundary "pore space–solid skeleton", and the second initial condition in (3.1.13).

Sometimes we will use the identity (3.1.14) in its differential form

$$\tilde{\rho}^\varepsilon \frac{\partial^2 \mathbf{w}^\varepsilon}{\partial t^2} = \nabla \cdot \big(\zeta \mathbb{P}^0 + (1-\zeta)\mathbb{P}\big), \tag{3.1.15}$$

and say that (3.1.15) and (3.1.11) are understood in the sense of distributions.

For the problem (3.1.4)–(3.1.13) we will find all the possible limiting regimes (homogenized equations) as $\varepsilon \searrow 0$.

To do that we suppose that the dimensionless parameters $\bar{\alpha}_\mu$ and $\bar{\alpha}_\lambda$ depend on the small parameter ε and the (finite or infinite) limits exist:

$$\lim_{\varepsilon\searrow 0} \bar{\alpha}_\mu(\varepsilon) = \mu_0, \ \lim_{\varepsilon\searrow 0} \frac{\bar{\alpha}_\mu}{\varepsilon^2} = \mu_1, \ \lim_{\varepsilon\searrow 0} \bar{\alpha}_\lambda(\varepsilon) = \lambda_0, \ \lim_{\varepsilon\searrow 0} \frac{\bar{\alpha}_\lambda}{\varepsilon^2} = \lambda_1.$$

Throughout this chapter we assume that

$$\int_{Q_T} \Big(|\nabla p_0(\mathbf{x}, t)|^2 + \Big|\nabla \frac{\partial p_0}{\partial t}(\mathbf{x}, t)\Big|^2 \Big) dxdt = \mathfrak{P}^2 < \infty.$$

In what follows, we denote as C_0 any constant depending only on domains Ω, Y and Y_f.

The derivation of all these limiting regimes is based upon on the following.

Theorem 3.1 *Let $\{\mathbf{w}^\varepsilon, p^\varepsilon\}$ be the weak solution of the problem (3.1.4)–(3.1.13). Then*

$$\begin{aligned} \max_{0\le t\le T} \int_Q \Big(&|\frac{\partial^2 \mathbf{w}^\varepsilon}{\partial t^2}|^2 + |p^\varepsilon|^2 + \bar{\alpha}_\mu\big(\zeta + (1-\zeta)\chi^\varepsilon\big)|\mathbb{D}(x, \frac{\partial \mathbf{w}^\varepsilon}{\partial t})|^2 \\ &+ \bar{\alpha}_\lambda(1-\zeta)(1-\chi^\varepsilon)|\mathbb{D}(x, \frac{\partial \mathbf{w}^\varepsilon}{\partial t})|^2 \Big) dx \leqslant C_0 \mathfrak{P}^2, \end{aligned} \tag{3.1.16}$$

where the constant C_0 is independent of the small parameter ε.

The proof of this theorem is straightforward. In fact, the estimate (3.1.16) for displacements $\mathbf{w}^\varepsilon$ follows from the energy identity

$$\frac{1}{2}\frac{d}{dt}\int_Q\left(\tilde{\rho}^\varepsilon\left|\frac{\partial^2\mathbf{w}^\varepsilon}{\partial t^2}\right|^2+\bar{\alpha}_\lambda(1-\zeta)(1-\chi^\varepsilon)\left|\mathbb{D}\left(x,\frac{\partial\mathbf{w}^\varepsilon}{\partial t}\right)\right|^2\right)dx$$

$$+\int_Q\bar{\alpha}_\mu\left(\zeta+(1-\zeta)\chi^\varepsilon\right)\left|\mathbb{D}\left(x,\frac{\partial^2\mathbf{w}^\varepsilon}{\partial t^2}\right)\right|^2dx=\int_Q\nabla\left(\frac{\partial p_0}{\partial t}\right)\cdot\frac{\partial^2\mathbf{w}^\varepsilon}{\partial t^2}dx.$$

In turn, the last relation is the result of differentiation of (3.1.15) with respect to time, multiplication by $\dfrac{\partial^2\mathbf{w}^\varepsilon}{\partial t^2}$, and integration by parts over domain Q.

The estimation of the pressure p^ε repeats the estimation of the pressure in Theorem 1.2 with some evident changes. So, the integral identity (3.1.14) and estimates (3.1.16) for displacements imply

$$\left|\int_{Q_T}p^\varepsilon\,\nabla\cdot\varphi\,dxdt\right|\leqslant C_0\,\mathfrak{P}\left(\int_{Q_T}|\nabla\varphi|^2dxdt\right)^{\frac{1}{2}}. \tag{3.1.17}$$

The difference from Theorem 1.2 is in the choice of the test function φ. Here we choose the test function φ from the same conditions

$$\nabla\cdot\varphi=p^\varepsilon,\ \textit{and}\ \int_{Q_T}|\nabla\varphi|^2dxdt\leqslant C_0\int_{Q_T}|p^\varepsilon|^2dxdt.$$

and decompose the function φ into the sum of two functions φ_0 and $\nabla\,\psi$ such that

$$\triangle\,\psi=p^\varepsilon,\ \mathbf{x}\in Q,\ \psi\Big|_{S_2}=0,\ \frac{\partial\psi}{\partial x_3}\Big|_{S_1}=0, \tag{3.1.18}$$

$$\nabla\cdot\varphi_0=0,\ \mathbf{x}\in Q,\ \varphi_0+\nabla\psi=0,\ \mathbf{x}\in S_2. \tag{3.1.19}$$

The difference in the choice of test functions is in the boundary condition for the function ψ on the part S^1. Instead of the homogeneous Dirichlet condition we put the homogeneous Neumann condition.

The desired smoothness of the solutions of the problems (3.1.18) and (3.1.19) follows from the structure of the boundary S. In fact, we may extend the solution φ outside of Q near some small neighborhood of $\{x_3=0\}$ and S_2 as an odd function, and then from the domain obtained into $\{x_3>0\}$ as an even function satisfying the Poisson equation, and use the local estimates in $W_2^2(Q')$ [60, 61].

3.2 A Slightly Viscous Liquid in an Extremely Elastic Skeleton

Throughout this section we assume that

$$\mu_0=0,\ \lambda_0=0. \tag{3.2.1}$$

3.2.1 Main Results

Theorem 3.2 *Let* $\{\mathbf{w}^{\varepsilon}, p^{\varepsilon}\}$ *be the weak solution of the problem (3.1.4)–(3.1.13) and*

$$\mu_1 = \lambda_1 = \infty.$$

Then the sequence $\{p^{\varepsilon}\}$ *converges weakly in* $L_2(Q_T)$ *as* $\varepsilon \to 0$ *to the solution* $p \in W_2^{1,0}(Q_T)$ *of the mixed boundary value problem*

$$\nabla \cdot \Big(\frac{1}{\rho(\mathbf{x})}\nabla p\Big) = 0, \quad \mathbf{x} \in Q, \ t > 0, \tag{3.2.2}$$

$$p(\mathbf{x}, t) = p_0(\mathbf{x}, t), \ \mathbf{x} \in S^1, \ t > 0, \tag{3.2.3}$$

$$\nabla p(\mathbf{x}, t) \cdot \mathbf{n}(\mathbf{x}) = 0, \ \mathbf{x} \in S^2, \ t > 0. \tag{3.2.4}$$

In (3.2.2)–(3.2.4)

$$\rho(\mathbf{x}) = \Big(\zeta(\mathbf{x}) + \big(1 - \zeta(\mathbf{x})\big)m\Big)\rho_f + \big(1 - \zeta(\mathbf{x})\big)(1 - m)\rho_s,$$

and $\mathbf{n}(\mathbf{x})$ *is the normal vector to the boundary* S^2 *at the point* $\mathbf{x} \in S^2$.

We refer to the problems (3.2.2)–(3.2.4) as the homogenized **model** $(\mathbb{HS})_1$.

Theorem 3.3 *Let* $\{\mathbf{w}^{\varepsilon}, p^{\varepsilon}\}$ *be the weak solution of the problems (3.1.4)–(3.1.13) and*

$$0 \leqslant \mu_1, \ \lambda_1 < \infty.$$

Then the sequence $\{p^{\varepsilon}\}$ *converges weakly in* $L_2(Q_T)$ *as* $\varepsilon \to 0$ *to the solution* $p \in W_2^{1,0}(Q_T)$ *of the mixed boundary value problem, consisting of boundary conditions (3.2.3) on the part* S^1 *of the boundary* S*, boundary condition*

$$\Big(\int_0^t \mathbb{B}(\mu_1, \lambda_1; \mathbf{x}, t - \tau) \cdot \nabla p(\mathbf{x}, \tau) d\tau\Big) \cdot \mathbf{n}(\mathbf{x}) = 0 \tag{3.2.5}$$

on the part S^2 *of the boundary* S*, and homogenized equation*

$$\nabla \cdot \Big(\int_0^t \mathbb{B}(\mu_1, \lambda_1; \mathbf{x}, t - \tau) \cdot \nabla p(\mathbf{x}, \tau) d\tau\Big) = 0, \quad \mathbf{x} \in Q, \ t > 0. \tag{3.2.6}$$

In (3.2.5) and (3.2.6) $\mathbb{B}(\mu_1, \lambda_1; \mathbf{x}, t)$ *is given below by formula (3.2.24) and* $\mathbf{n}(\mathbf{x})$ *is the normal vector to the boundary* S^2 *at the point* $\mathbf{x} \in S^2$.

We refer to the problems (3.2.3), (3.2.5) and (3.2.6) as the homogenized **model** $(\mathbb{HS})_2$.

Note that the Eq. (3.2.6) and the boundary condition (3.2.5) are a formal expression of the integral identity

$$\int_{Q_T}\Big(\int_0^t \mathbb{B}(\mu_1,\lambda_1;\mathbf{x},t-\tau)\cdot\nabla p(\mathbf{x},\tau)d\tau\Big)\cdot\nabla\xi\,dxdt=0 \tag{3.2.7}$$

for any smooth function ξ, vanishing at the part S^1 of the boundary S. As we have mentioned above, p satisfies (3.2.5) and (3.2.6) in the sense of distributions.

To formulate the following statements we need an additional construction. So, let $Q_f^\varepsilon = Q\backslash\overline{\Omega}_s^\varepsilon$, and

$$\mathbf{w}_f^\varepsilon = \mathbb{E}_{Q_f^\varepsilon}(\mathbf{w}^\varepsilon),\ \ \mathbf{w}_s^\varepsilon = \mathbb{E}_{\Omega_s^\varepsilon}(\mathbf{w}^\varepsilon),$$

where

$$\mathbb{E}_{Q_f^\varepsilon}:\mathbf{W}_2^1(Q_f^\varepsilon)\to\mathbf{W}_2^1(Q)$$

is an extension operator from Q_f^ε on Q, and

$$\mathbb{E}_{\Omega_s^\varepsilon}:\mathbf{W}_2^1(\Omega_s^\varepsilon)\to\mathbf{W}_2^1(Q)$$

is an extension operator from Ω_s^ε on Q, such that

$$\mathbf{w}_f^\varepsilon=\mathbf{w}^\varepsilon\ \ \text{in}\ \ Q_f^\varepsilon\times(0,T),\ \ \ \mathbf{w}_s^\varepsilon=\mathbf{w}^\varepsilon\ \text{in}\ \Omega_s^\varepsilon\times(0,T),$$

and

$$\begin{aligned}
&\int_Q|\mathbf{w}_f^\varepsilon|^2dx\leqslant C_0\int_{Q_f^\varepsilon}|\mathbf{w}^\varepsilon|^2dx,\ \ \int_Q|\mathbf{w}_s^\varepsilon|^2dx\leqslant C_0\int_{\Omega_s^\varepsilon}|\mathbf{w}^\varepsilon|^2dx,\\
&\int_Q|\mathbb{D}(x,\mathbf{w}_f^\varepsilon)|^2dx\leqslant C_0\int_{Q_f^\varepsilon}|\mathbb{D}(x,\mathbf{w}^\varepsilon)|^2dx,\\
&\int_Q|\mathbb{D}(x,\mathbf{w}_s^\varepsilon)|^2dx\leqslant C_0\int_{\Omega_s^\varepsilon}|\mathbb{D}(x,\mathbf{w}^\varepsilon)|^2dx.
\end{aligned} \tag{3.2.8}$$

(for more details see the extension lemma in Appendix B).

Theorem 3.4 *Let $\{\mathbf{w}^\varepsilon,\ p^\varepsilon\}$ be the weak solution of the problem (3.1.4)–(3.1.13),*

$$\mu_1=\infty,\ \ 0\leqslant\lambda_1<\infty,$$

and $\mathbf{w}_f^\varepsilon=\mathbb{E}_{Q_f^\varepsilon}(\mathbf{w}^\varepsilon)$.

Then there exists a subsequence of small parameters $\{\varepsilon>0\}$ such that the sequences $\{p^\varepsilon\}$, $\Big\{(1-\zeta)(1-\chi^\varepsilon)\dfrac{\partial\mathbf{w}^\varepsilon}{\partial t}\Big\}$, $\Big\{\dfrac{\partial\mathbf{w}_f^\varepsilon}{\partial t}\Big\}$, and $\Big\{\dfrac{\partial^2\mathbf{w}_f^\varepsilon}{\partial t^2}\Big\}$ converge weakly *in* $L_2(Q_T)$ *and* $\mathbf{L}_2(Q_T)$ *as* $\varepsilon \searrow 0$ *to the functions* $p \in W_2^{1,0}(Q_T)$, $\dfrac{\partial \mathbf{w}^{(s)}}{\partial t}$, $\dfrac{\partial \mathbf{w}_f}{\partial t}$, *and* $\dfrac{\partial^2 \mathbf{w}_f}{\partial t^2}$ *respectively and these limiting functions satisfy in the domain* Q_T *the system of homogenized equations consisting of the continuity equation*

$$\nabla \cdot \mathbf{v} = 0, \tag{3.2.9}$$

where

$$\mathbf{v} = -\frac{\zeta}{\rho_f}\int_0^t \nabla p(\mathbf{x},\tau)d\tau + (1-\zeta)\Big(m\frac{\partial \mathbf{w}_f}{\partial t} + \frac{\partial \mathbf{w}^{(s)}}{\partial t}\Big),$$

the momentum balance equation

$$(1-\zeta)\Big(m\rho_f\frac{\partial \mathbf{w}_f}{\partial t} + \rho_s\frac{\partial \mathbf{w}^{(s)}}{\partial t} + \int_0^t \nabla p(\mathbf{x},\tau)d\tau\Big) = 0, \tag{3.2.10}$$

for the liquid component, and the momentum balance equation

$$(1-\zeta)\Big(\frac{\partial \mathbf{w}^{(s)}}{\partial t} - (1-m)\frac{\partial \mathbf{w}_f}{\partial t}\Big) \tag{3.2.11}$$
$$= -(1-\zeta)\int_0^t \mathbb{B}^{(s)}(\infty,\lambda_1;t-\tau)\cdot\Big(\nabla p(\mathbf{x},\tau) + \rho_s\frac{\partial^2 \mathbf{w}_f}{\partial \tau^2}(\mathbf{x},\tau)\Big)d\tau$$
$$\tag{3.2.11}$$

for the solid component.

Equations (3.2.9)–(3.2.11) are supplemented with the homogeneous initial conditions

$$\mathbf{w}^{(s)}(\mathbf{x},0) = \mathbf{w}_f(\mathbf{x},0) = 0, \tag{3.2.12}$$

for displacements in the liquid and the solid components and boundary conditions (3.2.3) and

$$\mathbf{v}(\mathbf{x},t)\cdot\mathbf{n}(\mathbf{x}) = 0,\ \ \mathbf{x}\in S^2,\ \ t>0, \tag{3.2.13}$$

for the velocity $\mathbf{v}_f$ *and pressure* p.

In (3.2.11) the matrix $\mathbb{B}^{(s)}(\infty,\lambda_1;t)$ *is defined below by formulae (3.2.47) and (3.2.54) and the constant matrix* $\mathbb{B}^{(s)}(\infty,0;t) = \mathbb{B}^{(s)}(\infty,0)$ *is strictly positively definite.*

We refer to the problem (3.2.3), (3.2.9)–(3.2.13) as homogenized **model** $(\mathbb{HS})_3$.

Theorem 3.5 *Let* $\{\mathbf{w}^\varepsilon, p^\varepsilon\}$ *be the weak solution of the problem (3.1.4)–(3.1.13),*

$$\lambda_1 = \infty,\ \ 0 \leqslant \mu_1 < \infty,$$

and $\mathbf{w}_s^\varepsilon = \mathbb{E}_{\Omega_s^\varepsilon}(\mathbf{w}^\varepsilon)$.

Then there exists a subsequence of small parameters $\{\varepsilon > 0\}$ *such that the sequences* $\{p^\varepsilon\}$, $\Big\{\chi^\varepsilon \frac{\partial \mathbf{w}^\varepsilon}{\partial t}\Big\}$, $\Big\{(1-\zeta)\frac{\partial \mathbf{w}_s^\varepsilon}{\partial t}\Big\}$, *and* $\Big\{(1-\zeta)\frac{\partial^2 \mathbf{w}_s^\varepsilon}{\partial t^2}\Big\}$ *converge weakly in* $L_2(Q_T)$ *and* $\mathbf{L}_2(Q_T)$ *as* $\varepsilon \searrow 0$ *to the functions* $p \in W_2^{1,0}(Q_T)$, $\frac{\partial \mathbf{w}^{(f)}}{\partial t}$, $\frac{\partial \mathbf{w}_s}{\partial t}$, *and* $\frac{\partial^2 \mathbf{w}_s}{\partial t^2}$ *respectively and these limiting functions satisfy in the domain* Q_T *the system of homogenized equations consisting of the continuity equation*

$$\nabla \cdot \mathbf{v} = 0, \tag{3.2.14}$$

where

$$\mathbf{v} = -\frac{\zeta}{\rho_f}\int_0^t \nabla p(\mathbf{x},\tau)d\tau + (1-\zeta)\Big(\frac{\partial \mathbf{w}^{(f)}}{\partial t} + (1-m)\frac{\partial \mathbf{w}_s}{\partial t}\Big),$$

the momentum balance equation

$$(1-\zeta)\Big(\rho^{(f)}\frac{\partial \mathbf{w}^{(f)}}{\partial t} + (1-m)\rho_s\frac{\partial \mathbf{w}_s}{\partial t} + \int_0^t \nabla p(\mathbf{x},\tau)d\tau\Big) = 0, \tag{3.2.15}$$

for the solid component, and the momentum balance equation

$$\begin{aligned}&(1-\zeta)\Big(\frac{\partial \mathbf{w}^{(f)}}{\partial t} - m\frac{\partial \mathbf{w}_s}{\partial t}\Big)\\ &= -(1-\zeta)\int_0^t \mathbb{B}^{(f)}(\mu_1,\infty;t-\tau)\cdot\big(\nabla p(\mathbf{x},\tau) + \rho_f\frac{\partial^2 \mathbf{w}_s}{\partial \tau^2}(\mathbf{x},\tau)\big)d\tau\end{aligned} \tag{3.2.16}$$

for the liquid component.

Equations (3.2.14)–(3.2.16) are supplemented with the homogeneous initial conditions (3.2.12) for displacements $\mathbf{w}^{(f)}$ *and* $\mathbf{w}_s$ *in the liquid and the solid components, and boundary conditions (3.2.3) and (3.2.13) for the pressure* p *and the velocity* $\mathbf{v}$.

In (3.2.16) the matrix $\mathbb{B}^{(f)}(\mu_1,\infty;t)$ *is defined below by formulae (3.2.70) and (3.2.76) and the constant matrix* $\mathbb{B}^{(f)}(0,\infty;t) = \mathbb{B}^{(f)}(0,\infty)$ *is strictly positively definite.*

We refer to the problem (3.2.3), (3.2.12)–(3.2.16) as the homogenized **model** $(\mathbb{HS})_4$.

3.2.2 Proof of Theorem 3.2

By Theorem 3.1, the sequences $\{p^\varepsilon\}$, $\{\mathbf{w}^\varepsilon\}$, $\Big\{\frac{\partial^2 \mathbf{w}^\varepsilon}{\partial t^2}\Big\}$, and $\{\alpha(\varepsilon)\nabla \mathbf{w}^\varepsilon\}$ where $\alpha^2(\varepsilon) = \min\{\bar\alpha_\mu, \bar\alpha_\lambda\}$, are bounded in $L_2(Q_T)$ and $\mathbf{L}_2(Q_T)$. Hence there exists a subsequence of small parameters $\{\varepsilon > 0\}$ and functions p and $\mathbf{w}$ such that

$$p^{\varepsilon} \rightharpoonup p, \quad \mathbf{w}^{\varepsilon} \rightharpoonup \mathbf{w}, \quad \frac{\partial^2 \mathbf{w}^{\varepsilon}}{\partial t^2} \rightharpoonup \frac{\partial^2 \mathbf{w}}{\partial t^2} \tag{3.2.17}$$

weakly in $L_2(Q_T)$ and $\mathbf{L}_2(Q_T)$ as $\varepsilon \searrow 0$.

Note also that

$$\bar{\alpha}_{\lambda}(1-\zeta)\mathbb{D}(x, \mathbf{w}^{\varepsilon}) \to 0, \quad \bar{\alpha}_{\mu}\big(\zeta + (1-\zeta)\chi^{\varepsilon}\big)\mathbb{D}(x, \mathbf{w}^{\varepsilon}) \to 0 \tag{3.2.18}$$

strongly in $L_2(Q_T)$ as $\varepsilon \searrow 0$.

Relabeling if necessary, we assume that the sequences themselves converge.

By Nguetseng's theorem, there exist 1-periodic in $\mathbf{y}$ functions $P(\mathbf{x}, t, \mathbf{y})$ and $\mathbf{W}(\mathbf{x}, t, \mathbf{y})$ such that the sequences $\{p^{\varepsilon}\}$, $\{\mathbf{w}^{\varepsilon}\}$, and $\left\{\dfrac{\partial \mathbf{w}^{\varepsilon}}{\partial t}\right\}$ converge two-scale in $L_2(Q_T)$ and $\mathbf{L}_2(Q_T)$ to $P(\mathbf{x}, t, \mathbf{y})$, $\mathbf{W}(\mathbf{x}, t, \mathbf{y})$, and $\dfrac{\partial \mathbf{W}}{\partial t}$ respectively.

By supposition of the theorem

$$\lim_{\varepsilon \searrow 0} \frac{\alpha(\varepsilon)}{\varepsilon} = \infty.$$

Applying Lemma B.13 (see Appendix B) we conclude that

$$\mathbf{W}(\mathbf{x}, t, \mathbf{y}) = \mathbf{w}(\mathbf{x}, t). \tag{3.2.19}$$

Next, we prove the following

Lemma 3.1 *Under the conditions* $\mu_0 = 0$, $\lambda_0 = 0$ *the two-scale limit of the sequence* $\{p^{\varepsilon}\}$ *coincides with its weak limit:*

$$P(\mathbf{x}, t, \mathbf{y}) = p(\mathbf{x}, t). \tag{3.2.20}$$

Proof To prove the statement we fulfill the two-scale limit in (3.1.14) with test function $\varphi = \varepsilon h(\mathbf{x}, t)\varphi_0\left(\dfrac{\mathbf{x}}{\varepsilon}\right)$:

$$\int_{Q_T} h(\mathbf{x}, t)\Big(\int_Y P(\mathbf{x}, t, \mathbf{y})\nabla_y \cdot \varphi_0(\mathbf{y})dy\Big)dxdt = 0.$$

After reintegrating we arrive at

$$\nabla_y P(\mathbf{x}, t, \mathbf{y}) = 0, \ \ \mathbf{y} \in Y,$$

which is equivalent to (3.2.20).

Now we are ready to derive the homogenized momentum balance equation and the homogenized continuity equation. First, we pass to the limit as $\varepsilon \to 0$ in (3.1.14) with test function $\varphi = \varphi(\mathbf{x}, t)$, vanishing on the part S^2 of the boundary S, and at $t = T$:

$$\int_{Q_T}\Big(\rho(\mathbf{x})\frac{\partial \mathbf{w}}{\partial t}\cdot\frac{\partial \varphi}{\partial t}+p\,\nabla\cdot\varphi\Big)dxdt=\int_{Q_T}\nabla\cdot(\varphi\,p_0)dxdt, \tag{3.2.21}$$

where

$$\rho(\mathbf{x})=\Big(\zeta(\mathbf{x})+\big(1-\zeta(\mathbf{x})\big)m\Big)\rho_f+\big(1-\zeta(\mathbf{x})\big)(1-m)\rho_s.$$

The simple analysis of the integral identity (3.2.21) and estimates (3.1.16) show that $\nabla p\in\mathbf{L}_2(Q_T)$, and functions $\mathbf{w}$ and p satisfy the homogenized momentum balance equation

$$\rho(\mathbf{x})\frac{\partial^2\mathbf{w}}{\partial t^2}+\nabla p=0, \tag{3.2.22}$$

and boundary condition (3.2.3) in the usual sense.

Next, we rewrite the continuity Eq. (3.1.4) as an integral identity

$$\int_{Q_T}\mathbf{w}^\varepsilon\nabla\xi\,dxdt=0$$

for an arbitrary smooth function ξ, and then pass to the limit as $\varepsilon\to 0$:

$$\int_{Q_T}\mathbf{w}\nabla\xi\,dxdt=0.$$

Choosing now $\xi=\dfrac{\partial^2\zeta}{\partial t^2}$, where ζ vanishes at $t=T$, we may rewrite the last identity as

$$\int_{Q_T}\frac{\partial^2\mathbf{w}}{\partial t^2}\nabla\zeta\,dxdt=0. \tag{3.2.23}$$

The combination of (3.2.22) and (3.2.23) gives us the desired integral identity

$$\int_{Q_T}\Big(\frac{1}{\rho(\mathbf{x})}\nabla p\Big)\cdot\nabla\zeta\,dxdt=0, \tag{3.2.24}$$

which is equivalent to Eq. (3.2.2) in the domain Q and boundary condition (3.2.4) on the boundary S^2 in the sense of distributions.

3.2.3 Proof of Theorem 3.3

By Theorem 3.1, the sequences $\{p^\varepsilon\}$, $\{\mathbf{w}^\varepsilon\}$, $\left\{\dfrac{\partial^2\mathbf{w}^\varepsilon}{\partial t^2}\right\}$, $\{\varepsilon\nabla\mathbf{w}^\varepsilon\}$, and $\left\{\varepsilon\nabla\Big(\dfrac{\partial\mathbf{w}^\varepsilon}{\partial t}\Big)\right\}$ are bounded in $L_2(Q_T)$ and $\mathbf{L}_2(Q_T)$.

Hence there exists a subsequence of small parameters $\{\varepsilon > 0\}$ and functions p and $\mathbf{w}$ such that

$$p^{\varepsilon} \rightharpoonup p, \quad \mathbf{w}^{\varepsilon} \rightharpoonup \mathbf{w}, \quad \frac{\partial^2 \mathbf{w}^{\varepsilon}}{\partial t^2} \rightharpoonup \frac{\partial^2 \mathbf{w}}{\partial t^2}$$

weakly in $L_2(Q_T)$ and $\mathbf{L}_2(Q_T)$ as $\varepsilon \searrow 0$.

Owing to Nguetseng's theorem and Lemma 3.2, there exists a 1-periodic in $\mathbf{y}$ function $\mathbf{W}(\mathbf{x}, t, \mathbf{y})$ such that the sequences $\{p^{\varepsilon}\}$, $\{\mathbf{w}^{\varepsilon}\}$, $\left\{\dfrac{\partial^2 \mathbf{w}^{\varepsilon}}{\partial t^2}\right\}$, $\{\varepsilon \nabla \mathbf{w}^{\varepsilon}\}$, and $\left\{\varepsilon \nabla \Big(\dfrac{\partial \mathbf{w}^{\varepsilon}}{\partial t}\Big)\right\}$ converge two-scale in $L_2(Q_T)$ and $\mathbf{L}_2(Q_T)$ to $p(\mathbf{x}, t)$, $\mathbf{W}(\mathbf{x}, t, \mathbf{y})$, $\left\{\dfrac{\partial^2 \mathbf{W}}{\partial t^2}(\mathbf{x}, t, \mathbf{y})\right\}$, $\nabla_y \mathbf{W}(\mathbf{x}, t, \mathbf{y})$ and $\nabla_y \Big(\dfrac{\partial \mathbf{W}}{\partial t}\Big)(\mathbf{x}, t, \mathbf{y})$ respectively.

Nguetseng's theorem also guarantees that

$$\mathbf{W}, \ \frac{\partial^2 \mathbf{W}}{\partial t^2}, \ \nabla_y \mathbf{W}, \ \nabla_y \Big(\frac{\partial \mathbf{W}}{\partial t}\Big) \in \mathbf{L}_2(Q_T \times Y). \tag{3.2.25}$$

Lemma 3.2 *The limiting functions* $\mathbf{w}$ *and* $\mathbf{W}$ *satisfy the macroscopic continuity equation and boundary condition*

$$\nabla \cdot \mathbf{w} = 0, \ \ \mathbf{w} \cdot \mathbf{n} = 0 \tag{3.2.26}$$

in the domain Q_T *and at the boundary* S_T^2, *and microscopic continuity equation*

$$\nabla_y \cdot \mathbf{W} = 0 \tag{3.2.27}$$

in the domain Y_T *for almost all* $(\mathbf{x}, t) \in Q_T$.

The proof is straightforward (for details see Chap. 1 and proof of Theorem 3.1).

Lemma 3.3 *For almost all* $(\mathbf{x}, t) \in Q_T$ *the limiting functions* p *and* $\mathbf{W}$ *satisfy the microscopic momentum balance equation*

$$\begin{aligned}\bar{\rho}(\mathbf{x}, \mathbf{y}) \frac{\partial^2 \mathbf{W}}{\partial t^2} = &\nabla_y \cdot \Big(\mu_1 \bar{\chi}(\mathbf{x}, \mathbf{y}) \mathbb{D}(y, \frac{\partial \mathbf{W}}{\partial t}) \\ &+ \lambda_1 \big(1 - \bar{\chi}(\mathbf{x}, \mathbf{y})\big) \mathbb{D}(y, \mathbf{W}) - \Pi \, \mathbb{I}\Big) - \nabla p, \ \ \mathbf{y} \in Y, \ t > 0,\end{aligned} \tag{3.2.28}$$

completed with the homogeneous initial conditions

$$\mathbf{W}(\mathbf{x}, \mathbf{y}, 0) = \frac{\partial \mathbf{W}}{\partial t}(\mathbf{x}, \mathbf{y}, 0) = 0, \quad \mathbf{y} \in Y. \tag{3.2.29}$$

In (3.2.28)

$$\bar{\chi}(\mathbf{x}, \mathbf{y}) = \zeta(\mathbf{x}) + \big(1 - \zeta(\mathbf{x})\big) \chi(\mathbf{y}),$$

$$\bar{\rho}(\mathbf{x}, \mathbf{y}) = \rho_f \bar{\chi}(\mathbf{x}, \mathbf{y}) + \rho_s\big(1 - \bar{\chi}(\mathbf{x}, \mathbf{y})\big).$$

To prove this lemma we simply pass to the limit as $\varepsilon \to 0$ in (3.1.14) with test functions φ in the form $\varphi = h(\mathbf{x}, t)\varphi_0\Big(\dfrac{\mathbf{x}}{\varepsilon}\Big)$, where $\varphi_0(\mathbf{y})$ is 1-periodic in $\mathbf{y}$ function, solenoidal in Y, and arrive at the integral identity

$$\int_{Q_T} \big(\mathbf{a}(\mathbf{x}, t)h + p(\mathbf{x}, t) \cdot \nabla h\big)dxdt = 0,$$

where

$$\mathbf{a}(\mathbf{x}, t) = \int_Y \Big(\bar{\rho}\frac{\partial^2 \mathbf{W}}{\partial t^2} \cdot \varphi_0 + \Big(\mu_1\bar{\chi}\mathbb{D}\Big(y, \frac{\partial \mathbf{W}}{\partial t}\Big)$$
$$+ \lambda_1(1 - \bar{\chi})\mathbb{D}(y, \mathbf{W}) - \Pi\,\mathbb{I}\Big) : \mathbb{D}(y, \varphi_0)\Big)dy \in L_2(Q_T).$$

The last identity shows the validity of (3.2.28) with $\nabla p \in \mathbf{L}_2(Q_T)$.

As a final step we have to solve the periodic initial boundary value problem (3.2.27)–(3.2.29) and find $\mathbf{W}$ as an operator on ∇p. The desired Eq. (3.2.6) and boundary condition (3.2.5) are the result of substituting $\mathbf{w} = \langle \mathbf{W}\rangle_Y$ into the macroscopic continuity equation (3.2.26).

We look for the solution $\{\mathbf{W},\ \Pi\}$ of the problem (3.2.27)–(3.2.29) in the form

$$\mathbf{W}(\mathbf{x}, t, \mathbf{y}) = \sum_{i=1}^{3} \int_0^t \mathbf{W}^{(i)}(\mathbf{x}, \mathbf{y}, t - \tau)\frac{\partial p}{\partial x_i}(\mathbf{x}, \tau)d\tau,$$

$$\Pi(\mathbf{x}, t, \mathbf{y}) = \sum_{i=1}^{3} \int_0^t \Pi^{(i)}(\mathbf{x}, \mathbf{y}, t - \tau)\frac{\partial p}{\partial x_i}(\mathbf{x}, \tau)d\tau,$$

where $\{\mathbf{W}^{(i)},\ \Pi^{(i)}\}$, $i = 1, 2, 3$, are solutions to the following periodic initial boundary value problem

$$\bar{\rho}(\mathbf{x}, \mathbf{y})\frac{\partial^2 \mathbf{W}^{(i)}}{\partial t^2} = \nabla_y \cdot \Big(\mu_1\bar{\chi}(\mathbf{x}, \mathbf{y})\mathbb{D}\Big(y, \frac{\partial \mathbf{W}^{(i)}}{\partial t}\Big)$$
$$+ \lambda_1\big(1 - \bar{\chi}(\mathbf{x}, \mathbf{y})\big)\mathbb{D}(y, \mathbf{W}^{(i)}) - \Pi^{(i)}\,\mathbb{I}\Big),\ \ \nabla_y \cdot \mathbf{W}^{(i)} = 0, \tag{3.2.30}$$

in Y, for $t > 0$, completed with homogeneous initial conditions

$$\mathbf{W}^{(i)}(\mathbf{x}, \mathbf{y}, 0) = 0,\ \ \bar{\rho}(\mathbf{x}, \mathbf{y})\frac{\partial \mathbf{W}^{(i)}}{\partial t}(\mathbf{x}, \mathbf{y}, 0) = -\mathbf{e}_i, \quad \mathbf{y} \in Y. \tag{3.2.31}$$

The solvability of this problem is standard and is based upon the energy equality

$$\frac{1}{2}\int_Y\left(\bar{\rho}(\mathbf{x},\mathbf{y})\left|\frac{\partial \mathbf{W}^{(i)}}{\partial t}(\mathbf{x},t,\mathbf{y})\right|^2+\lambda_1\big(1-\bar{\chi}(\mathbf{x},\mathbf{y})\big)|\mathbb{D}(y,\mathbf{W}^{(i)})|^2\right)dy$$
$$+\int_0^t\int_Y \mu_1\bar{\chi}(\mathbf{x},\mathbf{y})\left|\mathbb{D}\left(y,\frac{\partial \mathbf{W}^{(i)}}{\partial t}\right)\right|^2 dyd\tau=\frac{1}{2}\int_Y\frac{1}{\bar{\rho}(\mathbf{x},\mathbf{y})}dy.$$

Thus,

$$\mathbf{W}=\int_0^t\left(\sum_{i=1}^3\mathbf{W}^{(i)}(\mathbf{x},\mathbf{y},t-\tau)\otimes\mathbf{e}_i\right)\cdot\nabla p(\mathbf{x},\tau)d\tau,$$

and

$$\mathbf{w}(\mathbf{x},t)=\int_0^t\mathbb{B}(\mu_1,\lambda_1;\mathbf{x},t-\tau)\cdot\nabla p(\mathbf{x},\tau)d\tau,$$

where

$$\mathbb{B}(\mu_1,\lambda_1;\mathbf{x},t)=\int_Y\sum_{i=1}^3\mathbf{W}^{(i)}(\mathbf{x},t,\mathbf{y})\otimes\mathbf{e}_i dy. \tag{3.2.32}$$

To end the proof we have to show that $\nabla p\in\mathbf{L}_2(Q_T)$, and the function p satisfies the boundary condition (3.2.3). In fact, the passage to the limit as $\varepsilon\to 0$ in (3.1.14) with test functions $\varphi=\varphi(\mathbf{x},t)$ result in the identity

$$\int_{Q_T}\left(\left(\int_Y\bar{\rho}(\mathbf{x},\mathbf{y})\frac{\partial^2\mathbf{W}}{\partial t^2}dy\right)\cdot\varphi-p\nabla\cdot\varphi\right)dxdt=-\int_{Q_T}\nabla\cdot(\varphi p_0)dxdt,$$

which together with (3.2.25) prove the statement.

3.2.4 Proof of Theorem 3.4

3.2.4.1 The Case $\lambda_1>0$.

Estimates (3.2.8) and (3.1.16) provide the boundedness of sequences $\{p^\varepsilon\}$, $\{\mathbf{w}_f^\varepsilon\}$, $\left\{\dfrac{\partial^2\mathbf{w}_f^\varepsilon}{\partial t^2}\right\}$, $\{\alpha_\lambda\nabla\mathbf{w}^\varepsilon\}$, $\{(1-\zeta)(1-\chi^\varepsilon)\mathbf{w}^\varepsilon\}$, and $\left\{(1-\zeta)(1-\chi^\varepsilon)\dfrac{\partial^2\mathbf{w}^\varepsilon}{\partial t^2}\right\}$ in $L_2(Q_T)$ and $\mathbf{L}_2(Q_T)$.

Hence there exists a subsequence of small parameters $\{\varepsilon>0\}$ and functions p, $\mathbf{w}_f$, and $\mathbf{w}^{(s)}$, such that

$$p^{\varepsilon} \rightharpoonup p, \ \mathbf{w}^{\varepsilon}_{f} \rightharpoonup \mathbf{w}_{f}, \ \frac{\partial^2 \mathbf{w}^{\varepsilon}_{f}}{\partial t^2} \rightharpoonup \frac{\partial^2 \mathbf{w}_{f}}{\partial t^2},$$

$$(1-\zeta)(1-\chi^{\varepsilon})\mathbf{w}^{\varepsilon} \rightharpoonup \mathbf{w}^{(s)}, \ (1-\zeta)(1-\chi^{\varepsilon})\frac{\partial^2 \mathbf{w}^{\varepsilon}}{\partial t^2} \rightharpoonup \frac{\partial^2 \mathbf{w}^{(s)}}{\partial t^2} \tag{3.2.33}$$

weakly in $L_2(Q_T)$ and $\mathbf{L}_2(Q_T)$ as $\varepsilon \searrow 0$.

In particular, the weak passage to the limit as $\varepsilon \searrow 0$ in integral identities

$$\int_Q \Big(\mathbf{w}^{\varepsilon}_{f} \cdot \frac{\partial \varphi}{\partial t} + \frac{\partial \mathbf{w}^{\varepsilon}_{f}}{\partial t} \cdot \varphi\Big) dxdt = 0,$$

$$\int_Q (1-\zeta)(1-\chi^{\varepsilon})\Big(\mathbf{w}^{\varepsilon} \cdot \frac{\partial \varphi}{\partial t} + \frac{\partial \mathbf{w}^{\varepsilon}}{\partial t} \cdot \varphi\Big) dxdt = 0,$$

for any smooth functions $\varphi(\mathbf{x}, t)$, such that $\varphi(\mathbf{x}, T) = 0$, results in the initial conditions (3.2.12).

By Nguetseng's theorem and Lemma 3.1, there exists a 1-periodic in $\mathbf{y}$ function $\mathbf{W}(\mathbf{x}, t, \mathbf{y})$ such that the sequences $\{p^{\varepsilon}\}$, $\{\mathbf{w}^{\varepsilon}_{f}\}$, $\{\mathbf{w}^{\varepsilon}\}$, $\{(1-\chi^{\varepsilon})\mathbf{w}^{\varepsilon}\}$, $\left\{\frac{\partial^2 \mathbf{w}^{\varepsilon}}{\partial t^2}\right\}$, and $\{\varepsilon \nabla \mathbf{w}^{\varepsilon}\}$ converge two-scale in $L_2(Q_T)$ and $\mathbf{L}_2(Q_T)$ to $p(\mathbf{x}, t)$, $\mathbf{w}_f(\mathbf{x}, t)$, $\mathbf{W}(\mathbf{x}, t, \mathbf{y})$, $\big(1-\chi(\mathbf{y})\big)\mathbf{W}(\mathbf{x}, t, \mathbf{y})$, $\frac{\partial^2 \mathbf{W}}{\partial t^2}(\mathbf{x}, t, \mathbf{y})$, and $\nabla_y \mathbf{W}(\mathbf{x}, t, \mathbf{y})$ respectively.

Nguetseng's theorem also guarantees that

$$\mathbf{W}, \ \frac{\partial^2 \mathbf{W}}{\partial t^2}, \ \nabla_y \mathbf{W} \in \mathbf{L}_2(Q_T \times Y) \tag{3.2.34}$$

and

$$\mathbf{W}(\mathbf{x}, t, \mathbf{y}) = \chi(\mathbf{y})\mathbf{w}_f(\mathbf{x}, t) + \big(1-\chi(\mathbf{y})\big)\mathbf{W}(\mathbf{x}, t, \mathbf{y}). \tag{3.2.35}$$

Lemma 3.4 *The limiting functions* $\mathbf{v} = \frac{\partial \mathbf{w}}{\partial t}$ *and* $\mathbf{W}$ *satisfy the homogenized continuity equation (3.2.9) in the domain* Q_T *and boundary condition (3.2.13) on the boundary* S^2_T *for the velocity*

$$\mathbf{v} = \big(\zeta + (1-\zeta)m\big)\frac{\partial \mathbf{w}_f}{\partial t} + (1-\zeta)\frac{\partial \mathbf{w}^{(s)}}{\partial t}, \tag{3.2.36}$$

and microscopic continuity equation (3.2.27) in the domain Y_T *for almost all* $(\mathbf{x}, t) \in Q_T$.

The proof of this lemma repeats the proof of Lemma 3.3 if we note, that the weak limit of the sequence of functions $\frac{\partial \mathbf{w}^{\varepsilon}}{\partial t}$ in the form

$$\frac{\partial \mathbf{w}^{\varepsilon}}{\partial t} = \big(\zeta + (1-\zeta)m\big)\frac{\partial \mathbf{w}_f^{\varepsilon}}{\partial t} + (1-\zeta)(1-\chi^{\varepsilon})\frac{\partial \mathbf{w}^{\varepsilon}}{\partial t}$$

results in (3.2.36).

Lemma 3.5 *The limiting functions p, $\mathbf{w}_f$ and $\mathbf{w}^{(s)}$ satisfy the integral identity*

$$\int_{Q_T}\Bigg(\bigg(\rho_f\big(\zeta + (1-\zeta)m\big)\frac{\partial^2 \mathbf{w}_f}{\partial t^2} + \rho_s(1-\zeta)\frac{\partial^2 \mathbf{w}^{(s)}}{\partial t^2}\bigg)\cdot\varphi - p\,\nabla\cdot\varphi\Bigg)dxdt$$
$$= -\int_{Q_T}\nabla\cdot(\varphi\, p_0)dxdt \tag{3.2.37}$$

for all smooth functions φ, such that $\varphi(\mathbf{x},t) = 0$ on the boundary S_T^2.

The proof of this lemma is straightforward. We simply pass to the limit as $\varepsilon \to 0$ in (3.1.14) with test functions $\varphi = \varphi(\mathbf{x},t)$.

The integral identity (3.2.37) evidently results in the inclusion $\nabla p \in \mathbf{L}_2(Q_T)$, the boundary condition (3.2.3), and the momentum balance equation for the liquid component in the form

$$\rho_f\big(\zeta + (1-\zeta)m\big)\frac{\partial \mathbf{w}_f}{\partial t} + \rho_s(1-\zeta)\frac{\partial \mathbf{w}^{(s)}}{\partial t} = -\int_0^t \nabla p(\mathbf{x},\tau)d\tau.$$

Multiplying the last equation by ζ and $(1-\zeta)$ we obtain (3.2.10) and

$$\zeta\rho_f\frac{\partial \mathbf{w}_f}{\partial t} = -\zeta\int_0^t \nabla p(\mathbf{x},\tau)d\tau.$$

This last formula and (3.2.36) give us

$$\mathbf{v} = -\frac{\zeta}{\rho_f}\int_0^t \nabla p(\mathbf{x},\tau)d\tau + (1-\zeta)\Big(m\frac{\partial \mathbf{w}_f}{\partial t} + \frac{\partial \mathbf{w}^{(s)}}{\partial t}\Big). \tag{3.2.38}$$

To derive the momentum balance equation for the solid component we simply pass to the limit as $\varepsilon \to 0$ in (3.1.14) with test functions $\varphi = h(\mathbf{x},t)\varphi_0\Big(\dfrac{\mathbf{x}}{\varepsilon},t\Big)$, where $h(\mathbf{x},t)$ is smooth and finite in Ω_T, and the 1-periodic in $\mathbf{y}$ smooth function $\varphi_0(\mathbf{y})$ is divergence free and finite in Y_s. If $\mathbf{W}^{(s)} = \big(1-\chi(\mathbf{y})\big)\mathbf{W}$, then the pair $\{\mathbf{W}^{(s)}, \Pi^{(s)}\}$ satisfies the equation

$$\rho_s\frac{\partial^2 \mathbf{W}^{(s)}}{\partial t^2} = \frac{\lambda_1}{2}\Delta_y \mathbf{W}^{(s)} - \nabla_y \Pi^{(s)} - \nabla\, p \tag{3.2.39}$$

in the domain $Y_s \times (0,T)$ and initial conditions

$$\mathbf{W}^{(s)}(\mathbf{x},0,\mathbf{y})=\frac{\partial \mathbf{W}^{(s)}}{\partial t}(\mathbf{x},0,\mathbf{y})=0,\ \ \mathbf{y}\in Y_s \tag{3.2.40}$$

for almost all $\mathbf{x}\in\Omega_T$.

Conditions (3.2.34) and formula (3.2.35) provide the boundary condition

$$\mathbf{W}^{(s)}(\mathbf{x},t,\mathbf{y})=\mathbf{w}_f(\mathbf{x},t),\ \ (\mathbf{y},t)\in\gamma\times(0,T) \tag{3.2.41}$$

for almost all $\mathbf{x}\in\Omega_T$.

Therefore, the solution $\{\mathbf{W}^{(s)},\ \Pi^{(s)}\}$ to the periodic initial boundary value problem (3.2.27), (3.2.39)–(3.2.41) has the form

$$\mathbf{W}^{(s)}=\mathbf{w}_f(\mathbf{x},t)+\sum_{i=1}^{3}\int_0^t \mathbf{W}_i^{(s)}(\mathbf{y},t-\tau)\Big(\frac{\partial p}{\partial x_i}(\mathbf{x},\tau)+\rho_s\frac{\partial^2 w_{f,i}}{\partial t^2}(\mathbf{x},\tau)\Big)d\tau,$$

$$\Pi^{(s)}(\mathbf{x},t,\mathbf{y})=\sum_{i=1}^{3}\int_0^t \Pi_i^{(s)}(\mathbf{y},t-\tau)\Big(\frac{\partial p}{\partial x_i}(\mathbf{x},\tau)+\rho_s\frac{\partial^2 w_{f,i}}{\partial t^2}(\mathbf{x},\tau)\Big)d\tau,$$

where $\mathbf{w}_f=(w_{f,1},w_{f,2},w_{f,3})$ and $\{\mathbf{W}_i^{(s)},\ \Pi_i^{(s)}\}$, $i=1,2,3$, are solutions to the following periodic initial boundary value problem

$$\rho_s\frac{\partial^2 \mathbf{W}_i^{(s)}}{\partial t^2}=\frac{\lambda_1}{2}\Delta_y\mathbf{W}_i^{(s)}-\nabla_y\Pi_i^{(s)},\ \ (\mathbf{y},t)\in Y_s\times(0,T), \tag{3.2.42}$$

$$\nabla_y\cdot\mathbf{W}_i^{(s)}(\mathbf{y},t)=0,\ \ (\mathbf{y},t)\in Y_s\times(0,T), \tag{3.2.43}$$

$$\mathbf{W}_i^{(s)}(\mathbf{y},0)=0,\ \ \rho_s\frac{\partial \mathbf{W}_i^{(s)}}{\partial t}(\mathbf{y},0)=\mathbf{e}_i,\ \ \mathbf{y}\in Y_s, \tag{3.2.44}$$

$$\mathbf{W}_i^{(s)}(\mathbf{y},t)=0,\ \ (\mathbf{y},t)\in\gamma\times(0,T), \tag{3.2.45}$$

for almost all $\mathbf{x}\in\Omega_T$.

The correctness of the problem (3.2.42)–(3.2.45) follows from the energy equality

$$\int_{Y_s}\Big(\rho_s\Big|\frac{\partial \mathbf{W}_i^{(s)}}{\partial t}(\mathbf{y},t)\Big|^2+\frac{\lambda_1}{2}|\nabla\mathbf{W}_i^{(s)}(\mathbf{y},t)|^2\Big)dy=\frac{(1-m)}{\rho_s}.$$

We recall that the problem (3.2.42)–(3.2.45) for solenoidal functions $\mathbf{W}_s^{(i)}$, vanishing at γ and $t=0$, is understood as integral identity

$$\int_0^T \int_{Y_s} \Big(\rho_s \frac{\partial \mathbf{W}_i^{(s)}}{\partial t} \cdot \frac{\partial \varphi}{\partial t} - \lambda_1 \nabla \mathbf{W}_i^{(s)} : \nabla \varphi \Big) dydt = \int_{Y_s} \mathbf{e}_i \cdot \varphi(\mathbf{y}, 0) dy$$

for any solenoidal 1-periodic smooth function φ, vanishing at γ and $t = T$. By definition

$$\frac{\partial \mathbf{w}^{(s)}}{\partial t}(\mathbf{x}, t) = \int_{Y_s} \frac{\partial \mathbf{W}^{(s)}}{\partial t}(\mathbf{x}, t, \mathbf{y}) dy$$
$$= (1-m)\frac{\partial \mathbf{w}_f}{\partial t} - \int_0^t \Big(\sum_{i=1}^{3} \big(\int_{Y_s} \frac{\partial \mathbf{W}_i^{(s)}}{\partial t}(\mathbf{y}, t-\tau) dy\big) \otimes \mathbf{e}_i \Big) \cdot \big(\nabla p(\mathbf{x}, \tau)$$
$$+ \rho_s \frac{\partial^2 \mathbf{w}_f}{\partial \tau^2}(\mathbf{x}, \tau)\big) d\tau = (1-m)\frac{\partial \mathbf{w}_f}{\partial t}$$
$$- \int_0^t \mathbb{B}^{(s)}(\infty, \lambda_1; t-\tau) \cdot \big(\nabla p(\mathbf{x}, \tau) + \rho_s \frac{\partial^2 \mathbf{w}_f}{\partial \tau^2}(\mathbf{x}, \tau)\big) d\tau, \quad (3.2.46)$$

where

$$\mathbb{B}^{(s)}(\infty, \lambda_1; t) = \sum_{i=1}^{3} \Big(\int_{Y_s} \frac{\partial \mathbf{W}_i^{(s)}}{\partial t}(\mathbf{y}, t) dy \Big) \otimes \mathbf{e}_i. \quad (3.2.47)$$

3.2.4.2 The Case $\lambda_1 = 0$.

For this case we may repeat everything as for the previous case $\lambda_1 > 0$, except for:

(1) two-scale convergence of the sequence $\{\varepsilon \nabla \mathbf{w}^{\varepsilon}\}$ to the function $\nabla \mathbf{W} \in \mathbf{L}_2 (Q_T \times Y)$, and
(2) derivation of the momentum balance equation for the solid component.

For $\lambda_1 = 0$ the microscopic momentum balance equation for the solid component has the form

$$\rho_s \frac{\partial^2 \mathbf{W}^{(s)}}{\partial t^2} = -\nabla_y \Pi^{(s)} - \nabla p. \quad (3.2.48)$$

Instead of condition (3.2.41) on the boundary γ one has there a condition

$$\big(\mathbf{W}^{(s)}(\mathbf{x}, t, \mathbf{y}) - \mathbf{w}_f(\mathbf{x}, t)\big) \cdot \mathbf{n}(\mathbf{y}) = 0, \quad (3.2.49)$$

which is a consequence of the microscopic continuity equation (3.2.27) and the representation

$$\mathbf{W}(\mathbf{x}, t, \mathbf{y}) = \chi(\mathbf{y}) \mathbf{w}_f(\mathbf{x}, t) + \big(1 - \chi(\mathbf{y})\big) \mathbf{W}^{(s)}(\mathbf{x}, t, \mathbf{y}), \ \ \mathbf{y} \in Y. \quad (3.2.50)$$

To solve (3.2.48) we apply to this equation the operation $\nabla_y \cdot$ and use again (3.2.27):

$$0 = \nabla_y \cdot \left(\rho_s \frac{\partial^2 \mathbf{W}^{(s)}}{\partial t^2}\right) = -\nabla_y \cdot (\nabla_y \Pi^{(s)}). \tag{3.2.51}$$

The boundary condition (3.2.49) and Eq. (3.2.48) provide the boundary condition on the boundary γ for the pressure $\Pi^{(s)}$:

$$\nabla_y \Pi^{(s)} \cdot \mathbf{n}(\mathbf{y}) = -\left(\nabla p + \rho_s \frac{\partial^2 \mathbf{w}_f}{\partial t^2}\right) \cdot \mathbf{n}(\mathbf{y}). \tag{3.2.52}$$

Let

$$\Pi^{(s)} = -\left(\sum_{i=1}^{3} \Pi_i^{(s)}(\mathbf{y})\mathbf{e}_i\right) \cdot \left(\nabla p + \rho_s \frac{\partial^2 \mathbf{w}_f}{\partial t^2}\right),$$

where $\Pi_i^{(s)}$, $i = 1, 2, 3$, are solutions to the periodic boundary value problems

$$\triangle_y \Pi_i^{(s)} = 0, \ \mathbf{y} \in Y_s, \ (\nabla_y \Pi_i^{(s)} - \mathbf{e}_i) \cdot \mathbf{n}(\mathbf{y}) = 0. \ \mathbf{y} \in \gamma. \tag{3.2.53}$$

Then

$$\nabla_y \Pi^{(s)} = -\left(\sum_{i=1}^{3} \nabla_y \Pi_i^{(s)} \otimes \mathbf{e}_i\right) \cdot \left(\nabla p + \rho_s \frac{\partial^2 \mathbf{w}_f}{\partial t^2}\right).$$

After integration (3.2.48) over domain Y_s we arrive at the desired momentum balance Eq. (3.2.11) for the solid component with

$$\rho_s \mathbb{B}^{(s)}(\infty, 0) = (1 - m)\mathbb{I} - \left(\sum_{i=1}^{3} \int_{Y_s} \nabla_y \Pi_i^{(s)}(\mathbf{y}) dy \otimes \mathbf{e}_i\right). \tag{3.2.54}$$

To prove that $\mathbb{B}^{(s)}(\infty, 0; t)$ is symmetric and strictly positively definite, we use the definition of the solution to the problem (3.2.53) with test function $\Pi_j^{(s)}$

$$\int_{Y_s} \nabla \Pi_j^{(s)} \cdot \nabla \Pi_i^{(s)} dy - \int_{Y_s} \frac{\partial \Pi_i^{(s)}}{\partial y_j} dy = 0. \tag{3.2.55}$$

Let $\zeta = (\zeta_1, \zeta_2, \zeta_3)$ and $\eta = (\eta_1, \eta_2, \eta_3)$ be arbitrary constant vectors and

$$\mathbf{z}_\zeta = \sum_{i=1}^{3} \Pi_i^{(s)} \zeta_i, \quad \mathbf{z}_\eta = \sum_{i=1}^{3} \Pi_i^{(s)} \eta_i.$$

Then (3.2.55) is equivalent to

$$\int_{Y_s} \nabla \mathbf{z}_\zeta \cdot \nabla \mathbf{z}_\eta dy - \int_{Y_s} \nabla \mathbf{z}_\zeta \cdot \eta dy = 0.$$

On the other hand

$$\rho_s\big(\mathbb{B}^{(s)}(\infty,0)\cdot\eta\big)\cdot\zeta = \int_{Y_s} \eta\cdot\zeta dy - \int_{Y_s} \nabla \mathbf{z}_\eta\cdot\zeta dy = 0.$$

The combination of the last two relations result in the equality

$$\rho_s\big(\mathbb{B}^{(s)}(\infty,0)\cdot\eta\big)\cdot\zeta = \int_{Y_s} (\eta - \mathbf{z}_\zeta)\cdot(\zeta - \mathbf{z}_\eta)dy,$$

which proves the last statement of the theorem.

3.2.5 *Proof of Theorem 3.5*

The proof of this theorem repeats the proof of the previous theorem with evident symmetric changes. Therefore, we only formulate the main results, omitting all proofs.

3.2.5.1 The Case $\mu_1 > 0$.

The sequences $\{p^\varepsilon\}$, $\{\mathbf{w}^\varepsilon\}$, $\Big\{\dfrac{\partial^2\mathbf{w}^\varepsilon}{\partial t^2}\Big\}$, $\{(1-\zeta)\mathbf{w}_s^\varepsilon\}$, $\Big\{(1-\zeta)\dfrac{\partial^2\mathbf{w}_s^\varepsilon}{\partial t^2}\Big\}$, $\Big\{\alpha_\mu\nabla\Big(\dfrac{\partial\mathbf{w}^\varepsilon}{\partial t}\Big)\Big\}$, $\{\chi^\varepsilon\mathbf{w}^\varepsilon\}$, and $\Big\{\chi^\varepsilon\dfrac{\partial^2\mathbf{w}^\varepsilon}{\partial t^2}\Big\}$ are bounded in $L_2(Q_T)$ and $\mathbf{L}_2(Q_T)$.

Hence there exists a subsequence of small parameters $\{\varepsilon > 0\}$ and functions p, $\mathbf{w}_s$, and $\mathbf{w}^{(f)}$, such that

$$\begin{aligned} &p^\varepsilon \rightharpoonup p,\ \mathbf{w}^\varepsilon \rightharpoonup \mathbf{w},\ \frac{\partial^2\mathbf{w}^\varepsilon}{\partial t^2} \rightharpoonup \frac{\partial^2\mathbf{w}}{\partial t^2},\ (1-\zeta)\mathbf{w}_s^\varepsilon \rightharpoonup \mathbf{w}_s,\\ &(1-\zeta)\frac{\partial^2\mathbf{w}_s^\varepsilon}{\partial t^2} \rightharpoonup \frac{\partial^2\mathbf{w}_s}{\partial t^2},\ \chi^\varepsilon\mathbf{w}^\varepsilon \rightharpoonup \mathbf{w}^{(f)},\ \chi^\varepsilon\frac{\partial^2\mathbf{w}^\varepsilon}{\partial t^2} \rightharpoonup \frac{\partial^2\mathbf{w}^{(f)}}{\partial t^2} \end{aligned} \tag{3.2.56}$$

weakly in $L_2(Q_T)$ and $\mathbf{L}_2(Q_T)$ as $\varepsilon \searrow 0$.

As in the previous subsubsection we conclude that functions $\mathbf{w}^{(f)}$ and $\mathbf{w}_s$ satisfy the initial condition (3.2.12).

Next, there exists a 1-periodic in $\mathbf{y}$ function $\mathbf{W}(\mathbf{x},t,\mathbf{y})$ such that the sequences $\{p^\varepsilon\}$, $\{\mathbf{w}_s^\varepsilon\}$, $\{\mathbf{w}^\varepsilon\}$, $\{\chi^\varepsilon\mathbf{w}^\varepsilon\}$, $\Big\{\dfrac{\partial^2\mathbf{w}^\varepsilon}{\partial t^2}\Big\}$, and $\Big\{\varepsilon\nabla\Big(\dfrac{\partial\mathbf{w}^\varepsilon}{\partial t}\Big)\Big\}$ converge two-scale in $L_2(Q_T)$

and $\mathbf{L}_2(Q_T)$ to $p(\mathbf{x},t)$, $\mathbf{w}_s(\mathbf{x},t)$, $\mathbf{W}(\mathbf{x},t,\mathbf{y})$, $\chi(\mathbf{y}))\mathbf{W}(\mathbf{x},t,\mathbf{y})$, $\dfrac{\partial^2\mathbf{W}}{\partial t^2}(\mathbf{x},t,\mathbf{y})$, and $\nabla_y\Big(\dfrac{\partial W}{\partial t}\Big)(\mathbf{x},t,\mathbf{y})$ respectively, and

$$\mathbf{W},\ \frac{\partial^2\mathbf{W}}{\partial t^2},\ \nabla_y\left(\frac{\partial\mathbf{W}}{\partial t}\right)\in\mathbf{L}_2(Q_T\times Y), \tag{3.2.57}$$

$$\frac{\partial\mathbf{W}}{\partial t}(\mathbf{x},t,\mathbf{y})=\big(1-\chi(\mathbf{y})\big)\frac{\partial\mathbf{w}_s}{\partial t}(\mathbf{x},t)+\chi(\mathbf{y})\frac{\partial\mathbf{W}}{\partial t}(\mathbf{x},t,\mathbf{y}). \tag{3.2.58}$$

Lemma 3.6 *The limiting functions* $\mathbf{v}=\dfrac{\partial\mathbf{w}}{\partial t}$ *and* $\mathbf{W}$ *satisfy the homogenized continuity equation (3.2.14) in the domain* Q_T *and boundary condition (3.2.13) on the boundary* S^2_T *for the velocity*

$$\mathbf{v}=\zeta\frac{\partial\mathbf{w}}{\partial t}+(1-\zeta)\Big(\frac{\partial\mathbf{w}^{(f)}}{\partial t}+(1-m)\frac{\partial\mathbf{w}_s}{\partial t}\Big), \tag{3.2.59}$$

and microscopic continuity equation (3.2.27) in the domain Y_T *for almost all* $(\mathbf{x},t)\in Q_T$.

Lemma 3.7 *The limiting functions* p, $\mathbf{w}$, $\mathbf{w}_s$ *and* $\mathbf{w}^{(f)}$ *satisfy the integral identity*

$$\int_{Q_T}\Big(\big(\rho_f\big(\zeta\frac{\partial^2\mathbf{w}}{\partial t^2}+(1-\zeta)\frac{\partial^2\mathbf{w}^{(f)}}{\partial t^2}\big)+\rho_s(1-\zeta)(1-m)\frac{\partial^2\mathbf{w}_s}{\partial t^2}\big)\cdot\varphi-p\,\nabla\cdot\varphi\Big)dxdt$$
$$=-\int_{Q_T}\nabla\cdot(\varphi\, p_0)dxdt \tag{3.2.60}$$

for all smooth functions φ, *such that* $\varphi(\mathbf{x},t)=0$ *on the boundary* S^2_T.

The integral identity (3.2.60) results in the inclusion $\nabla p\in\mathbf{L}_2(Q_T)$, the boundary condition (3.2.3), and the momentum balance equation for the solid component in the form

$$\rho_f\Big(\zeta\frac{\partial\mathbf{w}}{\partial t}+(1-\zeta)\frac{\partial\mathbf{w}^{(f)}}{\partial t}\Big)+\rho_s(1-\zeta)(1-m)\frac{\partial\mathbf{w}_s}{\partial t}=-\int_0^t\nabla p(\mathbf{x},\tau)d\tau.$$

Multiplying the last equation by ζ and $(1-\zeta)$ we obtain (3.2.15) and

$$\zeta\rho_f\frac{\partial\mathbf{w}}{\partial t}=-\zeta\int_0^t\nabla p(\mathbf{x},\tau)d\tau.$$

This last formula and (3.2.59) give us

$$\mathbf{v} = -\frac{\zeta}{\rho_f}\int_0^t \nabla p(\mathbf{x}, \tau)d\tau + (1-\zeta)\left((1-m)\frac{\partial \mathbf{w}_s}{\partial t} + \frac{\partial \mathbf{w}^{(f)}}{\partial t}\right). \tag{3.2.61}$$

To derive the momentum balance equation for the liquid component we simply pass to the limit as $\varepsilon \to 0$ in (3.1.14) with test functions $\varphi = h(\mathbf{x}, t)\varphi_0\left(\frac{\mathbf{x}}{\varepsilon}, t\right)$, where $h(\mathbf{x}, t)$ are smooth and finite in Ω_T, and 1-periodic in $\mathbf{y}$ smooth functions $\varphi_0(\mathbf{y})$ are divergence free and finite in Y_f. If we put $\mathbf{W}^{(f)} = \chi(\mathbf{y})\mathbf{W}$, then the limiting integral identity for the pair $\{\mathbf{W}^{(f)},\ \Pi^{(f)}\}$ is equivalent to the differential equation

$$\rho_f\frac{\partial^2 \mathbf{W}^{(f)}}{\partial t^2} = \frac{\mu_1}{2}\Delta_y\left(\frac{\partial \mathbf{W}^{(f)}}{\partial t}\right) - \nabla_y \Pi^{(f)} - \nabla p \tag{3.2.62}$$

in the domain $Y_f \times (0, T)$ and initial conditions

$$\mathbf{W}^{(f)}(\mathbf{x}, 0, \mathbf{y}) = \frac{\partial \mathbf{W}^{(f)}}{\partial t}(\mathbf{x}, 0, \mathbf{y}) = 0,\ \ \mathbf{y} \in Y_f \tag{3.2.63}$$

for almost all $\mathbf{x} \in \Omega_T$. Relations (3.2.57) and (3.2.58) imply the boundary condition

$$\mathbf{W}^{(f)}(\mathbf{x}, t, \mathbf{y}) = \mathbf{w}_s(\mathbf{x}, t),\ \ (\mathbf{y}, t) \in \gamma \times (0, T) \tag{3.2.64}$$

for almost all $\mathbf{x} \in \Omega_T$.

Therefore, the solution $\{\mathbf{W}^{(f)},\ \Pi^{(f)}\}$ to the periodic initial boundary value problem (3.2.27), (3.2.62)–(3.2.64) has the form

$$\mathbf{W}^{(f)} = \mathbf{w}_s(\mathbf{x}, t) + \sum_{i=1}^{3}\int_0^t \mathbf{W}_i^{(f)}(\mathbf{y}, t-\tau)\left(\frac{\partial p}{\partial x_i}(\mathbf{x}, \tau) + \rho_f\frac{\partial^2 w_{s,i}}{\partial \tau^2}(\mathbf{x}, \tau)\right)d\tau,$$

$$\Pi^{(f)}(\mathbf{x}, t, \mathbf{y}) = \sum_{i=1}^{3}\int_0^t \Pi_i^{(f)}(\mathbf{y}, t-\tau)\left(\frac{\partial p}{\partial x_i}(\mathbf{x}, \tau) + \rho_f\frac{\partial^2 w_{s,i}}{\partial \tau^2}(\mathbf{x}, \tau)\right)d\tau,$$

where $\mathbf{w}_s = (w_{s,1}, w_{s,2}, w_{s,3})$ and $\{\mathbf{W}_i^{(f)},\ \Pi_i^{(f)}\}$, $i = 1, 2, 3$, are solutions to the following periodic initial boundary value problem

$$\rho_f\frac{\partial^2 \mathbf{W}_i^{(f)}}{\partial t^2} = \frac{\mu_1}{2}\Delta_y\left(\frac{\partial \mathbf{W}_i^{(f)}}{\partial t}\right) - \nabla_y \Pi_i^{(f)},\ \ (\mathbf{y}, t) \in Y_f \times (0, T), \tag{3.2.65}$$

$$\nabla_y \cdot \mathbf{W}_i^{(f)}(\mathbf{y}, t) = 0,\ \ (\mathbf{y}, t) \in Y_f \times (0, T), \tag{3.2.66}$$

$$\mathbf{W}_i^{(f)}(\mathbf{y}, 0) = 0,\ \ \rho_f\frac{\partial \mathbf{W}_i^{(f)}}{\partial t}(\mathbf{y}, 0) = \mathbf{e}_i,\ \ \mathbf{y} \in Y_f, \tag{3.2.67}$$

$$\mathbf{W}_i^{(f)}(\mathbf{y}, t) = 0, \quad (\mathbf{y}, t) \in \gamma \times (0, T), \tag{3.2.68}$$

for almost all $\mathbf{x} \in \Omega_T$.

By definition

$$\frac{\partial \mathbf{w}^{(f)}}{\partial t}(\mathbf{x}, t) = \int_{Y_f} \frac{\partial \mathbf{W}^{(f)}}{\partial t}(\mathbf{x}, t, \mathbf{y}) dy \tag{3.2.69}$$

$$= m\frac{\partial \mathbf{w}_s}{\partial t} - \int_0^t \Big(\sum_{i=1}^3 \Big(\int_{Y_f} \frac{\partial \mathbf{W}_i^{(f)}}{\partial t}(\mathbf{y}, t-\tau) dy\Big) \otimes \mathbf{e}_i\Big) \cdot \Big(\nabla p(\mathbf{x}, \tau)$$
$$+ \rho_f \frac{\partial^2 \mathbf{w}_s}{\partial \tau^2}(\mathbf{x}, \tau)\Big) d\tau = m\frac{\partial \mathbf{w}_s}{\partial t}$$
$$- \int_0^t \mathbb{B}^{(f)}(\mu_1, \infty; t-\tau) \cdot \Big(\nabla p(\mathbf{x}, \tau) + \rho_f \frac{\partial^2 \mathbf{w}_s}{\partial \tau^2}(\mathbf{x}, \tau)\Big) d\tau, \tag{3.2.69}$$

where

$$\mathbb{B}^{(f)}(\mu_1, \infty; t) = \sum_{i=1}^3 \Big(\int_{Y_f} \frac{\partial \mathbf{W}_i^{(f)}}{\partial t}(\mathbf{y}, t) dy\Big) \otimes \mathbf{e}_i. \tag{3.2.70}$$

3.2.5.2 The Case $\mu_1 = 0$.

For $\mu_1 = 0$ the microscopic momentum balance equation for the liquid component has the form

$$\rho_f \frac{\partial^2 \mathbf{W}^{(f)}}{\partial t^2} = -\nabla_y \Pi^{(f)} - \nabla p. \tag{3.2.71}$$

Instead of condition (3.2.64) on the boundary γ one has there a condition

$$\big(\mathbf{W}^{(f)}(\mathbf{x}, t, \mathbf{y}) - \mathbf{w}_s(\mathbf{x}, t)\big) \cdot \mathbf{n}(\mathbf{y}) = 0, \tag{3.2.72}$$

which is a consequence of the microscopic continuity equation (3.2.27) and the representation

$$\mathbf{W}(\mathbf{x}, t, \mathbf{y}) = \chi(\mathbf{y})\mathbf{W}^{(f)}(\mathbf{x}, t, \mathbf{y}) + \big(1 - \chi(\mathbf{y})\big)\mathbf{w}_s(\mathbf{x}, t), \quad \mathbf{y} \in Y. \tag{3.2.73}$$

To solve (3.2.71) we apply to this equation the operation $\nabla_y \cdot$ and use again (3.2.27):

$$0 = \nabla_y \cdot \Big(\rho_f \frac{\partial^2 \mathbf{W}^{(f)}}{\partial t^2}\Big) = -\nabla_y \cdot (\nabla_y \Pi^{(f)}). \tag{3.2.74}$$

The boundary condition Eqs. (3.2.72) and (3.2.71) provide the boundary condition

$$\nabla_y \Pi^{(f)} \cdot \mathbf{n}(\mathbf{y}) = -\left(\nabla p + \rho_f \frac{\partial^2 \mathbf{w}_s}{\partial t^2}\right) \cdot \mathbf{n}(\mathbf{y}) \tag{3.2.75}$$

on the boundary γ for the pressure $\Pi^{(f)}$.

Let

$$\Pi^{(f)} = -\left(\sum_{i=1}^{3} \Pi_i^{(f)}(\mathbf{y})\mathbf{e}_i\right) \cdot \left(\nabla p + \rho_f \frac{\partial^2 \mathbf{w}_s}{\partial t^2}\right),$$

where $\Pi_i^{(f)}$, $i = 1, 2, 3$, are solutions to the periodic boundary value problems

$$\triangle_y \Pi_i^{(f)} = 0, \ \mathbf{y} \in Y_f, \ (\nabla_y \Pi_i^{(f)} - \mathbf{e}_i) \cdot \mathbf{n}(\mathbf{y}) = 0, \ \mathbf{y} \in \gamma.$$

Then

$$\nabla_y \Pi^{(f)} = -\left(\sum_{i=1}^{3} \nabla_y \Pi_i^{(f)} \otimes \mathbf{e}_i\right) \cdot \left(\nabla p + \rho_f \frac{\partial^2 \mathbf{w}_s}{\partial t^2}\right).$$

After integration (3.2.71) over domain Y_f we arrive at the desired momentum balance equation (3.2.16) for the liquid component with

$$\rho_f \mathbb{B}^{(f)}(0, \infty) = m\mathbb{I} - \left(\sum_{i=1}^{3} \int_{Y_f} \nabla_y \Pi_i^{(f)}(\mathbf{y}) dy \otimes \mathbf{e}_i\right). \tag{3.2.76}$$

The proof of the last statement of the theorem repeats the proof of the same statement in Theorem 3.4.

3.3 A Viscous Liquid in an Extremely Elastic Skeleton

Throughout this section we assume that

$$\bar{\alpha}_\mu = \mu_0, \ 0 < \mu_0 < \infty, \ \lambda_0 = 0. \tag{3.3.1}$$

3.3.1 Main Results

Theorem 3.6 *Let $\{\mathbf{w}^\varepsilon, p^\varepsilon\}$ be the weak solution of the problem (3.1.4)–(3.1.13),*

$$\lambda_1 = \infty,$$

and $\mathbf{v}_f^\varepsilon = \mathbb{E}_{Q_f^\varepsilon}\left(\frac{\partial \mathbf{w}^\varepsilon}{\partial t}\right)$. Then

(1) *the sequence* $\{\mathbf{v}_f^\varepsilon\}$ *converges weakly in* $\mathbf{W}_2^{1,0}(Q_T)$ *to the function* $\mathbf{v}_f$, *sequences* $\left\{\dfrac{\partial \mathbf{w}^\varepsilon}{\partial t}\right\}$ *and* $\{p^\varepsilon\}$ *converge weakly in* $\mathbf{L}_2(Q_T)$ *and* $L_2(Q_T)$ *to functions* $\mathbf{v} = \dfrac{\partial \mathbf{w}}{\partial t} = \mathbf{v}_f$ *and* p *respectively;*

(2) *limiting functions* $\mathbf{v}_f$ *and* p *solve the system of homogenized equations in the domain* Q_T, *consisting of the continuity equation*

$$\nabla \cdot \mathbf{v}_f = 0, \tag{3.3.2}$$

and the homogenized momentum balance equation

$$\rho(\mathbf{x}) \frac{\partial \mathbf{v}_f}{\partial t} = \nabla \cdot \widehat{\mathbb{P}}^f_0, \tag{3.3.3}$$

$$\widehat{\mathbb{P}}^f_0 = \mu_0\big(\zeta \mathbb{D}(x, \mathbf{v}_f) + (1-\zeta)\, \mathfrak{N}^f_0 : \mathbb{D}(x, \mathbf{v}_f)\big) - p\,\mathbb{I},$$

completed with the boundary conditions

$$\widehat{\mathbb{P}}^f_0 \cdot \mathbf{e}_3 = -p_0 \mathbf{e}_3, \;\; \mathbf{x} \in S^1, \tag{3.3.4}$$

$$\mathbf{v}_f(\mathbf{x}, t) = 0, \;\; \mathbf{x} \in S^2, \tag{3.3.5}$$

for $t \in (0, T)$, *and initial condition*

$$\mathbf{v}_f(\mathbf{x}, 0) = 0, \;\; \mathbf{x} \in Q; \tag{3.3.6}$$

(3) *the problem (3.3.2)–(3.3.6) has a unique solution.*
In (3.3.3)

$$\rho(\mathbf{x}) = \Big(\zeta(\mathbf{x}) + \big(1 - \zeta(\mathbf{x})\big)m\Big)\rho_f + \big(1 - \zeta(\mathbf{x})\big)(1-m)\rho_s,$$

and the symmetric strictly positively definite constant fourth-rank tensor $\mathfrak{N}^f_0$ *is given below by (3.3.22).*

We refer to the problem (3.3.2)–(3.3.6) as the homogenized **model** $(\mathbb{HS})_5$.

Theorem 3.7 *Let* $\{\mathbf{w}^\varepsilon, p^\varepsilon\}$ *be the weak solution of the problem (3.1.4)–(3.1.13),*

$$0 \leqslant \lambda_1 < \infty,$$

and $\mathbf{v}_f^\varepsilon = \mathbb{E}_{Q_f^\varepsilon}\left(\dfrac{\partial \mathbf{w}^\varepsilon}{\partial t}\right)$. *Then*

(1) the sequence $\{\mathbf{v}_f^\varepsilon\}$ *converges weakly in* $W_2^{1,0}(Q_T)$ *to the function* $\mathbf{v}_f$, *sequences* $\left\{\dfrac{\partial \mathbf{v}_f^\varepsilon}{\partial t}\right\}$, $\left\{(1-\zeta)(1-\chi^\varepsilon)\dfrac{\partial \mathbf{w}^\varepsilon}{\partial t}\right\}$, $\left\{(1-\zeta)(1-\chi^\varepsilon)\dfrac{\partial^2 \mathbf{w}^\varepsilon}{\partial t^2}\right\}$, $\{p^\varepsilon\}$, *and* $\{(1-\zeta)(1-\chi^\varepsilon)p^\varepsilon\}$ *converge weakly in* $\mathbf{L}_2(Q_T)$ *and* $L_2(Q_T)$ *to functions* $\dfrac{\partial \mathbf{v}_f}{\partial t}$, $\mathbf{v}^{(s)}$, $\dfrac{\partial \mathbf{v}^{(s)}}{\partial t}$, p, *and* $(1-m)p_s \in W_2^{1,0}(\Omega_T)$ *respectively;*

(2) limiting functions $\mathbf{v}_f$, $\mathbf{v}^{(s)}$, p, *and* p_s *solve the system of homogenized equations in the domain* Q_T, *consisting of the continuity equation*

$$\nabla\cdot\big((\zeta+(1-\zeta)m)\mathbf{v}_f+\mathbf{v}^{(s)}\big)=0, \tag{3.3.7}$$

the homogenized momentum balance equation

$$\rho_f\big(\zeta+m(1-\zeta)\big)\frac{\partial \mathbf{v}_f}{\partial t}+\rho_s\frac{\partial \mathbf{v}^{(s)}}{\partial t}=\nabla\cdot\widehat{\mathbb{P}^f}, \tag{3.3.8}$$

$$\widehat{\mathbb{P}^f}=\mu_0\big(\zeta\,\mathbb{D}(x,\mathbf{v}_f)+(1-\zeta)\,\mathfrak{N}_1^f:\mathbb{D}(x,\mathbf{v}_f)\big)-\big(\zeta\,p+(1-\zeta)\,p_s\big)\mathbb{I},$$

for the liquid component, and the homogenized momentum balance equation

$$\int_0^t\mathbb{B}^{(s)}(\infty,\lambda_1;t-\tau)\cdot\Big(\nabla p_s(\mathbf{x},\tau)+(1-\zeta)\rho_s\frac{\partial \mathbf{v}_f}{\partial \tau}(\mathbf{x},\tau)\Big)d\tau$$
$$=-\big(\mathbf{v}^{(s)}-(1-m)(1-\zeta)\mathbf{v}_f\big) \tag{3.3.9}$$

for the solid component, completed with the boundary and initial conditions (3.3.5)–(3.3.6) for the liquid component, the boundary condition

$$\mathbf{v}^{(s)}\cdot\mathbf{n}=0,\ \ \mathbf{x}\in S^2,\ \ t\in(0,T) \tag{3.3.10}$$

for the solid component, and the boundary condition

$$\widehat{\mathbb{P}^f}\cdot\mathbf{e}_3=-p_0\mathbf{e}_3 \tag{3.3.11}$$

on the boundary S^1 *for* $t\in(0,T)$ *for the liquid and solid components.*

In (3.3.8)–(3.3.11) the symmetric strictly positively definite constant fourth-rank tensor $\mathfrak{N}_1^f$ *is given below by (3.3.25), the matrix* $\mathbb{B}^{(s)}(\infty,\lambda_1;t)$ *is defined in Theorem 3.4, and* $\mathbf{n}$ *is the normal vector to the boundary* S^2.

We refer to the problem (3.3.4)–(3.3.11) as the homogenized **model** $(\mathbb{HS})_6$.

3.3.2 Proof of Theorem 3.6

By Theorem 3.1 and the properties of the extension operator $\mathbb{E}_{Q_f^\varepsilon}$ the sequences $\{p^\varepsilon\}$, $\{(1-\chi^\varepsilon)p^\varepsilon\}$, $\{\mathbf{v}_f^\varepsilon\}$, $\{\mathbb{D}(x,\mathbf{v}_f^\varepsilon)\}$, $\left\{\frac{\partial \mathbf{w}^\varepsilon}{\partial t}\right\}$, $\left\{\frac{\partial \mathbf{v}_f^\varepsilon}{\partial t}\right\}$, $\left\{\mathbb{D}(x,\frac{\partial \mathbf{v}_f^\varepsilon}{\partial t})\right\}$, $\left\{\frac{\partial^2 \mathbf{w}^\varepsilon}{\partial t^2}\right\}$, and $\left\{\bar{\alpha}_\lambda\nabla(\frac{\partial \mathbf{w}^\varepsilon}{\partial t})\right\}$ are bounded in $L_2(Q_T)$ and $\mathbf{L}_2(Q_T)$.

Hence there exists a subsequence of small parameters $\{\varepsilon > 0\}$ and functions p, p_s, $\mathbf{v}$, and $\mathbf{v}_f$ such that

$$p^\varepsilon \rightharpoonup p,\ (1-\zeta)(1-\chi^\varepsilon)p^\varepsilon \rightharpoonup (1-\zeta)(1-m)p_s,$$

$$\mathbf{v}^\varepsilon = \frac{\partial \mathbf{w}^\varepsilon}{\partial t} \rightharpoonup \mathbf{v},\ \frac{\partial^2 \mathbf{w}^\varepsilon}{\partial t^2} \rightharpoonup \frac{\partial \mathbf{v}}{\partial t}$$

weakly in $L_2(Q_T)$ and $\mathbf{L}_2(Q_T)$ as $\varepsilon \searrow 0$, and

$$\mathbf{v}_f^\varepsilon \rightharpoonup \mathbf{v}_f,\ \frac{\partial \mathbf{v}_f^\varepsilon}{\partial t} \rightharpoonup \frac{\partial \mathbf{v}_f}{\partial t}$$

weakly in $\mathbf{W}_2^{1,0}(Q_T)$ as $\varepsilon \searrow 0$.

Just as in Theorem 1.3 we conclude that $\mathbf{v}_f$ satisfies the boundary condition (3.3.5). Note also that

$$\bar{\alpha}_\lambda(1-\zeta)\mathbb{D}(x,\mathbf{w}^\varepsilon) \to 0$$

strongly in $\mathbf{L}_2(Q_T)$ as $\varepsilon \searrow 0$.

Relabeling if necessary, we assume that the sequences themselves converge.

By Nguetseng's theorem there exist 1-periodic in $\mathbf{y}$ functions $P(\mathbf{x},t,\mathbf{y})$, $\mathbf{V}(\mathbf{x},t,\mathbf{y})$, and $\mathbf{V}_f(\mathbf{x},t,\mathbf{y})$, such that

$$P \in L_2(Q_T \times Y),\ \mathbf{V},\ \mathbf{V}_f \in \mathbf{L}_2(Q_T \times Y),$$
$$\nabla\mathbf{V},\ \nabla\mathbf{V}_f,\ \frac{\partial \mathbf{V}}{\partial t},\ \frac{\partial \mathbf{V}_f}{\partial t} \in \mathbf{L}_2(Q_T \times Y),$$

and that the sequences $\{p^\varepsilon\}$, $\{\mathbf{v}^\varepsilon\}$, $\{\mathbf{v}_f^\varepsilon\}$, $\{\varepsilon\mathbb{D}(x,\mathbf{v}^\varepsilon)\}$, and $\{\mathbb{D}(x,\mathbf{v}_f^\varepsilon)\}$ converge two-scale in $L_2(Q_T)$ and $\mathbf{L}_2(Q_T)$ to $P(\mathbf{x},t,\mathbf{y})$, $\mathbf{V}(\mathbf{x},t,\mathbf{y})$, $\mathbf{v}_f(\mathbf{x},t)$, $\mathbb{D}(y,\mathbf{V})$, and $\mathbb{D}(x,\mathbf{v}_f)+\mathbb{D}(y,\mathbf{V}_f)$ respectively.

Lemma 3.8 *The limiting functions $\mathbf{v}_f$, $\mathbf{v}$, and $\mathbf{V}_f$ satisfy the macroscopic and microscopic continuity equations*

$$\nabla\cdot\mathbf{v} = 0,\ (\mathbf{x},t) \in Q_T,\ \mathbf{v}\cdot\mathbf{n} = 0,\ (\mathbf{x},t) \in S_T^2, \tag{3.3.12}$$

$$\chi(\mathbf{y})(\nabla\cdot\mathbf{v}_f + \nabla_y\cdot\mathbf{V}_f) = 0,\ (\mathbf{x},t) \in \Omega_T,\ \mathbf{y} \in Y, \tag{3.3.13}$$

and

$$\nabla_y \cdot \mathbf{V} = 0, \ (\mathbf{x}, t) \in \Omega_T, \ \mathbf{y} \in Y, \tag{3.3.14}$$

where

$$\mathbf{V}(\mathbf{x}, t, \mathbf{y}) = \big(\zeta + \chi(\mathbf{y})(1-\zeta)\big)\mathbf{v}_f + (1-\zeta)\big(1-\chi(\mathbf{y})\big)\mathbf{V}(\mathbf{x}, t, \mathbf{y}), \tag{3.3.15}$$

$$\mathbf{v} = \langle \mathbf{V} \rangle_Y = \big(\zeta + m\,(1-\zeta)\big)\mathbf{v}_f + (1-\zeta)\langle \mathbf{V} \rangle_{Y_s},$$

and $\mathbf{n}$ *is a unit normal vector to the boundary* S^2.

The proof of (3.3.12) repeats the proof of (1.3.14) in Chap. 1.
Equation (3.3.13) is a result of a two-scale limit in the equality

$$\chi^{\varepsilon}(\mathbf{x})\nabla \cdot \mathbf{v}_f^{\varepsilon} = 0$$

for $(\mathbf{x}, t) \in \Omega_T$.
Equation (3.3.14) is a simple consequence of a two-scale limit in the continuity equation (3.1.4) in its integral form:

$$\int_{Q_T} \varepsilon\,\mathbf{v}^{\varepsilon} \cdot \nabla \left(h_0(\mathbf{x}, t) h\left(\frac{\mathbf{x}}{\varepsilon}\right)\right) dxdt = 0.$$

Finally, the relation (3.3.15) is a result of two-scale limit in the equality

$$\mathbf{v}^{\varepsilon} = \big(\zeta + \chi^{\varepsilon}(1-\zeta)\big)\mathbf{v}_f^{\varepsilon} + (1-\chi^{\varepsilon})(1-\zeta)\mathbf{v}^{\varepsilon}.$$

Lemma 3.9 *The following equality holds true*

$$(1-\zeta)P(\mathbf{x}, t, \mathbf{y}) = (1-\zeta)\Big(P_f(\mathbf{x}, t, \mathbf{y}) + \big(1-\chi(\mathbf{y})\big)\,p_s(\mathbf{x}, t)\Big), \tag{3.3.16}$$

where $(1-\zeta)P_f = (1-\zeta)\chi(\mathbf{y})P(\mathbf{x}, t, \mathbf{y})$.

The proof of this lemma repeats the proof of Lemma 1.4 with evident symmetric changes.

Lemma 3.10 *The limiting functions* $\mathbf{v}_f$, p, $\mathbf{V}_f$, *and* P *satisfy the macroscopic momentum balance equation*

$$\int_{Q_T} \Big(\rho_f(\zeta + m(1-\zeta))\frac{\partial \mathbf{v}_f}{\partial t} + \rho_s(1-\zeta)\Big\langle\frac{\partial \mathbf{V}}{\partial t}\Big\rangle_{Y_s} \cdot \varphi + \big(\mu_0\big((\zeta + m(1-\zeta))\,\mathbb{D}(x, \mathbf{v}_f)$$
$$+ (1-\zeta)\langle\mathbb{D}(y, \mathbf{V}_f)\rangle_{Y_f}\big) - p\,\mathbb{I}\big) : \mathbb{D}(x, \varphi)\Big)dxdt = \int_{Q_T} \nabla \cdot (\varphi\, p_0)dxdt \qquad (3.3.17)$$

in the domain Q_T and the microscopic momentum balance equation

$$\nabla_y \cdot \Big(\chi\big(\mu_0\big(\mathbb{D}(x, \mathbf{v}_f) + \mathbb{D}(y, \mathbf{V}_f)\big) - (P_f - p_s)\mathbb{I}\big)\Big) = 0 \qquad (3.3.18)$$

in the domain Y for almost all $(\mathbf{x}, t) \in \Omega_T$.

Proof Equation (3.3.17) follow from (3.1.14) after the two-scale limit with test functions $\varphi = \varphi(\mathbf{x}, t)$. Equation (3.3.18) follows from (3.1.14) after the two-scale limit with test functions $\varphi = \varepsilon h(\mathbf{x}, t)\varphi_0\left(\frac{\mathbf{x}}{\varepsilon}\right)$, where h is finite in Ω.

Lemma B.13 and the boundedness of the sequence $\left\{\bar{\alpha}_\lambda \nabla\left(\frac{\partial \mathbf{w}^\varepsilon}{\partial t}\right)\right\}$ in $\mathbf{L}_2(Q_T)$ result

$$\mathbf{V}(\mathbf{x}, t, \mathbf{y}) = \mathbf{v}(\mathbf{x}, t).$$

Applying the two-scale limit to the equality $\chi^\varepsilon(\mathbf{v}^\varepsilon - \mathbf{v}_f^\varepsilon) = 0$, we arrive at

$$\chi(\mathbf{y})\big(\mathbf{v}(\mathbf{x}, t) - \mathbf{v}_f(\mathbf{x}, t)\big) = 0,$$

or $\mathbf{v}(\mathbf{x}, t) = \mathbf{v}_f(\mathbf{x}, t)$. Therefore, the function $\mathbf{v}_f$ satisfies the continuity equation (3.3.2) and the continuity equation (3.3.13) takes the form

$$\chi(\mathbf{y})\nabla_y \cdot \mathbf{V}_f = 0, \;\; (\mathbf{x}, t) \in \Omega_T, \; \mathbf{y} \in Y, \qquad (3.3.19)$$

while the macroscopic momentum balance equation (3.3.17) becomes

$$\int_{Q_T} \Big(\rho(\mathbf{x})\frac{\partial \mathbf{v}_f}{\partial t} \cdot \varphi - \nabla \cdot (\varphi\, p_0) + \big(\mu_0\big((\zeta + m(1-\zeta))\,\mathbb{D}(x, \mathbf{v}_f)$$
$$+ (1-\zeta)\langle\mathbb{D}(y, \mathbf{V}_f)\rangle_{Y_f}\big) - p\,\mathbb{I}\big) : \mathbb{D}(x, \varphi)\Big)dxdt = 0,$$

which is equivalent to the macroscopic momentum balance equation

$$\rho(\mathbf{x})\frac{\partial \mathbf{v}_f}{\partial t} = \nabla \cdot \Big(\mu_0\big((\zeta + m(1-\zeta))\,\mathbb{D}(x, \mathbf{v}_f)$$
$$+ (1-\zeta)\langle\mathbb{D}(y, \mathbf{V}_f)\rangle_{Y_f}\big) - p\,\mathbb{I}\Big). \qquad (3.3.20)$$

in the differential form, the boundary condition (3.3.4), and the initial condition (3.3.6).

Lemma 3.11 *The limiting functions $\mathbf{v}_f$ and p satisfy in the domain Q_T the homogenized momentum balance equation (3.3.3).*

Proof We simply repeat the proof of Theorem 1.3 with symmetric change of the domain Y_f onto domain Y_s. In fact, to find $\langle\mathbb{D}(y, \mathbf{V}_f)\rangle_{Y_f}$ in the domain Ω_T we look for the 1-periodic solution $\mathbf{V}_f,\ P_f$ of the system (3.3.18) and (3.3.19) in the form

$$\mathbf{V}_f = \sum_{i,j=1}^{3} \mathbf{V}^{(ij)}(\mathbf{y}) D_{ij}(\mathbf{x}, t),$$

$$P_f - p_s = \mu_0 \sum_{i,j=1}^{3} P^{(ij)}(\mathbf{y}) D_{ij}(\mathbf{x}, t),$$

where

$$D_{ij}(\mathbf{x}, t) = \frac{1}{2}\left(\frac{\partial v_{f,i}}{\partial x_j}(\mathbf{x}, t) + \frac{\partial v_{f,j}}{\partial x_i}(\mathbf{x}, t)\right), \quad \mathbf{v}_f = (v_{f,1}, v_{f,2}, v_{f,3}),$$

$$\mathbb{D}(x, \mathbf{v}_f) = \sum_{i,j=1}^{3} D_{ij}(\mathbf{x}, t)\, \mathbb{J}^{ij},$$

and

$$\left.\begin{array}{l} \nabla_y \cdot \left(\chi\left(\mathbb{D}(y, \mathbf{V}^{(ij)}) + \mathbb{J}^{ij} - P^{(ij)}\mathbb{I}\right)\right) = 0, \ \mathbf{y} \in Y, \\ \chi \nabla_y \cdot \mathbf{V}^{(ij)} = 0, \ \langle \mathbf{V}^{(ij)} \rangle_{Y_f} = 0, \quad \mathbf{y} \in Y. \end{array}\right\} \tag{3.3.21}$$

Then

$$\mathfrak{N}_0^f = (1-m)\mathbb{J} + \sum_{i,j=1}^{3} \langle \mathbb{D}(y, \mathbf{V}^{(ij)}) \rangle_{Y_f} \otimes \mathbb{J}^{ij}. \tag{3.3.22}$$

Recall that the homogenized momentum balance equation (3.3.3) is understood as an integral identity

$$\int_{Q_T} \left(\rho(\mathbf{x})\frac{\partial \mathbf{v}_f}{\partial t} \cdot \varphi + \left(\mu_0\big(\zeta\mathbb{J} + (1-\zeta)\,\mathfrak{N}_0^f\big) : \mathbb{D}(x, \mathbf{v}_f) - p\,\mathbb{I}\right) : \mathbb{D}(x, \varphi)\right) dxdt$$
$$= \int_{Q_T} \nabla \cdot (\varphi\, p_0) dxdt$$

for all functions $\varphi \in \mathbf{W}_2^{1,0}(Q_T)$, such that $\varphi(\mathbf{x}, t) = 0$ on the boundary S_T^2, and $\varphi(\mathbf{x}, T) = 0,\ \mathbf{x} \in Q$.

Therefore the function $\mathbf{v}_f$ satisfies the boundary and initial conditions (3.3.4) and (3.3.6). Finally, the solution $\mathbf{v}_f$ vanishes on the boundary S_T^2 for the same reason, as in Theorem 1.3.

The uniqueness of the problem (3.3.2)–(3.3.6) follows from the energy equality

$$\frac{1}{2}\frac{d}{dt}\int_Q \rho|\mathbf{v}_f|^2 dx + \mu_0 \int_Q \Big(\big(\zeta\mathbb{J} + (1-\zeta)\,\mathfrak{N}_0^f\big) : \mathbb{D}(x,\mathbf{v}_f)\Big) : \mathbb{D}(x,\mathbf{v}_f)dx = 0$$

for the homogeneous problem and the properties of the tensor $\mathfrak{N}_0^f$.

3.3.3 *Proof of Theorem 3.7*

3.3.3.1 The Case $\lambda_1 > 0$.

As in the previous subsection we conclude that the sequences $\{p^\varepsilon\}$, $\{\mathbf{v}_f^\varepsilon\}$, $\{\mathbb{D}(x,\mathbf{v}_f^\varepsilon)\}$ $\left\{\dfrac{\partial\mathbf{w}^\varepsilon}{\partial t}\right\}$, and $\left\{\varepsilon\nabla(\dfrac{\partial\mathbf{w}^\varepsilon}{\partial t})\right\}$ are bounded in $L_2(Q_T)$ and $\mathbf{L}_2(Q_T)$.

Hence there exists a subsequence of small parameters $\{\varepsilon > 0\}$ and functions p, p_s, $\mathbf{v}$, and $\mathbf{v}_f$ such that

$$p^\varepsilon \rightharpoonup p,\ (1-\zeta)(1-\chi^\varepsilon)p^\varepsilon \rightharpoonup (1-\zeta)(1-m)p_s,$$

$$\mathbf{v}^\varepsilon = \frac{\partial\mathbf{w}^\varepsilon}{\partial t} \rightharpoonup \mathbf{v},\quad \frac{\partial^2\mathbf{w}^\varepsilon}{\partial t^2} \rightharpoonup \frac{\partial\mathbf{v}}{\partial t}$$

weakly in $L_2(Q_T)$ and $\mathbf{L}_2(Q_T)$ as $\varepsilon \searrow 0$, and

$$\mathbf{v}_f^\varepsilon \rightharpoonup \mathbf{v}_f,\quad \frac{\partial\mathbf{v}_f^\varepsilon}{\partial t} \rightharpoonup \frac{\partial\mathbf{v}_f}{\partial t}$$

weakly in $\mathbf{W}_2^{1,0}(Q_T)$.

As in Theorem 1.3 we conclude that $\mathbf{v}_f$ satisfies the boundary condition (3.3.5).

Note also that

$$\bar{\alpha}_\lambda(1-\zeta)\mathbb{D}(x,\mathbf{w}^\varepsilon) \to 0 \tag{3.3.23}$$

strongly in $\mathbf{L}_2(Q_T)$ as $\varepsilon \searrow 0$.

At the same time there exist 1-periodic in $\mathbf{y}$ functions $P(\mathbf{x},t,\mathbf{y})$ from $L_2(Q_T\times Y)$, and $\mathbf{V}(\mathbf{x},t,\mathbf{y})$, $\mathbf{V}_f(\mathbf{x},t,\mathbf{y})$ from $\mathbf{L}_2\big(Q; W_2^{1,0}(Y_T)\big)$, such that $\dfrac{\partial\mathbf{V}}{\partial t}$, $\dfrac{\partial\mathbf{V}_f}{\partial t} \in \mathbf{L}_2(Q_T)$ and the sequences $\{p^\varepsilon\}$, $\{\mathbf{v}^\varepsilon\}$, $\{\mathbf{v}_f^\varepsilon\}$, $\{\varepsilon\nabla\mathbf{v}^\varepsilon\}$ and $\{\nabla\mathbf{v}_f^\varepsilon\}$ converge two-scale in $L_2(Q_T)$ and $\mathbf{L}_2(Q_T)$ to $P(\mathbf{x},t,\mathbf{y})$, $\mathbf{V}(\mathbf{x},t,\mathbf{y})$, $\mathbf{v}_f(\mathbf{x},t)$, $\nabla_y\mathbf{V}(\mathbf{x},t,\mathbf{y})$, and $\nabla\,\mathbf{v}_f(\mathbf{x},t)+\nabla_y\mathbf{V}_f(\mathbf{x},t,\mathbf{y})$ respectively.

For all these functions the statements of Lemma 3.8, Lemma 3.9, Lemma 3.10, and the boundary condition (3.3.4) hold true.

To find $\langle \mathbb{D}(y, \mathbf{V}_f)\rangle_{Y_f}$ in (3.3.17) we have to solve the system (3.3.13) and (3.3.18) in the domain Y for almost all $(\mathbf{x}, t) \in \Omega_T$. This system is similar to the system (1.3.31), (1.3.33), where we simply have to change the domain Y_s onto domain Y_f.

Therefore

$$\mathbf{V}_f = \sum_{i,j=1}^{3} \mathbf{V}^{(ij)}(\mathbf{y}) D_{ij}(\mathbf{x}, t) + \mathbf{V}_0(\mathbf{y})\big(\nabla \cdot \mathbf{v}_f(\mathbf{x}, t)\big),$$

$$P_f - p_s = \mu_0 \sum_{i,j=1}^{3} P^{(ij)}(\mathbf{y}) D_{ij}(\mathbf{x}, t) + \lambda_0 P_0(\mathbf{y})\big(\nabla \cdot \mathbf{v}_f(\mathbf{x}, t)\big),$$

where

$$\left.\begin{array}{l} \nabla_y \cdot \Big(\chi \left(\mathbb{D}(y, \mathbf{V}_0) - P_0 \mathbb{I}\right)\Big) = 0, \\ \chi\left(\nabla_y \cdot \mathbf{V}_0 + 1\right) = 0, \quad \langle \mathbf{V}_0 \rangle_{Y_f} = 0, \quad \mathbf{y} \in Y, \end{array}\right\} \tag{3.3.24}$$

and

$$\begin{aligned} \widehat{\mathbb{P}}^f &= \mu_0\big((\zeta + m(1-\zeta))\, \mathbb{D}(x, \mathbf{v}_f) + (1-\zeta)\langle \mathbb{D}(y, \mathbf{V}_f)\rangle_{Y_f}\big) - p\,\mathbb{I} \\ &= \zeta\left(\mu_0 \mathbb{D}(x, \mathbf{v}_f) - p\,\mathbb{I}\right) \\ &\quad + (1-\zeta)\Big(\mu_0\big(m\mathbb{D}(x, \mathbf{v}_f) + \langle \mathbb{D}(y, \mathbf{V}_f)\rangle_{Y_f}\big) - \langle \chi_f P + (1-\chi) p_s\rangle_Y)\mathbb{I}\Big) \\ &= \zeta\left(\mu_0 \mathbb{D}(x, \mathbf{v}_f) - p\,\mathbb{I}\right) \\ &\quad + (1-\zeta)\Big(\mu_0\big(\mathfrak{N}_0^f + \langle \mathbb{D}(y, \mathbf{V}_0)\rangle_{Y_f} \otimes \mathbb{I}\big) : \mathbb{D}(x, \mathbf{v}_f) - (p_s + \langle P - p_s\rangle_Y \mathbb{I}\Big) \\ &= \zeta\left(\mu_0 \mathbb{D}(x, \mathbf{v}_f) - p\,\mathbb{I}\right) + (1-\zeta)\big(\mu_0 \mathfrak{N}_1^f : \mathbb{D}(x, \mathbf{v}_f) - p_s \mathbb{I}\big), \end{aligned}$$

where

$$\mathfrak{N}_1^f = \mathfrak{N}_0^f + \langle \mathbb{D}(y, \mathbf{V}_0)\rangle_{Y_f} \otimes \mathbb{I} - \left\langle \sum_{i,j=1}^{3} P^{(ij)} \right\rangle_{Y_f} \mathbb{I} \otimes \mathbb{J}^{ij} - \langle P_0 \rangle_{Y_f} \mathbb{I} \otimes \mathbb{I}. \tag{3.3.25}$$

All properties of the tensor $\mathfrak{N}_1^f$ are the same as those of the tensor $\mathfrak{N}_1^s$ in Theorem 1.4.

It is easy to see that the sequence $\{(1-\zeta)(1-\chi^\varepsilon)\mathbf{v}^\varepsilon\}$ converges two-scale in $\mathbf{L}_2(Q_T)$ to the function $\mathbf{V}^{(s)} = \big(1-\zeta(\mathbf{x})\big)\big(1-\chi(\mathbf{y})\big)\mathbf{V}$, and weakly in $\mathbf{L}_2(Q_T)$ to the function $\mathbf{v}^{(s)} = (1-\zeta)\langle \mathbf{V}\rangle_{Y_s}$. Therefore, the limiting functions $\mathbf{v}_f$, $\mathbf{v}^{(s)}$, and p_s satisfy the homogenized momentum balance equation (3.3.8), boundary and initial conditions (3.3.4)–(3.3.6) and continuity equation (3.3.7).

To derive the homogenized momentum balance equation (3.3.9) for the solid component we pass to the limit in (3.1.14) as $\varepsilon \searrow 0$ with test functions $\varphi =$

$h(\mathbf{x},t)\varphi_0\left(\dfrac{\mathbf{x}}{\varepsilon}\right)$, where $h(\mathbf{x},t)$ is finite in Ω for all $t\in(0,T)$, and 1-periodic smooth function $\varphi_0(\mathbf{y})$ is divergence free and finite in Y_s:

$$\int_{\Omega_T}\big(a\,h+p_s\,\mathbf{b}\cdot\nabla\,h\big)dxdt=0,$$

where

$$a(\mathbf{x},t)=\int_{Y_s}\left(\rho_s\frac{\partial\mathbf{V}^{(s)}}{\partial t}\cdot\varphi_0+\mu_1\mathbb{D}(y,\mathbf{V}^{(s)}):\mathbb{D}(y,\varphi_0)\right)dy\in L_2(\Omega_T),$$

$$\mathbf{b}=-\int_{Y_s}\varphi_0 dy=const.$$

Just as in the proof of Theorem 1.1 we conclude that

$$\nabla\,p_f\in\mathbf{L}_2(\Omega_T).$$

One has for the function

$$\mathbf{W}^{(s)}=\int_0^t\mathbf{V}^{(s)}(\mathbf{y},\mathbf{x},\tau)d\tau$$

the microscopic momentum balance equation

$$\rho_s\frac{\partial^2\mathbf{W}^{(s)}}{\partial t^2}=\frac{\lambda_1}{2}\triangle\,\mathbf{W}^{(s)}-P^{(s)}-\nabla\,p_s(\mathbf{x},t),\ \ \mathbf{y}\in Y_s,\ t\in(0,T)\tag{3.3.26}$$

and initial conditions

$$\mathbf{W}^{(s)}(\mathbf{y},\mathbf{x},0)=\frac{\partial\mathbf{W}^{(s)}}{\partial t}(\mathbf{y},\mathbf{x},0)=0,\ \ \mathbf{y}\in Y_s,\tag{3.3.27}$$

for almost all $(\mathbf{x},t)\in\Omega_T$.

We complete (3.3.26) and (3.3.27) with continuity equation (3.3.14), boundary condition

$$\mathbf{W}^{(s)}(\mathbf{y},\mathbf{x},t)=\int_0^t\mathbf{v}_f(\mathbf{x},\tau)d\tau,\ \ \mathbf{y}\in\gamma,\ (\mathbf{x},t)\in\Omega_T,\tag{3.3.28}$$

which is a consequence of (3.3.15) and regularity condition

$$\mathbf{V},\ \nabla_y\mathbf{V}\in\mathbf{L}_2(Q_T\times Y).$$

Let us look for the solution of the obtained system in the form

$$\mathbf{W}^{(s)} = \int_0^t \mathbf{v}_f(\mathbf{x}, \tau)d\tau + \sum_{i=1}^{3} \int_0^t \mathbf{W}_i^{(s)}(\mathbf{y}, t-\tau)z_i(\mathbf{x}, \tau)d\tau,$$

$$P^{(s)} = \sum_{i=1}^{3} \int_0^t \Pi_i^{(s)}(\mathbf{y}, t-\tau)z_i(\mathbf{x}, \tau)d\tau,$$

where

$$\mathbf{z} = \nabla p_s(\mathbf{x}, t) + \rho_f \frac{\partial \mathbf{v}_f}{\partial t}(\mathbf{x}, t), \ \ \mathbf{z} = \sum_{i=1}^{3} z_i(\mathbf{x}, t)\mathbf{e}_i.$$

Then for $i = 1, 2, 3$ periodic in variable $\mathbf{y}$ functions $\mathbf{W}_i^{(s)}$, $\Pi_i^{(s)}$ satisfy an initial boundary-value problem

$$\left.\begin{array}{l}
\rho_s \dfrac{\partial^2 \mathbf{W}_i^{(s)}}{\partial t^2} = \dfrac{\lambda_1}{2} \triangle \mathbf{W}_i^{(s)} - \Pi_i^{(s)}, \ \ (\mathbf{y}, t) \in Y_s \times (0, T), \\
\nabla_y \cdot \mathbf{W}_i^{(s)} = 0, \ \ (\mathbf{y}, t) \in Y_s \times (0, T), \\
\mathbf{W}_i^{(s)}(\mathbf{y}, t) = 0, \ \ (\mathbf{y}, t) \in \gamma \times (0, T), \\
\mathbf{W}_i^{(s)}(\mathbf{y}, 0) = 0, \ \ \rho_s \dfrac{\partial \mathbf{W}_i^{(s)}}{\partial t}(\mathbf{y}, 0) = -\mathbf{e}_i, \ \ \mathbf{y} \in Y_s.
\end{array}\right\} \tag{3.3.29}$$

The problem (3.3.29) has been already studied in the proof of Theorem 3.4.

Thus,

$$\mathbf{V}^{(s)} = \mathbf{v}_f(\mathbf{x}, t) + \sum_{i=1}^{3} \int_0^t \frac{\partial \mathbf{W}_s^{(i)}}{\partial t}(\mathbf{y}, t-\tau)z_i(\mathbf{x}, \tau)d\tau,$$

and for function $\mathbf{v}^{(s)}$ the Eq. (3.3.9) holds true. In particular, this equation provides the boundary condition

$$\mathbf{v}^{(s)}(\mathbf{x}, 0) = 0, \ \ \mathbf{x} \in \Omega,$$

which, in turn, together with (3.3.17) guarantee the initial condition (3.3.6).

The boundary condition (3.3.10) follows from (3.3.5) and (3.3.12).

3.3.3.2 The Case $\lambda_1 = 0$.

The proof of this case completely repeats the proof of the previous case $\lambda_1 > 0$, except for the derivation of the homogenized momentum balance equation for the solid component. Here we only have to repeat the proof of Theorem 3.4 for a similar case $\lambda_1 = 0$.

3.4 A Slightly Viscous Liquid in an Elastic Skeleton

Throughout this section we assume that

$$\bar{\alpha}_\lambda = \lambda_0, \;\; 0 < \lambda_0 < \infty, \;\; \mu_0 = 0. \tag{3.4.1}$$

3.4.1 Main Results

Theorem 3.8 *Let $\{\mathbf{w}^\varepsilon, p^\varepsilon\}$ be the weak solution of the problem (3.1.4)–(3.1.13),*

$$\mu_1 = \infty,$$

and $\mathbf{w}_s^\varepsilon = \mathbb{E}_{\Omega_s^\varepsilon}\big(\mathbf{w}^\varepsilon\big)$. Then

(1) the sequence $\{\mathbf{w}_s^\varepsilon\}$ converges weakly in $\mathbf{W}_2^{1,0}(Q_T)$ to the function $\mathbf{w}_s$, sequences $\left\{\dfrac{\partial \mathbf{w}^\varepsilon}{\partial t}\right\}$ and $\{p^\varepsilon\}$ converge weakly in $\mathbf{L}_2(Q_T)$ and $L_2(Q_T)$ to functions $\mathbf{v} = \dfrac{\partial \mathbf{w}}{\partial t}$ and p respectively and

$$(1-\zeta)\left(\mathbf{v} - \frac{\partial \mathbf{w}_s}{\partial t}\right) = 0; \tag{3.4.2}$$

(2) limiting functions $\mathbf{v}$, $\mathbf{w}_s$, and p solve the system of homogenized equations in the domain Q_T, consisting of the continuity equation

$$\nabla \cdot \mathbf{v} = 0, \tag{3.4.3}$$

and the homogenized momentum balance equation

$$\rho_f\,\zeta \frac{\partial \mathbf{v}}{\partial t} + \hat{\rho}\,(1-\zeta)\frac{\partial^2 \mathbf{w}_s}{\partial t^2} = \nabla \cdot \widehat{\mathbb{P}}_0^s, \tag{3.4.4}$$

$$\widehat{\mathbb{P}}_0^s = \lambda_0(1-\zeta)\,\mathfrak{N}_0^s : \mathbb{D}(x, \mathbf{w}_s)\big) - p\,\mathbb{I},$$

completed with the boundary conditions

$$\widehat{\mathbb{P}}_0^s \cdot \mathbf{e}_3 = -p_0\mathbf{e}_3, \;\; \mathbf{x} \in S^1, \tag{3.4.5}$$

$$\mathbf{w}_s(\mathbf{x}, t) = 0, \;\; \mathbf{x} \in S^2, \tag{3.4.6}$$

for $t \in (0, T)$, and initial conditions

$$\zeta\mathbf{v}(\mathbf{x}, 0) = (1-\zeta)\mathbf{w}_s(\mathbf{x}, 0) = (1-\zeta)\frac{\partial \mathbf{w}_s}{\partial t}(\mathbf{x}, 0) = 0, \;\; \mathbf{x} \in Q; \tag{3.4.7}$$

(3) the problem (3.4.2)–(3.4.7) has a unique solution.
In (3.4.4)

$$\hat{\rho} = m\,\rho_f + (1-m)\,\rho_s,$$

and the symmetric strictly positively definite constant fourth-rank tensor $\mathfrak{N}^s_0$ is defined in Theorem 1.3.

We refer to the problem (3.4.2)–(3.4.7) as the homogenized **model** $(\mathbb{HS})_7$.

Theorem 3.9 *Let $\{\mathbf{w}^\varepsilon,\ p^\varepsilon\}$ be the weak solution of the problem (3.1.4)–(3.1.13),*

$$0 \leqslant \mu_1 < \infty,$$

and $\mathbf{w}^\varepsilon_s = \mathbb{E}_{\Omega^\varepsilon_s}(\mathbf{w}^\varepsilon)$.
Then

(1) the sequence $\{\mathbf{w}^\varepsilon_s\}$ converges weakly in $\mathbf{W}^{1,0}_2(Q_T)$ to the function $\mathbf{w}_s$, sequences $\{(\zeta + (1-\zeta)\chi^\varepsilon)\mathbf{w}^\varepsilon\}$, $\{p^\varepsilon\}$, and $\{(1-\zeta)\chi^\varepsilon p^\varepsilon\}$ converge weakly in $\mathbf{L}_2(Q_T)$ and $L_2(Q_T)$ to functions $\mathbf{w}^{(f)}$, p, and $m\,p_f$ respectively, and $\big(\zeta\, p + (1-\zeta)p_f\big) \in W^{1,0}_2(Q_T)$.

(2) limiting functions $\mathbf{w}_s$, $\mathbf{w}^{(f)}$, and p_f solve the system of homogenized equations in the domain Q_T, consisting of the continuity equation

$$\nabla\cdot\big(\mathbf{w}^{(f)} + (1-m)(1-\zeta)\mathbf{w}_s\big) = 0, \tag{3.4.8}$$

the homogenized momentum balance equation

$$\rho_f \frac{\partial^2 \mathbf{w}^{(f)}}{\partial t^2} + \rho_s(1-\zeta)\frac{\partial^2 \mathbf{w}_s}{\partial t^2} = \nabla\cdot\widehat{\mathbb{P}}^s, \tag{3.4.9}$$

$$\widehat{\mathbb{P}}^s = \lambda_0\,(1-\zeta)\mathfrak{N}^s_1 : \mathbb{D}(x,\mathbf{w}_s)\big) - \big(\zeta\, p + (1-\zeta)p_f\big)\mathbb{I}$$

for the solid component, and the homogenized momentum balance equation

$$(1-\zeta)\Big(\frac{\partial \mathbf{w}^{(f)}}{\partial t} - m\,\frac{\partial \mathbf{w}_s}{\partial t}\Big) = -\int_0^t \mathbb{B}^{(f)}(\mu_1,\infty;t-\tau)\cdot\big(\nabla p_f(\mathbf{x},\tau)$$
$$+\,\rho_f(1-\zeta)\frac{\partial^2 \mathbf{w}_s}{\partial \tau^2}(\mathbf{x},\tau)\big)d\tau \tag{3.4.10}$$

for the liquid component, completed with the boundary and initial conditions (3.4.6)–(3.4.7) for the solid component, and boundary and initial conditions

$$\mathbf{w}^{(f)}\cdot\mathbf{n} = 0,\ \ \mathbf{x}\in S^2,\ \ t\in(0,T), \tag{3.4.11}$$

$$\mathbf{w}^{(f)}(\mathbf{x},0) = \frac{\partial \mathbf{w}^{(f)}}{\partial t}(\mathbf{x},0) = 0,\ \ \mathbf{x}\in Q \tag{3.4.12}$$

for the liquid component, and the boundary condition

$$\widehat{\mathbb{P}}^s \cdot \mathbf{e}_3 = -p_0\mathbf{e}_3, \tag{3.4.13}$$

on the boundary S^1 for $t \in (0,T)$ for the solid and liquid components.

In (3.4.9)–(3.4.11) the symmetric strictly positively definite constant fourth-rank tensor $\mathfrak{N}_1^s$ is defined in Theorem 1.4, the matrix $\mathbb{B}^{(f)}(\mu_1, \infty; t)$ is defined in Theorem 3.4, and $\mathbf{n}$ is the normal vector to the boundary S^2.

We refer to the problem (3.4.4)–(3.4.12) as the homogenized **model** $(\mathbb{HS})_8$.

3.4.2 Proof of Theorem 3.8

By Theorem 3.1 and the properties of the extension operator $\mathbb{E}_{\Omega_s^\varepsilon}$ the sequences $\{p^\varepsilon\}, \{(1-\zeta)\mathbf{w}_s^\varepsilon\}, \{(1-\zeta)\mathbb{D}(x, \mathbf{w}_s^\varepsilon)\}\left\{\dfrac{\partial \mathbf{w}^\varepsilon}{\partial t}\right\}, \left\{(1-\zeta)\dfrac{\partial \mathbf{w}_s^\varepsilon}{\partial t}\right\}, \left\{(1-\zeta)\mathbb{D}(x, \dfrac{\partial \mathbf{w}_s^\varepsilon}{\partial t})\right\}$, $\left\{\dfrac{\partial^2 \mathbf{w}^\varepsilon}{\partial t^2}\right\}$, and $\left\{\bar{\alpha}_\mu \nabla(\dfrac{\partial \mathbf{w}^\varepsilon}{\partial t})\right\}$ are bounded in $L_2(Q_T)$ and $\mathbf{L}_2(Q_T)$.

Hence there exists a subsequence of small parameters $\{\varepsilon > 0\}$ and functions p, $\mathbf{w}_s$, and $\mathbf{w}$ such that

$$p^\varepsilon \rightharpoonup p, \ \mathbf{v}^\varepsilon = \frac{\partial \mathbf{w}^\varepsilon}{\partial t} \rightharpoonup \frac{\partial \mathbf{w}}{\partial t} = \mathbf{v}, \ \frac{\partial \mathbf{v}^\varepsilon}{\partial t} \rightharpoonup \frac{\partial \mathbf{v}}{\partial t}$$

weakly in $L_2(Q_T)$ and $\mathbf{L}_2(Q_T)$ as $\varepsilon \searrow 0$, and

$$\mathbf{w}_s^\varepsilon \rightharpoonup \mathbf{w}_s \ \frac{\partial \mathbf{w}_s^\varepsilon}{\partial t} \rightharpoonup \frac{\partial \mathbf{w}_s}{\partial t}$$

weakly in $\mathbf{W}_2^{1,0}(\Omega_T)$ as $\varepsilon \searrow 0$.

As in Theorem 1.3 we conclude that $\mathbf{w}_s$ satisfies the boundary condition (3.4.6).

Note also that

$$\bar{\alpha}_\mu \mathbb{D}(x, \mathbf{w}^\varepsilon) \to 0$$

strongly in $\mathbf{L}_2(Q_T)$ as $\varepsilon \searrow 0$.

Relabeling if necessary, we assume that the sequences themselves converge.

By Nguetseng's theorem, there exist 1-periodic in $\mathbf{y}$ functions $P(\mathbf{x}, t, \mathbf{y})$, $\mathbf{U}(\mathbf{x}, t, \mathbf{y})$, and $\mathbf{V}(\mathbf{x}, t, \mathbf{y})$ such that

(1) $P \in L_2(Q_T \times Y)$, $\mathbf{U}, \ \mathbf{V}, \ \nabla_y \mathbf{U}, \ \nabla_y \mathbf{V}, \ \dfrac{\partial \mathbf{V}}{\partial t} \in \mathbf{L}_2(Q_T \times Y)$;

(2) the sequences $\{p^\varepsilon\}$, $\{\mathbf{v}^\varepsilon\}$, $\{\mathbf{w}_s^\varepsilon\}$, $\{\varepsilon\nabla\mathbf{v}^\varepsilon\}$, and $\{\nabla \mathbf{w}_s^\varepsilon\}$ converge two-scale in $L_2(Q_T)$ and $\mathbf{L}_2(Q_T)$ to P, $\mathbf{V}$, $\mathbf{w}_s(\mathbf{x}, t)$, $\nabla_y \mathbf{V}$, and $\big(\nabla \mathbf{w}_s + \nabla_y \mathbf{U}\big)$ respectively.

Lemma 3.12 *The limiting functions* $\mathbf{w}_s$, $\mathbf{v}$, *and* $\mathbf{w}_s$ *satisfy the macroscopic and microscopic continuity equations*

$$\nabla \cdot \mathbf{v} = 0, \ \ (\mathbf{x}, t) \in Q_T, \tag{3.4.14}$$

$$\big(1 - \chi(\mathbf{y})\big)(\nabla \cdot \mathbf{w}_s + \nabla_y \cdot \mathbf{U}) = 0, \ \ (\mathbf{x}, t) \in \Omega_T, \ \mathbf{y} \in Y, \tag{3.4.15}$$

and

$$\nabla_y \cdot \mathbf{V} = 0, \ \ (\mathbf{x}, t) \in \Omega_T, \ \mathbf{y} \in Y, \tag{3.4.16}$$

where

$$\mathbf{V}(\mathbf{x}, t, \mathbf{y}) = \big(\zeta + \chi(\mathbf{y})(1 - \zeta)\big)\mathbf{V}(\mathbf{x}, t, \mathbf{y}) + \big(1 - \chi(\mathbf{y})\big)\frac{\partial \mathbf{w}_s}{\partial t}(\mathbf{x}, t), \tag{3.4.17}$$

and

$$\mathbf{v} = \langle \mathbf{V} \rangle_Y = \zeta \mathbf{v} + (1 - \zeta)\langle \mathbf{V} \rangle_{Y_f} + (1 - m)\frac{\partial \mathbf{w}_s}{\partial t}(\mathbf{x}, t). \tag{3.4.18}$$

Lemma 3.13 *The following equality holds true*

$$(1 - \zeta)P(\mathbf{x}, t, \mathbf{y}) = (1 - \zeta)\big(P_s(\mathbf{x}, t, \mathbf{y}) + \chi(\mathbf{y})\, p_f(\mathbf{x}, t)\big), \tag{3.4.19}$$

where $P_s = \big(1 - \chi(\mathbf{y})\big)P(\mathbf{x}, t, \mathbf{y})$.

The proofs of these lemmas repeat proofs of Lemmas 3.8 and 1.4 respectively, with evident symmetric changes.

Lemma 3.14 *The limiting functions* $\mathbf{w}_s$, p, $\mathbf{V}_f$, *and* P *satisfy the macroscopic momentum balance equation*

$$\begin{aligned} &\rho_f\big(\zeta\frac{\partial \mathbf{v}}{\partial t} + (1 - \zeta)\frac{\partial}{\partial t}\langle \mathbf{V} \rangle_{Y_f}\big) + \rho_s(1 - m)\frac{\partial^2 \mathbf{w}_s}{\partial t^2} \\ &\quad = \nabla \cdot \Big(\lambda_0\big((1 - m)\,\mathbb{D}(x, \mathbf{w}_s) + \langle \mathbb{D}(y, \mathbf{U})\rangle_{Y_s}\big) - p\,\mathbb{I}\Big), \end{aligned} \tag{3.4.20}$$

in the domain Q_T, *the boundary condition (3.4.5), the initial condition (3.4.7) for the function* $\mathbf{v}$, *and the microscopic momentum balance equation*

$$\nabla_y \cdot \Big((1 - \chi)\lambda_0\big(\mathbb{D}(x, \mathbf{w}_s) + \mathbb{D}(y, \mathbf{U})\big) - \big(P_s + \chi\, p_f\big)\mathbb{I}\Big) = 0 \tag{3.4.21}$$

in the domain Y *for almost all* $(\mathbf{x}, t) \in \Omega_T$.

Proof Equation (3.4.20) follows from (3.1.14) after two-scale limit with test functions $\varphi = \varphi(\mathbf{x}, t)$:

$$\int_{Q_T}\Big(-\Big(\rho_f\big(\zeta\mathbf{v}+(1-\zeta)\langle\mathbf{V}\rangle_{Y_f}\big)+\rho_s(1-m)\frac{\partial\mathbf{w}_s}{\partial t}\Big)\cdot\frac{\partial\varphi}{\partial t} \tag{3.4.22}$$

$$+\big(\lambda_0\big((1-m)\,\mathbb{D}(x,\mathbf{w}_s)+\langle\mathbb{D}(y,\mathbf{U})\rangle_{Y_s}\big)-p\,\mathbb{I}\big):\mathbb{D}(x,\varphi)\Big)dxdt$$

$$=-\int_{Q_T}\nabla\cdot(\varphi\,p_0)dxdt. \tag{3.4.22}$$

This last integral identity in the form

$$\int_{Q_T}\Big(-\rho_f\,\mathbf{v}\cdot\frac{\partial\varphi}{\partial t}-p\,\nabla\cdot\varphi+\nabla\cdot(\varphi\,p_0)\Big)dxdt=0$$

for the finite in Ω^0 functions φ provides the boundary condition (3.4.5), initial condition (3.4.7) for the function $\mathbf{v}$, and an estimate

$$\int_{Q_T}\zeta|\nabla p(\mathbf{x},t)|^2dxdt\leqslant C_0\mathfrak{P}^2. \tag{3.4.23}$$

Equation (3.4.21) follows from (3.1.14) after the two-scale limit with test functions $\varphi=\varepsilon h(\mathbf{x},t)\varphi_0\left(\frac{\mathbf{x}}{\varepsilon}\right)$, where h is finite in Ω.

Lemma B.13 and the boundedness of the sequence $\left\{\bar{\alpha}_\mu\nabla\Big(\frac{\partial\mathbf{w}^\varepsilon}{\partial t}\Big)\right\}$ in $\mathbf{L}_2(Q_T)$ result in

$$\mathbf{V}(\mathbf{x},t,\mathbf{y})=\mathbf{v}(\mathbf{x},t). \tag{3.4.24}$$

Applying the two-scale limit to the equality

$$(1-\zeta)(1-\chi^\varepsilon)\Big(\mathbf{v}^\varepsilon-\frac{\partial\mathbf{w}_s^\varepsilon}{\partial t}\Big)=0,$$

we arrive at (3.4.2). Therefore, the function $\mathbf{v}$ satisfies the continuity equation (3.4.3) and the continuity equation (3.4.15) takes the form

$$\big(1-\chi(\mathbf{y})\big)\nabla_y\cdot\mathbf{U}=0,\ \ (\mathbf{x},t)\in\Omega_T,\ \mathbf{y}\in Y, \tag{3.4.25}$$

while the macroscopic momentum balance equation (3.4.22) becomes

$$\int_{Q_T}\Big(-\Big(\rho_f\zeta\mathbf{v}+\rho_s\frac{\partial\mathbf{w}_s}{\partial t}\Big)\cdot\frac{\partial\varphi}{\partial t}+\int_{Q_T}\nabla\cdot(\varphi\,p_0)dxdt \tag{3.4.26}$$

$$+\big(\lambda_0\big((1-m)\,\mathbb{D}(x,\mathbf{w}_s)+\langle\mathbb{D}(y,\mathbf{U})\rangle_{Y_s}\big)-p\,\mathbb{I}\big):\mathbb{D}(x,\varphi)\Big)dxdt=0, \tag{3.4.26}$$

which is equivalent to the differential equation

$$\rho_f\zeta\frac{\partial \mathbf{v}}{\partial t}+\rho_s\frac{\partial^2 \mathbf{w}_s}{\partial t^2}=\nabla\cdot\Big(\lambda_0\big((1-m)\mathbb{D}(x,\mathbf{w}_s)+\langle\mathbb{D}(y,\mathbf{U})\rangle_{Y_s}\big)-p\,\mathbb{I}\Big) \tag{3.4.27}$$

and initial condition (3.4.7) for the function $\mathbf{w}_s$.

Lemma 3.15 *The limiting functions* $\mathbf{v}$, $\mathbf{w}_s$, *and* p *satisfy in the domain* Q_T *the homogenized momentum balance equation (3.4.4).*

The proof of this lemma repeats the proof of the corresponding statement in Theorem 1.3.

The uniqueness of the problem (3.4.2)–(3.4.7) follows from the energy equality

$$\frac{1}{2}\frac{d}{dt}\int_{\Omega}\left((1-\zeta)\lambda_0\big(\mathfrak{N}_0^s:\mathbb{D}(x,\mathbf{w}_s)\big):\mathbb{D}(x,\mathbf{w}_s)+\rho_f\zeta|\mathbf{v}|^2+\hat{\rho}(1-\zeta)\left|\frac{\partial \mathbf{w}_s}{\partial t}\right|^2\right)dx=0$$

for the solution $\mathbf{w}_s$ and $\mathbf{v}$ of the homogeneous ($p_0=0$) problem. This equality is a result of multiplying equation (3.4.4) by $\mathbf{v}$ and integrating by parts over domain Q with the use of (3.4.2).

3.4.3 *Proof of Theorem 3.9*

3.4.3.1 The Case $\mu_1 > 0$.

As in the previous subsection we conclude that the sequences $\{p^\varepsilon\}$, $\{(1-\zeta)\chi^\varepsilon p^\varepsilon\}$, $\{\mathbf{w}_s^\varepsilon\}$, $\{\mathbb{D}(x,\mathbf{w}_s^\varepsilon)\}$, $\left\{\frac{\partial \mathbf{w}^\varepsilon}{\partial t}\right\}$, $\left\{\frac{\partial \mathbf{w}_s^\varepsilon}{\partial t}\right\}$, $\left\{\mathbb{D}\Big(x,\frac{\partial \mathbf{w}_s^\varepsilon}{\partial t}\Big)\right\}$ $\left\{\frac{\partial^2 \mathbf{w}^\varepsilon}{\partial t^2}\right\}$, and $\left\{\bar{\alpha}_\mu\nabla\Big(\frac{\partial \mathbf{w}^\varepsilon}{\partial t}\Big)\right\}$ are bounded in $L_2(Q_T)$ and $\mathbf{L}_2(Q_T)$.

Hence there exists a subsequence of small parameters $\{\varepsilon>0\}$ and functions p, p_f, $\mathbf{w}_s$, and $\mathbf{v}^{(f)}$ such that

$$p^\varepsilon\rightharpoonup p,\ (1-\zeta)\chi^\varepsilon p^\varepsilon\rightharpoonup(1-\zeta)(1-m)p_f,$$

$$\big(\zeta+(1-\zeta)\chi^\varepsilon\big)\frac{\partial \mathbf{w}^\varepsilon}{\partial t}\rightharpoonup\mathbf{v}^{(f)},\ \big(\zeta+(1-\zeta)\chi^\varepsilon\big)\frac{\partial^2 \mathbf{w}^\varepsilon}{\partial t^2}\rightharpoonup\frac{\partial \mathbf{v}^{(f)}}{\partial t}$$

weakly in $L_2(Q_T)$ and $\mathbf{L}_2(Q_T)$ as $\varepsilon\searrow 0$, and

$$\mathbf{w}_s^\varepsilon\rightharpoonup\mathbf{w}_s,\quad\frac{\partial \mathbf{w}_s^\varepsilon}{\partial t}\rightharpoonup\frac{\partial \mathbf{w}_s}{\partial t}$$

weakly in $\mathbf{W}_2^{1,0}(\Omega_T)$, and $\mathbf{w}_s\in\overset{\circ}{\mathbf{W}}{}_2^{1,0}(\Omega_T)$.

We also have that

$$\bar{\alpha}_\mu \mathbb{D}(x, \mathbf{w}^\varepsilon) \to 0 \tag{3.4.28}$$

strongly in $\mathbf{L}_2(Q_T)$ as $\varepsilon \searrow 0$.

At the same time there exist 1-periodic in $\mathbf{y}$ functions $P(\mathbf{x}, t, \mathbf{y})$, $\mathbf{U}(\mathbf{x}, t, \mathbf{y})$, and $\mathbf{V}(\mathbf{x}, t, \mathbf{y})$ such that

(1) $P \in L_2(Q_T \times Y), \mathbf{U},\ \mathbf{V},\ \nabla_y \mathbf{U},\ \nabla_y \mathbf{V},\ \dfrac{\partial \mathbf{V}}{\partial t} \in \mathbf{L}_2(Q_T \times Y)$;
(2) the sequences $\{p^\varepsilon\}, \{\mathbf{v}^\varepsilon\}, \{\mathbf{w}_s^\varepsilon\}, \{\varepsilon \nabla \mathbf{v}^\varepsilon\}$, and $\{\nabla \mathbf{w}_s^\varepsilon\}$ converge two-scale in $L_2(Q_T)$ and $\mathbf{L}_2(Q_T)$ to P, $\mathbf{V}$, $\mathbf{w}_s(\mathbf{x}, t)$, $\nabla_y \mathbf{V}$, and $\big(\nabla\, \mathbf{w}_s + \nabla_y \mathbf{U}\big)$ respectively.

For all these functions hold true statements of Lemma 3.12, Lemma 3.13, and Lemma 3.14.

To find $\langle \mathbb{D}(y, \mathbf{U})\rangle_{Y_s}$ in (3.4.20) we have to solve the system (3.4.15) and (3.4.21) in the domain Y for almost all $(\mathbf{x}, t) \in \Omega_T$. This system is already studied in Theorem 1.4. Therefore the functions $\mathbf{w}_s$, $\mathbf{w}^{(f)}$, and p_f satisfy the continuity equation (3.4.8), the homogenized momentum balance equation (3.4.9), and boundary conditions (3.4.5) and (3.4.11).

In fact, the boundary condition (3.4.11) follows from the continuity equation in the form

$$\int_{Q_T} \mathbf{w} \cdot \nabla \psi \, dx dt = 0 \tag{3.4.29}$$

for any smooth functions ψ and the boundary condition (3.4.6). The validity of the boundary condition (3.4.5) has been shown in Theorem 3.8.

It is easy to see that the sequence $\left\{\big(\zeta + \chi^\varepsilon(1 - \zeta)\big)\dfrac{\partial \mathbf{w}^\varepsilon}{\partial t}\right\}$ converges two-scale in $\mathbf{L}_2(Q_T)$ to the function $\mathbf{V}^{(f)} = \big(\zeta + (1 - \zeta)\chi(\mathbf{y})\big)\mathbf{V}$.

To derive the homogenized momentum balance equation (3.4.10) for the liquid component we pass to the limit in (3.1.14) as $\varepsilon \searrow 0$ with test functions $\varphi = h(\mathbf{x}, t)\varphi_0\Big(\dfrac{\mathbf{x}}{\varepsilon}\Big)$, where $h(\mathbf{x}, t)$ is finite in Q for all $t \in (0, T)$, and 1-periodic smooth function $\varphi_0(\mathbf{y})$ is divergence free and finite in Y_f:

$$\int_{Q_T} \Big(a\, h + \big(\zeta\, p + (1 - \zeta) p_f\big)\, \mathbf{b} \cdot \nabla\, h\Big) dx dt = 0,$$

where

$$a(\mathbf{x}, t) = \int_{Y_f} \Big(\rho_f \frac{\partial \mathbf{V}}{\partial t} \cdot \varphi_0 + \mu_1 \mathbb{D}(y, \mathbf{V}) : \mathbb{D}(y, \varphi_0)\Big) dy \in L_2(Q_T),$$

$$\mathbf{b} = -\int_{Y_f} \varphi_0 dy = const.$$

Just as in the proof of Theorem 1.1 we conclude that

$$\nabla\left(\zeta\, p + (1-\zeta)p_f\right) \in L_2(Q_T).$$

For $\mathbf{x} \in \Omega$ one has for the function $\mathbf{V}^{(f)}$ the microscopic momentum balance equation

$$\rho_f \frac{\partial \mathbf{V}^{(f)}}{\partial t} = \frac{\mu_1}{2} \triangle \mathbf{V}^{(f)} - \nabla_y \Pi^{(f)} - \nabla\, p_f \tag{3.4.30}$$

in the domain $Y_f \times (0, T)$ and initial conditions

$$\mathbf{V}^{(f)}(\mathbf{x}, 0, \mathbf{y}) = 0, \;\; \mathbf{y} \in Y_f \tag{3.4.31}$$

for almost all $\mathbf{x} \in \Omega_T$.

The relation (3.4.17) and the smoothness of the function $\mathbf{V}$: $\mathbf{V}, \nabla_y \mathbf{V} \in \mathbf{L}_2$ $(\Omega_T \times Y)$, imply the boundary condition

$$\mathbf{V}^{(f)}(\mathbf{x}, t, \mathbf{y}) = \frac{\partial \mathbf{w}_s}{\partial t}(\mathbf{x}, t), \;\; (\mathbf{y}, t) \in \gamma \times (0, T) \tag{3.4.32}$$

for almost all $\mathbf{x} \in \Omega_T$.

We complete (3.4.30)–(3.4.32) with continuity equation (3.4.16) in the form

$$\nabla_y \cdot \mathbf{V}^{(f)} = 0, \;\; \mathbf{y} \in Y_f, \; (\mathbf{x}, t) \in \Omega_T. \tag{3.4.33}$$

This problem has been already studied in the proof of the Theorem 3.5. Therefore, the limiting functions $\mathbf{w}_s$, $\mathbf{w}^{(f)}$, and p_f satisfy the homogenized momentum balance equation (3.4.10).

3.4.3.2 The Case $\mu_1 = 0$.

The proof of this case completely repeats the proof of the previous case $\mu_1 > 0$, except for the derivation of the homogenized momentum balance equation for the liquid component. Here we only have to repeat the proof of Theorem 3.5 for a similar case $\mu_1 = 0$.

3.5 A Viscous Liquid in an Elastic Skeleton

Throughout this section we assume that

$$\bar{\alpha}_\lambda = \lambda_0, \;\; \bar{\alpha}_\mu = \mu_0, \;\; 0 < \lambda_0, \;\; \mu_0 < \infty. \tag{3.5.1}$$

3.5.1 Main Results

Theorem 3.10 *Let $\{\mathbf{w}^{\varepsilon}, p^{\varepsilon}\}$ be the weak solution of the problem (3.1.4)–(3.1.13). Then*

(1) sequences $\{\mathbf{w}^{\varepsilon}\}$ and $\left\{\dfrac{\partial \mathbf{w}^{\varepsilon}}{\partial t}\right\}$ converge weakly in $\overset{\circ}{\mathbf{W}}{}_2^{1,0}(\Omega_T)$ to the functions $\mathbf{w}$ and $\mathbf{v}$ respectively, sequences $\left\{\dfrac{\partial^2 \mathbf{w}^{\varepsilon}}{\partial t^2}\right\}$ and $\{p^{\varepsilon}\}$ converge weakly in $\mathbf{L}_2(Q_T)$ and $L_2(Q_T)$ to functions $\dfrac{\partial^2 \mathbf{w}}{\partial t^2}$ and p respectively;

(2) limiting functions $\mathbf{w}$, $\mathbf{v}$, and p solve the system of homogenized equations in the domain Q_T, consisting of the continuity equation

$$\nabla \cdot \mathbf{w} = 0, \tag{3.5.2}$$

and the homogenized momentum balance equation

$$\rho(\mathbf{x})\frac{\partial^2 \mathbf{w}}{\partial t^2} + \nabla p = \nabla \cdot \widehat{\mathbb{P}}, \tag{3.5.3}$$

$$\begin{aligned}\widehat{\mathbb{P}} &= \zeta\, \mu_0\, \mathbb{D}(x, \mathbf{v}) \\ &\quad + (1-\zeta)\left(\mathfrak{N}_1 : \mathbb{D}(x, \mathbf{v}) + \mathfrak{N}_2 : \mathbb{D}(x, \mathbf{w}) + \int_0^t \mathfrak{N}_3(t-\tau) : \mathbb{D}(x, \mathbf{w}(\mathbf{x}, \tau))d\tau,\right.\end{aligned}$$

completed with the boundary and initial conditions

$$\widehat{\mathbb{P}} \cdot \mathbf{e}_3 = -p_0 \mathbf{e}_3, \ \ \mathbf{x} \in S_T^1, \tag{3.5.4}$$

$$\mathbf{w}(\mathbf{x}, t) = 0, \ \ \mathbf{x} \in S_T^2, \tag{3.5.5}$$

$$\mathbf{w}(\mathbf{x}, 0) = \mathbf{v}(\mathbf{x}, 0) = 0, \ \ \mathbf{x} \in Q; \tag{3.5.6}$$

(3) the problem (3.5.2)–(3.5.6) has a unique solution.
In (3.5.3)

$$\rho(\mathbf{x}) = \Big(\zeta(\mathbf{x}) + \big(1-\zeta(\mathbf{x})\big)m\Big)\rho_f + \big(1-\zeta(\mathbf{x})\big)(1-m)\rho_s,$$

and fourth-rank tensors $\mathfrak{N}_1$, $\mathfrak{N}_1$, and $\mathfrak{N}_3(t)$ are defined in Theorem 1.11.

We refer to the problem (3.5.2)–(3.5.6) as the homogenized **model** $(\mathbb{HS})_9$.

3.5.2 Proof of Theorem 3.10

By Theorem 3.1 the sequences $\{p^\varepsilon\}$, $\{\mathbf{w}^\varepsilon\}$, $\left\{\dfrac{\partial \mathbf{w}^\varepsilon}{\partial t}\right\}$, $\left\{\dfrac{\partial^2 \mathbf{w}^\varepsilon}{\partial t^2}\right\}$, $\{\mathbb{D}(x, \mathbf{w}^\varepsilon)\}$, and $\left\{\mathbb{D}\Big(x, \dfrac{\partial \mathbf{w}^\varepsilon}{\partial t}\Big)\right\}$ are bounded in $L_2(Q_T)$ and $\mathbf{L}_2(Q_T)$.

Hence there exists a subsequence of small parameters $\{\varepsilon > 0\}$ and functions p, $\mathbf{w}$, and $\mathbf{v}$ such that

$$p^\varepsilon \rightharpoonup p, \quad \frac{\partial^2 \mathbf{w}^\varepsilon}{\partial t^2} \rightharpoonup \frac{\partial^2 \mathbf{w}}{\partial t^2}$$

weakly in $L_2(Q_T)$ and $\mathbf{L}_2(Q_T)$ as $\varepsilon \searrow 0$, and

$$\mathbf{w}^\varepsilon \rightharpoonup \mathbf{w} \quad \frac{\partial \mathbf{w}^\varepsilon}{\partial t} \rightharpoonup \mathbf{v} = \frac{\partial \mathbf{w}}{\partial t}$$

weakly in $\overset{\circ}{\mathbf{W}}{}_2^{1,0}(\Omega_T)$ as $\varepsilon \searrow 0$.

Relabeling if necessary, we assume that the sequences themselves converge.

By Nguetseng's theorem, there exist functions $P(\mathbf{x}, t, \mathbf{y})$ from $L_2(Q_T \times Y)$, and $\mathbf{W}(\mathbf{x}, t, \mathbf{y})$ from $\mathbf{L}_2\big(Q_T; \mathbf{W}_2^1(Y)\big)$ that are 1-periodic in $\mathbf{y}$ and satisfy the condition that the sequences $\{p^\varepsilon\}$, $\{\mathbf{w}^\varepsilon\}$, $\left\{\dfrac{\partial \mathbf{w}^\varepsilon}{\partial t}\right\}$, $\{\nabla \mathbf{w}^\varepsilon\}$, and $\left\{\nabla\Big(\dfrac{\partial \mathbf{w}^\varepsilon}{\partial t}\Big)\right\}$ converge two-scale in $L_2(Q_T)$ and $\mathbf{L}_2(Q_T)$ to P, $\mathbf{w}(\mathbf{x}, t)$, $\mathbf{v}(\mathbf{x}, t)$, $\big(\nabla\, \mathbf{w} + \nabla_y \mathbf{W}\big)$, and $\left(\nabla\, \mathbf{v} + \nabla_y \left(\dfrac{\partial \mathbf{W}}{\partial t}\right)\right)$ respectively.

Lemma 3.16 *The limiting functions* $\mathbf{w}$ *and* $\mathbf{W}$ *satisfy the macroscopic and microscopic continuity equations*

$$\nabla \cdot \mathbf{w} = 0, \ \ (\mathbf{x}, t) \in Q_T, \tag{3.5.7}$$

$$\nabla_y \cdot \mathbf{W} = 0, \ \ (\mathbf{x}, t) \in Q_T, \ \mathbf{y} \in Y. \tag{3.5.8}$$

The proof of this lemma is straightforward (see also the proof of Theorem 1.11).

Lemma 3.17 *The limiting functions* $\mathbf{w}$, p, $\mathbf{W}$, *and* P *satisfy the macroscopic momentum balance equation*

$$\begin{aligned}
&\rho(\mathbf{x})\frac{\partial^2 \mathbf{w}}{\partial t^2} + \nabla p - \nabla \cdot \Big(\zeta\, \mu_0\, \mathbb{D}(x, \mathbf{v})\Big) \\
&= \nabla \cdot \big((1-\zeta)\big(\mu_0\, m\, \mathbb{D}(x, \mathbf{v}) + \lambda_0\, (1-m)\, \mathbb{D}(x, \mathbf{w})\big)\big) \\
&\quad + \nabla \cdot \left((1-\zeta)\left(\mu_0 \left\langle \mathbb{D}\left(y, \frac{\partial \mathbf{W}}{\partial t}\right)\right\rangle_{Y_f}\right) + \lambda_0\, \langle \mathbb{D}(y, \mathbf{W})\rangle_{Y_s}\right),
\end{aligned} \tag{3.5.9}$$

in the domain Q_T, the boundary condition (3.5.4), the initial condition (3.5.6), and the microscopic momentum balance equation

$$\nabla_y \cdot \Big(\mu_0\, \chi(\mathbf{y})\, \mathbb{D}\Big(y, \frac{\partial \mathbf{W}}{\partial t}\Big) + \lambda_0\, \big(1 - \chi(\mathbf{y})\big)\mathbb{D}(y, \mathbf{W}) - P\mathbb{I}\Big)$$
$$= \nabla_y \cdot \Big(\mu_0\, \chi(\mathbf{y})\, \mathbb{D}(x, \mathbf{v}) + \lambda_0\, \big(1 - \chi(\mathbf{y})\big)\, \mathbb{D}(x, \mathbf{w})\Big). \tag{3.5.10}$$

in the domain Y for almost all $(\mathbf{x}, t) \in \Omega_T$.

Proof Equation (3.5.9) follows from (3.1.14) after the two-scale limit with test functions $\varphi = \varphi(\mathbf{x}, t)$:

$$\int_{Q_T} \Big(\frac{\partial \mathbf{w}}{\partial t} \cdot \frac{\partial \varphi}{\partial t} - \nabla \cdot (\varphi\, p_0) + \big(p\,\mathbb{I} - \zeta\, \mu_0\, \mathbb{D}(x, \mathbf{v})\big) : \mathbb{D}(x, \varphi)\Big) dxdt$$
$$= \int_{\Omega_T} \big(\mu_0\, m\, \mathbb{D}(x, \mathbf{v}) + \lambda_0\, (1 - m)\, \mathbb{D}(x, \mathbf{w})\big) : \mathbb{D}(x, \varphi) dxdt$$
$$+ \int_{\Omega_T} \big(\mu_0\, \langle \mathbb{D}(y, \frac{\partial \mathbf{W}}{\partial t})\rangle_{Y_f}\big) + \lambda_0\, \langle \mathbb{D}(y, \mathbf{W})\rangle_{Y_s}\big) : \mathbb{D}(x, \varphi) dxdt. \tag{3.5.11}$$

This last integral identity provides the boundary condition (3.5.4) and initial condition (3.5.6).

Equation (3.5.10) follows from (3.1.14) after the two-scale limit with test functions $\varphi = \varepsilon h(\mathbf{x}, t)\varphi_0\Big(\dfrac{\mathbf{x}}{\varepsilon}\Big)$, where h is finite in Ω.

To derive the homogenized momentum balance equation (3.5.6) we simply have to solve the periodic problem (3.5.8), (3.5.10) in the domain Y_T, to calculate terms $\Big\langle \mathbb{D}\Big(y, \dfrac{\partial \mathbf{W}}{\partial t}\Big)\Big\rangle_{Y_f}$ and $\langle \mathbb{D}(y, \mathbf{W})\rangle_{Y_s}$, and substitute these expressions in (3.5.9). But all these steps have already been taken in the proof of Theorem 1.11.

The uniqueness of the problem (3.5.2)–(3.4.6) follows from the energy equality

$$\frac{1}{2} \int_\Omega \rho(\mathbf{x})|\mathbf{v}(\mathbf{x}, t_0)|^2 dx + \int_0^{t_0} \int_\Omega \big(\mathfrak{N}_1 : \mathbb{D}(x, \mathbf{v})\big) : \mathbb{D}(x, \mathbf{v}) dxdt$$
$$= -\frac{1}{2} \int_\Omega \Big(\mathfrak{N}_2 : \mathbb{D}\big(x, \mathbf{w}(\mathbf{x}, t_0)\big)\Big) : \mathbb{D}(x, \mathbf{w}(\mathbf{x}, t_0)) dx$$
$$- \int_0^{t_0} \int_\Omega \mathbb{D}\big(x, \mathbf{v}(\mathbf{x}, t)\big) : \Big(\int_0^t \mathfrak{N}_3(t - \tau) : \mathbb{D}\big(x, \mathbf{w}(\mathbf{x}, \tau)\big) d\tau\Big) dxdt. \tag{3.5.12}$$

for the solution $\mathbf{v} = \dfrac{\partial \mathbf{w}}{\partial t}$ of the homogeneous ($p_0 = 0$) problem.

In fact, the representation

$$\int_{\Omega} |\mathbb{D}(x, \mathbf{w}(\mathbf{x}, t)|^2 dx = 2 \int_0^t \int_{\Omega} \mathbb{D}(x, \mathbf{w}(\mathbf{x}, \tau)) : \mathbb{D}(x, \mathbf{v}(\mathbf{x}, \tau)) dx d\tau$$

implies

$$\begin{aligned} \int_{\Omega} |\mathbb{D}(x, \mathbf{w}(\mathbf{x}, t_0)|^2 dx &\leqslant 4t_0 \int_0^{t_0} \int_{\Omega} |\mathbb{D}(x, \mathbf{v})|^2 dx dt \\ &\leqslant 4t_* \int_0^{t_0} \int_{\Omega} |\mathbb{D}(x, \mathbf{v})|^2 dx dt \end{aligned} \tag{3.5.13}$$

for $t_0 < t_*$.

Therefore, (3.5.12), (3.5.13), and the strict positive definiteness of $\mathfrak{N}_1$ result in

$$\int_0^{t_0} \int_{\Omega} \left|\mathbb{D}(x, \mathbf{v})\right|^2 dx dt \leqslant t_* C_0 \int_0^{t_0} \int_{\Omega} |\mathbb{D}(x, \mathbf{v})|^2 dx dt$$

for $t_0 < t_*$. Choosing $t_* < \frac{1}{2} C_0$ we obtain $\mathbf{v} = 0$ in Ω_{t_*}. Repeating once more we will prove our statement for a finite number of steps.

Chapter 4
Double Porosity Models for a Liquid Filtration

The liquid domain Ω_f^δ, which is a subdomain of a bounded domain Ω with a Lipschitz continuous boundary $S = \partial\Omega$, is defined in the following way. Let K be a unit cube, $K = Z_f \cup Z_s \cup \gamma_c$, where Z_f and Z_s are open sets, the common boundary $\gamma_c = \partial Z_f \cap \partial Z_s$ is a Lipschitz continuous surface, and a periodic repetition in $\mathbb{R}^3$ of the domain Z_s is a connected domain with a Lipschitz continuous boundary. The elementary cell Z_f models a fracture space Ω_c^δ : the domain Ω_c^δ is the intersection of the cube Ω with a periodic repetition in $\mathbb{R}^3$ of the elementary cell δZ_f. In the same way we define the pore space Ω_p^ε: $K = Y_f \cup Y_s \cup \gamma_p$, γ_p is a Lipschitz continuous surface, a periodic repetition in $\mathbb{R}^3$ of the domain Y_s is a connected domain with a Lipschitz continuous boundary, and Ω_p^ε is the intersection of $\Omega \backslash \Omega_c^\delta$ with a periodic repetition in $\mathbb{R}^3$ of the elementary cell εY_f. Finally, we put $\Omega_f^\delta = \Omega_p^\varepsilon \cup \Omega_c^\delta$, $\Omega_s^\delta = \Omega \backslash \overline{\Omega_f^\delta}$ is the solid skeleton, and $\Gamma^\delta = \partial\Omega_s^\delta \cap \partial\Omega_f^\delta$ is the "solid skeleton–liquid domain" interface.

We also may characterize liquid and solid domains using indicator functions in Ω. Let $\varsigma(\mathbf{x})$ be the indicator function of the domain Ω in $\mathbb{R}^3$, that is $\varsigma(\mathbf{x}) = 1$ if $\mathbf{x} \in \Omega$ and $\varsigma(\mathbf{x}) = 0$ if $\mathbf{x} \in \mathbb{R}^3 \backslash \Omega$. Let also $\chi_p(\mathbf{y})$ be the 1-periodic extension of the indicator function of the domain Y_f in K and $\chi_c(\mathbf{z})$ be the 1-periodic extension of the indicator function of the domain Z_f in K. Then $\chi_c^\delta(\mathbf{x}) = \varsigma(\mathbf{x})\chi_c(\frac{\mathbf{x}}{\delta})$ stands for the indicator function of the domain Ω_c^δ (fracture space), $\chi_p^\delta(\mathbf{x}) = \varsigma(\mathbf{x})(1 - \chi_c(\frac{\mathbf{x}}{\delta}))\chi_p(\frac{\mathbf{x}}{\varepsilon})$ stands for the indicator function of the domain Ω_p^ε (pore space) and

$$\tilde{\chi}^\delta(\mathbf{x}) = \varsigma(\mathbf{x})\tilde{\chi}\left(\frac{\mathbf{x}}{\varepsilon}, \frac{\mathbf{x}}{\delta}\right), \quad \tilde{\chi}(\mathbf{y}, \mathbf{z}) = \chi_c(\mathbf{z}) + \big(1 - \chi_c(\mathbf{z})\big)\chi_p(\mathbf{y})$$

stands for the indicator function of the liquid domain Ω_f^δ.

Let us call such a geometry a *double porosity geometry* and corresponding mathematical model a *double porosity model*. In this chapter we consider the motion of an incompressible liquid in an incompressible elastic skeleton, and the motion of a compressible liquid in a compressible elastic skeleton (Figs. 4.1, 4.2 and 4.3).

A. Meirmanov, *Mathematical Models for Poroelastic Flows*,
Atlantis Studies in Differential Equations 1, DOI: 10.2991/978-94-6239-015-7_4,

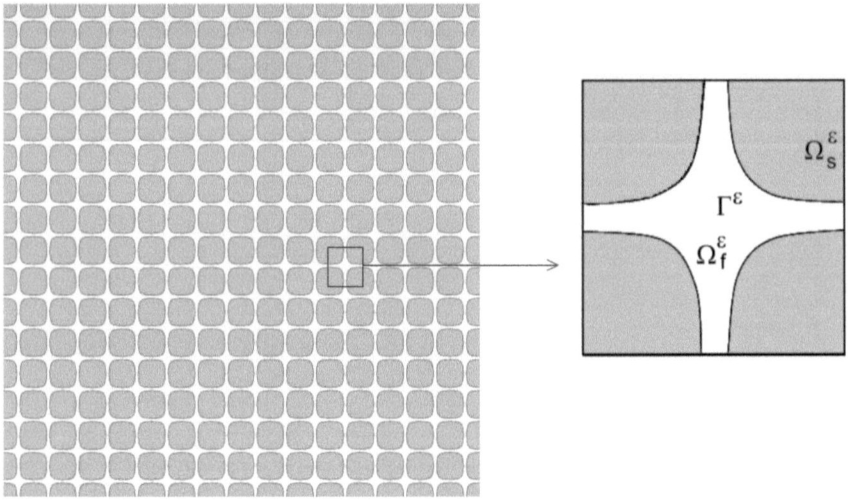

Fig. 4.1 Single porosity geometry

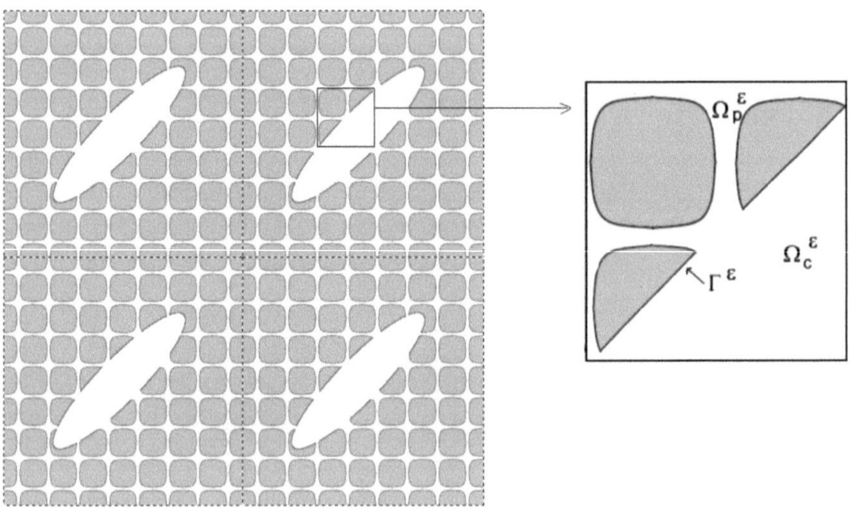

Fig. 4.2 Double porosity geometry: isolated fractures

We consider the model $\mathbb{M}_{15}$ as the basic mathematical model at the microscopic level for an incompressible liquid in an incompressible elastic skeleton:

$$\nabla \cdot \mathbf{w} = 0, \ (\mathbf{x}, t) \in \Omega_T = \Omega \times (0, T), \tag{4.0.1}$$

$$\nabla \cdot \mathbb{P} + \rho^{\delta} \mathbf{F} = 0, \ \ (\mathbf{x}, t) \in \Omega_T, \tag{4.0.2}$$

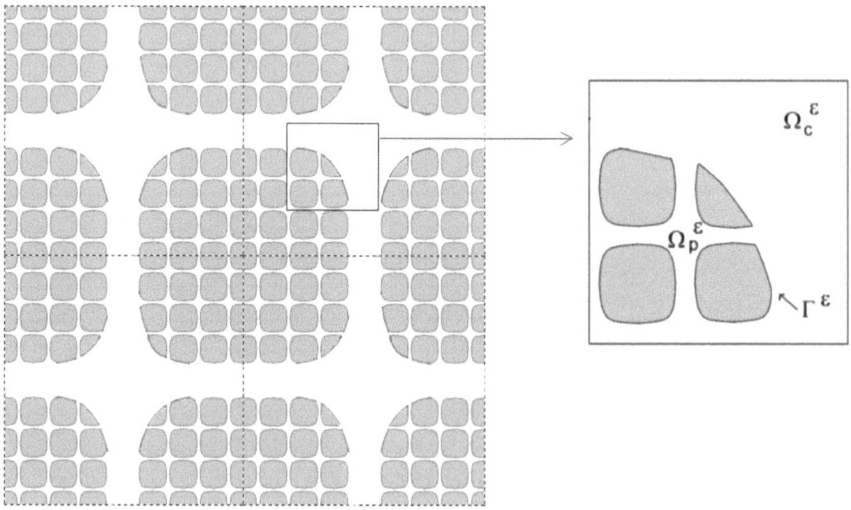

Fig. 4.3 Double porosity geometry: connected fracture space

$$\mathbb{P} = \tilde{\chi}^{\delta} \alpha_{\mu} \mathbb{D}\Big(x, \frac{\partial \mathbf{w}}{\partial t}\Big) + (1 - \tilde{\chi}^{\delta}) \lambda_0 \, \mathbb{D}(x, \mathbf{w}) - q \, \mathbb{I}, \tag{4.0.3}$$

$$\mathbf{w}(\mathbf{x}, t) = 0, \ \ (\mathbf{x}, t) \in \Omega_T, \tag{4.0.4}$$

$$\int_{\Omega} q(\mathbf{x}, t) dx = 0, \ t \in (0, T), \ \ \tilde{\chi}^{\delta} \, \mathbf{w}(\mathbf{x}, 0) = 0, \ \ \mathbf{x} \in \Omega, \tag{4.0.5}$$

where

$$\rho^{\delta} = \rho_f \, \tilde{\chi}^{\delta} + \rho_s \, (1 - \tilde{\chi}^{\delta}).$$

For the motion of a compressible liquid in a compressible elastic skeleton as a basic mathematical model at the microscopic level we consider the model $\mathbb{M}_{14}$

$$\frac{1}{\alpha_q^{\delta}} q + \nabla \cdot \mathbf{w} = 0, \ \ (\mathbf{x}, t) \in \Omega_T, \tag{4.0.6}$$

$$\nabla \cdot \mathbb{P} + \rho^{\delta} \mathbf{F} = 0, \ \ (\mathbf{x}, t) \in \Omega_T, \tag{4.0.7}$$

$$\mathbb{P} = \tilde{\chi}^{\delta} \alpha_{\mu} \mathbb{D}\Big(x, \frac{\partial \mathbf{w}}{\partial t}\Big) + (1 - \tilde{\chi}^{\delta}) \lambda_0 \, \mathbb{D}(x, \mathbf{w}) - q \, \mathbb{I}, \tag{4.0.8}$$

$$\mathbf{w}(\mathbf{x}, t) = 0, \ \ \mathbf{x} \in S, \ \ t \in (0, T), \tag{4.0.9}$$

$$\tilde{\chi}^{\delta} \, \mathbf{w}(\mathbf{x}, 0) = 0, \ \ \mathbf{x} \in \Omega, \tag{4.0.10}$$

where

$$\alpha_q^\delta = c_f^2\,\tilde{\chi}^{\,\delta} + c_s^2\,(1-\tilde{\chi}^{\,\delta}).$$

4.1 Main Results

Definition 4.1 We say that the pair of functions $\{\mathbf{w}^{\,\delta},\ q^{\,\delta}\}$, such that

$$\mathbf{w}^\delta \in \overset{\circ}{\mathbf{W}}{}_2^{1,0}(\Omega_T),\quad \frac{\partial \mathbf{w}^{\,\delta}}{\partial t} \in \mathbf{L}_2\big((0,T);\overset{\circ}{\mathbf{W}}{}_2^1(\Omega_f^\delta)\big),\quad q^{\,\delta} \in L_2\big(\Omega_T\big),$$

is a generalized solution to the problem (4.0.1)–(4.0.5), if it satisfies the continuity equation (4.0.1) in the usual sense almost everywhere in $\Omega \times (0,T)$, normalization and initial conditions (4.0.5), and the integral identity

$$\int_0^T\!\!\int_\Omega \left(\alpha_\mu \tilde{\chi}^\delta \mathbb{D}\left(x, \frac{\partial \mathbf{w}^\delta}{\partial t}\right) + \lambda_0\,(1-\tilde{\chi}^\delta)\mathbb{D}(x,\mathbf{w}^\delta)\right) : \mathbb{D}(x,\varphi)dxdt \tag{4.1.1}$$
$$= \int_0^T\!\!\int_\Omega \big(\rho^\delta \mathbf{F}\cdot\varphi + q^\delta\,(\nabla\cdot\varphi)\big)\,dxdt$$

for any vector-functions $\varphi \in \mathbf{L}_2\big((0,T);\overset{\circ}{\mathbf{W}}{}_2^1(\Omega)\big)$.

The homogeneous boundary condition (4.0.4) is already included into the corresponding functional space.

Definition 4.2 We say, that the pair of functions $\{\mathbf{w}^{\,\delta},\ q^{\,\delta}\}$, such that

$$\mathbf{w}^\delta \in \overset{\circ}{\mathbf{W}}{}_2^{1,0}(\Omega_T),\quad \frac{\partial \mathbf{w}^{\,\delta}}{\partial t} \in \mathbf{L}_2\big((0,T);\overset{\circ}{\mathbf{W}}{}_2^1(\Omega_f^\delta)\big),\quad q^{\,\delta} \in L_2\big(\Omega_T\big),$$

is a generalized solution to the problem (4.0.6)–(4.0.10), if it satisfies the continuity equation (4.0.6) in the usual sense almost everywhere in $\Omega \times (0,T)$, the initial condition (4.0.10), and the integral identity (4.1.1) for any vector-functions $\varphi \in \mathbf{L}_2\big((0,T);\overset{\circ}{\mathbf{W}}{}_2^1(\Omega)\big)$.

As before, the homogeneous boundary condition (4.0.9) is already included into the corresponding functional space.

Let

$$\lim_{\delta\searrow 0}\alpha_\mu(\delta) = 0,\quad \lim_{\delta\searrow 0}\frac{\alpha_\mu}{\delta^2} = \mu_1,\quad \lim_{\delta\searrow 0}\frac{\alpha_\mu}{\varepsilon^2} = \mu_2,$$

Throughout this chapter we impose Assumption 1.1, and suppose that

$$0 < \lambda_0,\ c_f^2,\ c_s^2 < \infty,\ 1 < r < \infty,$$

$$\int_0^T \int_\Omega \left(|\mathbf{F}|^2 + \left| \frac{\partial \mathbf{F}}{\partial t} \right|^2 + \left| \frac{\partial^2 \mathbf{F}}{\partial t^2} \right|^2 \right) dxdt = F^2 < \infty.$$

To formulate the existence and uniqueness results we need to extend the function $\mathbf{w}^\delta$ from Ω_s^δ to Ω: $\mathbf{w}_s^\varepsilon = \mathbb{E}_{\Omega_s^\delta}(\mathbf{w}^\varepsilon)$ (for the definition of this extension see Appendix B):

$$\int_\Omega |\mathbf{w}_s^\delta|^2 dx \le C \int_{\Omega_s^\delta} |\mathbf{w}^\delta|^2 dx, \quad \int_\Omega |\mathbb{D}(x, \mathbf{w}_s^\delta)|^2 dx \le C \int_{\Omega_s^\delta} |\mathbb{D}(x, \mathbf{w}^\delta)|^2 dx,$$

where C is independent of δ and t.

Theorem 4.1 *There exists a sufficiently small $\delta_0 > 0$, such that for any $0 < \delta < \delta_0$ and for any $0 < t < T$ problems (4.0.1)–(4.0.5) and (4.0.6)–(4.0.10) have unique generalized solutions and the following estimates hold true*

$$\int_\Omega |\mathbf{w}^\delta(\mathbf{x}, t)|^2 dx + \alpha_\mu \int_{\Omega_f^\delta} \left|\mathbb{D}\big(x, \mathbf{w}^\delta(\mathbf{x}, t)\big)\right|^2 dx + \int_{\Omega_s^\delta} \left|\mathbb{D}\big(x, \mathbf{w}_s^\delta(\mathbf{x}, t)\big)\right|^2 dx \leqslant CF^2, \tag{4.1.2}$$

$$\int_\Omega \left|\mathbf{v}^\delta(\mathbf{x}, t)\right|^2 dx + \alpha_\mu \int_{\Omega_f^\delta} \left|\mathbb{D}\big(x, \mathbf{v}^\delta(\mathbf{x}, t)\big)\right|^2 dx + \int_{\Omega_s^\delta} \left|\mathbb{D}\left(x, \frac{\partial \mathbf{w}_s^\delta}{\partial t}(\mathbf{x}, t)\right)\right|^2 dx \leqslant CF^2, \tag{4.1.3}$$

$$\int_\Omega \left|q^\delta(\mathbf{x}, t)\right|^2 dx \leqslant CF^2, \tag{4.1.4}$$

$$\frac{\alpha_\mu}{\varepsilon^2} \int_{\Omega_p^\varepsilon} \left|(\mathbf{w}^\delta - \mathbf{w}_s^\delta)(\mathbf{x}, t)\right|^2 dx + \frac{\alpha_\mu}{\delta^2} \int_{\Omega_c^\delta} \left|(\mathbf{w}^\delta - \mathbf{w}_s^\delta)(\mathbf{x}, t)\right|^2 dx \leqslant CF^2, \tag{4.1.5}$$

$$\frac{\alpha_\mu}{\varepsilon^2} \int_{\Omega_p^\varepsilon} \left|\left(\mathbf{v}^\delta - \frac{\partial \mathbf{w}_s^\delta}{\partial t}\right)(\mathbf{x}, t)\right|^2 dx + \frac{\alpha_\mu}{\delta^2} \int_{\Omega_c^\delta} \left|\left(\mathbf{v}^\delta - \frac{\partial \mathbf{w}_s^\delta}{\partial t}\right)(\mathbf{x}, t)\right|^2 dx \leqslant CF^2, \tag{4.1.6}$$

where $\mathbf{v}^\delta = \dfrac{\partial \mathbf{w}^\delta}{\partial t}$ *and* C *is independent of* δ *and* t.

Theorem 4.2 *Under the conditions of Theorem 4.1 let* $\{\mathbf{w}^\delta,\ q^\delta\}$ *be a solution to the problem (4.0.1)–(4.0.5) and* $\mu_2 = \infty$, *or the fracture space is disconnected (isolated fractures).*

Then there exist a subsequence of small parameters $\{\delta > 0\}$, *and functions* $\mathbf{v}_p \in \mathbf{L}_2(\Omega_T)$*—the limiting velocity of the liquid in pores,* $\mathbf{v}_c \in \mathbf{L}_2(\Omega_T)$*—the limiting velocity of the liquid in fractures,* $\mathbf{w}_s \in \overset{\circ}{\mathbf{W}}{}_2^{1,0}(\Omega_T)$*—the limiting displacements of the solid skeleton, and* $q_f \in L_2(\Omega_T)$*—the limiting pressure in the liquid, such that the sequences* $\left\{\chi_p^\varepsilon \dfrac{\partial \mathbf{w}^\delta}{\partial t}\right\}$, $\left\{\chi_c^\delta \dfrac{\partial \mathbf{w}^\delta}{\partial t}\right\}$, *and* $\{\tilde{\chi}^\delta q^\delta\}$ *converge weakly in* $\mathbf{L}_2(\Omega_T)$ *and* $L_2(\Omega_T)$ *to the functions* $\mathbf{v}_p$, $\mathbf{v}_c$, *and* mq_f, *respectively as* $\delta \searrow 0$.

At the same time the sequence $\{\mathbf{w}_s^\delta\}$ *converges weakly in* $\overset{\circ}{\mathbf{W}}{}_2^{1,0}(\Omega_T)$ *to the function* $\mathbf{w}_s$ *as* $\delta \searrow 0$.

The limiting functions $\mathbf{v}_p$, $\mathbf{v}_c$, $\mathbf{w}_s$, *and* q_f *satisfy the following relations*

$$\mathbf{v}_p = (1 - m_c)\, m_p \frac{\partial \mathbf{w}_s}{\partial t}, \quad \mathbf{v}_c = m_c \frac{\partial \mathbf{w}_s}{\partial t},$$
$$\mathbf{v} \equiv \mathbf{v}_c + \mathbf{v}_p + (1 - m)\frac{\partial \mathbf{w}_s}{\partial t} = \frac{\partial \mathbf{w}_s}{\partial t}, \tag{4.1.7}$$

and the anisotropic Lamé's system

$$\nabla \cdot \mathbf{v} = 0, \tag{4.1.8}$$

$$\lambda_0 \nabla \cdot \left(\mathfrak{B}_0^{(s)} : \mathbb{D}(\mathbf{w}_s)\right) - \nabla q_f + \hat{\rho}\mathbf{F} = 0, \tag{4.1.9}$$

in $\Omega \times (0, T)$ *with homogeneous normalization and boundary conditions*

$$\int_\Omega q_f(\mathbf{x}, t)dt = 0, \quad \mathbf{w}_s(\mathbf{x}, t) = 0, \quad \mathbf{x} \in S,\ t \geqslant 0. \tag{4.1.10}$$

The fourth-rank constant tensor $\mathfrak{B}_0^{(s)}$ *is defined below by formula (4.3.36),* $\hat{\rho} = m\,\rho_f + (1 - m)\,\rho_s$, $m = \int_K \int_K \tilde{\chi} dydz$*—the porosity of the liquid domain,* $m_p = \int_K \chi_p dy$*—the porosity of the pore space, and* $m_c = \int_K \chi_c dz$*—the porosity of the fracture space,* $m = m_c + (1 - m_c)m_p$. *The tensor* $\mathfrak{B}_0^{(s)}$ *is symmetric, strictly positively definite, and depends only on the geometry of the solid cells* Y_s *and* Z_s.

We refer to the problem (4.1.7)–(4.1.10) as the homogenized **model** $(\mathbb{DPF})_1$.

Theorem 4.3 *Under the conditions of Theorem 4.1 let* $\{\mathbf{w}^\delta,\ q^\delta\}$ *be a solution to the problem (4.0.1)–(4.0.5),* $\mu_2 < \infty$, *and the fracture space be connected.*

Then there exist a subsequence of small parameters $\{\delta > 0\}$, *and functions* $\mathbf{v}_p \in \mathbf{L}_2(\Omega_T)$—*the limiting velocity of the liquid in pores,* $\mathbf{v}_c \in \mathbf{L}_2(\Omega_T)$—*the limiting velocity of the liquid in fractures,* $\mathbf{w}_s \in \overset{\circ}{\mathbf{W}}{}_2^{1,0}(\Omega_T)$—*the limiting displacements of the solid skeleton, and* $q_f \in L_2(\Omega_T)$—*the limiting pressure in the liquid, such that the sequences* $\left\{\chi_p^{\varepsilon}\,\dfrac{\partial \mathbf{w}^{\delta}}{\partial t}\right\}$, $\left\{\chi_c^{\delta}\,\dfrac{\partial \mathbf{w}^{\delta}}{\partial t}\right\}$, *and* $\{\tilde{\chi}^{\delta}\,q^{\delta}\}$ *converge weakly in* $\mathbf{L}_2(\Omega_T)$ *and* $L_2(\Omega_T)$ *to the functions* $\mathbf{v}_p$, $\mathbf{v}_c$, *and* mq_f, *respectively as* $\delta \searrow 0$.

At the same time the sequence $\{\mathbf{w}_s^{\delta}\}$ *converges weakly in* $\overset{\circ}{\mathbf{W}}{}_2^{1,0}(\Omega_T)$ *to the function* $\mathbf{w}_s$ *as* $\delta \searrow 0$.

The limiting functions $\mathbf{v}_p$, $\mathbf{v}_c$, $\mathbf{w}_s$, *and* q_f *satisfy in* Ω_T *the following relations*

$$\mathbf{v}_p = (1-m_c)\,m_p\,\frac{\partial \mathbf{w}_s}{\partial t}, \quad \mathbf{v} = \mathbf{v}_c + (1-m_c)\frac{\partial \mathbf{w}_s}{\partial t}, \tag{4.1.11}$$

Equations (4.1.8), (4.1.9), Darcy's law in the form

$$\mathbf{v}_c = m_c\,\frac{\partial \mathbf{w}_s}{\partial t} + \frac{1}{\mu_2}\mathbb{B}^{(c)}\big(\rho_f\mathbf{F} - \nabla q_f\big), \quad \mathbf{x} \in \Omega, \tag{4.1.12}$$

normalization and boundary conditions (4.1.10), and boundary condition

$$\mathbf{v}\cdot\mathbf{n} = 0, \quad \mathbf{x} \in S, \tag{4.1.13}$$

where $\mathbf{n}$ *is a unit normal vector to the boundary* S *at* $\mathbf{x} \in S$. *In (4.1.12) the strictly positively definite constant matrix* $\mathbb{B}^{(c)}$, *is defined below by formula (4.3.18) and depends only on the geometry of the liquid cell* Z_f. *The fourth-rank constant tensor* $\mathfrak{B}_0^{(s)}$ *is defined below by formula (4.3.36),* $\hat{\rho} = m\,\rho_f + (1-m)\,\rho_s$, $m = \int_K\int_K \tilde{\chi}\,dydz$—*the porosity of the liquid domain,* $m_p = \int_K \chi_p dy$—*the porosity of the pore space, and* $m_c = \int_K \chi_c dz$—*the porosity of the fracture space,* $m = m_c + (1-m_c)m_p$. *The tensor* $\mathfrak{B}_0^{(s)}$ *is symmetric, strictly positively definite, and depends only on the geometry of the solid cells* Y_s *and* Z_s.

We refer to the problem (4.1.8)–(4.1.13) as the homogenized **model** $(\mathbb{DPF})_2$.

We obtain the double-porosity model for the absolutely rigid solid skeleton from the limit of solutions to the model $(\mathbb{DPF})_2$ as $\lambda_0 \to \infty$. In this case the liquid in pores is blocked and unmoving.

Theorem 4.4 *Let* $\mu_2 < \infty$, $\lambda_0 = n$ *and* $\mathbf{w}_s^{(n)}$, $\mathbf{v}_c^{(n)}$ *and* $q_f^{(n)}$ *be a solution to the problem (4.1.8)–(4.1.13). Then there exists a subsequence* $\{n_k\}$, *such that the sequence* $\left\{\dfrac{\partial \mathbf{w}_s^{(n_k)}}{\partial t}\right\}$ *converges strongly in* $\overset{\circ}{\mathbf{W}}{}_2^{1,0}(\Omega_T)$ *to zero as* $n_k \to \infty$, *the sequence* $\{\mathbf{v}_p^{(n_k)}\}$ *converges strongly in* $\mathbf{L}_2(\Omega_T)$ *to zero as* $n_k \to \infty$, *and sequences* $\{\mathbf{v}_c^{(n_k)}\}$ *and*

$\{q_f^{(n_k)}\}$ *converge weakly in* $\mathbf{L}_2(\Omega_T)$ *and* $L_2(\Omega_T)$ *to functions* $\mathbf{v}_c$, *and* $q_f \in W_2^{1,0}(\Omega_T)$ *respectively as* $n_k \to \infty$.

These limiting functions solve the problem

$$\frac{\partial \mathbf{w}_s}{\partial t} = \mathbf{v}_p = 0, \quad \mathbf{v}_c = \frac{1}{\mu_2}\mathbb{B}^{(c)}\big(\rho_f \mathbf{F} - \nabla q_f\big), \quad \mathbf{x} \in \Omega, \tag{4.1.14}$$

$$\nabla \cdot \mathbf{v}_c = 0, \ \mathbf{x} \in \Omega, \quad \mathbf{v}_c \cdot \mathbf{n} = 0, \ \mathbf{x} \in S. \tag{4.1.15}$$

We refer to the problem (4.1.14) and (4.1.15) as the homogenized **model** $(\mathbb{DPF})_3$.

Theorem 4.5 *Under the conditions of Theorem 4.1 let* $\{\mathbf{w}^\delta,\ q^\delta\}$ *be a solution to the problem (4.0.6)–(4.0.10) and* $\mu_2 = \infty$, *or the fracture space is disconnected (isolated fractures).*

Then there exist a subsequence of small parameters $\{\delta > 0\}$, *and functions* $\mathbf{v}_p \in \mathbf{L}_2(\Omega_T)$*—the limiting velocity of the liquid in pores,* $\mathbf{v}_c \in \mathbf{L}_2(\Omega_T)$*—the limiting velocity of the liquid in fractures,* $\mathbf{w}_s \in \overset{\circ}{\mathbf{W}}{}_2^{1,0}(\Omega_T)$*—the limiting displacements of the solid skeleton,* $q_s \in L_2(\Omega_T)$*—the limiting pressure in the solid skeleton, and* $q_f \in L_2(\Omega_T)$*—the limiting pressure in the liquid, such that the sequences* $\left\{\chi_p^\varepsilon \dfrac{\partial \mathbf{w}^\delta}{\partial t}\right\}$, $\left\{\chi_c^\delta \dfrac{\partial \mathbf{w}^\delta}{\partial t}\right\}$, $\{(1-\tilde{\chi}^\delta)\, q^\delta\}$ *and* $\{\tilde{\chi}^\delta\, q^\delta\}$ *converge weakly in* $\mathbf{L}_2(\Omega_T)$ *and* $L_2(\Omega_T)$ *to the functions* $\mathbf{v}_p$, $\mathbf{v}_c$, q_s, *and* mq_f, *respectively as* $\delta \searrow 0$.

At the same time the sequence $\{\mathbf{w}_s^\delta\}$ *converges weakly in* $\overset{\circ}{\mathbf{W}}{}_2^{1,0}(\Omega_T)$ *to the function* $\mathbf{w}_s$ *as* $\delta \searrow 0$.

The limiting functions $\mathbf{v}_p$, $\mathbf{v}_c$, $\mathbf{w}_s$, q_s, *and* q_f *satisfy in* $\Omega \times (0,T)$ *relations (4.1.7) and the anisotropic Lamé's system*

$$\nabla \cdot \Big(\lambda_0\, \mathfrak{B}^{(s)} : \mathbb{D}(x, \mathbf{w}_s) + q_f\, \mathbb{C}^{(s)}\Big) + \hat{\rho}\mathbf{F} = 0, \tag{4.1.16}$$

$$\frac{1}{c_s^2}\, q_s + \frac{a^{(s)}}{\lambda_0}\, q_f + \mathbb{A}^{(s)} : \mathbb{D}(x, \mathbf{w}_s) = 0, \tag{4.1.17}$$

$$\frac{1}{c_f^2}\frac{\partial q_f}{\partial t} + \frac{1}{c_s^2}\frac{\partial q_s}{\partial t} + \nabla \cdot \mathbf{v} = 0, \tag{4.1.18}$$

completed with the homogeneous boundary condition

$$\mathbf{w}_s(\mathbf{x}, t) = 0, \quad \mathbf{x} \in S. \tag{4.1.19}$$

In (4.1.16) and (4.1.17) the symmetric, strictly positively definite fourth-rank constant tensor $\mathfrak{B}^{(s)}$, *constant matrices* $\mathbb{C}^{(s)}$ *and* $\mathbb{A}^{(s)}$, *and constant* $a^{(s)}$ *are defined below by formulae (4.5.23)–(4.5.26),* $\hat{\rho} = m\,\rho_f + (1-m)\,\rho_s$, $m = \int_K \int_K \chi\, dy dz$*—the* *porosity of the liquid domain,* $m_p = \int_K \chi_p dy$*—the porosity of the pore space, and* $m_c = \int_K \chi_c dz$*—the porosity of the fracture space. The tensor* $\mathbb{A}^{(s)}$ *is , and depends only on the geometry of the solid cells* Y_s *and* Z_s*. The tensor* $\mathfrak{B}^{(s)}$*, matrices* $\mathbb{C}^{(s)}$ *and* $\mathbb{A}^{(s)}$*, and the constant* $a^{(s)}$ *depend only on the geometry of the solid cells* Y_s *and* Z_s*, and criterion* $\beta = \dfrac{c_s^2}{\lambda_0}$.

We refer to the problem (4.1.7), (4.1.16)–(4.1.19) as the homogenized **model** $(\mathbb{DPF})_4$.

Theorem 4.6 *Under the conditions of Theorem 4.1 let* $\{\mathbf{w}^\delta,\ q^\delta\}$ *be a solution to the problem (4.0.6)–(4.0.10),* $\mu_2 < \infty$*, and the fracture space is connected.*

Then there exist a subsequence of small parameters $\{\delta > 0\}$*, and functions* $\mathbf{v}_p \in \mathbf{L}_2(\Omega_T)$*—the limiting velocity of the liquid in pores,* $\mathbf{v}_c \in \mathbf{L}_2(\Omega_T)$*—the limiting velocity of the liquid in fractures,* $\mathbf{w}_s \in \overset{\circ}{\mathbf{W}}{}^{1,0}_2(\Omega_T)$*—the limiting displacements of the solid skeleton,* $q_s \in L_2(\Omega_T)$*—the limiting pressure in the solid skeleton, and* $q_f \in L_2(\Omega_T)$*—the limiting pressure in the liquid, such that the sequences* $\left\{\chi_p^\varepsilon \dfrac{\partial \mathbf{w}^\delta}{\partial t}\right\}$*,* $\left\{\chi_c^\delta \dfrac{\partial \mathbf{w}^\delta}{\partial t}\right\}$*,* $\{(1-\tilde{\chi}^\delta)\, q^\delta\}$ *and* $\{\tilde{\chi}^\delta\, q^\delta\}$ *converge weakly in* $\mathbf{L}_2(\Omega_T)$ *and* $L_2(\Omega_T)$ *as* $\delta \searrow 0$ *to the functions* $\mathbf{v}_p$*,* $\mathbf{v}_c$*,* q_s*, and* mq_f *respectively.*

At the same time the sequence $\{\mathbf{w}_s^\delta\}$ *converges weakly in* $\overset{\circ}{\mathbf{W}}{}^{1,0}_2(\Omega_T)$ *as* $\delta \searrow 0$ *to the function* $\mathbf{w}_s$ *as* $\delta \searrow 0$.

The limiting functions $\mathbf{v}_p$*,* $\mathbf{v}_c$*,* $\mathbf{w}_s$*,* q_s*, and* q_f *satisfy in* $\Omega \times (0,T)$ *relations (4.1.11), equations (4.1.16)–(4.1.17), Darcy's law (4.1.12), and boundary conditions (4.1.13) and (4.1.19).*

We refer to the problem (4.1.11)–(4.1.13), (4.1.16), (4.1.17), (4.1.19) as the homogenized **model** $(\mathbb{DPF})_5$.

4.2 Proof of Theorem 4.1

The only nonstandard element here is a proof of estimates (4.1.5) and (4.1.6), which are the basis of other estimates.

To prove (4.1.2) we choose as a test function in (4.1.1) the function $h(\tau)\dfrac{\partial \mathbf{w}^\delta}{\partial \tau}(\mathbf{x},\tau)$, where $h(\tau) = 1,\ \tau \in (0,t)$ and $h(\tau) = 0,\ \tau \in [t,T)$:

$$\alpha_\mu \int_0^t \int_\Omega \chi^\delta \left|\mathbb{D}\left(x, \frac{\partial \mathbf{w}^\delta}{\partial t}(\mathbf{x},\tau)\right)\right|^2 dxd\tau$$
$$+\frac{1}{2}\lambda_0 \int_\Omega (1-\chi^\delta)\left|\mathbb{D}\big(x, \mathbf{w}^\delta(\mathbf{x},t)\big)\right|^2 dx = \int_0^t \int_\Omega \rho^\delta \mathbf{F}\cdot \frac{\partial \mathbf{w}^\delta}{\partial t} dxd\tau.$$

Passing the time derivative from $\dfrac{\partial \mathbf{w}^\delta}{\partial t}$ to $\mathbf{F}$ in the right-hand side integral, applying after that to this integral the Hölder inequality and the evident estimate

$$\int_\Omega \chi^\delta \big|\mathbb{D}\big(x, \mathbf{w}^\delta(\mathbf{x}, t)\big)\big|^2 dx \leq C \int_0^t \int_\Omega \chi^\delta \left|\mathbb{D}\left(x, \frac{\partial \mathbf{w}^\delta}{\partial t}(\mathbf{x}, \tau)\right)\right|^2 dx d\tau,$$

and using the equality

$$\int_\Omega (1-\chi^\delta)\big|\mathbb{D}\big(x, \mathbf{w}^\delta(\mathbf{x}, t)\big)\big|^2 dx = \int_\Omega (1-\chi^\delta)\big|\mathbb{D}\big(x, \mathbf{w}_s^\delta(\mathbf{x}, t)\big)\big|^2 dx,$$

we arrive at

$$\begin{aligned} J(t) \equiv \alpha_\mu \int_\Omega \chi^\delta \big|\mathbb{D}\big(x, \mathbf{w}^\delta(\mathbf{x}, t)\big)\big|^2 dx + \lambda_0 \int_\Omega (1-\chi^\delta)\big|\mathbb{D}\big(x, \mathbf{w}_s^\delta(\mathbf{x}, t)\big)\big|^2 dx \\ \leqslant CF^2 + \int_0^t \int_\Omega |\mathbf{w}^\delta(\mathbf{x}, \tau)|^2 dx d\tau. \end{aligned} \tag{4.2.1}$$

Next we put $\mathbf{w}_0^\delta = \mathbf{w}^\delta - \mathbf{w}_s^\delta$. By construction $\mathbf{w}_0^\delta \in \overset{\circ}{\mathbf{W}}{}_2^1 (\Omega_f^\delta)$.

To estimate the integral

$$I_f^\delta = \int_{\Omega_f^\delta} |\mathbf{w}_0^\delta|^2 dx$$

we divide it by two parts:

$$I_f^\delta = I_p^\varepsilon + I_c^\delta, \quad I_p^\varepsilon = \int_{\Omega_p^\varepsilon} |\mathbf{w}_0^\delta|^2 dx, \quad I_c^\delta = \int_{\Omega_c^\delta} |\mathbf{w}_0^\delta|^2 dx.$$

Let $G_p^{(\mathbf{k})}$, where $\mathbf{k} = (k_1, k_2, k_3) \in \mathbb{Z}^3$, be the intersection of Ω_p^ε with a set $\{\mathbf{x} : \mathbf{x} = \varepsilon(\mathbf{y} + \mathbf{k}),\ \mathbf{y} \in K\}$. Then $\Omega_p^\varepsilon = \cup_{\mathbf{k}\in\mathbb{Z}^3} G_p^{(\mathbf{k})}$ and

$$I_p^\varepsilon = \sum_{\mathbf{k}\in\mathbb{Z}^3} I_p^\varepsilon(\mathbf{k}), \quad I_p^\varepsilon(\mathbf{k}) = \int_{G_p^{(\mathbf{k})}} |\mathbf{w}_0^\delta|^2 dx.$$

next we divide the sets $G_p^{(\mathbf{k})}$ by two groups:

(1) the set $G_p^{(\mathbf{k})}$, $\mathbf{k} \in \mathbb{Z}_0$, has no intersection with the boundary between pore and fracture spaces, and
(2) the set $G_p^{(\mathbf{k})}$, $\mathbf{k} \in \mathbb{Z}_1$, has an intersection with the boundary between pore and fracture spaces.

For the first group in each integral I_p^ε we change variable by $\mathbf{x} = \varepsilon(\mathbf{y} - \mathbf{y}_\mathbf{k})$, then apply the Friedrichs-Poincaré inequality and finally return to the original variables:

$$\int_{G_p^{(\mathbf{k})}} |\mathbf{w}_0^\delta|^2 dx = \varepsilon^3 \int_{Y^{(\mathbf{k})}} |\overline{\mathbf{w}}_{0,\mathbf{k}}^\delta|^2 dy$$
$$\leqslant \varepsilon^3 C \int_{Y^{(\mathbf{k})}} |\mathbb{D}(y, \overline{\mathbf{w}}_{0,\mathbf{k}}{}^\delta)|^2 dy = \varepsilon^2 C \int_{G_p^{(\mathbf{k})}} |\mathbb{D}(x, \mathbf{w}_0^\delta)|^2 dx,$$

and

$$\sum_{\mathbf{k}\in\mathbb{Z}_0} \int_{G_p^{(\mathbf{k})}} |\mathbf{w}_0^\delta|^2 dx \leqslant \varepsilon^2\, C \int_{\Omega_f^\delta} |\mathbb{D}(x, \mathbf{w}_0^\delta)|^2 dx.$$

Here $\overline{\mathbf{w}}_{0,\mathbf{k}}{}^\delta(\mathbf{y}, t) = \mathbf{w}_0^\delta(\mathbf{x}, t)$ and C is a constant in the Friedrichs-Poincaré inequality for the domain $Y_f \subset K$.

All sets $G_p^{(\mathbf{k})}$ from the second group are in the εN—neighborhood of the set $\cup_{\mathbf{k}\in\mathbb{Z}_0} G_p^{(\mathbf{k})}$, where $N \in \mathbb{Z}$ is an integer bound for the Lipschitz norm for γ_c. Therefore, for each $G_p^{(\mathbf{k}')}$ from the second group there is at least one $G_p^{(\mathbf{k}'')}$ from the first group such that

$$\int_{G_p^{(\mathbf{k}')}} |\mathbf{w}_0^\delta|^2 dx \leqslant C(N) \int_{G_p^{(\mathbf{k}'')}} |\mathbf{w}_0^\delta|^2 dx,$$

and each set of the first group is repeated no more than N times.

Thus,

$$\sum_{\mathbf{k}\in\mathbb{Z}_1} \int_{G_p^{(\mathbf{k})}} |\mathbf{w}_0^\delta|^2 dx \leqslant C(N)\, N \sum_{\mathbf{k}\in\mathbb{Z}_0} \int_{G_p^{(\mathbf{k})}} |\mathbf{w}_0^\delta|^2 dx,$$

$$I_p^\varepsilon \leqslant \varepsilon^2\, C \sum_{\mathbf{k}\in\mathbb{Z}^3} \int_{G_p^{(\mathbf{k})}} |\mathbb{D}_x(\mathbf{w}_0^\delta)|^2 dx \leqslant \varepsilon^2\, C \int_{\Omega_f^\delta} |\mathbb{D}(x, \mathbf{w}_0^\delta)|^2 dx. \tag{4.2.2}$$

In the same way we show that

$$I_c^\delta \leq \delta^2\, C \int_{\Omega_f^\delta} |\mathbb{D}(x, \mathbf{w}_0^\delta)|^2 dx. \tag{4.2.3}$$

In fact, as before we again divide the integral I_c^δ into the sum of integrals over domains $G_c^{(\mathbf{k})}$ and make a change of variables:

$$\mathbf{x} = \delta \mathbf{z}, \quad \mathbf{w}_0^\delta(\mathbf{x}, t) = \tilde{\mathbf{w}}_0^\delta(\mathbf{z}, t), \qquad \int_{G_c^{(\mathbf{k})}} |\mathbf{w}_0^\delta|^2 dx = \delta^3 \int_{Z^{(\mathbf{k})}} |\tilde{\mathbf{w}}_0^\delta|^2 dz.$$

For integrals over domains $G_c^{(\mathbf{k})}$ we use the Friedrichs-Poincaré inequality, based on the fact that the function $\tilde{\mathbf{w}}_0^\delta$ vanishes on the some periodic $\big($with period $\dfrac{\varepsilon}{\delta}\big)$ part of the boundary $\partial G_c^{(\mathbf{k})}$ with a strictly positive measure, which is bounded from below independently of δ.

Thus,

$$I_f^\delta \leqslant C(\varepsilon^2+\delta^2)\int_{\Omega_f^\delta}|\mathbb{D}(x,\mathbf{w}_0^\delta)|^2dx \leq C\left(\frac{\varepsilon^2}{\alpha_\mu}+\frac{\delta^2}{\alpha_\mu}\right)\alpha_\mu\int_{\Omega_f^\delta}|\mathbb{D}(x,\mathbf{w}^\delta)|^2dx$$
$$+\,C(\varepsilon^2+\delta^2)\int_{\Omega_f^\delta}|\mathbb{D}(x,\mathbf{w}_s^\delta)|^2dx \leq C\,J(t),$$

and

$$\int_{\Omega_f^\delta}|\mathbf{w}^\delta|^2dx \leq \int_{\Omega_f^\delta}|\mathbf{w}_0^\delta|^2dx+\int_{\Omega_f^\delta}|\mathbf{w}_s^\delta|^2dx \leq C\big(J(t)+\int_{\Omega_s^\delta}|\mathbf{w}^\delta|^2dx\big).$$

To estimate the integral

$$I_s^\delta=\int_{\Omega_s^\delta}|\mathbf{w}^\delta|^2dx$$

we use the Friedrichs-Poincaré inequality and the properties of the extension $\mathbf{w}_s$:

$$I_s^\delta \leq \int_{\Omega}|\mathbf{w}_s^\delta|^2dx \leq C\int_{\Omega}|\mathbb{D}(x,\mathbf{w}_s^\delta)|^2dx \leq C\lambda_0\int_{\Omega_s^\delta}|\mathbb{D}(x,\mathbf{w}^\delta)|^2dx \leq CJ(t).$$

Gathering everything together we have

$$\int_{\Omega}|\mathbf{w}^\delta|^2dx \leq CJ(t).$$

Estimate (4.1.2) follows now from (4.2.1) and Gronwall's inequality. The same estimate (4.1.2) together with (4.2.2) and (4.2.3) result (4.1.5).

To prove estimates (4.1.3) and (4.1.6) we simply repeat once more for the "time derivative" of identity (4.1.1) and $\dfrac{\partial^2\mathbf{w}^\delta}{\partial t^2}$.

Estimate (4.1.4) for an incompressible liquid is a simple consequence of (4.1.2), (4.1.3), (4.1.5), and (4.1.6). For a compressible liquid this estimate follows from the same estimates (4.1.2), (4.1.3), (4.1.5), and (4.1.6) and the continuity equation (4.0.6).

4.3 Proofs of Theorems 4.2 and 4.3

4.3.1 Weak and Three-Scale Limits of Sequences of Displacements, Velocities and Pressure

First, we define the velocity of the liquid in pores as $\mathbf{v}_p^\varepsilon=\chi_p^\varepsilon\dfrac{\partial\mathbf{w}^\delta}{\partial t}$, the velocity of the liquid in fractures as $\mathbf{v}_c^\delta=\chi_c^\delta\dfrac{\partial\mathbf{w}^\delta}{\partial t}$ and the velocity of the solid skeleton as

$\mathbf{v}_s^\delta = \dfrac{\partial \mathbf{w}_s^\delta}{\partial t}$. By definition

$$\mathbf{v}^\delta = \mathbf{v}_p^\varepsilon + \mathbf{v}_c^\delta + (1 - \tilde{\chi}^\delta)\mathbf{v}_s^\delta. \tag{4.3.1}$$

On the strength of Theorem 4.1, the sequences $\{\tilde{\chi}^\delta q^\delta\}$, $\{(1-\tilde{\chi}^\delta)q^\delta\}$, $\{\mathbf{v}^\delta\}$, $\{\mathbf{v}_p^\varepsilon\}$, $\{\mathbf{v}_c^\delta\}$, $\{\mathbf{w}_s^\delta\}$, $\{\mathbf{v}_s^\delta\}$, and $\{\mathbb{D}(x, \mathbf{w}_s^\delta)\}$ are bounded in $L_2(\Omega_T)$ and $\mathbf{L}_2(\Omega_T)$.

Hence there exists a subsequence of small parameters $\{\delta > 0\}$ and functions $q_f,\, q_s \in L_2(\Omega_T)$, $\mathbf{v}, \mathbf{v}_p, \mathbf{v}_c, \mathbf{v}_s \in \mathbf{L}_2(\Omega_T)$ and $\mathbf{w}_s \in \overset{\circ}{\mathbf{W}}{}_2^{1,0}(\Omega_T)$ such that

$$\begin{aligned}
&\tilde{\chi}^\delta q^\delta \rightharpoonup m q_f, \quad (1-\tilde{\chi}^\delta)\, q^\delta \rightharpoonup q_s, \quad \mathbf{v}^\delta \rightharpoonup \mathbf{v}, \quad \mathbf{v}_p^\varepsilon \rightharpoonup \mathbf{v}_p, \quad \mathbf{v}_c^\delta \rightharpoonup \mathbf{v}_c, \\
&\mathbf{v}_s^\delta \rightharpoonup \mathbf{v}_s, \quad \mathbf{w}_s^\delta \rightharpoonup \mathbf{w}_s, \quad \mathbb{D}(x, \mathbf{w}_s^\delta) \rightharpoonup \mathbb{D}(x, \mathbf{w}_s)
\end{aligned} \tag{4.3.2}$$

weakly in $L_2(\Omega_T)$ and $\mathbf{L}_2(\Omega_T)$ as $\delta \searrow 0$.

Note also that

$$\tilde{\chi}^\delta\, \alpha_\mu \mathbb{D}(x, \mathbf{v}^\delta) \to 0 \tag{4.3.3}$$

strongly in $\mathbf{L}_2(\Omega_T)$ as $\delta \searrow 0$.

Next we apply the method of reiterated homogenization (see [6, 70] and Appendix B): there exist functions $Q_f(\mathbf{x}, t, \mathbf{y}, \mathbf{z})$, $Q_s(\mathbf{x}, t, \mathbf{y}, \mathbf{z})$, $\mathbf{V}(\mathbf{x}, t, \mathbf{y}, \mathbf{z})$, $\mathbf{V}_c(\mathbf{x}, t, \mathbf{y}, \mathbf{z})$, $\mathbf{U}_c(\mathbf{x}, t, \mathbf{z})$, and $\mathbf{U}_p(\mathbf{x}, t, \mathbf{y}, \mathbf{z})$ that are 1-periodic in $\mathbf{y}$ and $\mathbf{z}$ and satisfy the condition that the sequences $\{\tilde{\chi}^\delta q^\delta\}$, $\{(1-\tilde{\chi}^\delta)\, q^\delta\}$, $\{\mathbf{V}^\delta\}$, $\{\mathbf{V}_c^\delta\}$, and $\{\mathbb{D}(x, \mathbf{w}_s^\delta)\}$ three-scale converge (up to some subsequences) to $Q_f(\mathbf{x}, t, \mathbf{y}, \mathbf{z})$, $Q_s(\mathbf{x}, t, \mathbf{y}, \mathbf{z})$, $\mathbf{V}(\mathbf{x}, t, \mathbf{y}, \mathbf{z})$, $\mathbf{V}_C(\mathbf{x}, t, \mathbf{y}, \mathbf{z})$, and $\mathbb{D}(x, \mathbf{w}_s) + \mathbb{D}\big(z, \mathbf{U}_c(\mathbf{x}, t, \mathbf{z})\big) + \mathbb{D}\big(y, \mathbf{U}_p(\mathbf{x}, t, \mathbf{y}, \mathbf{z})\big)$, respectively.

The sequence $\{\mathbf{w}_s^\delta\}$ three-scale converges to the function $\mathbf{w}_s(\mathbf{x}, t)$.

Relabeling if necessary, we assume that the sequences themselves converge.

Remember that *three-scale convergence* of the sequence $\{\pi^\delta\}$ to the function $\Pi(\mathbf{x}, t, \mathbf{y}, \mathbf{z})$ means the convergence of integrals

$$\begin{aligned}
&\int_{\Omega_T} \pi^\delta(\mathbf{x}, t)\, \varphi\left(\mathbf{x}, t, \frac{\mathbf{x}}{\varepsilon}, \frac{\mathbf{x}}{\delta}\right) dxdt \\
&\quad \to \int_{\Omega_T} \int_Y \int_Z \Pi(\mathbf{x}, t, \mathbf{y}, \mathbf{z})\, \varphi(\mathbf{x}, t, \mathbf{y}, \mathbf{z}) dzdydxdt,
\end{aligned}$$

for any smooth 1-periodic in $\mathbf{y}$ and $\mathbf{z}$ function $\varphi(\mathbf{x}, t, \mathbf{y}, \mathbf{z})$.

By definition the function

$$\pi(\mathbf{x}, t) = \langle\langle \Pi \rangle_y\rangle_z,$$

where

$$\langle \Pi \rangle_y = \int_K \Pi(\mathbf{x}, t, \mathbf{y}, \mathbf{z}) dy, \quad \langle \Pi \rangle_z = \int_K \Pi(\mathbf{x}, t, \mathbf{y}, \mathbf{z}) dz,$$

is a weak limit in $L_2(\Omega_T)$ of the sequence $\{\pi^\delta\}$.

4.3.2 Macro- and Microscopic Equations

We start the proof of the theorem from the macro- and microscopic equations related to the liquid motion and to the continuity equation.

Lemma 4.1 *For almost all* $(\mathbf{x}, t) \in \Omega_T$, $\mathbf{y} \in Y$ *and* $\mathbf{z} \in Z$, *the weak and three-scale limits of the sequences* $\{\tilde{\chi}^\delta q^\delta\}$, $\{(1-\tilde{\chi}^\delta) q^\delta\}$, $\{\mathbf{v}^\delta\}$, $\{\mathbf{v}_c^\delta\}$, $\{\mathbf{v}_p^\delta\}$, *and* $\{\mathbf{w}_s^\delta\}$ *satisfy the relations*

$$Q_f = q_f(\mathbf{x}, t)\tilde{\chi}(\mathbf{y}, \mathbf{z}), \quad Q_s = Q_s\big(1 - \tilde{\chi}(\mathbf{y}, \mathbf{z})\big), \tag{4.3.4}$$

$$\mathbf{v}_p = (1 - m_c)\, m_p\, \mathbf{v}_s, \quad \mathbf{v} = \mathbf{v}_c + (1 - m_c)\mathbf{v}_s, \tag{4.3.5}$$

$$\nabla \cdot \mathbf{v}(\mathbf{x}, t) = 0, \;\; (\mathbf{x}, t) \in \Omega_T, \;\; \mathbf{v}(\mathbf{x}, t) \cdot \mathbf{n}(\mathbf{x}) = 0, \;\; (\mathbf{x}, t) \in S_T, \tag{4.3.6}$$

$$(1 - \tilde{\chi})\big(\nabla \cdot \mathbf{w}_s + \nabla_z \cdot \mathbf{U_c} + \nabla_y \cdot \mathbf{U_p}\big) = 0, \tag{4.3.7}$$

where

$$\tilde{\chi}(\mathbf{y}, \mathbf{z}) = \chi_c(\mathbf{z}) + \big(1 - \chi_c(\mathbf{z})\big)\chi_p(\mathbf{y}),$$

$\mathbf{n}(\mathbf{x})$ *is a normal vector to S at* $\mathbf{x} \in S$, $m = \langle\langle\tilde{\chi}\rangle_y\rangle_z$—*the porosity of the liquid domain,* $m_p = \langle\chi_p\rangle_y$—*the porosity of the pore space, and* $m_c = \langle\chi_c\rangle_z$—*the porosity of the fracture space.*

Proof By the properties of three-scale convergence one has equalities

$$Q_f = \tilde{\chi}\, Q_f, \quad Q_s = (1 - \tilde{\chi})\, Q_s.$$

Choosing in (4.1.1) a test function in the form $\varphi = \varepsilon\, h_0(t)\, h(\mathbf{x})\, \psi\big(\frac{\mathbf{x}}{\varepsilon}, \frac{\mathbf{x}}{\delta}\big)$, where $\psi(\mathbf{y}, \mathbf{z})$ is finite in $Y_f \times Z_f$, and passing to the limit as $\delta \searrow 0$ we arrive at

$$\tilde{\chi}(\mathbf{y}, \mathbf{z})\, \nabla_y Q_f = 0, \quad \text{or} \quad Q_f = \tilde{\chi}(\mathbf{y}, \mathbf{z})\, Q_f(\mathbf{x}, t, \mathbf{z}).$$

Now we repeat once more with $\varphi = \delta\, h_0(t)\, h(\mathbf{x})\, \psi\big(\frac{\mathbf{x}}{\varepsilon}, \frac{\mathbf{x}}{\delta}\big)$, where $\psi(\mathbf{y}, \mathbf{z})$ is finite in $Y_f \times Z_f$, and get

$$\tilde{\chi}(\mathbf{y}, \mathbf{z})\, \nabla_z Q_f = 0, \quad \text{or} \quad Q_f = \tilde{\chi}(\mathbf{y}, \mathbf{z})\, Q_f(\mathbf{x}, t),$$

which results in (4.3.4).

Equalities (4.3.5) are a simple consequence of (4.3.1), (4.1.6) and the properties of three-scale convergence.

The continuity equation and boundary condition in (4.3.6) follow from the continuity equation (4.0.1) in the form

$$\int_{\Omega} \mathbf{v}^{\delta} \cdot \nabla \psi dx = 0, \tag{4.3.8}$$

which holds true for any smooth functions ψ, after passing in (4.3.8) to the limit as $\delta \searrow 0$.

The three-scale limit in continuity equation (4.0.1) in the form

$$(1 - \tilde{\chi}^{\delta}) \nabla \cdot \mathbf{v}_s^{\delta} = 0$$

results in the continuity equation (4.3.6).

Lemma 4.2 *Let* $\widetilde{\mathbf{V}} = \langle \mathbf{V}_c \rangle_{\mathcal{Y}}$. *If* $\mu_2 = \infty$, *then*

$$\widetilde{\mathbf{V}} = \mathbf{V}_c = \mathbf{v_s}(\mathbf{x}, t)\, \chi_c(\mathbf{z}), \quad \mathbf{v_c} = m_c \mathbf{v_s}. \tag{4.3.9}$$

If $\mu_2 < \infty$, *then for almost every* $(\mathbf{x}, t) \in \Omega_T$ *the function* $\widetilde{\mathbf{V}}$ *is a 1-periodic in* $\mathbf{z}$ *solution to the Stokes system*

$$-\frac{\mu_2}{2} \Delta_z \widetilde{\mathbf{V}} = -\nabla_z \widetilde{\Pi} - \nabla q_f + \rho_f \mathbf{F}, \tag{4.3.10}$$

$$\nabla_z \cdot \widetilde{\mathbf{V}} = 0, \tag{4.3.11}$$

in the domain Z_f, *such that*

$$\widetilde{\mathbf{V}}(\mathbf{x}, t, \mathbf{z}) = \mathbf{v_s}(\mathbf{x}, t), \quad \mathbf{z} \in \gamma_c. \tag{4.3.12}$$

Proof Firstly we derive the continuity equation (4.3.11). To do that we put $\psi = \delta \psi_0\big(\mathbf{x}, \frac{\mathbf{x}}{\delta}\big)$ in the integral identity (4.3.8), pass to the limit as $\delta \searrow 0$, and get identity

$$\int_{\Omega} \int_{Z_f} \widetilde{\mathbf{V}} \cdot \nabla_z \psi_0(\mathbf{x}, \mathbf{z}) dx dz = 0,$$

which is obviously equivalent to (4.3.11).

If $\mu_2 = \infty$, then (4.3.9) follows from estimate (4.1.6).

Let now $\mu_2 < \infty$. If we choose in the integral identity (4.1.1) a test function φ in the form $\varphi = h_0(t) h_1(\mathbf{x})\, \psi\big(\frac{\mathbf{x}}{\delta}\big)$, where $\operatorname{supp} h_1 \subset \Omega$, $\operatorname{supp} \psi(\mathbf{z}) \subset Z_f$, $\nabla_z \cdot \psi = 0$, and pass to the limit as $\delta \searrow 0$, we arrive at

$$\int_{\Omega} \int_{Z_f} \big(h_1 \mu_2 \widetilde{\mathbf{V}} \cdot (\nabla_z \cdot \mathbb{D}_z(\psi)) + q_f (\nabla h_1 \cdot \psi) + \rho_f (\mathbf{F} \cdot \psi) h_1\big) dx dz = 0.$$

In the same way as in Chap. 1 we may show that $\nabla q_f \in \mathbf{L}_2(\Omega_T)$.

The desired equation (4.3.10) follows from the last identity, if we pass derivatives from the test function to $\widetilde{\mathbf{V}}$ and take into account (4.3.11). The term $\nabla_z \widetilde{\Pi}$ appears due to condition $\nabla_z \cdot \psi = 0$.

Finally, the boundary condition (4.3.12) follows from the representation

$$\langle \mathbf{V} \rangle_y = \widetilde{\mathbf{V}} + \big(1 - \chi_c(\mathbf{z})\big)\, \mathbf{v_s}(\mathbf{x}, t),$$

and inclusion $\langle \mathbf{V} \rangle_y \in \mathbf{W}_2^1(Z)$ for almost every $(\mathbf{x}, t) \in \Omega_T$.

Now we derive macro- and microscopic equations for the solid motion. Let

$$\tilde{q}_f = \frac{1}{\lambda_0} q_f, \quad \tilde{Q}_s = \left(\frac{1}{\lambda_0} Q_s - \tilde{q}_f\right)(1 - \tilde{\chi}), \quad \tilde{q}_s = \langle\langle \tilde{Q}_s \rangle_{Z_s} \rangle_{Y_s},$$

where

$$\langle \Psi \rangle_{Z_s} = \langle (1 - \chi_c) \Psi \rangle_z, \quad \langle \Phi \rangle_{Y_s} = \langle (1 - \chi_p) \Psi \rangle_y.$$

Then

$$\frac{1}{\lambda_0}(q_f + q_s) = \frac{1}{\lambda_0} \langle\langle (1 - \tilde{\chi})(Q_f + Q_s) \rangle_z \rangle_y = \langle\langle \tilde{q}_f + \tilde{Q}_s \rangle_z \rangle_y = \tilde{q}_f + \tilde{q}_s$$

Lemma 4.3 *Functions* $\mathbf{w}_s$, $\mathbf{U}_c$, $\mathbf{U}_p$, $\tilde{q}_f$, *and* $\tilde{q}_s$ *satisfy in* Ω_T *the macroscopic equation*

$$\nabla \cdot \Big((1 - m) \mathbb{D}(x, \mathbf{w}_s) + (1 - m_p) \langle \mathbb{D}(z, \mathbf{U}_c) \rangle_{Z_s} + \langle \langle \mathbb{D}(y, \mathbf{U}_p) - \tilde{Q}_s \rangle_{Z_s} \rangle_{Y_s} - \tilde{q}_f\, \mathbb{I} \Big) + \frac{\hat{\rho}}{\lambda_0} \mathbf{F} = 0, \tag{4.3.13}$$

where

$$\hat{\rho} = m\, \rho_f + (1 - m)\, \rho_s \quad \text{and} \quad \tilde{q} = \tilde{q}_f + \tilde{q}_s.$$

To prove this lemma we put in (4.1.1) $\varphi = h_0(t)\mathbf{h}_1(\mathbf{x})$, where $\mathbf{h}$ is finite in Ω, and pass to the limit as $\delta \searrow 0$, taking into account (4.3.3).

Lemma 4.4 *Functions* $\mathbf{w}_s$, $\mathbf{U}_c$, $\mathbf{U}_p$, *and* $\tilde{Q}_s$ *satisfy in* Z_s *and almost everywhere in* Ω_T *the microscopic equation*

$$\nabla_z \cdot \Big((1 - \chi_c) \big((1 - m_p) \big(\mathbb{D}(x, \mathbf{w}_s) + \mathbb{D}(z, \mathbf{U}_c) \big) \tag{4.3.14}$$

$$+ \langle \mathbb{D}(y, \mathbf{U}_p) - \tilde{Q}_s\, \mathbb{I} \rangle_{Y_s} \big) \Big) = 0. \tag{4.3.14}$$

To prove this lemma we put in (4.1.1) $\varphi = \delta h_0(t) h_1(\mathbf{x}) \varphi_0\big(\frac{\mathbf{x}}{\delta}\big)$, where h_1 is finite in Ω, pass to the limit as $\delta \searrow 0$, and use the equality

$$(1-\tilde{\chi}) = (1-\chi_p)(1-\chi_c).$$

Lemma 4.5 *Functions* $\mathbf{w}_s$, $\mathbf{U}_c$, $\mathbf{U}_p$, *and* $\tilde{Q}_s$ *satisfy in* Y_s *and almost everywhere in* $\Omega_T \times Z_s$ *the microscopic equation*

$$\nabla_y \cdot \Big((1-\chi_p)\big(\mathbb{D}(x,\mathbf{w}_s) + \mathbb{D}(z,\mathbf{U}_c) + \mathbb{D}(y,\mathbf{U}_p) - \tilde{Q}_s\mathbb{I}\big)\Big) = 0. \tag{4.3.15}$$

To prove the lemma we put in (4.1.1) $\varphi = \varepsilon h_0(t) h_1(\mathbf{x})\,\varphi_0\big(\frac{\mathbf{x}}{\delta}\big)\,\varphi_1\big(\frac{\mathbf{x}}{\varepsilon}\big)$, where h_1 is finite in Ω, and pass to the limit as $\delta \searrow 0$.

4.3.3 Homogenized Equations

The derivation of homogenized equations is quite standard (see previous chapters). For the liquid motion we solve the microscopic system (4.3.9)–(4.3.12), find $\widetilde{\mathbf{V}}$ as an operator on ∇q_f and $\dfrac{\partial \mathbf{w}_s}{\partial t}$, and then use the relation $\mathbf{v}_c = \langle \widetilde{\mathbf{V}} \rangle_{Z_f}$. That is, holds true the following lemma.

Lemma 4.6 *Let* $\mu_1 < \infty$. *Then functions* $\mathbf{v}_c$, $\mathbf{v}_s$, $\mathbf{v} = \mathbf{v}_c + (1-m_c)\,\mathbf{v}_s$, *and* q_f *satisfy in the domain* Ω_T *the usual Darcy system of filtration*

$$\mathbf{v}_c = m_c\,\mathbf{v}_s + \frac{1}{\mu_2}\mathbb{B}^{(c)}\big(\rho_f \mathbf{F} - \nabla q_f\big), \quad (\mathbf{x},t) \in \Omega_T, \tag{4.3.16}$$

$$\nabla \cdot \mathbf{v} = 0, \quad \mathbf{x} \in \Omega, \quad \mathbf{v}\cdot\mathbf{n} = 0, \quad \mathbf{x} \in S,\ t > 0, \tag{4.3.17}$$

where $\mathbf{n}$ *is a unit normal vector to the boundary* S *at* $\mathbf{x} \in S$.

If the fracture space is connected, then the strictly positively definite constant matrix $\mathbb{B}^{(c)}$, *is defined by the formula*

$$\mathbb{B}^{(c)} = \frac{1}{\mu_1}\sum_{i=1}^{3} \langle \mathbf{V}^i \rangle_{Z_f} \otimes \mathbf{e}_i. \tag{4.3.18}$$

In (4.3.18) functions $\mathbf{V}^i(\mathbf{z})$, $i = 1, 2, 3$, *are solutions to the periodic boundary-value problems*

$$\left.\begin{array}{l} -\frac{1}{2}\triangle_z \mathbf{V}^i + \nabla \Pi^i = \mathbf{e}_i, \quad \nabla_y \cdot \mathbf{V}^i = 0, \quad \mathbf{z} \in Z_f, \\ \mathbf{V}^i = 0, \quad \mathbf{z} \in \gamma_c, \end{array}\right\} \tag{4.3.19}$$

where $\mathbf{e}_i$, $i = 1, 2, 3$, *are the standard Cartesian basis vectors.*

If the fracture space is disconnected (isolated fractures), then the unique solution to the problem (4.3.19) is $\mathbf{V}^i = 0$, $i = 1, 2, 3$, $\mathbb{B}^{(c)} = 0$, *and*

$$\mathbf{v}_c = m_c\,\mathbf{v}_s.$$

The same procedure is applied to the solid motion. First, we solve the microscopic equation (4.3.15) coupled with a continuity equation (4.3.7), find $\mathbf{U}_p$ as an operator on $\mathbb{D}(z,\mathbf{U}_c)$ and $\mathbb{D}(x,\mathbf{w}_s)$, and substitute the result into equation (4.3.14). Next, we solve the obtained microscopic equation and find $\mathbf{U}_c$ as an operator on $\mathbb{D}(x,\mathbf{w}_s)$. Finally, we substitute expressions $\mathbf{U}_p$ and $\mathbf{U}_c$ as operators on $\mathbb{D}(x,\mathbf{w}_s)$ into macroscopic equation (4.3.13) and arrive at the desired homogenized equation for the function $\mathbf{w}_s$.

Lemma 4.7 *Functions $\mathbf{w}_s$ and $\mathbf{U}_c$ satisfy in Z_s the microscopic equation*

$$\nabla_z\cdot\Big((1-\chi_c)\,\mathfrak{A}_0^{(c)}:\big(\mathbb{D}(x,\mathbf{w}_s)+\mathbb{D}(z,\mathbf{U}_c)\big)\Big)=0, \tag{4.3.20}$$

where fourth-rank constant tensor $\mathfrak{A}_0^c$ is defined below by formula (4.3.23).

Proof Let

$$D^{ij}=\frac{1}{2}\left(\frac{\partial w_s^i}{\partial x_j}+\frac{\partial w_s^j}{\partial x_i}\right),\quad d=\nabla\cdot\mathbf{w}_s,\quad \mathbf{w}_s=(w_s^1,\,w_s^2,\,w_s^3),$$

$$D^{ij}_{(c)}=\frac{1}{2}\left(\frac{\partial U_c^i}{\partial z_j}+\frac{\partial U_c^j}{\partial z_i}\right),\quad d_{(c)}=\nabla_z\cdot\mathbf{U}_c,\quad \mathbf{U}_c=(U_c^1,\,U_c^2,\,U_c^3),$$

$$D^{ij}_{(p)}=D^{ij}+D^{ij}_{(c)},\quad d_{(p)}=d+d_{(c)}.$$

As usual, Eq. (4.3.20) follows from the microscopic equations (4.3.14) after we insert into (4.3.14) the expression

$$\langle\mathbb{D}(y,\mathbf{U}_p)\rangle_{Y_s}-\langle\tilde{Q}_s\rangle_{Y_s}\mathbb{I}=\mathfrak{C}_0^{(p)}:\big(\mathbb{D}(x,\mathbf{w}_s)+\mathbb{D}(z,\mathbf{U}_c)\big).$$

To find the exact form of this last expression we look for the solution $\mathbf{U}_p$ to the system of microscopic equations (4.3.15) and (4.3.7) in the form

$$\mathbf{U}_p=\sum_{i,j=1}^{3}\mathbf{U}_p^{ij}(\mathbf{y})\,D^{ij}_{(p)}+\mathbf{U}_p^0(\mathbf{y})\,d_{(p)},\quad \tilde{Q}_s=\sum_{i,j=1}^{3}Q_p^{ij}(\mathbf{y})\,D^{ij}_{(p)}+Q_p^0(\mathbf{y})\,d_{(p)}$$

and arrive at the following periodic boundary-value problems in Y_s:

$$\left.\begin{aligned}&\nabla_y\cdot\Big((1-\chi_p)\big(\mathbb{D}(y,\mathbf{U}_p^{ij})+\mathbb{J}^{ij}-Q_p^{ij}\mathbb{I}\big)\Big)=0,\quad \mathbf{y}\in Y,\\&\nabla_y\cdot\mathbf{U}_p^{ij}=0,\quad \langle\mathbf{U}_p^{ij}\rangle_{Y_s}=0,\quad \mathbf{y}\in Y_s,\end{aligned}\right\} \tag{4.3.21}$$

$$\left.\begin{aligned} \nabla_y \cdot \Big((1-\chi_p)\big(\mathbb{D}(y, \mathbf{U}_p^0) - Q_p^0\,\mathbb{I}\big)\Big) = 0, \quad \mathbf{y} \in Y, \\ \nabla_y \cdot \mathbf{U}_p^0 + 1 = 0, \quad \langle \mathbf{U}_p^0 \rangle_{Y_s} = 0, \quad \mathbf{y} \in Y_s. \end{aligned}\right\} \tag{4.3.22}$$

In (4.3.21)

$$\mathbb{J}^{ij} = \frac{1}{2}(\mathbb{I}^{ij} + \mathbb{I}^{ji}) = \frac{1}{2}(\mathbf{e}_i \otimes \mathbf{e}_j + \mathbf{e}_i \otimes \mathbf{e}_i).$$

Problems (4.3.21) and (4.3.22) are understood in the sense of distributions. For example, the first equation in (4.3.21) is equivalent to the integral identity

$$\int_Y (1-\chi_p)\big(\mathbb{D}(y, \mathbf{U}_p^{ij}) + \mathbb{J}^{ij} - Q_p^{ij}\mathbb{I}\big) : \mathbb{D}(y, \varphi)dy = 0$$

for any smooth and periodic in $\mathbf{y}$ function $\varphi(\mathbf{y})$.

The solvability of the problem (4.3.21) directly follows from the a priori estimate

$$\int_{Y_s} |\nabla \mathbf{U}_p^{ij}|^2 dy \leqslant C,$$

and the this is a consequence of the energy identity

$$\int_{Y_s} \Big(\mathbb{D}(y, \mathbf{U}_p^{ij}) : \mathbb{D}(y, \mathbf{U}_p^{ij}) + \mathbb{J}^{ij} : \mathbb{D}(y, \mathbf{U}_p^{ij})\Big)dy = 0.$$

To solve the problem (4.3.22) we use a 1-periodic function $\mathbf{V}_0 \in \mathbf{W}_2^1(Y_s)$ such that

$$\nabla_y \cdot \mathbf{V}_0 + 1 = 0, \quad \mathbf{y} \in Y_s$$

(see proof of Theorem 1.4).

Then the solvability of the problem (4.3.22) follows from the energy equality

$$\int_{Y_s} \Big(\mathbb{D}(y, \mathbf{U}_p^0) : \big(\mathbb{D}(y, \mathbf{U}_p^0) - \mathbb{D}(y, \mathbf{V}_0)\big)\Big)dy = 0,$$

which is a result of substituting the test function $\varphi = (\mathbf{U}_p^0 - \mathbf{V}_0)$ into the corresponding to the first equation in (4.3.22) integral identity.

Thus,

$$\begin{aligned} \langle \mathbb{D}(y, \mathbf{U}_p) \rangle_{Y_s} - \langle \widetilde{Q}_s \rangle_{Y_s}\mathbb{I} &= \sum_{i,j=1}^{3} \langle \mathbb{D}(y, \mathbf{U}_p^{ij}) \rangle_{Y_s} D_{(p)}^{ij} + \langle \mathbb{D}(y, \mathbf{U}_p^0) \rangle_{Y_s} d_{(p)} \\ &\quad - \Big(\sum_{i,j=1}^{3} \langle Q_p^{ij} \rangle_{Y_s} D_{(p)}^{ij}\Big)\mathbb{I} - \Big(\langle Q_p^0 \rangle_{Y_s} d^{(p)}\Big)\mathbb{I} \\ &= \sum_{i,j=1}^{3} \Big(\langle \mathbb{D}(y, \mathbf{U}_p^{ij}) \rangle_{Y_s} - \langle Q_p^{ij} \rangle_{Y_s}\mathbb{I}\Big) D_{(p)}^{ij} \end{aligned}$$

$$
\begin{aligned}
&+ \Big(\langle \mathbb{D}(y, \mathbf{U}_p^0) \rangle_{Y_s} - \langle Q_p^0 \rangle_{Y_s} \mathbb{I} \Big)\, d_{(p)} \\
&= \sum_{i,j=1}^{3} \Big(\langle \mathbb{D}(y, \mathbf{U}_p^{ij}) \rangle_{Y_s} \otimes \mathbb{J}^{ij} - \langle Q_p^{ij} \rangle_{Y_s} \mathbb{I} \otimes \mathbb{J}^{ij} \Big) \\
&: \Big(\mathbb{D}(x, \mathbf{w}_s) + \mathbb{D}(z, \mathbf{U}_c) \Big) \\
&+ \Big(\langle \mathbb{D}(y, \mathbf{U}_p^0) \rangle_{Y_s} \otimes \mathbb{I} - \langle Q_p^0 \rangle_{Y_s} \mathbb{I} \otimes \mathbb{I} \Big) : \Big(\mathbb{D}(x, \mathbf{w}_s) + \mathbb{D}(z, \mathbf{U}_c) \Big) \\
&= \Big(\mathfrak{C}_{0,1}^{(p)} + \mathfrak{C}_{0,2}^{(p)} + \mathfrak{C}_{0,3}^{(p)} + \mathfrak{C}_{0,4}^{(p)} \Big) : \Big(\mathbb{D}(x, \mathbf{w}_s) + \mathbb{D}(z, \mathbf{U}_c) \Big) \\
&= \mathfrak{C}_0^{(p)} : \Big(\mathbb{D}(x, \mathbf{w}_s) + \mathbb{D}(z, \mathbf{U}_c) \Big),
\end{aligned}
$$

where

$$
\begin{aligned}
\mathfrak{A}_0^{(c)} &= (1 - m_p) \sum_{i,j=1}^{3} \mathbb{J}^{ij} \otimes \mathbb{J}^{ij} + \mathfrak{C}_0^{(p)} = (1 - m_p)\mathfrak{J} + \mathfrak{C}_0^{(p)}, \\
\mathfrak{J} &= \sum_{i,j=1}^{3} \mathbb{J}^{ij} \otimes \mathbb{J}^{ij}, \quad \mathfrak{C}_0^{(p)} = \mathfrak{C}_{0,1}^{(p)} + \mathfrak{C}_{0,2}^{(p)} + \mathfrak{C}_{0,3}^{(p)} + \mathfrak{C}_{0,4}^{(p)}, \\
\mathfrak{C}_{0,1}^{(p)} &= \sum_{i,j=1}^{3} \langle \mathbb{D}(y, \mathbf{U}_p^{ij}) \rangle_{Y_s} \otimes \mathbb{J}^{ij}, \quad \mathfrak{C}_{0,2}^{(p)} = \langle \mathbb{D}(y, \mathbf{U}_p^0) \rangle_{Y_s} \otimes \mathbb{I}, \\
\mathfrak{C}_{0,3}^{(p)} &= -\sum_{i,j=1}^{3} \langle Q_p^{ij} \rangle_{Y_s} \mathbb{I} \otimes \mathbb{J}^{ij}, \quad \mathfrak{C}_{0,4}^{(p)} = -\langle Q_p^0 \rangle_{Y_s} \mathbb{I} \otimes \mathbb{I}.
\end{aligned}
\tag{4.3.23}
$$

Lemma 4.8 *Tensors* $\mathfrak{A}_0^{(c)}$ *and* $\mathfrak{C}_0^{(p)}$ *are symmetric and the tensor* $\mathfrak{A}_0^{(c)}$ *is strictly positively definite, that is for any arbitrary symmetric matrices* $\zeta = (\zeta_{ij})$ *and* $\eta = (\eta_{ij})$

$$
\big(\mathfrak{A}_0^{(c)} : \zeta\big) : \eta = \big(\mathfrak{A}_0^{(c)} : \eta\big) : \zeta\,, \quad \textit{and} \quad \big(\mathfrak{A}_0^{(c)} : \zeta\big) : \zeta \geqslant \beta\,(\zeta : \zeta),
$$

where positive constant β *is independent of* ζ.

Proof To prove the lemma we need some properties of the tensor $\mathfrak{A}_0^{(c)}$, which follow from the equalities

$$
-\langle Q_p^0 \rangle_{Y_s} = \langle \mathbb{D}(y, \mathbf{U}_p^0) : \mathbb{D}(y, \mathbf{U}_p^0) \rangle_{Y_s}, \tag{4.3.24}
$$

$$
\langle \mathbb{D}(y, \mathbf{U}_p^{ij}) : \mathbb{D}(y, \mathbf{U}_p^0) \rangle_{Y_s} = 0, \tag{4.3.25}
$$

$$
\langle Q_p^{ij} \rangle_{Y_s} = -\langle \mathbb{D}(y, \mathbf{U}_p^0) : \mathbb{J}^{ij} \rangle_{Y_s}, \tag{4.3.26}
$$

$$
\langle \mathbb{D}(y, \mathbf{U}_p^{ij}) : \mathbb{D}(y, \mathbf{U}_p^{kl}) \rangle_{Y_s} + \langle \mathbb{J}^{ij} : \mathbb{D}(y, \mathbf{U}_p^{kl}) \rangle_{Y_s} = 0, \tag{4.3.27}
$$

for all $i,\ j,\ k,\ l = 1,\ 2,\ 3$.

Equation (4.3.24) is an integral identity, corresponding to the first equation in (4.3.22) with the test function $\mathbf{U}_p^0$.

Equation (4.3.25) is an integral identity, corresponding to the first equation in (4.3.22) with the test function $\mathbf{U}_p^{ij}$.

Equation (4.3.26) is an integral identity, corresponding to the first equation in (4.3.21) with the test function $\mathbf{U}_p^0$. Here we additionally took into account relations (4.3.25).

Finally, equation (4.3.27) is an integral identity, corresponding to the first equation in (4.3.21) with the test function $\mathbf{U}_p^{kl}$.

Next we put

$$\mathbf{Y}_\zeta = \sum_{i,j=1}^{3} \mathbf{U}_p^{ij}\zeta_{ij}, \quad \mathbf{Y}_\eta = \sum_{i,j=1}^{3} \mathbf{U}_p^{ij}\eta_{ij}, \quad \mathbf{Y}_\zeta^0 = \mathbf{u}_p^0 \operatorname{tr}\zeta, \quad \mathbf{Y}_\eta^0 = \mathbf{u}_p^0 \operatorname{tr}\eta.$$

Then

$$\mathfrak{C}_{0,1}^{(p)} : \zeta = \langle \mathbb{D}_y(\mathbf{Y}_\zeta)\rangle_{Y_s}, \quad \mathfrak{C}_{0,2}^{(p)} : \zeta = \langle \mathbb{D}_y(\mathbf{Y}_\zeta^0)\rangle_{Y_s},$$

and equations (4.3.24)–(4.3.27) take the form

$$\big(\mathfrak{C}_{0,4}^{(p)} : \zeta\big) : \eta = \langle \mathbb{D}_y(\mathbf{Y}_\zeta^0) : \mathbb{D}_y(\mathbf{Y}_\eta^0)\rangle_{Y_s}, \tag{4.3.28}$$

$$\langle \mathbb{D}_y(\mathbf{Y}_\eta) : \mathbb{D}_y(\mathbf{Y}_\zeta^0)\rangle_{Y_s} = 0, \tag{4.3.29}$$

$$\big(\mathfrak{C}_{0,3}^{(p)} : \zeta\big) : \eta = \big(\mathfrak{C}_{0,2}^{(p)} : \eta\big) : \zeta, \tag{4.3.30}$$

$$\big(\mathfrak{C}_{0,1}^{(p)} : \eta\big) : \zeta + \langle \mathbb{D}_y(\mathbf{Y}_\zeta) : \mathbb{D}_y(\mathbf{Y}_\eta)\rangle_{Y_s} = 0. \tag{4.3.31}$$

Therefore,

$$\begin{aligned}\big(\mathfrak{A}_0^{(c)} : \zeta\big) : \eta &= (1-m_p)\zeta : \eta + \big(\mathfrak{C}_0^{(p)} : \zeta\big) : \eta = \langle \mathbb{D}_y(\mathbf{Y}_\eta^0)\rangle_{Y_s} : \zeta \\ &\quad + \langle \mathbb{D}_y(\mathbf{Y}_\zeta^0)\rangle_{Y_s} : \eta + \eta : \langle \mathbb{D}_y(\mathbf{Y}_\zeta)\rangle_{Y_s} + \langle \mathbb{D}_y(\mathbf{Y}_\zeta^0) : \mathbb{D}_y(\mathbf{Y}_\eta^0)\rangle_{Y_s} \\ &\quad + (1-m_p)\zeta : \eta.\end{aligned}$$

Taking into account (4.3.29) and (4.3.31) we finally get

$$\begin{aligned}\big(\mathfrak{A}_0^{(c)} : \zeta\big) : \eta = {} & (1-m_p)\zeta : \eta + \langle \mathbb{D}(y, \mathbf{Y}_\zeta^0) : \mathbb{D}(y, \mathbf{y}_\eta^0)\rangle_{Y_s} \\ & + \langle \mathbb{D}(y, \mathbf{Y}_\eta^0)\rangle_{Y_s} : \zeta + \langle \mathbb{D}(y, \mathbf{Y}_\zeta^0)\rangle_{Y_s} : \eta + \langle \mathbb{D}(y, \mathbf{y}_\zeta) : \mathbb{D}(y, \mathbf{Y}_\eta)\rangle_{Y_s} \\ & + \zeta : \langle \mathbb{D}(y, \mathbf{Y}_\eta)\rangle_{Y_s} + \eta : \langle \mathbb{D}(y, \mathbf{Y}_\zeta)\rangle_{Y_s} \\ = {} & \Big\langle \big(\mathbb{D}(y, \mathbf{Y}_\zeta + \mathbf{y}_\zeta^0) + \zeta\big) : \big(\mathbb{D}(y, \mathbf{Y}_\eta + \mathbf{Y}_\eta^0) + \eta\big)\Big\rangle_{Y_s}.\end{aligned} \tag{4.3.32}$$

Equations (4.3.32) and (4.3.23) show that tensors $\mathfrak{A}_0^{(c)}$ and $\mathfrak{C}_0^{(p)}$ are symmetric:

$$\big(\mathfrak{A}_0^{(c)} : \zeta\big) : \eta = \big(\mathfrak{A}_0^{(c)} : \eta\big) : \zeta, \quad \big(\mathfrak{C}_0^{(p)} : \zeta\big) : \eta = -(1-m_p)\zeta : \zeta + \big(\mathfrak{A}_0^{(c)} : \zeta\big) : \eta.$$

In particular,

$$\big(\mathfrak{A}_0^{(c)} : \zeta\big) : \zeta = \big\langle \big(\mathbb{D}(y, \mathbf{Y}_\zeta + \mathbf{Y}_\zeta^0) + \zeta\big) : \big(\mathbb{D}(y, \mathbf{Y}_\zeta + \mathbf{Y}_\zeta^0) + \zeta\big)\big\rangle_{Y_s} > 0,$$

and $\mathfrak{A}_0^{(c)}$ is strictly positively definite.

In fact, if $\big(\mathfrak{A}_0^{(c)} : \zeta^0\big) : \zeta^0 = 0$ for some ζ^0, such that $\zeta^0 : \zeta^0 = 1$, then

$$\mathbb{D}(y, \mathbf{Y}_{\zeta^0} + \mathbf{Y}_{\zeta^0}^0) + \zeta^0 = 0.$$

The last equality is possible if and only if the periodic function $\mathbf{Y}_{\zeta^0} + \mathbf{Y}_{\zeta^0}^0$ is a linear one. But due to the geometry of the solid cell Y_s it is possible only if

$$\mathbf{Y}_{\zeta^0}(\mathbf{Y}) + \mathbf{Y}_{\zeta^0}^0(\mathbf{y}) = \mathbf{U}^0 = const.$$

On the other hand, the function

$$\mathbf{U}^0(\mathbf{y}) = \sum_{i,j=1}^{3} \mathbf{U}_p^{ij}(\mathbf{y})\zeta_{ij}^0 + \mathbf{u}_p^0(\mathbf{y}) \operatorname{tr}\zeta^0$$

is a solution to the problem

$$\left.\begin{array}{l} \nabla_y \cdot \Big((1-\chi_p)\big(\mathbb{D}(y, \mathbf{U}^0) + \zeta^0 - Q^0\mathbb{I}\big)\Big) = 0, \quad \mathbf{y} \in Y, \\ \nabla_y \cdot \mathbf{U}^0 + \operatorname{tr}\zeta^0 = 0, \quad \langle \mathbf{U}^0\rangle_{Y_s} = 0, \quad \mathbf{y} \in Y_s. \end{array}\right\}$$

Therefore $\zeta^0 = 0$, which contradicts the supposition.

Lemma 4.9 *Functions $\mathbf{w}_s$ and q_f satisfy in Ω_T the homogenized equation*

$$\nabla \cdot \Big(\lambda_0\, \mathfrak{B}_0^{(s)} : \mathbb{D}(x, \mathbf{w}_s) - q_f\mathbb{I}\Big) + \hat{\rho}\mathbf{F} = 0, \tag{4.3.33}$$

where fourth-rank constant tensor $\mathfrak{B}_0^{(s)}$ is defined below by formula (4.3.36).

Proof Following the standard scheme, we look for the solution to the microscopic equation (4.3.20) in the form

$$\mathbf{U}_c(\mathbf{x}, t, \mathbf{z}) = \sum_{i,j=1}^{3} \mathbf{U}_c^{ij}(\mathbf{z}) D_{ij}(\mathbf{x}, t),$$

where functions $\mathbf{U}_c^{ij}$ satisfy in Z the periodic boundary-value problem

$$\nabla_z \cdot \Big((1-\chi_c)\,\mathfrak{A}_0^{(c)} : \big(\mathbb{D}(z, \mathbf{U}_c^{ij}) + J^{ij}\big)\Big) = 0, \quad \langle \mathbf{U}_c^{ij}\rangle_{Z_s} = 0, \tag{4.3.34}$$

which is understood in the sense of distributions. Thus

$$\langle \mathbb{D}(z, \mathbf{U}_c)\rangle_{Z_s} = \Big(\sum_{i,j=1}^{3} \langle \mathbb{D}(z, \mathbf{U}_c^{ij})\rangle_{Z_s} \otimes \mathbb{J}^{ij}\Big) : \mathbb{D}(x, \mathbf{w}_s) = \mathfrak{C}^{(c)} : \mathbb{D}(x, \mathbf{w}_s),$$

$$\mathfrak{C}_0^{(c)} = \sum_{i,j=1}^{3} \langle \mathbb{D}(z, \mathbf{U}_c^{ij})\rangle_{Z_s} \otimes \mathbb{J}^{ij}, \tag{4.3.35}$$

and

$$\begin{aligned}
\Big\langle\big\langle\big(\mathbb{D}(y, \mathbf{U}_p) - \widetilde{Q}_s\mathbb{I}\big)\big\rangle_{Y_s}\Big\rangle_{Z_s} &= \mathfrak{C}_0^{(p)} : \big((1-m_c)\mathbb{D}(x, \mathbf{w}_s) + \langle \mathbb{D}(z, \mathbf{U}_c)\rangle_{Z_s}\big)\\
&= \mathfrak{C}_0^{(p)} : \big((1-m_c)\mathbb{D}(x, \mathbf{w}_s) + \mathfrak{C}_0^{(c)} : \mathbb{D}(x, \mathbf{w}_s)\big)\\
&= \mathfrak{C}_0^{(p)} : \Big(\big((1-m_c)\,\mathfrak{J} + \mathfrak{C}_0^{(c)}\big) : \mathbb{D}(x, \mathbf{w}_s)\Big)\\
&= \Big((1-m_c)\,\mathfrak{C}_0^{(p)} + \mathfrak{C}_0^{(p)} : \mathfrak{C}_0^{(c)}\Big) : \mathbb{D}(x, \mathbf{w}_s),
\end{aligned}$$

$$\begin{aligned}
\mathfrak{B}_0^{(s)} &= (1-m)\,\mathfrak{J} + (1-m_p)\,\mathfrak{C}_0^{(c)} + (1-m_c)\,\mathfrak{C}_0^{(p)} + \mathfrak{C}_0^{(p)} : \mathfrak{C}_0^{(c)}\\
&= (1-m)\,\mathfrak{J} + \big((1-m_p)\,\mathfrak{J} + \mathfrak{C}_0^{(p)}\big) : \mathfrak{C}_0^{(c)} + (1-m_c)\,\mathfrak{C}_0^{(p)}\\
&= (1-m)\,\mathfrak{J} + \mathfrak{A}_0^{(c)} : \mathfrak{C}_0^{(c)} + (1-m_c)\,\mathfrak{C}_0^{(p)}\\
&= (1-m_c)\,\big((1-m_p)\,\mathfrak{J} + \mathfrak{C}_0^{(p)}\big) + \mathfrak{A}_0^{(c)} : \mathfrak{C}_0^{(c)}\\
&= (1-m_c)\,\mathfrak{A}_0^{(c)} + \mathfrak{A}_0^{(c)} : \mathfrak{C}_0^{(c)} = \mathfrak{A}_0^{(c)} : \big((1-m_c)\,\mathfrak{J} + \mathfrak{C}_0^{(c)}\big),
\end{aligned}$$

where we have used equalities

$$(1-m) = (1-m_p)(1-m_c)$$

and

$$\mathfrak{J} : \mathfrak{A} = \mathfrak{A} : \mathfrak{J} = \mathfrak{A}$$

for any fourth-rank tensor $\mathfrak{A}$.

Finally

$$\mathfrak{B}_0^{(s)} = \mathfrak{A}_0^{(c)} : \big((1-m_c)\,\mathfrak{J} + \mathfrak{C}_0^{(c)}\big), \tag{4.3.36}$$

where $\mathfrak{C}_0^{(c)}$ is defined by (4.3.35).

Lemma 4.10 *The tensor $\mathfrak{B}_0^{(s)}$ is symmetric and strictly positively definite.*

Proof To prove the second statement of the Lemma we use the equality

$$\int_{Z_s} \big(\mathfrak{A}_0^{(c)} : \mathbb{D}(z, \mathbf{U}_c^{ij})\big) : \mathbb{D}(z, \mathbf{U}_c^{kl})dz + \int_{Z_s} \big(\mathfrak{A}_0^{(c)} : \mathbb{D}(z, \mathbf{J}^{ij})\big) : \mathbb{D}(z, \mathbf{U}_c^{kl})dz = 0, \tag{4.3.37}$$

which is simply the integral identity with the test function $\mathbf{U}_c^{kl}$, corresponding to the differential equation (4.3.34).

Let

$$\mathbf{Z}_\zeta = \sum_{i,j=1}^{3} \mathbf{U}_c^{ij}\zeta_{ij}, \quad \mathbf{Z}_\eta = \sum_{i,j=1}^{3} \mathbf{U}_c^{ij}\eta_{ij}.$$

Then (4.3.37) takes the form

$$\big\langle\big(\mathfrak{A}_0^{(c)} : \mathbb{D}(z, \mathbf{Z}_\zeta)\big) : \mathbb{D}(z, \mathbf{Z}_\eta)\big\rangle_{Z_s} + \big\langle\big(\mathfrak{A}_0^{(c)} : \mathbb{D}(z, \mathbf{Z}_\eta)\big) : \zeta\big\rangle_{Z_s} = 0. \tag{4.3.38}$$

Also note that by definition

$$\mathfrak{C}_0^{(c)} : \zeta = \langle \mathbb{D}(z, \mathbf{Z}_\zeta)\rangle_{Z_s}. \tag{4.3.39}$$

Relations (4.3.38) and (4.3.39) result in

$$\begin{aligned}
\big(\mathfrak{B}_0^{(s)} : \zeta\big) : \eta &= (1 - m_c)\big(\mathfrak{A}_0^{(c)} : \zeta\big) : \eta + \Big(\big(\mathfrak{A}_0^{(c)} : \mathfrak{C}_0^{(c)}\big) : \zeta\Big) : \eta \\
&= (1 - m_c)\big(\mathfrak{A}_0^{(c)} : \zeta\big) : \eta + \big(\mathfrak{A}_0^{(c)} : \langle \mathbb{D}(z, \mathbf{Z}_\zeta)\rangle_{Z_s}\big) : \eta \\
&= (1 - m_c)\big(\mathfrak{A}_0^{(c)} : \zeta\big) : \eta + \langle\big(\mathfrak{A}_0^{(c)} : \mathbb{D}(z, \mathbf{Z}_\zeta)\big) : \mathbb{D}(z, \mathbf{Z}_\eta)\rangle_{Z_s} \\
&\quad + \big\langle\big(\mathfrak{A}_0^{(c)} : \mathbb{D}(z, \mathbf{Z}_\eta)\big) : \zeta\rangle_{Z_s} + \big(\mathfrak{A}_0^{(c)} : \langle \mathbb{D}(z, \mathbf{Z}_\zeta)\rangle_{Z_s}\big) : \eta \\
&= \big\langle\big(\mathfrak{A}_0^{(c)} : \big(\mathbb{D}(z, \mathbf{Z}_\zeta) + \zeta\big)\big) : \big(\mathbb{D}(z, \mathbf{Z}_\eta) + \eta\big)\big\rangle_{Z_s},
\end{aligned}$$

which proves the symmetry of $\mathfrak{B}_0^{(s)}$.

In particular,

$$\big(\mathfrak{B}_0^{(s)} : \eta\big) : \eta = \Big\langle\Big(\mathfrak{A}_0^{(c)} : \big(\mathbb{D}(z, \mathbf{z}_\eta) + \eta\big)\Big) : \big(\mathbb{D}(z, \mathbf{z}_\eta) + \eta\big)\Big\rangle_{Z_s} > \beta\,(\eta : \eta).$$

4.4 Proof of Theorem 4.4

The homogenization procedure proves the existence at least one (weak) solution to the problem (4.1.8)–(4.1.13). But we also may prove the correctness (uniqueness and existence of the solution) of the problem (4.1.8)–(4.1.13) directly,using basic a priori estimates

$$n\int_0^t\int_\Omega |\nabla \mathbf{v}_s^{(n)}(\mathbf{x},\tau)|^2 dx d\tau + \frac{1}{\mu_2}\int_\Omega |\nabla q_f^{(n)}(\mathbf{x},t)|^2 dx \leqslant CF^2. \tag{4.4.1}$$

To derive (4.4.1) we multiply the continuity equation (4.1.8) by $\dfrac{\partial q_f^{(n)}}{\partial t}$, integrate by parts over domain Ω, express the velocity using the representation (4.1.11) and Darcy's law (4.1.12). Next we differentiate (4.1.9) with respect to time, multiply by $\mathbf{v}_s^{(n)}$, integrate by parts over domain Ω, and sum results:

$$\int_\Omega \left(\lambda_0\, \mathbb{D}(x,\mathbf{v}_s) : \left(\mathbb{A}^{(s)} : \mathbb{D}(x,\mathbf{v}_s)\right) + \frac{1}{\mu_2}\left(\mathbb{B}^{(c)} \cdot (\nabla\, q_f^{(n)})\right) \cdot \left(\nabla \frac{\partial q_f^{(n)}}{\partial t}\right)\right) dx$$

$$= \int_\Omega \left(\frac{\rho_f}{\mu_2}(\mathbb{B}^{(c)} \cdot \mathbf{F}) \cdot \left(\nabla \frac{\partial q_f^{(n)}}{\partial t}\right) + \hat{\rho}\frac{\partial \mathbf{F}}{\partial t} \cdot \mathbf{v}_s\right) dx. \tag{4.4.2}$$

Estimate (4.4.1) follows now from (4.4.2), and Hölder, Gronwall, and Korn's inequalities.

Finally, we apply the standard compactness results to choose the convergent subsequences $\{\mathbf{v}_c^{(n_k)}\}$ and $\{q_f^{(n_k)}\}$, and pass to the limit as $n_k \to \infty$ in integral identities, corresponding to (4.1.8) and (4.1.9).

The estimate (4.4.1) and the representation (4.1.11) also guarantee the strong convergence of $\{\mathbf{v}_s^{(n)}\}$ and $\{\mathbf{v}_p^{(n)}\}$ to zero as $n \to \infty$.

4.5 Proofs of Theorems 4.5 and 4.6

As in the proofs of Theorems 4.2 and 4.3 we conclude that there exists a subsequence of small parameters $\{\delta > 0\}$ and functions $q_f, q_s \in L_2(\Omega_T)$, $\mathbf{v}, \mathbf{v}_p, \mathbf{v}_c, \mathbf{v}_s \in \mathbf{L}_2(\Omega_T)$ and $\mathbf{w}_s \in \overset{\circ}{\mathbf{W}}{}_2^{1,0}(\Omega_T)$ such that

$$\tilde{\chi}^\delta q^\delta \rightharpoonup q_f, \quad (1-\tilde{\chi}^\delta) q^\delta \rightharpoonup q_s, \quad \mathbf{v}^\delta \rightharpoonup \mathbf{v}, \quad \mathbf{v}_p^\varepsilon \rightharpoonup \mathbf{v}_p, \quad \mathbf{v}_c^\delta \rightharpoonup \mathbf{v}_c,$$
$$\mathbf{v}_s^\delta \rightharpoonup \mathbf{v}_s, \quad \mathbf{w}_s^\delta \rightharpoonup \mathbf{w}_s, \quad \mathbb{D}(x,\mathbf{w}_s^\delta) \rightharpoonup \mathbb{D}(x,\mathbf{w}_s) \tag{4.5.1}$$

weakly in $L_2(\Omega_T)$ and $\mathbf{L}_2(\Omega_T)$ as $\delta \searrow 0$, and

$$\tilde{\chi}^\delta \alpha_\mu \mathbb{D}(x,\mathbf{v}^\delta) \to 0 \tag{4.5.2}$$

strongly in $\mathbf{L}_2(\Omega_T)$ as $\delta \searrow 0$.

For $\mu_2 = \infty$ relations (4.1.7) hold true, and for $\mu_2 < \infty$ relations (4.1.11) hold true.

At the same time the sequences $\{\tilde{\chi}^\delta q^\delta\}$, $\{(1-\tilde{\chi}^\delta) q^\delta\}$, $\{\mathbf{v}^\delta\}$, $\{\mathbf{v}_c^\delta\}$, and $\{\mathbb{D}(x, \mathbf{w}_s^\delta)\}$ three-scale converge (up to some subsequences) to $Q_f(\mathbf{x}, t, \mathbf{y}, \mathbf{z}) = (1/m)\, q_f(\mathbf{x}, t)\, \tilde{\chi}(\mathbf{y}, \mathbf{z})$, $Q_s(\mathbf{x}, t, \mathbf{y}, \mathbf{z}) = (1-\tilde{\chi})Q_s$, $\mathbf{V}(\mathbf{x}, t, \mathbf{y}, \mathbf{z})$, $\mathbf{V}_c(\mathbf{x}, t, \mathbf{y}, \mathbf{z})$, and $\mathbb{D}(x, \mathbf{w}_s) + \mathbb{D}\big(z, \mathbf{U}_c(\mathbf{x}, t, \mathbf{z})\big) + \mathbb{D}\big(y, \mathbf{U}_p(\mathbf{x}, t, \mathbf{y}, \mathbf{z})\big)$ respectively, and the sequence $\{\mathbf{w}_s^\delta\}$ three-scale converges to the function $\mathbf{w}_s(\mathbf{x}, t)$.

These functions satisfy micro- and macroscopic equations (4.3.10)–(4.3.15), and additional micro- and macroscopic continuity equations and the boundary condition

$$\frac{1}{c_f^2}\frac{\partial q_f}{\partial t} + \frac{1}{c_s^2}\frac{\partial q_s}{\partial t} + \nabla \cdot \mathbf{v} = 0, \quad \mathbf{v} \cdot \mathbf{n} = 0,\, \mathbf{x} \in S, \tag{4.5.3}$$

$$\frac{1}{c_s^2} Q_s + (1-\chi)\big(\nabla \cdot \mathbf{w}_s + \nabla_z \cdot \mathbf{U_c} + \nabla_y \cdot \mathbf{U_p}\big) = 0. \tag{4.5.4}$$

The equation (4.5.3) is derived in the usual way (see Chaps. 1 and 2), and the Eq. (4.5.4) is just a three-scale limit in the continuity equation

$$(1-\tilde{\chi}^\varepsilon)\left(\frac{1}{c_s^2} q^\varepsilon + \nabla \cdot \mathbf{w}_s^\varepsilon\right) = 0$$

for the solid component.

Equations (4.3.10)–(4.3.12) result in Darcy's law (4.1.12) as in Theorem 4.2.

The dynamics equations for the solid component follow from Eqs. (4.3.13)–(4.3.15) and the Eq. (4.5.4).

Lemma 4.11 *For almost all* $(\mathbf{x}, t) \in \Omega_T$ *functions* $\mathbf{w}_s$, $\mathbf{U}_c$, *and* $\tilde{q}_f$ *satisfy in* Z_s *the microscopic equation*

$$\nabla_z \cdot \Big((1-\chi_c)\Big(\mathfrak{A}^{(c)} : \big(\mathbb{D}(x, \mathbf{w}_s) + \mathbb{D}(z, \mathbf{U}_c)\big) + \tilde{q}_f\, \mathbb{C}^{(c)}\Big)\Big) = 0, \tag{4.5.5}$$

where the constant forth rank tensor $\mathfrak{A}^{(c)}$ *and the constant matrix* $\mathbb{C}^{(c)}$ *are defined below by formulae (4.5.8) and (4.5.9).*

Proof Let

$$\mathbb{T}^{(p)} = \mathbb{D}(x, \mathbf{w}_s) + \mathbb{D}(z, \mathbf{U}_c),\;\; d^{(p)} = \nabla \cdot \mathbf{w}_s + \nabla_z \cdot \mathbf{U}_c + \frac{1}{\beta}\tilde{q}_f,$$

$$\beta = \frac{c_s^*}{\lambda_0},\;\; \mathbb{T}^{(p)} = \sum_{i,j=1}^{3} T_{ij}^{(p)} \mathbb{J}^{ij}.$$

As usual, Eq. (4.5.5) follows from the microscopic equations (4.3.14), after we insert into (4.3.14) the expression

$$\langle \mathbb{D}_y(\mathbf{U}_p) - \tilde{Q}_s \mathbb{I} \rangle_{Y_s} = \mathfrak{C}^{(p)} : \big(\mathbb{D}(x, \mathbf{w}_s) + \mathbb{D}(z, \mathbf{U}_c) \big) + \tilde{q}_f \, \mathbb{C}^{(c)}.$$

To find the exact form of this last expression we look for the solution $\mathbf{U}_p$ and $\tilde{Q}_s$ to the system of microscopic equations (4.3.15) and (4.5.4) in the form

$$-\tilde{Q}_s = \beta \, \nabla_y \cdot \mathbf{U_p} + \beta \, d^{\,(p)}, \mathbf{U}_p$$

$$= \sum_{i,j=1}^{3} \mathbf{V}_p^{(ij)}(\mathbf{y}) \, T_{ij}^{(p)} + \mathbf{V}_p^{(0)}(\mathbf{y}) \, d^{\,(p)},$$

and arrive at the following periodic boundary value problems in Y_s:

$$\nabla_y \cdot \Big((1 - \chi_p) \big((\mathbb{D}_y(\mathbf{V}_p^{(ij)}) + \mathbb{J}^{ij}) + \beta \, (\nabla_y \cdot \mathbf{V}_p^{(ij)}) \, \mathbb{I} \big) \Big) = 0, \tag{4.5.6}$$

$$\nabla_y \cdot \Big((1 - \chi_p) \big(\mathbb{D}_y(\mathbf{V}_p^{(0)}) + \beta \, (\nabla_y \cdot \mathbf{V}_p^{(0)} + 1) \mathbb{I} \big) \Big) = 0 \tag{4.5.7}$$

with the following normalization conditions

$$\langle \mathbf{V}_p^{(ij)} \rangle_{Y_s} = 0, \quad \langle \mathbf{V}_p^{(0)} \rangle_{Y_s} = 0.$$

Equations (4.5.6) and (4.5.7) are understood in the sense of distributions. For example, the Eq. (4.5.6) is equivalent to the integral identity

$$\int_Y (1 - \chi_p) \big((\mathbb{D}(y, \mathbf{V}_p^{(ij)}) + \mathbb{J}^{ij}) + \beta \, (\nabla_y \cdot \mathbf{V}_p^{(ij)}) \, \mathbb{I} \big) : \mathbb{D}(y, \varphi) dy = 0$$

for any smooth periodic in $\mathbf{y}$ function $\varphi(\mathbf{y})$.

The solvability of (4.5.6) and (4.5.7) is already discussed in previous sections. Thus,

$$\langle \mathbb{D}(y, \mathbf{U}_p) - \tilde{Q}_s \mathbb{I} \rangle_{Y_s} = \sum_{i,j=1}^{3} \langle \mathbb{D}(y, \mathbf{V}_p^{(ij)}) \rangle_{Y_s} T_{ij}^{(p)} + \langle \mathbb{D}(y, \mathbf{V}_p^{(0)}) \rangle_{Y_s} d^{(p)}$$

$$+ \beta \Big(\sum_{i,j=1}^{3} \langle \nabla \cdot \mathbf{V}_p^{(ij)} \rangle_{Y_s} T_{ij}^{(p)} \Big) \mathbb{I} + d^{(p)} \, \beta \, \langle \nabla_y \cdot \mathbf{V}_p^{(0)} + 1 \rangle_{Y_s} \mathbb{I}$$

$$= \sum_{i,j=1}^{3} \Big(\langle (\mathbb{D}(y, \mathbf{V}_p^{(ij)}) + \beta \, \nabla \cdot \mathbf{V}_p^{(ij)} \mathbb{I} \rangle_{Y_s} \Big) T_{ij}^{(p)}$$

$$+ d^{(p)} \, \langle \mathbb{D}(y, \mathbf{V}_p^{(0)}) + \beta \, (\nabla_y \cdot \mathbf{V}_p^{(0)} + 1) \mathbb{I} \rangle_{Y_s}$$

$$
\begin{aligned}
&= \Big(\sum_{i,j=1}^{3} \langle (\mathbb{D}(y, \mathbf{V}_p^{(ij)}) + \beta\, \nabla \cdot \mathbf{V}_p^{(ij)} \mathbb{I} \rangle_{Y_s} \otimes \mathbb{J}^{ij} \Big) : \mathbb{T}^{(p)} \\
&+ \Big(\langle \mathbb{D}(y, \mathbf{V}_p^{(0)}) + \beta\, (\nabla_y \cdot \mathbf{V}_p^{(0)} + 1) \mathbb{I} \rangle_{Y_s} \otimes \mathbb{I} \Big) : \mathbb{T}^{(p)} \\
&+ \frac{1}{\beta} \tilde{q}_f \, \langle \mathbb{D}(y, \mathbf{V}_p^{(0)}) + \beta\, (\nabla_y \cdot \mathbf{V}_p^{(0)} + 1) \mathbb{I} \rangle_{Y_s} \\
&= \mathfrak{C}^{(p)} : \mathbb{T}^{(p)} + \tilde{q}_f \, \mathbb{C}^{(c)},
\end{aligned}
$$

where

$$
\mathfrak{A}^{(c)} = (1 - m_p) \mathfrak{J} + \mathfrak{C}^{(p)}, \tag{4.5.8}
$$

$$
\mathbb{C}^{(c)} = \frac{1}{\beta} \langle \mathbb{D}(y, \mathbf{V}_p^{(0)}) + \beta\, (\nabla_y \cdot \mathbf{V}_p^{(0)} + 1) \mathbb{I} \rangle_{Y_s}, \tag{4.5.9}
$$

and

$$
\mathfrak{C}^{(p)} = \sum_{i,j=1}^{3} \langle (\mathbb{D}(y, \mathbf{V}_p^{(ij)}) + \beta\, \nabla \cdot \mathbf{V}_p^{(ij)} \mathbb{I} \rangle_{Y_s} \otimes \mathbb{J}^{ij} + \beta\, \mathbb{C}^{(c)} \otimes \mathbb{I}.
$$

Lemma 4.12 *Tensor $\mathfrak{A}^{(c)}$ is symmetric and is strictly positively definite, that is for any arbitrary symmetric matrices $\zeta = (\zeta_{ij})$ and $\eta = (\eta_{ij})$*

$$
\big(\mathfrak{A}^{(c)} : \zeta \big) : \eta = \big(\mathfrak{A}^{(c)} : \eta \big) : \zeta \,, \quad \textit{and} \quad \big(\mathfrak{A}^{(c)} : \zeta \big) : \zeta \geqslant \alpha \, (\zeta : \zeta),
$$

where positive constant α is independent of ζ.

Proof To prove the lemma we need some properties of the tensor $\mathfrak{A}_0^{(c)}$, which follow from the equalities

$$
-\beta \, \langle \nabla_y \cdot \mathbf{V}_p^{(0)} \rangle_{Y_s} = \langle \mathbb{D}_y(\mathbf{V}_p^{(0)}) : \mathbb{D}_y(\mathbf{V}_p^{(0)}) \rangle_{Y_s} + \beta \langle (\nabla_y \cdot \mathbf{V}_p^{(0)})(\nabla_y \cdot \mathbf{V}_p^{(0)}) \rangle_{Y_s}, \tag{4.5.10}
$$

$$
-\beta \, \langle \nabla_y \cdot \mathbf{V}_p^{(ij)} \rangle_{Y_s} = \langle \mathbb{D}_y(\mathbf{V}_p^{(ij)}) : \mathbb{D}_y(\mathbf{V}_p^{(0)}) \rangle_{Y_s} + \beta \langle (\nabla_y \cdot \mathbf{V}_p^{(0)})(\nabla_y \cdot \mathbf{V}_p^{(ij)}) \rangle_{Y_s}, \tag{4.5.11}
$$

$$
- \langle \mathbb{D}_y(\mathbf{V}_p^{(0)}) : \mathbb{J}^{ij} \rangle_{Y_s} = \langle \mathbb{D}_y(\mathbf{V}_p^{(ij)}) : \mathbb{D}_y(\mathbf{V}_p^{(0)}) \rangle_{Y_s} + \beta \langle (\nabla_y \cdot \mathbf{V}_p^{(0)})(\nabla_y \cdot \mathbf{V}_p^{(ij)}) \rangle_{Y_s}, \tag{4.5.12}
$$

$$
- \langle \mathbb{D}_y(\mathbf{V}_p^{(kl)}) : \mathbb{J}^{ij} \rangle_{Y_s} = \langle \mathbb{D}_y(\mathbf{V}_p^{(ij)}) : \mathbb{D}_y(\mathbf{V}_p^{(kl)}) \rangle_{Y_s} + \beta \langle (\nabla_y \cdot \mathbf{V}_p^{(kl)})(\nabla_y \cdot \mathbf{V}_p^{(ij)}) \rangle_{Y_s}, \tag{4.5.13}
$$

for all $i, \, j, \, k, \, l = 1, \, 2, \, 3$.

Equation (4.5.10) is an integral identity, corresponding to the Eq. (4.5.7) with the test function $\mathbf{V}_p^{(0)}$.

Equation (4.5.11) is an integral identity, corresponding to the Eq. (4.5.7) with the test function $\mathbf{V}_p^{(ij)}$.

Equation (4.5.12) is an integral identity, corresponding to the Eq. (4.5.6) with the test function $\mathbf{V}_p^{(0)}$.

Finally, Eq. (4.5.13) is an integral identity, corresponding to the Eq. (4.5.6) with the test function $\mathbf{V}_p^{(kl)}$.

Next we put

$$\mathbf{Y}_\zeta = \sum_{i,j=1}^{3} \mathbf{V}_p^{(ij)} \zeta_{ij}, \quad \mathbf{Y}_\eta = \sum_{i,j=1}^{3} \mathbf{V}_p^{(ij)} \eta_{ij},$$

$$\mathbf{Y}_\zeta^0 = \mathbf{V}_p^{(0)} \operatorname{tr} \zeta, \quad \mathbf{Y}_\eta^0 = \mathbf{V}_p^{(0)} \operatorname{tr} \eta.$$

Then equations (4.5.10)–(4.5.13) transform to

$$\beta \langle \nabla_y \cdot \mathbf{Y}_\eta^0 \rangle_{Y_s} \operatorname{tr} \zeta + \langle \mathbb{D}_y(Y_\zeta^0) : \mathbb{D}_y(Y_\eta^0) \rangle_{Y_s} + \beta \langle (\nabla_y \cdot \mathbf{Y}_\zeta^0)(\nabla_y \cdot \mathbf{Y}_\eta^0) \rangle_{Y_s} = 0, \tag{4.5.14}$$

$$\beta \langle \nabla_y \cdot \mathbf{Y}_\eta \rangle_{Y_s} \operatorname{tr} \zeta + \langle \mathbb{D}_y(\mathbf{Y}_\zeta) : \mathbb{D}_y(\mathbf{Y}_\eta^0) \rangle_{Y_s} + \beta \langle (\nabla_y \cdot \mathbf{Y}_\zeta)(\nabla_y \cdot \mathbf{Y}_\eta^0) \rangle_{Y_s} = 0, \tag{4.5.15}$$

$$\langle \mathbb{D}_y(\mathbf{Y}_\eta^0) \rangle_{Y_s} : \zeta + \langle \mathbb{D}_y(\mathbf{Y}_\zeta) : \mathbb{D}_y(\mathbf{Y}_\eta^0) \rangle_{Y_s} + \beta \langle (\nabla_y \cdot \mathbf{Y}_\zeta)(\nabla_y \cdot \mathbf{Y}_\eta^0) \rangle_{Y_s} = 0, \tag{4.5.16}$$

$$\langle \mathbb{D}_y(\mathbf{Y}_\eta) \rangle_{Y_s} : \zeta + \langle \mathbb{D}_y(\mathbf{Y}_\zeta) : \mathbb{D}_y(\mathbf{Y}_\eta) \rangle_{Y_s} + \beta \langle (\nabla_y \cdot \mathbf{Y}_\zeta)(\nabla_y \cdot \mathbf{Y}_\eta) \rangle_{Y_s} = 0. \tag{4.5.17}$$

Thus,

$$\begin{aligned}
\big(\mathfrak{A}^{(c)} : \zeta\big) : \eta &= (1 - m_p)\zeta : \eta + \big(\mathfrak{C}^{(p)} : \zeta\big) : \eta \\
&= (1 - m_p)\zeta : \eta + \langle \mathbb{D}(y, \mathbf{Y}_\zeta) \rangle_{Y_s} : \eta + \beta \langle \nabla_y \cdot \mathbf{Y}_\zeta \rangle_{Y_s} \operatorname{tr} \eta \\
&\quad + \langle \mathbb{D}(y, \mathbf{Y}_\zeta^0) \rangle_{Y_s} : \eta + \beta \langle \nabla_y \cdot \mathbf{Y}_\zeta^0 \rangle_{Y_s} \operatorname{tr} \eta + \beta (1 - m_p) \operatorname{tr} \zeta \operatorname{tr} \eta.
\end{aligned}$$

Adding to the right hand side of the last equality the left hand side of (4.5.14)–(4.5.17) we finally get

$$\begin{aligned}
\big(\mathfrak{A}^{(c)} : \zeta\big) : \eta =& \big\langle \big(\mathbb{D}(y, \mathbf{Y}_\zeta + \mathbf{Y}_\zeta^0) + \zeta\big) : \big(\mathbb{D}(y, \mathbf{Y}_\eta + \mathbf{Y}_\eta^0) + \eta\big) \big\rangle_{Y_s} \\
&+ \beta \big\langle \big(\nabla_y \cdot (\mathbf{Y}_\zeta + \mathbf{Y}_\zeta^0) + \operatorname{tr} \zeta\big) \big(\nabla_y \cdot (\mathbf{Y}_\eta + \mathbf{Y}_\eta^0) + \operatorname{tr} \eta\big) \big\rangle_{Y_s}.
\end{aligned}$$

This representation proves the symmetry of $\mathfrak{A}^{(c)}$. In particular,

$$(\mathfrak{A}^{(c)} : \zeta) : \zeta = \big\langle \big(\mathbb{D}(y, \mathbf{Y}_\zeta + \mathbf{Y}_\zeta^0) + \zeta\big) : \big(\mathbb{D}(y, \mathbf{Y}_\zeta + \mathbf{Y}_\zeta^0) + \zeta\big)\big\rangle_{Y_s} + \beta\big\langle\big(\nabla_y \cdot (\mathbf{Y}_\zeta + \mathbf{Y}_\zeta^0) + \operatorname{tr}\zeta\big)\big(\nabla_y \cdot (\mathbf{Y}_\zeta + \mathbf{Y}_\zeta^0) + \operatorname{tr}\zeta\big)\big\rangle_{Y_s} > 0,$$

which proves that $\mathfrak{A}^{(c)}$ is strictly positively definite.

Lemma 4.13 *Functions $\mathbf{w}_s$ and q_f satisfy in Ω_T the homogenized momentum balance equation*

$$\nabla \cdot \Big(\lambda_0\, \mathfrak{B}^{(s)} : \mathbb{D}(\mathbf{w}_s) + q_f\, \mathbb{C}^{(s)}\Big) + \hat{\rho}\mathbf{F} = 0, \tag{4.5.18}$$

and the pressure q_s in the solid skeleton is defined by

$$-\frac{1}{c_s^*} q_s = \frac{1}{\lambda_0} a^{(s)}\, q_f + \mathbb{A}^{(s)} : \mathbb{D}(\mathbf{w}_s). \tag{4.5.19}$$

In (4.5.18) and (4.5.19) the constant forth rank tensor $\mathfrak{B}^{(s)}$, constant matrices $\mathbb{C}^{(s)}$ and $\mathbb{A}^{(s)}$, and the constant $a^{(s)}$ are defined below by (4.5.23)–(4.5.26). The tensor $\mathfrak{B}^{(s)}$, matrices $\mathbb{C}^{(s)}$ and $\mathbb{A}^{(s)}$, and the constant $a^{(s)}$ depend only on the geometry of the solid cells Y_s and Z_s, and criterion $\beta = \dfrac{c_s^2}{\lambda_0}$.

Proof As usual, we look for the solution to the microscopic equation (4.5.5) in the form

$$\mathbf{V}_c(\mathbf{x}, t, \mathbf{z}) = \sum_{i,j=1}^{3} \mathbf{V}_c^{(ij)}(\mathbf{z}) D^{ij}(\mathbf{x}, t) + \mathbf{V}_c^{(0)}(\mathbf{z}) \tilde{q}_f(\mathbf{x}, t),$$

where

$$D^{ij} = \frac{1}{2}\left(\frac{\partial w_s^i}{\partial x_j} + \frac{\partial w_s^j}{\partial x_i}\right), \quad \mathbf{w}_s = (w_s^1, w_s^2, w_s^3),$$

and functions $\mathbf{V}_c^{(ij)}$ and $\mathbf{V}_c^{(0)}$ satisfy in Z following periodic boundary value problems

$$\nabla_z \cdot \Big((1 - \chi_c)\, \mathfrak{A}^{(c)} : \big(\mathbb{D}(z, \mathbf{V}_c^{(ij)}) + J^{ij}\big)\Big) = 0, \quad \langle \mathbf{V}_c^{(ij)} \rangle_{Z_s} = 0, \tag{4.5.20}$$

$$\nabla_z \cdot \Big((1 - \chi_c)\big(\mathfrak{A}^{(c)} : \mathbb{D}(z, \mathbf{V}_c^{(0)}) + \mathbb{C}^{(c)}\big)\Big) = 0, \quad \langle \mathbf{V}_c^{(0)} \rangle_{Z_s} = 0, \tag{4.5.21}$$

which are understood in the sense of distributions.

Therefore,

$$\langle \mathbb{D}(z, \mathbf{U}_c) \rangle_{Z_s} = \Big(\sum_{i,j=1}^{3} \langle \mathbb{D}(z, \mathbf{V}_c^{(ij)}) \rangle_{Z_s} \otimes \mathbb{J}^{ij}\Big) : \mathbb{D}(x, \mathbf{w}_s)$$

$$+ \tilde{q}_f \langle \mathbb{D}(z, \mathbf{V}_c^{(0)}) \rangle_{Z_s} = \mathfrak{A}_1^{(c)} : \mathbb{D}(x, \mathbf{w}_s) + \tilde{q}_f \, \mathbb{C}_1^{(s)},$$

$$\mathfrak{A}_1^{(c)} = \sum_{i,j=1}^{3} \langle \mathbb{D}(z, \mathbf{V}_c^{(ij)}) \rangle_{Z_s} \otimes \mathbb{J}^{ij}, \quad \mathbb{C}_1^{(s)} = \langle \mathbb{D}(z, \mathbf{V}_c^{(0)}) \rangle_{Z_s}, \tag{4.5.22}$$

and

$$\begin{aligned}
\langle\langle (\mathbb{D}(y, \mathbf{U}_p) - \tilde{Q}_s \mathbb{I}) \rangle_{Y_s} \rangle_{Z_s} = & \, \mathfrak{C}^{(p)} : \big((1 - m_c) \mathbb{D}(x, \mathbf{w}_s) + \langle \mathbb{D}(z, \mathbf{V}_c) \rangle_{Z_s} \big) \\
& + \tilde{q}_f \, (1 - m_c) \mathbb{C}^{(c)} = \mathfrak{C}^{(p)} : \big((1 - m_c) \mathbb{D}(x, \mathbf{w}_s) \\
& + \mathfrak{A}_1^{(c)} : \mathbb{D}(x, \mathbf{w}_s) \big) \\
& + \tilde{q}_f \, \big(\mathfrak{C}^{(p)} : \mathbb{C}_1^{(s)} + (1 - m_c) \mathbb{C}^{(c)} \big) \\
= & \, \mathfrak{C}^{(p)} : \Big(\big((1 - m_c) \, \mathbb{J} + \mathfrak{A}_1^{(c)} \big) : \mathbb{D}(x, \mathbf{w}_s) \Big) \\
& + \tilde{q}_f \, \big(\mathfrak{C}^{(p)} : \mathbb{C}_1^{(s)} + (1 - m_c) \mathbb{C}^{(c)} \big) \\
= & \Big((1 - m_c) \, \mathfrak{C}^{(p)} + \mathfrak{C}^{(p)} : \mathfrak{A}_1^{(c)} \Big) : \mathbb{D}(x, \mathbf{w}_s) \\
& + \tilde{q}_f \, \big(\mathfrak{C}^{(p)} : \mathbb{C}_1^{(s)} + (1 - m_c) \mathbb{C}^{(c)} \big).
\end{aligned}$$

Coming back to the macroscopic equation (4.3.13) we see that

$$\begin{aligned}
\mathfrak{B}^{(s)} &= (1 - m) \, \mathfrak{J} + (1 - m_p) \, \mathfrak{A}_1^{(c)} + (1 - m_c) \, \mathfrak{C}^{(p)} + \mathfrak{C}^{(p)} : \mathfrak{A}_1^{(c)} \\
&= (1 - m) \, \mathfrak{J} + \big((1 - m_p) \, \mathfrak{J} + \mathfrak{C}^{(p)} \big) : \mathfrak{A}_1^{(c)} + (1 - m_c) \, \mathfrak{C}^{(p)} \\
&= (1 - m) \, \mathfrak{J} + \mathfrak{A}^{(c)} : \mathfrak{A}_1^{(c)} + (1 - m_c) \, \mathfrak{C}^{(p)} \\
&= (1 - m_c) \, \big((1 - m_p) \, \mathfrak{J} + \mathfrak{C}^{(p)} \big) + \mathfrak{A}^{(c)} : \mathfrak{A}_1^{(c)} \\
&= (1 - m_c) \, \mathfrak{A}^{(c)} + \mathfrak{A}^{(c)} : \mathfrak{A}_1^{(c)} = \mathfrak{A}^{(c)} : \big((1 - m_c) \, \mathfrak{J} + \mathfrak{A}_1^{(c)} \big),
\end{aligned}$$

where we have used equalities $(1 - m) = (1 - m_p)(1 - m_c)$ and $\mathfrak{J} : \mathfrak{A} = \mathfrak{A} : \mathfrak{J} = \mathfrak{A}$ for any forth rank tensor $\mathfrak{A}$.

Thus,

$$\mathfrak{B}^{(s)} = \mathfrak{A}^{(c)} : \big((1 - m_c) \, \mathfrak{J} + \mathfrak{A}_1^{(c)} \big). \tag{4.5.23}$$

In a completely analogous way, we obtain

$$\begin{aligned}
m \, \mathbb{C}^{(s)} &= (1 - m_p) \mathbb{C}_1^{(s)} + \mathfrak{C}^{(p)} : \mathbb{C}_1^{(s)} + (1 - m_c) \mathbb{C}^{(c)} - \mathbb{I} \\
&= \big((1 - m_p) \mathfrak{J} + \mathfrak{C}^{(p)} \big) : \mathbb{C}_1^{(s)} + (1 - m_c) \mathbb{C}^{(c)} - \mathbb{I} \\
&= \mathfrak{A}^{(c)} : \mathbb{C}_1^{(s)} + (1 - m_c) \mathbb{C}^{(c)} - \mathbb{I}.
\end{aligned}$$

Therefore,

$$m\,\mathbb{C}^{(s)} = \mathfrak{A}^{(c)} : \mathbb{C}_1^{(s)} + (1 - m_c)\mathbb{C}^{(c)} - \mathbb{I}. \tag{4.5.24}$$

To calculate the pressure in the solid component we use the macroscopic continuity equation for the solid component

$$-\frac{1}{c_s^2} q_s = (1 - m)\nabla \cdot \mathbf{w}_s + (1 - m_p)\langle \nabla_z \cdot \mathbf{U_c} \rangle_{Z_s} + \langle \langle \nabla_y \cdot \mathbf{U_p} \rangle_{Y_s} \rangle_{Z_s},$$

which follows from the microscopic continuity equation (4.5.4) after integration over Y_s and Z_s. Next we have

$$\langle \nabla_y \cdot \mathbf{U_p} \rangle_{Z_s} = \mathbb{A}^{(p)} : \big(\mathbb{D}(x, \mathbf{w}_s) + \mathbb{D}(z, \mathbf{U}_c)\big) + a^{(p)}\,\tilde{q}_f,$$

where

$$\mathbb{A}^{(p)} = \sum_{i,j=1}^{3} \langle \nabla_y \cdot \mathbf{v}_p^{(ij)} \rangle_{Y_s}\, \mathbb{J}^{ij} + \langle \nabla_y \cdot \mathbf{v}_p^{(0)} \rangle_{Y_s}\, \mathbb{I},$$

$$a^{(p)} = \frac{1}{\beta} \langle \nabla_y \cdot \mathbf{v}_p^{(0)} \rangle_{Y_s},$$

and

$$\langle \langle \nabla_y \cdot \mathbf{U_p} \rangle_{Y_s} \rangle_{Z_s} = \mathbb{A}^{(p)} : \Big((1 - m_c)\mathbb{D}(x, \mathbf{w}_s) + \langle \mathbb{D}(z, \mathbf{U_c}) \rangle_{Z_s}\Big) + (1 - m_c) a^{(p)}\,\tilde{q}_f.$$

Thus,

$$\begin{aligned} -\frac{1}{c_s^*} q_s &= (1 - m)\mathbb{I} : \mathbb{D}(x, \mathbf{w}_s) + (1 - m_c)\mathbb{A}^{(p)} : \mathbb{D}(x, \mathbf{w}_s) \\ &\quad + \mathbb{A}^{(p)} : \langle \mathbb{D}(z, \mathbf{U_c}) \rangle_{Z_s} + (1 - m_c) a^{(p)}\,\tilde{q}_f \\ &= (1 - m_c)\mathbb{A}^{(c)} : \mathbb{D}(x, \mathbf{w}_s) + \mathbb{A}^{(c)} : \langle \mathbb{D}(z, \mathbf{U_c}) \rangle_{Z_s} + a^{(c)}\,\tilde{q}_f \\ &= \mathbb{A}^{(c)} : \Big((1 - m_c)\mathbb{J} + \langle \mathbb{D}(z, \mathbf{U_c}) \rangle_{Z_s}\Big) + + a^{(c)}\,\tilde{q}_f, \end{aligned}$$

where

$$\mathbb{A}^{(c)} = \mathbb{A}^{(p)} + (1 - m_p)\mathbb{I}, \quad a^{(c)} = a^{(p)}(1 - m_c).$$

Since

$$\langle \mathbb{D}(z, \mathbf{U}_c) \rangle_{Z_s} = \mathfrak{A}_1^{(c)} : \mathbb{D}(x, \mathbf{w}_s) + \tilde{q}_f\, \mathbb{C}_1^{(s)},$$

then finally one has

$$-\frac{1}{c_s^*}q_s = \mathbb{A}^{(c)} : \Big((1-m_c)\mathbb{J} + \mathfrak{A}_1^{(c)}\Big) : \mathbb{D}(x, \mathbf{w}_s) + \big(\mathbb{A}^{(c)} : \mathbb{C}_1^{(s)} + a^{(c)}\big)\tilde{q}_f,$$

and

$$\mathbb{A}^{(s)} = \mathbb{A}^{(c)} : \big((1-m_c)\mathbb{J} + \mathfrak{A}_1^{(c)}\big), \tag{4.5.25}$$

$$a^{(s)} = \frac{1}{m}\big(a^{(c)} + \mathbb{A}^{(c)} : \mathbb{C}_1^{(s)}\big). \tag{4.5.26}$$

Lemma 4.14 *The tensor $\mathbb{B}^{(s)}$ is symmetric and strictly positively definite.*

Proof To prove the first statement of the Lemma we use the equality

$$\int_{Z_s} \big(\mathfrak{A}^{(c)} : (\mathbb{D}(z, \mathbf{V}_c^{(ij)}) + \mathbf{J}^{ij}\big) : \mathbb{D}(z, \mathbf{V}_c^{(kl)})dz = 0, \tag{4.5.27}$$

which is an integral identity corresponding to (4.5.20) with the test function $\mathbf{V}_c^{(kl)}$.
Let

$$\mathbf{Z}_\zeta = \sum_{i,j-1}^{3} \mathbf{V}_c^{(ij)}\zeta_{ij}, \quad \mathbf{Z}_\eta = \sum_{i,j=1}^{3} \mathbf{V}_c^{(ij)}\eta_{ij}.$$

Then (4.5.27) takes the form

$$\big\langle\big(\mathfrak{A}^{(c)} : \mathbb{D}(z, \mathbf{Z}_\zeta)\big) : \mathbb{D}(z, \mathbf{Z}_\eta)\big\rangle_{Z_s} + \big\langle\big(\mathfrak{A}^{(c)} : \mathbb{D}(z, \mathbf{Z}_\eta)\big) : \zeta\big\rangle_{Z_s} = 0. \tag{4.5.28}$$

Note that by definition (see (4.5.22))

$$\mathfrak{A}_1^{(c)} : \zeta = \langle \mathbb{D}(z, \mathbf{Z}_\zeta)\rangle_{Z_s}. \tag{4.5.29}$$

Relations (4.5.28) and (4.5.29) lead to the following chain of equalities

$$\begin{aligned}
\big(\mathfrak{B}^{(s)} : \zeta\big) : \eta &= (1-m_c)\big(\mathfrak{A}^{(c)} : \zeta\big) : \eta + \Big(\big(\mathfrak{A}^{(c)} : \mathfrak{A}_1^{(c)}\big) : \zeta\Big) : \eta \\
&= (1-m_c)\big(\mathfrak{A}^{(c)} : \zeta\big) : \eta + \big(\mathfrak{A}^{(c)} : \langle \mathbb{D}(z, \mathbf{Z}_\zeta)\rangle_{Z_s}\big) : \eta \\
&= (1-m_c)\big(\mathfrak{A}^{(c)} : \zeta\big) : \eta + \langle\big(\mathfrak{A}^{(c)} : \mathbb{D}(z, \mathbf{Z}_\zeta)\big) : \mathbb{D}(z, \mathbf{Z}_\eta)\rangle_{Z_s} \\
&\quad + \langle\big(\mathfrak{A}^{(c)} : \mathbb{D}(z, \mathbf{Z}_\eta)\big) : \zeta\rangle_{Z_s} + \big(\mathfrak{A}^{(c)} : \langle \mathbb{D}(z, \mathbf{Z}_\zeta)\rangle_{Z_s}\big) : \eta \\
&= \big\langle\big(\mathfrak{A}^{(c)} : \big(\mathbb{D}(z, \mathbf{Z}_\zeta) + \zeta\big)\big) : \big(\mathbb{D}(z, \mathbf{Z}_\eta) + \eta\big)\big\rangle_{Z_s},
\end{aligned}$$

which proves the symmetry of $\mathfrak{B}^{(s)}$.
In particular,

$$\big(\mathfrak{B}^{(s)} : \eta\big) : \eta = \big\langle\big(\mathfrak{A}^{(c)} : \big(\mathbb{D}(z, \mathbf{Z}_\eta) + \eta\big)\big) : \big(\mathbb{D}(z, \mathbf{Z}_\eta) + \eta\big)\big\rangle_{Z_s} > 0.$$

Chapter 5
Filtration in Composite Incompressible Media

5.1 Filtration from a Reservoir into a Porous Medium

5.1.1 The Problem Statements and Main Results

The problem in its simplest setting is modeled by two domains Ω_0 and Ω having a common boundary S^0 (Fig. 5.1). The domain Ω^0 models a reservoir and is occupied by liquid, and the domain Ω models a porous medium. The motion of the liquid in Ω^0 for $t > 0$ is governed by the non-stationary Stokes system

$$\nabla \cdot \mathbf{w} = 0, \tag{5.1.1}$$

$$\tau_0 \rho_f \frac{\partial^2 \mathbf{w}}{\partial t^2} = \nabla \cdot \mathbb{P}_f + \rho_f \mathbf{e}, \quad \mathbb{P}_f = \alpha_\mu \mathbb{D}\left(x, \frac{\partial \mathbf{w}}{\partial t}\right) - p\,\mathbb{I}, \tag{5.1.2}$$

and the joint motion of the poroelastic media in Ω for $t > 0$ is governed by the model $\mathbb{M}_{24}$ consisting of the continuity equation (5.1.1) and the momentum balance equation

$$\tau_0 \rho^\varepsilon \frac{\partial^2 \mathbf{w}}{\partial t^2} = \nabla \cdot \mathbb{P} + \rho^\varepsilon \mathbf{e}, \tag{5.1.3}$$

where

$$\mathbb{P} = \chi^\varepsilon \alpha_\mu \mathbb{D}\left(x, \frac{\partial \mathbf{w}}{\partial t}\right) + (1 - \chi^\varepsilon)\lambda_0 \mathbb{D}(x, \mathbf{w}) - p\,\mathbb{I}, \tag{5.1.4}$$

and $\rho^\varepsilon = \rho_f \chi^\varepsilon + \rho_s(1 - \chi^\varepsilon)$.

On the common boundary $S^0 = \partial\Omega \cap \partial\Omega^0$ for $t > 0$ the continuity conditions

$$\lim_{\substack{\mathbf{x} \to \mathbf{x}^0 \\ \mathbf{x} \in \Omega^0}} \mathbf{w}(\mathbf{x}, t) = \lim_{\substack{\mathbf{x} \to \mathbf{x}^0 \\ \mathbf{x} \in \Omega}} \mathbf{w}(\mathbf{x}, t), \tag{5.1.5}$$

A. Meirmanov, *Mathematical Models for Poroelastic Flows*,
Atlantis Studies in Differential Equations 1, DOI: 10.2991/978-94-6239-015-7_5,

Fig. 5.1 Filtration from reservoir

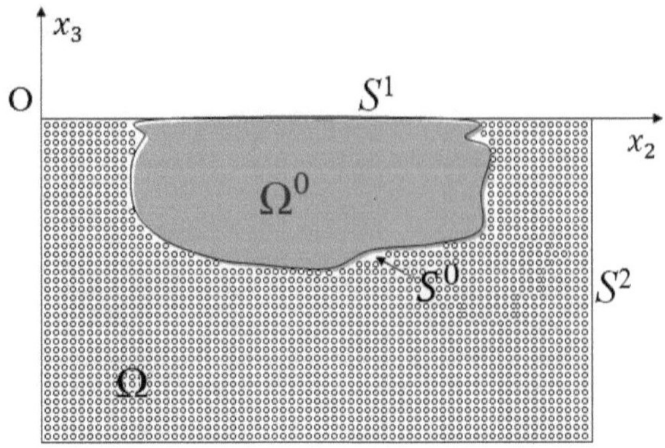

$$\lim_{\substack{\mathbf{x}\to\mathbf{x}^0\\ \mathbf{x}\in\Omega^0}} \mathbb{P}_f(\mathbf{x},t)\cdot\mathbf{n}(\mathbf{x}^0) = \lim_{\substack{\mathbf{x}\to\mathbf{x}^0\\ \mathbf{x}\in\Omega}} \mathbb{P}(\mathbf{x},t)\cdot\mathbf{n}(\mathbf{x}^0), \tag{5.1.6}$$

hold true for displacements and for normal tensions. Here $\mathbf{n}(\mathbf{x}^0)$ is a normal vector to the boundary S^0 at $\mathbf{x}^0 \in S^0$.

We complete the problem with the Neumann boundary condition

$$\mathbb{P}_f(\mathbf{x},t)\cdot\mathbf{n} = -p^0(\mathbf{x},t)\mathbf{n}, \tag{5.1.7}$$

on the part S^1 of the outer boundary S of the domain $Q = \Omega^0 \cup S^0 \cup \Omega$ (which is also the part of the boundary $\partial\Omega^0$) for $t > 0$, the Dirichlet boundary condition

$$\mathbf{w}(\mathbf{x},t) = 0 \tag{5.1.8}$$

on the part $S^2 = S\backslash\overline{S^1}$ of the outer boundary S for $t > 0$, and initial conditions

$$\mathbf{w}(\mathbf{x},0) = \frac{\partial\mathbf{w}}{\partial t}(\mathbf{x},0) = 0, \ \ \mathbf{x}\in Q. \tag{5.1.9}$$

In (5.1.1)–(5.1.9) the characteristic function $\chi^\varepsilon(\mathbf{x})$ of the domain Ω^ε_f is given by the expression

$$\chi^\varepsilon(\mathbf{x}) = \varsigma(\mathbf{x})\chi\left(\frac{\mathbf{x}}{\varepsilon}\right),$$

where $\varsigma(\mathbf{x})$ is the characteristic function of the domain Ω, $\chi(\mathbf{y})$ is the characteristic function of the liquid cell Y_f in the unit cube Y, and $\mathbf{e}$ is a unit vector in the direction of gravity.

Let, as usual,

$$\lim_{\varepsilon\searrow 0}\alpha_\mu(\varepsilon) = \mu_0, \quad \lim_{\varepsilon\searrow 0}\frac{\alpha_\mu}{\varepsilon^2} = \mu_1.$$

We use the problem (5.1.1)–(5.1.9) to derive the desired homogenized problem for the case $\mu_0 = 0$ and $0 < \mu_1 < \infty$ and will do it in two steps. First we fix τ_0 and pass to the limit as $\varepsilon \searrow 0$, and after that we pass to the limit as $\tau_0 \searrow 0$.

For the case $0 < \mu_0 < \infty$ we will use the model $\mathbb{M}_{15}$, or, more precisely, the formal limit as $\tau_0 \searrow 0$ of the model (5.1.1)–(5.1.9), which consists of relations (5.1.1), (5.1.4)–(5.1.8), completed with the momentum balance equation

$$\nabla \cdot \mathbb{P}_f + \rho_f \mathbf{e} = 0 \tag{5.1.10}$$

in the domain Ω^0 for $t > 0$, the momentum balance equation

$$\nabla \cdot \mathbb{P} + \rho^{\varepsilon} \mathbf{e} = 0 \tag{5.1.11}$$

in the domain Ω for $t > 0$, and the initial condition

$$\mathbf{w}(\mathbf{x}, 0) = 0, \ \ \mathbf{x} \in \Omega \cup \Omega^{\varepsilon}_f. \tag{5.1.12}$$

Throughout this section we impose Assumptions 0.1, 1.2, and 3.1 for the structure of the pore space, defined by the characteristic function $\chi(\mathbf{y})$. We also assume that S^1 is a part of the plane $\{x_3 = 0\}$, $\mathbf{e} = -\mathbf{e}_3$, and that the domain Q is a subset of the half-space $\{x_3 < 0\}$. Moreover we suppose that S^2 is a C^2—smooth surface and in some small neighborhood of the plane $\{x_3 - 0\}$ it is represented by the equation $\Phi(x_1, x_2) = 0$.

The given function p^0 is supposed to be smooth:

$$\int_{Q_T} \Big(|\nabla p^0(\mathbf{x}, t)|^2 + \Big|\nabla \Big(\frac{\partial p^0}{\partial t}\Big)(\mathbf{x}, t)\Big|^2 \Big) dxdt = \mathfrak{P}^2 < \infty. \tag{5.1.13}$$

Definition 5.1 We say that the pair of functions $\{\mathbf{w}^{\varepsilon}, \ p^{\varepsilon}\}$, such that

$$p^{\varepsilon} \in L_2(Q_T), \ \mathbf{w}^{\varepsilon}, \ \frac{\partial \mathbf{w}^{\varepsilon}}{\partial t}, \ (\zeta + (1-\zeta)\chi^{\varepsilon})\nabla \frac{\partial \mathbf{w}^{\varepsilon}}{\partial t}, \ \nabla \mathbf{w}^{\varepsilon} \in \mathbf{L}_2(Q_T),$$

is a weak solution of the problem (5.1.1)–(5.1.9), if it satisfies the continuity equation (5.1.1) almost everywhere in $Q_T = Q \times (0, T)$, the boundary condition (5.1.8), the initial condition (5.1.9) for the function $\mathbf{w}^{\varepsilon}$, and the integral identity

$$\begin{aligned} &\int_{Q_T} \Big(-\tau_0 \tilde{\rho}^{\varepsilon} \frac{\partial \mathbf{w}^{\varepsilon}}{\partial t} \cdot \frac{\partial \varphi}{\partial t} + \big(\zeta \mathbb{P}_f + (1-\zeta)\mathbb{P}\big) : \mathbb{D}(x, \varphi) \Big) dxdt \\ &\quad = \int_{Q_T} \big(\tilde{\rho}^{\varepsilon} \mathbf{e} \cdot \varphi - \nabla \cdot (\varphi\, p^0)\big) dxdt \end{aligned} \tag{5.1.14}$$

for all smooth functions φ, such that $\varphi(\mathbf{x}, t) = 0$ at the boundary S^2_T, and $\varphi(\mathbf{x}, T) = 0$, $\mathbf{x} \in Q$.

In (5.1.14) $\tilde{\rho}^{\varepsilon} = \big(\zeta + (1-\zeta)\chi^{\varepsilon}\big)\rho_f + (1-\zeta)(1-\chi^{\varepsilon})\rho_s$ and $\zeta = \zeta(\mathbf{x})$ is the characteristic function of the domain Ω^0 in Q.

The identity (5.1.14) obviously contains Eqs. (5.1.2) and (5.1.3), and boundary conditions (5.1.6) and (5.1.7).

Definition 5.2 We say that the pair of functions $\{\mathbf{w}^{\varepsilon},\ p^{\varepsilon}\}$, such that

$$p^{\varepsilon} \in L_2(Q_T),\ \mathbf{w}^{\varepsilon},\ \mathbb{D}(x,\mathbf{w}),\ \big(\zeta + (1-\zeta)\chi^{\varepsilon}\big)\mathbb{D}\left(x, \frac{\partial \mathbf{w}^{\varepsilon}}{\partial t}\right) \in \mathbf{L}_2(Q_T),$$

is a weak solution of the problem (5.1.1), (5.1.4)–(5.1.8), (5.1.10)–(5.1.12), if it satisfies the continuity equation (5.1.1) almost everywhere in Q_T, the boundary condition (5.1.8), the initial condition (5.1.12), and the integral identity

$$\int_{Q_T} \Big(\big(\zeta\mathbb{P}_f + (1-\zeta)\mathbb{P}\big) : \mathbb{D}(x,\varphi) + \nabla\cdot(\varphi\, p^0) - \tilde{\rho}^{\varepsilon}\mathbf{e}\cdot\varphi\Big)dxdt = 0 \quad (5.1.15)$$

for all smooth functions φ, such that $\varphi(\mathbf{x},t) = 0$ at the boundary S^2_T.

Theorem 5.1 *Let*

$$p^0 = p^0(t). \quad (5.1.16)$$

Then for all $\varepsilon > 0$ *and for an arbitrary time interval* $[0,T]$ *there exists a unique generalized solution of problem (5.1.1)–(5.1.9) and*

$$\max_{0\le t\le T}\int_Q \Big(\tau_0^2\Big|\frac{\partial^2 \mathbf{w}^{\varepsilon}}{\partial t^2}\Big|^2 + \tau_0\Big|\frac{\partial \mathbf{w}^{\varepsilon}}{\partial t}\Big|^2 + \lambda_0(1-\zeta)(1-\chi^{\varepsilon})\Big|\mathbb{D}(x,\mathbf{w}^{\varepsilon})\Big|^2\Big)dx$$
$$+ \int_{Q_T}\Big(|p^{\varepsilon}|^2 + \alpha_{\mu}\big(\zeta + (1-\zeta)\chi^{\varepsilon}\big)\Big|\mathbb{D}(x,\frac{\partial \mathbf{w}^{\varepsilon}}{\partial t})\Big|^2\Big)dxdt \leqslant C_0, \quad (5.1.17)$$

where here and in what follows, we denote as C_0 *any constant independent of* τ_0 *and* ε.

Theorem 5.2 *Under condition (5.1.13) for all* $\varepsilon > 0$ *and for an arbitrary time interval* $[0,T]$ *there exists a unique generalized solution of problem (5.1.1), (5.1.4)–(5.1.8), (5.1.10)–(5.1.12) and*

$$\int_{Q_T}\Big(\mid p^{\varepsilon}\mid^2 + \alpha_{\mu}\big(\zeta + (1-\zeta)\chi^{\varepsilon}\big)\Big|\mathbb{D}\Big(x,\frac{\partial \mathbf{w}^{\varepsilon}}{\partial t}\Big)\Big|^2\Big)dxdt$$
$$+ \lambda_0 \max_{0\le t\le T}\int_{\Omega}(1-\chi^{\varepsilon}) \mid \mathbb{D}(x,\mathbf{w}^{\varepsilon}) \mid^2 dx \leqslant C_0(\mathfrak{P}^2 + 1). \quad (5.1.18)$$

Theorem 5.3 *Under the conditions of Theorem 5.1 let*

$$\mu_0 = 0,\ 0 < \lambda_0,\ \tau_0 < \infty,\ \mu_1 = \infty,$$

$\{\mathbf{w}^\varepsilon,\ p^\varepsilon\}$ be the weak solution of the problem (5.1.1)–(5.1.9) and $\mathbf{w}_s^\varepsilon = \mathbb{E}_{\Omega_s^\varepsilon}(\mathbf{w}^\varepsilon)$ be an extension (3.2.8) from the domain Ω_s^ε onto the domain Ω.

Then sequences $\{\mathbf{w}^\varepsilon\}$, $\left\{\dfrac{\partial \mathbf{w}^\varepsilon}{\partial t}\right\}$, $\left\{(1-\zeta)\dfrac{\partial^2 \mathbf{w}_s^\varepsilon}{\partial t^2}\right\}$ and $\{p^\varepsilon\}$ converge weakly in $\mathbf{L}_2(Q_T)$ and $L_2(Q_T)$ to the functions $\mathbf{W}$, $\mathbf{v}$, $(1-\zeta)\dfrac{\partial^2 \mathbf{w}_s}{\partial t^2}$, and p respectively as $\varepsilon \to 0$.

At the same time the sequence $\{\mathbf{w}_s^\varepsilon\}$ converges weakly in $\mathbf{W}_2^{1,0}(\Omega_T)$ to the function $\mathbf{w}_s$ as $\varepsilon \to 0$.

The limiting pressure p and the limiting velocity $\mathbf{v}$ of the liquid satisfy the system

$$\tau_0 \rho_f \frac{\partial \mathbf{v}}{\partial t} + \nabla p = \rho_f\,\mathbf{e},\quad \nabla \cdot \mathbf{v} = 0 \tag{5.1.19}$$

in the domain Ω_0 for $t > 0$.

In the domain Ω for $t > 0$ limiting functions $\mathbf{w}_s$ and p solve the homogenized system, consisting of the homogenized momentum balance equation

$$\tau_0 \hat{\rho} \frac{\partial^2 \mathbf{w}_s}{\partial t^2} = \nabla \cdot \mathbb{P}_0^{(s)} + \hat{\rho}\,\mathbf{e}, \tag{5.1.20}$$

$$\mathbb{P}_0^{(s)} = \lambda_0\, \mathfrak{N}_0^s : \mathbb{D}(x, \mathbf{w}_s) - p\,\mathbb{I},$$

and the continuity equation

$$\nabla \cdot \mathbf{w}_s = 0. \tag{5.1.21}$$

The problem is completed with the continuity conditions

$$\lim_{\substack{\mathbf{x} \to \mathbf{x}^0 \in S^0 \\ \mathbf{x} \in \Omega}} \mathbf{w}_s(\mathbf{x}, t) \cdot \mathbf{n}(\mathbf{x}^0) = \lim_{\substack{\mathbf{x} \to \mathbf{x}^0 \in S^0 \\ \mathbf{x} \in \Omega_0}} \mathbf{w}(\mathbf{x}, t) \cdot \mathbf{n}(\mathbf{x}^0), \tag{5.1.22}$$

$$\lim_{\substack{\mathbf{x} \to \mathbf{x}^0 \in S^0 \\ \mathbf{x} \in \Omega}} \mathbb{P}_0^{(s)}(\mathbf{x}, t) \cdot \mathbf{n}(\mathbf{x}^0) = - \lim_{\substack{\mathbf{x} \to \mathbf{x}^0 \in S^0 \\ \mathbf{x} \in \Omega_0}} p(\mathbf{x}, t) \cdot \mathbf{n}(\mathbf{x}^0) \tag{5.1.23}$$

on the common boundary S^0, the homogeneous boundary condition

$$\mathbf{w}_s = 0 \tag{5.1.24}$$

on the part S^2 of the outer boundary S, the boundary condition

$$p(\mathbf{x},t) = p_0(t) \tag{5.1.25}$$

on the part $S_0^1 = S^1 \cap \overline{\Omega}_0$ *of the outer boundary* S, *and the boundary condition*

$$\mathbb{P}_0^{(s)}(\mathbf{x},t)\cdot\mathbf{n}(\mathbf{x}^0) = -p_0(t)\mathbf{n}(\mathbf{x}^0) \tag{5.1.26}$$

on the part $S_1^1 = S^1 \cap \overline{\Omega}$ *of the outer boundary* S.

In (5.1.20)–(5.1.26) $\mathbf{n}(\mathbf{x}^0)$ *is a unit normal to* S^0 *(or* S_1^1*) at* $\mathbf{x}^0 \in S^0$ *(or* S_1^1*),*

$$\hat{\rho} = m\,\rho_f + (1-m)\,\rho_s, \quad m = \int_Y \chi(\mathbf{y})dy,$$

and the symmetric strictly positively definite constant fourth-rank tensor $\mathfrak{N}_0^s$ *is given by (1.2.35) (see Theorem 1.3 of Chap. 1).*

We refer to the problem (5.1.19)–(5.1.26) as the homogenized **model** $(\mathbb{FCM})_1$.

Theorem 5.4 *Under the conditions of Theorem 5.3 let* $\tau_0 = \dfrac{1}{n}$, *and* $p^{(n)}$, $\mathbf{w}^{(n)}$, *and* $\mathbf{w}_s^{(n)}$ *be the weak solution of the problem (5.1.19)–(5.1.26).*

Then the sequence $\{p^{(n)}\}$ *converges weakly in* $L_2(Q_T)$ *to the function* p, *and the sequence* $\{\mathbf{w}_s^{(n)}\}$ *converges weakly in* $\mathbf{W}_2^{1,0}(\Omega_T)$ *to the function* $\mathbf{w}_s$ *as* $n \to \infty$.

The limiting pressure p *of the liquid in the domain* Ω_0 *coincides for* $t > 0$ *with the hydrostatic pressure*

$$p(\mathbf{x},t) = p^0(t) - \rho_f x_3 \equiv p_0(\mathbf{x},t). \tag{5.1.27}$$

The limiting functions solve the homogenized system in the domain Ω *for* $t > 0$, *consisting of the homogenized momentum balance equation*

$$\nabla\cdot\mathbb{P}_0^{(s)} + \hat{\rho}\,\mathbf{e} = 0, \tag{5.1.28}$$

and the continuity equation (5.1.21).

The problem is completed with the boundary condition (5.1.24) on the part S^2 *of the outer boundary* S, *the boundary condition (5.1.26) on the part* $S_1^1 = S^1 \cap \overline{\Omega}$ *of the outer boundary* S, *and the boundary condition*

$$\lim_{\substack{\mathbf{x}\to\mathbf{x}^0\\ \mathbf{x}\in\Omega}} \mathbb{P}_0^{(s)}(\mathbf{x},t)\cdot\mathbf{n}(\mathbf{x}^0) = -p_0(\mathbf{x}^0,t)\cdot\mathbf{n}(\mathbf{x}^0) \tag{5.1.29}$$

on the common boundary S^0.

In (5.1.26)–(5.1.29) $\mathbf{n}(\mathbf{x}^0)$ *is a unit normal to* S^0 *(or* S_1^1*) at* $\mathbf{x}^0 \in S^0$ *(or* S_1^1*),*

$$\hat{\rho} = m\,\rho_f + (1-m)\,\rho_s, \quad m = \int_Y \chi(\mathbf{y})dy,$$

and the symmetric strictly positively definite constant fourth-rank tensor $\mathfrak{N}_0^s$ *is given by (1.2.35) (see Theorem 1.3 of Chap. 1).*

We refer to the problem (5.1.24), (5.1.26)–(5.1.29) as the homogenized **model** $(\mathbb{FCM})_2$.

Theorem 5.5 *Under the conditions of Theorem 5.1 let*

$$\mu_0 = 0, \; 0 < \lambda_0, \; \mu_1, \; \tau_0 < \infty,$$

$\{\mathbf{w}^\varepsilon, \, p^\varepsilon\}$ *be the weak solution of the problem* (5.1.1)–(5.1.9) *and* $\mathbf{w}_s^\varepsilon = \mathbb{E}_{\Omega_s^\varepsilon}(\mathbf{w}^\varepsilon)$ *be an extension* (1.2.9) *from the domain* Ω_s^ε *onto the domain* Ω.

Then sequences $\{(1-\zeta)\chi^\varepsilon p^\varepsilon\}$, $\{p^\varepsilon\}$, $\{\mathbf{w}^\varepsilon\}$, $\{\zeta \dfrac{\partial \mathbf{w}^\varepsilon}{\partial t}\}$, $\left\{\zeta \dfrac{\partial^2 \mathbf{w}^\varepsilon}{\partial t^2}\right\}$, $\left\{(1-\zeta)\dfrac{\partial^2 \mathbf{w}_s^\varepsilon}{\partial t^2}\right\}$, $\{(1-\zeta)\chi^\varepsilon \mathbf{w}^\varepsilon\}$, $\left\{(1-\zeta)\chi^\varepsilon \dfrac{\partial \mathbf{w}^\varepsilon}{\partial t}\right\}$ *and* $\left\{(1-\zeta)\chi^\varepsilon \dfrac{\partial^2 \mathbf{w}^\varepsilon}{\partial t^2}\right\}$ *converge weakly in* $L_2(Q_T)$ *and* $\mathbf{L}_2(Q_T)$ *to the functions* $(1-\zeta)mp_f \in W_2^{1,0}(\Omega_T)$, p, $\mathbf{w}$, $\zeta\mathbf{v}$, $\zeta\dfrac{\partial \mathbf{v}}{\partial t}$, $(1-\zeta)\dfrac{\partial^2 \mathbf{w}_s}{\partial t^2}$, $(1-\zeta)\mathbf{w}^{(f)}$, $(1-\zeta)\dfrac{\partial \mathbf{w}^{(f)}}{\partial t}$ *and* $(1-\zeta)\dfrac{\partial^2 \mathbf{w}^{(f)}}{\partial t^2}$ *respectively as* $\varepsilon \to 0$, *and the sequence* $\{\mathbf{w}_s^\varepsilon\}$ *converges weakly in* $\mathbf{W}_2^{1,0}(\Omega_T)$ *to the function* $\mathbf{w}_s$ *as* $\varepsilon \to 0$.

The limiting pressure p and the limiting velocity $\mathbf{v}$ *of the liquid in the domain* Ω_0 *satisfy in* Ω_0 *for* $t > 0$ *the system (5.1.19).*

In the domain Ω *for* $t > 0$ *limiting functions solve the homogenized system, consisting of the continuity equation*

$$\nabla \cdot \left(\mathbf{w}^{(f)} + (1-m)\,\mathbf{w}_s\right) = 0, \tag{5.1.30}$$

the homogenized momentum balance equation

$$\nabla \cdot \mathbb{P}_1^{(s)} + \hat{\rho}\,\mathbf{e} = 0, \tag{5.1.31}$$

$$\mathbb{P}_1^{(s)} = \lambda_0\,\mathfrak{N}_1^s : \mathbb{D}(x, \mathbf{w}_s) - p_f\,\mathbb{I}$$

for the solid component, and the momentum balance equation

$$\frac{\partial \mathbf{w}^{(f)}}{\partial t} - m\,\frac{\partial \mathbf{w}_s}{\partial t} = \int_0^t \mathbb{B}^{(f)}(\tau_0; t-\tau)\cdot\left(-\nabla\, p_f\,(\mathbf{x}, \tau) - \rho_f \frac{\partial^2 \mathbf{w}_s}{\partial \tau^2}(\mathbf{x}, \tau) + \rho_f \mathbf{e}\right) d\tau \tag{5.1.32}$$

for the liquid component, completed with the continuity conditions

$$\lim_{\substack{\mathbf{x} \to \mathbf{x}^0 \in S^0 \\ \mathbf{x} \in \Omega}} \mathbb{P}_1^{(s)}(\mathbf{x}, t)\cdot \mathbf{n}(\mathbf{x}^0) = -\lim_{\substack{\mathbf{x} \to \mathbf{x}^0 \in S^0 \\ \mathbf{x} \in \Omega_0}} p(\mathbf{x}, t)\cdot \mathbf{n}(\mathbf{x}^0), \tag{5.1.33}$$

$$\lim_{\substack{\mathbf{x}\to\mathbf{x}^0\in S^0\\ \mathbf{x}\in\Omega}}\big(\mathbf{w}^f+(1-m)\mathbf{w}_s\big)(\mathbf{x},t)\cdot\mathbf{n}(\mathbf{x}^0)=\lim_{\substack{\mathbf{x}\to\mathbf{x}^0\in S^0\\ \mathbf{x}\in\Omega_0}}\mathbf{w}(\mathbf{x},t)\cdot\mathbf{n}(\mathbf{x}^0),\tag{5.1.34}$$

$$\lim_{\substack{\mathbf{x}\to\mathbf{x}^0\in S^0\\ \mathbf{x}\in\Omega}}p_f(\mathbf{x},t)=\lim_{\substack{\mathbf{x}\to\mathbf{x}^0\in S^0\\ \mathbf{x}\in\Omega_0}}p(\mathbf{x},t)\tag{5.1.35}$$

on the common boundary S^0, boundary conditions (5.1.24), (5.1.25), the boundary condition

$$\mathbf{w}^{(f)}(\mathbf{x}^0,t)\cdot\mathbf{n}(\mathbf{x}^0)=0\tag{5.1.36}$$

on the part S^2 of the outer boundary S, and boundary conditions

$$\mathbb{P}_1^{(s)}(\mathbf{x},t)\cdot\mathbf{n}(\mathbf{x}^0)=-p_0(t)\mathbf{n}(\mathbf{x}^0),\tag{5.1.37}$$

$$p_f(\mathbf{x},t)=p^0(t)\tag{5.1.38}$$

on the part S_1^1 of the outer boundary S.

In (5.1.30)–(5.1.38) $\mathbf{n}(\mathbf{x}^0)$ is a unit normal to S^0 (or S^2) at $\mathbf{x}^0\in S^0$ (or S^2),

$$\hat{\rho}=m\,\rho_f+(1-m)\,\rho_s,\quad m=\int_Y\chi(\mathbf{y})dy,$$

the symmetric matrix $\mathbb{B}^{(f)}(\tau_0;t)$ is given below by formula (5.1.70), and the symmetric strictly positively definite constant fourth-rank tensor $\mathfrak{N}_1^s$ is given by (1.2.38) (see Theorem 1.4 of Chap. 1).

We refer to the problem (5.1.19), (5.1.24), (5.1.25), (5.1.30)–(5.1.38) as the homogenized **model** $(\mathbb{FCM})_3$.

Theorem 5.6 *Under the conditions of Theorem 5.5 let $\tau_0=\dfrac{1}{n}$, and $p^{(n)}$, $\mathbf{w}_s^{(n)}$, $\mathbf{w}^{(f,n)}$, $p^{(n)}$, $\pi_f{}^{(n)}$ be a solution of the model $(\mathbb{FCM})_3$.*

Then sequences $\{p^{(n)}\}$, $\{(1-\zeta)\pi_f{}^{(n)}\}$, and $\{(1-\zeta)\mathbf{w}^{(f,n)}\}$ converge weakly in $L_2(Q_T)$ and $\mathbf{L}_2(Q_T)$ to the functions p, $(1-\zeta)\pi_f$, and $(1-\zeta)\mathbf{w}^{(f)}$ respectively as $n\to\infty$, and the sequence $\{\mathbf{w}_s^{(n)}\}$ converges weakly in $\mathbf{W}_2^{1,0}(\Omega_T)$ to the function $\mathbf{W}_s$ as $n\to\infty$.

The limiting pressure p of the liquid in the domain Ω_0 coincides for $t>0$ with the hydrostatic pressure (5.1.27).

The limiting functions $\mathbf{w}_s$, π_f, and $\mathbf{w}^{(f)}$ solve in the domain Ω for $t>0$ the homogenized system, consisting of the homogenized momentum balance equation (5.1.31), the continuity equation (5.1.30), and Darcy's law

$$\mathbf{w}^{(f)}=\frac{1}{\mu_1}\mathbb{B}\cdot\big(-\nabla\pi_f+t\,\rho_f\mathbf{e}\big),\tag{5.1.39}$$

for the liquid component, completed with the boundary conditions (5.1.24), (5.1.33), (5.1.36), (5.1.37), and the boundary condition

$$\pi_f(\mathbf{x}, t) = \int_0^t p_0(\mathbf{x}, \tau)d\tau \tag{5.1.40}$$

on the common boundary S^0 *and on the part* S_1^1 *of the outer boundary* S.

In (5.1.39), (5.1.40)

$$\pi_f(\mathbf{x}, t) = \int_0^t p(\mathbf{x}, \tau)d\tau, \quad m = \int_Y \chi(\mathbf{y})dy,$$

the symmetric strictly positive definite constant matrix $\mathbb{B}$ *is given by (1.1.27) (see Theorem 1.1 of Chap. 1).*

We refer to the problem (5.1.24), (5.1.25), (5.1.33), (5.1.36), (5.1.37), (5.1.39), (5.1.40) as the homogenized **model** $(\mathbb{FCM})_4$.

Theorem 5.7 *Under the conditions of Theorem 5.4 let* $\{\mathbf{w}_s^{(k)}, \mathbf{w}^{(f,k)}, \pi_f{}^{(k)}\}$ *be a solution of the model* $(\mathbb{FCM})_4$ *with* $\lambda_0 = k$.

Then sequences $\{\pi_f{}^{(k)}\}$, *and* $\{\mathbf{w}^{(f,k)}\}$ *converge weakly in* $L_2(Q_T)$ *and* $\mathbf{L}_2(Q_T)$ *to the functions* π_f, *and* $\mathbf{w}^{(f)}$ *respectively as* $k \to \infty$, *and the sequence* $\{\mathbf{w}_s^{(k)}\}$ *converges strongly in* $\mathbf{L}_2(\Omega_T)$ *to zero.*

In the domain Ω *for* $t > 0$ *limiting functions solve the homogenized system, consisting of the continuity equation*

$$\nabla \cdot \mathbf{w}^{(f)} = 0 \tag{5.1.41}$$

and Darcy's law

$$\mathbf{w}^{(f)} = \frac{1}{\mu_1}\mathbb{B} \cdot \big(- \nabla\pi_f + t\,\rho_f\mathbf{e}\big), \tag{5.1.42}$$

for the liquid component, completed with the boundary conditions (5.1.36) and (5.1.40).

We refer to the problem (5.1.36), (5.1.40)–(5.1.42) as the homogenized **model** $(\mathbb{FCM})_5$.

Theorem 5.8 *Under the condition of Theorem 5.2 let*

$$\alpha_\mu = \mu_0, \; 0 < \mu_0, \; \lambda_0 < \infty$$

and $\{\mathbf{w}^\varepsilon, \, p^\varepsilon\}$ *be the weak solution of the problem (5.1.1), (5.1.4)–(5.1.8), (5.1.10)–(5.1.12).*

Then the sequence $\{p^\varepsilon\}$ *converges weakly in* $L_2(Q_T)$ *as* $\varepsilon \to 0$ *to the function* p *and the sequence* $\{\mathbf{w}^\varepsilon\}$ *converges weakly in* $\mathbf{W}_2^{1,0}(Q_T)$ *as* $\varepsilon \to 0$ *to the func-*

tion $\mathbf{w}$. *The limiting functions solve the homogenized system, consisting of the Stokes equations*

$$\nabla \cdot \mathbf{w} = 0, \tag{5.1.43}$$

$$\nabla \cdot \Big(\mu_0 \mathbb{D}\Big(x, \frac{\partial \mathbf{w}}{\partial t}\Big) - p\,\mathbb{I}\Big) + \rho_f\,\mathbf{e} = 0 \tag{5.1.44}$$

in the domain Ω^0 *for* $t > 0$, *the continuity equation* (5.1.39) *and the homogenized momentum balance equation*

$$\nabla \cdot \widehat{\mathbb{P}} + \hat{\rho}\,\mathbf{e} = 0 \tag{5.1.45}$$

in the domain Ω *for* $t > 0$, *where*

$$\widehat{\mathbb{P}} = -p\,\mathbb{I} + \mathfrak{N}_1 : \mathbb{D}\Big(x, \frac{\partial \mathbf{w}}{\partial t}\Big) + \mathfrak{N}_2 : \mathbb{D}(x, \mathbf{w}) + \int_0^t \mathfrak{N}_3(t-\tau) : \mathbb{D}(x, \mathbf{w}(\mathbf{x}, \tau))d\tau.$$

The problem is completed with the continuity condition for normal tensions

$$\lim_{\substack{\mathbf{x}\to\mathbf{x}^0\\ \mathbf{x}\in\Omega^0}} \Big(\mu_0 \mathbb{D}\Big(x, \frac{\partial \mathbf{w}}{\partial t}(\mathbf{x}, t)\Big) - p(\mathbf{x}, t)\,\mathbb{I}\Big) \cdot \mathbf{n}(\mathbf{x}^0) = \lim_{\substack{\mathbf{x}\to\mathbf{x}^0\\ \mathbf{x}\in\Omega}} \widehat{\mathbb{P}}(\mathbf{x}, t) \cdot \mathbf{n}(\mathbf{x}^0) \tag{5.1.46}$$

on the common boundary S^0, *the Neumann boundary condition*

$$\Big(\mu_0 \mathbb{D}\Big(x, \frac{\partial \mathbf{w}}{\partial t}(\mathbf{x}, t)\Big) - p(\mathbf{x}, t)\,\mathbb{I}\Big) \cdot \mathbf{n} = -p^0(\mathbf{x}, t)\mathbf{n}, \tag{5.1.47}$$

on the part S_0^1 *of the outer boundary* S, *the Neumann boundary condition*

$$\widehat{\mathbb{P}} \cdot \mathbf{n} = -p^0(\mathbf{x}, t)\mathbf{n}, \tag{5.1.48}$$

on the part S_1^1 *of the outer boundary* S, *the Dirichlet boundary condition*

$$\mathbf{w}(\mathbf{x}, t) = 0 \tag{5.1.49}$$

on the part S^2 *of the outer boundary* S, *and the initial condition*

$$\mathbf{w}(\mathbf{x}, 0) = 0, \;\; \mathbf{x} \in \Omega. \tag{5.1.50}$$

Fourth-rank tensors $\mathfrak{N}_1$, $\mathfrak{N}_2$, $\mathfrak{N}_3(t)$ *are given by formulae* (1.4.30) *(see Theorem 1.11 of Chap. 1) and the symmetric tensor* $\mathfrak{N}_1$ *is strictly positively definite.*

We refer to the problem (5.1.43)–(5.1.50) as the homogenized **model** $(\mathbb{FCM})_6$.

5.1.2 Proof of Theorem 5.1

The proof of this theorem repeats the proofs of similar theorems in the previous chapters and is based on the energy equalities

$$\frac{1}{2}\int_{Q}\Big(\tau_0\tilde{\rho}^{\varepsilon}\Big|\frac{\partial\mathbf{w}^{\varepsilon}}{\partial t}(\mathbf{x},t)\Big|^2+\lambda_0(1-\zeta)(1-\chi^{\varepsilon})\big|\mathbb{D}\big(x,\mathbf{w}^{\varepsilon}(\mathbf{x},t)\big)\big|^2\Big)dx$$
$$+\ \alpha_\mu\int_0^t\int_Q\big(\zeta+(1-\zeta)\chi^{\varepsilon}\big)\Big|\mathbb{D}\Big(x,\frac{\partial\mathbf{w}^{\varepsilon}}{\partial\tau}(\mathbf{x},\tau)\Big)\Big|^2dxd\tau$$
$$=\int_0^t\int_Q\tilde{\rho}^{\varepsilon}\mathbf{e}\cdot\frac{\partial\mathbf{w}^{\varepsilon}}{\partial t}(\mathbf{x},\tau)dxd\tau, \tag{5.1.51}$$

and

$$\frac{1}{2}\int_{Q}\Big(\tau_0\tilde{\rho}^{\varepsilon}\Big|\frac{\partial^2\mathbf{w}^{\varepsilon}}{\partial t^2}(\mathbf{x},t)\Big|^2+\lambda_0(1-\zeta)(1-\chi^{\varepsilon})\Big|\mathbb{D}\Big(x,\frac{\partial\mathbf{w}^{\varepsilon}}{\partial t}(\mathbf{x},t)\Big)\Big|^2\Big)dx$$
$$+\alpha_\mu\int_0^t\int_Q\big(\zeta+(1-\zeta)\chi^{\varepsilon}\big)\Big|\mathbb{D}\Big(x,\frac{\partial^2\mathbf{w}^{\varepsilon}}{\partial\tau^2}(\mathbf{x},\tau)\Big)\Big|^2dxd\tau$$
$$=\frac{1}{2}\int_Q\tau_0\tilde{\rho}^{\varepsilon}\Big|\frac{\partial^2\mathbf{w}^{\varepsilon}}{\partial t^2}\Big|^2(\mathbf{x},0)dx=I_0. \tag{5.1.52}$$

We may use, for example, Galerkin's method. This method shows that for any $t\geqslant 0$ and any divergent free function $\varphi\in W_2^1(Q)$, vanishing at $\mathbf{x}\in S^2$, the equality

$$\int_Q\tau_0\tilde{\rho}^{\varepsilon}\frac{\partial^2\mathbf{w}^{\varepsilon}}{\partial t^2}(\mathbf{x},t)\cdot\varphi(\mathbf{x})dx+\int_Q\big(\zeta\mathbb{P}_f+(1-\zeta)\mathbb{P}\big)(\mathbf{x},t):\mathbb{D}\big(x,\varphi(\mathbf{x})\big)dx$$
$$=\int_Q\tilde{\rho}^{\varepsilon}\mathbf{e}\cdot\varphi(\mathbf{x})dx$$

holds true.

For $t=0$ $\mathbb{P}_f+(1-\zeta)\mathbb{P}=0$, and therefore

$$\int_Q\tau_0\tilde{\rho}^{\varepsilon}\frac{\partial^2\mathbf{w}^{\varepsilon}}{\partial t^2}(\mathbf{x},0)\cdot\varphi(\mathbf{x})dx=\int_Q\tilde{\rho}^{\varepsilon}\mathbf{e}\cdot\varphi(\mathbf{x})dx.$$

In particular, Galerkin's method states that $\dfrac{\partial^2\mathbf{w}^{\varepsilon}}{\partial t^2}(\mathbf{x},0)$ is a divergent free function in Q. Therefore, for

$$\varphi(\mathbf{x})=\frac{\partial^2\mathbf{w}^{\varepsilon}}{\partial t^2}(\mathbf{x},0)$$

$$\int_Q \tau_0\tilde{\rho}^{\,\varepsilon}\Big|\frac{\partial^2 \mathbf{w}^\varepsilon}{\partial t^2}(\mathbf{x},0)\Big|^2 dx = \int_Q \tilde{\rho}^{\,\varepsilon}\mathbf{e}\cdot\frac{\partial^2 \mathbf{w}^\varepsilon}{\partial t^2}(\mathbf{x},0)dx,$$

which implies

$$\int_Q \tau_0\tilde{\rho}^{\,\varepsilon}\Big|\frac{\partial^2 \mathbf{w}^\varepsilon}{\partial t^2}(\mathbf{x},0)\Big|^2 dx \leqslant \frac{C_0}{\tau_0}.$$

The last relation and (5.1.52) provide an estimate of the time derivative $\dfrac{\partial^2 \mathbf{w}^\varepsilon}{\partial t^2}$ in (5.1.17).

To estimate the right-hand side of (5.1.51) we use representations

$$\tilde{\rho}^{\,\varepsilon} = \rho_f + (1-\zeta)(1-\chi^\varepsilon)(\rho_s-\rho_f), \quad \mathbf{e} = -\nabla x_3,$$

integration by parts and the continuity equation (5.1.1)

$$\rho_f\int_Q \mathbf{e}\cdot\mathbf{w}^\varepsilon dx = -\rho_f\int_Q (\nabla x_3)\cdot\mathbf{w}^\varepsilon dx = 0.$$

So,

$$I = \int_0^t\int_Q \tilde{\rho}^\varepsilon\mathbf{e}\cdot\frac{\partial \mathbf{w}^\varepsilon}{\partial t}dxd\tau = -\rho_f\int_Q(\nabla x_3)\cdot\mathbf{w}^\varepsilon dx + (\rho_s-\rho_f)\int_\Omega(1-\chi^\varepsilon)\mathbf{e}\cdot\mathbf{w}^\varepsilon dx$$
$$= (\rho_s-\rho_f)\int_\Omega(1-\chi^\varepsilon)\mathbf{e}\cdot\mathbf{w}^\varepsilon dx.$$

Next we apply the Hölder inequality

$$I \leqslant (\rho_s-\rho_f)\Big(\int_\Omega dx\Big)^{\frac{1}{2}}\Big(\int_\Omega(1-\chi^\varepsilon)|\mathbf{w}^\varepsilon|^2dx\Big)^{\frac{1}{2}}$$
$$\leqslant \frac{(\rho_s-\rho_f)^2}{2\delta}|\Omega| + \frac{\delta}{2}\int_\Omega(1-\chi^\varepsilon)|\mathbf{w}^\varepsilon|^2dx$$

and extension $\mathbf{w}_s^\varepsilon = \mathbb{E}_{\Omega_s^\varepsilon}(\mathbf{w}^\varepsilon)$ (see (3.2.8) and Appendix B, Lemma B.9) from the domain Ω_s^ε onto the domain Q with Friedrichs–Poincaré's inequality:

$$I_1 = \int_\Omega(1-\chi^\varepsilon)|\mathbf{w}^\varepsilon|^2dx = \int_\Omega(1-\chi^\varepsilon)|\mathbf{w}_s^\varepsilon|^2dx \leqslant C\int_\Omega(1-\chi^\varepsilon)|\nabla\,\mathbf{w}_s^\varepsilon|^2dx,$$

and Korn's inequality

$$\int_\Omega(1-\chi^\varepsilon)|\nabla\,\mathbf{w}_s^\varepsilon|^2dx \leqslant C\int_\Omega(1-\chi^\varepsilon)|\mathbb{D}(x,\mathbf{w}_s^\varepsilon)|^2dx$$
$$= C\int_Q(1-\chi^\varepsilon)(1-\zeta)|\mathbb{D}(x,\mathbf{w}^\varepsilon)|^2dx.$$

Finally one has

$$I \leqslant \frac{(\rho_s - \rho_f)^2}{2\delta}|\Omega| + C\frac{\delta}{2}\int_Q (1-\chi^\varepsilon)(1-\zeta)|\mathbb{D}(x,\mathbf{w}^\varepsilon)|^2 dx,$$

which together with (5.1.52) prove (5.1.17).

The pressure p is estimated in the same way as in Chap. 3 (Theorem 3.1).

5.1.3 Proof of Theorem 5.2

As above, the proof of this theorem repeats the proofs of similar theorems in the previous chapters. Estimates (5.1.18) for displacements are based on the energy equality

$$\begin{aligned}
&\frac{\lambda_0}{2}\int_\Omega (1-\chi^\varepsilon)\Big|\mathbb{D}\big(x,\mathbf{w}^\varepsilon(\mathbf{x},t)\big)\Big|^2 dx \\
&\quad + \mu_0\int_0^t\int_Q \big(\zeta + (1-\zeta)\chi^\varepsilon\big)\Big|\mathbb{D}\Big(x,\frac{\partial \mathbf{w}^\varepsilon}{\partial \tau}(\mathbf{x},\tau)\Big)\Big|^2 dxd\tau \\
&\quad = \int_0^t\int_Q \big(\tilde{\rho}^\varepsilon \mathbf{e} - \nabla p^0\big)\cdot\frac{\partial \mathbf{w}^\varepsilon}{\partial t}(\mathbf{x},\tau)dxd\tau.
\end{aligned}$$

The estimation of the pressure p^ε is the same as in Theorem 3.1.

5.1.4 Proof of Theorem 5.3

On the basis of estimates (5.1.17), the results of Chap. 1, and Lemma B.5.1 ($\mu_1 = \infty$) we conclude that for $\varepsilon \to 0$

$$p^\varepsilon \to p \text{ weakly in } L_2(Q_T),$$

$$\mathbf{w}^\varepsilon \to \mathbf{w}(\mathbf{x},t) \text{ weakly and two-scale in } \mathbf{L}_2(Q_T),$$

$$\mathbf{w}_s^\varepsilon \to \mathbf{w}_s(\mathbf{x},t) \text{ weakly and two-scale in } \mathbf{L}_2(Q_T),$$

$$\frac{\partial \mathbf{w}^\varepsilon}{\partial t} \to \frac{\partial \mathbf{w}}{\partial t}(\mathbf{x},t) \text{ weakly and two-scale in } \mathbf{L}_2(Q_T),$$

$$\frac{\partial \mathbf{w}_s^\varepsilon}{\partial t} \to \frac{\partial \mathbf{w}_s}{\partial t}(\mathbf{x},t) \text{ weakly and two-scale in } \mathbf{L}_2(\Omega_T),$$

$$\frac{\partial^2 \mathbf{w}^\varepsilon}{\partial t^2} \to \frac{\partial \mathbf{w}^2}{\partial t^2}(\mathbf{x}, t) \text{ weakly and two-scale in } \mathbf{L}_2(Q_T),$$

$$\frac{\partial^2 \mathbf{w}_s^\varepsilon}{\partial t^2} \to \frac{\partial \mathbf{w}_s^2}{\partial t^2}(\mathbf{x}, t) \text{ weakly and two-scale in } \mathbf{L}_2(\Omega_T),$$

$$\mathbf{w}(\mathbf{x}, t) = \mathbf{w}_s(\mathbf{x}, t), \ \mathbf{x} \in \Omega, \ t > 0,$$

$$\mathbb{D}\big(x, \mathbf{w}_s^\varepsilon\big) \to \mathbb{D}(x, \mathbf{w}_s) + \mathbb{D}\big(y, \mathbf{U}(\mathbf{x}, t, \mathbf{y})\big) \text{ two-scale in } \mathbf{L}_2(\Omega_T).$$

Passing to the limit as $\varepsilon \to 0$ in (5.1.14) with test functions $\varphi = \varphi(\mathbf{x}, t)$, vanishing at $t = T$ and at S^2, we arrive at the microscopic momentum balance equation in the form of the integral identity

$$\int_{Q_T} \nabla \cdot (\varphi\, p^0) dx dt - Q_T \int_{\Omega_T} \Big(\tau_0 \hat{\rho} \frac{\partial \mathbf{w}_s}{\partial t} \cdot \frac{\partial \varphi}{\partial t} + \hat{\rho}\, \mathbf{e} \cdot \varphi\Big) dx dt$$
$$+ \int_{\Omega_T} \Big(\lambda_0\big((1-m)\mathbb{D}(x, \mathbf{w}_s) + \langle \mathbb{D}(x, \mathbf{U})\rangle_{Y_s}\big) - p\,\mathbb{I}\Big) : \mathbb{D}(x, \varphi) dx dt$$
$$= \int_{\Omega_T^0} \Big(\tau_0 \rho_f \frac{\partial \mathbf{w}}{\partial t} \cdot \frac{\partial \varphi}{\partial t} + \rho_f\, \mathbf{e} \cdot \varphi + p\,(\nabla \cdot \varphi)\Big) dx dt.$$

In Theorem 1.3 of Chap. 1 we have shown that

$$(1-m)\mathbb{D}(x, \mathbf{w}_s) + \langle \mathbb{D}(x, \mathbf{U})\rangle_{Y_s} = \mathfrak{N}_0^s : \mathbb{D}(x, \mathbf{w}_s).$$

Therefore, the last identity takes the form

$$\int_{Q_T} \nabla \cdot (\varphi\, p^0) dx dt - \int_{\Omega_T} \Big(\tau_0 \hat{\rho} \frac{\partial \mathbf{w}_s}{\partial t} \cdot \frac{\partial \varphi}{\partial t} + \hat{\rho}\, \mathbf{e} \cdot \varphi\Big) dx dt$$
$$+ \int_{\Omega_T} \big(\lambda_0 \mathfrak{N}_0^s : \mathbb{D}(x, \mathbf{w}_s) - p\,\mathbb{I}\big) : \mathbb{D}(x, \varphi) dx dt$$
$$= \int_{\Omega_T^0} \Big(\tau_0 \rho_f \frac{\partial \mathbf{w}}{\partial t} \cdot \frac{\partial \varphi}{\partial t} + \rho_f\, \mathbf{e} \cdot \varphi + p\,(\nabla \cdot \varphi)\Big) dx dt. \tag{5.1.53}$$

The continuity equation (5.1.1) for $\mathbf{w}^\varepsilon$ is evidently transformed into the continuity equation in the form of the integral identity

$$\int_{Q_T} \nabla \xi \cdot \mathbf{w} dx dt = 0 \tag{5.1.54}$$

for any smooth function ξ vanishing at $S^1 \times (0, T)$. This identity implies the continuity equation in (5.1.19), the continuity equation (5.1.21), and the continuity condition (5.1.22) on the common boundary S^0.

In turn, the integral identity (5.1.53) implies the dynamic equation in (5.1.19), the dynamic equation (5.1.20), the continuity condition (5.1.23) on the common boundary S^0, the boundary condition (5.1.26) on the part S_1^1 of the outer boundary S, and the boundary condition (5.1.25) on the part S_0^1 of the outer boundary S. The boundary condition (5.1.24) is a consequence of the corresponding boundary condition (5.1.8) (see Chap. 1).

5.1.5 *Proof of Theorem 5.4*

Note that estimates (5.1.17) are still valid for the functions $p^{(n)}$ and $\mathbf{w}_s^{(n)}$. Then (5.1.17) and (5.1.19) imply the inclusion $\nabla\, p^{(n)} \in \mathbf{L}_2(\Omega_T^0)$, and estimates

$$\max_{0\le t\le T}\int_Q \Big(\frac{1}{n^2}\zeta\Big|\frac{\partial^2 \mathbf{w}^{(n)}}{\partial t^2}\Big|^2 + \frac{1}{n^2}(1-\zeta)\Big|\frac{\partial^2 \mathbf{w}_s^{(n)}}{\partial t^2}\Big|^2\Big)dx$$
$$+\max_{0\le t\le T}\int_Q \Big(\frac{1}{n}\zeta\Big|\frac{\partial \mathbf{w}^{(n)}}{\partial t}\Big|^2 + \frac{1}{n}(1-\zeta)\Big|\frac{\partial \mathbf{w}_s^{(n)}}{\partial t}\Big|^2 + \lambda_0(1-\zeta)|\mathbb{D}(x,\mathbf{w}_s^{(n)})|^2\Big)dx$$
$$+\int_0^T\int_Q \Big(|p^{(n)}|^2 + \zeta\,|\nabla\, p^{(n)}|^2\Big)dxdt \leqslant C_0. \tag{5.1.55}$$

Coming back to (5.1.53) in the form

$$\int_{Q_T} \nabla\cdot(\varphi\, p^0)dxdt - \int_{\Omega_T}\Big(\frac{1}{n}\hat{\rho}\frac{\partial \mathbf{w}_s^{(n)}}{\partial t}\cdot\frac{\partial \varphi}{\partial t} + \hat{\rho}\,\mathbf{e}\cdot\varphi\Big)dxdt$$
$$+\int_{\Omega_T}\Big(\lambda_0\mathfrak{N}_0^s : \mathbb{D}(x,\mathbf{w}_s^{(n)}) - p^{(n)}\,\mathbb{I}\Big) : \mathbb{D}(x,\varphi)dxdt$$
$$=\int_{\Omega_T^0}\Big(\frac{1}{n}\rho_f\frac{\partial \mathbf{w}^{(n)}}{\partial t}\cdot\frac{\partial \varphi}{\partial t} + \rho_f\,\mathbf{e}\cdot\varphi + p^{(n)}\,(\nabla\cdot\varphi)\Big)dxdt \tag{5.1.56}$$

we conclude that

$$p^{(n)}(\mathbf{x},t) = p^0(t),\ \mathbf{x}\in S_0^1,\ t>0 \tag{5.1.57}$$

as a trace of function from $W_2^{1,0}(\Omega_T^0)$.

Due to estimates (5.1.55) there exists a subsequence of n (still denoted for simplicity by n) such that

$$\frac{1}{n}\hat{\rho}\frac{\partial \mathbf{w}_s^{(n)}}{\partial t} \to 0 \text{ strongly in } \mathbf{L}_2(\Omega_T),$$

$$\frac{1}{n}\rho_f\frac{\partial \mathbf{w}^{(n)}}{\partial t} \to 0 \text{ strongly in } \mathbf{L}_2(\Omega_T^0),$$

$$\nabla p^{(n)} \rightharpoonup \nabla p_0, \; p^{(n)} \rightharpoonup p_0 \text{ weakly in } \mathbf{L}_2(\Omega_T^0) \text{ and } L_2(\Omega_T^0),$$

and

$$p^{(n)} \rightharpoonup p, \; \mathbf{w}_s^{(n)} \rightharpoonup \mathbf{w}_s, \; \nabla \mathbf{w}_s^{(n)} \to \nabla \mathbf{w}_s \text{ weakly in } L_2(\Omega_T) \text{ and } \mathbf{L}_2(\Omega_T)$$

as $n \to \infty$.

The limit in (5.1.56) and in (5.1.54) as $n \to \infty$ results in the integral identity

$$\int_{Q_T} \nabla \cdot (\varphi\, p^0) dxdt - \int_{\Omega_T^0} \big(\rho_f\, \mathbf{e} \cdot \varphi + p_0\, (\nabla \cdot \varphi)\big) dxdt$$
$$+ \int_{\Omega_T} \Big(\big(\lambda_0 \mathfrak{N}_0^s : \mathbb{D}(x, \mathbf{w}_s) - p\, \mathbb{I}\big) : \mathbb{D}(x, \varphi) - \hat{\rho}\, \mathbf{e} \cdot \varphi\Big) dxdt = 0, \quad (5.1.58)$$

and the continuity equation (5.1.21).

Integral identity (5.1.58) obviously implies (5.1.26)–(5.1.29).

The validity of the boundary condition (5.1.24) has been proved earlier in Chap. 1.

5.1.6 Proof of Theorem 5.5

On the basis of estimates (5.1.17), results of Chap. 1, and Nguetseng's theorem we conclude that for $\varepsilon \to 0$

$$p^{\varepsilon} \rightharpoonup p(\mathbf{x}, t) \text{ weakly } L_2(Q_T),$$

$$(1 - \zeta) p^{\varepsilon} \chi^{\varepsilon} \to (1 - \zeta) p_f(\mathbf{x}, t) \chi(\mathbf{y}) \text{ two-scale in } L_2(Q_T),$$

$$\frac{\partial \mathbf{w}^{\varepsilon}}{\partial t} \rightharpoonup \mathbf{v} \text{ weakly in } \mathbf{L}_2(Q_T),$$

$$\frac{\partial^2 \mathbf{w}^{\varepsilon}}{\partial t^2} \rightharpoonup \frac{\partial \mathbf{v}}{\partial t}(\mathbf{x}, t) \text{ weakly in } \mathbf{L}_2(Q_T),$$

$$\frac{\partial \mathbf{w}^{\varepsilon}}{\partial t} \to \mathbf{V}(\mathbf{x}, t, \mathbf{y}) \text{ two-scale in } \mathbf{L}_2(Q_T),$$

$$\frac{\partial^2 \mathbf{w}^{\varepsilon}}{\partial t^2} \to \frac{\partial \mathbf{V}}{\partial t}(\mathbf{x}, t, \mathbf{y}) \text{ two-scale in } \mathbf{L}_2(Q_T),$$

$$\varepsilon \mathbb{D}\Big(x, \frac{\partial \mathbf{w}^{\varepsilon}}{\partial t}\Big) \to \mathbb{D}\big(y, \mathbf{V}(\mathbf{x}, t, \mathbf{y})\big) \text{ two-scale in } \mathbf{L}_2(Q_T),$$

$$\mathbf{w}_s^{\varepsilon} \to \mathbf{w}_s(\mathbf{x}, t) \text{ weakly and two-scale in } \mathbf{L}_2(\Omega_T),$$

$$\frac{\partial \mathbf{w}_s^\varepsilon}{\partial t} \to \frac{\partial \mathbf{w}_s}{\partial t}(\mathbf{x}, t) \text{ weakly and two-scale in } \mathbf{L}_2(\Omega_T),$$

$$\frac{\partial^2 \mathbf{w}_s^\varepsilon}{\partial t^2} \to \frac{\partial^2 \mathbf{w}_s}{\partial t^2}(\mathbf{x}, t) \text{ weakly and two-scale in } \mathbf{L}_2(\Omega_T),$$

$$\mathbb{D}\big(x, \mathbf{w}_s^\varepsilon\big) \to \mathbb{D}(x, \mathbf{w}_s) + \mathbb{D}\big(y, \mathbf{U}(\mathbf{x}, t, \mathbf{y})\big) \text{ two-scale in } \mathbf{L}_2(\Omega_T).$$

It is easy to see that

$$\mathbf{V} = \zeta\, \mathbf{V} + (1-\zeta)\Big(\chi(\mathbf{y})\, \mathbf{V} + \big(1-\chi(\mathbf{y})\big)\frac{\partial \mathbf{w}_s}{\partial t}\Big), \tag{5.1.59}$$

and

$$\int_{Q_T}\int_Y \Big(\tau_0|\mathbf{V}|^2 + \tau_0^2\Big|\frac{\partial \mathbf{V}}{\partial t}\Big|^2 + |\mathbb{D}(y, \mathbf{V})|^2\Big)dxdt \leqslant C_0, \tag{5.1.60}$$

where C_0 is independent of τ_0.

Moreover, we have the equality $\zeta\, \mathbf{V} = \zeta\, \mathbf{v}(\mathbf{x}, t)$, but this fact is not useful to us.

The two-scale limit in (5.1.14) as $\varepsilon \to 0$ with test functions $\varphi = \varphi(\mathbf{x}, t)$, vanishing at $t = T$ and at S^2 results in

$$\int_{Q_T} \nabla\cdot(\varphi\, p^0)dxdt - \int_{\Omega_T}\Big(\tau_0\Big(\rho_f\frac{\partial \mathbf{w}^f}{\partial t} + \rho_s(1-m)\frac{\partial \mathbf{w}_s}{\partial t}\Big)\cdot\frac{\partial \varphi}{\partial t} + \hat{\rho}\,\mathbf{e}\cdot\varphi\Big)dxdt$$
$$+\int_{\Omega_T}\Big(\lambda_0\big((1-m)\mathbb{D}(x, \mathbf{w}_s) + \langle\mathbb{D}(x, \mathbf{U})\rangle_{Y_s}\big) - p\,\mathbb{I}\Big) : \mathbb{D}(x, \varphi)dxdt$$
$$= \int_{\Omega_T^0}\Big(\tau_0\rho_f\mathbf{v}\cdot\frac{\partial \varphi}{\partial t} + \rho_f\,\mathbf{e}\cdot\varphi + p\,(\nabla\cdot\varphi)\Big)dxdt.$$

Earlier in Chap. 1 (Theorem 1.4) it was shown that

$$\lambda_0\big((1-m)\mathbb{D}(x, \mathbf{w}_s) + \langle\mathbb{D}(x, \mathbf{U})\rangle_{Y_s}\big) - p\,\mathbb{I} = \lambda_0\mathfrak{N}_1^s : \mathbb{D}(x, \mathbf{w}_s^{(n)}) - p_f\,\mathbb{I}.$$

Therefore,

$$\int_{Q_T} \nabla\cdot(\varphi\, p^0)dxdt - \int_{\Omega_T}\Big(\tau_0\Big(\rho_f\mathbf{v} + \rho_s(1-m)\frac{\partial \mathbf{w}_s}{\partial t}\Big)\cdot\frac{\partial \varphi}{\partial t} + \hat{\rho}\,\mathbf{e}\cdot\varphi\Big)dxdt$$
$$+\int_{\Omega_T}\Big(\lambda_0\mathfrak{N}_1^s : \mathbb{D}(x, \mathbf{w}_s) - p_f\,\mathbb{I}\Big) : \mathbb{D}(x, \varphi)dxdt$$
$$= \int_{\Omega_T^0}\Big(\tau_0\rho_f\frac{\partial \mathbf{w}}{\partial t}\cdot\frac{\partial \varphi}{\partial t} + \rho_f\,\mathbf{e}\cdot\varphi + p\,(\nabla\cdot\varphi)\Big)dxdt. \tag{5.1.61}$$

The continuity equation (5.1.1) for $\mathbf{w}^\varepsilon$ after taking the limit as $\varepsilon \to 0$ is transformed into the macroscopic continuity equation in the form of the integral identity

$$\int_{\Omega_T} \nabla\xi \cdot \Big(\zeta\, \mathbf{v} + (1-\zeta)\Big(\frac{\partial \mathbf{w}^f}{\partial t} + (1-m)\frac{\partial \mathbf{w}_s}{\partial t}\Big)\Big) dxdt = 0 \tag{5.1.62}$$

and the microscopic continuity equation

$$\nabla_y \cdot \mathbf{V} = 0,\ \mathbf{y} \in Y,\ \mathbf{x} \in Q,\ t > 0 \tag{5.1.63}$$

(for details see Chap. 1).

Integral identities (5.1.61) and (5.1.62) imply differential equations (5.1.19), (5.1.30), and (5.1.31), continuity condition (5.1.23), the continuity condition (5.1.33) on the common boundary S^0, and boundary conditions (5.1.25) and (5.1.37) on the outer boundary S.

The validity of the boundary condition (5.1.24) has been proved earlier in Chap. 1. So, we have only to derive the dynamic equation for the liquid component in Ω. To do that we consider the integral identity (5.1.14) in the form

$$\int_{Q_T} \Big(-\tau_0\, \tilde{\rho}^\varepsilon \frac{\partial \mathbf{w}^\varepsilon}{\partial t} \cdot \frac{\partial \varphi}{\partial t} + \big(\zeta \widetilde{\mathbb{P}}_f + (1-\zeta)\widetilde{\mathbb{P}}\big) : \mathbb{D}(x,\varphi)\Big) dxdt$$
$$= \int_{Q_T} \big(\tilde{\rho}^\varepsilon \mathbf{e}\cdot\varphi + (p^\varepsilon - p^0)\nabla\cdot\varphi\big) dxdt,$$
$$\widetilde{\mathbb{P}}_f = \mathbb{P}_f - p^\varepsilon\,\mathbb{I},\ \widetilde{\mathbb{P}} = \mathbb{P} - p^\varepsilon\,\mathbb{I},$$

with test functions $\varphi = h(\mathbf{x},t)\varphi_0\left(\frac{\mathbf{x}}{\varepsilon}\right)$, where h vanishes at S^2, and $\varphi_0(\mathbf{y})$ is a 1-periodic in $\mathbf{y}$ function, such that $\nabla_y \cdot \varphi_0 = 0$ for $\mathbf{y} \in Y$, and $supp\,\varphi_0 \subset Y_f$. One has

$$\int_{Q_T} \Big(h\, \tau_0 \rho_f \frac{\partial^2 \mathbf{w}^\varepsilon}{\partial t^2} \cdot \varphi_0 + h\frac{\alpha_\mu}{\varepsilon^2}\Big(\varepsilon \mathbb{D}\Big(x, \frac{\partial \mathbf{w}^\varepsilon}{\partial t}\Big)\Big) : \mathbb{D}(y,\varphi_0)$$
$$- \big(\zeta\,(p^\varepsilon - p^0) + (1-\zeta)\,\chi^\varepsilon\,(p^\varepsilon - p^0)\big)\nabla h\cdot\varphi_0 - h\,\rho_f \mathbf{e}\cdot\varphi_0\Big) dxdt$$
$$= -\frac{\varepsilon}{2}\int_{Q_T} \frac{\alpha_\mu}{\varepsilon^2}\Big(\varepsilon \mathbb{D}\Big(x, \frac{\partial \mathbf{w}^\varepsilon}{\partial t}\Big)\Big) : (\nabla h \otimes \varphi_0 + \varphi_0 \otimes \nabla h) dxdt.$$

The limit in the last identity as $\varepsilon \to 0$ results in

$$\int_{Q_T} \Big(h\Big(\tau_0 \rho_f \Big\langle \frac{\partial \mathbf{V}}{\partial t}\cdot\varphi_0\Big\rangle_Y + h\,\mu_1 \langle \mathbb{D}(y,\mathbf{V}) : \mathbb{D}(y,\varphi_0)\rangle_Y - \rho_f \mathbf{e}\cdot\langle\varphi_0\rangle_Y\Big)$$
$$- \big(\zeta\,(p - p^0) + (1-\zeta)\,(p_f - p^0)\big)\,\langle\varphi_0\rangle_Y\cdot\nabla h\Big) dxdt = 0, \tag{5.1.64}$$

or

$$\int_{Q_T} \big(h\,a_\varphi + \tilde{p}\,(\mathbf{b}_\varphi \cdot \nabla h)\big)dxdt = 0. \tag{5.1.65}$$

In (5.1.65)

$$a_\varphi = \tau_0 \rho_f \left\langle \frac{\partial \mathbf{V}}{\partial t} \cdot \varphi_0 \right\rangle_Y + h\,\mu_1 \langle \mathbb{D}(y, \mathbf{V}) : \mathbb{D}(y, \varphi_0)\rangle_Y - \rho_f \mathbf{e} \cdot \langle \varphi_0 \rangle_Y,$$

where $a_\varphi \in L_2\big((0, T); L_2(Q)\big)$ due to (5.1.56), and

$$\tilde{p} = \zeta\,(p - p^0) + (1 - \zeta)\,(p_f - p^0), \;\; \mathbf{b}_\varphi = -\langle \varphi_0 \rangle_Y.$$

In the same way as in Chap. 1 and on the base of Lemma B.15, choosing in (5.1.65) φ_0 from conditions $\langle \varphi_0 \rangle_Y = \mathbf{e}_i$, $i = 1, 2, 3$ ($\{\mathbf{e}_1, \mathbf{e}_2, \mathbf{e}_3\}$ is a standard Cartesian basis), we conclude that the function $\tilde{p}^{(n)}$ belongs to $\mathbf{w}_2^{1,0}(Q_T)$,

$$\int_{Q_T} |\nabla\, \tilde{p}^{(n)}|^2 dxdt \leqslant C_0, \tag{5.1.66}$$

where C_0 does not depend on τ_0, and the function p_f satisfies boundary condition (5.1.34).

Estimate (5.1.66) obviously implies the continuity condition (5.1.35) on the common boundary S^0.

For *supp* $h \subset \Omega$ (5.1.64) is equivalent to the differential equation

$$\tau_0 \rho_f \frac{\partial \mathbf{V}}{\partial t} = \frac{\mu_1}{2} \triangle_y \mathbf{V} - \nabla_y \Pi - \nabla p_f + \rho_f \mathbf{c} \tag{5.1.67}$$

in Y_f for $t > 0$, which we complete with continuity equation (5.1.63), the boundary condition

$$\mathbf{V}(\mathbf{x}, t, \mathbf{y}) = \frac{\partial \mathbf{w}_s}{\partial t}(\mathbf{x}, t), \;\; \mathbf{y} \in \gamma, \tag{5.1.68}$$

(see (5.1.59)), and the initial condition

$$\mathbf{V}(\mathbf{y}, 0) = 0, \;\; \mathbf{y} \in Y_f. \tag{5.1.69}$$

The problem (5.1.63), (5.1.67)–(5.1.69) for $\tau_0 = 1$ has been solved in Chap. 3 (see proof of Theorem 3.5). Therefore, we simply formulate the result.

Lemma 5.1 *For almost all* $\mathbf{x} \in \Omega$ *the function* $\dfrac{\partial \mathbf{w}^{(f)}}{\partial t} = \langle \mathbf{V} \rangle_{Y_f}$ *satisfies equation* (5.1.32), *where*

$$\mathbb{B}^{(f)}(\tau_0; t) = \sum_{i=1}^{3} \left(\int_{Y_f} \mathbf{V}_i^{(f)}(\mathbf{y}, t) dy \right) \otimes \mathbf{e}_i, \tag{5.1.70}$$

and $\mathbf{V}_i^{(f)}$, $i = 1, 2, 3$, *are solutions to the following periodic initial boundary value problem*

$$\tau_0\rho_f\frac{\partial\mathbf{V}_i^{(f)}}{\partial t} = \frac{\mu_1}{2}\Delta_y\mathbf{V}_i^{(f)} - \nabla_y\Pi_i^{(f)},\ (\mathbf{y}, t) \in Y_f \times (0, T), \tag{5.1.71}$$

$$\nabla_y \cdot \mathbf{V}_i^{(f)}(\mathbf{y}, t) = 0,\ (\mathbf{y}, t) \in Y_f,\ t > 0, \tag{5.1.72}$$

$$\tau_0\rho_f\mathbf{V}_i^{(f)}(\mathbf{y}, 0) = \mathbf{e}_i,\ \mathbf{y} \in Y_f, \tag{5.1.73}$$

$$\mathbf{V}_i^{(f)}(\mathbf{y}, t) = 0,\ \mathbf{y} \in \gamma,\ t > 0. \tag{5.1.74}$$

5.1.7 Proof of Theorem 5.6

As in the proof of Theorem 5.4 we use estimates (5.1.55), (5.1.66), the integral identity (5.1.61) to conclude that there exists a subsequence of n (still denoted for simplicity by n) such that

$$\frac{1}{n}\hat{\rho}\frac{\partial\mathbf{w}_s^{(n)}}{\partial t} \to 0 \text{ strongly in } \mathbf{L}_2(\Omega_T),$$

$$\frac{1}{n}\rho_f\frac{\partial\mathbf{w}^{(n)}}{\partial t} \to 0 \text{ strongly in } \mathbf{L}_2(\Omega_T^0),$$

$$\nabla p^{(n)} \rightharpoonup \nabla p_0,\ p^{(n)} \rightharpoonup p_0 \text{ weakly in } \mathbf{L}_2(\Omega_T^0) \text{ and } L_2(\Omega_T^0),$$

$$\nabla p_f^{(n)} \rightharpoonup \nabla p_f,\ p_f^{(n)} \rightharpoonup p_f \text{ weakly in } \mathbf{L}_2(\Omega_T^0) \text{ and } L_2(\Omega_T),$$

$$\mathbf{w}_s^{(n)} \rightharpoonup \mathbf{w}_s,\ \nabla\mathbf{w}_s^{(n)} \rightharpoonup \nabla\mathbf{w}_s \text{ weakly in } \mathbf{L}_2(\Omega_T)$$

as $n \to \infty$.

The limiting functions are bounded

$$\int_{\Omega_T}\big(|\nabla p_f|^2 + |\mathbf{w}_s|^2 + |\mathbb{D}(x, \mathbf{w}_s)|^2\big)dxdt \leqslant C_0, \tag{5.1.75}$$

and satisfy the integral identity

$$\int_{\Omega_T}\Big(\big(\lambda_0\mathfrak{N}_1^s : \mathbb{D}(x, \mathbf{w}_s) - (p_f - p^0)\,\mathbb{I}\big) : \mathbb{D}(x, \varphi) - \hat{\rho}\,\mathbf{e}\cdot\varphi\Big)dxdt$$

$$= \int_{\Omega_T^0} \big(\rho_f \mathbf{e} \cdot \varphi + (p - p^0)(\nabla \cdot \varphi)\big) dx dt. \tag{5.1.76}$$

which results in the dynamic equation (5.1.31), the relation (5.1.27), and boundary conditions (5.1.33) and (5.1.37).

The boundary condition (5.1.35), which implies (5.1.40) on the common boundary S^0, follows from (5.1.66), (5.1.75), and the boundary condition (5.1.35) for $p_f^{(n)}$.

The validity of the boundary condition (5.1.24) has been proved earlier in Chap. 1.

To derive the continuity equation (5.1.30) and Darcy's law (5.1.39) we somehow have to pass to the limit as $n \to \infty$ in equations (5.1.30) and (5.1.32). Unfortunately we have no uniform estimates with respect to n not only for $\dfrac{\partial \mathbf{w}^{(f,n)}}{\partial t}$ and $\dfrac{\partial \mathbf{w}_s^{(n)}}{\partial t}$, but neither for $\mathbf{w}^{(f,n)}$.

Let us try to find estimates using estimate (5.1.66) for pressure and estimates (5.1.60) for two-scale limits in the form

$$\int_{Q_T} \int_Y \Big(\frac{1}{n}|\mathbf{V}^{(n)}|^2 + \frac{1}{n^2}\Big|\frac{\partial \mathbf{V}^{(n)}}{\partial t}\Big|^2 + \Big|\mathbb{D}\big(y, \mathbf{V}^{(n)}\big)\Big|^2\Big) dx dt \leqslant C_0. \tag{5.1.77}$$

The problem (5.1.63), (5.1.67)–(5.1.69) can be rewritten for functions

$$\mathbf{w}^{(n)}(\mathbf{x}, t, \mathbf{y}) = \int_0^t \mathbf{V}^{(n)}(\mathbf{x}, \mathbf{y}, \tau) d\tau,$$

$$P^{(n)}(\mathbf{x}, t, \mathbf{y}) = \int_0^t \Pi^{(n)}(\mathbf{x}, \mathbf{y}, \tau) d\tau, \text{ and } \pi_f^{(n)}(\mathbf{x}, t) = \int_0^t p_f^{(n)}(\mathbf{x}, \tau) d\tau$$

as

$$\left.\begin{array}{l} \dfrac{1}{n}\rho_f \mathbf{V}^{(n)} = \dfrac{\mu_1}{2}\triangle_y \mathbf{W}^{(n)} - \nabla_y P^{(n)} - \nabla \pi_f^{(n)} + t\,\rho_f \mathbf{e}, \\ \nabla \cdot \mathbf{W}^{(n)} = 0, \quad \mathbf{y} \in Y_f; \\ \mathbf{W}^{(n)}(\mathbf{x}, t, \mathbf{y}) = \mathbf{w}_s^{(n)}(\mathbf{x}, t), \quad \mathbf{y} \in \gamma. \end{array}\right\} \tag{5.1.78}$$

The standard procedure results in the equality

$$\frac{\mu_1}{2} \int_{Y_f} |\nabla(\mathbf{W}^{(n)} - \mathbf{w}_s^{(n)})|^2 dy$$
$$= \int_{Y_f} \Big(t\,\rho_f \mathbf{e} - \nabla \pi_f^{(n)} - \frac{1}{n}\rho_f \mathbf{V}^{(n)}\Big) \cdot (\mathbf{W}^{(n)} - \mathbf{w}_s^{(n)}) dy,$$

and, successively, the a priori estimate

$$\int_{\Omega_T} \int_{Y_f} \big(|\mathbf{W}^{(n)} - \mathbf{w}_s^{(n)}|^2 + |\nabla(\mathbf{W}^{(n)} - \mathbf{w}_s^{(n)})|^2\big) dy dx dt$$

$$\leqslant C_0 \int_{\Omega_T} \Big(1 + \int_{Y_f} \Big|\frac{1}{n}\mathbf{V}^{(n)}\Big|^2 dydxdt\Big) \leqslant C_0. \tag{5.1.79}$$

This last estimate in the usual way implies the uniform estimate

$$\int_{\Omega_T} \int_{Y_f} |P^{(n)}|^2 dydxdt) \leqslant C_0 \tag{5.1.80}$$

for the pressure $P^{(n)}$.

Now we may extract convergent subsequences (for simplicity keeping the same notations)

$$p^{(n)} \rightharpoonup p = p_0(\mathbf{x}, t) \text{ weakly in } W_2^{1,0}(\Omega_T^0),$$

$$p_f^{(n)} \rightharpoonup p_f \text{ weakly in } W_2^{1,0}(\Omega_T),$$

$$\tilde{p}^{(n)} \rightharpoonup \tilde{p} \text{ weakly in } W_2^{1,0}(Q_T),$$

$$\tilde{\pi}^{(n)} \rightharpoonup \tilde{\pi} \text{ weakly in } W_2^{1,0}(Q_T)),$$

$$\frac{1}{n}\frac{\partial \mathbf{w}_s^{(n)}}{\partial t} \to 0 \text{ strongly in } \mathbf{L}_2(\Omega_T),$$

$$\mathbf{w}_s^{(n)} \to \mathbf{w}_s^{(n)} \text{ weakly in } \mathbf{W}_2^{1,0}(\Omega_T),$$

$$\frac{1}{n}\mathbf{V}^{(n)} \to 0 \text{ strongly in } \mathbf{L}_2(\Omega_T \times Y),$$

$$\mathbf{W}^{(n)} \to \mathbf{W},\ \nabla_y \mathbf{W}^{(n)} \to \nabla_y \mathbf{W} \text{ weakly in } \mathbf{L}_2\big(\Omega_T \times Y_f\big),$$

$$P^{(n)} \to P \text{ weakly in } L_2\big(\Omega_T \times Y_f\big),$$

$$\mathbf{w}^{(f,n)} = \langle \mathbf{W}^{(n)} \rangle_{Y_f} \rightharpoonup \mathbf{w}^f = \langle \mathbf{W} \rangle_{Y_f} \text{ weakly in } L_2(\Omega_T)$$

as $n \to \infty$.

Here

$$\tilde{\pi}^{(n)}(\mathbf{x}, t) = \int_0^t \tilde{p}^{(n)}(\mathbf{x}, \tau)d\tau,\ \tilde{\pi}(\mathbf{x}, t) = \int_0^t \tilde{p}(\mathbf{x}, \tau)d\tau,$$

$$\tilde{p}^{(n)} = \zeta p^{(n)} + (1-\zeta)p_f^{(n)},\ \ \tilde{p} = \zeta p + (1-\zeta)p_f,$$

and $p_0(\mathbf{x}, t)$ is the hydrostatic pressure defined by (5.1.27).

The limit as $n \to \infty$ in the continuity equation (5.1.30) in its integral form

$$\int_{\Omega_T} \nabla\xi \cdot \left(\mathbf{w}^{(f,n)} + (1-m)\,\mathbf{w}_s^{(n)}\right) dx dt = 0$$

with test functions ξ vanishing in Ω_0 results in the same continuity equation (5.1.30) for the limiting functions $\mathbf{w}^{(f)}$ and w_s, and the boundary condition (5.1.36).

The microscopic periodic boundary value problem, defining Darcy's law (5.1.39) (see proofs of Theorems 1.1 and 1.4 in Chap. 1), follows from (5.1.78) after taking the limit $n \to \infty$:

$$\left.\begin{array}{l} \frac{\mu_1}{2}\triangle_y \mathbf{W} - \nabla_y P - \nabla \pi_f + t\,\rho_f \mathbf{e} = 0, \\ \nabla \cdot \mathbf{W} = 0 \;\; \mathbf{y} \in Y_f; \\ \mathbf{W}(\mathbf{x}, t, \mathbf{y}) = \mathbf{w}_s(\mathbf{x}, t), \;\; \mathbf{y} \in \gamma. \end{array}\right\} \tag{5.1.81}$$

As usual, we must fulfill the limiting procedure in the corresponding integral identity.

Finally, the boundary condition (5.1.40) on the part S_1^1 of the outer boundary S is a consequence of the convergence

$$\tilde{\pi}^{(n)} \rightharpoonup \tilde{\pi} \;\text{ weakly in } W_2^{1,0}(Q_T)$$

and the corresponding boundary condition for $\tilde{\pi}$.

5.1.8 Proof of Theorem 5.7

Let

$$\bar{p}^{(k)} = p_f^{(k)} - p_0, \;\; \bar{\pi}^{(k)} = \pi_f^{(k)} - \pi_0$$

and

$$\pi_0(\mathbf{x}, t) = \int_0^t p_0(\mathbf{x}, \tau) d\tau.$$

Firstly we find estimates for the solution $\{\mathbf{w}_s^{(k)}, \mathbf{v}^{(k)},\; p_f{}^{(k)}\}$, independent of k. To do that we rewrite Eqs. (5.1.31) and (5.1.39) as

$$-\,k\,\nabla \cdot \left(\mathfrak{N}_1^s : \mathbb{D}(x, \mathbf{w}_s^{(k)})\right) + \nabla \bar{p}^{(k)} = \hat{\rho}\,\mathbf{e} - \nabla\, p_0 \equiv \mathbf{F}_s, \tag{5.1.82}$$

$$\mathbf{w}^{(f,k)} - m\,\mathbf{w}_s^{(k)} + \frac{1}{\mu_1}\mathbb{B} \cdot (\nabla \bar{\pi}^{(k)}) = \frac{1}{\mu_1}\mathbb{B} \cdot (t\,\rho_f \mathbf{e} - \nabla\, \pi_0) \equiv \mathbf{F}_l. \tag{5.1.83}$$

Next we multiply (5.1.82) by $\mathbf{w}_s^{(k)}$, (5.1.83) by $\nabla \bar{p}^{(k)}$, integrate by parts over Ω, sum results, and take into account (5.1.30) in its form of the integral identity:

$$\int_{\Omega}\big(\mathbf{w}^{(f,k)}+(1-m)\mathbf{w}_s^{(k)}\big)\cdot\nabla\bar{p}^{(k)}dx=0.$$

One has

$$k\int_{\Omega}(\mathfrak{N}_1^s:\mathbb{D}(x,\mathbf{w}_s^{(k)})):\mathbb{D}(x,\mathbf{w}_s^{(k)})dx+\frac{1}{2\mu_1}\frac{d}{dt}\int_{\Omega}\nabla\bar{\pi}^{(k)}\cdot\big(\mathbb{B}\cdot(\nabla\bar{\pi}^{(k)})\big)dx$$
$$=\int_{\Omega}\mathbf{F}_s\cdot\mathbf{w}_s^{(k)}dx+\frac{d}{dt}\int_{\Omega}\mathbf{F}_l\cdot\nabla\bar{\pi}^{(k)}dx-\int_{\Omega}\frac{\partial\mathbf{F}_l}{\partial t}\cdot\nabla\bar{\pi}^{(k)}dx. \quad (5.1.84)$$

Note that all integrals over boundaries disappear due to the choice of functions $\bar{\pi}^{(k)}$ and $\mathbf{u}^{(k)}$ and corresponding boundary conditions.

The last relation and the imbedding theorem [61] imply estimates

$$k\int_{\Omega_T}\Big(|\mathbf{w}_s^{(k)}(\mathbf{x},t)|^2+\Big|\mathbb{D}\big(x,\mathbf{w}_s^{(k)}(\mathbf{x},t)\big)\Big|^2\Big)dxdt$$
$$+\max_{0\le t\le T}\int_{\Omega}|\nabla\pi^{(k)}(\mathbf{x},t)|^2dx\leqslant C_0. \quad (5.1.85)$$

The derivation of (5.1.85) is quite formal because we do not have any information about the existence of $\nabla p^{(k)}$.

For a rigorous proof we must use mollifiers

$$u_{(h)}(\mathbf{x},t)=\frac{1}{h^3}\int_{\mathbb{R}^3}\eta\Big(\frac{|\mathbf{x}-\mathbf{y}|}{h}\Big)u(\mathbf{y},t)dy$$

(see [3, 61]) with some smooth and finite kernel η, and instead of (5.1.83) we consider the corresponding integral identity

$$\int_{\Omega}\Big(-\nabla\xi\cdot\mathbf{w}_s^{(k)}+\frac{1}{\mu_1}\nabla\xi\cdot\mathbb{B}\cdot(\nabla\bar{\pi}^{(k)})\Big)dx=\int_{\Omega}\nabla\xi\cdot\mathbf{F}_l dx, \quad (5.1.86)$$

where we have used the continuity equation (5.1.30) in its form as an integral identity.

To obtain the desired estimates we choose in (5.1.86) ξ as $\xi=\psi_{(h)}$, pass the smoothing from the test function ψ to the cofactors, and put $\psi=(\bar{p}^{(k)})_{(h)}$:

$$\int_{\Omega}\Big(-\nabla(\bar{p}^{(k)})_{(h)}\cdot(\mathbf{w}_s^{(k)})_{(h)}+\frac{1}{\mu_1}\nabla(\bar{p}^{(k)})_{(h)}\cdot\mathbb{B}\cdot\nabla(\bar{\pi}^{(k)})_{(h)}\Big)dx$$
$$=\int_{\Omega}\nabla(\bar{p}^{(k)})_{(h)}\cdot(\mathbf{F}_l)_{(h)}dx.$$

The last identity is easily transformed to

$$\int_0^t\int_{\Omega}(\bar{p}^{(k)})_{(h)}\,\nabla\cdot(\mathbf{w}_s^{(k)})_{(h)}dxd\tau+\frac{1}{2\mu_1}\int_{\Omega}\nabla(\bar{\pi}^{(k)})_{(h)}\cdot\mathbb{B}\cdot\nabla(\bar{\pi}^{(k)})_{(h)}dx$$

$$= \int_{\Omega} \nabla(\bar{\pi}^{(k)})_{(h)} \cdot (\mathbf{F}_l)_{(h)} dx - \int_0^t \int_{\Omega} \nabla(\bar{\pi}^{(k)})_{(h)} \cdot \left(\frac{\partial \mathbf{F}_l}{\partial t}\right)_{(h)} dx d\tau,$$

and, after the limit as $h \to 0$, to the equality

$$\int_0^t \int_{\Omega} \tilde{p}^{(k)} \nabla \cdot \mathbf{w}_s^{(k)} dx d\tau + \frac{1}{2\mu_1} \int_{\Omega} \nabla \bar{\pi}^{(k)} \cdot \big(\mathbb{B} \cdot \nabla \bar{\pi}^{(k)}\big) dx$$
$$= \int_{\Omega} \nabla \bar{\pi}^{(k)} \cdot \mathbf{F}_l dx - \int_0^t \int_{\Omega} \nabla \bar{\pi}^{(k)} \cdot \frac{\partial \mathbf{F}_l}{\partial t} dx d\tau.$$

This relation and the evident consequence of (5.1.82)

$$k \int_0^t \int_{\Omega} \big(\mathfrak{N}_1^s : \mathbb{D}(x, \mathbf{w}_s^{(k)})\big) : \mathbb{D}(x, \mathbf{w}_s^{(k)}) dx \tau - \int_0^t \int_{\Omega} \tilde{p}^{(k)} \nabla \cdot \mathbf{w}_s^{(k)} dx d\tau$$
$$= \int_0^t \int_{\Omega} \mathbf{F}_s \cdot \mathbf{w}_s^{(k)} dx \tau$$

result in (5.1.85).

Estimates (5.1.85) and Darcy's law (5.1.39) imply the estimate

$$\int_{\Omega_T} |\mathbf{w}^{(f,k)}|^2 dx dt \leqslant C_0.$$

On the basis of the last estimate and estimates (5.1.85) we may choose the convergent subsequences

$$\mathbf{w}_s^{(k_n)} \to 0 \text{ strongly in } \mathbf{W}_2^{1,0}(\Omega_T),$$

$$\mathbf{w}^{(f,k_n)} \rightharpoonup \mathbf{w}^{(f)} \text{ weakly in } \mathbf{L}_2(\Omega_T),$$

$$\pi_f^{(k_n)} \rightharpoonup \pi_f \text{ weakly in } L_2(\Omega_T)$$

as $k_n \to \infty$.

Darcy's law (5.1.83) in the form

$$\nabla \bar{\pi}^{(k)} = \mu_1 \mathbb{B}^{-1} \cdot \big(\mathbf{F}_l - \mathbf{w}^{(f,k_n)} + m\, \mathbf{w}_s^{(k_n)}\big)$$

provides that

$$\nabla \pi_f^{(k_n)} \rightharpoonup \nabla \pi_f \text{ weakly in } \mathbf{L}_2(\Omega_T)$$

as $k_n \to \infty$.

After taking the limit in (5.1.39) and in the continuity equation (5.1.30) in the form of the integral identity

$$\int_{\Omega_T} \big(\mathbf{w}^{(f,k_n)} + (1-m)\mathbf{w}_s^{(k_n)}\big) \cdot \nabla \xi dxdt = 0$$

we arrive at (5.1.36), (5.1.40)–(5.1.42).

5.1.9 Proof of Theorem 5.8

The main part of the proof of this Theorem repeats the proof of Theorem 1.11 of Chap 1. Let

$$\mathbf{v}^\varepsilon = \mathbb{E}_{\Omega_f^\varepsilon}\Big(\frac{\partial \mathbf{w}^\varepsilon}{\partial t}\Big)$$

be an extension of $\dfrac{\partial \mathbf{w}^\varepsilon}{\partial t}$ from $\Omega_f^\varepsilon \cup \Omega_0$ onto Q (see Theorem 1.11).

Estimates (5.1.18) provides the existence of convergent subsequences (for simplicity still denoted by ε), such that

$$p^\varepsilon \rightharpoonup p(\mathbf{x}, t) \text{ weakly in in } L_2(Q_T),$$

$$p^\varepsilon \to P(\mathbf{x}, t, \mathbf{y}) \text{ two-scale in } L_2(Q_T),$$

$$\mathbf{w}^\varepsilon \to \mathbf{w}(\mathbf{x}, t) \text{ two-scale in } \mathbf{L}_2(Q_T),$$

$$\mathbf{v}^\varepsilon \to \mathbf{v}(\mathbf{x}, t) = \frac{\partial \mathbf{w}}{\partial t}(\mathbf{x}, t) \text{ two-scale in } \mathbf{L}_2(Q_T),$$

$$\mathbb{D}\big(x, \mathbf{w}^\varepsilon\big) \to \mathbb{D}\big(x, \mathbf{w}\big) + \mathbb{D}\big(y, \mathbf{W}(\mathbf{x}, t, \mathbf{y})\big) \text{ two-scale in } \mathbf{L}_2(Q_T),$$

$$\mathbb{D}\big(x, \mathbf{v}^\varepsilon\big) \to \mathbb{D}\big(x, \frac{\partial \mathbf{w}}{\partial t}\big) + \mathbb{D}\Big(y, \frac{\partial \mathbf{W}}{\partial t}\Big) \text{ two-scale in } \mathbf{L}_2(Q_T).$$

The two-scale limit in (5.1.15) with test functions $\varphi = \varphi(\mathbf{x}, t)$ results in the integral identity

$$\begin{aligned}\int_{Q_T} \Bigg(\Bigg(\mu_0 \mathbb{D}\Big(x, \frac{\partial \mathbf{w}}{\partial t}\Big) - p\,\mathbb{I} + (1-\zeta)\Big(\mu_0 \Big\langle \mathbb{D}\Big(y, \frac{\partial \mathbf{W}}{\partial t}\Big)\Big\rangle_{Y_f} + \lambda_0 \mathbb{D}(x, \mathbf{w}) \\ + \lambda_0 \langle \mathbb{D}(y, \mathbf{W})\rangle_{Y_s}\Big)\Bigg) : \mathbb{D}(x, \varphi) + \nabla \cdot (\varphi\, p^0) \\ - \big(\zeta \rho_f + (1-\zeta)\hat{\rho}\big)\mathbf{e} \cdot \varphi\Bigg) dxdt = 0. \qquad (5.1.87)\end{aligned}$$

Due to Theorem 1.11

$$(1-\zeta)\Big(\mu_0\Big(\mathbb{D}(x,\frac{\partial \mathbf{w}}{\partial t})+\Big\langle\mathbb{D}\Big(y,\frac{\partial \mathbf{W}}{\partial t}\Big)\Big\rangle_{Y_f}\Big)+\lambda_0(\mathbb{D}(x,\mathbf{w})+\langle\mathbb{D}(y,\mathbf{w})\rangle_{Y_s})-p\,\mathbb{I}\Big)=\widehat{\mathbb{P}}.$$

Therefore, (5.1.87) transforms to

$$\int_{Q_T}\Big(\Big(\zeta\mu_0\mathbb{D}\Big(x,\frac{\partial \mathbf{w}}{\partial t}\Big)-p\,\mathbb{I}+(1-\zeta)\widehat{\mathbb{P}}\Big):\mathbb{D}(x,\varphi)$$
$$+\nabla\cdot(\varphi\,p^0)-\big(\zeta\rho_f+(1-\zeta)\hat{\rho}\big)\mathbf{e}\cdot\varphi\Big)dxdt=0. \qquad (5.1.88)$$

The continuity equation (5.1.1) after taking the limit as $\varepsilon\to 0$ does not change its form:

$$\nabla\cdot\mathbf{w}=0. \qquad (5.1.89)$$

In the usual way one may show that the integral identity (5.1.88) is equivalent to the Eqs. (5.1.44) and (5.1.45), and the boundary conditions (5.1.46)–(5.1.48). The boundary condition (5.1.49) follows from the integral identity

$$\int_{\Omega_T}\big(\mathbf{w}^\varepsilon(\nabla\cdot\varphi)+\nabla\mathbf{w}^\varepsilon\cdot\varphi\big)dxdt=0$$

for any smooth function φ, vanishing at S^0, after taking the limit as $\varepsilon\to 0$.

Finally, the initial condition (5.1.50) follows from the integral identity

$$\int_{\Omega_T} m\left(\frac{\partial \mathbf{w}}{\partial t}\cdot\varphi+\mathbf{w}\cdot\frac{\partial\varphi}{\partial t}\right)dxdt=0,$$

which holds true for any smooth function $\varphi=\varphi(\mathbf{x},t)$ vanishing at $t=T$. The last identity is a result of the two-scale limit as $\varepsilon\to 0$ in

$$\int_{\Omega_T}\chi^\varepsilon\left(\mathbf{v}^\varepsilon\cdot\varphi+\mathbf{w}^\varepsilon\cdot\frac{\partial\varphi}{\partial t}\right)dxdt=0.$$

5.2 Filtration in Two Different Poroelastic Media

5.2.1 Statement of the Problems

This section is devoted to the joint motion in the domain $Q=\Omega_0\cup S^0\cup\Omega$ of two different poroelastic media in Ω_0 and Ω respectively (Fig. 5.2). We suppose that Ω_0 and Ω have a common boundary S^0.

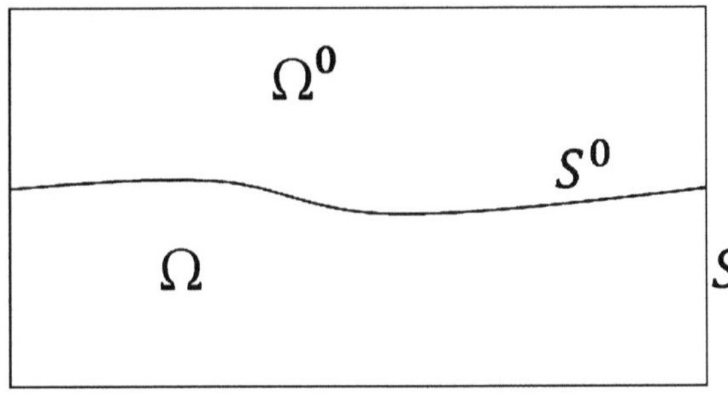

Fig. 5.2 Two different poroelastic media

In the domain Ω^0 for $t > 0$ the motion of the medium is described by the model $\mathbb{M}_{15}$

$$\nabla \cdot \mathbf{w} = 0, \tag{5.2.1}$$

$$\nabla \cdot \mathbb{P}^0 + \rho^{0,\varepsilon}\mathbf{F} = 0, \tag{5.2.2}$$

where

$$\mathbb{P}^0 = \chi_0^\varepsilon \alpha_\mu \mathbb{D}\Big(x, \frac{\partial \mathbf{w}}{\partial t}\Big) + (1 - \chi_0^\varepsilon)\lambda_0^0 \mathbb{D}(x, \mathbf{w}) - p\,\mathbb{I}.$$

The motion in Ω for $t > 0$ is also governed by the model $\mathbb{M}_{15}$, consisting of the continuity equation (5.2.1) and the momentum balance equation

$$\nabla \cdot \mathbb{P} + \rho^\varepsilon \mathbf{F} = 0, \tag{5.2.3}$$

where

$$\mathbb{P} = \chi^\varepsilon \alpha_\mu \mathbb{D}\Big(x, \frac{\partial \mathbf{w}}{\partial t}\Big) + (1 - \chi^\varepsilon)\lambda_0 \mathbb{D}(x, \mathbf{w}) - p\,\mathbb{I}.$$

On the common boundary $S^0 = \partial\Omega \cap \partial\Omega^0$ for $t > 0$ the continuity conditions

$$\lim_{\substack{\mathbf{x} \to \mathbf{x}^0 \\ \mathbf{x} \in \Omega^0}} \mathbf{w}(\mathbf{x}, t) = \lim_{\substack{\mathbf{x} \to \mathbf{x}^0 \\ \mathbf{x} \in \Omega}} \mathbf{w}(\mathbf{x}, t), \tag{5.2.4}$$

$$\lim_{\substack{\mathbf{x} \to \mathbf{x}^0 \\ \mathbf{x} \in \Omega^0}} \mathbb{P}^0(\mathbf{x}, t) \cdot \mathbf{n}(\mathbf{x}^0) = \lim_{\substack{\mathbf{x} \to \mathbf{x}^0 \\ \mathbf{x} \in \Omega}} \mathbb{P}(\mathbf{x}, t) \cdot \mathbf{n}(\mathbf{x}^0) \tag{5.2.5}$$

hold true for displacements and normal tensions. Here $n(\mathbf{x}^0)$ is a normal vector to the boundary S^0 at $\mathbf{x}^0 \in S^0$.

The problem is concluded with the Dirichlet boundary condition

$$\mathbf{w}(\mathbf{x}, t) = 0 \tag{5.2.6}$$

on the outer boundary $S = \partial Q$ for $t > 0$, the initial condition

$$\hat{\chi}^{\varepsilon}(\mathbf{x})\mathbf{w}(\mathbf{x},0)=0,\ \ \mathbf{x}\in Q, \tag{5.2.7}$$

and the normalization condition

$$\int_{Q}\frac{\hat{\chi}^{\varepsilon}(\mathbf{x})}{\hat{m}(\mathbf{x})}\,p\,(\mathbf{x},t)\,dx=0. \tag{5.2.8}$$

In (5.2.1)–(5.2.8) $\mathbf{F}$ is a given density of distributed mass forces,

$$\hat{\chi}^{\varepsilon}(\mathbf{x})=\zeta(\mathbf{x})\chi_0^{\varepsilon}(\mathbf{x})+\big(1-\zeta(\mathbf{x})\big)\chi^{\varepsilon}(\mathbf{x})$$

is the characteristic function of the liquid domain $Q_f^{\varepsilon}=\Omega_f^{0,\varepsilon}\cup\Omega_f^{\varepsilon}$,

$$\hat{m}=\zeta(\mathbf{x})m_0+\big(1-\zeta(\mathbf{x})\big)m,\ \ m=\int_Y\chi(\mathbf{y})dy,\ \ m_0=\int_Y\chi_0(\mathbf{y})dy,$$

$$\rho^{0,\varepsilon}=\rho_f\chi_0^{\varepsilon}+\rho_s^0(1-\chi_0^{\varepsilon}),\quad \rho^{\varepsilon}=\rho_f\chi^{\varepsilon}+\rho_s(1-\chi^{\varepsilon}).$$

As usual, λ_0^0 and λ_0 are dimensionless Lamé's constants of the solid component in Ω^0 and Ω respectively, ρ_s^0 and ρ_s are dimensionless densities of the solid component in Ω^0 and Ω respectively, $\chi^{\varepsilon}(\mathbf{x})$ is the characteristic function of the liquid domain Ω_f^{ε}, $\chi_0^{\varepsilon}(\mathbf{x})$ is the characteristic function of the liquid domain $\Omega_f^{0,\varepsilon}$,

$$\chi^{\varepsilon}(\mathbf{x})=\big(1-\zeta(\mathbf{x})\big)\chi\left(\frac{\mathbf{x}}{\varepsilon}\right),\quad \chi_0^{\varepsilon}(\mathbf{x})=\zeta(\mathbf{x})\chi_0\left(\frac{\mathbf{x}}{\varepsilon}\right).$$

Finally, $\zeta(\mathbf{x})$ is the characteristic function of the domain Ω^0 in Q, $\chi(\mathbf{y})$ is the characteristic function of the domain Y_f in the unit cube Y, and $\chi_0(\mathbf{y})$ is the characteristic function of the domain Y_f^0 in the unit cube Y.

Here we consider three different cases, when

(I) $\mu_0=0,\ 0<\mu_1<\infty,\ \lambda_0=\infty$ (absolutely rigid solid skeleton);
(II) $\mu_0=0,\ 0<\mu_1\leqslant\infty,\ 0<\lambda_0<\infty$ (slightly viscous liquid in an elastic solid skeleton);
(III) $0<\mu_0,\ \lambda_0<\infty$ (viscous liquid in an elastic solid skeleton),

where

$$\mu_0=\lim_{\varepsilon\searrow 0}\alpha_{\mu}(\varepsilon),\quad \mu_1=\lim_{\varepsilon\searrow 0}\frac{\alpha_{\mu}}{\varepsilon^2}.$$

We will obtain the homogenized model for the first case as a limit in the homogenized model for the second case with $\mu_1<\infty$, when $\lambda_0\to\infty$.

Throughout this section we impose Assumptions 0.1, 1.2, and 3.1 for the pore spaces, defined by characteristic functions $\chi(\mathbf{y})$ and $\chi_0(\mathbf{y})$, and suppose that

$$\max_{0<t<T}\int_Q|\mathbf{F}(\mathbf{x},t)|^2dx=\mathfrak{P}^2<\infty.$$

5.2.2 Main Results

It is clear, what type of homogenized equations we will obtain in each domain Ω^0 and Ω for $t > 0$. All these equations have already been described in Chap. 1. The main problem here is the continuity conditions on the common boundary S^0. In turn these conditions depend on the structures of the corresponding pore spaces, or, on the functions $\chi_0(\mathbf{y})$ and $\chi(\mathbf{y})$. For the sake of simplicity, we consider only two different cases (Figs. 5.3, 5.4, 5.5 and 5.6).

So,

(1) for the **first structure** of the common pore space elementary liquid domains Y_f^0 and Y_f have a nonempty intersection in Y:

$$Y_f^0 \cap Y_f \neq \emptyset; \tag{5.2.9}$$

(2) for the **second structure** of the common pore space elementary liquid domains Y_f^0 and Y_f have an empty intersection in Y:

$$Y_f^0 \cap Y_f = \emptyset; \tag{5.2.10}$$

(3) for **both structures** of the common pore space elementary solid domains Y_s^0 and Y_s have a nonempty intersection in Y:

$$Y_s^0 \cap Y_s \neq \emptyset. \tag{5.2.11}$$

We say that for the first structure the **common pore space is connected**, and for the second structure the **common pore space is disconnected**.

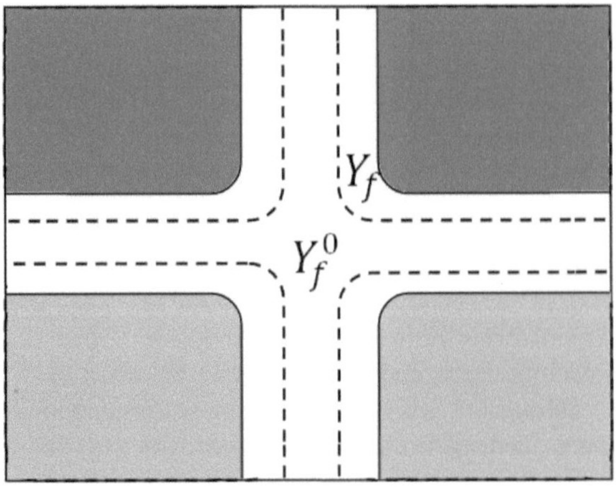

Fig. 5.3 Connected common pore space

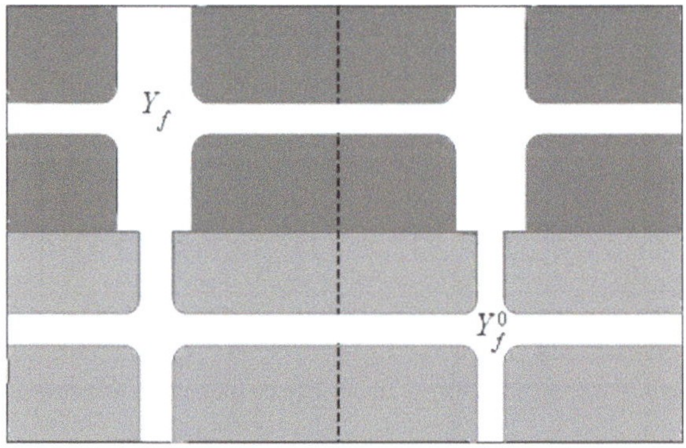

Fig. 5.4 Connected common pore space

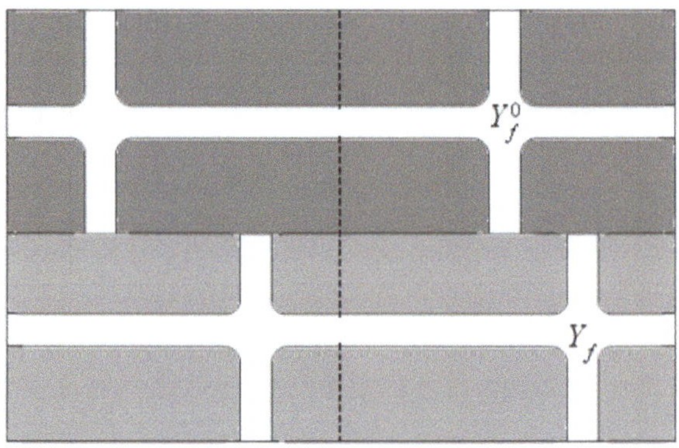

Fig. 5.5 Disconnected common pore space

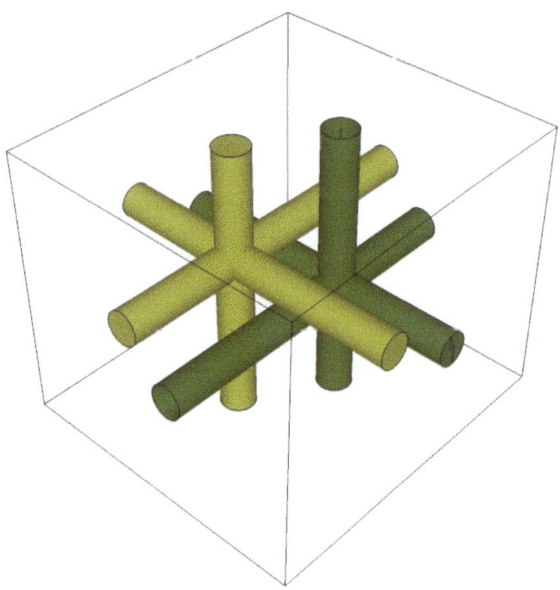

Fig. 5.6 Disconnected common pore space

To formulate the next statement we introduce the common liquid and solid domains Q_f^ε and Q_s^ε as

$$Q_f^\varepsilon = \Omega_f^{0,\varepsilon} \cup \Omega_f^\varepsilon, \quad Q_s^\varepsilon = \Omega_s^{0,\varepsilon} \cup \Omega_s^\varepsilon,$$

where $\Omega_f^{0,\varepsilon}$, $\Omega_s^{0,\varepsilon}$, Ω_f^ε, and Ω_s^ε are liquid and solid domains in Ω^0 and Ω respectively.

Lemma 5.2 *Let* $\mathbf{w}^\varepsilon \in \mathbf{L}_2\big((0,T); \mathbf{W}_2^1(Q_s^\varepsilon)\big)$. *Then for the first and second structures of the pore space in* Q *there exists an extension operator*

$$E_{Q_s^\varepsilon} : \mathbf{L}_2\big((0,T); \mathbf{W}_2^1(Q_s^\varepsilon)\big) \to \mathbf{L}_2\big((0,T); \mathbf{W}_2^1(Q)\big), \\ \mathbf{w}_s^\varepsilon = E_{Q_s^\varepsilon}(\mathbf{w}^\varepsilon), \tag{5.2.12}$$

such that

$$\big(1 - \hat{\chi}^\varepsilon(\mathbf{x})\big)\big(\mathbf{w}_s^\varepsilon(\mathbf{x},t) - \mathbf{w}^\varepsilon(\mathbf{x},t)\big) = 0, \ \mathbf{x} \in Q, \ t > 0, \tag{5.2.13}$$

and

$$\int_Q |\mathbf{w}_s^\varepsilon|^2 dx \leqslant C_0 \int_{Q_s^\varepsilon} |\mathbf{w}^\varepsilon|^2 dx, \\ \int_Q |\mathbb{D}(x,\mathbf{w}_s^\varepsilon)|^2 dx \leqslant C_0 \int_{Q_s^\varepsilon} |\mathbb{D}(x,\mathbf{w}^\varepsilon)|^2 dx, \tag{5.2.14}$$

where C_0 *is independent of* ε *and* $t \in (0,T)$.

Definition 5.3 We say that the pair of functions $\{\mathbf{w}^\varepsilon, \ p^\varepsilon\}$, such that

$$\mathbf{w}^\varepsilon \in \overset{\circ}{\mathbf{W}}{}_2^{1,0}(G_T), \ p^\varepsilon \in L_2(G_T), \ G_T = Q \times (0,T),$$

is a weak solution of the problem (5.2.1)–(5.2.8), if it satisfies the continuity equation (5.2.1) almost everywhere in G_T, the normalization condition (5.2.8), and the integral identity

$$\int_{G_T} \Big(-\alpha_\mu \hat{\chi}^\varepsilon \mathbb{D}(x,\mathbf{w}^\varepsilon) : \mathbb{D}\Big(x, \frac{\partial \varphi}{\partial t}\Big) + (1-\hat{\chi}^\varepsilon)\lambda(\mathbf{x})\mathbb{D}(x,\mathbf{w}^\varepsilon) : \mathbb{D}(x,\varphi)\Big) dxdt \\ = \int_{G_T} \big(p^\varepsilon(\nabla \cdot \varphi) + \hat{\rho}^\varepsilon \mathbf{F} \cdot \varphi\big) dxdt \tag{5.2.15}$$

for all functions φ vanishing at $t = T$, such that $\varphi, \ \dfrac{\partial \varphi}{\partial t} \in \overset{\circ}{\mathbf{W}}{}_2^{1,0}(G_T)$.

In (5.2.15)

$$\lambda(\mathbf{x}) = \lambda_0^0 \,\zeta(\mathbf{x}) + \lambda_0 \left(1 - \zeta(\mathbf{x})\right).$$

Theorem 5.9 *For all $\varepsilon > 0$ and for an arbitrary time interval $[0, T]$ there exists a unique generalized solution of problem (5.2.1)–(5.2.8) and*

$$\max_{0<t<T} \int_Q \hat{\chi}^{\varepsilon}\Big(\alpha_\mu \Big|\mathbb{D}\big(x, \mathbf{w}^\varepsilon(\mathbf{x}, t)\big)\Big|^2 + \frac{\alpha_\mu}{\varepsilon^2}|\mathbf{w}^\varepsilon(\mathbf{x}, t) - \mathbf{w}_s^\varepsilon(\mathbf{x}, t)|^2\Big)dx$$
$$+ \int_{G_T} \Big(|\pi^{\varepsilon}|^2 + \lambda(\mathbf{x})|\mathbb{D}(x, \mathbf{w}^\varepsilon)|^2\Big)dxdt \leqslant C_0 \mathfrak{P}^2, \qquad (5.2.16)$$

where C_0 is independent of ε, λ_0^0, and λ_0 for $\lambda_0 > 1$, $\lambda_0^0 > 1$, and

$$\pi^{\varepsilon}(\mathbf{x}, t) = \int_0^t p^\varepsilon(\mathbf{x}, \tau)d\tau.$$

Theorem 5.10 *Let $\mu_0 = 0$, $\mu_1 = \infty$, $0 < \lambda_0$, $\lambda_0^0 < \infty$, $\{\mathbf{w}^\varepsilon,\ p^{\varepsilon}\}$ be the weak solution of the problem (5.2.1)–(5.2.8),*

$$\pi^{\varepsilon}(\mathbf{x}, t) = \int_0^t p^\varepsilon(\mathbf{x}, \tau)d\tau,$$

and $\mathbf{w}_s^\varepsilon = \mathbb{E}_{Q_s^\varepsilon}(\mathbf{w}^\varepsilon)$ be an extension from the domain Q_s^ε onto the domain Q.

Then up to some subsequences the sequences $\{\mathbf{w}^\varepsilon\}$ and $\{\hat{\chi}^{\varepsilon}\pi^{\varepsilon}\}$ converge weakly in $\mathbf{L}_2(G_T)$ and $L_2(G_T)$ as $\varepsilon \to 0$ to the functions $\mathbf{w}_s$ and $\hat{m}\,\pi_f$ respectively, and the sequence $\{\mathbf{w}_s^\varepsilon\}$ converges weakly in $\mathbf{W}_2^{1,0}(G_T))$ as $\varepsilon \to 0$ to the function $\mathbf{w}_s$.

The limiting functions $\mathbf{w}_s$ and π_f, where $\dfrac{\partial \pi_f}{\partial t} \in L_2(G_T)$, solve in the domain Ω^0 for $t > 0$ the homogenized system, consisting of the homogenized momentum balance equation

$$\nabla \cdot \mathbb{P}_0^{(s,0)} + \hat{\rho}^{\,0}\,\mathbf{F} = 0, \qquad (5.2.17)$$

$$\mathbb{P}_0^{(s,0)} = \lambda_0^0\, \mathfrak{N}_0^{s,0} : \mathbb{D}(x, \mathbf{w}_s) - p_f\, \mathbb{I},$$

and the continuity equation

$$\nabla \cdot \mathbf{w}_s = 0. \qquad (5.2.18)$$

In the domain Ω for $t > 0$ the limiting functions solve the homogenized system, consisting of the homogenized momentum balance equation

$$\nabla \cdot \mathbb{P}_0^{(s)} + \hat{\rho}\,\mathbf{F} = 0, \qquad (5.2.19)$$

$$\mathbb{P}_0^{(s)} = \lambda_0\, \mathfrak{N}_0^{s} : \mathbb{D}(x, \mathbf{w}_s) - p_f\, \mathbb{I},$$

and the continuity equation (5.2.18).

The problem is completed with the normalization condition

$$\int_Q p_f(\mathbf{x}, t)dx = 0, \tag{5.2.20}$$

the boundary condition

$$\mathbf{w}_s = 0 \tag{5.2.21}$$

on the outer boundary S for $t > 0$, and the continuity conditions

$$\lim_{\substack{\mathbf{x} \to \mathbf{x}^0 \\ \mathbf{x} \in \Omega}} \mathbf{w}_s(\mathbf{x}, t) = \lim_{\substack{\mathbf{x} \to \mathbf{x}^0 \\ \mathbf{x} \in \Omega^0}} \mathbf{w}_s(\mathbf{x}, t), \ \mathbf{x}^0 \in S^0, \tag{5.2.22}$$

$$\lim_{\substack{\mathbf{x} \to \mathbf{x}^0 \\ \mathbf{x} \in \Omega}} \mathbb{P}_0^{(s)}(\mathbf{x}, t) \cdot \mathbf{n}(\mathbf{x}^0) = \lim_{\substack{\mathbf{x} \to \mathbf{x}^0 \\ \mathbf{x} \in \Omega^0}} \mathbb{P}_0^{(s,0)}(\mathbf{x}, t) \cdot \mathbf{n}(\mathbf{x}^0), \ \mathbf{x}^0 \in S^0 \tag{5.2.23}$$

on the common boundary S^0 for $t > 0$.

In (5.2.17)–(5.2.23) $\mathbf{n}(\mathbf{x}^0)$ is a unit normal to S^0 at $\mathbf{x}^0 \in S^0$,

$$p_f = \frac{\partial \pi_f}{\partial t}, \quad \hat{m} = \zeta\, m_0 + (1 - \zeta)m,$$

$$\hat{\rho}^0 = m_0\, \rho_f + (1 - m_0)\, \rho_s^0, \quad m_0 = \int_Y \chi_0 dy,$$

$$\hat{\rho} = m\, \rho_f + (1 - m)\, \rho_s, \quad m = \int_Y \chi dy,$$

the symmetric strictly positively definite constant fourth-rank tensors $\mathfrak{N}_0^{s,0}$ and $\mathfrak{N}_0^s$ are given by (5.2.54).

We refer to the problem (5.2.17)–(5.2.23) as the homogenized **model** $(\mathbb{FCM})_7$.

Theorem 5.11 *Let $\mu_0 = 0$, $0 < \mu_1, \lambda_0 < \infty$, $\{\mathbf{w}^\varepsilon, \ p^\varepsilon\}$ be the weak solution of the problem (5.2.1)–(5.2.8), and $\mathbf{w}_s^\varepsilon = \mathbb{E}_{Q_s^\varepsilon}(\mathbf{w}^\varepsilon)$ be an extension from the domain Q_s^ε onto the domain Q, and*

$$\pi^\varepsilon(\mathbf{x}, t) = \int_0^t p^\varepsilon(\mathbf{x}, \tau)d\tau,$$

Then up to some subsequences the sequences $\{\hat{\chi}^\varepsilon \pi^\varepsilon\}$, and $\{\hat{\chi}^\varepsilon \mathbf{w}^\varepsilon\}$ converge weakly in $L_2(G_T)$ and $\mathbf{L}_2(G_T)$ as $\varepsilon \to 0$ to the functions $\hat{m}\, \pi_f$ and $\mathbf{w}^{(f)}$ respectively, and the sequence $\{\mathbf{w}_s^\varepsilon\}$ converges weakly in $\mathbf{W}_2^1(Q))$ as $\varepsilon \to 0$ to the function $\mathbf{w}_s$.

In the domain Ω^0 for $t > 0$ limiting functions $\mathbf{w}^{(f)}$, $\mathbf{w}_s$, and π_f, where $\nabla \pi_f^{(\delta)} \in \mathbf{L}_2(\Omega_T)$, $\dfrac{\partial \pi_f^{(\delta)}}{\partial t} \in L_2(\Omega_T)$, solve the homogenized system, consisting of the continuity equation

$$\nabla \cdot \left(\mathbf{w}^{(f)} + (1 - m_0)\,\mathbf{w}_s\right) = 0, \tag{5.2.24}$$

the homogenized momentum balance equation

$$\nabla \cdot \mathbb{P}_1^{(s,0)} + \hat{\rho}^{\,0}\,\mathbf{F} = 0, \tag{5.2.25}$$

$$\mathbb{P}_1^{(s,0)} = \lambda_0^0\,\mathfrak{N}_1^{s,0} : \mathbb{D}(x, \mathbf{w}_s) - p_f\,\mathbb{I}$$

for the solid component, and Darcy's law in the form

$$\mathbf{w}^{(f)} = m_0\,\mathbf{w}_s + \frac{1}{\mu_1}\mathbb{B}^0 \cdot \left(- \nabla\,\pi_f + \rho_f \int_0^t \mathbf{F}(\mathbf{x}, \tau) d\tau \right) \tag{5.2.26}$$

for the liquid component.

In the domain Ω for $t > 0$ limiting functions solve the homogenized system, consisting of the continuity equation

$$\nabla \cdot \left(\mathbf{w}^{(f)} + (1 - m)\,\mathbf{w}_s\right) = 0, \tag{5.2.27}$$

the homogenized momentum balance equation

$$\nabla \cdot \mathbb{P}_1^{(s)} + \hat{\rho}\,\mathbf{F} = 0, \tag{5.2.28}$$

$$\mathbb{P}_1^{(s)} = \lambda_0\,\mathfrak{N}_1^{s} : \mathbb{D}(x, \mathbf{w}_s) - p_f\,\mathbb{I}$$

for the solid component, and Darcy's law in the form

$$\mathbf{w}^{(f)} = m\,\mathbf{w}_s + \frac{1}{\mu_1}\mathbb{B} \cdot \left(- \nabla\,\pi_f + \rho_f \int_0^t \mathbf{F}(\mathbf{x}, \tau) d\tau \right) \tag{5.2.29}$$

for the liquid component.

The problem is completed with the normalization condition

$$\int_Q \pi_f(\mathbf{x}, t) dx = 0, \tag{5.2.30}$$

the boundary condition (5.2.21) for the solid displacements $\mathbf{w}_s$ and the boundary condition

$$\mathbf{w}^{(f)}(\mathbf{x}, t) \cdot \mathbf{n}(\mathbf{x}) = 0 \tag{5.2.31}$$

for the liquid displacements on the outer boundary S for $t > 0$, and continuity conditions

$$\lim_{\substack{\mathbf{x} \to \mathbf{x}^0 \\ \mathbf{x} \in \Omega}} \mathbf{w}_s(\mathbf{x}, t) = \lim_{\substack{\mathbf{x} \to \mathbf{x}^0 \\ \mathbf{x} \in \Omega^0}} \mathbf{w}_s(\mathbf{x}, t), \;\; \mathbf{x}^0 \in S^0, \tag{5.2.32}$$

$$\lim_{\substack{\mathbf{x}\to\mathbf{x}^0\\ \mathbf{x}\in\Omega}} \mathbb{P}_1^{(s)}(\mathbf{x},t)\cdot\mathbf{n}(\mathbf{x}^0) = \lim_{\substack{\mathbf{x}\to\mathbf{x}^0\\ \mathbf{x}\in\Omega^0}} \mathbb{P}_1^{(s,0)}(\mathbf{x},t)\cdot\mathbf{n}(\mathbf{x}^0),\ \mathbf{x}^0\in S^0, \tag{5.2.33}$$

and

$$\lim_{\substack{\mathbf{x}\to\mathbf{x}^0\\ \mathbf{x}\in\Omega}} \mathbf{n}(\mathbf{x}^0)\cdot\big(\mathbf{w}^{(f)} + (1-m)\,\mathbf{w}_s\big)(\mathbf{x},t)$$
$$= \lim_{\substack{\mathbf{x}\to\mathbf{x}^0\\ \mathbf{x}\in\Omega^0}} \mathbf{n}(\mathbf{x}^0)\cdot\big(\mathbf{w}^{(f)} + (1-m_0)\,\mathbf{w}_s\big)(\mathbf{x},t),\ \mathbf{x}^0\in S^0 \tag{5.2.34}$$

on the common boundary S^0 for $t > 0$.

Finally, the last missing continuity condition on S^0 depends on the structure of the common pore space, namely, if the common pore space is connected (as in the case of the first structure), then

$$\lim_{\substack{\mathbf{x}\to\mathbf{x}^0\\ \mathbf{x}\in\Omega}} \pi_f(\mathbf{x},t) = \lim_{\substack{\mathbf{x}\to\mathbf{x}^0\\ \mathbf{x}\in\Omega^0}} \pi_f(\mathbf{x},t),\ \mathbf{x}^0\in S^0, \tag{5.2.35}$$

and if the common pore space is disconnected (as in the case of the second structure), then

$$\lim_{\substack{\mathbf{x}\to\mathbf{x}^0\\ \mathbf{x}\in\Omega}} \mathbf{n}(\mathbf{x}^0)\cdot\big(\mathbf{w}^{(f)} - m\,\mathbf{w}_s\big)(\mathbf{x},t) = 0,\ \mathbf{x}^0\in S^0. \tag{5.2.36}$$

Conditions (5.2.34) and (5.2.36) result in

$$\lim_{\substack{\mathbf{x}\to\mathbf{x}^0\\ \mathbf{x}\in\Omega^0}} \mathbf{n}(\mathbf{x}^0)\cdot\big(\mathbf{w}^{(f)} - m_0\,\mathbf{w}_s\big)(\mathbf{x},t) = 0,\ \mathbf{x}^0\in S^0. \tag{5.2.37}$$

In (5.2.24)–(5.2.37) $\mathbf{n}(\mathbf{x}^0)$ is a unit normal to S^0 at $\mathbf{x}^0 \in S^0$,

$$p_f = \frac{\partial \pi_f}{\partial t},\quad \hat{m} = \zeta\, m_0 + (1-\zeta)m,$$
$$\hat{\rho}^0 = m_0\,\rho_f + (1-m_0)\,\rho_s^0,\quad m_0 = \int_Y \chi_0 dy,$$
$$\hat{\rho} = m\,\rho_f + (1-m)\,\rho_s,\quad m = \int_Y \chi dy,$$

the symmetric strictly positively definite constant fourth-rank tensor $\mathfrak{N}_1^{s,0}$ is given by (1.2.38) for the pore space with the characteristic function $\chi_0(\mathbf{y})$, and the symmetric strictly positively definite constant fourth-rank tensor $\mathfrak{N}_1^s$ is given by (1.2.38) for the pore space with the characteristic function $\chi(\mathbf{y})$ (see Theorem 1.4 of Chap. 1), the symmetric strictly positive definite constant matrix $\mathbb{B}^0$ is given by (1.1.27) for *the pore space with the characteristic function $\chi_0(\mathbf{y})$, and the symmetric strictly positive definite constant matrix $\mathbb{B}$ is given by (1.1.27) for the pore space with the characteristic function $\chi(\mathbf{y})$ (see Theorem 1.1 of Chap. 1).*

We refer to the problem (5.2.21), (5.2.24)–(5.2.35) for the first structure of the common pore space as the homogenized **model** $(\mathbb{FCM})_8$, and to the problem (5.2.21), (5.2.24)–(5.2.34), (5.2.36), (5.2.37) for the second structure of the common pore space as the homogenized **model** $(\mathbb{FCM})_9$.

Theorem 5.12 *Under the conditions of Theorem 5.9 let $\{\mathbf{w}_s^{(k)}, \mathbf{w}^{(f,k)}, \pi_f^{(k)}\}$ be the weak solution of the model $(\mathbb{FCM})_8$ with $\lambda_0^0 = \lambda_0 = k$.*

Then up to some subsequences sequences $\{\pi_f^{(k)}\}$, and $\{\mathbf{w}^{(f,k)}\}$ converge weakly in $L_2(G_T)$ and $\mathbf{L}_2(G_T)$ as $k \to \infty$ to the functions π_f and $\mathbf{w}^{(f)}$ respectively, and the sequence $\{\mathbf{w}_s^{(k)}\}$ converges strongly in $\mathbf{L}_2(G_T)$ to zero.

In the domain Ω^0 for $t > 0$ limiting functions solve the homogenized system, consisting of the continuity equation

$$\nabla \cdot \mathbf{w}^{(f)} = 0 \tag{5.2.38}$$

and Darcy's law

$$\mathbf{w}^{(f)} = \frac{1}{\mu_1}\mathbb{B}^0 \cdot \Big(-\nabla \pi_f + \rho_f \int_0^t \mathbf{F}(\mathbf{x}, \tau) d\tau \Big), \tag{5.2.39}$$

and in the domain Ω for $t > 0$ limiting functions solve the homogenized system, consisting of the continuity equation (5.2.38) and Darcy's law

$$\mathbf{w}^{(f)} = \frac{1}{\mu_1}\mathbb{B} \cdot \Big(-\nabla \pi_f + \rho_f \int_0^t \mathbf{F}(\mathbf{x}, \tau) d\tau \Big), \tag{5.2.40}$$

completed with the boundary condition (5.2.31) for the liquid velocity on the outer boundary S for $t > 0$, the normalization condition (5.2.30), and the continuity conditions (5.2.35) and

$$\lim_{\substack{\mathbf{x} \to \mathbf{x}^0 \\ \mathbf{x} \in \Omega}} \mathbf{n}(\mathbf{x}^0) \cdot \mathbf{w}^{(f)}(\mathbf{x}, t) = \lim_{\substack{\mathbf{x} \to \mathbf{x}^0 \\ \mathbf{x} \in \Omega^0}} \mathbf{n}(\mathbf{x}^0) \cdot \mathbf{w}^{(f)}(\mathbf{x}, t), \ \mathbf{x}^0 \in S^0 \tag{5.2.41}$$

on the common boundary S^0 for $t > 0$.

The symmetric strictly positive definite constant matrix $\mathbb{B}^0$ is given by (1.1.27) for the pore space with the characteristic function $\chi_0(\mathbf{y})$, and the symmetric strictly positive definite constant matrix $\mathbb{B}$ is given by (1.1.27) for the pore space with the characteristic function $\chi(\mathbf{y})$ (see Theorem 1.1 of Chap. 1).

We refer to the problem (5.2.30), (5.2.31), (5.2.35), (5.2.38)–(5.2.41) as the homogenized **model** $(\mathbb{FCM})_{10}$.

Theorem 5.13 *Under the conditions of Theorem 5.9 let* $\{\mathbf{w}_s^{(k)},\ \mathbf{w}^{(f,k)},\ \pi_f^{(k)}\}$ *be the weak solution of the model* $(\mathbb{FCM})_9$ *with* $\lambda_0^0 = \lambda_0 = k$.

Then, up to some subsequences sequences $\{\pi_f^{(k)}\}$, *and* $\{\mathbf{w}^{(f,k)}\}$ *converge weakly in* $L_2(G_T)$ *and* $\mathbf{L}_2(G_T)$ *as* $k \to \infty$ *to the functions* π_f *and* $\mathbf{w}^{(f)}$ *respectively, and the sequence* $\{\mathbf{w}_s^{(k)}\}$ *converges strongly in* $\mathbf{L}_2(G_T)$ *to zero.*

In the domain Ω^0 *for* $t > 0$ *limiting functions solve the homogenized system, consisting of the continuity equation (5.2.38) and Darcy's law (5.2.39), and in the domain* Ω *for* $t > 0$ *limiting functions solve the homogenized system, consisting of the continuity equation (5.2.38) and Darcy's law (5.2.40), completed with the boundary condition (5.2.31) for the liquid velocity on the outer boundary* S *for* $t > 0$, *the normalization condition (5.2.30), and the continuity conditions (5.2.41) and*

$$\lim_{\substack{\mathbf{x}\to\mathbf{x}^0\\ \mathbf{x}\in\Omega}} \mathbf{n}(\mathbf{x}^0)\cdot\mathbf{w}^{(f)}(\mathbf{x},t) = 0,\ \ \mathbf{x}^0 \in S^0 \tag{5.2.42}$$

on the common boundary S^0 *for* $t > 0$.

The symmetric strictly positive definite constant matrix $\mathbb{B}^0$ *is given by (1.1.27) for the pore space with the characteristic function* $\chi_0(\mathbf{y})$, *and the symmetric strictly positive definite constant matrix* $\mathbb{B}$ *is given by (1.1.27) for the pore space with the characteristic function* $\chi(\mathbf{y})$ *(see Theorem 1.1 of Chap. 1).*

We refer to the problem (5.2.30), (5.2.31), (5.2.38)–(5.2.42) as the homogenized **model** $(\mathbb{FCM})_{11}$.

To consider the following case we change the setting of the problem at the microscopic level, namely, instead of the normalization condition (5.2.8) we consider the normalization condition

$$\int_Q p^\varepsilon(\mathbf{x},t)\,dx = 0. \tag{5.2.43}$$

The proof of the solvability of (5.2.1)–(5.2.7), (5.2.43) and the derivation of the a priori estimates repeat exactly the proof of the solvability and the derivation of the a priori estimates of (5.2.1)–(5.2.8).

Theorem 5.14 *Let*

$$\alpha_\mu = \mu_0,\ 0 < \mu_0,\ \lambda_0,\ \lambda_0^0 < \infty,$$

$$\int_{G_T} (1-\chi^\varepsilon)\Big|\frac{\partial \mathbf{F}}{\partial t}\Big|^2 dxdt = \mathfrak{P}_1^2 < \infty,$$

and $\{\mathbf{w}^\varepsilon,\ p^\varepsilon\}$ *be the weak solution of the problem (5.2.1)–(5.2.7), (5.2.43).*

Then up to some subsequences the sequence $\{p^{\varepsilon}\}$ converges weakly in $L_2(G_T)$ as $\varepsilon \to 0$ to the function p, and the sequence $\{\mathbf{w}^{\varepsilon}\}$ converges weakly in $\mathbf{W}_2^{1,0}(G_T))$ as $\varepsilon \to 0$ to the function **w**.

The limiting functions solve the homogenized system, consisting of the continuity equation

$$\nabla \cdot \mathbf{w} = 0, \tag{5.2.44}$$

and the homogenized momentum balance equation

$$\begin{aligned}
&\nabla \cdot \widehat{\mathbb{P}}^0 + \hat{\rho}\,\mathbf{F} = 0,\\
&\widehat{\mathbb{P}}^0 = -p\,\mathbb{I} + \mathfrak{N}_1^0 : \mathbb{D}(x, \frac{\partial \mathbf{w}}{\partial t}) + \mathfrak{N}_2^0 : \mathbb{D}(x, \mathbf{w})\\
&\quad + \int_0^t \mathfrak{N}_3^0(t-\tau) : \mathbb{D}(x, \mathbf{w}(\mathbf{x}, \tau))d\tau
\end{aligned} \tag{5.2.45}$$

in the domain Ω^0 for $t > 0$, the continuity equation (5.2.44) and the homogenized momentum balance equation

$$\begin{aligned}
&\nabla \cdot \widehat{\mathbb{P}} + \hat{\rho}\,\mathbf{F} = 0,\\
&\widehat{\mathbb{P}} = -p\,\mathbb{I} + \mathfrak{N}_1 : \mathbb{D}\Big(x, \frac{\partial \mathbf{w}}{\partial t}\Big) + \mathfrak{N}_2 : \mathbb{D}(x, \mathbf{w})\\
&\quad + \int_0^t \mathfrak{N}_3(t-\tau) : \mathbb{D}(x, \mathbf{w}(\mathbf{x}, \tau))d\tau
\end{aligned} \tag{5.2.46}$$

in the domain Ω for $t > 0$.

The problem is completed with the normalization condition

$$\int_Q p\,(\mathbf{x}, t)\,dx = 0, \tag{5.2.47}$$

the continuity condition for normal tensions

$$\lim_{\substack{\mathbf{x} \to \mathbf{x}^0\\ \mathbf{x} \in \Omega^0}} \widehat{\mathbb{P}}^0(\mathbf{x}, t) \cdot \mathbf{n}(\mathbf{x}^0) = \lim_{\substack{\mathbf{x} \to \mathbf{x}^0\\ \mathbf{x} \in \Omega}} \widehat{\mathbb{P}}(\mathbf{x}, t) \cdot \mathbf{n}(\mathbf{x}^0) \tag{5.2.48}$$

on the common boundary S^0, the Dirichlet boundary condition

$$\mathbf{w}(\mathbf{x}, t) = 0 \tag{5.2.49}$$

the outer boundary S, and the initial condition

$$\mathbf{w}(\mathbf{x}, 0) = 0, \;\; \mathbf{x} \in Q. \tag{5.2.50}$$

Fourth-rank tensors $\mathfrak{N}_1^0$, $\mathfrak{N}_2^0$, $\mathfrak{N}_3^0(t)$ are given by formulae (1.4.30) for criteria μ_0 and λ_0^0, and the pore space with the characteristic function $\chi_0(\mathbf{y})$, and fourth-rank tensors $\mathfrak{N}_1$, $\mathfrak{N}_2$, $\mathfrak{N}_3(t)$ are given by formulae (1.4.30) for criteria μ_0 and λ_0, and the pore space with the characteristic function $\chi(\mathbf{y})$ (see Theorem 1.11 of Chap. 1). The symmetric tensors $\mathfrak{N}_1^0$ and $\mathfrak{N}_1$ are strictly positively definite.

We refer to the problem (5.2.44)–(5.2.50) as the homogenized **model** $(\mathbb{FCM})_{12}$.

5.2.3 Proof of Lemma 5.2

For all elementary cells which have no intersection with the common boundary S^0 we construct the extension in the same way, as in the corresponding extension lemma in Appendix B. So, we have to consider more precisely only cells $\varepsilon\tilde{Y}^{(k)}$ with nonempty intersections with S^0. To simplify the proof we consider only two different cases: (1) $Y_f^0 \subset Y_f$ (first structure), and (2) $Y_f^0 \cap Y_f = \emptyset$ (second structure). All these cubes have two parts. The first part belongs to the domain Ω^0 and has the pore space defined by $\chi_0(\mathbf{y})$, and the second part belongs to Ω and has the pore space defined by $\chi(\mathbf{y})$. For the first structure of the common pore space (see Fig. 5.4) the pore space in the whole cube has the same structure as in other cubes in Ω^0 and in Ω. So we may use the same method of extension as in that extension lemma. For the second structure of the common pore space (see Fig. 5.5) one has two disconnected sets, but it is again possible to apply the same extension method as before.

5.2.4 Proof of Theorem 5.9

The proof of this theorem repeats the proofs of similar theorems in the previous chapters (see, for example, the proof of Theorem 1.2). In fact, setting in (5.2.15) $\varphi(\mathbf{x},\tau) = h(\tau)\mathbf{w}^\varepsilon(\mathbf{x},\tau)$, where $h(\tau) = 1$ for $0 < \tau < t$ and $h(\tau) = 0$ for $t < \tau < T$ we first obtain that

$$\alpha_\mu \int_{Q_f^\varepsilon} \big|\mathbb{D}\big(x, \mathbf{w}^\varepsilon(\mathbf{x},t)\big)\big|^2 dx + \min(\lambda_0^0, \lambda_0) \int_0^t \int_{Q_s^\varepsilon} \big|\mathbb{D}\big(x, \mathbf{w}^\varepsilon(\mathbf{x},\tau)\big)\big|^2 dx d\tau \leqslant C_0 \mathfrak{P}^2, \tag{5.2.51}$$

$$\int_0^T \int_Q |\mathbf{w}^\varepsilon(\mathbf{x},t) - \mathbf{w}_s^\varepsilon(\mathbf{x},t)|^2 dx dt z$$

$$\leqslant \frac{C_0}{\mu_1}\alpha_\mu \int_0^T \int_{Q_f^\varepsilon} \left|\mathbb{D}\big(x, \mathbf{w}^\varepsilon(\mathbf{x}, t)\big) - \mathbb{D}\big(x, \mathbf{w}_s^\varepsilon(\mathbf{x}, t)\big)\right|^2 dx, \qquad (5.2.52)$$

where $\mathbf{w}_s^\varepsilon$ is an extension of $\mathbf{w}^\varepsilon$ from the solid part Q_s^ε onto the liquid part Q_f^ε:

$$\int_Q |\mathbf{w}_s^\varepsilon(\mathbf{x}, t)|^2 dx \leqslant C_0 \int_{Q_s^\varepsilon} |\mathbf{w}^\varepsilon(\mathbf{x}, t)|^2 dx,$$

$$\int_Q |\mathbf{w}_s^\varepsilon(\mathbf{x}, t)|^2 dx \leqslant C_0 \int_Q \left|\mathbb{D}\big(x, \mathbf{w}_s^\varepsilon(\mathbf{x}, t)\big)\right|^2 dx$$

$$\leqslant C_0 \int_{Q_s^\varepsilon} \left|\mathbb{D}\big(x, \mathbf{w}^\varepsilon(\mathbf{x}, t)\big)\right|^2 dx, \qquad (5.2.53)$$

$$Q_s^\varepsilon = \{\mathbf{x} \in Q : \hat{\chi}^\varepsilon(\mathbf{x}) = 0\}, \quad Q_f^\varepsilon = \{\mathbf{x} \in Q : \hat{\chi}^\varepsilon(\mathbf{x}) = 1\}.$$

In (5.2.51) and (5.2.52) C_0 depends only on the domain Q, and the geometry of pore spaces in Ω^0 and Ω, and does not depend on ε, and in (5.2.51) C_0 additionally depends on $\min\{\lambda_0^0, \lambda_0, 1\}$.

5.2.5 Proof of Theorem 5.10

On the basis of estimates (5.2.16) and in the same way as in Chap. 1 we conclude that for $\varepsilon \to 0$

$$\pi^\varepsilon \to \pi_f(\mathbf{x}, t)\hat{\chi}(\mathbf{y}) + \big(1 - \hat{\chi}(\mathbf{y})\big)\Pi_s(\mathbf{x}, t, \mathbf{y}) \text{ two-scale in } L_2(G_T),$$

$$\hat{\chi}^\varepsilon \pi^\varepsilon \rightharpoonup \hat{m}\,\pi_f \text{ weakly in } L_2(G_T),$$

$$\mathbf{w}_s^\varepsilon \to \mathbf{w}_s(\mathbf{x}, t) \text{ weakly and two-scale in } \mathbf{L}_2(G_T),$$

$$\mathbf{w}^\varepsilon \to \mathbf{w} = \mathbf{w}_s(\mathbf{x}, t) \text{ weakly and two-scale in } \mathbf{L}_2(G_T),$$

$$\mathbb{D}\big(x, \mathbf{w}_s^\varepsilon\big) \to \mathbb{D}\big(x, \mathbf{w}_s(\mathbf{x}, t)\big) + \mathbb{D}\big(y, \hat{\mathbf{U}}(\mathbf{x}, t, \mathbf{y})\big) \text{ two-scale in } \mathbf{L}_2(G_T).$$

The continuity condition (5.2.22) on the common boundary S^0 is a consequence of the smoothness of $\mathbf{w}_s$.

The weak limit in the continuity equation (5.2.1) in its integral form

$$\int_{G_T} \mathbf{w}^\varepsilon \cdot \nabla h dx dt = 0$$

for arbitrary smooth functions $h = h(\mathbf{x}, t)$ results in the continuity equation (5.2.18) in Q for $t > 0$.

As we have shown in Chap. 1 (Theorems 1.3 and 1.4)

$$p_f = \frac{\partial \pi_f}{\partial t} \in L_2(G_T), \quad P_s = \frac{\partial \Pi_s}{\partial t} \in L_2(Y_s \times G_T).$$

Passing to the limit as $\varepsilon \to 0$ in (5.2.15) with test functions $\varphi = \varphi(\mathbf{x}, t)$, vanishing at $t = T$ and at S, we arrive at the microscopic momentum balance equation in the form of the integral identity

$$\int_{G_T} \Big(\zeta\, \lambda_0^0\big((1-m_0)\mathbb{D}(x, \mathbf{w}_s) + \langle \mathbb{D}(x, \hat{\mathbf{U}})\rangle_{Y_s^0} - \langle (P_s - p_f)\rangle_{Y_s^0}\big) : \mathbb{D}(x, \varphi)$$
$$+ (1-\zeta)\lambda_0\big((1-m)\mathbb{D}(x, \mathbf{w}_s) + \langle \mathbb{D}(x, \hat{\mathbf{U}})\rangle_{Y_s} - \langle (P_s - p_f)\rangle_{Y_s}\big) : \mathbb{D}(x, \varphi)\Big) dxdt$$
$$= \int_{G_T} \Big(p_f\, (\nabla \cdot \varphi) + \big(\zeta \hat{\rho}^0 + (1-\zeta)\hat{\rho}\big)\mathbf{F} \cdot \varphi\Big) dxdt = 0.$$

for arbitrary smooth function φ vanishing at S.

The function $\hat{\mathbf{U}}$ is defined separately in each domain Ω_T^0 and Ω_T (see Theorem 1.3 of Chap. 1):

$$\hat{\mathbf{U}} = \mathbf{U}^0(\mathbf{x}, t, \mathbf{y}), \quad \text{for } \mathbf{x} \in \Omega^0, \quad \hat{\mathbf{U}} = \mathbf{U}(\mathbf{x}, t, \mathbf{y}), \quad \text{for } \mathbf{x} \in \Omega,$$

and

$$\mathbb{D}(x, \mathbf{w}_s) + \langle \mathbb{D}(x, \mathbf{U}^0)\rangle_{Y_s^0} - \langle (P_s - p_f)\rangle_{Y_s^0} = \mathfrak{N}_0^{s,0} : \mathbb{D}(x, \mathbf{w}_s)$$
$$= \Big(\mathfrak{N}^{s,0} - \Big\langle \sum_{i,j=1}^{3} P_0^{(ij,0)} \Big\rangle_{Y_s^0} \mathbb{I} \otimes \mathbb{J}^{ij}\Big) : \mathbb{D}(x, \mathbf{w}_s),$$
$$\mathbb{D}(x, \mathbf{w}_s) + \langle \mathbb{D}(x, \mathbf{U})\rangle_{Y_s} - \langle (P_s - p_f)\rangle_{Y_s} = \mathfrak{N}_0^{s} : \mathbb{D}(x, \mathbf{w}_s)$$
$$= \Big(\mathfrak{N}^{s} - \Big\langle \sum_{i,j=1}^{3} P_0^{(ij)} \Big\rangle_{Y_s} \mathbb{I} \otimes \mathbb{J}^{ij}\Big) : \mathbb{D}(x, \mathbf{w}_s). \tag{5.2.54}$$

Here $\mathfrak{N}^{s,0}$ and $\mathfrak{N}^{s}$ are given by (1.2.35) for the pore space with characteristic functions $\chi_0(\mathbf{y})$ and $\chi(\mathbf{y})$, and $P_0^{(ij,0)}$ and $P_0^{(ij)}$ are solutions of the system (1.2.34) in domains Y_s and Y_s^0 respectively.

Therefore the last identity takes the form

$$\int_{G_T} \Big(\zeta\, \big((\lambda_0^0 \mathfrak{N}_0^{s,0} : \mathbb{D}(x, \mathbf{w}_s) - p_f\, \mathbb{I}) : \mathbb{D}(x, \varphi) - \hat{\rho}^0 \mathbf{F} \cdot \varphi\big)$$
$$+ (1-\zeta)\big((\lambda_0 \mathfrak{N}_0^{s} : \mathbb{D}(x, \mathbf{w}_s) - p_f\, \mathbb{I}) : \mathbb{D}(x, \varphi) - \hat{\rho}\mathbf{F} \cdot \varphi\big)\Big) dxdt = 0. \tag{5.2.55}$$

This identity implies Eqs. (5.2.17) and (5.2.19) in Ω_T^0 and Ω_T respectively, and the continuity condition (5.2.23) on the common boundary S^0.

The weak limit in the normalization condition (5.2.8) in the form

$$0 = -\int_0^T \frac{dh}{dt}(t)\Big(\int_\Omega \frac{\hat{\chi}^\varepsilon}{\hat{m}}\pi^\varepsilon dx\Big)dt \to -\int_0^T \frac{dh}{dt}(t)\Big(\int_\Omega \pi_f\, dx\Big)dt$$
$$= \int_0^T h(t)\Big(\int_\Omega p_f\, dx\Big)dt = 0$$

for arbitrary smooth finite on $(0, T)$ functions $h(t)$ results in the normalization condition (5.2.20).

The boundary condition (5.2.21) on the outer boundary S follows from the Lemma B.14 in Appendix B.

Finally, the continuity condition (5.2.22) on the common boundary S^0 is a consequence of the inclusion $\mathbf{w}_s \in \mathbf{W}_2^{1,0}(G_T)$.

5.2.6 Proof of Theorem 5.11

As in the proofs of Theorems 5.5, 5.6, and 5.10 of this Chapter, and Theorem 1.4 of Chap. 1, we conclude that for $\varepsilon \to 0$

$$\hat{\chi}^\varepsilon \pi^\varepsilon \rightharpoonup \pi(\mathbf{x}, t) \text{ weakly in } L_2(G_T),$$

$$\pi^\varepsilon \to \hat{\chi}(\mathbf{x}, \mathbf{y})\,\pi_f + \big(1 - \hat{\chi}(\mathbf{x}, \mathbf{y})\big)\Pi_s(\mathbf{x}, t, \mathbf{y}) \text{ two-scale in } L_2(G_T),$$

$$\hat{\chi}^\varepsilon \mathbf{w}^\varepsilon \rightharpoonup \mathbf{w}^{(f)} \text{ weakly in } \mathbf{L}_2(G_T),$$

$$\hat{\chi}^\varepsilon \mathbf{w}^\varepsilon \to \hat{\chi}(\mathbf{x}, \mathbf{y})\hat{\mathbf{W}}(\mathbf{x}, t, \mathbf{y}) \text{ two-scale in } \mathbf{L}_2(G_T), \quad \mathbf{w}^{(f)} = \langle \hat{\chi}\hat{\mathbf{W}}\rangle_Y,$$

$$\mathbf{w}_s^\varepsilon \to \mathbf{w}_s(\mathbf{x}, t) \text{ weakly and two-scale in } \mathbf{L}_2(G_T),$$

$$\mathbb{D}\big(x, \mathbf{w}_s^\varepsilon\big) \to \mathbb{D}\big(x, \mathbf{w}_s(\mathbf{x}, t)\big) + \mathbb{D}\big(y, \hat{\mathbf{U}}(\mathbf{x}, t, \mathbf{y})\big) \text{ two-scale in } \mathbf{L}_2(G_T),$$

$$\hat{\mathbf{W}} = \zeta \mathbf{W}^0(\mathbf{x}, t, \mathbf{y}) + (1 - \zeta)\mathbf{W}(\mathbf{x}, t, \mathbf{y}), \quad \hat{\mathbf{U}} = \zeta \mathbf{U}^0(\mathbf{x}, t, \mathbf{y}) + (1 - \zeta)\mathbf{U}(\mathbf{x}, t, \mathbf{y}),$$

$$\hat{\chi}(\mathbf{x}, \mathbf{y}) = \zeta(\mathbf{x})\chi_0(\mathbf{y}) + \big(1 - \zeta(\mathbf{x})\big)\chi(\mathbf{y}),$$

$$p = \frac{\partial \pi}{\partial t} \in L_2(G_T), \; p_f = \frac{\partial \pi_f}{\partial t} \in L_2(G_T),$$

and the integral identities

$$\int_{G_T}\big(\zeta(\mathbf{w}^{(f)}+(1-m_0)\mathbf{w}_s)+(1-\zeta)(\mathbf{w}^{(f)}+(1-m)\mathbf{w}_s)\big)\cdot\nabla h dxdt=0, \quad (5.2.56)$$

$$\int_{G_T}\Big(\zeta\,\lambda_0^0\big(\mathbb{D}(x,\mathbf{w}_s)+\langle\mathbb{D}(x,\mathbf{U})\rangle_{Y_s^0}-p\,\mathbb{I}\big):\mathbb{D}(x,\varphi)-\big(\zeta\hat{\rho}^0+(1-\zeta)\hat{\rho}\big)\mathbf{F}\cdot\varphi$$
$$+\ (1-\zeta)\lambda_0\big(\mathbb{D}(x,\mathbf{w}_s)+\langle\mathbb{D}(x,\mathbf{U})\rangle_{Y_s}-p\,\mathbb{I}\big):\mathbb{D}(x,\varphi)\Big)dxdt=0. \quad (5.2.57)$$

hold true for any smooth functions $h=h(\mathbf{x},t)$ and $\varphi=\varphi(\mathbf{x},t)$, vanishing at S.

Next, following the proofs of Theorem 1.4 and Theorem 5.8 we use representations

$$\mathbb{D}(x,\mathbf{w}_s)+\langle\mathbb{D}(x,\mathbf{U}^0)\rangle_{Y_s^0}-p\,\mathbb{I}=\mathfrak{N}_1^{s,0}:\mathbb{D}(x,\mathbf{w}_s)-p_f\,\mathbb{I}$$

for $\mathbf{x}\in\Omega^0$, and

$$\mathbb{D}(x,\mathbf{w}_s)+\langle\mathbb{D}(x,\mathbf{U})\rangle_{Y_s}-p\,\mathbb{I}=\mathfrak{N}_1^{s}:\mathbb{D}(x,\mathbf{w}_s)-p_f\,\mathbb{I},$$

for $\mathbf{x}\in\Omega$, and arrive at the integral identity

$$\int_{G_T}\Big(\zeta\,\big((\lambda_0^0\mathfrak{N}_1^{s,0}:\mathbb{D}(x,\mathbf{w}_s)-p_f\,\mathbb{I}):\mathbb{D}(x,\varphi)-\hat{\rho}^0\mathbf{F}\cdot\varphi\big)$$
$$+(1-\zeta)\big((\lambda_0\mathfrak{N}_1^{s}:\mathbb{D}(x,\mathbf{w}_s)-p_f\,\mathbb{I}):\mathbb{D}(x,\varphi)-\hat{\rho}\mathbf{F}\cdot\varphi\big)\Big)dxdt=0. \quad (5.2.58)$$

This identity implies Eqs. (5.2.25) and (5.2.28) in Ω_T^0 and Ω_T respectively, and the continuity condition (5.2.33) on the common boundary S^0. The boundary condition (5.2.21) on the outer boundary S follows from the Lemma B.14 in Appendix B, the continuity condition (5.2.32) on the common boundary S^0 is a consequence of the inclusion $\mathbf{w}_s\in\mathbf{W}_2^{1,0}(G_T)$.

The integral identity (5.2.56) implies continuity equations (5.2.24) and (5.2.27) in Ω_T^0 and Ω_T respectively, the continuity condition (5.2.34) on the common boundary S^0, and boundary condition (5.2.31) on the outer boundary S.

To derive Darcy's laws (5.2.26) and (5.2.29) in Ω_T^0 and Ω_T respectively, and prove that $\nabla\pi_f\in\mathbf{L}_2(G_T)$, we pass to the limit as $\varepsilon\to 0$ in (5.2.15) with test functions

$$\varphi=\int_0^t h(\mathbf{x},\tau)d\tau\,\varphi_0\Big(\frac{\mathbf{x}}{\varepsilon}\Big),$$

where

(1) h is finite in Ω^0 and $\varphi_0(\mathbf{y})$ is divergent free and finite in Y_f^0, and
(2) h is finite in Ω and $\varphi_0(\mathbf{y})$ is divergent free and finite in Y_f.

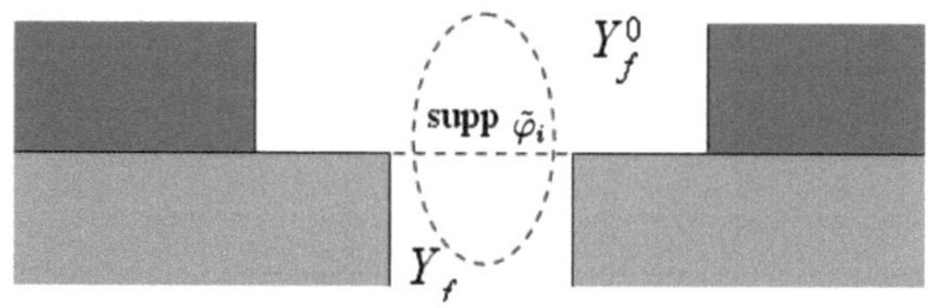

Fig. 5.7 Continuity of the pressure on S^0

This procedure has been done in Chap. 1 (proofs of Theorems 1.3 and 1.4) and together with microscopic continuity equations

$$\nabla_y \cdot \mathbf{W}^0(\mathbf{x}, t, \mathbf{y}) = 0, \ \mathbf{y} \in Y_f^0, \quad \nabla_y \cdot \mathbf{W}(\mathbf{x}, t, \mathbf{y}) = 0, \ \mathbf{y} \in Y_f^0$$

defines (5.2.26) and (5.2.29).

The missing continuity condition (5.2.35) or (5.2.36) on S^0 depends on the structure of the common pore space. For the first structure there exist divergent free and finite in $Y_f \cap Y_f^0$ smooth functions $\tilde{\varphi}_i(\mathbf{y})$ (Fig. 5.7), such that

$$\langle \tilde{\varphi}_i \rangle_Y = \mathbf{e}_i, \ \ i = 1, 2, 3,$$

where $\{\mathbf{e}_1, \ \mathbf{e}_1, \ \mathbf{e}_1\}$ is a standard Cartesian basis. The existence of such functions is proved in Lemma B.15 of Appendix B.

Next we put in (5.2.15)

$$\varphi = \int_0^t h(\mathbf{x}, \tau) d\tau \, \tilde{\varphi}_i \left(\frac{\mathbf{x}}{\varepsilon} \right),$$

where $h(\mathbf{x}^0, t) \neq 0$ for $\mathbf{x}^0 \in S^0$ and is $h(\mathbf{x}, t)$ vanishes outside of some small neighborhood of $\mathbf{x}^0$, and pass to the limit as $\varepsilon \to 0$:

$$\begin{aligned}
&\int_{\Omega_T} h \left(\mu_1 \langle \mathbb{D}(y, \mathbf{W}) : \mathbb{D}(y, \tilde{\varphi}_i) \rangle_Y - \rho_f \left(\int_0^t \mathbf{F} d\tau \right) \cdot \langle \tilde{\varphi}_i \rangle_Y \right) dx dt \\
&\quad + \int_{\Omega_T^0} h \left(\mu_1 \langle \mathbb{D}(y, \mathbf{W}^0) : \mathbb{D}(y, \tilde{\varphi}_i) \rangle_Y - \rho_f \left(\int_0^t \mathbf{F} d\tau \right) \cdot \langle \tilde{\varphi}_i \rangle_Y \right) dx dt \\
&\quad - \int_{\Omega_T} \pi_f \, (\nabla h \cdot \mathbf{e}_i) dx dt - \int_{\Omega_T^0} \pi_f \, (\nabla h \cdot \mathbf{e}_i) dx dt \\
&\quad = \int_{\Omega_T} \left(h \, P_i + \frac{\partial h}{\partial x_i} \pi_f \right) dx dt + \int_{\Omega_T^0} \left(h \, P_i^0 + \frac{\partial h}{\partial x_i} \pi_f \right) dx dt = 0, \ \ i = 1, 2, 3,
\end{aligned} \tag{5.2.59}$$

where

$$P_i^0 = -\mu_1 \langle \mathbb{D}(y, \mathbf{W}^0) : \mathbb{D}(y, \tilde{\varphi}_i) \rangle_Y + \rho_f \left(\int_0^t \mathbf{F} d\tau \right) \cdot \langle \tilde{\varphi}_i \rangle_Y,$$

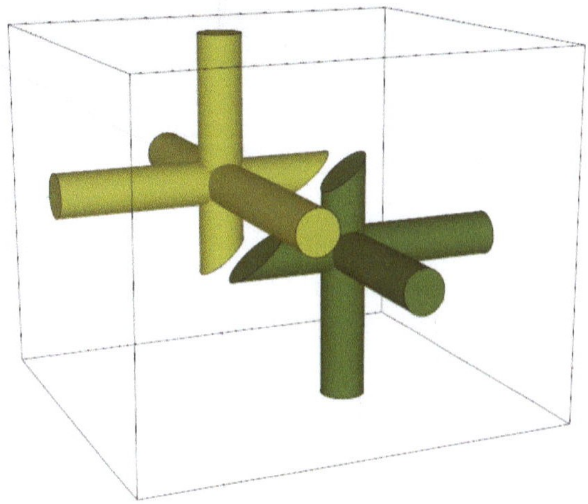

Fig. 5.8 Disconnected pore space

$$P_i = -\mu_1 \langle \mathbb{D}(y, \mathbf{W}) : \mathbb{D}(y, \tilde{\varphi}_i) \rangle_Y + \rho_f \left(\int_0^t \mathbf{F} d\tau \right) \cdot \langle \tilde{\varphi}_i \rangle_Y .$$

Due to estimates (5.2.16) and the two-scale convergence results (see Appendix B) we have

$$\pi_f,\ P_i^0 \in L_2(\Omega_T^0),\ \pi_f,\ P_i \in L_2(\Omega_T),\ i = 1, 2, 3.$$

Therefore $\pi_f \in W_2^{1,0}(G_T)$ and function π_f satisfies the continuity condition (5.2.35) on the common boundary S^0.

Now, let the common pore space be disconnected (the second structure). The function $\tilde{\mathbf{w}}^\varepsilon = \mathbf{w}^\varepsilon - \mathbf{w}_s^\varepsilon$ is identically equal to zero in the solid domain Q_s^ε due to the properties of the extension $\mathbf{w}_s^\varepsilon$.

Moreover, by supposition on a structure of the common pore space we have the equality

$$\tilde{\mathbf{w}}^\varepsilon = 0$$

on the common boundary S^0 (see Figs. 5.5 and 5.8).

Let

$$\omega = \{\mathbf{x} \in \Omega : |\mathbf{x} - \mathbf{x}^0| < \delta\}, \quad \omega^\varepsilon = \{\mathbf{x} \in \Omega_f^\varepsilon : |\mathbf{x} - \mathbf{x}^0| < \delta\}$$

for $\mathbf{x}^0 \in S^0$ and sufficiently small positive δ, and $\xi(\mathbf{x}, t)$ be a smooth function, such that

$$\xi(\mathbf{x}^0, t) \neq 0, \quad supp\, \xi \subset \omega^\varepsilon .$$

By the choice of the function ξ and the domain ω^ε

$$\tilde{\mathbf{w}}^\varepsilon(\mathbf{x}, t) = 0 \ \text{ for } \mathbf{x} \in \partial\omega^\varepsilon .$$

Therefore

$$\int_0^T\int_\omega \chi^\varepsilon\,\xi\,\nabla\cdot\mathbf{w}_s^\varepsilon dxdt = \int_0^T\int_{\omega^\varepsilon}\xi\,\nabla\cdot\mathbf{w}_s^\varepsilon dxdt$$
$$= -\int_0^T\int_{\omega^\varepsilon}\xi\,\nabla\cdot\tilde{\mathbf{w}}^\varepsilon dxdt = \int_0^T\int_{\omega^\varepsilon}\nabla\xi\cdot\tilde{\mathbf{w}}^\varepsilon dxdt$$
$$= \int_0^T\int_\omega \nabla\xi\cdot\big(\chi^\varepsilon\,\mathbf{w}^\varepsilon - \chi^\varepsilon\,\mathbf{w}_s^\varepsilon\big)dxdt.$$

Passing to the two-scale limit in the last identity as $\varepsilon\to 0$ we arrive at

$$\int_0^T\int_\omega \xi\,\big(m\,\nabla\cdot\mathbf{w}_s + \langle\nabla_y\cdot\mathbf{W}\rangle_{Y_f}\big)dxdt$$
$$= \int_0^T\int_\omega \nabla\xi\cdot\big(\mathbf{w}^{(f)} - m\,\mathbf{w}_s\big)dxdt. \tag{5.2.60}$$

The arbitrary choice of ξ, the condition $\xi(\mathbf{x}^0,t)\neq 0$ for $\mathbf{x}^0\in S^0$, and the identity (5.2.60) imply (5.2.36).

The validity of the normalization condition (5.2.20) for p_f is proved in the the previous theorem. By definition

$$\pi_f(\mathbf{x},t) = \int_0^t p_f(\mathbf{x},\tau)d\tau.$$

Therefore,

$$\int_Q \pi_f(\mathbf{x},t)dx = \int_0^t\Big(\int_Q p_f(\mathbf{x},\tau)dx\Big)d\tau = 0,$$

which proves (5.2.30).

5.2.7 *Proof of Theorem 5.12*

Note that estimates (5.2.16) are still valid for the solutions of $(\mathbb{FCM})_6$. These estimates are independent of $\lambda_0 = k$. But the problem itself also possesses such estimates and we will try to obtain them.

First we use continuity equations in the form

$$\nabla\cdot\mathbf{w}^{(f,k)} - m_0\,\nabla\cdot\mathbf{w}_s^{(k)} = -\nabla\cdot\mathbf{w}_s^{(k)},\quad (\mathbf{x},t)\in\Omega_T^0,$$
$$\nabla\cdot\mathbf{w}^{(f,k)} - m\,\nabla\cdot\mathbf{w}_s^{(k)} = -\nabla\cdot\mathbf{w}_s^{(k)},\quad (\mathbf{x},t)\in\Omega_T,$$

and rewrite the problem as two integral identities

$$\int_Q \Big(k\big(\tilde{\mathfrak{N}} : \mathbb{D}(x, \mathbf{w}_s^{(k)})\big) : \mathbb{D}(x, \varphi) - p_f^{(k)} \nabla \cdot \varphi\Big) dx = \int_Q \tilde{\rho} \mathbf{F} \cdot \varphi dx, \qquad (5.2.61)$$

$$\int_Q \big((\tilde{\mathbb{B}} \cdot \nabla \pi_f^{(k)}) \cdot \nabla \xi + \xi \, \nabla \cdot \mathbf{w}_s^{(k)}\big) dx = \int_Q \rho_f \nabla \xi \cdot \tilde{\mathbb{B}} \cdot \Big(\int_0^t \mathbf{F}(\mathbf{x}, \tau) d\tau\Big) dx \quad (5.2.62)$$

for any functions $\varphi \in \overset{\circ}{\mathbf{W}}{}_2^{1,0}(G_T)$ and $\xi \in W_2^{1,0}(G_T)$.

In (5.2.61), (5.2.62)

$$\pi_f^{(k)}(\mathbf{x}, t) = \int_0^t p_f^{(k)}(\mathbf{x}, \tau) d\tau,$$
$$\tilde{\mathfrak{N}} = \zeta \, \mathfrak{N}_1^{s,0} + (1 - \zeta) \, \mathfrak{N}_1^s,$$
$$\tilde{\mathbb{B}} = \zeta \, \mathbb{B}^0 + (1 - \zeta) \, \mathbb{B}.$$

Suppose for the moment that $\dfrac{\partial}{\partial t}(\nabla \pi_f^{(k)}) \in \mathbf{L}_2(G_T)$.

If we put $\varphi = \mathbf{w}_s^{(k)}$ in (5.2.61) and $\xi = p_f^{(k)}$ in (5.2.62), and then sum results, we obtain

$$\begin{aligned} k \int_Q \big(\tilde{\mathfrak{N}} : \mathbb{D}(x, \mathbf{w}_s^{(k)})\big) : \mathbb{D}(x, \mathbf{w}_s^{(k)}) dx + \frac{1}{2} \frac{d}{dt} \int_Q \nabla \pi_f^{(k)} \cdot \big(\tilde{\mathbb{B}} \cdot \nabla \pi_f^{(k)}\big) dx \\ = \int_Q \tilde{\rho} \mathbf{F} \cdot \mathbf{w}_s^{(k)} dx + \frac{d}{dt} \int_Q \rho_f \nabla \pi_f^{(k)} \cdot \tilde{\mathbb{B}} \cdot \Big(\int_0^t \mathbf{F}(\mathbf{x}, \tau) d\tau\Big) dx \\ - \int_Q \rho_f \nabla \pi_f^{(k)} \cdot \tilde{\mathbb{B}} \cdot \mathbf{F} dx. \end{aligned} \qquad (5.2.63)$$

This equality and the properties of tensors $\mathfrak{N}_1^{s,0}$ and $\mathfrak{N}_1^s$ and matrices $\mathbb{B}^0$ and $\mathbb{B}$ imply the desired estimates

$$k \int_{G_T} |\mathbb{D}(x, \mathbf{w}_s^{(k)})|^2 dx dt + \max_{0<t<T} \int_Q |\nabla \pi_f^{(k)}(\mathbf{x}, t)|^2 dx \leqslant C_0 \mathfrak{P}^2, \qquad (5.2.64)$$

where C_0 is independent of ε and k for $k > 1$.

The rigorous derivation of (5.2.63) requires the use of mollifiers and one may find such types of estimates for parabolic equations in [61].

Estimates (5.2.64), and Eqs. (5.2.26) and (5.2.29) show that there exist subsequences $\{\mathbf{w}_s^{(k_n)}\}$, $\{\pi_f^{(k_n)}\}$, and $\{\mathbf{w}^{(f,k_n)}\}$, such that

$$\mathbf{w}_s^{(k_n)} \to 0 \text{ strongly in } \overset{\circ}{\mathbf{W}}{}_2^{1,0}(G_T),$$

$$\pi_f^{(k_n)} \to \pi_f \text{ weakly in } W_2^{1,0}(G_T),$$

and

$$\mathbf{w}^{(f,k_n)} \to \mathbf{w}^{(f)} \text{ weakly in } \mathbf{L}_2(G_T)$$

as $k_n \to \infty$, and the limiting functions obviously solve the problem (5.2.30), (5.2.31), (5.2.35), (5.2.38)–(5.2.41).

In fact, Eqs. (5.2.39), (5.2.40) are the direct result of the limit in Eqs. (5.2.26), (5.2.29). The continuity equation (5.2.38) in Q, the continuity condition (5.2.41), and the boundary condition (5.2.31) follow from the continuity equation (5.2.27) in its integral form

$$\int_{G_T} \nabla \xi \cdot \Big(\mathbf{w}^{(f,k_n)} + \big(\zeta(1-m_0) + (1-\zeta)(1-m)\big)\mathbf{w}_s^{(k_n)}\Big) dx dt = 0,$$

which holds true for any smooth function $\xi(\mathbf{x}, t)$.

The continuity condition (5.2.35) follows from the smoothness of the function π_f.

As before, the weak limit as $k_n \to \infty$ in the normalization condition (5.2.30) for $\pi_f^{(k_n)}$ in the form

$$0 = \int_0^T h(t)\Big(\int_\Omega \pi_f^{(k_n)} dx\Big) dt \to \int_0^T h(t)\Big(\int_\Omega \pi_f \, dx\Big) dt = 0$$

for arbitrary smooth functions $h(t)$ results in the normalization condition (5.2.30) for the function π_f.

5.2.8 *Proof of Theorem 5.13*

The proof of this theorem is similar to the proof of the previous one. To explain our ideas on how to obtain estimates independent of k, we again use the formal method under the supposition $\nabla p_f^{(k)} \in \mathbf{L}_2(G_T)$.

Firstly we rewrite (5.2.58) as

$$\int_Q \Big(\zeta\, \big((k\,\mathfrak{N}_1^{s,0} : \mathbb{D}(x, \mathbf{w}_s) - p_f\, \mathbb{I}) : \mathbb{D}(x, \varphi) - \hat{\rho}^0 \mathbf{F} \cdot \varphi\big)$$
$$+ (1-\zeta)\big((k\,\mathfrak{N}_1^s : \mathbb{D}(x, \mathbf{w}_s) - p_f\, \mathbb{I}) : \mathbb{D}(x, \varphi) - \hat{\rho}\mathbf{F} \cdot \varphi\big)\Big) dx = 0 \quad (5.2.65)$$

and put $\varphi = \mathbf{w}_s^{(k)}$.

Then we multiply (5.2.24) and (5.2.27) by $p_f^{(k)}$, integrate over domain Ω^0 and Ω respectively, and sum the results with (5.2.65):

$$0 = k \int_Q \Big(\zeta\, \mathbb{D}(x, \mathbf{w}_s^{(k)}) : (\mathfrak{N}_1^{s,0} : \mathbb{D}(x, \mathbf{w}_s^{(k)}))$$
$$+ (1-\zeta)\mathbb{D}(x, \mathbf{w}_s^{(k)}) : (\mathfrak{N}_1^{s,0} : \mathbb{D}(x, \mathbf{w}_s^{(k)}))\Big) dx - \int_Q \hat{\rho}^0 \mathbf{F} \cdot \mathbf{w}_s^{(k)} dx$$
$$+ \int_{\Omega^0} p_f \nabla \cdot (\mathbf{w}^{(f,k)} - m_0 \mathbf{w}_s^{(k)}) dx + \int_\Omega p_f \nabla \cdot (\mathbf{w}^{(f,k)} - m \mathbf{w}_s^{(k)}) dx$$
$$= I_1 + I_2 + I_3 + I_4 = 0.$$

The last two integrals I_3 and I_4 we rewrite using integration by parts, boundary conditions (5.2.21), (5.2.31), (5.2.36), and (5.2.37), and Eqs. (5.2.26) and (5.2.29) as

$$I_3 = \frac{1}{\mu_1} \int_{\Omega^0} \nabla p_f \cdot \mathbb{B}^0 \cdot \Big(\nabla \pi_f - \rho_f \int_0^t \mathbf{F}(\mathbf{x}, \tau) d\tau \Big) dx,$$

$$I_4 = \frac{1}{\mu_1} \int_{\Omega} \nabla p_f \cdot \mathbb{B} \cdot \Big(\nabla \pi_f - \rho_f \int_0^t \mathbf{F}(\mathbf{x}, \tau) d\tau \Big) dx.$$

Gathering all together we arrive at (5.2.63).

The rest of the proof repeats the proof of Theorem 5.12. The slight difference here is only in the derivation of the boundary condition (5.2.42). To prove that we rewrite the continuity equation (5.2.27) in Ω as

$$\nabla \cdot (\mathbf{w}^{(f,k)} - m \mathbf{w}_s^{(k)}) + \mathbf{w}_s^{(k)} = 0,$$

multiply by an arbitrary smooth function ξ and integrate by parts over domain Ω_T using boundary conditions (5.2.21), (5.2.31), and (5.2.36):

$$\int_\Omega \Big(-\nabla \xi \cdot (\mathbf{w}^{(f,k)} - m \mathbf{w}_s^{(k)}) + \xi \nabla \cdot \mathbf{w}_s^{(k)} \Big) dx dt = 0.$$

The limit as $k \to \infty$ in the last identity results in

$$\int_\Omega \nabla \xi \cdot \mathbf{w}^{(f)} dx dt = 0,$$

which is equivalent to the continuity equation (5.2.38) and the boundary condition (5.2.42).

5.2.9 *Proof of Theorem 5.14*

The new supposition of the theorem, that

$$\int_{\Omega_T} (1 - \chi^\varepsilon) \Big| \frac{\partial \mathbf{F}}{\partial t} \Big|^2 dx dt = \mathfrak{P}_1^2 < \infty,$$

permits us to get estimates

$$\int_{\Omega_T}\Big(|\mathbf{w}^{\varepsilon}|^2+|\mathbb{D}(x,\mathbf{v}^{\varepsilon})|^2\Big)dxdt$$
$$+\max_{0<t<T}\int_{\Omega}\Big(|\mathbb{D}(x,\mathbf{w}^{\varepsilon})|^2+|p^{\varepsilon}|^2\Big)dx\leqslant C_0(\mathfrak{P}^2+\mathfrak{P}_1^2),\tag{5.2.66}$$

where $\mathbf{v}^{\varepsilon}=\mathbb{E}_{Q_f^{\varepsilon}}\Big(\frac{\partial\mathbf{w}^{\varepsilon}}{\partial t}\Big)$ is an extension of $\frac{\partial\mathbf{w}^{\varepsilon}}{\partial t}$ from the liquid domain Q_f^{ε} onto Q (see the proof of Theorem 1.9).

On the basis of these estimates we conclude that for $\varepsilon\to 0$

$$p^{\varepsilon}\rightharpoonup p \text{ weakly in } L_2(G_T),$$

$$\mathbf{w}^{\varepsilon}\to\mathbf{w}(\mathbf{x},t) \text{ weakly and two-scale in } \mathbf{L}_2(G_T),$$

$$\mathbf{v}^{\varepsilon}\to\mathbf{v}(\mathbf{x},t)=\frac{\partial\mathbf{w}}{\partial t} \text{ weakly in } \overset{\circ}{\mathbf{W}}{}_2^{1,0}(G_T) \text{ and two-scale in } \mathbf{L}_2(G_T),$$

$$\mathbb{D}\big(x,\mathbf{w}^{\varepsilon}\big)\to\mathbb{D}\big(x,\mathbf{w}\big)+\mathbb{D}\big(y,\hat{\mathbf{W}}\big) \text{ two-scale in } \mathbf{L}_2(G_T),$$

$$\mathbb{D}\Big(x,\mathbf{v}^{\varepsilon}\Big)\to\mathbb{D}\Big(x,\mathbf{v}\Big)+\mathbb{D}\Big(y,\hat{\mathbf{V}}\Big)=\mathbb{D}\Big(x,\frac{\partial\mathbf{w}}{\partial t}\Big)$$
$$+\mathbb{D}\Big(y,\frac{\partial\hat{\mathbf{W}}}{\partial t}\Big) \text{ two-scale in } \mathbf{L}_2(G_T),$$

$$\hat{\mathbf{W}}=\zeta\,\mathbf{W}^0+(1-\zeta)\,\mathbf{W},$$

where functions $\mathbf{W}^0$ and $\mathbf{W}$ are defined separately in $\Omega_T^0\times Y$ and $\Omega_T\times Y$ (for details see the proof of Theorem 1.11). The same theorem says that limiting functions satisfy Eqs. (5.2.44)–(5.2.46), the boundary condition (5.2.49), and the initial condition (5.2.50). The validity of the normalization condition (5.2.47) is a consequence of the weak convergence of $\{p^{\varepsilon}\}$ and the normalization condition (5.2.43).

Finally, the boundary condition (5.2.48) follows from the integral identity (5.2.15) after taking the two-scale limit as $\varepsilon\to 0$ with test functions $\varphi=\varphi(\mathbf{x},t)$:

$$\int_{\Omega_T^0}\Big(\mu_0\Big(m_0\mathbb{D}\Big(x,\frac{\partial\mathbf{w}}{\partial t}\Big)+\Big\langle\mathbb{D}\Big(y,\frac{\partial\mathbf{W}^0}{\partial t}\Big)\Big\rangle_{Y_f^0}\Big)+\lambda_0\big((1-m_0)\mathbb{D}(x,\mathbf{w})$$
$$+\langle\mathbb{D}(y,\mathbf{W}^0)\rangle_{Y_s^0}\big)-p\,\mathbb{I}\Big):\mathbb{D}(x,\varphi)dxdt-\int_{\Omega_T^0}\hat{\rho}^0\mathbf{F}\cdot\varphi dxdt$$
$$+\int_{\Omega_T}\Big(\mu_0\Big(m\mathbb{D}\Big(x,\frac{\partial\mathbf{w}}{\partial t}\Big)+\Big\langle\mathbb{D}(y,\frac{\partial\mathbf{W}}{\partial t})\Big\rangle_{Y_f}\Big)+\lambda_0\big((1-m)\mathbb{D}(x,\mathbf{w})$$

$$+\langle\mathbb{D}(y,\mathbf{W})\rangle_{Y_s}\big) - p\,\mathbb{I}\Big) : \mathbb{D}(x,\varphi)dxdt - \int_{\Omega_T} \hat{\rho}\mathbf{F}\cdot\varphi dxdt = 0. \qquad (5.2.67)$$

In fact, by Theorem 1.11

$$\widehat{\mathbb{P}}^0 = -p\,\mathbb{I} + \mu_0\Big(m_0\mathbb{D}\Big(x,\frac{\partial \mathbf{w}}{\partial t}\Big) + \Big\langle\mathbb{D}\Big(y,\frac{\partial \mathbf{W}^0}{\partial t}\Big)\Big\rangle_{Y_f^0}\Big)$$
$$+\lambda_0\big((1-m_0)\mathbb{D}(x,\mathbf{w}) + \langle\mathbb{D}(y,\mathbf{W}^0)\rangle_{Y_s^0}\big),$$

$$\widehat{\mathbb{P}} = -p\,\mathbb{I} + \mu_0\Big(m\mathbb{D}\Big(x,\frac{\partial \mathbf{W}}{\partial t}\Big) + \Big\langle\mathbb{D}\Big(y,\frac{\partial \mathbf{w}}{\partial t}\Big)\Big\rangle_{Y_f}\Big)$$
$$+\lambda_0\big((1-m)\mathbb{D}(x,\mathbf{w}) + \langle\mathbb{D}(y,\mathbf{W})\rangle_{Y_s}\big),$$

and the integral identity (5.2.67) takes the form

$$\int_{G_T}\big(\zeta\widehat{\mathbb{P}}^0 + (1-\zeta)\widehat{\mathbb{P}}\big) : \mathbb{D}(x,\varphi)dxdt = \int_{G_T}\big(\zeta\hat{\rho}^0 + (1-\zeta)\hat{\rho}\big)\mathbf{F}\cdot\varphi dxdt,$$

which obviously implies the boundary condition (5.2.48).

5.3 Filtration in Poroelastic Media with a Variable Structure

In this section we will try to model nonperiodic poroelastic media with a variable structure in the domain Ω, described for $t > 0$ by the model $\mathbb{M}_{15}$ with the characteristic function $\chi_0(\mathbf{x})$ of the liquid domain Ω_f. To do this we use the standard procedure of approximation of variable coefficients by means of step functions.

Suppose that for some small positive δ

$$\chi_0(\mathbf{x}) = \chi_n\left(\frac{\mathbf{x}}{\delta}\right), \quad \lambda_0 = \lambda_0^n \;\; \rho_s = \rho_s^n, \;\; \text{for } \mathbf{x} \in K_n^{(\delta)},$$

where $\chi_n(\mathbf{y})$ is a 1-periodic in $\mathbf{y}$ function,

$$\lambda_0^n = const., \;\; \rho_s^n = const., \;\; \Omega = \bigcup_{n=1}^{N} K_n^{(\delta)},$$

and for $\delta > 0$ the cube $K_n^{(\delta)}$ is an intersection of the domain Ω with the cube $\delta\,K$, $K = [0,1]^3 \subset \mathbb{R}^3$, $Int K_n^{(\delta)} \cap Int K_m^{(\delta)} = \emptyset$ for $m \neq n$.

Let

$$\chi^{(\delta)}(\mathbf{x},\mathbf{y})=\chi_n(\mathbf{y}),\ \lambda_0^{(\delta)}(\mathbf{x})=\lambda_0^n,\ \rho_s^{(\delta)}(\mathbf{x})=\rho_s^n\ \text{ for } \mathbf{x}\in K_n^{(\delta)}$$

be step functions in the variable $\mathbf{x}$. Then $\chi^{(\delta)}(\mathbf{x},\mathbf{y})$ is a 1-periodic function in the variable $\mathbf{y}$.

Now, as usual, we consider in the domain Ω for $t>0$ the problem

$$\nabla\cdot\mathbf{w}^{\delta,\varepsilon}=0, \tag{5.3.1}$$

$$\nabla\cdot\Big(\chi^{\delta,\varepsilon}\alpha_\mu\mathbb{D}\Big(x,\frac{\partial\mathbf{w}^{\delta,\varepsilon}}{\partial t}\Big)+(1-\chi^{\delta,\varepsilon})\lambda_0^{(\delta)}\mathbb{D}(x,\mathbf{w}^{\delta,\varepsilon})-p^{\delta,\varepsilon}\,\mathbb{I}\Big)+\rho^{\delta,\varepsilon}\mathbf{F}=0, \tag{5.3.2}$$

with the characteristic function

$$\chi^{\delta,\varepsilon}(\mathbf{x})=\chi^{(\delta)}\Big(\mathbf{x},\frac{\mathbf{x}}{\varepsilon}\Big)$$

of the pore space $\Omega_f^{\delta,\varepsilon}$, the solid density $\rho_s^{(\delta)}(\mathbf{x})$, and the elasticity coefficient $\lambda_0^{(\delta)}(\mathbf{x})$, depending on the variable $\mathbf{x}\in\Omega$ and the small parameter $\varepsilon<\delta$.

The problem is completed with the boundary condition

$$\mathbf{w}^{\delta,\varepsilon}(\mathbf{x},0)=0,\ \ \mathbf{x}\in S=\partial\Omega,\ t>0, \tag{5.3.3}$$

and initial and normalization conditions

$$\chi^{\delta,\varepsilon}(\mathbf{x})\mathbf{w}^{\delta,\varepsilon}(\mathbf{x},t)=0,\ \ \mathbf{x}\in\Omega, \tag{5.3.4}$$

$$\int_\Omega\frac{\chi^{\delta,\varepsilon}(\mathbf{x})}{m^{(\delta)}(\mathbf{x})}\,p^{\delta,\varepsilon}(\mathbf{x},t)dx=0,\ \ m^{(\delta)}(\mathbf{x})=\int_Y\chi^{(\delta)}(\mathbf{x},\mathbf{y})dy. \tag{5.3.5}$$

In (5.3.2)

$$\rho^{\delta,\varepsilon}(\mathbf{x})=\chi^{\delta,\varepsilon}(\mathbf{x})\rho_f+\big(1-\chi^{\delta,\varepsilon}(\mathbf{x})\big)\rho_s^{(\delta)}(\mathbf{x}).$$

Differential equation (5.3.2) is understood as an integral identity

$$\int_0^T\int_\Omega\Big(-\alpha_\mu\chi^{\delta,\varepsilon}\mathbb{D}(x,\mathbf{w}^{\delta,\varepsilon}):\mathbb{D}\Big(x,\frac{\partial\varphi}{\partial t}\Big)+\lambda_0^{(\delta)}(1-\chi^{\delta,\varepsilon})\mathbb{D}(x,\mathbf{w}^{\delta,\varepsilon}):\mathbb{D}(x,\varphi)\Big)dxdt$$
$$=\int_0^T\int_\Omega\big(p^{\delta,\varepsilon}(\nabla\cdot\varphi)+\rho^{\delta,\varepsilon}\mathbf{F}\cdot\varphi\big)dxdt \tag{5.3.6}$$

for all functions φ vanishing at $t=T$, such that $\varphi,\ \dfrac{\partial\varphi}{\partial t}\in\overset{\circ}{\mathbf{W}}{}_2^{1,0}(G_T)$.

Throughout this section we impose Assumptions 0.1, 1.2, and 3.1 for the structures, defined by characteristic functions $\chi_n(\mathbf{y})$, completed with an additional supposition.

Assumption 5.1 Let $\Omega = \bigcup_{n=1}^{N} K_n^{(\delta)}$, where $K_n^{(\delta)}$ is an intersection of Ω with the cube $\delta\, K$, $K = [0, 1]^3 \subset \mathbb{R}^3$, $Int\, K_n^{(\delta)} \cap Int\, K_m^{(\delta)} = \emptyset$ for $m \neq n$, and

$$\chi^{(\delta)}(\mathbf{x}, \mathbf{y}) = \chi_n(\mathbf{y}), \quad \text{for } \mathbf{x} \in K_n^{(\delta)}$$

be a characteristic function of the pore space in Ω.

Then the common pore space in Ω is connected (see previous section), that is, for any $K_n^{(\delta)}$ and $K_m^{(\delta)}$, having a common boundary,

$$Y_f^{(n)} \cap Y_f^{(m)} \neq \emptyset, \quad Y_s^{(n)} \cap Y_s^{(m)} \neq \emptyset,$$

where $Y_f^{(n)}$ and $Y_f^{(m)}$ are elementary liquid domains and $Y_s^{(n)}$ and $Y_s^{(m)}$ are elementary solid domains, defined by characteristic functions $\chi_n(\mathbf{y})$ and $\chi_m(\mathbf{y})$ respectively.

Next we introduce an extension

$$\mathbf{w}_s^{\delta,\varepsilon}(\mathbf{x}, t) = \mathbb{E}_{\Omega_s^\varepsilon}^{(\delta)}\left(\mathbf{w}^{\delta,\varepsilon}\right)$$

from the solid part

$$\Omega_s^\varepsilon = \left\{\mathbf{x} \in \Omega : \chi^{(\delta)}\left(\mathbf{x}, \frac{\mathbf{x}}{\varepsilon}\right) = 0\right\}$$

of the domain Ω onto the whole domain Ω, with the following properties:

$$\left(1 - \chi^{\delta,\varepsilon}(\mathbf{x})\right)\left(\mathbf{w}^{\delta,\varepsilon}(\mathbf{x}, t) - \mathbf{w}_s^{\delta,\varepsilon}(\mathbf{x}, t)\right) = 0, \quad \mathbf{x} \in \Omega, \ t \in (0, T),$$

and

$$\int_\Omega \left|\mathbf{w}_s^{\delta,\varepsilon}(\mathbf{x}, t)\right|^2 dx \leqslant C_0 \int_{\Omega_s^\varepsilon} \left|\mathbf{w}^{\delta,\varepsilon}(\mathbf{x}, t)\right|^2 dx,$$

$$\int_\Omega \left|\mathbb{D}\left(x, \mathbf{w}_s^{\delta,\varepsilon}(\mathbf{x}, t)\right)\right|^2 dx \leqslant C_0 \int_{\Omega_s^\varepsilon} \left|\mathbb{D}\left(x, \mathbf{w}^{\delta,\varepsilon}(\mathbf{x}, t)\right)\right|^2 dx, \quad t \in (0, T),$$

where C_0 is independent of ε, δ, and $t \in (0, T)$.

The existence of such an extension for domains Ω_s^ε with a non- periodic structure is proved as well as the existence of the extension (1.2.9) for domains Ω_s^ε with periodic structure.

Under these assumptions for solutions $\{\mathbf{w}^{\delta,\varepsilon},\ p^{\delta,\varepsilon}\}$ of the problem (5.3.1)–(5.3.5) all statements of the previous section hold true, which we reformulate as the following theorems.

Theorem 5.15 *For all* $\varepsilon > 0$ *and for arbitrary time interval* $[0, T]$ *there exists a unique generalized solution of problem (5.3.1)–(5.3.5) and*

$$\max_{0<t<T}\int_{\Omega}\chi^{\delta,\varepsilon}\Big(\alpha_\mu|\mathbb{D}\big(x,\mathbf{w}^{\delta,\varepsilon}(\mathbf{x},t)\big)|^2+\frac{\alpha_\mu}{\varepsilon^2}|\mathbf{w}^{\delta,\varepsilon}(\mathbf{x},t)-\mathbf{w}_s^{\delta,\varepsilon}(\mathbf{x},t)|^2\Big)dx$$
$$+\int_0^T\int_{\Omega}\Big(|\pi^{\delta,\varepsilon}|^2+\lambda_0^{\delta}|\mathbb{D}(x,\mathbf{w}^{\delta,\varepsilon})|^2\Big)dxdt\leqslant C_0\mathfrak{P}^2, \tag{5.3.7}$$

where C_0 is independent of ε, λ_0^{δ} for $\lambda_0^{(\delta)}>\lambda^-$, and

$$\pi^{\delta,\varepsilon}(\mathbf{x},t)=\int_0^t p^{\delta,\varepsilon}(\mathbf{x},\tau)d\tau,$$

$$\mathfrak{P}^2=\max_{0<t<T}\int_{\Omega}|\mathbf{F}(\mathbf{x},t)|^2dx<\infty.$$

Theorem 5.16 *Let*

$$\mu_0=0,\ \ \mu_1=\infty,\ \ 0<\lambda^-<\lambda_0^{(\delta)}(\mathbf{x})<\lambda^+<\infty,$$

$\{\mathbf{w}^{\delta,\varepsilon},\ p^{\delta,\varepsilon}\}$ be the weak solution of the problem (5.3.1)–(5.3.5),

$$\pi^{\delta,\varepsilon}(\mathbf{x},t)=\int_0^t p^{\delta,\varepsilon}(\mathbf{x},\tau)d\tau,$$

and $\mathbf{w}_s^{\delta,\varepsilon}=\mathbb{E}_{Q_s^{\varepsilon}}\big(\mathbf{w}^{\delta,\varepsilon}\big)$ be an extension from the domain $\Omega_s^{\delta,\varepsilon}=\{\mathbf{x}\in\Omega:\chi^{\delta,\varepsilon}(\mathbf{x})=0\}$ onto the domain Ω.

Then the sequences $\{\mathbf{w}^{\delta,\varepsilon}\}$ and $\{\chi^{\delta,\varepsilon}\pi^{\delta,\varepsilon}\}$ converge weakly in $\mathbf{L}_2(\Omega_T)$ and $L_2(\Omega_T)$ as $\varepsilon\to0$ to the functions $\mathbf{w}_s^{(\delta)}$ and $m^{(\delta)}\,\pi_f^{(\delta)}$ respectively, and the sequence $\{\mathbf{w}_s^{\delta,\varepsilon}\}$ converges weakly in $W_2^{1,0}(\Omega_T)$ as $\varepsilon\to0$ to the function $\mathbf{w}_s^{(\delta)}$.

The limiting functions solve in the domain Ω for $t>0$ the homogenized system, consisting of the homogenized momentum balance equation

$$\nabla\cdot\mathbb{P}_1^{(\delta)}(\mathbf{x})+\hat{\rho}^{(\delta)}(\mathbf{x})\,\mathbf{F}=0, \tag{5.3.8}$$

$$\mathbb{P}_1^{(\delta)}(\mathbf{x})=\lambda_0^{(\delta)}(\mathbf{x})\,\mathfrak{N}_1^{s,\delta}(\mathbf{x}):\mathbb{D}(x,\mathbf{w}_s^{(\delta)})-p_f^{(\delta)}\,\mathbb{I}, \tag{5.3.9}$$

and the continuity equation

$$\nabla\cdot\mathbf{w}_s^{(\delta)}=0. \tag{5.3.10}$$

The problem is completed with the normalization condition

$$\int_{\Omega}p_f^{(\delta)}(\mathbf{x},t)dx=0 \tag{5.3.11}$$

and the boundary condition

$$\mathbf{w}_s^{(\delta)} = 0 \tag{5.3.12}$$

on the outer boundary S for $t > 0$.

In (5.3.8), (5.3.9)

$$p_f^{(\delta)} = \frac{\partial \pi_f^{(\delta)}}{\partial t}, \ \hat{\rho}^{(\delta)}(\mathbf{x}) = m^{(\delta)}(\mathbf{x})\,\rho_f + \big(1 - m^{(\delta)}(\mathbf{x})\big)\,\rho_s^{(\delta)}(\mathbf{x}),$$

the symmetric strictly positively definite fourth-rank tensor $\mathfrak{N}_1^{s,\delta}(\mathbf{x})$ is given at point $\mathbf{x} \in \Omega$ by (1.2.38) for the pore space with the characteristic function $\chi^{(\delta)}(\mathbf{x}, \mathbf{y})$ (see Theorem 1.4 of Chap. 1).

We refer to the problem (5.3.8)–(5.3.12) as the homogenized **model** $(\mathbb{FCM})_{13}^{(\delta)}$.

Theorem 5.17 *Let*

$$\mu_0 = 0, \ 0 < 0 < \lambda^- < \mu_1, \ \lambda_0^{(\delta)}(\mathbf{x}) < \lambda^+ < \infty,$$

$\{\mathbf{w}^{\delta,\varepsilon}, \ p^{\delta,\varepsilon}\}$ be the weak solution of the problem (5.3.1)–(5.3.5), $\mathbf{w}_s^{\delta,\varepsilon} = \mathbb{E}_{\Omega_s^\varepsilon}\big(\mathbf{w}^{\delta,\varepsilon}\big)$ be an extension from the domain $\Omega_s^\varepsilon = \{\mathbf{x} \in \Omega : \chi^{\delta,\varepsilon}(\mathbf{x}) = 0\}$ onto the domain Ω, and

$$\pi^{\delta,\varepsilon}(\mathbf{x}, t) = \int_0^t p^{\delta,\varepsilon}(\mathbf{x}, \tau)d\tau.$$

Then the sequences $\{\chi^{\delta,\varepsilon}\,\pi^{\delta,\varepsilon}\}$, and $\{\chi^{\delta,\varepsilon}\,\mathbf{w}^{\delta,\varepsilon}\}$ converge weakly in $L_2(\Omega_T)$ and $\mathbf{L}_2(\Omega_T)$ as $\varepsilon \to 0$ to the functions $m^{(\delta)}\,\pi_f^{(\delta)}$ and $\mathbf{w}^{(\delta,f)}$ respectively, and the sequence $\{\mathbf{w}_s^{\delta,\varepsilon}\}$ converges weakly in $\mathbf{W}_2^{1,0}(\Omega_T)$ as $\varepsilon \to 0$ to the function $\mathbf{w}_s^{(\delta)}$.

The limiting functions $\pi_f^{(\delta)}$, $\mathbf{w}^{(\delta,f)}$, and $\mathbf{w}_s^{(\delta)}$, where $\nabla \pi_f^{(\delta)} \in \mathbf{L}_2(\Omega_T)$, $\dfrac{\partial \pi_f^{(\delta)}}{\partial t} \in L_2(\Omega_T)$, solve in the domain Ω for $t > 0$ the homogenized system, consisting of the continuity equation

$$\nabla \cdot \big(\mathbf{w}^{(\delta,f)} + \big(1 - m^{(\delta)}(\mathbf{x})\big)\,\mathbf{w}_s^{(\delta)}\big) = 0, \tag{5.3.13}$$

the homogenized momentum balance equation

$$\nabla \cdot \mathbb{P}_1^{(\delta)}(\mathbf{x}) + \hat{\rho}^{(\delta)}(\mathbf{x})\,\mathbf{F} = 0, \tag{5.3.14}$$

$$\mathbb{P}_1^{(\delta)}(\mathbf{x}) = \lambda_0^{(\delta)}(\mathbf{x})\,\mathfrak{N}_1^{s,\delta}(\mathbf{x}) : \mathbb{D}(x, \mathbf{w}_s^{(\delta)}) - p_f^{(\delta)}\,\mathbb{I} \tag{5.3.15}$$

for the solid component, and Darcy's law in the form

$$\mathbf{w}^{(\delta,f)} = m^{(\delta)}(\mathbf{x})\,\mathbf{w}_s^{(\delta)} + \frac{1}{\mu_1}\mathbb{B}^{(\delta)}(\mathbf{x}) \cdot \Big(- \nabla \pi_f^{(\delta)} + \rho_f \int_0^t \mathbf{F}(\mathbf{x}, \tau)d\tau \Big) \tag{5.3.16}$$

for the liquid component.

The problem is completed with the normalization condition (5.3.11), the boundary condition (5.3.12) for the solid displacements $\mathbf{w}_s^{(\delta)}$, *and the boundary condition*

$$\mathbf{w}^{(\delta,f)}(\mathbf{x},t)\cdot\mathbf{n}(\mathbf{x}) = 0 \tag{5.3.17}$$

for the liquid displacements on the outer boundary S *for* $t > 0$.

In (5.3.13)–(5.3.17) $\mathbf{n}(\mathbf{x})$ *is a unit normal to* S *at* $\mathbf{x} \in S$,

$$p_f^{(\delta)} = \frac{\partial \pi_f^{(\delta)}}{\partial t},\ \hat{\rho}^{(\delta)}(\mathbf{x}) = m^{(\delta)}(\mathbf{x})\,\rho_f + \big(1 - m^{(\delta)}(\mathbf{x})\big)\,\rho_s^{(\delta)}(\mathbf{x}),$$

the symmetric strictly positively definite fourth-rank tensor $\mathfrak{N}_1^{s,\delta}(\mathbf{x})$ *is given for almost all points* $\mathbf{x} \in \Omega$ *by (1.2.38) for the pore space with the characteristic function* $\chi^{(\delta)}(\mathbf{x},\mathbf{y})$ *(see Theorem 1.4 of Chap. 1), the symmetric strictly positive definite matrix* $\mathbb{B}^{(\delta)}(\mathbf{x})$ *is given for almost all points* $\mathbf{x} \in \Omega$ *by (1.1.27) for the pore space with the characteristic function* $\chi^{(\delta)}(\mathbf{x},\mathbf{y})$ *(see Theorem 1.1 of Chap. 1).*

We refer to the problem (5.3.11), (5.3.12)–(5.3.17) as a homogenized **model** $(\mathbb{FCM})_{14}^{(\delta)}$.

Theorem 5.18 *Under the conditions of Theorem 5.17 let* $\{\mathbf{w}_s^{(\delta,k)}, \mathbf{w}^{(\delta,f,k)}, \pi_f^{(\delta,k)}\}$ *be a weak solution of the model* $(\mathbb{FCM})_{10}^{(\delta)}$ *with* $\lambda_0^{(\delta)} = k$.

Then the sequences $\{\pi_f^{(\delta,k)}\}$ *and* $\{\mathbf{w}^{(\delta,f,k)}\}$ *converge weakly in* $L_2(\Omega_T)$ *and* $\mathbf{L}_2(\Omega_T)$ *as* $k \to \infty$ *to the functions* $\pi_f^{(\delta)}$ *and* $\mathbf{w}^{(\delta,f)}$ *respectively, and the sequence* $\{\mathbf{w}_s^{(\delta,k)}\}$ *converges strongly in* $\mathbf{L}_2(\Omega_T)$ *to zero.*

The limiting functions solve in the domain Ω *for* $t > 0$ *the homogenized system, consisting of the continuity equation*

$$\nabla\cdot\mathbf{w}^{(\delta,f)} = 0 \tag{5.3.18}$$

and Darcy's law

$$\mathbf{w}^{(\delta,f)} = \frac{1}{\mu_1}\mathbb{B}^{(\delta)}(\mathbf{x})\cdot\Big(-\nabla\pi_f^{(\delta)} + \rho_f\int_0^t \mathbf{F}(\mathbf{x},\tau)d\tau\Big), \tag{5.3.19}$$

completed with the boundary condition (5.3.17) for the liquid velocity on the outer boundary S *for* $t > 0$, *and the normalization condition (5.3.11).*

The symmetric strictly positively definite matrix $\mathbb{B}^{(\delta)}(\mathbf{x})$ *is given for almost all points* $\mathbf{x} \in \Omega$ *by (1.1.27) for the pore space with the characteristic function* $\chi^{(\delta)}(\mathbf{x},\mathbf{y})$ *(see Theorem 1.1 of Chap. 1).*

We refer to the problem (5.3.11), (5.3.17)–(5.3.19) as the homogenized **model** $(\mathbb{FCM})_{15}^{(\delta)}$.

To consider the following case we change the setting of the problem at the microscopic level, namely, instead of the normalization condition (5.3.5) we consider the normalization condition

$$\int_\Omega p^{\delta,\varepsilon}(\mathbf{x},t)\,dx = 0. \tag{5.3.20}$$

The proof of the solvability of the problem (5.3.1)–(5.3.4), (5.3.20) and the derivation of the a priori estimates repeat exactly the proof of the solvability and the derivation of the a priori estimates of the problem (5.3.1)–(5.3.5).

Theorem 5.19 *Let*

$$\alpha_\mu = \mu_0,\ 0 < \lambda^- < \mu_0,\ \lambda_0^{(\delta)}(\mathbf{x}) < \lambda^+ < \infty,$$

$$\int_0^T\int_\Omega (1-\chi^\varepsilon)\left|\frac{\partial \mathbf{F}}{\partial t}\right|^2 dxdt = \mathfrak{P}_1^2 < \infty,$$

and $\{\mathbf{w}^{\delta,\varepsilon},\ p^{\delta,\varepsilon}\}$ *be the weak solution of the problem (5.3.1)–(5.3.4), (5.3.20).*

Then the sequence $\{p^{\delta,\varepsilon}\}$ *converges weakly in* $L_2(\Omega_T)$ *as* $\varepsilon \to 0$ *to the function* $p^{(\delta)}$ *and the sequence* $\{\mathbf{w}^{\delta,\varepsilon}\}$ *converges weakly in* $\mathbf{W}_2^{1,0}(\Omega_T)$ *as* $\varepsilon \to 0$ *to the function* $\mathbf{w}^{(\delta)}$.

The limiting functions solve in the domain Ω *for* $t > 0$ *the homogenized system, consisting of the continuity equation*

$$\nabla \cdot \mathbf{w}^{(\delta)} = 0, \tag{5.3.21}$$

and the homogenized momentum balance equation

$$\nabla \cdot \widehat{\mathbb{P}}^{(\delta)}(\mathbf{x}) + \hat{\rho}^{(\delta)}(\mathbf{x})\,\mathbf{F} = 0, \tag{5.3.22}$$

$$\begin{aligned}\widehat{\mathbb{P}}^{(\delta)}(\mathbf{x}) = &-p^{(\delta)}\,\mathbb{I} + \mathfrak{N}_1^{(\delta)}(\mathbf{x}) : \mathbb{D}\Big(x, \frac{\partial \mathbf{w}^{(\delta)}}{\partial t}\Big) + \mathfrak{N}_2^{(\delta)}(\mathbf{x}) : \mathbb{D}(x, \mathbf{w}^{(\delta)})\\ &+ \int_0^t \mathfrak{N}_3^{(\delta)}(\mathbf{x}, t-\tau) : \mathbb{D}\big(x, \mathbf{w}^{(\delta)}(\mathbf{x},\tau)\big)d\tau.\end{aligned} \tag{5.3.23}$$

The problem is completed with the normalization condition

$$\int_\Omega p^{(\delta)}(\mathbf{x},t)\,dx = 0, \tag{5.3.24}$$

the Dirichlet boundary condition

$$\mathbf{w}^{(\delta)}(\mathbf{x},t) = 0 \tag{5.3.25}$$

at the outer boundary S, and the initial condition

$$\mathbf{w}^{(\delta)}(\mathbf{x},0)=0,\ \ \mathbf{x}\in\Omega. \tag{5.3.26}$$

In (5.3.21)–(5.3.26)

$$\hat{\rho}^{(\delta)}(\mathbf{x})=m^{(\delta)}(\mathbf{x})\,\rho_f+\big(1-m^{(\delta)}(\mathbf{x})\big)\,\rho_s^{(\delta)}(\mathbf{x}),$$

fourth-rank tensors $\mathfrak{N}_1^{(\delta)}(\mathbf{x})$, $\mathfrak{N}_2^{(\delta)}(\mathbf{x})$, *and* $\mathfrak{N}_3^{(\delta)}(\mathbf{x},t)$ *are given for almost all points* $\mathbf{x}\in\Omega$ *by formulae (1.4.30) for criteria* μ_0 *and* $\lambda_0^{(\delta)}(\mathbf{x})$, *and the pore space with the characteristic function* $\chi^{(\delta)}(\mathbf{x},\mathbf{y})$ *(see Theorem 1.11 of Chap. 1). The symmetric tensor* $\mathfrak{N}_1^{(\delta)}$ *is strictly positively definite.*

We refer to the problem (5.3.21)–(5.3.26) as the homogenized **model** $(\mathbb{FCM})_{16}^{(\delta)}$.

Now we may complete the construction of mathematical models with variable properties of the medium by the following

Assumption 5.2 Under the conditions of Assumption 5.1 let $\{\chi^{(\delta)}(\mathbf{x},\mathbf{y})\}$ be a sequence of characteristic functions, which approximately describe the pore space in Ω.

Then there exists a function $\chi(\mathbf{x},\mathbf{y})$m 1-periodic in the variable $\mathbf{y}$, such that the sequence $\{\chi^{(\delta)}\}$ converges uniformly in $\Omega\times Y$ as $\delta\to 0$ to the function $\chi(\mathbf{x},\mathbf{y})$.

Let $Y_f(\mathbf{x})$ and $Y_s(\mathbf{x})$ be elementary liquid and solid cells in Y, defined by the characteristic function $\chi(\mathbf{x},\mathbf{y})$. Then due to Assumptions 5.1 and 5.2 for any $\mathbf{x}_0\in\Omega$ there exists $\gamma>0$, such that

$$Y_f(\mathbf{x})\cap Y_f(\mathbf{x}_0)\neq\emptyset,\quad Y_s(\mathbf{x})\cap Y_s(\mathbf{x}_0)\neq\emptyset,\quad \forall\,\mathbf{x}:\ |\mathbf{x}-\mathbf{x}_0|<\gamma. \tag{5.3.27}$$

We additionally suppose that

$$\lambda_0^{(\delta)}(\mathbf{x})\to\lambda_0(\mathbf{x}),\ \ \rho_s^{(\delta)}(\mathbf{x})\to\rho_s(\mathbf{x})\ \text{in}\ L_2(\Omega) \tag{5.3.28}$$

as $\delta\to 0$. These last assumptions permit us to pass to the limit as $\delta\to 0$ in the mathematical models $(\mathbb{FCM})_{13}^{(\delta)}-(\mathbb{FCM})_{16}^{(\delta)}$.

So, the following statements hold true.

Theorem 5.20 *Under Assumptions 5.1 and 5.2 and conditions (5.3.28) let* $\{\mathbf{w}_s^{(\delta)},\mathbf{w}^{(\delta,f)},\pi_f^{(\delta)}\}$ *be a solution of the model* $(\mathbb{FCM})_{13}^{(\delta)}$.

Then the sequences $\{\mathbf{w}^{(\delta,f)}\}$ *and* $\{\pi_f^{(\delta)}\}$ *converge weakly in* $\mathbf{L}_2(\Omega_T)$ *and* $L_2(\Omega_T)$ *as* $\delta\to 0$ *to the functions* $\mathbf{w}_s$ *and* π_f *respectively, and the sequence* $\{\mathbf{w}_s^{(\delta)}\}$ *converges weakly in* $\mathbf{W}_2^{1,0}(\Omega_T)$ *as* $\delta\to 0$ *to the function* $\mathbf{w}_s$.

The limiting functions solve in the domain Ω *for* $t>0$ *the homogenized system, consisting of the homogenized momentum balance equation*

$$\nabla \cdot \mathbb{P}_1(\mathbf{x}) + \hat{\rho}(\mathbf{x})\,\mathbf{F} = 0, \tag{5.3.29}$$

$$\mathbb{P}_1(\mathbf{x}) = \lambda_0(\mathbf{x})\,\mathfrak{N}_1^s(\mathbf{x}) : \mathbb{D}(x, \mathbf{w}_s) - p_f\,\mathbb{I}, \tag{5.3.30}$$

and the continuity equation

$$\nabla \cdot \mathbf{w}_s = 0. \tag{5.3.31}$$

The problem is completed with the normalization condition

$$\int_\Omega p_f(\mathbf{x}, t)dx = 0 \tag{5.3.32}$$

and the boundary condition

$$\mathbf{w}_s = 0 \tag{5.3.33}$$

on the outer boundary S for $t > 0$.

In (5.3.29)–(5.3.33)

$$p_f = \frac{\partial \pi_f}{\partial t},\ \ \hat{\rho}(\mathbf{x}) = m(\mathbf{x})\,\rho_f + \big(1 - m(\mathbf{x})\big)\,\rho_s(\mathbf{x}),\ \ m(\mathbf{x}) = \int_Y \chi(\mathbf{x}, \mathbf{y})dy,$$

the symmetric strictly positively definite fourth-rank tensor $\mathfrak{N}_1^s(\mathbf{x})$ is given at point $\mathbf{x} \in \Omega$ by (1.2.38) for the pore space with the characteristic function $\chi(\mathbf{x}, \mathbf{y})$ (see Theorem 1.4 of Chap. 1).

We refer to the problem (5.3.29)–(5.3.33) as the homogenized **model** $(\mathbb{FCM})_{13}^{(0)}$.

Theorem 5.21 *Under Assumptions 5.1 and 5.2 and conditions (5.3.28) let $\{\mathbf{w}_s^{(\delta)}, \mathbf{w}^{(\delta,f)}, \pi_f^{(\delta)}\}$ be a solution of the model $(\mathbb{FCM})_{14}^{(\delta)}$.*

Then the sequences $\{\mathbf{w}^{(\delta,f)}\}$ and $\{\pi_f^{(\delta)}\}$ converge weakly in $\mathbf{L}_2(\Omega_T)$ and $L_2(\Omega_T)$ as $\delta \to 0$ to the functions $\mathbf{w}^f$ and π_f respectively, and the sequence $\{\mathbf{w}_s^{(\delta)}\}$ converges weakly in $\mathbf{W}_2^{1,0}(\Omega_T)$ as $\delta \to 0$ to the function $\mathbf{w}_s$.

The limiting functions solve in the domain Ω for $t > 0$ the homogenized system, consisting of the continuity equation

$$\nabla \cdot \big(\mathbf{w}^f + \big(1 - m(\mathbf{x})\big)\,\mathbf{w}_s\big) = 0, \tag{5.3.34}$$

the homogenized momentum balance equation

$$\nabla \cdot \mathbb{P}_1(\mathbf{x}) + \hat{\rho}(\mathbf{x})\,\mathbf{F} = 0, \tag{5.3.35}$$

$$\mathbb{P}_1(\mathbf{x}) = \lambda_0(\mathbf{x})\,\mathfrak{N}_1^s(\mathbf{x}) : \mathbb{D}(x, \mathbf{w}_s) - p_f\,\mathbb{I} \tag{5.3.36}$$

for the solid component, and Darcy's law in the form

$$\mathbf{w}^f = m(\mathbf{x})\,\mathbf{w}_s + \frac{1}{\mu_1}\mathbb{B}(\mathbf{x}) \cdot \Big(- \nabla \pi_f + \rho_f \int_0^t \mathbf{F}(\mathbf{x}, \tau) d\tau \Big) \tag{5.3.37}$$

for the liquid component.

The problem is completed with the normalization condition (5.3.32), the boundary condition (5.3.33) for the solid displacements $\mathbf{w}_s$*, and the boundary condition*

$$\mathbf{w}^f(\mathbf{x}, t) \cdot \mathbf{n}(\mathbf{x}) = 0 \tag{5.3.38}$$

for the liquid displacements on the outer boundary S *for* $t > 0$.

In (5.3.34)–(5.3.38) $\mathbf{n}(\mathbf{x})$ *is a unit normal to* S *at* $\mathbf{x} \in S$,

$$p_f = \frac{\partial \pi_f}{\partial t}, \quad \hat{\rho}(\mathbf{x}) = m(\mathbf{x})\,\rho_f + \big(1 - m(\mathbf{x})\big)\,\rho_s(\mathbf{x}), \quad m(\mathbf{x}) = \int_Y \chi(\mathbf{x}, \mathbf{y}) dy,$$

the symmetric strictly positively definite fourth-rank tensor $\mathfrak{N}_1^s(\mathbf{x})$ *is given for almost all points* $\mathbf{x} \in \Omega$ *by (1.2.38) for the pore space with the characteristic function* $\chi(\mathbf{x}, \mathbf{y})$ *(see Theorem 1.4 of Chap. 1), the symmetric strictly positively definite matrix* $\mathbb{B}(\mathbf{x})$ *is given for almost all points* $\mathbf{x} \in \Omega$ *by (1.1.27) for the pore space with the characteristic function* $\chi(\mathbf{x}, \mathbf{y})$ *(see Theorem 1.1 of Chap. 1).*

We refer to the problem (5.3.32), (5.3.33)–(5.3.38) as the homogenized **model** $(\mathbb{FCM})_{14}^{(0)}$.

Theorem 5.22 *Under the conditions of Theorem 5.21 let* $\{\mathbf{w}_s^{(k)}, \mathbf{w}^{(f,k)}, \pi_f^{(k)}\}$ *be the weak solution of the model* $(\mathbb{FCM})_{14}^{(0)}$ *with* $\lambda_0(\mathbf{x}) = k$.

Then the sequences $\{\pi_f^{(k)}\}$ *and* $\{\mathbf{w}^{(f,k)}\}$ *converge weakly in* $L_2(\Omega_T)$ *and* $\mathbf{L}_2(\Omega_T)$ *as* $k \to \infty$ *to the functions* π_f *and* $\mathbf{w}^f$ *respectively, and the sequence* $\{\mathbf{w}_s^{(k)}\}$ *converges strongly in* $\mathbf{L}_2(\Omega_T)$ *to zero.*

The limiting functions solve in the domain Ω *for* $t > 0$ *the homogenized system, consisting of the continuity equation*

$$\nabla \cdot \mathbf{w}^f = 0 \tag{5.3.39}$$

and Darcy's law

$$\mathbf{w}^f = \frac{1}{\mu_1}\mathbb{B}(\mathbf{x}) \cdot \Big(- \nabla \pi_f + \rho_f \int_0^t \mathbf{F}(\mathbf{x}, \tau) d\tau \Big), \tag{5.3.40}$$

completed with the boundary condition (5.3.38) for the liquid velocity on the outer boundary S *for* $t > 0$*, and the normalization condition (5.3.32).*

The symmetric strictly positively definite matrix $\mathbb{B}(\mathbf{x})$ *is given for almost all points* $\mathbf{x} \in \Omega$ *by (1.1.27) for the pore space with the characteristic function* $\chi(\mathbf{x}, \mathbf{y})$ *(see Theorem 1.1 of Chap. 1).*

We refer to the problem (5.3.32), (5.3.38)–(5.3.40) as the homogenized **model** $(\mathbb{FCM})^{(0)}_{15}$.

Theorem 5.23 *Under Assumptions 5.1 and 5.2 and conditions (5.3.28) let* $\{\mathbf{w}^{(\delta)}, p^{(\delta)}\}$ *be a solution of the model* $(\mathbb{FCM})^{(\delta)}_{16}$.

Then the sequence $\{p^{(\delta)}\}$ *converges weakly in* $L_2(\Omega_T)$ *as* $\delta \to 0$ *to the function* p *and the sequence* $\{\mathbf{w}^{(\delta)}\}$ *converges weakly in* $\mathbf{w}^{1,0}_2(\Omega_T)$ *as* $\delta \to 0$ *to the function* $\mathbf{w}$.

The limiting functions solve in the domain Ω *for* $t > 0$ *the homogenized system, consisting of the continuity equation*

$$\nabla \cdot \mathbf{w} = 0, \tag{5.3.41}$$

and the homogenized momentum balance equation

$$\nabla \cdot \widehat{\mathbb{P}}(\mathbf{x}) + \hat{\rho}(\mathbf{x})\,\mathbf{F} = 0, \tag{5.3.42}$$

$$\begin{aligned}\widehat{\mathbb{P}}(\mathbf{x}) = & -p\,\mathbb{I} + \mathfrak{N}_1(\mathbf{x}) : \mathbb{D}\Big(x, \frac{\partial \mathbf{w}}{\partial t}\Big) + \mathfrak{N}_2(\mathbf{x}) : \mathbb{D}(x, \mathbf{w}) \\ & + \int_0^t \mathfrak{N}_3(\mathbf{x}, t-\tau) : \mathbb{D}(x, \mathbf{w}(\mathbf{x}, \tau))d\tau. \end{aligned} \tag{5.3.43}$$

The problem is completed with the normalization condition

$$\int_\Omega p\,(\mathbf{x}, t)\,dx = 0, \tag{5.3.44}$$

the Dirichlet boundary condition

$$\mathbf{w}(\mathbf{x}, t) = 0 \tag{5.3.45}$$

at the outer boundary S, and the initial condition

$$\mathbf{w}(\mathbf{x}, 0) = 0,\ \ \mathbf{x} \in \Omega. \tag{5.3.46}$$

In (5.3.41)–(5.3.46)

$$\hat{\rho}(\mathbf{x}) = m(\mathbf{x})\,\rho_f + \big(1 - m(\mathbf{x})\big)\,\rho_s(\mathbf{x}),\ \ m(\mathbf{x}) = \int_Y \chi(\mathbf{x}, \mathbf{y})dy,$$

fourth-rank tensors $\mathfrak{N}_1(\mathbf{x})$, $\mathfrak{N}_2(\mathbf{x})$, *and* $\mathfrak{N}_3(\mathbf{x}, t)$ *are given for almost all points* $\mathbf{x} \in \Omega$ *by formulae (1.4.30) for criteria* μ_0 *and* $\lambda_0(\mathbf{x})$, *and the pore space with the characteristic function* $\chi(\mathbf{x}, \mathbf{y})$ *(see Theorem 1.11 of Chap. 1). The symmetric tensor* $\mathfrak{N}_1$ *is strictly positively definite.*

We refer to the problem (5.3.41)–(5.3.46) as the homogenized **model** $(\mathbb{FCM})^{(0)}_{16}$.

To prove all these statements we only have to show that tensors $\mathfrak{N}_1^{s,\delta}(\mathbf{x})$, $\mathfrak{N}_1^{(\delta)}(\mathbf{x})$, $\mathfrak{N}_2^{(\delta)}(\mathbf{x})$, and $\mathfrak{N}_3^{(\delta)}(\mathbf{x},t)$, and matrices $\mathbb{B}^{(\delta)}(\mathbf{x})$ continuously depend on δ as $\delta \to 0$.

The proof of this fact is quite standard and we do it schematically only for the tensor $\mathfrak{N}_1^{s,\delta}(\mathbf{x})$.

Lemma 5.3 *The tensor $\mathfrak{N}_1^{s,\delta}(\mathbf{x})$ is a continuous with respect to parameter δ.*

Proof To prove the statement we only have to show the continuity of the solution $\{\mathbf{U}_\delta^{(ij)},\ P_\delta^{(ij)}\}$ of the problem

$$\left.\begin{aligned} &\nabla_y \cdot \Big((1-\chi^{(\delta)})\big(\mathbb{D}(y,\mathbf{U}^{(\delta)}) + \mathbb{J} - P^{(\delta)}\mathbb{I}\big)\Big) = 0,\ \ \mathbf{y}\in Y,\\ &(1-\chi^{(\delta)})\nabla_y\cdot\mathbf{U}^{(\delta)} = 0,\ \ \mathbf{y}\in Y,\ \ \langle\mathbf{U}^{(\delta)}\rangle_{Y_s^{(\delta)}} = 0 \end{aligned}\right\} \tag{5.3.47}$$

with respect to δ, if the characteristic function $\big(1-\chi^{(\delta)}(\mathbf{y})\big)$ of the solid cell $Y_s^{(\delta)}$ is a continuous with respect to δ.

Without loss of generality we may assume that functions $\mathbf{U}^{(\delta)}$ and $P^{(\delta)}$ are defined in Y,

$$\int_Y \big(|\nabla_y\mathbf{U}^{(\delta)}|^2 + |P^{(\delta)}|^2\big)dy \leqslant C_0, \tag{5.3.48}$$

and that

$$\nabla_y\cdot\mathbf{U}^{(\delta)} = 0,\ \ \mathbf{y}\in Y. \tag{5.3.49}$$

Then the difference $\mathbf{U} = \mathbf{U}^{(\delta_1)} - \mathbf{U}^{(\delta_2)}$, $P = P^{(\delta_1)} - P^{(\delta_2)}$ is a solution of the integral identity

$$\begin{aligned} &\int_Y (1-\chi^{(\delta_1)})\big(\mathbb{D}(y,\mathbf{U}) - P\mathbb{I}\big) : \mathbb{D}(y,\varphi)dy\\ &\quad = \int_Y (\chi^{(\delta_1)} - \chi^{(\delta_2)})\mathbb{D}(y,\mathbf{U}_2) : \mathbb{D}(y,\varphi)dy, \end{aligned} \tag{5.3.50}$$

completed with the continuity equation

$$\nabla_y\cdot\mathbf{U} = 0,\ \ \mathbf{y}\in Y. \tag{5.3.51}$$

Setting in (5.3.50) $\varphi = \mathbf{U}$ one has the equality

$$\int_Y (1-\chi^{(\delta_1)})|\mathbb{D}(y,\mathbf{U})|^2 dy = \int_Y (\chi^{(\delta_1)} - \chi^{(\delta_2)})\mathbb{D}(y,\mathbf{U}_2) : \mathbb{D}(y,\mathbf{U})dy,$$

which provides the first estimate

$$\int_Y |\nabla_y\mathbf{U}|^2 dy \leqslant C_0 \max_{\mathbf{y}\in Y} |\chi^{(\delta_1)}(\mathbf{y}) - \chi^{(\delta_2)}(\mathbf{y})|^2. \tag{5.3.52}$$

Let now $\varphi_0 \in \overset{\circ}{\mathbf{W}}{}^1_2(Y)$ be found from the condition

$$\nabla_y \cdot \varphi_0 = P, \ \mathbf{y} \in Y. \tag{5.3.53}$$

Such a choice is always possible (see [59]) and

$$\int_Y |\nabla_y \varphi_0|^2 dy \leqslant C_0 \int_Y |P|^2 dy \leqslant C_1. \tag{5.3.54}$$

Setting in (5.3.50) $\varphi = \varphi_0$ we arrive at the second estimate

$$\int_Y |P|^2 dy \leqslant C_0 \max_{\mathbf{y}\in Y} |\chi^{(\delta_1)}(\mathbf{y}) - \chi^{(\delta_2)}(\mathbf{y})|^2. \tag{5.3.55}$$

Note that we have started with the mathematical problem at the microscopic level (5.3.1)–(5.3.5), depending on two small parameters ε and δ, then first pass to the limit as $\varepsilon \to 0$, and after that pass to the limit as $\delta \to 0$.

For such limiting procedures with two independent small parameters pure mathematics requires proofs for the full diagram. It means that now we must first pass to the limit as $\delta \to 0$, and after that pass to the limit as $\varepsilon \to 0$. The limiting model will be correct, if in both cases we get the same result.

The limit as $\delta \to 0$ in (5.3.1)–(5.3.5) obviously results in the problem

$$\nabla \cdot \mathbf{w}^\varepsilon = 0, \tag{5.3.56}$$

$$\nabla \cdot \Big(\chi^\varepsilon \alpha_\mu \mathbb{D}\Big(x, \frac{\partial \mathbf{w}^\varepsilon}{\partial t}\Big) + (1-\chi^\varepsilon)\lambda_0(\mathbf{x})\mathbb{D}(x, \mathbf{w}^\varepsilon) - p^\varepsilon\, \mathbb{I}\Big) + \rho^\varepsilon(\mathbf{x})\mathbf{F} = 0, \tag{5.3.57}$$

$$\mathbf{w}^\varepsilon(\mathbf{x}, 0) = 0, \ \ \mathbf{x} \in S = \partial\Omega, \ t > 0, \tag{5.3.58}$$

$$\chi^\varepsilon(\mathbf{x})\mathbf{w}^\varepsilon(\mathbf{x}, t) = 0, \ \ \mathbf{x} \in \Omega, \tag{5.3.59}$$

$$\int_\Omega \frac{\chi^\varepsilon(\mathbf{x})}{m(\mathbf{x})}\, p^\varepsilon(\mathbf{x}, t) dx = 0, \tag{5.3.60}$$

where

$$\rho^\varepsilon(\mathbf{x}) = \chi^\varepsilon(\mathbf{x})\rho_f + \big(1 - \chi^\varepsilon(\mathbf{x})\big)\rho_s(\mathbf{x}), \ m(\mathbf{x}) = \int_Y \chi(\mathbf{x}, \mathbf{y}) dy,$$

with the limiting characteristic function

$$\chi^\varepsilon(\mathbf{x}) = \chi\left(\mathbf{x}, \frac{\mathbf{x}}{\varepsilon}\right)$$

of the pore space.

The correctness of the problem (5.3.56)–(5.3.60) is proved in a way similar to the the correctness of the problem (5.3.1)–(5.3.5).

As above, we introduce an extension

$$\mathbf{w}_s^\varepsilon(\mathbf{x},t) = \mathbb{E}^{(*)}_{\Omega_s^\varepsilon}\big(\mathbf{w}^\varepsilon\big) \tag{5.3.61}$$

from the solid part

$$\Omega_s^\varepsilon = \Big\{\mathbf{x}\in\Omega : \chi\Big(\mathbf{x},\frac{\mathbf{x}}{\varepsilon}\Big)=0\Big\}$$

of the domain Ω onto the whole domain Ω, with the following properties:

$$\big(1-\chi^\varepsilon(\mathbf{x})\big)\big(\mathbf{w}^\varepsilon(\mathbf{x},t)-\mathbf{w}_s^\varepsilon(\mathbf{x},t)\big)=0,\quad \mathbf{x}\in\Omega,\ t\in(0,T),$$

and

$$\begin{aligned}
&\int_\Omega |\mathbf{w}_s^\varepsilon(\mathbf{x},t)|^2dx \leqslant C_0\int_{\Omega_s^\varepsilon}|\mathbf{w}^\varepsilon(\mathbf{x},t)|^2dx,\\
&\int_\Omega \Big|\mathbb{D}\big(x,\mathbf{w}_s^\varepsilon(\mathbf{x},t)\big)\Big|^2dx \leqslant C_0\int_{\Omega_s^\varepsilon}\Big|\mathbb{D}\big(x,\mathbf{w}^\varepsilon(\mathbf{x},t)\big)\Big|^2dx,\quad t\in(0,T),
\end{aligned}\tag{5.3.62}$$

where C_0 is independent of ε and $t\in(0,T)$.

The existence of such an extension for domains Ω_s^ε with a non-periodic structure might be proved as well as the existence of the extension (1.2.9) for domains Ω_s^ε with a periodic structure.

So, the following theorem holds true.

Theorem 5.24 *Under Assumptions 5.1 and 5.2 for all $\varepsilon>0$ and for an arbitrary time interval $[0,T]$ there exists a unique generalized solution of problem (5.3.56)–(5.3.60) and*

$$\begin{aligned}
&\max_{0<t<T}\int_\Omega \chi^\varepsilon\Big(\alpha_\mu\Big|\mathbb{D}\big(x,\mathbf{w}^\varepsilon(\mathbf{x},t)\big)\Big|^2+\frac{\alpha_\mu}{\varepsilon^2}|\mathbf{w}^\varepsilon(\mathbf{x},t)-\mathbf{w}_s^\varepsilon(\mathbf{x},t)|^2\Big)dx\\
&\quad+\int_0^T\int_\Omega\Big(|\pi^\varepsilon|^2+\lambda_0(\mathbf{x})|\mathbb{D}(x,\mathbf{w}_s^\varepsilon)|^2\Big)dxdt\leqslant C_0\mathfrak{P}^2,
\end{aligned}\tag{5.3.63}$$

where C_0 is independent of ε, $\lambda_0(\mathbf{x})$ for $\lambda_0(\mathbf{x})>\lambda^-$, and

$$\pi^\varepsilon(\mathbf{x},t)=\int_0^t p^\varepsilon(\mathbf{x},\tau)d\tau,$$

$$\mathfrak{P}^2=\max_{0<t<T}\int_\Omega|\mathbf{F}(\mathbf{x},t)|^2dx<\infty.$$

The limit as $\varepsilon \to 0$ in (5.3.56)–(5.3.60) does not cause any difficulties and one formulates the following theorems.

Theorem 5.25 *Under Assumptions 5.1 and 5.2 let*

$$\mu_0 = 0, \ \mu_1 = \infty, \ 0 < \lambda^- < \lambda_0(\mathbf{x}) < \lambda^+ < \infty,$$

$\{\mathbf{w}^\varepsilon, \ p^\varepsilon\}$ *be the weak solution of the problem (5.3.56)–(5.3.60),*

$$\pi^\varepsilon(\mathbf{x}, t) = \int_0^t p^\varepsilon(\mathbf{x}, \tau)d\tau,$$

and $\mathbf{w}_s^\varepsilon = \mathbb{E}^{(*)}_{Q_s^\varepsilon}(\mathbf{w}^\varepsilon)$ *be an extension (5.3.61).*

Then the sequences $\{\mathbf{w}^\varepsilon\}$ *and* $\{\chi^\varepsilon \pi^{\delta,\varepsilon}\}$ *converge weakly in* $\mathbf{L}_2(\Omega_T)$ *and* $L_2(\Omega_T)$ *as* $\varepsilon \to 0$ *to the functions* $\mathbf{w}_s$ *and* $m(\mathbf{x})\pi_f$ *respectively, and the sequence* $\{\mathbf{w}_s^\varepsilon\}$ *converges weakly in* $\mathbf{W}_2^{1,0}(\Omega_T)$ *as* $\varepsilon \to 0$ *to the function* $\mathbf{w}_s$.

The pair of functions $\{\mathbf{w}_s, \ \pi_f\}$ *solves in the domain* Ω *for* $t > 0$ *the problem* $(\mathbb{FCM})^{(0)}_{13}$.

Theorem 5.26 *Under Assumptions 5.1 and 5.2 let*

$$\mu_0 = 0, \ 0 < \lambda^- < \mu_1, \ \lambda_0(\mathbf{x}) < \lambda^+ < \infty,$$

$\{\mathbf{w}^\varepsilon, \ p^\varepsilon\}$ *be the weak solution of the problem (5.3.56)–(5.3.60),* $\mathbf{w}_s^\varepsilon = \mathbb{E}^{(*)}_{Q_s^\varepsilon}(\mathbf{w}^\varepsilon)$ *be an extension (5.3.61), and*

$$\pi^\varepsilon(\mathbf{x}, t) = \int_0^t p^\varepsilon(\mathbf{x}, \tau)d\tau.$$

Then the sequences $\{\chi^\varepsilon \pi^\varepsilon\}$, *and* $\{\chi^\varepsilon \mathbf{w}^\varepsilon\}$ *converge weakly in* $L_2(\Omega_T)$ *and* $\mathbf{L}_2(\Omega_T)$ *as* $\varepsilon \to 0$ *to the functions* $m(\mathbf{x})\pi_f$ *and* $\mathbf{w}^{(f)}$ *respectively, and the sequence* $\{\mathbf{w}_s^\varepsilon\}$ *converges weakly in* $\mathbf{W}_2^{1,0}(\Omega_T)$ *as* $\varepsilon \to 0$ *to the function* $\mathbf{w}_s$.

The triple of functions $\{\mathbf{w}_s, \ \mathbf{w}^{(f)}, \ \pi_f\}$, *where* $\pi_f \in W_2^{1,0}(\Omega_T)$, $\dfrac{\partial \pi_f}{\partial t} \in L_2(\Omega_T)$, *solves in the domain* Ω *for* $t > 0$ *the problem* $(\mathbb{FCM})^{(0)}_{14}$.

Theorem 5.27 *Under the conditions of Theorem 5.26 let* $\{\mathbf{w}_s^{(k)}, \ \mathbf{w}^{(f,k)}, \ \pi_f^{(k)}\}$ *be the weak solution of the model* $(\mathbb{FCM})^{(0)}_{14}$ *with* $\lambda_0(\mathbf{x}) = k$.

Then the sequences $\{\pi_f^{(k)}\}$ *and* $\{\mathbf{w}^{(f,k)}\}$ *converge weakly in* $L_2(\Omega_T)$ *and* $\mathbf{L}_2(\Omega_T)$ *as* $k \to \infty$ *to the functions* π_f *and* $\mathbf{w}^{(f)}$ *respectively, and the sequence* $\{\mathbf{w}_s^{(k)}\}$ *converges strongly in* $\mathbf{L}_2(\Omega_T)$ *to zero.*

The pair of functions $\{\mathbf{w}^{(f)}, \ \pi_f\}$ *solves in the domain* Ω *for* $t > 0$ *the problem* $(\mathbb{FCM})^{(0)}_{15}$.

For the case

$$0 < \lambda^{-} < \mu_0,\ \lambda_0(\mathbf{x}) < \lambda^{+} < \infty, \tag{5.3.64}$$

instead of the problem (5.3.56)–(5.3.60) we consider the problem (5.3.56)–(5.3.59) with the normalization condition

$$\int_{\Omega} p^{\varepsilon}(\mathbf{x}, t)dx = 0. \tag{5.3.65}$$

Theorem 5.28 *Under Assumptions 5.1 and 5.2 for all $\varepsilon > 0$ and for arbitrary time interval $[0, T]$ there exists a unique generalized solution $\{\mathbf{w}^{\varepsilon},\ p^{\varepsilon}\}$ of problem (5.3.56)–(5.3.59), (5.3.65) and*

$$\max_{0<t<T} \lambda_0 \int_{\Omega} (1-\chi^{\varepsilon})(\mathbf{x})\Big|\mathbb{D}\big(x, \mathbf{w}^{\varepsilon}(\mathbf{x}, t)\big)\Big|^2 dx \\ + \int_0^T \int_{\Omega} \Big(|p^{\varepsilon}|^2 + \alpha_{\mu}\Big|\mathbb{D}\Big(x, \frac{\partial \mathbf{w}^{\varepsilon}}{\partial t}\Big)\Big|^2\Big)dxdt \leqslant C_0\mathfrak{P}^2, \tag{5.3.66}$$

where C_0 is independent of ε and

$$\mathfrak{P}^2 = \max_{0<t<T} \int_{\Omega} |\mathbf{F}(\mathbf{x}, t)|^2 dx < \infty.$$

The sequence $\{p^{\varepsilon}\}$ converges weakly in $L_2(\Omega_T)$ as $\varepsilon \to 0$ to the function p and the sequence $\{\mathbf{w}^{\varepsilon}\}$ converges weakly in $\mathbf{W}_2^{1,0}(\Omega_T)$ as $\varepsilon \to 0$ to the function $\mathbf{w}$.

The pair of functions $\{\mathbf{w},\ p\}$ solves in the domain Ω for $t > 0$ the problem $(\mathbb{FCM})_{16}^{(0)}$.

Proofs of these theorems in its main points repeats the proofs of the similar theorems above. That is why we prove only Theorem 5.26 to outline the differences.

5.3.1 Proof of Theorem 5.26

On the basis of estimates (5.3.66) we conclude that for $\varepsilon \to 0$

$$\chi^{\varepsilon}\pi^{\varepsilon} \rightharpoonup m(\mathbf{x})\pi_f(\mathbf{x}, t) \ \text{ weakly in } L_2(\Omega_T),$$

$$\pi^{\varepsilon} \to \Pi = \chi(\mathbf{x}, \mathbf{y})\pi_f + \big(1-\chi(\mathbf{x}, \mathbf{y})\big)\Pi_s(\mathbf{x}, t, \mathbf{y}) \ \text{ two-scale in } L_2(\Omega_T),$$

$$\chi^{\varepsilon}\mathbf{w}^{\varepsilon} \to \chi(\mathbf{x}, \mathbf{y})\mathbf{W}(\mathbf{x}, t, \mathbf{y}) \ \text{ two-scale in } \mathbf{L}_2(\Omega_T),$$

$$\chi^{\varepsilon}\mathbf{w}^{\varepsilon} \rightharpoonup \mathbf{w}^{(f)}(\mathbf{x}, t) \ \text{ weakly in } \mathbf{L}_2(\Omega_T), \quad \mathbf{w}^{(f)} = \langle \mathbf{W}\rangle_{Y_f},$$

$$\mathbf{w}_s^\varepsilon \to \mathbf{w}_s(\mathbf{x},t) \text{ weakly and two-scale in } \mathbf{L}_2(\Omega_T),$$

$$\mathbb{D}\big(x,\mathbf{w}_s^\varepsilon\big) \to \mathbb{D}\big(x,\mathbf{w}_s(\mathbf{x},t)\big) + \mathbb{D}\big(y,\mathbf{U}(\mathbf{x},t,\mathbf{y})\big) \text{ two-scale in } \mathbf{L}_2(\Omega_T).$$

The properties of the function $\chi(\mathbf{x},\mathbf{y})$ admit the two-scale limit in the integrals

$$I^{(\varepsilon)} = \int_0^T \int_\Omega \chi^\varepsilon(\mathbf{x})\varphi\left(\mathbf{x},\frac{\mathbf{x}}{\varepsilon},t\right) u^\varepsilon(\mathbf{x},t)dxdt,$$

where $\varphi(\mathbf{x},t,\mathbf{y})$ is a smooth 1-periodic in $\mathbf{y}$ function, and $u^\varepsilon \to U(\mathbf{x},t,\mathbf{y})$ two-scale in $L_2(\Omega_T)$.

Indeed, by construction,

$$\begin{aligned} I^{(\varepsilon,\delta)} &= \int_0^T \int_\Omega \chi^{\delta,\varepsilon}(\mathbf{x})\varphi\left(\mathbf{x},\frac{\mathbf{x}}{\varepsilon},t\right) u^\varepsilon(\mathbf{x},t)dxdt \\ &\to \int_0^T \int_\Omega \int_Y \chi^{(\delta)}(\mathbf{x},\mathbf{y})\,\varphi(\mathbf{x},t,\mathbf{y})U(\mathbf{x},t,\mathbf{y})dydxdt = I^{(0,\delta)} \end{aligned}$$

as $\varepsilon \to 0$.

Therefore, if

$$I^{(0)} = \int_0^T \int_\Omega \int_Y \chi(\mathbf{x},\mathbf{y})\,\varphi(\mathbf{x},t,\mathbf{y})U(\mathbf{x},t,\mathbf{y})dydxdt,$$

then for any $\gamma > 0$ there exists $\varepsilon_0 = \varepsilon_0(\gamma)$, such that

$$|I^{(\varepsilon)} - I^{(0)}| = |I^{(\varepsilon)} - I^{(\varepsilon,\delta)}| + |I^{(\varepsilon,\delta)} - I^{(0,\delta)}| + |I^{(0,\delta)} - I^0| < \gamma$$

for any $\varepsilon < \varepsilon_0$.

We simply choose δ from conditions

$$|I^{(\varepsilon)} - I^{(\varepsilon,\delta)}| < \frac{\gamma}{3}, \quad |I^{(0,\delta)} - I^0| < \frac{\gamma}{3},$$

and after that for fixed δ we choose ε_0 from the condition

$$|I^{(\varepsilon,\delta)} - I^{(0,\delta)}| < \frac{\gamma}{3}$$

for any $\varepsilon < \varepsilon_0$.

Now we pass to the limit as $\varepsilon \to 0$ in the integral identity

$$\begin{aligned} \int_0^T \int_\Omega \Big(&\big(\chi^\varepsilon \alpha_\mu \mathbb{D}(x,\mathbf{w}^\varepsilon) - \pi^\varepsilon\,\mathbb{I}\big) : \mathbb{D}\left(x,\frac{\partial\varphi}{\partial t}\right) \\ &- (1-\chi^\varepsilon)\lambda_0(\mathbf{x})\mathbb{D}(x,\mathbf{w}_s^\varepsilon)\big) : \mathbb{D}(x,\varphi) + \rho^\varepsilon(\mathbf{x})\mathbf{F}\cdot\varphi\Big)dxdt = 0, \end{aligned} \tag{5.3.67}$$

which is equivalent to Eq. (5.3.57).

After the limit with test functions $\varphi = \varphi(\mathbf{x}, t)$ we arrive at the macroscopic momentum balance equation

$$\int_0^T \int_\Omega \Big(\lambda_0(\mathbf{x})\big((1 - m(\mathbf{x}))\mathbb{D}(x, \mathbf{w}_s) + \langle \mathbb{D}(y, \mathbf{U})\rangle_{Y_s(\mathbf{x})}\big) : \mathbb{D}(x, \varphi) \\ + \big(\pi_f + \langle (\Pi_f - \pi_f)\rangle_{Y_s(\mathbf{x})}\big)\nabla \cdot \Big(\frac{\partial \varphi}{\partial t}\Big) - \hat{\rho}(\mathbf{x})\mathbf{F} \cdot \varphi\Big) dx dt = 0, \qquad (5.3.68)$$

and taking in (5.3.67) $\varphi = \varepsilon h(\mathbf{x})\varphi_0\left(\frac{\mathbf{x}}{\varepsilon}, t\right)$, we obtain the microscopic momentum balance equation

$$\int_0^T \int_Y \lambda_0 (1 - \chi)\big(\mathbb{D}(x, \mathbf{w}_s) + \mathbb{D}(y, \mathbf{U}) : \mathbb{D}(y, \varphi_0) dy dt \\ = -\int_0^T \int_Y \Pi \nabla \cdot \Big(\frac{\partial \varphi_0}{\partial t}\Big) dy dt. \qquad (5.3.69)$$

The two-scale limit as $\varepsilon \to 0$ in the continuity equation (5.3.56) results the microscopic continuity equation

$$(1 - \chi)(\nabla \cdot \mathbf{w}_s + \nabla_y \cdot \mathbf{U}) = 0 \qquad (5.3.70)$$

for the solid component.

Just as in Theorem 1.4, we conclude that (5.3.68)–(5.3.70) imply the inclusion

$$p_f = \frac{\partial \pi_f}{\partial t} \in L_2 L_2(\Omega_T),$$

and differential equation (5.3.35).

The boundary condition (5.3.33) is a consequence of properties of the extension operator (5.3.61), and the validity of the normalization condition (5.3.32) is proved in the same way as in previous statements.

Next, after the two-scale limits as $\varepsilon \to 0$ in the continuity equation (5.3.56) in its form of the integral identity

$$\int_0^T \int_\Omega \big(\chi^\varepsilon \mathbf{w}^\varepsilon + (1 - \chi^\varepsilon)\mathbf{w}_s^\varepsilon\big) \cdot \nabla \xi dx dt = 0$$

with two different groups of test functions $\xi = \xi(\mathbf{x}, t)$ and $\xi = \varepsilon h(\mathbf{x}, t)\xi_0\Big(\frac{\mathbf{x}}{\varepsilon}\Big)$, $supp\, \xi_0 \subset Y_f(\mathbf{x})$, we arrive at macroscopic continuity equation in the form of an integral identity, which implies continuity equation (5.3.34) for the mixture, and boundary condition (5.3.38) for the liquid component, and the microscopic continuity equation

$$\nabla_y \cdot \mathbf{W} = 0, \quad \mathbf{y} \in Y_f(\mathbf{x}) \tag{5.3.71}$$

for the liquid component.

The last step is a limit as $\varepsilon \to 0$ in (5.3.67) with test functions $\varphi = h(\mathbf{x}, t)\varphi_0\left(\frac{\mathbf{x}}{\varepsilon}\right)$, where smooth function $\varphi_0(\mathbf{y})$ is 1-periodic in the variable $\mathbf{y}$, divergent free and has a finite support in $Y_f(\mathbf{x})$:

$$\int_0^T \int_\Omega \left(\frac{\partial h}{\partial t}\mu_1\langle \mathbb{D}(y, \mathbf{W}) : \mathbb{D}(y, \varphi_0)\rangle_{Y_f(\mathbf{x})} - \pi_f \langle \varphi_0\rangle_{Y_f(\mathbf{x})} \nabla \cdot \left(\frac{\partial h}{\partial t}\right) \right.$$
$$\left. - \frac{\partial h}{\partial t}\rho_f \int_0^t \mathbf{F}(\mathbf{x}, \tau)d\tau \cdot \langle \varphi_0\rangle_{Y_f(\mathbf{x})}\right) dxdt = 0. \tag{5.3.72}$$

Let $\mathbf{x}_0 \in \Omega$. By Assumptions 5.1 and 5.2 there exists some small $\delta > 0$, such that for any $\mathbf{x} \in \Omega$, $|\mathbf{x}_0 - \mathbf{x}| < \delta$, one has $Y_f(\mathbf{x}_0) \cap Y_f(\mathbf{x}) \neq \emptyset$.

So, we may choose functions $\varphi_{0,i}(\mathbf{y}),\ i = 1, 2, 3$, such that

$$\varphi_{0,i} \in \overset{\circ}{W}{}^1_2(Y_0),\ \nabla \cdot \varphi_{0,i} = 0,\ \langle \varphi_{0,i}\rangle_{Y_0} = \mathbf{e}_i,\ Y_0 \subset \cap_{|\mathbf{x}_0 - \mathbf{x}| < \delta} Y_f(\mathbf{x}),$$

where $\{\mathbf{e}_1, \mathbf{e}_2, \mathbf{e}_3\}$ is an orthogonal Cartesian basis. The existence of such functions follows from Lemma B.15 (see Appendix B).

Let $Q_0 = \{\mathbf{x} \in \Omega : |\mathbf{x}_0 - \mathbf{x}| < \delta\}$ and $h_0 \in L_2\big((0, T); \overset{\circ}{W}{}^1_2(Q_0)\big)$. Setting in (5.3.72) $\dfrac{\partial h}{\partial t} = h_0$ and $\varphi_0 = \varphi_{0,i}$ we obtain

$$\int_0^T \int_\Omega \left(h_0\, \Phi_i(\mathbf{x}) - \pi_f \frac{\partial h}{\partial x_i}\right) dxdt = 0, \tag{5.3.73}$$

where

$$\Phi_i(\mathbf{x}) = \mu_1\langle \mathbb{D}(y, \mathbf{W}) : \mathbb{D}(y, \varphi_{0,i})\rangle_{Y_f(\mathbf{x})} - \rho_f \int_0^t \mathbf{F}(\mathbf{x}, \tau)d\tau \cdot \mathbf{e}_i.$$

Nguetseng's theorem guarantees the inclusion

$$\mathbb{D}(y, \mathbf{W}) \in \mathbf{L}_2(\Omega_T \times Y).$$

Therefore

$$\frac{\partial \pi_f}{\partial x_i}(\mathbf{x}_0) = \Phi_i(\mathbf{x}_0),\ i = 1, 2, 3,$$

and

$$\nabla \pi_f = \big(\Phi_1, \Phi_2, \Phi_3\big) \in \mathbf{L}_2(\Omega_T).$$

after reintegrating (5.3.72) we arrive at the equation

$$\mu_1 \nabla_y \cdot \mathbb{D}(y, \mathbf{W}) - \nabla_y \Psi - \nabla \pi_f + \rho_f \int_0^t \mathbf{F}(\mathbf{x}, \tau) d\tau = 0 \tag{5.3.74}$$

in the domain $Y_f(\mathbf{x})$, which together with continuity equation (5.3.71) and the boundary condition

$$\mathbf{W}(\mathbf{x}, t, \mathbf{y}) = \mathbf{w}(\mathbf{x}, t), \quad \mathbf{y} \in \gamma(\mathbf{x}) = Y_f(\mathbf{x}) \cap Y_s(\mathbf{x}) \tag{5.3.75}$$

result in Darcy's law (5.3.37).

Chapter 6
Isothermal Liquid Filtration

The mathematical model $\mathbb{M}_{20}$ consists of the differential equations

$$\frac{1}{\bar{\alpha}_p^{\varepsilon}}\, p + \nabla \cdot \mathbf{w} = 0, \;\; (\mathbf{x}, t) \in \Omega_T = \Omega \times (0, T), \tag{6.0.1}$$

$$\rho^{\varepsilon} \frac{\partial^2 \mathbf{w}}{\partial t^2} = \nabla \cdot \mathbb{P} + \rho^{\varepsilon} \mathbf{F}, \;\; (\mathbf{x}, t) \in \Omega_T, \tag{6.0.2}$$

$$\mathbb{P} = \chi^{\varepsilon} \bar{\alpha}_{\mu} \mathbb{D}\Big(x, \frac{\partial \mathbf{w}}{\partial t}\Big) + (1 - \chi^{\varepsilon}) \bar{\alpha}_{\lambda} \mathbb{D}(x, \mathbf{w}) - \Big(p - \chi^{\varepsilon} \bar{\alpha}_{\nu} \nabla \cdot \frac{\partial \mathbf{w}}{\partial t}\Big) \mathbb{I}, \tag{6.0.3}$$

$$\mathbf{w}(\mathbf{x}, t) = 0, \;\; (\mathbf{x}, t) \in S_T = S \times (0, T), \tag{6.0.4}$$

$$\mathbf{w}(\mathbf{x}, 0) = \frac{\partial \mathbf{w}}{\partial t}(\mathbf{x}, 0) = 0, \mathbf{x} \in \Omega, \tag{6.0.5}$$

and the model $\mathbb{M}_{21}$ consists of the differential equations

$$\chi^{\varepsilon} \Big(\frac{1}{\bar{c}_f^2}\, p + \nabla \cdot \mathbf{w}\Big) = 0, (\mathbf{x}, t) \in \Omega_T, \tag{6.0.6}$$

$$\rho_f \frac{\partial^2 \mathbf{w}}{\partial t^2} = \chi^{\varepsilon} \Big(\nabla \cdot \mathbb{P} + \rho_f \mathbf{F}\Big), (\mathbf{x}, t) \in \Omega_T, \tag{6.0.7}$$

$$\mathbb{P} = \bar{\alpha}_{\mu} \mathbb{D}\Big(x, \frac{\partial \mathbf{w}}{\partial t}\Big) - \Big(p - \bar{\alpha}_{\nu} \nabla \cdot \frac{\partial \mathbf{w}}{\partial t}\Big) \mathbb{I}, \tag{6.0.8}$$

in the bounded domain $\Omega = \Omega_f^{\varepsilon} \cup \Gamma \cup \Omega_s^{\varepsilon} \subset \mathbb{R}^3$, $\Gamma^{\varepsilon} = \partial \Omega_f^{\varepsilon} \cap \partial \Omega_s^{\varepsilon}$, with a C^2 continuous boundary $S = \partial \Omega$ for $t \in (0, T)$.

A. Meirmanov, *Mathematical Models for Poroelastic Flows*,
Atlantis Studies in Differential Equations 1, DOI: 10.2991/978-94-6239-015-7_6,

Recall that in (6.0.1)–(6.0.8) the characteristic function $\chi^{\varepsilon}(\mathbf{x})$ of the domain Ω_f^{ε} is given by the expression

$$\chi^{\varepsilon}(\mathbf{x}) = \zeta(\mathbf{x})\chi\left(\frac{\mathbf{x}}{\varepsilon}\right), \tag{6.0.9}$$

where $\zeta(\mathbf{x})$ is the characteristic function of the domain Ω, $\chi(\mathbf{y})$ is the characteristic function of the domain Y_f, and

$$\bar{\alpha}_p^{\varepsilon} = \chi^{\varepsilon}\bar{c}_f^2 + (1-\chi^{\varepsilon})\bar{c}_s^2, \;\; \rho^{\varepsilon} = \chi^{\varepsilon}\rho_f + (1-\chi^{\varepsilon})\rho_s.$$

For the definition of $\bar{\alpha}_{\mu}$, $\bar{\alpha}_{\nu}$, $\bar{\alpha}_{\lambda}$, $\bar{c}_f$, and $\bar{c}_s$ see Appendix A.

As usual, the function $p_f^{\varepsilon} = \chi^{\varepsilon}p^{\varepsilon}$ stands for the liquid pressure, and the function $p_s^{\varepsilon} = (1-\chi^{\varepsilon})p^{\varepsilon}$ stands for the solid pressure.

We also assume that all dimensionless parameters depend on the small parameter ε and that the (finite or infinite) limits exist:

$$\lim_{\varepsilon\searrow 0}\bar{\alpha}_{\mu}(\varepsilon) = \mu_0, \quad \lim_{\varepsilon\searrow 0}\bar{\alpha}_{\nu}(\varepsilon) = \nu_0, \quad \lim_{\varepsilon\searrow 0}\bar{\alpha}_{\lambda}(\varepsilon) = \lambda_0,$$

$$\lim_{\varepsilon\searrow 0}\frac{\bar{\alpha}_{\mu}}{\varepsilon^2} = \mu_1, \quad \lim_{\varepsilon\searrow 0}\frac{\bar{\alpha}_{\lambda}}{\varepsilon^2} = \lambda_1.$$

In the following sections we will find homogenized equations of acoustics for

(I) a slightly viscous liquid in an absolutely rigid solid skeleton: $\mu_0 = 0, \; \lambda_0 = \infty$,
(II) a slightly viscous liquid in an extremely elastic solid skeleton: $\mu_0 = 0, \; \lambda_0 = 0$, and
(III) a slightly viscous liquid in an elastic solid skeleton: $\mu_0 = 0, \; 0 < \lambda_0 < \infty$.

Throughout this chapter it is assumed that

$$\int_{Q_T}\left(|\mathbf{F}(\mathbf{x},t)|^2 + \left|\frac{\partial \mathbf{F}}{\partial t}(\mathbf{x},t)\right|^2\right)dxdt = F^2 < \infty,$$

and that Assumptions 0.1, 1.1 and 3.1 hold true.

Definition 6.1 We say that the pair of functions $\{\mathbf{w}^{\varepsilon}, \; p^{\varepsilon}\}$ such that

$$\mathbf{w}^{\varepsilon} \in \overset{\circ}{\mathbf{W}}{}_2^{1,0}(\Omega_T), \;\; \chi^{\varepsilon}\nabla\frac{\partial \mathbf{w}^{\varepsilon}}{\partial t} \in \mathbf{L}_2(\Omega_T), \;\; p^{\varepsilon} \in L_2(\Omega_T),$$

is a weak solution of the problem (6.0.1)–(6.0.5), if it satisfies the continuity equation (6.0.1) almost everywhere in Ω_T, the first initial condition in (6.0.5), and the integral identity

$$\int_{\Omega_T}\left(-\rho^{\varepsilon}\frac{\partial \mathbf{w}^{\varepsilon}}{\partial t}\cdot\frac{\partial \varphi}{\partial t} + \mathbb{P} : \mathbb{D}(x,\varphi)\right)dxdt = -\int_{\Omega_T}\rho^{\varepsilon}\mathbf{F}\cdot\varphi dxdt \tag{6.0.10}$$

for all functions φ vanishing at $t = T$, $t = 0$ and S_T, such that $\varphi \in \overset{\circ}{\mathbf{W}}{}_2^{1,0}(\Omega_T)$, $\frac{\partial \varphi}{\partial t} \in \mathbf{L}_2(\Omega_T)$.

Theorem 6.1 *For all $\varepsilon > 0$ and for any time interval $[0, T]$ there exists a unique generalized solution of the problem (6.0.1)–(6.0.5) and*

$$\begin{aligned}
&\max_{0<t<T} \int_\Omega \Big(|p^\varepsilon(\mathbf{x}, t)|^2 + \Big| \frac{\partial \mathbf{w}^\varepsilon}{\partial t}(\mathbf{x}, t) \Big|^2 + (1 - \chi^\varepsilon) \bar{\alpha}_\lambda |\mathbb{D}(x, \mathbf{w}^\varepsilon)|^2 \Big) dx \\
&+ \max_{0<t<T} \int_\Omega \Big(\Big| \frac{\partial p^\varepsilon}{\partial t}(\mathbf{x}, t) \Big|^2 + \Big| \frac{\partial^2 \mathbf{w}^\varepsilon}{\partial t^2}(\mathbf{x}, t) \Big|^2 + (1 - \chi^\varepsilon) \bar{\alpha}_\lambda \Big| \mathbb{D}\Big(x, \frac{\partial \mathbf{w}^\varepsilon}{\partial t}\Big) \Big|^2 \Big) dx \\
&+ \int_{\Omega_T} \chi^\varepsilon \Big(\bar{\alpha}_\mu \Big| \mathbb{D}\Big(x, \frac{\partial \mathbf{w}^\varepsilon}{\partial t}\Big) \Big|^2 + \bar{\alpha}_\nu \Big| \nabla \cdot \frac{\partial \mathbf{w}^\varepsilon}{\partial t} \Big|^2 \Big) dx dt \\
&+ \int_{\Omega_T} \chi^\varepsilon \Big(\bar{\alpha}_\mu \Big| \mathbb{D}\Big(x, \frac{\partial^2 \mathbf{w}^\varepsilon}{\partial t^2}\Big) \Big|^2 + \bar{\alpha}_\nu \Big| \nabla \cdot \frac{\partial^2 \mathbf{w}^\varepsilon}{\partial t^2} \Big|^2 \Big) dx dt \leqslant C_0 F^2,
\end{aligned} \tag{6.0.11}$$

where here and in what follows, we denote as C_0 any constant depending only on domains Ω, Y and Y_f.

The proof of this theorem repeats the proofs of similar theorems in the previous chapters and is based on the energy equalities

$$\begin{aligned}
&\frac{1}{2} \frac{d}{dt} \int_\Omega \Big(\rho^\varepsilon \Big| \frac{\partial \mathbf{w}^\varepsilon}{\partial t} \Big|^2 + (1 - \chi^\varepsilon) \bar{\alpha}_\lambda \mathbb{D}(x, \mathbf{w}^\varepsilon) : \mathbb{D}(x, \mathbf{w}^\varepsilon) + \frac{1}{\bar{\alpha}_p^\varepsilon} |p^\varepsilon|^2 \Big) dx \\
&+ \int_\Omega \chi^\varepsilon \Big(\bar{\alpha}_\mu \mathbb{D}\Big(x, \frac{\partial \mathbf{w}^\varepsilon}{\partial t}\Big) : \mathbb{D}\Big(x, \frac{\partial \mathbf{w}^\varepsilon}{\partial t}\Big) + \bar{\alpha}_\nu \Big(\nabla \cdot \frac{\partial \mathbf{w}^\varepsilon}{\partial t} \Big)^2 \Big) dx \\
&= \int_\Omega \rho^\varepsilon \mathbf{F} \cdot \frac{\partial \mathbf{w}^\varepsilon}{\partial t} dx,
\end{aligned}$$

$$\begin{aligned}
&\frac{1}{2} \frac{d}{dt} \int_\Omega \Big(\rho^\varepsilon \Big| \frac{\partial^2 \mathbf{w}^\varepsilon}{\partial t^2} \Big|^2 + (1 - \chi^\varepsilon) \bar{\alpha}_\lambda \mathbb{D}\Big(x, \frac{\partial \mathbf{w}^\varepsilon}{\partial t}\Big) : \mathbb{D}\Big(x, \frac{\partial \mathbf{w}^\varepsilon}{\partial t}\Big) + \frac{1}{\bar{\alpha}_p^\varepsilon} \Big| \frac{\partial p^\varepsilon}{\partial t} \Big|^2 \Big) dx \\
&+ \int_\Omega \chi^\varepsilon \Big(\bar{\alpha}_\mu \mathbb{D}\Big(x, \frac{\partial^2 \mathbf{w}^\varepsilon}{\partial t^2}\Big) : \mathbb{D}\Big(x, \frac{\partial^2 \mathbf{w}^\varepsilon}{\partial t^2}\Big) + \bar{\alpha}_\nu \Big(\nabla \cdot \frac{\partial^2 \mathbf{w}^\varepsilon}{\partial t^2} \Big)^2 \Big) dx \\
&= \int_\Omega \rho^\varepsilon \frac{\partial \mathbf{F}}{\partial t} \cdot \frac{\partial^2 \mathbf{w}^\varepsilon}{\partial t^2} dx,
\end{aligned}$$

For example, the first equality follows from the Eq. (6.0.2), if we express the stress tensor $\mathbb{P}$ and the pressure p^ε there using state equations (6.0.3) and continuity equation (6.0.1), multiply the result by $\dfrac{\partial \mathbf{w}^\varepsilon}{\partial t}$ and integrate by parts.

6.1 A Compressible Slightly Viscous Liquid in an Absolutely Rigid Skeleton

In this section as a basic mathematical model at the microscopic level we consider the model $\mathbb{M}_{21}$ of the motion of a compressible liquid in an absolutely rigid solid skeleton, where

$$\mathbf{w}(\mathbf{x},t)=0,\quad p(\mathbf{x},t)=0,\quad \mathbf{x}\in\Omega_s^{\varepsilon}.$$

If we put $\mathbf{v}=\dfrac{\partial\mathbf{w}}{\partial t}$, then we may rewrite the last condition and Eqs. (6.0.6)–(6.0.8) in the form

$$\frac{1}{\bar{c}_f^2}\frac{\partial p}{\partial t}+\nabla\cdot\mathbf{v}=0, \mathbf{x}\in\Omega,\ t\in(0,T), \tag{6.1.1}$$

$$\rho_f\frac{\partial\mathbf{v}}{\partial t}=\nabla\cdot\mathbb{P}+\rho_f\mathbf{F}, \mathbf{x}\in\Omega_f^{\varepsilon},\ t\in(0,T), \tag{6.1.2}$$

$$\mathbb{P}=\bar{\alpha}_\mu\mathbb{D}\big(x,\mathbf{v}\big)+\big(\bar{\alpha}_\nu\nabla\cdot\mathbf{v}-p\big)\mathbb{I}, \tag{6.1.3}$$

$$\mathbf{v}(\mathbf{x},t)=0,\quad p(\mathbf{x},t)=0,\quad \mathbf{x}\in\Omega_s^{\varepsilon}\cup S,\ t\in(0,T), \tag{6.1.4}$$

$$\mathbf{v}(\mathbf{x},0)=0,\quad p(\mathbf{x},0)=0,\quad \mathbf{x}\in\Omega. \tag{6.1.5}$$

Throughout this section we assume that conditions

$$\mu_0=0,\ 0\leqslant\mu_1<\infty,\ 0<c_f<\infty,\ 0\leqslant\nu_0<\infty$$

hold true.

6.1.1 Statement of the Problem and Main Results

Definition 6.2 We say that the pair of functions $\{\mathbf{v}^{\varepsilon},\ p^{\varepsilon}\}$ such that

$$\mathbf{v}^{\varepsilon}\in\overset{\circ}{\mathbf{W}}{}_2^{1,0}(\Omega_T),\ p^{\varepsilon}\in L_2(\Omega_T),$$

is a weak solution of the problem (6.1.1)–(6.1.5), if it satisfies condition (6.1.4) and integral identities

$$\int_{\Omega_T} \chi^\varepsilon \Big(\bar{\alpha}_\mu \mathbb{D}(x, \mathbf{v}^\varepsilon) : \mathbb{D}(x, \varphi) + \big(\bar{\alpha}_\nu \nabla \cdot \mathbf{v}^\varepsilon - p^\varepsilon\big) \nabla \cdot \varphi\Big) dx dt$$
$$= \int_{\Omega_T} \chi^\varepsilon \rho_f \Big(\frac{\partial \varphi}{\partial t} \cdot \mathbf{v}^\varepsilon + \mathbf{F} \cdot \varphi\Big) dx dt, \tag{6.1.6}$$

$$\int_{\Omega_T} \Big(\nabla \xi \cdot \mathbf{v}^\varepsilon + \frac{1}{\bar{c}_f^2} \frac{\partial \xi}{\partial t} p^\varepsilon\Big) dx dt = 0, \tag{6.1.7}$$

for any smooth functions φ and ξ, such that φ and ξ satisfy condition (6.1.4), and conditions $\varphi(\mathbf{x}, T) = 0, \xi(\mathbf{x}, T) = 0$.

Theorem 6.2 *For all $\varepsilon > 0$ and for an arbitrary time interval $[0, T]$ there exists a unique generalized solution of problem (6.1.1)–(6.1.5) and*

$$\int_{\Omega_T} \Big(\bar{\alpha}_\mu |\nabla \mathbf{v}^\varepsilon|^2 + \bar{\alpha}_\nu |\nabla \cdot \mathbf{v}^\varepsilon|^2\Big) dx dt$$
$$+ \max_{0<t<T} \int_\Omega \Big(|p^\varepsilon(\mathbf{x}, t)|^2 + |\mathbf{v}^\varepsilon(\mathbf{x}, t)|^2\Big) dx \leqslant C_0 F^2. \tag{6.1.8}$$

The proof of this theorem repeats the proofs of similar theorems in the previous chapters and is based on the energy equality

$$\frac{1}{2} \frac{d}{dt} \int_\Omega \Big(\rho_f |\mathbf{v}^\varepsilon|^2 + \frac{1}{\bar{c}_f^2} |p^\varepsilon|^2\Big) dx + \int_\Omega \Big(\bar{\alpha}_\mu \mathbb{D}(x, \mathbf{v}^\varepsilon) : \mathbb{D}(x, \mathbf{v}) + \bar{\alpha}_\nu (\nabla \cdot \mathbf{v}^\varepsilon)^2\Big) dx$$
$$= \int_\Omega \rho_f \mathbf{F} \cdot \mathbf{v}^\varepsilon dx.$$

Theorem 6.3 *Let $\{\mathbf{v}^\varepsilon, p^\varepsilon\}$ be the weak solution of the problem (6.1.1)–(6.1.5), $q^\varepsilon = p^\varepsilon - \bar{\alpha}_\nu (\nabla \cdot \mathbf{v}^\varepsilon)$, and*

$$\mu_1 > 0, \quad \nu_0 \geqslant 0.$$

Then for $\nu_0 > 0$ the sequences $\{\mathbf{v}^\varepsilon\}$, $\{\nabla \cdot \mathbf{v}^\varepsilon\}$, $\{q^\varepsilon\}$ and $\{p^\varepsilon\}$ converge weakly in $\mathbf{L}_2(\Omega_T)$ and $L_2(\Omega_T)$ as $\varepsilon \leftarrow 0$ to functions $\mathbf{v}$, $\nabla \cdot \mathbf{v}$, q, and p respectively.

These limiting functions, where $q = p - \nu_0 (\nabla \cdot \mathbf{v}) \in W_2^{1,0}(\Omega_T)$, solve the homogenized system of equations in the domain Ω for $t \in (0, T)$, consisting of the continuity equation

$$\frac{1}{\bar{c}_f^2} \frac{\partial p}{\partial t} + \nabla \cdot \mathbf{v} = 0, \tag{6.1.9}$$

and the dynamic equation in the form

$$\mathbf{v} = \int_0^t \mathbb{B}^{(f)}(\mu_1, \infty; t - \tau) \cdot \Big(-\nabla\big(p - \nu_0 (\nabla \cdot \mathbf{v})\big) + \rho_f\Big)(\mathbf{x}, \tau) d\tau. \tag{6.1.10}$$

For $\nu_0 = 0$ the sequences $\{\mathbf{v}^\varepsilon\}$ and $\{p^\varepsilon\}$ converge weakly in $\mathbf{L}_2(\Omega_T)$ and $L_2(\Omega_T)$ as $\varepsilon \leftarrow 0$ to functions $\mathbf{v}$ and $p \in W_2^{1,0}(\Omega_T)$ respectively and these limiting functions solve the homogenized system of equations, consisting of the continuity equation (6.1.9) and the dynamic equation in the form

$$\mathbf{v} = \int_0^t \mathbb{B}^{(f)}(\mu_1, \infty; t-\tau) \cdot \big(-\nabla p + \rho_f \mathbf{F} \big)(\mathbf{x}, \tau) d\tau. \tag{6.1.11}$$

Equations (6.1.9), (6.1.10) and (6.1.9), (6.1.11) are completed with boundary and initial conditions

$$\mathbf{v}(\mathbf{x}, t) \cdot \mathbf{n}(\mathbf{x}) = 0, \;\; \mathbf{x} \in S, \; t \in (0, T), \tag{6.1.12}$$

$$p(\mathbf{x}, 0) = \nu_0 \frac{\partial p}{\partial t}(\mathbf{x}, 0) = 0, \mathbf{x} \in \Omega. \tag{6.1.13}$$

The matrix $\mathbb{B}^{(f)}(\mu_1, \infty; t)$ has been defined in Chap. 3 by the formula (3.2.70).

Theorem 6.4 *Let $\{\mathbf{v}^\varepsilon, \, p^\varepsilon\}$ be the weak solution of the problem (6.1.1)–(6.1.5), $q^\varepsilon = p^\varepsilon - \bar{\alpha}_\nu (\nabla \cdot \mathbf{v}^\varepsilon)$, and*

$$\mu_1 = 0, \quad \nu_0 \geqslant 0.$$

Then for $\nu_0 > 0$ the sequences $\{\mathbf{v}^\varepsilon\}$, $\{\nabla \cdot \mathbf{v}^\varepsilon\}$, $\{q^\varepsilon\}$, and $\{p^\varepsilon\}$ converge weakly in $\mathbf{L}_2(\Omega_T)$ and $L_2(\Omega_T)$ to functions $\mathbf{v}$, $\nabla \cdot \mathbf{v}$, q, and p respectively.

These limiting functions, where $q = p - \nu_0(\nabla \cdot \mathbf{v}) \in W_2^{1,0}(\Omega_T)$, solve the homogenized system of equations in the domain Ω for $t \in (0, T)$, consisting of the continuity equation (6.1.9) and the dynamic equation in the form

$$\frac{\partial \mathbf{v}}{\partial t} = \mathbb{B}^{(f)}(0, \infty) \cdot \Big(-\frac{1}{m} \nabla \Big(p - \nu_0 (\nabla \cdot \mathbf{v}) \Big) + \rho_f \mathbf{F} \Big). \tag{6.1.14}$$

For $\nu_0 = 0$ the sequences $\{\mathbf{v}^\varepsilon\}$ and $\{p^\varepsilon\}$ converge weakly in $\mathbf{L}_2(\Omega_T)$ and $L_2(\Omega_T)$ as $\varepsilon \to 0$ to functions $\mathbf{v}$ and $p \in W_2^{1,0}(\Omega_T)$ respectively and these limiting functions solve the homogenized system of equations, consisting of the continuity equation (6.1.9) and the dynamic equation in the form

$$\frac{\partial \mathbf{v}}{\partial t} = \mathbb{B}^{(f)}(0, \infty) \cdot \Big(-\frac{1}{m} \nabla p + \rho_f \mathbf{F} \Big). \tag{6.1.15}$$

Equations (6.1.9), (6.1.14) and (6.1.9), (6.1.15) are completed with boundary and initial conditions (6.1.12) and (6.1.13).

The symmetric and strictly positively definite constant matrix $\mathbb{B}^{(f)}(0, \infty)$ has been defined in Chap. 3 by formula (3.2.76).

Problems (6.1.9), (6.1.11)–(6.1.14) and (6.1.9), (6.1.12), (6.1.13), (6.1.15) have unique solutions.

We refer to these described problems as the homogenized **models** $(\mathbb{IA})_1$ $(\mu_1 > 0,\ \nu_0 > 0)$, $(\mathbb{IA})_2$ $(\mu_1 > 0,\ \nu_0 = 0)$, $(\mathbb{IA})_3$ $(\mu_1 = 0,\ \nu_0 > 0)$, and $(\mathbb{IA})_4$ $(\mu_1 = 0,\ \nu_0 = 0)$ of isothermal acoustics in an absolutely rigid body.

6.1.2 Proofs of Theorems 6.2–6.4

The main parts of all these proofs repeat the similar proofs from Chaps. 1 and 3. The only differences here are the macro- and microscopic equations.

So, we may assume, that the sequence $\{\mathbf{v}^\varepsilon\}$ converges two-scale and weakly in $\mathbf{L}_2(\Omega_T)$ to functions $\mathbf{V}(\mathbf{x}, t, \mathbf{y})$ and $\mathbf{v}(\mathbf{x}, t) = \langle \mathbf{V}\rangle_Y$ respectively, and the sequence $\{p^\varepsilon\}$ weakly converges in $L_2(\Omega_T)$ to function p. At the same time the sequence $\{\nabla\cdot\mathbf{v}^\varepsilon\}$ for $\nu_0 > 0$ weakly converges in $\mathbf{L}_2(\Omega_T)$ to the function $\nabla\cdot\mathbf{v}$ and the sequence $\{\bar{\alpha}_\nu\nabla\cdot\mathbf{v}^\varepsilon\}$ for $\nu_0 = 0$ converges strongly in $\mathbf{L}_2(\Omega_T)$ to zero.

The limiting functions satisfy the macroscopic continuity equation (6.1.9) in Ω_T.

For $\nu_0 > 0$ and $\mu_1 > 0$ the limiting functions satisfy the microscopic dynamic equation

$$\rho_f\frac{\partial\mathbf{V}}{\partial t} = \frac{\mu_1}{2}\triangle_y\mathbf{V} - \nabla_y\Pi + \nabla\big(-p + \nu_0(\nabla\cdot\mathbf{v})\big) + \rho_f\mathbf{F}. \tag{6.1.16}$$

For $\nu_0 = 0$ and $\mu_1 > 0$ the limiting functions satisfy the microscopic dynamic equation

$$\rho_f\frac{\partial\mathbf{V}}{\partial t} = \frac{\mu_1}{2}\triangle_y\mathbf{V} - \nabla_y\Pi - \nabla\ p + \rho_f\mathbf{F}. \tag{6.1.17}$$

For $\nu_0 > 0$ and $\mu_1 = 0$ the limiting functions satisfy the microscopic dynamic equation

$$\rho_f\frac{\partial\mathbf{V}}{\partial t} = \nabla_y\Pi + \nabla\big(-p + \nu_0(\nabla\cdot\mathbf{v})\big) + \rho_f\mathbf{F}, \tag{6.1.18}$$

Finally, for $\nu_0 = 0$ and $\mu_1 = 0$ the limiting functions satisfy the microscopic dynamic equation

$$\rho_f\frac{\partial\mathbf{V}}{\partial t} = \nabla_y\Pi - \nabla\ p + \rho_f\mathbf{F}, \tag{6.1.19}$$

For all cases the limiting functions satisfy the microscopic continuity equation

$$\nabla_y\cdot\mathbf{V} = 0 \tag{6.1.20}$$

in Y, and the condition

$$\mathbf{V}(\mathbf{x}, t, \mathbf{y}) = 0\ \text{ for } \mathbf{y}\in Y_s. \tag{6.1.21}$$

Recall that all equations are understood in the sense of distributions. For example, we must complete the microscopic continuity equation (6.1.20) with the boundary condition

$$[\mathbf{V}] \cdot \mathbf{n} = 0, \quad \mathbf{y} \in \gamma, \tag{6.1.22}$$

where $\mathbf{n}$ is a normal vector to the boundary γ. Relations (6.1.21) and (6.1.22) give us the condition

$$\mathbf{V} \cdot \mathbf{n} = 0, \quad \mathbf{y} \in \gamma, \tag{6.1.23}$$

which we will use as a boundary condition for the case $\mu_0 = 0$.

For the case $\mu_0 > 0$ condition (6.1.21) and the imbedding $\nabla \mathbf{V} \in \mathbf{L}_2(\Omega_T \times Y)$ result in

$$\mathbf{V}(\mathbf{x}, t, \mathbf{y}) = 0, \quad \mathbf{y} \in \gamma. \tag{6.1.24}$$

The problem (6.1.16), (6.1.20), (6.1.24), the problem (6.1.17), (6.1.20), (6.1.24), the problem (6.1.18), (6.1.20), (6.1.23), and the problem (6.1.19), (6.1.20), (6.1.23), completed with initial condition

$$\mathbf{V}(\mathbf{x}, \mathbf{y}, 0) = 0, \quad \mathbf{y} \in Y, \tag{6.1.25}$$

have been considered in Chap. 3 (proof of Theorem 3.5).

It is clear that the problems (6.1.9), (6.1.11)–(6.1.14) and (6.1.9), (6.1.12), (6.1.13), (6.1.15) are reduced to linear hyperbolic equations for the pressure. Therefore, the uniqueness of these problems follow from the properties of the matrix $\mathbb{B}^{(f)}(0, \infty)$.

6.2 A Compressible Slightly Viscous Liquid in a Compressible Extremely Elastic Skeleton

Throughout this section we assume that

$$\mu_0 = 0, \ \lambda_0 = 0. \tag{6.2.1}$$

6.2.1 Main Results

Theorem 6.5 *Let $\{\mathbf{v}^\varepsilon, p^\varepsilon\}$ be the weak solution of the problem (6.0.1)–(6.0.5) and*

$$\mu_1 = \lambda_1 = \infty.$$

Then the sequence $\{p^\varepsilon\}$ converges weakly in $L_2(\Omega_T)$ as $\varepsilon \to 0$ to the limiting pressure $p(\mathbf{x}, t)$ of the mixture, which satisfies the initial boundary-value problem

$$\frac{\hat{\rho}}{\tilde{c}^2}\frac{\partial^2 p}{\partial t^2} = \triangle\,\tilde{p} - \hat{\rho}\,\nabla\cdot\mathbf{F},\ \ \mathbf{x}\in\Omega,\ t>0, \tag{6.2.2}$$

$$\big(\nabla\,\tilde{p} - \hat{\rho}\mathbf{F}\big)\cdot\mathbf{n}(\mathbf{x}) = 0,\ \ \mathbf{x}\in S,\ t>0, \tag{6.2.3}$$

$$p(\mathbf{x},0) = \frac{\partial p}{\partial t}(\mathbf{x},0) = 0,\ \ \mathbf{x}\in\Omega. \tag{6.2.4}$$

In (6.2.2)–(6.2.4)

$$\tilde{p}(\mathbf{x},t) = p(\mathbf{x},t) + m\,\frac{\nu_0}{\bar{c}_f^2}\frac{\partial p}{\partial t}(\mathbf{x},t), \tag{6.2.5}$$

$$\hat{\rho} = m\rho_f + (1-m)\rho_s,\quad \frac{1}{\tilde{c}^2} = \frac{m}{\bar{c}_f^2} + \frac{(1-m)}{\bar{c}_s^2},$$

and $\mathbf{n}(\mathbf{x})$ *is the normal vector to the boundary* S *at the point* $\mathbf{x}\in S$.

For $\nu_0 = 0$ *the limiting pressure* p *of the mixture is given by formula*

$$p = \tilde{p}, \tag{6.2.6}$$

and satisfies the initial boundary-value problem

$$\frac{\hat{\rho}}{\tilde{c}^2}\frac{\partial^2 p}{\partial t^2} = \triangle\,p - \hat{\rho}\,(\nabla\cdot\mathbf{F}),\ \ \mathbf{x}\in\Omega,\ t>0, \tag{6.2.7}$$

$$\big(\nabla p - \hat{\rho}\mathbf{F}\big)\cdot\mathbf{n}(\mathbf{x}) = 0,\ \ \mathbf{x}\in S,\ t>0, \tag{6.2.8}$$

$$p(\mathbf{x},0) = \frac{\partial p}{\partial t}(\mathbf{x},0) = 0,\ \ \mathbf{x}\in\Omega \tag{6.2.9}$$

for the wave equation.

We refer to the problem (6.2.2)–(6.2.4) as the homogenized **model** $(\mathbb{IA})_5$ and to the problem (6.2.7)–(6.2.9) as the homogenized **model** $(\mathbb{IA})_6$.

Theorem 6.6 *Let* $\{\mathbf{w}^\varepsilon,\ p^\varepsilon\}$ *be the weak solution of the problem (6.0.1)–(6.0.5) and*

$$0\leqslant \mu_1,\ \lambda_1 < \infty.$$

Then the sequence $\{p_f^\varepsilon\}$ *converges weakly in* $L_2(\Omega_T)$ *as* $\varepsilon\to 0$ *to the function* p, *where* $p\in W_2^{1,0}(\Omega_T)$, $\nu_0\nabla\big(\frac{\partial p}{\partial t}\big)\in\mathbf{W}_2^{1,0}(\Omega_T)$, *and this limiting pressure* p *of the mixture satisfies the initial boundary-value problem consisting of the homogenized equation*

$$\nabla \cdot \int_0^t \left(\frac{\nu_0}{\bar{c}_f^2} \mathbb{B}_0^{(a)}(\mu_1, \lambda_1; t-\tau) \cdot \nabla \left(\frac{\partial p}{\partial t} \right) + \mathbb{B}^{(a)}(\mu_1, \lambda_1; t-\tau) \cdot \nabla p \right)(\mathbf{x}, \tau) d\tau$$
$$= -\frac{\hat{\rho}}{\tilde{c}^2} \frac{\partial p}{\partial t} - \nabla \cdot \mathbf{f} \tag{6.2.10}$$

in the domain Ω_T, the boundary condition

$$\int_0^t \left(\frac{\nu_0}{\bar{c}_f^2} \mathbb{B}_0^{(a)}(\mu_1, \lambda_1; t-\tau) \cdot \nabla \left(\frac{\partial p}{\partial t} \right) + \mathbb{B}^{(a)}(\mu_1, \lambda_1; t-\tau) \cdot \nabla p \right)(\mathbf{x}, \tau) d\tau \cdot \mathbf{n}(\mathbf{x})$$
$$= -\mathbf{f}(\mathbf{x}, t) \cdot \mathbf{n}(\mathbf{x}) \tag{6.2.11}$$

on the boundary S_T, and the initial condition

$$p\,(\mathbf{x}, 0) = (\mathbf{x}, 0) = 0, \ \ \mathbf{x} \in \Omega. \tag{6.2.12}$$

In (6.2.10), (6.2.11) matrices $\mathbb{B}^{(a)}(\mu_1, \lambda_1; t)$ and $\mathbb{B}_0^{(a)}(\mu_1, \lambda_1; t)$, and the function $\mathbf{f}(\mathbf{x}, t)$ are given below by formulae (6.2.40)–(6.2.42), and $\mathbf{n}(\mathbf{x})$ is the normal vector to the boundary S at the point $\mathbf{x} \in S$.

For $\nu_0 = 0$ the limiting pressure p of the mixture satisfies the homogenized equation

$$\frac{\hat{\rho}}{\tilde{c}^2} \frac{\partial p}{\partial t} + \nabla \cdot \int_0^t \mathbb{B}^{(a)}(\mu_1, \lambda_1; t-\tau) \cdot \nabla p(\mathbf{x}, \tau) d\tau + \nabla \cdot \mathbf{f} = 0 \tag{6.2.13}$$

in the domain Ω_T, the boundary condition

$$\int_0^t \mathbb{B}^{(a)}(\mu_1, \lambda_1; t-\tau) \cdot \nabla p\,(\mathbf{x}, \tau) d\tau \cdot \mathbf{n}(\mathbf{x}) = -\mathbf{f}\,(\mathbf{x}, t) \cdot \mathbf{n}(\mathbf{x}) \tag{6.2.14}$$

on the boundary S_T, and initial conditions (6.2.12).

We refer to the problem (6.2.10)–(6.2.12) as the homogenized **model** $(\mathbb{IA})_7$, and to the problem (6.2.12)–(6.2.14) as the homogenized **model** $(\mathbb{IA})_8$.

To formulate the following statements we consider extensions

$$\mathbf{w}_f^{\varepsilon} = \mathbb{E}_{\Omega_f^{\varepsilon}}(\mathbf{w}^{\varepsilon}), \ \text{ and } \mathbf{w}_s^{\varepsilon} = \mathbb{E}_{\Omega_s^{\varepsilon}}(\mathbf{w}^{\varepsilon}),$$

where

$$\mathbb{E}_{\Omega_f^{\varepsilon}} : \mathbf{W}_2^1(\Omega_f^{\varepsilon}) \to \mathbf{W}_2^1(\Omega)$$

is an extension operator from Ω_f^{ε} on Ω, and

$$\mathbb{E}_{\Omega_s^{\varepsilon}} : \mathbf{W}_2^1(\Omega_s^{\varepsilon}) \to \mathbf{W}_2^1(\Omega)$$

is an extension operator from Ω_s^ε on Ω, such that

$$\mathbf{w}_f^\varepsilon = \mathbf{w}^\varepsilon \ \text{ in } \ \Omega_f^\varepsilon \times (0,T), \quad \mathbf{w}_s^\varepsilon = \mathbf{w}^\varepsilon \text{ in } \Omega_s^\varepsilon \times (0,T),$$

and

$$\int_\Omega |\mathbf{w}_f^\varepsilon|^2 dx \leqslant C_0 \int_{\Omega_f^\varepsilon} |\mathbf{w}^\varepsilon|^2 dx, \quad \int_\Omega |\mathbf{w}_s^\varepsilon|^2 dx \leqslant C_0 \int_{\Omega_s^\varepsilon} |\mathbf{w}^\varepsilon|^2 dx,$$
$$\int_\Omega |\mathbb{D}(x,\mathbf{w}_f^\varepsilon)|^2 dx \leqslant C_0 \int_{Q_f^\varepsilon} |\mathbb{D}(x,\mathbf{w}^\varepsilon)|^2 dx,$$
$$\int_\Omega |\mathbb{D}(x,\mathbf{w}_s^\varepsilon)|^2 dx \leqslant C_0 \int_{\Omega_s^\varepsilon} |\mathbb{D}(x,\mathbf{w}^\varepsilon)|^2 dx. \qquad (6.2.15)$$

(for more details see Appendix B, Lemma B.9).

Theorem 6.7 *Let $\{\mathbf{w}^\varepsilon,\ p^\varepsilon\}$ be the weak solution of the problem (6.0.1)–(6.0.5),*

$$\mu_1 = \infty, \ 0 \leqslant \lambda_1 < \infty,$$

and $\mathbf{w}_f^\varepsilon = \mathbb{E}_{\Omega_f^\varepsilon}(\mathbf{w}^\varepsilon)$.

Then for $\nu_0 > 0$ there exists a subsequence of small parameters $\{\varepsilon > 0\}$ such that the sequences $\{p^\varepsilon\}$, $\{(1-\chi^\varepsilon)\mathbf{w}^\varepsilon\}$, and $\{\mathbf{w}_f^\varepsilon\}$, converge weakly in $L_2(\Omega_T)$ and $\mathbf{L}_2(\Omega_T)$ as $\varepsilon \searrow 0$ to the functions p, $\mathbf{w}^{(s)}$ and $\mathbf{w}_f$ respectively and these limiting functions, where $p \in W_2^{1,0}(\Omega_T)$, $\nu_0\nabla\big(\dfrac{\partial p}{\partial t}\big) \in \mathbf{W}_2^{1,0}(\Omega_T)$, satisfy in the domain Ω_T the system of homogenized equations consisting of the continuity equation

$$\left(\frac{m}{\bar{c}_f^2} + \frac{(1-m)}{\bar{c}_s^2}\right)\frac{\partial p}{\partial t} + \nabla\cdot\left(m\frac{\partial \mathbf{w}_f}{\partial t} + \frac{\partial \mathbf{w}^{(s)}}{\partial t}\right) = 0, \qquad (6.2.16)$$

the momentum balance equation

$$m\rho_f\frac{\partial \mathbf{w}_f}{\partial t} + \rho_s\frac{\partial \mathbf{w}^{(s)}}{\partial t} + \int_0^t \Big(-\hat{\rho}\mathbf{F} + \nabla\,\tilde{p}\Big)(\mathbf{x},\tau)d\tau = 0, \qquad (6.2.17)$$

for the liquid component, and the momentum balance equation

$$\frac{\partial \mathbf{w}^{(s)}}{\partial t} - (1-m)\frac{\partial \mathbf{w}_f}{\partial t}$$
$$= -\int_0^t \mathbb{B}^{(s)}(\infty,\lambda_1;t-\tau)\cdot\left(\nabla\,\tilde{p} + \rho_s\left(\frac{\partial^2 \mathbf{w}_f}{\partial \tau^2} - \mathbf{F}\right)\right)(\mathbf{x},\tau)d\tau \qquad (6.2.18)$$

for the solid component.

Equations (6.2.16)–(6.2.18) are supplemented with the homogeneous initial conditions

$$\mathbf{w}^{(s)}(\mathbf{x},0)=\mathbf{w}_f(\mathbf{x},0)=0, \tag{6.2.19}$$

for displacements in the liquid and the solid components and boundary condition

$$\Big(m\frac{\partial \mathbf{w}_f}{\partial t}(\mathbf{x},t)+\frac{\partial \mathbf{w}^{(s)}}{\partial t}(\mathbf{x},t)\Big)\cdot \mathbf{n}(\mathbf{x})=0,\ \ \mathbf{x}\in S,\ \ t>0. \tag{6.2.20}$$

For $\nu_0=0$ the limiting pressure p of the mixture and functions $\mathbf{w}^{(s)}$ and $\mathbf{w}_f$ satisfy in the domain Ω_T the system of homogenized equations consisting the continuity equation

$$\Big(\frac{m}{\bar{c}_f^2}+\frac{(1-m)}{\bar{c}_s^2}\Big)\frac{\partial p}{\partial t}+\nabla\cdot\Big(m\frac{\partial \mathbf{w}_f}{\partial t}+\frac{\partial \mathbf{w}^{(s)}}{\partial t}\Big)=0, \tag{6.2.21}$$

the momentum balance equation

$$m\rho_f\frac{\partial \mathbf{w}_f}{\partial t}+\rho_s\frac{\partial \mathbf{w}^{(s)}}{\partial t}+\int_0^t\big(-\hat{\rho}\mathbf{F}+\nabla p\big)(\mathbf{x},\tau)d\tau=0, \tag{6.2.22}$$

for the liquid component, the momentum balance equation

$$\frac{\partial \mathbf{w}^{(s)}}{\partial t}-(1-m)\frac{\partial \mathbf{w}_f}{\partial t} = -\int_0^t \mathbb{B}^{(s)}(\infty,\lambda_1;t-\tau)\cdot\Big(\nabla p+\rho_s\Big(\frac{\partial^2 \mathbf{w}_f}{\partial \tau^2}-\mathbf{F}\Big)\Big)(\mathbf{x},\tau)d\tau \tag{6.2.23}$$

for the solid component, and initial and boundary conditions (6.2.19) and (6.2.20).

In (6.2.17) the function $\tilde{p}$ is given by (6.2.5), in (6.2.18) and in (6.2.23) the matrix $\mathbb{B}^{(s)}(\infty,\lambda_1;t)$ has been defined in Chap. 3 by formulas (3.2.76) and (3.2.53).

We refer to the problem (6.2.16)–(6.2.20) as the homogenized **model** $(\mathbb{IA})_9$, and to the problem (6.2.19)–(6.2.23) as the homogenized **model** $(\mathbb{IA})_{10}$.

Theorem 6.8 *Let* $\{\mathbf{w}^{\varepsilon},\ p^{\varepsilon}\}$ *be the weak solution of the problem (6.0.1)–(6.0.5),*

$$\lambda_1 = \infty,\ 0 \leqslant \mu_1 < \infty,$$

and $\mathbf{w}_s^{\varepsilon} = \mathbb{E}_{\Omega_s^{\varepsilon}}\left(\mathbf{w}^{\varepsilon}\right)$.

Then for $\nu_0 > 0$ *there exists a subsequence of small parameters* $\{\varepsilon > 0\}$ *such that the sequences* $\{p^{\varepsilon}\}$, $\{\chi^{\varepsilon}\mathbf{w}^{\varepsilon}\}$, $\{\mathbf{w}_s^{\varepsilon}\}$ *converge weakly in* $L_2(\Omega_T)$ *as* $\varepsilon \searrow 0$ *to the functions* p, $\mathbf{w}^{(s)}$, $\mathbf{w}^{(f)}$, *and* w_s *respectively and these limiting functions, where* $p \in W_2^{1,0}(\Omega_T)$, $\nu_0\nabla\left(\dfrac{\partial p}{\partial t}\right) \in \mathbf{W}_2^{1,0}(\Omega_T)$, *satisfy in the domain* Ω_T *the system of homogenized equations consisting of the continuity equation*

$$\left(\frac{m}{\bar{c}_f^2} + \frac{(1-m)}{\bar{c}_s^2}\right)\frac{\partial p}{\partial t} + \nabla\cdot\left(\frac{\partial \mathbf{w}^{(f)}}{\partial t} + (1-m)\frac{\partial \mathbf{w}_s}{\partial t}\right) = 0, \tag{6.2.24}$$

the momentum balance equation

$$\rho_f\frac{\partial \mathbf{w}^{(f)}}{\partial t} + (1-m)\rho_s\frac{\partial \mathbf{w}_s}{\partial t} = \int_0^t \left(\hat{\rho}\mathbf{F} - \nabla\,\tilde{p}\right)(\mathbf{x},\tau)d\tau, \tag{6.2.25}$$

for the solid component, and the momentum balance equation

$$\frac{\partial \mathbf{w}^{(f)}}{\partial t} - m\,\frac{\partial \mathbf{w}_s}{\partial t} = -\int_0^t \mathbb{B}^{(f)}(\mu_1,\infty;t-\tau)\cdot\left(\nabla\,\tilde{p} + \rho_f\left(\frac{\partial^2 \mathbf{w}_s}{\partial \tau^2} - \mathbf{F}\right)\right)(\mathbf{x},\tau)d\tau \tag{6.2.26}$$

for the liquid component.

Equations (6.2.24)–(6.2.26) are supplemented with the homogeneous initial conditions

$$\mathbf{w}^{(f)}(\mathbf{x},0) = \mathbf{w}_s(\mathbf{x},0) = 0,\ \ \mathbf{x}\in\Omega \tag{6.2.27}$$

for displacements in the liquid and the solid components and boundary condition

$$\left(\frac{\partial \mathbf{w}^{(f)}}{\partial t} + (1-m)\frac{\partial \mathbf{w}_s}{\partial t}\right)\cdot\mathbf{n}(\mathbf{x}) = 0,\ \ \mathbf{x}\in S,\ t>0. \tag{6.2.28}$$

For $\nu_0 = 0$ *the limiting pressure* p *of the sequence* $\{p^{\varepsilon}\}$ *and functions* $\mathbf{w}^{(f)}$ *and* $\mathbf{w}_s$ *satisfy in the domain* Ω_T *the system of homogenized equations consisting the continuity equation*

$$\left(\frac{m}{\bar{c}_f^2} + \frac{(1-m)}{\bar{c}_s^2}\right)\frac{\partial p}{\partial t} + \nabla\cdot\left(\frac{\partial \mathbf{w}^{(f)}}{\partial t} + (1-m)\frac{\partial \mathbf{w}_s}{\partial t}\right) = 0, \tag{6.2.29}$$

the momentum balance equation

$$\rho_f \frac{\partial \mathbf{w}^{(f)}}{\partial t} + (1-m)\rho_s \frac{\partial \mathbf{w}_s}{\partial t} = \int_0^t \Big(\hat{\rho}\mathbf{F} - \nabla p\Big)(\mathbf{x},\tau)d\tau, \tag{6.2.30}$$

for the solid component, and the momentum balance equation

$$\frac{\partial \mathbf{w}^{(f)}}{\partial t} - m\,\frac{\partial \mathbf{w}_s}{\partial t}$$
$$= -\int_0^t \mathbb{B}^{(f)}(\mu_1,\infty;t-\tau)\cdot\left(\nabla p + \rho_f\left(\frac{\partial^2 \mathbf{w}_s}{\partial \tau^2} - \mathbf{F}\right)\right)(\mathbf{x},\tau)d\tau \tag{6.2.31}$$

for the liquid component, and initial and boundary conditions (6.2.27) and (6.2.28). In (6.2.26) and (6.2.31) the matrix $\mathbb{B}^{(f)}(\mu_1,\infty;t)$ has been defined in Chap. 3 by formula (3.2.70) and formula (3.2.76), and in (6.2.25) and (6.2.26) the function $\tilde{p}$ is given by (6.2.5).

We refer to the problem (6.2.24)–(6.2.28) as the homogenized **model** $(\mathbb{IA})_{11}$, and to the problem (6.2.27)–(6.2.31) as the homogenized **model** $(\mathbb{IA})_{12}$.

6.2.2 Proofs of Theorems 6.5–6.8

Proofs of Theorems 6.5–6.8 repeat the proofs of the corresponding Theorems in Chaps. 1 and 3 with evident changes.

In the same way as in Lemma 1.1. in Chap. 1 and Lemma 3.9 in Chap. 3 one may show that the sequences $\{p^\varepsilon\}$ and $\{\tilde{p}^\varepsilon\}$, where $\tilde{p}^\varepsilon = \chi^\varepsilon\left(\left(\frac{\nu_0}{\bar{c}_f^2}\right)\frac{\partial p^\varepsilon}{\partial t}\right) + p^\varepsilon$, converge two-scale in $L_2(\Omega_T)$ to functions $p(\mathbf{x},t)$ and $\tilde{P}(\mathbf{x},t,\mathbf{y})$ respectively and

$$\tilde{P}(\mathbf{x},t,\mathbf{y}) = p(\mathbf{x},t) + \frac{\nu_0}{\bar{c}_f^2}\chi(\mathbf{y})\frac{\partial p}{\partial t}(\mathbf{x},t).$$

Correspondingly, the sequences $\{p^\varepsilon\}$ and $\{\tilde{p}^\varepsilon\}$ converge weakly in $L_2(\Omega_T)$ to functions $p(\mathbf{x},t)$ and $\tilde{p}(\mathbf{x},t)$, where

$$\tilde{p}(\mathbf{x},t) = p(\mathbf{x},t) + m\,\frac{\nu_0}{\bar{c}_f^2}\frac{\partial p}{\partial t}(\mathbf{x},t).$$

The main differences here from Chap. 3 are the derivations of continuity equations and boundary conditions, which repeat the same procedure for compressible media in Chap. 1, and the derivation of the microscopic momentum balance equation for the case $0 \leqslant \mu_1,\ \lambda_1 < \infty$.

In general, one has the following limiting continuity equation

$$\Big(\frac{m}{\bar{c}_f^2} + \frac{(1-m)}{\bar{c}_s^2}\Big)\frac{\partial p}{\partial t} + \nabla \cdot \mathbf{v} = 0, \ \mathbf{x} \in \Omega, \ t > 0,$$

and boundary condition

$$\mathbf{v} \cdot \mathbf{n} = 0, \ \mathbf{x} \in S, \ t > 0.$$

For $\mu_1 = \infty$ and $\lambda_1 = \infty$, or $\mu_1 < \infty$ and $\lambda_1 < \infty$

$$\mathbf{v} = \frac{\partial \mathbf{w}}{\partial t}.$$

For $\mu_1 = \infty$ and $\lambda_1 < \infty$

$$\mathbf{v} = m\,\frac{\partial \mathbf{w}_f}{\partial t} + \frac{\partial \mathbf{w}^{(s)}}{\partial t}.$$

Finally, for $\mu_1 < \infty$ and $\lambda_1 = \infty$

$$\mathbf{v} = \frac{\partial \mathbf{w}^{(f)}}{\partial t} + \frac{\partial \mathbf{w}_s}{\partial t}.$$

6.2.2.1 Proof of Theorem 6.5

The continuity equation and boundary condition for this case have a form

$$\Big(\frac{m}{\bar{c}_f^2} + \frac{(1-m)}{\bar{c}_s^2}\Big)\frac{\partial p}{\partial t} + \nabla \cdot \frac{\partial \mathbf{w}}{\partial t} = 0, \ \mathbf{x} \in \Omega, \ t > 0,$$

$$\mathbf{w} \cdot \mathbf{n} = 0, \ \mathbf{x} \in S, \ t > 0.$$

The weak limit as $\varepsilon \to 0$ in the integral identity (6.0.10) results

$$\hat{\rho}\frac{\partial^2 \mathbf{w}}{\partial t^2} = -\nabla \tilde{p} + \hat{\rho}\mathbf{F}.$$

The combination of these relations give us models $(\mathbb{IA})_5$ and $(\mathbb{IA})_6$.

6.2.2.2 Proof of Theorem 6.6

As in the previous subsection the continuity equation and boundary condition for this case have the form

$$\left(\frac{1}{\bar{c}_f^2}+\frac{1}{\bar{c}_s^2}\right)\frac{\partial p}{\partial t}+\nabla\cdot\frac{\partial \mathbf{w}}{\partial t}=0,\ \mathbf{x}\in\Omega,\ t>0,$$

$$\mathbf{w}\cdot\mathbf{n}=0,\ \mathbf{x}\in S,\ t>0.$$

Next, in the usual way (see Lemma 3.2.3 in Chap. 3) we prove the inclusion $\nabla\,\tilde{P}\in \mathbf{L}_2(\Omega_T\times Y)$, that is $\nabla\,p\in\mathbf{L}_2(\Omega_T)$, $\nabla\left(\dfrac{\partial p}{\partial t}\right)\in\mathbf{L}_2(\Omega_T)$, and derive the microscopic momentum balance equation:

$$\rho(\mathbf{y})\frac{\partial^2\mathbf{W}}{\partial t^2}=\nabla_y\cdot\left(\mu_1\chi(\mathbf{y})\mathbb{D}\left(y,\frac{\partial \mathbf{W}}{\partial t}\right)+\lambda_1(1-\chi(\mathbf{y}))\mathbb{D}(y,\mathbf{W})-\Pi\,\mathbb{I}\right)$$
$$-\nabla\,\tilde{P}+\rho(\mathbf{y})\mathbf{F},\ \mathbf{y}\in Y,\ t>0,\qquad(6.2.32)$$

where

$$\rho(\mathbf{y})=\rho_f\,\chi(\mathbf{y})+\rho_s\left(1-\chi(\mathbf{y})\right),$$

and the microscopic continuity equation

$$\nabla_y\cdot\mathbf{W}=0,\quad \mathbf{y}\in Y.\qquad(6.2.33)$$

These equations are completed with homogeneous initial conditions

$$\mathbf{W}(\mathbf{x},\mathbf{y},0)=\frac{\partial \mathbf{W}}{\partial t}(\mathbf{x},\mathbf{y},0)=0,\quad \mathbf{y}\in Y.$$

We look for the periodic solution of the problem as a sum

$$\mathbf{W}(\mathbf{x},t,\mathbf{y})=\sum_{i=1}^{3}\int_0^t\mathbf{W}^{(i)}(\mathbf{y},t-\tau)\frac{\partial p}{\partial x_i}(\mathbf{x},\tau)d\tau$$
$$+\frac{\nu_0}{\bar{c}_f^2}\sum_{i=1}^{3}\int_0^t\mathbf{W}_0^{(i)}(\mathbf{y},t-\tau)\frac{\partial^2 p}{\partial x_i\partial\tau}(\mathbf{x},\tau)d\tau$$
$$+\sum_{i=1}^{3}\int_0^t\mathbf{W}_F^{(i)}(\mathbf{y},t-\tau)F_i(\mathbf{x},\tau)d\tau,$$

$$\Pi(\mathbf{x},t,\mathbf{y}) = \sum_{i=1}^{3}\int_0^t \Pi^{(i)}(\mathbf{y},t-\tau)\frac{\partial p}{\partial x_i}(\mathbf{x},\tau)d\tau + \sum_{i=1}^{3}\int_0^t \Pi_0^{(i)}(\mathbf{y},t-\tau)\frac{\partial^2 p}{\partial x_i\partial\tau}(\mathbf{x},\tau)d\tau + \sum_{i=1}^{3}\int_0^t \Pi_F^{(i)}(\mathbf{y},t-\tau)F_i(\mathbf{x},\tau)d\tau,$$

where

$$\mathbf{F}(\mathbf{x},t) = \big(F_1(\mathbf{x},t),\, F_2(\mathbf{x},t),\, F_3(\mathbf{x},t)\big).$$

In turn, the pairs $\{\mathbf{W}^{(i)},\,\Pi^{(i)}\}$, $\{\mathbf{W}^{(i)},\,\Pi^{(i)}\}$, and $\{\mathbf{W}_F^{(i)},\,\Pi_F^{(i)}\}$ for $i=1,2,3$ solve periodic initial boundary value problems in the domain Y, for $t>0$

$$\rho(\mathbf{y})\frac{\partial^2\mathbf{W}^{(i)}}{\partial t^2} = \nabla_y\cdot\Big(\mu_1\chi(\mathbf{y})\nabla_y\Big(\frac{\partial\mathbf{W}^{(i)}}{\partial t}\Big) + \lambda_1\big(1-\chi(\mathbf{y})\big)\nabla_y\mathbf{W}^{(i)} - \Pi^{(i)}\,\mathbb{I}\Big),\quad \nabla_y\cdot\mathbf{W}^{(i)}=0, \tag{6.2.34}$$

$$\mathbf{W}^{(i)}(\mathbf{y},0)=0,\quad \rho(\mathbf{y})\frac{\partial\mathbf{W}^{(i)}}{\partial t}(\mathbf{y},0) = -\mathbf{e}_i,\quad \mathbf{y}\in Y, \tag{6.2.35}$$

$$\rho(\mathbf{y})\frac{\partial^2\mathbf{W}_0^{(i)}}{\partial t^2} = \nabla_y\cdot\Big(\mu_1\chi(\mathbf{y})\nabla_y\Big(\frac{\partial\mathbf{W}_0^{(i)}}{\partial t}\Big) + \lambda_1\big(1-\chi(\mathbf{y})\big)\nabla_y\mathbf{W}_0^{(i)} - \Pi_0^{(i)}\,\mathbb{I}\Big),\quad \nabla_y\cdot\mathbf{W}_0^{(i)}=0, \tag{6.2.36}$$

$$\mathbf{W}_0^{(i)}(\mathbf{y},0)=0,\quad \rho(\mathbf{y})\frac{\partial\mathbf{W}_0^{(i)}}{\partial t}(\mathbf{y},0) = -\chi(\mathbf{y})\mathbf{e}_i,\quad \mathbf{y}\in Y, \tag{6.2.37}$$

and

$$\rho(\mathbf{y})\frac{\partial^2\mathbf{W}_F^{(i)}}{\partial t^2} = \nabla_y\cdot v\Big(\mu_1\chi(\mathbf{y})\nabla_y\Big(\frac{\partial\mathbf{W}_F^{(i)}}{\partial t}\Big) + \lambda_1\big(1-\chi(\mathbf{y})\big)\nabla_y\mathbf{W}_F^{(i)} - \Pi_F^{(i)}\,\mathbb{I}\Big),\quad \nabla_y\cdot\mathbf{W}_F^{(i)}=0, \tag{6.2.38}$$

$$\mathbf{W}_F^{(i)}(\mathbf{y},0)=0,\quad \frac{\partial\mathbf{W}_F^{(i)}}{\partial t}(\mathbf{y},0) = \mathbf{e}_i,\quad \mathbf{y}\in Y \tag{6.2.39}$$

respectively.

Thus,

$$
\begin{aligned}
\frac{\partial \mathbf{W}}{\partial t} &= \sum_{i=1}^{3} \int_0^t \frac{\partial \mathbf{W}^{(i)}}{\partial t}(\mathbf{y}, t-\tau) \frac{\partial p}{\partial x_i}(\mathbf{x}, \tau) d\tau \\
&+ \frac{\nu_0}{\bar{c}_f^2} \sum_{i=1}^{3} \int_0^t \frac{\partial \mathbf{W}_0^{(i)}}{\partial t}(\mathbf{y}, t-\tau) \frac{\partial^2 p}{\partial x_i \partial \tau}(\mathbf{x}, \tau) d\tau \\
&+ \sum_{i=1}^{3} \int_0^t \frac{\partial \mathbf{W}_F^{(i)}}{\partial t}(\mathbf{y}, t-\tau) F_i(\mathbf{x}, \tau) d\tau,
\end{aligned}
$$

and

$$
\begin{aligned}
\frac{\partial \mathbf{w}}{\partial t} &= \sum_{i=1}^{3} \int_0^t \left\langle \frac{\partial \mathbf{W}^{(i)}}{\partial t} \right\rangle_Y (t-\tau) \frac{\partial p}{\partial x_i}(\mathbf{x}, \tau) d\tau \\
&+ \frac{\nu_0}{\bar{c}_f^2} \sum_{i=1}^{3} \int_0^t \left\langle \frac{\partial \mathbf{W}_0^{(i)}}{\partial t} \right\rangle_Y (t-\tau) \frac{\partial^2 p}{\partial x_i \partial \tau}(\mathbf{x}, \tau) d\tau \\
&+ \sum_{i=1}^{3} \int_0^t \left\langle \frac{\partial \mathbf{W}_F^{(i)}}{\partial t} \right\rangle_Y (\mathbf{y}, t-\tau) F_i(\mathbf{x}, \tau) d\tau \\
&= \int_0^t \mathbb{B}^{(a)}(\mu_1, \lambda_1; t-\tau) \cdot \nabla p\,(\mathbf{x}, \tau) d\tau \\
&+ \frac{\nu_0}{\bar{c}_f^2} \int_0^t \mathbb{B}_0^{(a)}(\mu_1, \lambda_1; t-\tau) \cdot \nabla \frac{\partial p}{\partial \tau}(\mathbf{x}, \tau) d\tau + \mathbf{f}(\mathbf{x}, t),
\end{aligned}
$$

where

$$
\mathbb{B}^{(a)}(\mu_1, \lambda_1; t) = \sum_{i=1}^{3} \left\langle \frac{\partial \mathbf{W}^{(i)}}{\partial t} \right\rangle_Y (t) \otimes \mathbf{e}_i, \tag{6.2.40}
$$

$$
\mathbb{B}_0^{(a)}(\mu_1, \lambda_1; t) = \sum_{i=1}^{3} \left\langle \frac{\partial \mathbf{W}_0^{(i)}}{\partial t} \right\rangle_Y (t) \otimes \mathbf{e}_i, \tag{6.2.41}
$$

$$
\mathbf{f}(\mathbf{x}, t) = \sum_{i=1}^{3} \int_0^t \left\langle \frac{\partial \mathbf{W}_F^{(i)}}{\partial t} \right\rangle_Y (\mathbf{y}, t-\tau) F_i(\mathbf{x}, \tau) d\tau. \tag{6.2.42}
$$

As before, the combination of all the relations obtained result in models $(\mathbb{IA})_7$ and $(\mathbb{IA})_8$.

6.2.2.3 Proof of Theorem 6.7

For this case the velocity of the mixture is given by the expression

$$\mathbf{v} = m\frac{\partial \mathbf{w}_f}{\partial t} + \frac{\partial \mathbf{w}^{(s)}}{\partial t},$$

and the continuity equation in the domain Ω for $t > 0$ takes the form

$$\left(\frac{m}{\bar{c}_f^2} + \frac{(1-m)}{\bar{c}_s^2}\right)\frac{\partial p}{\partial t} + \nabla \cdot \left(m\frac{\partial \mathbf{w}_f}{\partial t} + \frac{\partial \mathbf{w}^{(s)}}{\partial t}\right) = 0.$$

The weak limit as $\varepsilon \to 0$ in the integral identity (6.0.10) results in the momentum balance equation

$$m\rho_f\frac{\partial^2 \mathbf{w}_f}{\partial t^2} + \rho_s\frac{\partial^2 \mathbf{w}^{(s)}}{\partial t^2} = -\nabla\,\tilde{p} + \hat{\rho}\mathbf{F}$$

for the liquid component.

Finally, the derivation of the momentum balance equation for the solid component begins with the microscopic system of equations

$$\rho_s\frac{\partial^2 \mathbf{W}^{(s)}}{\partial t^2} = \frac{\lambda_1}{2}\Delta_y \mathbf{W}^{(s)} - \nabla_y \Pi^{(s)} - \nabla\, p, \quad \nabla \cdot \mathbf{W}^{(s)} = 0,$$

in the domain Y_s, and repeats the same procedure as in the proof of Theorem 3.4 of Chap. 3.

6.2.2.4 Proof of Theorem 6.8

Here for the velocity of the mixture one has

$$\mathbf{v} = \frac{\partial \mathbf{w}^{(\mathbf{f})}}{\partial t} + \frac{\partial \mathbf{w}_s}{\partial t},$$

and the continuity equation takes the form

$$\left(\frac{m}{\bar{c}_f^2} + \frac{(1-m)}{\bar{c}_s^2}\right)\frac{\partial p}{\partial t} + \nabla \cdot \left(\frac{\partial \mathbf{w}^{(f)}}{\partial t} + \frac{\partial \mathbf{w}_s}{\partial t}\right) = 0.$$

The weak limit as $\varepsilon \to 0$ in the integral identity (6.0.10) results in the momentum balance equation

$$\rho_f \frac{\partial^2 \mathbf{w}^{(f)}}{\partial t^2} + (1-m)\rho_s \frac{\partial^2 \mathbf{w}_s}{\partial t^2} = -\nabla\, \tilde{p} + \hat{\rho}\mathbf{F}$$

for the solid component.

The derivation of the momentum balance equation for the liquid component begins with the microscopic system of equations

$$\rho_f \frac{\partial^2 \mathbf{W}^{(f)}}{\partial t^2} = \frac{\mu_1}{2}\Delta_y \mathbf{W}^{(f)} - \nabla_y \Pi^{(f)} - \nabla\, \tilde{p}, \quad \nabla \cdot \mathbf{W}^{(f)} = 0,$$

in the domain Y_f, and repeats the same procedure in the proof of Theorem 3.5 of Chap. 3.

6.3 A Compressible Slightly Viscous Liquid in a Compressible Elastic Skeleton

Throughout this section we assume that

$$\mu_0 = 0, \; 0 < \lambda_0 < \infty. \tag{6.3.1}$$

6.3.1 Main Results

Theorem 6.9 *Let* $\{\mathbf{w}^\varepsilon, p^\varepsilon\}$ *be the weak solution of the problem (6.1.1)–(6.1.5),* $\mathbf{w}_s^\varepsilon = \mathbb{E}_{\Omega_s^\varepsilon}(\mathbf{w}^\varepsilon)$ *be an extension 1.2.9, and* $\mu_1 = \infty$ *or* $\mu_1 < \infty$ *but the pore space be disconnected.*

Then

(1) *for all* $\nu_0 > 0$ *the sequences* $\{\mathbf{w}^\varepsilon\}$, $\{\nabla\cdot\mathbf{w}^\varepsilon\}$, $\left\{\chi^\varepsilon p^\varepsilon\right\}$, $\left\{\chi^\varepsilon \frac{\partial p^\varepsilon}{\partial t}\right\}$, *and* $\{q^\varepsilon\}$, *where* $q^\varepsilon = \chi^\varepsilon\left(p^\varepsilon - \alpha_\nu \nabla\cdot\left(\frac{\partial \mathbf{w}^\varepsilon}{\partial t}\right)\right)$, *converge weakly in* $\mathbf{L}_2(\Omega_T)$ *and* $L_2(\Omega_T)$ *(up to some subsequences) to functions* $\mathbf{w}$, $\nabla\cdot\mathbf{w}$, $\frac{\partial p_f}{\partial t}$, *and* $m\, q = m\left(p_f + \frac{\nu_0}{c_f^2}\frac{\partial p_f}{\partial t}\right)$ *respectively;*

(2) *for* $\nu_0 = 0$ *the sequences* $\{\mathbf{w}^\varepsilon\}$, $\{\nabla\cdot\mathbf{w}^\varepsilon\}$ *and* $\{\chi^\varepsilon p^\varepsilon\}$ *converge weakly in* $\mathbf{L}_2(\Omega_T)$ *and* $L_2(\Omega_T)$ *(up to some subsequences) to functions* $\mathbf{w}$, $\nabla\cdot\mathbf{w}$, *and* mp_f *respectively;*

(3) *for all* $\nu_0 \geqslant 0$ *the sequence* $\{\mathbf{w}_s^\varepsilon\}$ *converges weakly in* $\mathbf{W}_2^{1,0}(\Omega_T)$ *to the function* $\mathbf{w}_s = \mathbf{w}$;

(4) *limiting functions solve the system of homogenized equations in the domain* Ω_T, *consisting of the homogenized continuity equation*

$$\frac{1}{c_f^2} p_f + m\, \nabla \cdot \mathbf{w}_s = \mathbb{C}_0^s : \mathbb{D}(x, \mathbf{w}_s) + \frac{c_0^s}{\lambda_0} q, \tag{6.3.2}$$

the state equation

$$q = p_f + \frac{\nu_0}{c_f^2} \frac{\partial p_f}{\partial t} \quad (q = p_f \text{ for } \nu_0 = 0), \tag{6.3.3}$$

and the homogenized momentum balance equation

$$\hat{\rho} \frac{\partial^2 \mathbf{w}_s}{\partial t^2} = \nabla \cdot \left(\lambda_0\, \mathfrak{N}_2^s : \mathbb{D}(x, \mathbf{w}_s) - q\, \mathbb{C}_1^s\right) + \hat{\rho} \mathbf{F}, \tag{6.3.4}$$

completed with homogeneous boundary and initial conditions

$$\mathbf{w}_s(\mathbf{x}, t) = 0, \;\; \mathbf{x} \in S, \;\; t \in (0, T), \tag{6.3.5}$$

$$\mathbf{w}_s(\mathbf{x}, 0) = \frac{\partial \mathbf{w}_s}{\partial t}(\mathbf{x}, 0) = 0, \;\; \mathbf{x} \in \Omega; \tag{6.3.6}$$

(5) *there exists $\lambda_* > 0$, such that for all $\lambda_0 > \lambda_*$ the problem (6.3.2)–(6.3.6) has a unique solution.*

In (6.3.2), (6.3.4)

$$\hat{\rho} = m\, \rho_f + (1 - m)\, \rho_s,$$

the symmetric strictly positively definite constant fourth-rank tensor $\mathfrak{N}_2^s$, matrices $\mathbb{C}_0^s$ and $\mathbb{C}_1^s$, and the constant c_0^s are given in Chap. 1 by formulas (1.3.26), (1.3.27) and 1.3.31 *and do not depend on λ_0.*

We refer to the problem (6.3.2)–(6.3.6) with $\nu_0 = 0$ as the homogenized **model** $(\mathbb{IA})_{13}$ and to the problem (6.3.2)–(6.3.6) with $\nu_0 > 0$ as the homogenized **model** $(\mathbb{IA})_{14}$.

Theorem 6.10 *Let $\{\mathbf{w}^\varepsilon,\, p^\varepsilon\}$ be the weak solution of the problem (6.1.1)–(6.1.5), $\mathbf{w}_s^\varepsilon = \mathbb{E}_{\Omega_s^\varepsilon}(\mathbf{w}^\varepsilon)$ be an extension (1.2.9), the pore space be connected, and $\mu_1 < \infty$. Then*

(1) *for all $\nu_0 > 0$ the sequences $\{\mathbf{w}^\varepsilon\}$, $\{\chi^\varepsilon \mathbf{w}^\varepsilon\}$, $\{\nabla \cdot \mathbf{w}^\varepsilon\}$, $\{\chi^\varepsilon\, p^\varepsilon\}$, $\left\{\chi^\varepsilon \dfrac{\partial p^\varepsilon}{\partial t}\right\}$, and $\{q^\varepsilon\}$, where $q^\varepsilon = \chi^\varepsilon\left(p^\varepsilon - \bar{\alpha}_\nu \nabla \cdot \left(\dfrac{\partial \mathbf{w}^\varepsilon}{\partial t}\right)\right)$, converge weakly in $\mathbf{L}_2(\Omega_T)$ and $L_2(\Omega_T)$ (up to some subsequences) to functions $\mathbf{w}$, $\mathbf{w}^{(f)}$, $\nabla \cdot \mathbf{w}$, $m \dfrac{\partial p_f}{\partial t}$, and $m\, q = m\left(p_f + \dfrac{\nu_0}{c_f^2} \dfrac{\partial p_f}{\partial t}\right) \in W_2^{1,0}(\Omega_T)$ respectively;*

(2) *for* $\nu_0 = 0$ *the sequences* $\{\mathbf{w}^\varepsilon\}$, $\{\chi^\varepsilon \mathbf{w}^\varepsilon\}$, $\{\nabla \cdot \mathbf{w}^\varepsilon\}$, *and* $\{\chi^\varepsilon p^\varepsilon\}$ *converge weakly in* $\mathbf{L}_2(\Omega_T)$ *and* $L_2(\Omega_T)$ *(up to some subsequences) to functions* $\mathbf{w}$, $\mathbf{w}^{(f)}$, $\nabla \cdot \mathbf{w}$, *and* $m p_f \in W_2^{1,0}(\Omega_T)$ *respectively;*

(3) *for all* $\nu_0 \geqslant 0$ *the sequence* $\{\mathbf{w}_s^\varepsilon\}$ *converges weakly in* $\mathbf{W}_2^{1,0}(\Omega_T)$ *to the function* $\mathbf{w}_s$*;*

(4) *limiting functions solve the system of homogenized equations in the domain* Ω_T*, consisting of the homogenized continuity equation*

$$\frac{m}{\bar{c}_f^2} p_f + \nabla \cdot \mathbf{w}^{(f)} = \mathbb{C}_0^s : \mathbb{D}(x, \mathbf{w}_s) + \frac{c_0^s}{\lambda_0} q, \tag{6.3.7}$$

the state equation (6.3.3), the homogenized momentum balance equation

$$\rho_f \frac{\partial^2 \mathbf{w}^{(f)}}{\partial t^2} + \rho_s (1-m) \frac{\partial^2 \mathbf{w}_s}{\partial t^2} = \nabla \cdot \left(\lambda_0\, \mathfrak{N}_2^s : \mathbb{D}(x, \mathbf{w}_s) - q\, \mathbb{C}_1^s\right) + \hat{\rho} \mathbf{F}, \tag{6.3.8}$$

for the solid component, the boundary and initial conditions (6.3.5) and (6.3.6), the homogenized momentum balance equation

$$-\int_0^t \mathbb{B}^{(f)}(\mu_1, \infty; t-\tau) \cdot \left(\nabla \left(p_f + \frac{\nu_0}{\bar{c}_f^2} \frac{\partial p_f}{\partial t}\right) + \rho_f \left(\frac{\partial^2 \mathbf{w}_s}{\partial \tau^2} - \mathbf{F}\right)\right)(\mathbf{x}, \tau) d\tau$$
$$= \frac{\partial \mathbf{w}^{(f)}}{\partial t} - m \frac{\partial \mathbf{w}_s}{\partial t} \tag{6.3.9}$$

for the liquid component, and homogeneous boundary condition

$$\mathbf{w}^{(f)}(\mathbf{x}, t) \cdot \mathbf{n}(\mathbf{x}) = 0, \;\; \mathbf{x} \in S, \;\; t \in (0, T) \tag{6.3.10}$$

for displacements $\mathbf{w}^{(f)}$ *of the liquid component.*

In (6.3.9) and (6.3.10) $\mathbf{n}$ *is the normal vector to the boundary S, the matrix* $\mathbb{B}^{(f)}(\mu_1, \infty; t)$ *is given in the proof of Theorem 3.4 of Chap. 3 (see formulae (3.2.70) and (3.2.76), and the constant matrix* $\mathbb{B}^{(s)}(\infty, 0; t) = \mathbb{B}^{(s)}(\infty, 0)$ *is strictly positively definite.*

We refer to the problem (6.3.3)–(6.3.10) with $\nu_0 = 0$ as the homogenized **model** $(\mathbb{IA})_{15}$ and to the problem (6.3.3)–(6.3.10) with $\nu_0 > 0$ as the homogenized **model** $(\mathbb{IA})_{16}$.

Theorem 6.11 *The solution* $\mathbf{w}_s$ *to the model* $(\mathbb{IA})_{13}$ *satisfies the homogenized equation*

$$\hat{\rho} \frac{\partial^2 \mathbf{w}_s}{\partial t^2} = \nabla \cdot \left(\lambda_0\, \mathfrak{N}_3^s : \mathbb{D}(x, \mathbf{w}_s)\right) + \hat{\rho} \mathbf{F} \tag{6.3.11}$$

in the domain Ω_T*, completed with the boundary and initial conditions (6.3.5) and (6.3.6).*

The symmetric strictly positive definite constant fourth-rank tensor $\mathfrak{N}_3^s$ is given in Chap. 1 by formula (1.3.39) and does not depend on λ_0.

6.3.2 Proofs of Theorems 6.9–6.11

As in the previous subsection, the proofs of these theorems repeat corresponding proofs of Theorems 1.6–1.8 in Chap. 1 and the proof of Theorem 6.8 of this chapter.

For example, the proof of Theorem 6.9 differs from the proof of Theorem 1.6 only in the form of the macroscopic momentum balance equation:

$$\hat{\rho}\frac{\partial^2 \mathbf{w}_s}{\partial t^2} = \nabla \cdot \Big(\lambda_0\, \mathfrak{N}^{(0)} : \big((1-m)\,\mathbb{D}(x, \mathbf{w}_s) + \langle \mathbb{D}(y, \mathbf{U})\rangle_{Y_s}\big)\Big) - \nabla q + \hat{\rho}\mathbf{F}.$$

The proof of Theorem 6.10 repeats the proofs of Theorem 6.9 and Theorem 1.7, where instead of Darcy's law (1.3.16) we use the homogenized momentum balance equation (6.2.26) for the liquid component.

Chapter 7
Non-isothermal Acoustics in Poroelastic Media

We consider the model $\mathbb{M}_{19}$ under the condition $\bar{\alpha}_\nu = 0$:

$$\frac{1}{\bar{\alpha}_p^\varepsilon}\, p + \nabla \cdot \mathbf{w} = 0, \ \ (\mathbf{x}, t) \in \Omega_T = \Omega \times (0, T), \tag{7.0.1}$$

$$\rho^\varepsilon \frac{\partial^2 \mathbf{w}}{\partial t^2} = \nabla \cdot \mathbb{P} + \rho^\varepsilon \mathbf{F}, \ \ (\mathbf{x}, t) \in \Omega_T, \tag{7.0.2}$$

$$\mathbb{P} = \chi^\varepsilon \bar{\alpha}_\mu \mathbb{D}\Big(x, \frac{\partial \mathbf{w}}{\partial t}\Big) + (1 - \chi^\varepsilon)\bar{\alpha}_\lambda \mathbb{D}(x, \mathbf{w}) - \big(p + \alpha_\vartheta^\varepsilon\, \vartheta\big)\mathbb{I}; \tag{7.0.3}$$

$$\eta_0^\varepsilon \frac{\partial \vartheta}{\partial t} = \nabla \cdot (\alpha_\varkappa^\varepsilon \nabla \vartheta) - \bar{\gamma}_0\, \alpha_\vartheta^\varepsilon\, \nabla \cdot \frac{\partial \mathbf{w}}{\partial t}, \ \ (\mathbf{x}, t) \in \Omega_T, \tag{7.0.4}$$

$$\mathbf{w}(\mathbf{x}, t) = 0, \ \ \vartheta(\mathbf{x}, t) = 0, \ \ (\mathbf{x}, t) \in S_T = S \times (0, T), \tag{7.0.5}$$

$$\mathbf{w}(\mathbf{x}, 0) = \frac{\partial \mathbf{w}}{\partial t}(\mathbf{x}, 0) = 0, \ \ \vartheta(\mathbf{x}, 0) = \vartheta_0(\mathbf{x}), \ \ \mathbf{x} \in \Omega, \tag{7.0.6}$$

in the bounded domain $\Omega = \Omega_f^\varepsilon \cup \Gamma \cup \Omega_s^\varepsilon \subset \mathbb{R}^3$, $\Gamma^\varepsilon = \partial \Omega_f^\varepsilon \cap \partial \Omega_s^\varepsilon$, with a C^2 continuous boundary $S = \partial \Omega$ for $t \in (0, T)$.

This model is derived in Appendix A.

Recall that in (7.0.1)–(7.0.6) the characteristic function $\chi^\varepsilon(\mathbf{x})$ of the domain Ω_f^ε is given by the expression

$$\chi^\varepsilon(\mathbf{x}) = \varsigma_0(\mathbf{x}) \chi\left(\frac{\mathbf{x}}{\varepsilon}\right),$$

where $\varsigma_0(\mathbf{x})$ is the characteristic function of the domain Ω, $\chi(\mathbf{y})$ is the characteristic function of the domain Y_f, and

$$\bar{\alpha}_p^\varepsilon = \chi^\varepsilon \bar{c}_f^{2} + (1 - \chi^\varepsilon)\bar{c}_s^{2}, \ \ \rho^\varepsilon = \chi^\varepsilon \rho_f + (1 - \chi^\varepsilon)\rho_s, \ \ \alpha_\vartheta^\varepsilon = \chi^\varepsilon \beta_f + (1 - \chi^\varepsilon)\beta_s,$$

A. Meirmanov, *Mathematical Models for Poroelastic Flows*,
Atlantis Studies in Differential Equations 1, DOI: 10.2991/978-94-6239-015-7_7,

$$\eta_0^\varepsilon = \chi^\varepsilon c_{p,f} + (1-\chi^\varepsilon)c_{p,s}, \ \alpha_\varkappa^\varepsilon = \chi^\varepsilon \alpha_{\varkappa,f} + (1-\chi^\varepsilon)\varkappa_s.$$

For the definition of $\bar{\gamma}_0$, $\bar{\alpha}_p^\varepsilon$, $\bar{\alpha}_\mu$, $\bar{\alpha}_\lambda$, η_0^ε, $\alpha_\vartheta^\varepsilon$, and $\alpha_\varkappa^\varepsilon$ see Appendix A.

As usual, the function $p_f^\varepsilon = \chi^\varepsilon p^\varepsilon$ stands for the liquid pressure, and the function $p_s^\varepsilon = (1-\chi^\varepsilon)p^\varepsilon$ stands for the solid pressure.

We also assume that all dimensionless parameters depend on the small parameter ε and the (finite or infinite) limits exist:

$$\lim_{\varepsilon\searrow 0} \bar{\alpha}_\mu(\varepsilon) = \mu_0, \quad \lim_{\varepsilon\searrow 0} \bar{\alpha}_\lambda(\varepsilon) = \lambda_0, \quad \lim_{\varepsilon\searrow 0} \bar{\alpha}_{\varkappa,f}(\varepsilon) = \varkappa_f,$$

$$\lim_{\varepsilon\searrow 0} \frac{\bar{\alpha}_\mu}{\varepsilon^2} = \mu_1, \quad \lim_{\varepsilon\searrow 0} \frac{\bar{\alpha}_\lambda}{\varepsilon^2} = \lambda_1, \quad \lim_{\varepsilon\searrow 0} \frac{\bar{\alpha}_{\varkappa,f}}{\varepsilon^2} = \varkappa_1.$$

Throughout this chapter it is assumed that

$$\int_{Q_T} \left(|\mathbf{F}(\mathbf{x},t)|^2 + \left|\frac{\partial \mathbf{F}}{\partial t}(\mathbf{x},t)\right|^2 \right) dxdt = F^2 < \infty,$$

$$0 < c_{p,f}, \ c_{p,s}, \ \bar{c}_f, \ \bar{c}_s, \ \varkappa_s, \ \bar{\gamma}_0 < \infty,$$

and that Assumptions 0.1, 1.1 and 3.1 hold true.

Definition 7.1 We say that the triple of functions $\{\mathbf{w}^\varepsilon, \ p^\varepsilon, \ \vartheta^\varepsilon\}$ such that

$$\mathbf{w}^\varepsilon \in \overset{\circ}{\mathbf{W}}{}_2^{1,0}(\Omega_T), \ \frac{\partial \mathbf{w}^\varepsilon}{\partial t} \in \mathbf{L}_2(\Omega_T), \ p^\varepsilon \in L_2(\Omega_T), \ \vartheta^\varepsilon \in \overset{\circ}{W}{}_2^{1,0}(\Omega_T)$$

is a weak solution of the problem (7.0.1)–(7.0.6), if it satisfies the continuity equation (7.0.1) almost everywhere in Ω_T, the first initial condition in (7.0.6)

$$\mathbf{w}^\varepsilon(\mathbf{x},0) = 0, \ \mathbf{x} \in \Omega,$$

and integral identities

$$\int_{\Omega_T} \Big(-\rho^\varepsilon \frac{\partial \mathbf{w}^\varepsilon}{\partial t} \cdot \frac{\partial \varphi}{\partial t} + \mathbb{P} : \mathbb{D}(x,\varphi) \Big) dxdt = -\int_{\Omega_T} \rho^\varepsilon \mathbf{F} \cdot \varphi dxdt, \tag{7.0.7}$$

$$\int_{\Omega_T} \Big(-\eta_0^\varepsilon \vartheta^\varepsilon \frac{\partial \psi}{\partial t} + \alpha_\varkappa^\varepsilon \nabla \vartheta^\varepsilon \cdot \nabla \psi - \bar{\gamma}_0 \alpha_\vartheta^\varepsilon (\nabla \cdot \mathbf{w}) \frac{\partial \psi}{\partial t} \Big) dxdt = \int_{\Omega_T} \eta_0^\varepsilon \vartheta_0(\mathbf{x}) \psi(\mathbf{x},0) dxdt \tag{7.0.8}$$

for all functions φ and ψ, such that $\varphi \in \overset{\circ}{\mathbf{W}}{}_2^{1,0}(\Omega_T)$, $\dfrac{\partial \varphi}{\partial t} \in \mathbf{L}_2(\Omega_T)$, $\psi \in \overset{\circ}{W}{}_2^{1,0}(\Omega_T)$, $\dfrac{\partial \psi}{\partial t} \in L_2(\Omega_T)$ and $\varphi(\mathbf{x},T) = 0$, $\psi(\mathbf{x},T) = 0$ for $\mathbf{x} \in \Omega$.

Theorem 7.1 *For all $\varepsilon > 0$ and for an arbitrary time interval $[0, T]$ there exists a unique generalized solution of problem (7.0.1)–(7.0.6) and*

$$\max_{0<t<T}\int_{\Omega}\Big(|p^{\varepsilon}(\mathbf{x},t)|^2+|\vartheta^{\varepsilon}(\mathbf{x},t)|^2+\Big|\frac{\partial \mathbf{w}^{\varepsilon}}{\partial t}(\mathbf{x},t)\Big|^2+(1-\chi^{\varepsilon})\bar{\alpha}_{\lambda}|\mathbb{D}(x,\mathbf{w}^{\varepsilon})|^2\Big)dx$$
$$+\max_{0<t<T}\int_{\Omega}\Big(\Big|\frac{\partial p^{\varepsilon}}{\partial t}(\mathbf{x},t)\Big|^2+\Big|\frac{\partial \vartheta^{\varepsilon}}{\partial t}(\mathbf{x},t)\Big|^2+\Big|\frac{\partial^2 \mathbf{w}^{\varepsilon}}{\partial t^2}(\mathbf{x},t)\Big|^2+(1-\chi^{\varepsilon})\bar{\alpha}_{\lambda}\Big|\mathbb{D}\Big(x,\frac{\partial \mathbf{w}^{\varepsilon}}{\partial t}\Big)\Big|^2\Big)dx$$
$$+\int_{\Omega_T}\chi^{\varepsilon}\Big(\bar{\alpha}_{\mu}\Big(\Big|\mathbb{D}\Big(x,\frac{\partial \mathbf{w}^{\varepsilon}}{\partial t}\Big)\Big|^2+\Big|\mathbb{D}\Big(x,\frac{\partial^2 \mathbf{w}^{\varepsilon}}{\partial t^2}\Big)\Big|^2\Big)+\bar{\alpha}_{\varkappa,f}\Big(|\nabla\vartheta^{\varepsilon}|^2+\Big|\nabla\frac{\partial\vartheta^{\varepsilon}}{\partial t}\Big|^2\Big)\Big)dxdt$$
$$+\int_{\Omega_T}(1-\chi^{\varepsilon})\Big(|\nabla\vartheta^{\varepsilon}|^2+\Big|\nabla\frac{\partial\vartheta^{\varepsilon}}{\partial t}\Big|^2\Big)dxdt\leqslant C_0F^2, \qquad (7.0.9)$$

where here and in what follows, we denote as C_0 any constant depending only on domains Ω, Y and Y_f.

The proof of this theorem repeats the proofs of similar theorems in the previous chapters and is based on the energy equalities

$$\frac{1}{2}\frac{d}{dt}\int_{\Omega}\Big(\rho^{\varepsilon}\Big|\frac{\partial \mathbf{w}^{\varepsilon}}{\partial t}\Big|^2+(1-\chi^{\varepsilon})\bar{\alpha}_{\lambda}\,|\mathbb{D}(x,\mathbf{w}^{\varepsilon})|^2+\frac{1}{\bar{\alpha}_p^{\varepsilon}}|p^{\varepsilon}|^2+\frac{\eta_0^{\varepsilon}}{\bar{\gamma}_0}|\vartheta^{\varepsilon}|^2\Big)dx$$
$$+\int_{\Omega}\Big(\chi^{\varepsilon}\,\bar{\alpha}_{\mu}\,\Big|\mathbb{D}\Big(x,\frac{\partial \mathbf{w}^{\varepsilon}}{\partial t}\Big)\Big|^2+\frac{\alpha_{\varkappa}^{\varepsilon}}{\bar{\gamma}_0}|\nabla\vartheta^{\varepsilon}|^2\Big)dx=\int_{\Omega}\rho^{\varepsilon}\mathbf{F}\cdot\frac{\partial \mathbf{w}^{\varepsilon}}{\partial t}dx,$$

$$\frac{1}{2}\frac{d}{dt}\int_{\Omega}\Big(\rho^{\varepsilon}\Big|\frac{\partial^2 \mathbf{w}^{\varepsilon}}{\partial t^2}\Big|^2+(1-\chi^{\varepsilon})\bar{\alpha}_{\lambda}\,\Big|\mathbb{D}\Big(x,\frac{\partial \mathbf{w}^{\varepsilon}}{\partial t}\Big)\Big|^2+\frac{1}{\bar{\alpha}_p^{\varepsilon}}\Big|\frac{\partial p^{\varepsilon}}{\partial t}\Big|^2+\frac{\eta_0^{\varepsilon}}{\bar{\gamma}_0}\Big|\frac{\partial\vartheta^{\varepsilon}}{\partial t}\Big|^2\Big)dx$$
$$+\int_{\Omega}\Big(\chi^{\varepsilon}\,\bar{\alpha}_{\mu}\,\Big|\mathbb{D}\Big(x,\frac{\partial^2 \mathbf{w}^{\varepsilon}}{\partial t^2}\Big)\Big|^2+\frac{\alpha_{\varkappa}^{\varepsilon}}{\bar{\gamma}_0}\Big|\nabla\frac{\partial\vartheta^{\varepsilon}}{\partial t}\Big|^2\Big)dx=\int_{\Omega}\rho^{\varepsilon}\frac{\partial \mathbf{F}}{\partial t}\cdot\frac{\partial^2 \mathbf{w}^{\varepsilon}}{\partial t^2}dx,$$

For example, the first equality follows from the Eq. (7.0.2), if we multiply it by $\dfrac{\partial \mathbf{w}^{\varepsilon}}{\partial t}$, integrate by parts, express the stress tensor $\mathbb{P}$ and $\nabla\cdot\mathbf{w}$ there using state equations (7.0.3) and continuity equation (7.0.1), then multiply the Eq. (7.0.4) by $\dfrac{\vartheta^{\varepsilon}}{\bar{\gamma}_0}$, integrate by parts and sum with the previous result.

7.1 A Slightly Viscous Liquid in an Extremely Elastic Solid Skeleton

Throughout this section we assume that

$$\mu_0=0,\;\;\lambda_0=0. \qquad (7.1.1)$$

7.1.1 Main Results

Theorem 7.2 *Let $\{\mathbf{v}^{\varepsilon},\ p^{\varepsilon},\ \vartheta^{\varepsilon}\}$ be a weak solution of the problem (7.0.1)–(7.0.6) and*

$$0 < \varkappa_f < \infty, \ \ \mu_1 = \lambda_1 = \infty.$$

Then the sequences $\{\mathbf{w}^{\varepsilon}\}$ and $\{p^{\varepsilon}\}$ converge as $\varepsilon \to 0$ weakly and two-scale in $\mathbf{L}_2(\Omega_T)$ and $L_2(\Omega_T)$ to the displacements $\mathbf{w}$ and the pressure $p \in W_2^{1,0}(\Omega_T)$ of the mixture correspondingly, and the sequence $\{\vartheta^{\varepsilon}\}$ converges as $\varepsilon \to 0$ weakly in $\overset{\circ}{W}{}_2^{1,0}(\Omega_T)$ and two-scale in $L_2(\Omega_T)$ to the temperature $\vartheta(\mathbf{x},t)$ of the mixture.

These limiting functions satisfy in the domain Ω_T the system of homogenized differential equations

$$\hat{\rho}\,\frac{\partial^2 \mathbf{w}}{\partial t^2} = \nabla\,(p + \hat{\beta}\,\vartheta) + \hat{\rho}\,\mathbf{F}, \tag{7.1.2}$$

$$\frac{1}{\tilde{c}^2}\,p + \nabla\cdot\mathbf{w} = 0, \tag{7.1.3}$$

$$\hat{c}_p\frac{\partial \vartheta}{\partial t} - \hat{\gamma}\,\frac{\partial p}{\partial t} = \nabla\cdot\big(\mathbb{B}^{\vartheta}\cdot\nabla\vartheta\big), \tag{7.1.4}$$

completed with boundary and initial conditions

$$\mathbf{w}(\mathbf{x},t)\cdot\mathbf{n}(\mathbf{x}) = 0, \ \ (\mathbf{x},t)\in S_T, \tag{7.1.5}$$

$$\vartheta(\mathbf{x},t) = 0, \ \ (\mathbf{x},t)\in S_T, \tag{7.1.6}$$

$$\mathbf{w}(\mathbf{x},0) = \frac{\partial \mathbf{w}}{\partial t}(\mathbf{x},0) = 0, \ \ \mathbf{x}\in\Omega, \tag{7.1.7}$$

$$\vartheta(\mathbf{x},0) = \vartheta_0(\mathbf{x}), \ \ \mathbf{x}\in\Omega. \tag{7.1.8}$$

In (7.1.2)–(7.1.8)

$$\hat{\rho} = m\rho_f + (1-m)\,\rho_s, \ \ \frac{1}{\tilde{c}^2} = \frac{m}{\bar{c}_f^2} + \frac{(1-m)}{\bar{c}_s^2},$$

$$\hat{\beta} = m\beta_f + (1-m)\,\beta_s, \ \ \hat{c_p} = mc_{p,f} + (1-m)\,c_{p,s},$$

$$m = \int_Y \chi(\mathbf{y})dy, \ \ \hat{\gamma} = m\gamma_f + (1-m)\,\gamma_s,$$

$$\gamma_f = \bar{\gamma}_0\,\frac{\beta_f}{\bar{c}_f^2}, \ \ \gamma_s = \bar{\gamma}_0\,\frac{\beta_s}{\bar{c}_s^2},$$

the matrix $\mathbb{B}^{\vartheta}$ *is defined in Chap. 2 (see Theorem 2.1), and* $\mathbf{n}(\mathbf{x})$ *is the normal vector to the boundary* S *at the point* $\mathbf{x} \in S$.

We refer to the problem (7.1.2)–(7.1.8) as the homogenized **model** $(\mathbb{NIA})_1$.
To formulate the following statements we consider an extension

$$\vartheta_s^{\varepsilon} = E_{\Omega_s^{\varepsilon}}(\vartheta^{\varepsilon}), \tag{7.1.9}$$

where

$$E_{\Omega_s^{\varepsilon}} : W_2^1(\Omega_s^{\varepsilon}) \to W_2^1(\Omega)$$

is an extension operator from Ω_s^{ε} on Ω, such that

$$\vartheta_s^{\varepsilon} = \vartheta^{\varepsilon} \text{ in } \Omega_s^{\varepsilon} \times (0, T),$$

and

$$\int_{\Omega} |\vartheta_s^{\varepsilon}|^2 dx \leqslant C_0 \int_{\Omega_s^{\varepsilon}} |\vartheta^{\varepsilon}|^2 dx, \quad \int_{\Omega} |\nabla \vartheta_s^{\varepsilon}|^2 dx \leqslant C_0 \int_{\Omega_s^{\varepsilon}} |\nabla \vartheta^{\varepsilon}|^2 dx. \tag{7.1.10}$$

(for more details see Appendix B, Lemma B.4.1).

Theorem 7.3 *Let* $\{\mathbf{v}^{\varepsilon},\ p^{\varepsilon},\ \vartheta^{\varepsilon}\}$ *be a weak solution of the problem (7.0.1)–(7.0.6),* $\vartheta_s^{\varepsilon} = E_{\Omega_s^{\varepsilon}}(\vartheta^{\varepsilon})$ *be an extension (7.1.9), and*

$$\varkappa_f = 0,\ 0 < \varkappa_1 < \infty,\ \ \mu_1 = \lambda_1 = \infty.$$

Then the sequence $\{p^{\varepsilon}\}$ *converges weakly and two-scale in* $L_2(\Omega_T)$ *as* $\varepsilon \to 0$ *to the pressure of the mixture* $p(\mathbf{x}, t)$*, the sequences* $\{\mathbf{w}^{\varepsilon}\}$ *and* $\{\chi^{\varepsilon}\,\vartheta^{\varepsilon}\}$ *converge weakly in* $\mathbf{L}_2(\Omega_T)$ *and* $L_2(\Omega_T)$ *as* $\varepsilon \to 0$ *to the displacements* $\mathbf{w}(\mathbf{x}, t)$ *of the mixture and the liquid temperature* $\vartheta^{(f)}(\mathbf{x}, t)$ *correspondingly, and the sequence* $\{\vartheta_s^{\varepsilon}\}$ *converges weakly in* $\overset{\circ}{W}{}_2^{1,0}(\Omega_T)$ *and two-scale in* $L_2(\Omega_T)$ *as* $\varepsilon \to 0$ *to the temperature* $\vartheta_s(\mathbf{x}, t)$ *of the solid component.*

These limiting functions satisfy in the domain Ω_T *the homogenized momentum balance equation*

$$\hat{\rho}\,\frac{\partial^2 \mathbf{w}}{\partial t^2} = \nabla\left(p + \beta_f\,\vartheta^{(f)} + \beta_s(1-m)\vartheta_s\right) + \hat{\rho}\,\mathbf{F}, \tag{7.1.11}$$

the homogenized continuity equation (7.1.3), and the homogenized heat equation

$$c_{p,f}\frac{\partial \vartheta^{(f)}}{\partial t} + c_{p,s}(1-m)\frac{\partial \vartheta_s}{\partial t} - \hat{\gamma}\,\frac{\partial p}{\partial t} = \nabla\cdot\left(\varkappa_s\,\mathbb{B}_s^{\vartheta}\cdot\nabla\vartheta_s\right), \tag{7.1.12}$$

completed with the state equation

$$\vartheta^{(f)}(\mathbf{x},t) = m\vartheta_s(\mathbf{x},t) + \int_0^t a^{(f)}(t-\tau)\left(\gamma_f \frac{\partial p}{\partial \tau} - c_{p,f}\frac{\partial \vartheta_s}{\partial \tau}\right)(\mathbf{x},\tau)d\tau, \quad (7.1.13)$$

and boundary and initial conditions (7.1.5), (7.1.7), and

$$\vartheta_s(\mathbf{x},t) = 0, \ \ (\mathbf{x},t) \in S_T, \quad (7.1.14)$$

$$\vartheta_s(\mathbf{x},0) = \vartheta_0(\mathbf{x}), \ \ \mathbf{x} \in \Omega. \quad (7.1.15)$$

In (7.1.11)–(7.1.15)

$$\hat{\rho} = m\rho_f + (1-m)\,\rho_s, \ \ m = \int_Y \chi(\mathbf{y})dy,$$

$$\hat{\gamma} = m\gamma_f + (1-m)\,\gamma_s, \ \ \gamma_f = \bar{\gamma}_0 \frac{\beta_f}{c_f^2\, c_{p,f}}, \ \ \gamma_s = \bar{\gamma}_0\, \frac{\beta_s}{c_s^2},$$

the matrix $\mathbb{B}_s^{\vartheta}$ and the function $a^{(f)}(t)$ are defined below by (7.1.39) and (7.1.45), and $\mathbf{n}(\mathbf{x})$ is the normal vector to the boundary S at the point $\mathbf{x} \in S$.

We refer to the problem (7.1.3), (7.1.11)–(7.1.15) as the homogenized **model** $(\mathbb{NIA})_2$.

Theorem 7.4 *Let $\{\mathbf{w}^{\varepsilon},\ p^{\varepsilon},\ \vartheta^{\varepsilon}\}$ be a weak solution of the problem (7.0.1)–(7.0.6) and*

$$0 < \varkappa_f < \infty, \ \ 0 \leqslant \mu_1, \ \lambda_1 < \infty.$$

Then the sequence $\{p^{\varepsilon}\}$ converges weakly and two-scale in $L_2(\Omega_T)$ as $\varepsilon \to 0$ to the pressure $p(\mathbf{x},t)$ of the mixture, the sequence $\{\mathbf{w}^{\varepsilon}\}$ converges weakly in $\mathbf{L}_2(\Omega_T)$ as $\varepsilon \to 0$ to the displacements $\mathbf{w}(\mathbf{x},t)$ of the mixture, and the sequence $\{\vartheta^{\varepsilon}\}$ converges weakly in $\overset{\circ}{W}{}_2^{1,0}(\Omega_T)$ and two-scale in $L_2(\Omega_T)$ as $\varepsilon \to 0$ to the temperature $\vartheta(\mathbf{x},t)$ of the mixture.

These limiting functions, where $p \in W_2^{1,0}(\Omega_T)$ and $\dfrac{\partial p}{\partial t} \in L_2(\Omega_T)$, satisfy in the domain Ω_T homogenized momentum balance equation

$$\nabla \cdot \int_0^t \mathbb{B}^{(a)}(\mu_1,\lambda_1; t-\tau)\cdot\nabla\big(p + \hat{\beta}\,\vartheta\big)(\mathbf{x},\tau)d\tau = -\frac{\hat{\rho}}{\bar{c}^2}\frac{\partial p}{\partial t} - \nabla\cdot\mathbf{f}, \quad (7.1.16)$$

and the homogenized heat equation (7.1.4), completed with boundary and initial conditions (7.1.6), (7.1.8), and

$$\int_0^t \mathbb{B}^{(a)}(\mu_1,\lambda_1; t-\tau)\cdot\nabla\big(p + \hat{\beta}\,\vartheta\big)(\mathbf{x},\tau)d\tau\cdot\mathbf{n}(\mathbf{x}) = -\mathbf{f}(\mathbf{x},t)\cdot\mathbf{n}(\mathbf{x}), \quad (7.1.17)$$

$$p(\mathbf{x},0)=0,\ \ \mathbf{x}\in\Omega. \tag{7.1.18}$$

In (7.1.16), (7.1.17)

$$\hat{\rho}=m\,\rho_f+(1-m)\,\rho_s,\quad \frac{1}{\bar{c}^2}=\frac{m}{\bar{c}_f^2}+\frac{(1-m)}{\bar{c}_s^2},$$

$$\hat{\beta}=m\,\beta_f+(1-m)\,\beta_s,\ \ m=\int_Y\chi(\mathbf{y})dy,$$

the matrix $\mathbb{B}^{(a)}(\mu_1,\lambda_1;t)$ *and the function* $\mathbf{f}(\mathbf{x},t)$ *have been defined in Chap. 6 by formulae (6.2.40)–(6.2.42), and* $\mathbf{n}(\mathbf{x})$ *is the normal vector to the boundary S at the point* $\mathbf{x}\in S$.

We refer to the problem (7.1.4), (7.1.6), (7.1.8), (7.1.16)–(7.1.18) as the homogenized **model** $(\mathbb{NIA})_3$.

Theorem 7.5 *Let* $\{\mathbf{w}^\varepsilon,\ p^\varepsilon,\ \vartheta^\varepsilon\}$ *be a weak solution of the problem (7.0.1)–(7.0.6),* $\vartheta_s^\varepsilon=E_{\Omega_s^\varepsilon}(\vartheta^\varepsilon)$ *be an extension (7.1.9), and*

$$\varkappa_f=0,\ \ \varkappa_1=\infty,\ \ 0\leqslant\mu_1,\ \lambda_1<\infty.$$

Then the sequence $\{p^\varepsilon\}$ *converges weakly and two-scale in* $L_2(\Omega_T)$ *as* $\varepsilon\to 0$ *to the pressure* $p(\mathbf{x},t)$ *of the mixture, the sequence* $\{\mathbf{w}^\varepsilon\}$ *converges weakly in* $\mathbf{L}_2(\Omega_T)$ *as* $\varepsilon\to 0$ *to the displacements* $\mathbf{w}(\mathbf{x},t)$ *of the mixture. At the same time the sequence* $\{\vartheta_s^\varepsilon\}$ *converges weakly in* $\overset{\circ}{W}{}_2^{1,0}(\Omega_T)$ *and two-scale in* $L_2(\Omega_T)$ *as* $\varepsilon\to 0$ *to the temperature* $\vartheta_s(\mathbf{x},t)$ *of the solid component, and the sequence* $\{\chi^\varepsilon\,\vartheta^\varepsilon\}$ *converges weakly and two-scale in* $L_2(\Omega_T)$ *as* $\varepsilon\to 0$ *to the liquid temperature* $\vartheta^{(f)}(\mathbf{x},t)=m\vartheta_s(\mathbf{x},t)$.

These limiting functions, where $p\in W_2^{1,0}(\Omega_T)$ *and* $\dfrac{\partial p}{\partial t}\in L_2(\Omega_T)$*, satisfy in the domain* Ω_T *the homogenized momentum balance equation (7.1.16) and the homogenized heat equation*

$$\hat{c}_p\frac{\partial\vartheta_s}{\partial t}-\hat{\gamma}\,\frac{\partial p}{\partial t}=\nabla\cdot\big(\varkappa_s\,\mathbb{B}_s^\vartheta\cdot\nabla\vartheta_s\big), \tag{7.1.19}$$

completed with boundary and initial conditions (7.1.14), (7.1.15), (7.1.17), and (7.1.18).

In (7.1.19)

$$\hat{c_p}=mc_{p,f}+(1-m)\,c_{p,s},\ \ m=\int_Y\chi(\mathbf{y})dy,$$

$$\hat{\gamma}=m\gamma_f+(1-m)\,\gamma_s,\ \ \gamma_f=\bar{\gamma}_0\,\frac{\beta_f}{c_f^2},\ \ \gamma_s=\bar{\gamma}_0\,\frac{\beta_s}{c_s^2},$$

and the matrix $\mathbb{B}^{\vartheta}_{s}$ *is defined below by (7.1.39).*

We refer to the problem (7.1.14)–(7.1.19) as the homogenized **model** $(\mathbb{NIA})_4$. To formulate the following statements we consider extensions

$$\mathbf{w}^{\varepsilon}_{f} = \mathbb{E}_{\Omega^{\varepsilon}_{f}}\big(\mathbf{w}^{\varepsilon}\big), \text{ and } \mathbf{w}^{\varepsilon}_{s} = \mathbb{E}_{\Omega^{\varepsilon}_{s}}\big(\mathbf{w}^{\varepsilon}\big), \tag{7.1.20}$$

where

$$\mathbb{E}_{\Omega^{\varepsilon}_{f}} : \mathbf{W}^{1}_{2}(\Omega^{\varepsilon}_{f}) \to \mathbf{W}^{1}_{2}(\Omega)$$

is an extension operator from Ω^{ε}_{f} on Ω, and

$$\mathbb{E}_{\Omega^{\varepsilon}_{s}} : \mathbf{W}^{1}_{2}(\Omega^{\varepsilon}_{s}) \to \mathbf{W}^{1}_{2}(\Omega)$$

is an extension operator from Ω^{ε}_{s} on Ω, such that

$$\mathbf{w}^{\varepsilon}_{f} = \mathbf{w}^{\varepsilon} \text{ in } \Omega^{\varepsilon}_{f} \times (0, T), \quad \mathbf{w}^{\varepsilon}_{s} = \mathbf{w}^{\varepsilon} \text{ in } \Omega^{\varepsilon}_{s} \times (0, T),$$

and

$$\int_{\Omega} |\mathbf{w}^{\varepsilon}_{f}|^2 dx \leqslant C_0 \int_{\Omega^{\varepsilon}_{f}} |\mathbf{w}^{\varepsilon}|^2 dx, \quad \int_{\Omega} |\mathbf{w}^{\varepsilon}_{s}|^2 dx \leqslant C_0 \int_{\Omega^{\varepsilon}_{s}} |\mathbf{w}^{\varepsilon}|^2 dx,$$
$$\int_{\Omega} |\mathbb{D}(x, \mathbf{w}^{\varepsilon}_{f})|^2 dx \leqslant C_0 \int_{Q^{\varepsilon}_{f}} |\mathbb{D}(x, \mathbf{w}^{\varepsilon})|^2 dx,$$
$$\int_{\Omega} |\mathbb{D}(x, \mathbf{w}^{\varepsilon}_{s})|^2 dx \leqslant C_0 \int_{\Omega^{\varepsilon}_{s}} |\mathbb{D}(x, \mathbf{w}^{\varepsilon})|^2 dx.$$

(for more details see Appendix B, Lemma B.9).

Theorem 7.6 *Let* $\{\mathbf{w}^{\varepsilon},\ p^{\varepsilon},\ \vartheta^{\varepsilon}\}$ *be a weak solution of the problems (7.0.1)–(7.0.6),*

$$0 < \varkappa_f < \infty, \ \mu_1 = \infty, \ 0 \leqslant \lambda_1 < \infty,$$

and $\mathbf{w}^{\varepsilon}_{f} = \mathbb{E}_{\Omega^{\varepsilon}_{f}}\big(\mathbf{w}^{\varepsilon}\big)$ *be an extension (7.1.20).*

Then the sequences $\{p^{\varepsilon}\}$ *and* $\{\mathbf{w}^{\varepsilon}_{f}\}$*, converge weakly and two-scale in* $L_2(\Omega_T)$ *and* $\mathbf{L}_2(\Omega_T)$ *as* $\varepsilon \searrow 0$ *to the pressure* $p(\mathbf{x}, t)$ *and the displacements* $\mathbf{w}_f(\mathbf{x}, t)$ *of the liquid component respectively. The sequence* $\{(1 - \chi^{\varepsilon})\mathbf{w}^{\varepsilon}\}$ *converges weakly in* $\mathbf{L}_2(\Omega_T)$ *as* $\varepsilon \searrow 0$ *to the displacements* $\mathbf{w}^{(s)}$ *of the solid component, and the sequence* $\{\vartheta^{\varepsilon}\}$ *converges weakly in* $\overset{\circ}{W}{}^{1,0}_{2}(\Omega_T)$ *and two-scale in* $L_2(\Omega_T)$ *as* $\varepsilon \to 0$ *to the temperature of the mixture* $\vartheta(\mathbf{x}, t)$.

These limiting functions, where $p \in W^{1,0}_{2}(\Omega_T)$ *and* $\dfrac{\partial p}{\partial t} \in L_2(\Omega_T)$*, satisfy in the domain* Ω_T *the system of homogenized differential equations, consisting of the continuity equation*

$$\frac{1}{\bar{c}^2}\frac{\partial p}{\partial t} + \nabla \cdot \left(m\frac{\partial \mathbf{w}_f}{\partial t} + \frac{\partial \mathbf{w}^{(s)}}{\partial t} \right) = 0, \tag{7.1.21}$$

the momentum balance equation

$$m\rho_f \frac{\partial \mathbf{w}_f}{\partial t} + \rho_s \frac{\partial \mathbf{w}^{(s)}}{\partial t} + \int_0^t \big(-\hat{\rho}\mathbf{F} + \nabla\, (p + \hat{\beta}\,\vartheta)\big)(\mathbf{x}, \tau)d\tau = 0 \tag{7.1.22}$$

for the liquid component, the momentum balance equation

$$\frac{\partial \mathbf{w}^{(s)}}{\partial t} - (1-m)\frac{\partial \mathbf{w}_f}{\partial t} \tag{7.1.23}$$

$$= -\int_0^t \mathbb{B}^{(s)}(\infty, \lambda_1; t-\tau) \cdot \left(\nabla (p + \beta_s \vartheta) + \rho_s \left(\frac{\partial^2 \mathbf{w}_f}{\partial \tau^2} - \mathbf{F} \right) \right)(\mathbf{x}, \tau)d\tau \tag{7.1.23}$$

for the solid component, and the heat equation (7.1.4).

Equations (7.1.4), (7.1.21)–(7.1.23) are supplemented with boundary and initial conditions (7.1.6), (7.1.8), and initial conditions

$$\mathbf{w}^{(s)}(\mathbf{x}, 0) = \mathbf{w}_f(\mathbf{x}, 0) = 0 \tag{7.1.24}$$

for displacements in the liquid and the solid components, and the boundary condition

$$\left(m\frac{\partial \mathbf{w}_f}{\partial t}(\mathbf{x}, t) + \frac{\partial \mathbf{w}^{(s)}}{\partial t}(\mathbf{x}, t) \right) \cdot \mathbf{n}(\mathbf{x}) = 0, \;\; \mathbf{x} \in S, \;\; t > 0. \tag{7.1.25}$$

In (7.1.21)–(7.1.25)

$$\hat{\rho} = m\rho_f + (1-m)\,\rho_s, \quad \frac{1}{\bar{c}^2} = \frac{m}{\bar{c}_f^2} + \frac{(1-m)}{\bar{c}_s^2},$$

$$\hat{\beta} = m\,\beta_f + (1-m)\,\beta_s, \;\; m = \int_Y \chi(\mathbf{y})dy,$$

and the matrix $\mathbb{B}^{(s)}(\infty, \lambda_1; t)$ has been defined in Chap. 3 by formulae (3.2.47) and (3.2.54).

We refer to the problem (7.1.4), (7.1.6), (7.1.8), (7.1.21)–(7.1.25) as the homogenized **model** $(\mathbb{NIA})_5$.

Theorem 7.7 *Let $\{\mathbf{w}^\varepsilon, \; p^\varepsilon, \; \vartheta^\varepsilon\}$ be a weak solution of the problem (7.0.1)–(7.0.6),*

$$\varkappa_f = 0, \;\; 0 < \varkappa_1 < \infty, \;\; \mu_1 = \infty, \;\; 0 \leqslant \lambda_1 < \infty,$$

$\vartheta_s^\varepsilon = E_{\Omega_s^\varepsilon}(\vartheta^\varepsilon)$ be an extension (7.1.9), and $\mathbf{w}_f^\varepsilon = \mathbb{E}_{\Omega_f^\varepsilon}\left(\mathbf{w}^\varepsilon\right)$ be an extension (7.1.20).

Then the sequences $\{p^\varepsilon\}$ and $\{\mathbf{w}_f^\varepsilon\}$, converge weakly and two-scale in $L_2(\Omega_T)$ as $\varepsilon \searrow 0$ and $\mathbf{L}_2(\Omega_T)$ to the pressure $p(\mathbf{x},t)$ and the displacements $\mathbf{w}_f(\mathbf{x},t)$ of the liquid component respectively, the sequences $\{(1-\chi^\varepsilon)\mathbf{w}^\varepsilon\}$ and $\{\chi^\varepsilon\,\vartheta^\varepsilon\}$ converge weakly in $\mathbf{L}_2(\Omega_T)$ and $L_2(\Omega_T)$ as $\varepsilon \searrow 0$ to the displacements $\mathbf{w}^{(s)}$ of the solid component and the liquid temperature $\vartheta^{(f)}(\mathbf{x},t)$, and the sequence $\{\vartheta_s^\varepsilon\}$ converges weakly in $\overset{\circ}{W}{}_2^{1,0}(\Omega_T)$ and two-scale in $L_2(\Omega_T)$ as $\varepsilon \to 0$ to the limiting temperature $\vartheta_s(\mathbf{x},t)$ of the solid component.

These limiting functions, where $p \in W_2^{1,0}(\Omega_T)$ and $\dfrac{\partial p}{\partial t} \in L_2(\Omega_T)$, satisfy in the domain Ω_T the system of homogenized differential equations, consisting of the continuity equation (7.1.21), the momentum balance equation

$$m\rho_f\frac{\partial \mathbf{w}_f}{\partial t} + \rho_s\frac{\partial \mathbf{w}^{(s)}}{\partial t}$$
$$= -\int_0^t \Big(-\hat{\rho}\mathbf{F} + \nabla\left(p + \beta_f\,\vartheta^{(f)} + (1-m)\beta_s\,\vartheta_s\right)\Big)(\mathbf{x},\tau)d\tau \qquad (7.1.26)$$

for the liquid component, the momentum balance equation

$$\frac{\partial \mathbf{w}^{(s)}}{\partial t} - (1-m)\frac{\partial \mathbf{w}_f}{\partial t}$$
$$= -\int_0^t \mathbb{B}^{(s)}(\infty,\lambda_1;t-\tau)\cdot\left(\nabla(p+\beta_s\vartheta_s) + \rho_s\left(\frac{\partial^2\mathbf{w}_f}{\partial\tau^2} - \mathbf{F}\right)\right)(\mathbf{x},\tau)d\tau \qquad (7.1.27)$$

for the solid component, and the heat equation (7.1.12).

Equations (7.1.10), (7.1.21), (7.1.26) and (7.1.27) are supplemented with boundary and initial conditions (7.1.14), (7.1.15), (7.1.24) and (7.1.25).

In (7.1.26) and (7.1.27)

$$\hat{\rho} = m\rho_f + (1-m)\,\rho_s, \quad m = \int_Y \chi(\mathbf{y})d\mathbf{y},$$

and the matrix $\mathbb{B}^{(s)}(\infty,\lambda_1;t)$ has been defined in Chap. 3 by formulae (3.2.47) and (3.2.54).

We refer to the problem (7.1.10), (7.1.14), (7.1.15), (7.1.21), (7.1.24)–(7.1.27) as the homogenized **model** $(\mathbb{NIA})_6$.

Theorem 7.8 *Let $\{\mathbf{w}^\varepsilon,\ p^\varepsilon,\ \vartheta^\varepsilon\}$ be a weak solution of the problem (7.0.1)–(7.0.6),*

$$0 < \varkappa_f < \infty,\ \lambda_1 = \infty,\ 0 \leqslant \mu_1 < \infty,$$

and $\mathbf{w}_s^\varepsilon = \mathbb{E}_{\Omega_s^\varepsilon}(\mathbf{w}^\varepsilon)$ *be an extension (7.1.20).*

Then the sequences $\{p^\varepsilon\}$ *and* $\{\mathbf{w}_s^\varepsilon\}$, *converge weakly and two-scale in* $L_2(\Omega_T)$ *and* $\mathbf{L}_2(\Omega_T)$ *as* $\varepsilon \searrow 0$ *to the pressure* $p(\mathbf{x}, t)$ *of the mixture and the displacements* $\mathbf{w}_s(\mathbf{x}, t)$ *of the solid component respectively, the sequence* $\{\chi^\varepsilon \mathbf{w}^\varepsilon\}$ *converges weakly in* $\mathbf{L}_2(\Omega_T)$ *as* $\varepsilon \searrow 0$ *to the displacements* $\mathbf{w}^{(f)}$ *of the liquid component, and the sequence* $\{\vartheta^\varepsilon\}$ *converges weakly in* $\overset{\circ}{W}{}_2^{1,0}(\Omega_T)$ *and two scale in* $L_2(\Omega_T)$ *as* $\varepsilon \to 0$ *to the temperature* $\vartheta(\mathbf{x}, t)$ *of the mixture.*

These limiting functions, where $p \in W_2^{1,0}(\Omega_T)$ *and* $\dfrac{\partial p}{\partial t} \in L_2(\Omega_T)$, *satisfy in the domain* Ω_T *the system of homogenized differential equations, consisting of the continuity equation*

$$\frac{1}{\bar{c}^2}\frac{\partial p}{\partial t} + \nabla \cdot \left(\frac{\partial \mathbf{w}^{(f)}}{\partial t} + (1-m)\frac{\partial \mathbf{w}_s}{\partial t}\right) = 0, \tag{7.1.28}$$

the momentum balance equation

$$\rho_f \frac{\partial \mathbf{w}^{(f)}}{\partial t} + (1-m)\rho_s \frac{\partial \mathbf{w}_s}{\partial t} = \int_0^t \Big(\hat{\rho}\mathbf{F} - \nabla\big(p + \hat{\beta}\,\vartheta\big)\Big)(\mathbf{x}, \tau)d\tau \tag{7.1.29}$$

for the solid component, the momentum balance equation

$$\frac{\partial \mathbf{w}^{(f)}}{\partial t} - m\frac{\partial \mathbf{w}_s}{\partial t}$$
$$= -\int_0^t \mathbb{B}^{(f)}(\mu_1, \infty; t-\tau) \cdot \left(\nabla\left(p + \beta_f\,\vartheta\right) + \rho_f\left(\frac{\partial^2 \mathbf{w}_s}{\partial \tau^2} - \mathbf{F}\right)\right)(\mathbf{x}, \tau)d\tau \tag{7.1.30}$$

for the liquid component, and the heat equation (7.1.4).

Equations (7.1.4), (7.1.28)–(7.1.30) are supplemented with the boundary and initial conditions (7.1.6), (7.1.8), initial conditions

$$\mathbf{w}^{(f)}(\mathbf{x}, 0) = \mathbf{w}_s(\mathbf{x}, 0) = 0, \ \ \mathbf{x} \in \Omega \tag{7.1.31}$$

for displacements in the liquid and the solid components, and the boundary condition

$$\left(\frac{\partial \mathbf{w}^{(f)}}{\partial t} + (1-m)\frac{\partial \mathbf{w}_s}{\partial t}\right) \cdot \mathbf{n}(\mathbf{x}) = 0, \ \ \mathbf{x} \in S, \ \ t > 0. \tag{7.1.32}$$

In (7.1.28)–(7.1.30)

$$\hat{\rho} = m\rho_f + (1-m)\,\rho_s, \quad \frac{1}{\bar{c}^2} = \frac{m}{\bar{c}_f^2} + \frac{(1-m)}{\bar{c}_s^2},$$

$$\hat{\beta} = m\,\beta_f + (1-m)\,\beta_s, \;\; m = \int_Y \chi(\mathbf{y})dy,$$

and the matrix $\mathbb{B}^{(f)}(\mu_1, \infty; t)$ *has been defined in Chap. 3 by formulae (3.2.70) and (3.2.76).*

We refer to the problem (7.1.4), (7.1.6), (7.1.8), (7.1.28)–(7.1.32) as the homogenized **model** $(\mathbb{NIA})_7$.

Theorem 7.9 *Let* $\{\mathbf{w}^\varepsilon,\; p^\varepsilon,\; \vartheta^\varepsilon\}$ *be a weak solution of the problem (7.0.1)–(7.0.6),*

$$\varkappa_f = 0, \;\; \varkappa_1 = \infty, \;\; \lambda_1 = \infty, \;\; 0 \leqslant \mu_1 < \infty,$$

$\vartheta^\varepsilon_s = E_{\Omega^\varepsilon_s}(\vartheta^\varepsilon)$ *be an extension (7.1.9), and* $\mathbf{w}^\varepsilon_s = \mathbb{E}_{\Omega^\varepsilon_s}\left(\mathbf{w}^\varepsilon\right)$ *be an extension (7.1.20).*

Then the sequences $\{p^\varepsilon\}$ *and* $\{\mathbf{w}^\varepsilon_s\}$*, converge weakly and two-scale in* $L_2(\Omega_T)$ *and* $\mathbf{L}_2(\Omega_T)$ *as* $\varepsilon \searrow 0$ *to the pressure* $p(\mathbf{x}, t)$ *of the mixture and the displacements* $\mathbf{w}_s(\mathbf{x}, t)$ *of the solid component respectively, and the sequence* $\{\chi^\varepsilon\,\mathbf{w}^\varepsilon\}$ *converges weakly in* $\mathbf{L}_2(\Omega_T)$ *as* $\varepsilon \searrow 0$ *to the displacements* $\mathbf{w}^{(f)}$ *of the liquid component. At the same time the sequence* $\{\vartheta^\varepsilon_s\}$ *converges weakly in* $\overset{\circ}{W}{}^{1,0}_2(\Omega_T)$ *and two-scale in* $L_2(\Omega_T)$ *as* $\varepsilon \to 0$ *to the limiting temperature* $\vartheta_s(\mathbf{x}, t)$ *of the solid component, and the sequence* $\{\chi^\varepsilon\,\vartheta^\varepsilon\}$ *converges weakly and two-scale in* $L_2(\Omega_T)$ *as* $\varepsilon \to 0$ *to the liquid temperature* $\vartheta^{(f)}(\mathbf{x}, t) = m\vartheta_s(\mathbf{x}, t)$.

These limiting functions, where $p \in W^{1,0}_2(\Omega_T)$ *and* $\dfrac{\partial p}{\partial t} \in L_2(\Omega_T)$*, satisfy in the domain* Ω_T *the system of homogenized differential equations (7.1.19), (7.1.28)–(7.1.30), and boundary and initial conditions (7.1.14), (7.1.15), (7.1.31) and (7.1.32).*

We refer to the problem (7.1.14), (7.1.15), (7.1.19), (7.1.28)–(7.1.32) as the homogenized **model** $(\mathbb{NIA})_8$.

7.1.2 Proofs of Theorems 7.2–7.9

Proofs of Theorems 7.2–7.9 repeat the proofs of corresponding Theorems in Chaps. 1–3 and 6. Dynamic equations are derived in the same way as in Chap. 6. Theorems 7.2–7.3 correspond to Theorem 6.5, Theorems 7.4–7.5 correspond to Theorem 6.6, Theorems 7.6–7.7 correspond to Theorem 6.7, and Theorems 7.8–7.9 correspond to Theorem 6.8.

The basic integral identity (7.1.4) of this chapter differs from the basic identity (6.0.10) in Chap. 6 in only one term

$$\int_{\Omega_T} \alpha^\varepsilon_\vartheta\, \vartheta^\varepsilon\, \nabla \cdot \varphi dxdt.$$

Therefore, the limiting procedure does not cause any difficulties.

The really new features here are the homogenized heat equations (7.1.4), (7.1.12) and (7.1.19).

For all these equations we start with the integral identity (7.0.8) in the form

$$\int_{\Omega_T}\Big(-\eta_0^{\varepsilon}\,\vartheta^{\varepsilon}\,\frac{\partial\psi}{\partial t}+\alpha_{\varkappa}^{\varepsilon}\,\nabla\vartheta^{\varepsilon}\cdot\nabla\psi+\gamma^{\varepsilon}\,p^{\varepsilon}\,\frac{\partial\psi}{\partial t}\Big)dxdt$$
$$=\int_{\Omega_T}\eta_0^{\varepsilon}\vartheta_0(\mathbf{x})\psi(\mathbf{x},0)dxdt, \tag{7.1.33}$$

where

$$\gamma^{\varepsilon}=\chi^{\varepsilon}\gamma_f+(1-\chi^{\varepsilon})\gamma_s,\ \ \gamma_f=\bar{\gamma}_0\,\frac{\beta_f}{c_f^2},\ \ \gamma_s=\bar{\gamma}_0\,\frac{\beta_s}{c_s^2}.$$

If $0<\varkappa_f<\infty$, then estimates (7.0.9) and (7.1.33) result in the macroscopic equation

$$\hat{c}_p\frac{\partial\vartheta}{\partial t}-\hat{\gamma}\,\frac{\partial p}{\partial t}=\nabla\cdot\big\langle\big(\varkappa_f\chi+\varkappa_s(1-\chi)\big)\big(\nabla\,\vartheta+\nabla_y\Theta\big)\big\rangle_Y+\hat{\rho}\,\mathbf{F},$$

the microscopic equation

$$\nabla_y\cdot\Big(\big(\varkappa_f\chi+\varkappa_s(1-\chi)\big)\big(\nabla\,\vartheta+\nabla_y\Theta\big)\Big)=0,$$

and, finally, the homogenized equation (7.1.4) (for details see Chap. 2).

Here $\hat{\gamma}=m\gamma_f+(1-m)\gamma_s$, and $\nabla\,\vartheta(\mathbf{x},t)+\nabla_y\Theta(\mathbf{x},t,\mathbf{y})$ is a two-scale limit of the sequence $\{\nabla\,\vartheta^{\varepsilon}\}$.

If $\varkappa_f=0$ and $0<\varkappa_1<\infty$, then estimates (7.0.9) and (7.1.33) result the macroscopic equation

$$c_{p,f}\frac{\partial\vartheta^{(f)}}{\partial t}+c_{p,s}(1-m)\frac{\partial\vartheta_s}{\partial t}-\hat{\gamma}\,\frac{\partial p}{\partial t}$$
$$=\nabla\cdot\big\langle\big(\varkappa_s(1-\chi)\big)\big(\nabla\,\vartheta_s+\nabla_y\Theta_s\big)\big\rangle_Y+\hat{\rho}\,\mathbf{F} \tag{7.1.34}$$

and microscopic equations

$$\nabla_y\cdot\Big(\big(\varkappa_s(1-\chi)\big)\big(\nabla\,\vartheta_s+\nabla_y\Theta_s\big)\Big)=0, \tag{7.1.35}$$

$$c_{p,f}\,\frac{\partial\,\Theta^{(f)}}{\partial t}=\varkappa_1\triangle_y\,\Theta^{(f)}+\gamma_f\,\frac{\partial p}{\partial t}, \tag{7.1.36}$$

were $\nabla\,\vartheta_s(\mathbf{x},t)+\nabla_y\Theta_s(\mathbf{x},t,\mathbf{y})$ is a two-scale limit of the sequence $\{\nabla\,\vartheta_s^{\varepsilon}\}$, and $\Theta^{(f)}(\mathbf{x},t,\mathbf{y})$ is a two-scale limit of the sequence $\{\chi^{\varepsilon}\,\vartheta^{\varepsilon}\}$.

Recall, that Eq. (7.1.35) is understood in the sense of distributions, and is equivalent to the integral identity

$$\int_Y \Big(((1-\chi))\big(\nabla \vartheta_s + \nabla_y \Theta_s)\big)\Big) : \nabla \varphi \, dy = 0$$

for any smooth 1-periodic function $\varphi(\mathbf{y})$. It means that

$$\triangle_y \Theta_s = 0, \;\; \mathbf{y} \in Y_s, \;\; (\nabla \vartheta_s + \nabla_y \Theta_s) \cdot \mathbf{n}(\mathbf{y}) = 0, \;\; \mathbf{y} \in \gamma. \tag{7.1.37}$$

Here $\mathbf{n}(\mathbf{y})$ is a normal vector to the boundary γ at $\mathbf{y} \in \gamma$.

We look for the solution to (7.1.37) in the form

$$\Theta_s(\mathbf{x}, t, \mathbf{y}) = \sum_{i=1}^{3} \Theta_s^{(i)}(\mathbf{y}) \frac{\partial \vartheta_s}{\partial x_i}(\mathbf{x}, t),$$

where

$$\triangle_y \Theta_s^{(i)} = 0, \;\; \mathbf{y} \in Y_s, \;\; (\mathbf{e}_i + \nabla_y \Theta_s^{(i)}) \cdot \mathbf{n}(\mathbf{y}) = 0, \;\; \mathbf{y} \in \gamma. \tag{7.1.38}$$

Then

$$\nabla_y \Theta_s = \left(\sum_{i=1}^{3} \nabla_y \Theta_s^{(i)} \otimes \mathbf{e}_i \right) \cdot \nabla \vartheta_s,$$

and

$$\mathbb{B}_s^{\vartheta} = (1-m)\mathbb{I} + \sum_{i=1}^{3} \langle \nabla_y \Theta_s^{(i)} \rangle_{Y_s} \otimes \mathbf{e}_i. \tag{7.1.39}$$

We consider the microscopic equation (7.1.35) in Y_f. Therefore it must be completed with the boundary condition

$$\Theta^{(f)}(\mathbf{x}, t, \mathbf{y}) = \vartheta_s(\mathbf{x}, t), \;\; \mathbf{y} \in \gamma, \tag{7.1.40}$$

and the initial condition

$$\Theta^{(f)}(\mathbf{x}, \mathbf{y}, 0) = \vartheta_s(\mathbf{x}, 0), \;\; \mathbf{y} \in Y_f. \tag{7.1.41}$$

The boundary condition (7.1.40) is a consequence of the equality

$$\Theta(\mathbf{x}, t, \mathbf{y}) = \Theta^{(f)}(\mathbf{x}, t, \mathbf{y})\chi(\mathbf{y}) + \vartheta_s(\mathbf{x}, t)\big(1 - \chi(\mathbf{y})\big),$$

and the inclusion $\Theta^{(f)} \in L_2\big(\Omega_T; W_2^1(Y)\big)$.

The initial condition (7.1.41) is a consequence of the equality

$$\begin{aligned} &\Theta^{(f)}(\mathbf{x}, \mathbf{y}, 0)\chi(\mathbf{y}) + \vartheta_s(\mathbf{x}, 0)\big(1 - \chi(\mathbf{y})\big) \\ &\qquad = \vartheta_0(\mathbf{x})\chi(\mathbf{y}) + \vartheta_0(\mathbf{x})\big(1 - \chi(\mathbf{y})\big), \end{aligned} \tag{7.1.42}$$

which follows from the integral identity

$$\int_{\Omega_T}\left(\frac{\partial\vartheta^\varepsilon}{\partial t}\varphi+\vartheta^\varepsilon\frac{\partial\varphi}{\partial t}\right)dxdt=-\int_\Omega\vartheta_0(\mathbf{x})\,\varphi\left(\mathbf{x},\frac{\mathbf{x}}{\varepsilon},0\right)dx$$

for any smooth 1-periodic in $\mathbf{y}$ functions $\varphi(\mathbf{x},t,\mathbf{y})$, such that $\varphi(\mathbf{x},\mathbf{y},T)=0$, after taking a limit as $\varepsilon\to 0$.

The same equality (7.1.42) provides the initial condition (7.1.15).

To solve the problem (7.1.36), (7.1.40), (7.1.41) we use a representation

$$\Theta^{(f)}(\mathbf{x},t,\mathbf{y})=\vartheta_s(\mathbf{x},t)+\int_0^t\bar{\Theta}(\mathbf{y},t-\tau)\left(\gamma_f\frac{\partial p}{\partial\tau}-c_{p,f}\frac{\partial\vartheta_s}{\partial\tau}\right)(\mathbf{x},\tau)d\tau,$$

where the function $\bar{\Theta}(\mathbf{y},t)$ is a solution to the periodic initial boundary-value problem

$$c_{p,f}\frac{\partial\bar{\Theta}}{\partial t}=\varkappa_1\triangle_y\bar{\Theta},\ \ \mathbf{y}\in Y_f \tag{7.1.43}$$

$$\bar{\Theta}(\mathbf{y},t)=0,\ \ \mathbf{y}\in\gamma,\ \ \bar{\Theta}(\mathbf{y},0)=\frac{1}{c_{p,f}},\ \ \mathbf{y}\in Y_f. \tag{7.1.44}$$

Thus,

$$a^{(f)}(t)=\langle\bar{\Theta}(.,t)\rangle_{Y_f}. \tag{7.1.45}$$

The Eq. (7.1.19) is a consequence of (7.1.12) for the case $\vartheta^{(f)}=m\vartheta_s$.

7.2 A Slightly Viscous Liquid in an Elastic Skeleton

Throughout this section we assume that

$$\mu_0=0,\ \ 0<\lambda_0<\infty. \tag{7.2.1}$$

7.2.1 Main Results

Theorem 7.10 *Let $\{\mathbf{w}^\varepsilon,\ p^\varepsilon,\ \vartheta^\varepsilon\}$ be a weak solution of the problem (7.0.1)–(7.0.6), $\mathbf{w}_s^\varepsilon=\mathbb{E}_{\Omega_s^\varepsilon}(\mathbf{w}^\varepsilon)$ be an extension (7.1.20), and*

$$\mu_1=\infty,\ \ 0<\varkappa_f<\infty.$$

Then up to some subsequences the sequences $\{\mathbf{w}^{\varepsilon}\}$ *and* $\{\chi^{\varepsilon} p^{\varepsilon}\}$ *converge weakly and two-scale in* $\mathbf{L}_2(\Omega_T)$ *and* $L_2(\Omega_T)$ *to functions* $\mathbf{w}(\mathbf{x}, t)$ *and* $p_f(\mathbf{x}, t)$ *respectively.*

At the same time the sequence $\{\mathbf{w}_s^{\varepsilon}\}$ *converges weakly in* $\mathbf{W}_2^{1,0}(\Omega_T)$ *and two-scale in* $\mathbf{L}_2(\Omega_T)$ *to the function* $\mathbf{w}_s(\mathbf{x}, t)$, $\mathbf{w}_s \in \overset{\circ}{\mathbf{W}}{}_2^{1,0}(\Omega_T)$, *and the sequence* $\{\vartheta^{\varepsilon}\}$ *converges weakly in* $\overset{\circ}{W}{}_2^{1,0}(\Omega_T)$ *and two-scale in* $L_2(\Omega_T)$ *to the function* $\vartheta(\mathbf{x}, t)$.

Limiting functions solve the system of homogenized equations in the domain Ω_T, *consisting of the homogenized continuity equation*

$$\frac{1}{c_f^2} p_f + m \nabla \cdot \mathbf{w}_s = \mathbb{C}_0^s : \mathbb{D}(x, \mathbf{w}_s) + \frac{c_0^s}{\lambda_0} q, \tag{7.2.2}$$

the state equation

$$q = p_f + m\,(\beta_f - \beta_s)\,\vartheta, \tag{7.2.3}$$

the homogenized momentum balance equation

$$\hat{\rho} \frac{\partial^2 \mathbf{w}_s}{\partial t^2} = \nabla \cdot \left(\lambda_0\, \mathfrak{N}_2^s : \mathbb{D}(x, \mathbf{w}_s) - q\, \mathbb{C}_1^s - \beta_s\, \vartheta\, \mathbb{I}\right) + \hat{\rho} \mathbf{F}, \tag{7.2.4}$$

and the homogenized heat equation

$$\hat{c}_p \frac{\partial \vartheta}{\partial t} = \nabla \cdot \left(\mathbb{B}^{\vartheta} \cdot \nabla \vartheta\right) + \mathbb{C}^{\vartheta} : \mathbb{D}\left(x, \frac{\partial \mathbf{w}_s}{\partial t}\right) + \frac{c_0^s}{\lambda_0} \frac{\partial q}{\partial t}. \tag{7.2.5}$$

The system is completed with boundary and initial conditions

$$\mathbf{w}_s(\mathbf{x}, t) = 0, \ \mathbf{x} \in S, \ t \in (0, T), \tag{7.2.6}$$

$$\vartheta(\mathbf{x}, t) = 0, \ \mathbf{x} \in S, \ t \in (0, T), \tag{7.2.7}$$

$$\mathbf{w}_s(\mathbf{x}, 0) = \frac{\partial \mathbf{w}_s}{\partial t}(\mathbf{x}, 0) = 0, \ \mathbf{x} \in \Omega, \tag{7.2.8}$$

$$\hat{c}_p\big(\vartheta(\mathbf{x}, 0) - \vartheta_0(\mathbf{x})\big) = \frac{c_0^s}{\lambda_0} q, \ \mathbf{x} \in \Omega. \tag{7.2.9}$$

In (7.2.2)–(7.2.9)

$$\hat{\rho} = m\rho_f + (1 - m)\,\rho_s, \ \hat{c_p} = m c_{p,f} + (1 - m)\, c_{p,s}, \ m = \int_Y \chi(\mathbf{y}) dy,$$

the symmetric strictly positively definite constant fourth-rank tensor $\mathfrak{N}_2^s$, *matrices* $\mathbb{C}_0^s$ *and* $\mathbb{C}_1^s$, *and the constant* c_0^s *are given in Chap. 1 by formulae (1.3.26), (1.3.27)* *and (1.3.31). The symmetric strictly positively definite constant matrix $\mathbb{B}^{\vartheta}$ and the matrix $\mathbb{C}^{\vartheta}$ are defined in Chap. 1 in Theorems 2.1 and 2.4.*

We refer to the problems (7.2.2)–(7.2.9) as the homogenized **model** $(\mathbb{NIA})_9$.

Theorem 7.11 *Let $\{\mathbf{w}^{\varepsilon},\ p^{\varepsilon},\ \vartheta^{\varepsilon}\}$ be a weak solution of the problems (7.0.1)–(7.0.6),*

$$\mu_1 = \infty,\quad \varkappa_f = 0,\quad \varkappa_1 = \infty,$$

$\vartheta_s^{\varepsilon} = E_{\Omega_s^{\varepsilon}}(\vartheta^{\varepsilon})$ be an extension (7.1.9), and $\mathbf{w}_s^{\varepsilon} = \mathbb{E}_{\Omega_s^{\varepsilon}}\big(\mathbf{w}^{\varepsilon}\big)$ be an extension (7.1.20).

Then up to some subsequences the sequences $\{\mathbf{w}^{\varepsilon}\}$ and $\{\chi^{\varepsilon}\,\vartheta^{\varepsilon}\}$ converge weakly and two-scale in $\mathbf{L}_2(\Omega_T)$ and $L_2(\Omega_T)$ to the displacements of the mixture $\mathbf{w}(\mathbf{x},t)$ and the liquid temperature $\vartheta^{(f)}(\mathbf{x},t)$ respectively.

The sequence$\{\chi^{\varepsilon}\,p^{\varepsilon}\}$ converge weakly and two-scale in $L_2(\Omega_T)$ to the liquid pressure $p_f(\mathbf{x},t)$ and the function $p_f(\mathbf{x},t)\chi(\mathbf{y})$ respectively.

At the same time the sequence $\{\mathbf{w}_s^{\varepsilon}\}$ converges weakly in $\mathbf{W}_2^{1,0}(\Omega_T)$ and two-scale in $\mathbf{L}_2(\Omega_T)$ to the solid displacements $\mathbf{w}_s(\mathbf{x},t)$ and the sequence $\{\vartheta_s^{\varepsilon}\}$ converges weakly in $W_2^{1,0}(\Omega_T)$ and two-scale in $L_2(\Omega_T)$ to the solid temperature $\vartheta_s(\mathbf{x},t)$.

Limiting functions, where $\mathbf{w}_s(\mathbf{x},t) = \mathbf{w}(\mathbf{x},t)$, $\mathbf{w}_s \in \overset{\circ}{\mathbf{W}}{}_2^{1,0}(\Omega_T)$, and $\vartheta^{(f)}(\mathbf{x},t) = m\vartheta_s(\mathbf{x},t)$, $\vartheta_s \in \overset{\circ}{W}{}_2^{1,0}(\Omega_T)$, solve the system of homogenized equations in the domain Ω_T, consisting of the homogenized continuity equation (7.2.2), the state equation

$$q = p_f + m\,(\beta_f - \beta_s)\,\vartheta_s, \tag{7.2.10}$$

the homogenized momentum balance equation

$$\hat{\rho}\frac{\partial^2 \mathbf{w}_s}{\partial t^2} = \nabla\cdot\Big(\lambda_0\,\mathfrak{N}_2^s : \mathbb{D}(x,\mathbf{w}_s) - q\,\mathbb{C}_1^s - \beta_s\,\vartheta_s\,\mathbb{I}\Big) + \hat{\rho}\mathbf{F}, \tag{7.2.11}$$

and the homogenized heat equation

$$\hat{c}_p\,\frac{\partial \vartheta_s}{\partial t} = \nabla\cdot\big(\mathbb{B}_s^{\vartheta}\cdot\nabla\,\vartheta_s\big) + \mathbb{C}^{\vartheta} : \mathbb{D}\left(x,\frac{\partial \mathbf{w}_s}{\partial t}\right) + \frac{c_0^s}{\lambda_0}\frac{\partial q}{\partial t}. \tag{7.2.12}$$

The system is completed with boundary and initial conditions (7.2.6), (7.2.8) and

$$\vartheta_s(\mathbf{x},t) = 0,\ \ \mathbf{x}\in S,\ \ t\in(0,T), \tag{7.2.13}$$

$$\hat{c}_p\,\big(\vartheta_s(\mathbf{x},0) - \vartheta_0(\mathbf{x})\big) = \frac{c_0^s}{\lambda_0}\,q,\ \ \mathbf{x}\in\Omega. \tag{7.2.14}$$

In (7.2.11)–(7.2.14)

$$\hat{\rho} = m\rho_f + (1-m)\,\rho_s,\ \ \hat{c_p} = mc_{p,f} + (1-m)\,c_{p,s},\ \ m = \int_Y \chi(\mathbf{y})dy,$$

the symmetric strictly positively definite constant fourth-rank tensor $\mathfrak{N}^s_2$, *the matrix* $\mathbb{C}^s_0$ *and the constant* c^s_0 *are given in Chap. 1 by formulae (1.3.26), (1.3.27) and (1.3.31). The symmetric strictly positively definite constant matrix* $\mathbb{B}^{\vartheta}_s$ *and the matrix* $\mathbb{C}^{\vartheta}$ *are defined in this chapter in Theorem 7.3, and in Chap. 1 in Theorem 2.4.*

We refer to the problem (7.2.2), (7.2.6), (7.2.8) (7.2.10)–(7.2.14) as the homogenized **model** $(\mathbb{NIA})_{10}$.

Theorem 7.12 *Let* $\{\mathbf{w}^{\varepsilon},\ p^{\varepsilon},\ \vartheta^{\varepsilon}\}$ *be a weak solution of the problems (7.0.1)–(7.0.6),* $\mathbf{w}^{\varepsilon}_s = \mathbb{E}_{\Omega^{\varepsilon}_s}\big(\mathbf{w}^{\varepsilon}\big)$ *be an extension (7.1.20),* $q^{\varepsilon} = \chi^{\varepsilon}\big(p^{\varepsilon} + (\beta_f - \beta_s)\vartheta^{\varepsilon}\big)$, *and*

$$\mu_1 < \infty,\ \ 0 < \varkappa_f < \infty.$$

Then up to some subsequences the sequence $\{\mathbf{w}^{\varepsilon}_s\}$ *converges weakly in* $\mathbf{W}^{1,0}_2(\Omega_T)$ *and two-scale in* $\mathbf{L}_2(\Omega_T)$ *to the solid displacements* $\mathbf{w}_s(\mathbf{x},t)$, *and the sequence* $\{\vartheta^{\varepsilon}\}$ *converges weakly in* $\overset{\circ}{W}{}^{1,0}_2(\Omega_T)$ *and two-scale in* $L_2(\Omega_T)$ *to the temperature of the mixture* $\vartheta(\mathbf{x},t)$.

At the same time the sequence $\{\chi^{\varepsilon} p^{\varepsilon}\}$ *converges weakly and two-scale in* $L_2(\Omega_T)$ *to the liquid pressure* $p_f(\mathbf{x},t)$ *and the function* $p_f(\mathbf{x},t)\chi(\mathbf{y})$ *respectively, the sequence* $\{\chi^{\varepsilon}\mathbf{w}^{\varepsilon}\}$ *converges weakly in* $\mathbf{L}_2(\Omega_T)$ *to the liquid displacements* $\mathbf{w}^{(f)}$, *and the sequence* $\{q^{\varepsilon}\}$ *converges weakly and two-scale in* $L_2(\Omega_T)$ *to the functions* $mq(\mathbf{x},t)$ *and* $q(\mathbf{x},t)\chi(\mathbf{y})$ *respectively.*

The limiting functions, where $\mathbf{w}_s \in \overset{\circ}{\mathbf{W}}{}^{1,0}_2(\Omega_T)$ *and* $q = (p_f + (\beta_f - \beta_s)\vartheta) \in W^{1,0}_2(\Omega_T)$, *solve the system of homogenized equations in the domain* Ω_T, *consisting of the homogenized continuity equation*

$$\frac{1}{c_f^2}p_f + \nabla\cdot\mathbf{w}^{(f)} = \mathbb{C}^s_0 : \mathbb{D}(x,\mathbf{w}_s) + \frac{c^s_0}{\lambda_0}q, \tag{7.2.15}$$

the state equation (7.2.3), the homogenized momentum balance equation

$$\rho_f\frac{\partial^2\mathbf{w}^{(f)}}{\partial t^2} + \rho_s\frac{\partial^2\mathbf{w}_s}{\partial t^2} = \nabla\cdot\big(\lambda_0\,\mathfrak{N}^s_2 : \mathbb{D}(x,\mathbf{w}_s) - q\,\mathbb{C}^s_1 - \beta_s\,\vartheta\,\mathbb{I}\big) + \hat{\rho}\mathbf{F}, \tag{7.2.16}$$

for the solid component with boundary and initial conditions (7.2.6) and (7.2.8), the homogenized heat equation (7.2.5) with boundary and initial conditions (7.2.7) and (7.2.9), and the homogenized momentum balance equation

$$-\int_0^t \mathbb{B}^{(f)}(\mu_1,\infty;t-\tau)\cdot\left(\nabla q + \rho_f\left(\frac{\partial^2\mathbf{w}_s}{\partial\tau^2} - \mathbf{F}\right)\right)(\mathbf{x},\tau)d\tau = \frac{\partial\mathbf{w}^{(f)}}{\partial t} - m\,\frac{\partial\mathbf{w}_s}{\partial t} \tag{7.2.17}$$

for the liquid component with homogeneous boundary and initial conditions

$$\mathbf{w}^{(f)}(\mathbf{x},t)\cdot\mathbf{n}(\mathbf{x})=0,\;\;\mathbf{x}\in S,\;\;t\in(0,T),\tag{7.2.18}$$

$$\mathbf{w}^{(f)}(\mathbf{x},0)=0,\;\;\mathbf{x}\in\Omega.\tag{7.2.19}$$

In (7.2.15)–(7.2.18) the symmetric strictly positively definite constant fourth-rank tensor $\mathfrak{N}_2^s$, *matrices* $\mathbb{C}_0^s$ *and* $\mathbb{C}_1^s$, *and the constant* c_0^s *are given in Chap. 1 by formulae (1.3.26), (1.3.27) and (1.3.31). the matrix* $\mathbb{B}^{(f)}(\mu_1,\infty;t)$ *is defined by (3.2.70) and (3.2.76) in Chap. 3 and* $\mathbf{n}$ *is the normal vector to the boundary S.*

We refer to the problem (7.2.3), (7.2.5)–(7.2.9), (7.2.15)–(7.2.19) as the homogenized **model** $(\mathbb{NIA})_{11}$.

Theorem 7.13 *Let* $\{\mathbf{w}^{\varepsilon},\ p^{\varepsilon},\ \vartheta^{\varepsilon}\}$ *be a weak solution of the problem (7.0.1)–(7.0.6),*

$$\mu_1<\infty,\;\;\varkappa_f=0,\;\;\varkappa_1=\infty,$$

$\mathbf{w}_s^{\varepsilon}=\mathbb{E}_{\Omega_s^{\varepsilon}}\big(\mathbf{w}^{\varepsilon}\big)$ *be an extension (7.1.20), and* $\vartheta_s^{\varepsilon}=E_{\Omega_s^{\varepsilon}}(\vartheta^{\varepsilon})$ *be an extension (7.1.9).*

Then up to some subsequences the sequence $\{\chi^{\varepsilon}p^{\varepsilon}\}$ *converges weakly and two-scale in* $L_2(\Omega_T)$ *to the liquid pressure* $p_f(\mathbf{x},t)$ *and the function* $p_f(\mathbf{x},t)\chi(\mathbf{y})$ *respectively, the sequence* $\{\chi^{\varepsilon}\,\vartheta^{\varepsilon}\}$ *converges weakly and two-scale in* $L_2(\Omega_T)$ *to the liquid temperature* $\vartheta^{(f)}(\mathbf{x},t)$, *and the sequence* $\{\chi^{\varepsilon}\,\mathbf{w}^{\varepsilon}\}$ *converges weakly in* $\mathbf{L}_2(\Omega_T)$ *to the liquid displacements* $\mathbf{w}^{(f)}(\mathbf{x},t)$.

At the same time the sequence $\{\mathbf{w}_s^{\varepsilon}\}$ *converges weakly in* $\mathbf{W}_2^{1,0}(\Omega_T)$ *and two-scale in* $\mathbf{L}_2(\Omega_T)$ *to the solid displacements* $\mathbf{w}_s(\mathbf{x},t)$, *the sequence* $\{\vartheta_s^{\varepsilon}\}$ *converges weakly in* $W_2^{1,0}(\Omega_T)$ *and two- scale in* $L_2(\Omega_T)$ *to the solid temperature* $\vartheta^{(f)}(\mathbf{x},t)$, *and* $\mathbf{w}_s(\mathbf{x},t)=\mathbf{w}(\mathbf{x},t)$, $\vartheta^{(f)}(\mathbf{x},t)=m\vartheta_s(\mathbf{x},t)$.

The limiting functions, where $\mathbf{w}_s\in\overset{\circ}{\mathbf{W}}{}_2^{1,0}(\Omega_T)$ *and* $\vartheta_s\in\overset{\circ}{W}{}_2^{1,0}(\Omega_T)$, *solve the system of homogenized equations in the domain* Ω_T, *consisting of the homogenized continuity equation (7.2.15), the state equation (7.2.10), the homogenized momentum balance equation*

$$\rho_f\frac{\partial^2\mathbf{w}^{(f)}}{\partial t^2}+\rho_s\frac{\partial^2\mathbf{w}_s}{\partial t^2}=\nabla\cdot\big(\lambda_0\,\mathfrak{N}_2^s:\mathbb{D}(x,\mathbf{w}_s)-q\,\mathbb{C}_1^s-\beta_s\,\vartheta_s\,\mathbb{I}\big)+\hat{\rho}\mathbf{F},\tag{7.2.20}$$

for the solid component with boundary and initial conditions (7.2.6) and (7.2.8), the homogenized heat equation (7.2.12) with boundary and initial conditions (7.2.13) and (7.2.14), and the homogenized momentum balance equation (7.2.17) for the liquid component with homogeneous boundary and initial conditions (7.2.18) and (7.2.19).

In (7.2.20) the symmetric strictly positively definite constant fourth-rank tensor $\mathfrak{N}_2^s$ *and matrix* $\mathbb{C}_1^s$, *and the constant* c_0^s *are given in Chap. 1 by formulae (1.3.26) and (1.3.27).*

We refer to the problem (7.2.6), (7.2.8), (7.2.10), (7.2.12)–(7.2.14), (7.2.15), (7.2.17)–(7.2.20) as the homogenized **model** $(\mathbb{NIA})_{12}$.

The proofs of these statements repeats the proofs of similar statements in Chaps. 1, 2, and 6.

Chapter 8
Isothermal Acoustics in Composite Media

We restrict ourself to a simple situation, when the domain Q is a unit cube: $Q = (0,1)\times(0,1)\times(0,1)$, the poroelastic medium occupies the domain $\Omega = (0,1)\times(0,1)\times(0,a),\ 0 < a < 1$ and the domain $\Omega^{(s)}$ ($\Omega^{(f)}$, or G) is an open complement to Ω:

$$Q = \Omega \cup \Omega^{(s)} \cup S^{(0)}, \quad S^{(0)} = \partial\Omega \cap \partial\Omega^{(s)},$$

or

$$Q = \Omega \cup \Omega^{(f)} \cup S^{(0)}, \quad S^{(0)} = \partial\Omega \cap \partial\Omega^{(f)},$$

or

$$Q = \Omega \cup G \cup S^{(0)}, \quad S^{(0)} = \partial\Omega \cap \partial G.$$

As in the previous chapter, we describe the motion of the mixture in the domain Ω for $t > 0$ by the system

$$\left(\frac{\chi^{\varepsilon}}{\bar{c}_f^2} + \frac{1-\chi^{\varepsilon}}{\bar{c}_s^2}\right) p + \nabla\cdot\mathbf{w} = 0, \tag{8.0.1}$$

$$\big(\rho_f\chi^{\varepsilon} + (1-\chi^{\varepsilon})\rho_s\big)\frac{\partial^2\mathbf{w}}{\partial t^2} = \nabla\cdot\mathbb{P} + \rho^{\varepsilon}\mathbf{F}, \tag{8.0.2}$$

$$\mathbb{P} = \chi^{\varepsilon}\bar{\alpha}_{\mu}\mathbb{D}\left(x, \frac{\partial\mathbf{w}}{\partial t}\right) + (1-\chi^{\varepsilon})\bar{\alpha}_{\lambda}\mathbb{D}(x,\mathbf{w}) - p\,\mathbb{I}, \tag{8.0.3}$$

The motion of an elastic body in the domain $\Omega^{(s)}$ for $t > 0$ is governed by Lamé's system

$$\frac{1}{\left(\bar{c}_s^{(0)}\right)^2}\,p + \nabla\cdot\mathbf{w} = 0, \tag{8.0.4}$$

A. Meirmanov, *Mathematical Models for Poroelastic Flows*,
Atlantis Studies in Differential Equations 1, DOI: 10.2991/978-94-6239-015-7_8,

$$\rho_s^{(0)} \frac{\partial^2 \mathbf{w}}{\partial t^2} = \nabla \cdot \mathbb{P}^{(s)} + \rho_s^{(0)} \mathbf{F}, \tag{8.0.5}$$

$$\mathbb{P}^{(s)} = \bar{\alpha}_\lambda^{(0)} \mathbb{D}(x, \mathbf{w}) - p\, \mathbb{I}, \tag{8.0.6}$$

where $\bar{\alpha}_\lambda^{(0)}$ and $\bar{c}_s^{(0)}$ are dimensionless Lamé's constants for the elastic body in $\Omega^{(s)}$.

The motion of the liquid in the domain $\Omega^{(f)}$ for $t > 0$ is described by the Stokes system

$$\frac{1}{\bar{c}_f^2}\, p + \nabla \cdot \mathbf{w} = 0, \tag{8.0.7}$$

$$\rho_f \frac{\partial^2 \mathbf{w}}{\partial t^2} = \nabla \cdot \mathbb{P}^{(f)} + \rho_f \mathbf{F}, \tag{8.0.8}$$

$$\mathbb{P}^{(f)} = \bar{\alpha}_\mu \mathbb{D}\left(x, \frac{\partial \mathbf{w}}{\partial t}\right) - p\, \mathbb{I}. \tag{8.0.9}$$

Finally, the motion of the mixture in the domain G for $t > 0$ is described by the system

$$\left(\frac{\chi_0^\varepsilon}{\bar{c}_f^2} + \frac{1 - \chi_0^\varepsilon}{\left(\bar{c}_s^{(0)}\right)^2}\right) p + \nabla \cdot \mathbf{w} = 0, \tag{8.0.10}$$

$$\left(\rho_f \chi_0^\varepsilon + (1 - \chi_0^\varepsilon) \rho_s^{(0)}\right) \frac{\partial^2 \mathbf{w}}{\partial t^2} = \nabla \cdot \mathbb{P}^{(0)} + \rho^\varepsilon \mathbf{F}, \tag{8.0.11}$$

$$\mathbb{P}^{(0)} = \chi_0^\varepsilon \bar{\alpha}_\mu \mathbb{D}\left(x, \frac{\partial \mathbf{w}}{\partial t}\right) + (1 - \chi_0^\varepsilon) \bar{\alpha}_\lambda^{(0)} \mathbb{D}(x, \mathbf{w}) - p\, \mathbb{I}, \tag{8.0.12}$$

where χ_0^ε is the characteristic function of the liquid domain G_f^ε in G:

$$\chi_0^\varepsilon(\mathbf{x}) = \chi_0\left(\frac{\mathbf{x}}{\varepsilon}\right),$$

For the first and third configurations the elastic properties of the solid material in $\Omega^{(s)}$ and Ω (G_s^ε and Ω) might be different, while in all cases the liquid must be the same.

On the common boundary $S^{(0)}$ the usual continuity conditions for displacements hold true:

$$\lim_{\substack{\mathbf{x} \to \mathbf{x}^0 \\ \mathbf{x} \in \Omega^{(s)}}} \mathbf{w}(\mathbf{x}, t) = \lim_{\substack{\mathbf{x} \to \mathbf{x}^0 \\ \mathbf{x} \in \Omega}} \mathbf{w}(\mathbf{x}, t), \tag{8.0.13}$$

and for the normal component of the momentum

$$\lim_{\substack{\mathbf{x}\to\mathbf{x}^0\\ \mathbf{x}\in\Omega^{(s)}}} \mathbb{P}^{(s)}(\mathbf{x},t)\cdot\mathbf{n}(\mathbf{x}^0) = \lim_{\substack{\mathbf{x}\to\mathbf{x}^0\\ \mathbf{x}\in\Omega}} \mathbb{P}(\mathbf{x},t)\cdot\mathbf{n}(\mathbf{x}^0), \tag{8.0.14}$$

for the first structure,

$$\lim_{\substack{\mathbf{x}\to\mathbf{x}^0\\ \mathbf{x}\in\Omega^{(f)}}} \mathbb{P}^{(f)}(\mathbf{x},t)\cdot\mathbf{n}(\mathbf{x}^0) = \lim_{\substack{\mathbf{x}\to\mathbf{x}^0\\ \mathbf{x}\in\Omega}} \mathbb{P}(\mathbf{x},t)\cdot\mathbf{n}(\mathbf{x}^0) \tag{8.0.15}$$

for the second structure, and

$$\lim_{\substack{\mathbf{x}\to\mathbf{x}^0\\ \mathbf{x}\in G}} \mathbb{P}^{(0)}(\mathbf{x},t)\cdot\mathbf{n}(\mathbf{x}^0) = \lim_{\substack{\mathbf{x}\to\mathbf{x}^0\\ \mathbf{x}\in\Omega}} \mathbb{P}(\mathbf{x},t)\cdot\mathbf{n}(\mathbf{x}^0) \tag{8.0.16}$$

for the third structure.

To complete the problems we impose a homogeneous boundary condition

$$\mathbf{w}(\mathbf{x},t)=0,\ \ (\mathbf{x},t)\in S_T=S\times(0,T), \tag{8.0.17}$$

on the boundary $S=\partial Q$, and homogeneous initial conditions

$$\mathbf{w}(\mathbf{x},0)=\frac{\partial\mathbf{w}}{\partial t}(\mathbf{x},0)=0,\ \ \mathbf{x}\in Q. \tag{8.0.18}$$

As before, we assume that

$$\int_{Q_T}\left(\left|\mathbf{F}(\mathbf{x},t)\right|^2+\left|\frac{\partial\mathbf{F}}{\partial t}(\mathbf{x},t)\right|^2\right)dxdt=F^2<\infty,$$

and Assumption 0.1, Assumption 1.2 and Assumption 3.1 hold true. It is also assumed that all dimensionless parameters depend on the small parameter ε and the (finite or infinite) limits exist:

$$\lim_{\varepsilon\searrow 0}\bar{\alpha}_\mu(\varepsilon)=\mu_0,\quad \lim_{\varepsilon\searrow 0}\bar{\alpha}_\lambda(\varepsilon)=\lambda_0,\quad \lim_{\varepsilon\searrow 0}\bar{\alpha}_\lambda^{(0)}(\varepsilon)=\lambda_0^{(0)},$$

$$\lim_{\varepsilon\searrow 0}\frac{\bar{\alpha}_\mu}{\varepsilon^2}=\mu_1,\quad \lim_{\varepsilon\searrow 0}\frac{\bar{\alpha}_\lambda}{\varepsilon^2}=\lambda_1,\quad \lim_{\varepsilon\searrow 0}\frac{\bar{\alpha}_\lambda^{(0)}}{\varepsilon^2}=\lambda_1^{(0)}.$$

Throughout this chapter we assume that

$$\mu_0=0.$$

In the usual way one may define weak solutions to the problem (I) (relations (8.0.1)–(8.0.6), (8.0.13), (8.0.14), (8.0.17), (8.0.18)), to the problem (II) (relations

(8.0.1)–(8.0.3), (8.0.7)–(8.0.9), (8.0.13), (8.0.15), (8.0.17), (8.0.18)) and to the problem (III) (relations (8.0.1)–(8.0.3), (8.0.10)–(8.0.13), (8.0.16)–(8.0.18)) by means of integral identities.

Let $\zeta(\mathbf{x})$ be the characteristic function of the domain Ω and

$$\rho^{\varepsilon}_{(s)} = (1-\zeta)\rho^{(0)}_{s} + \zeta\left(\rho_f\chi^{\varepsilon} + (1-\chi^{\varepsilon})\rho_s\right),$$

$$\rho^{\varepsilon}_{(f)} = (1-\zeta)\rho_f + \zeta\left(\rho_f\chi^{\varepsilon} + (1-\chi^{\varepsilon})\rho_s\right),$$

$$\rho^{\varepsilon}_{(0)} = (1-\zeta)\left(\rho_f\chi^{\varepsilon}_0 + (1-\chi^{\varepsilon}_0)\rho^{(0)}_{s}\right) + \zeta\left(\rho_f\chi^{\varepsilon} + (1-\chi^{\varepsilon})\rho_s\right).$$

Definition 8.1 We say that the pair of functions $\{\mathbf{w}^{\varepsilon},\ p^{\varepsilon}\}$ such that

$$\mathbf{w}^{\varepsilon} \in \overset{\circ}{\mathbf{W}}{}^{1,1}_2(Q_T),\ p^{\varepsilon} \in L_2(Q_T),$$

is a weak solution of the problem (I), if it satisfies the continuity equation

$$\left((1-\zeta)\frac{1}{\left(\bar{c}^{(0)}_s\right)^2} + \zeta\left(\frac{\chi^{\varepsilon}}{\bar{c}^2_f} + \frac{1-\chi^{\varepsilon}}{\bar{c}^2_s}\right)\right) p^{\varepsilon} + \nabla\cdot\mathbf{w}^{\varepsilon} = 0, \tag{8.0.19}$$

almost everywhere in Q_T, and the integral identity

$$\int_{Q_T}\rho^{\varepsilon}_{(s)}\left(\frac{\partial\mathbf{w}^{\varepsilon}}{\partial t}\cdot\frac{\partial\boldsymbol{\varphi}}{\partial t} + \mathbf{F}\cdot\boldsymbol{\varphi}\right)dxdt = \int_{Q_T}\left(\zeta\mathbb{P} + (1-\zeta)\mathbb{P}^{(s)}\right) : \mathbb{D}(x,\boldsymbol{\varphi})dxdt \tag{8.0.20}$$

for all functions $\boldsymbol{\varphi}$, such that $\boldsymbol{\varphi} \in \overset{\circ}{\mathbf{W}}{}^{1,0}_2(Q_T)$, $\dfrac{\partial\boldsymbol{\varphi}}{\partial t} \in \mathbf{L}_2(\Omega_T)$ and $\boldsymbol{\varphi}(\mathbf{x},T) = 0$ for $\mathbf{x} \in Q$.

Definition 8.2 We say that the pair of functions $\{\mathbf{w}^{\varepsilon},\ p^{\varepsilon}\}$ such that

$$\mathbf{w}^{\varepsilon} \in \overset{\circ}{\mathbf{W}}{}^{1,1}_2(Q_T),\ p^{\varepsilon} \in L_2(Q_T),$$

is a weak solution of the problem (II), if it satisfies the continuity equation

$$\left((1-\zeta)\frac{1}{\bar{c}^2_f} + \zeta\left(\frac{\chi^{\varepsilon}}{\bar{c}^2_f} + \frac{1-\chi^{\varepsilon}}{\bar{c}^2_s}\right)\right) p^{\varepsilon} + \nabla\cdot\mathbf{w}^{\varepsilon} = 0, \tag{8.0.21}$$

almost everywhere in Q_T, and the integral identity

$$\int_{Q_T} \rho^{\varepsilon}_{(f)} \left(\frac{\partial \mathbf{w}^{\varepsilon}}{\partial t} \cdot \frac{\partial \boldsymbol{\varphi}}{\partial t} + \mathbf{F} \cdot \boldsymbol{\varphi} \right) dx dt$$
$$= \int_{Q_T} \left(\zeta \mathbb{P} + (1 - \zeta) \mathbb{P}^{(f)} \right) : \mathbb{D}(x, \boldsymbol{\varphi}) dx dt \qquad (8.0.22)$$

for all functions $\boldsymbol{\varphi}$, such that $\boldsymbol{\varphi} \in \overset{\circ}{\mathbf{W}}{}^{1,0}_2(Q_T)$, $\dfrac{\partial \boldsymbol{\varphi}}{\partial t} \in \mathbf{L}_2(\Omega_T)$ and $\boldsymbol{\varphi}(\mathbf{x}, T) = 0$ for $\mathbf{x} \in Q$.

Definition 8.3 We say that the pair of functions $\{\mathbf{w}^{\varepsilon},\ p^{\varepsilon}\}$ such that

$$\mathbf{w}^{\varepsilon} \in \overset{\circ}{\mathbf{W}}{}^{1,1}_2(Q_T),\ p^{\varepsilon} \in L_2(Q_T),$$

is a weak solution of the problem (III), if it satisfies the continuity equation

$$\left((1 - \zeta) \left(\frac{\chi_0^{\varepsilon}}{\bar{c}_f^2} + \frac{1 - \chi_0^{\varepsilon}}{\left(\bar{c}_s^{(0)}\right)^2} \right) + \zeta \left(\frac{\chi^{\varepsilon}}{\bar{c}_f^2} + \frac{1 - \chi^{\varepsilon}}{\bar{c}_s^2} \right) \right) p^{\varepsilon} + \nabla \cdot \mathbf{w}^{\varepsilon} = 0, \quad (8.0.23)$$

almost everywhere in Q_T, and the integral identity

$$\int_{Q_T} \rho^{\varepsilon}_{(0)} \left(\frac{\partial \mathbf{w}^{\varepsilon}}{\partial t} \cdot \frac{\partial \boldsymbol{\varphi}}{\partial t} + \mathbf{F} \cdot \boldsymbol{\varphi} \right) dx dt$$
$$= \int_{Q_T} \left(\zeta \mathbb{P} + (1 - \zeta) \mathbb{P}^{(0)} \right) : \mathbb{D}(x, \boldsymbol{\varphi}) dx dt \qquad (8.0.24)$$

for all functions $\boldsymbol{\varphi}$, such that $\varphi \in \overset{\circ}{\mathbf{W}}{}^{1,0}_2(Q_T)$, $\dfrac{\partial \boldsymbol{\varphi}}{\partial t} \in \mathbf{L}_2(\Omega_T)$ and $\boldsymbol{\varphi}(\mathbf{x}, T) = 0$ for $\mathbf{x} \in Q$.

Theorem 8.1 *For all $\varepsilon > 0$ and for an arbitrary time interval $[0, T]$ there exists a unique generalized solution $\{\mathbf{w}^{\varepsilon},\ p^{\varepsilon}\}$ of the problem (I) and*

$$\max_{0<t<T} \int_{\Omega} \left(|p^{\varepsilon}(\mathbf{x}, t)|^2 + \left| \frac{\partial \mathbf{w}^{\varepsilon}}{\partial t}(\mathbf{x}, t) \right|^2 + (1 - \chi^{\varepsilon}) \bar{\alpha}_{\lambda} \left| \mathbb{D}(x, \mathbf{w}^{\varepsilon}) \right|^2 \right) dx$$
$$+ \max_{0<t<T} \int_{\Omega^{(s)}} \left(|p^{\varepsilon}(\mathbf{x}, t)|^2 + \left| \frac{\partial \mathbf{w}^{\varepsilon}}{\partial t}(\mathbf{x}, t) \right|^2 + (1 - \chi^{\varepsilon}) \bar{\alpha}_{\lambda}^{(0)} \left| \mathbb{D}(x, \mathbf{w}^{\varepsilon}) \right|^2 \right) dx$$
$$+ \max_{0<t<T} \int_{\Omega} \left(\left| \frac{\partial p^{\varepsilon}}{\partial t}(\mathbf{x}, t) \right|^2 + \left| \frac{\partial^2 \mathbf{w}^{\varepsilon}}{\partial t^2}(\mathbf{x}, t) \right|^2 + (1 - \chi^{\varepsilon}) \bar{\alpha}_{\lambda} \left| \mathbb{D}\left(x, \frac{\partial \mathbf{w}^{\varepsilon}}{\partial t}\right) \right|^2 \right) dx$$
$$+ \max_{0<t<T} \int_{\Omega^{(s)}} \left(\left| \frac{\partial p^{\varepsilon}}{\partial t}(\mathbf{x}, t) \right|^2 + \left| \frac{\partial^2 \mathbf{w}^{\varepsilon}}{\partial t^2}(\mathbf{x}, t) \right|^2 + (1 - \chi^{\varepsilon}) \bar{\alpha}_{\lambda}^{(0)} \left| \mathbb{D}\left(x, \frac{\partial \mathbf{w}^{\varepsilon}}{\partial t}\right) \right|^2 \right) dx$$

$$+\int_{\Omega_T} \chi^{\varepsilon} \bar{\alpha}_{\mu} \left(\left| \mathbb{D}\left(x, \frac{\partial \mathbf{w}^{\varepsilon}}{\partial t}\right) \right|^2 + \left| \mathbb{D}\left(x, \frac{\partial^2 \mathbf{w}^{\varepsilon}}{\partial t^2}\right) \right|^2 \right) dxdt \leqslant C_0 F^2, \qquad (8.0.25)$$

where the constant C_0 is independent of the small parameter ε and the criteria $\bar{\alpha}_{\lambda}$, $\bar{\alpha}_{\lambda}^{(0)}$, $\bar{\alpha}_{\mu}$.

The proof of this theorem repeats the proofs of similar theorems in the previous chapters and is based on the energy equalities

$$\frac{1}{2}\frac{d}{dt}\int_{\Omega}\left(\rho^{\varepsilon}\left|\frac{\partial \mathbf{w}^{\varepsilon}}{\partial t}\right|^2 + (1-\chi^{\varepsilon})\bar{\alpha}_{\lambda}\mathbb{D}(x,\mathbf{w}^{\varepsilon}) : \mathbb{D}(x,\mathbf{w}^{\varepsilon}) + \frac{1}{\bar{\alpha}_p^{\varepsilon}}|p^{\varepsilon}|^2\right)dx$$

$$+\frac{1}{2}\frac{d}{dt}\int_{\Omega^{(s)}}\left(\rho_s\left|\frac{\partial \mathbf{w}^{\varepsilon}}{\partial t}\right|^2 + (1-\chi^{\varepsilon})\bar{\alpha}_{\lambda}^{(0)}\mathbb{D}(x,\mathbf{w}^{\varepsilon}) : \mathbb{D}(x,\mathbf{w}^{\varepsilon}) + \frac{1}{\left(\bar{c}_s^{(0)}\right)^2}|p^{\varepsilon}|^2\right)dx$$

$$+\int_{\Omega}\chi^{\varepsilon}\left(\bar{\alpha}_{\mu}\mathbb{D}\left(x,\frac{\partial \mathbf{w}^{\varepsilon}}{\partial t}\right) : \mathbb{D}\left(x,\frac{\partial \mathbf{w}^{\varepsilon}}{\partial t}\right)\right)dx = \int_{Q}\tilde{\rho}^{\varepsilon}\mathbf{F}\cdot\frac{\partial \mathbf{w}^{\varepsilon}}{\partial t}dx,$$

$$\frac{1}{2}\frac{d}{dt}\int_{\Omega}\left(\rho^{\varepsilon}\left|\frac{\partial^2 \mathbf{w}^{\varepsilon}}{\partial t^2}\right|^2 + (1-\chi^{\varepsilon})\bar{\alpha}_{\lambda}\mathbb{D}\left(x,\frac{\partial \mathbf{w}^{\varepsilon}}{\partial t}\right) : \mathbb{D}\left(x,\frac{\partial \mathbf{w}^{\varepsilon}}{\partial t}\right) + \frac{1}{\bar{\alpha}_p^{\varepsilon}}\left|\frac{\partial p^{\varepsilon}}{\partial t}\right|^2\right)dx$$

$$+\frac{1}{2}\frac{d}{dt}\int_{\Omega^{(s)}}\left(\rho_s\left|\frac{\partial^2 \mathbf{w}^{\varepsilon}}{\partial t^2}\right|^2 + (1-\chi^{\varepsilon})\bar{\alpha}_{\lambda}^{(0)}\mathbb{D}\left(x,\frac{\partial \mathbf{w}^{\varepsilon}}{\partial t}\right) : \mathbb{D}\left(x,\frac{\partial \mathbf{w}^{\varepsilon}}{\partial t}\right) + \frac{1}{\left(\bar{c}_s^{(0)}\right)^2}\left|\frac{\partial p^{\varepsilon}}{\partial t}\right|^2\right)dx$$

$$+\int_{\Omega}\chi^{\varepsilon}\left(\bar{\alpha}_{\mu}\mathbb{D}\left(x,\frac{\partial^2 \mathbf{w}^{\varepsilon}}{\partial t^2}\right) : \mathbb{D}\left(x,\frac{\partial^2 \mathbf{w}^{\varepsilon}}{\partial t^2}\right)\right)dx = \int_{Q}\tilde{\rho}^{\varepsilon}\frac{\partial \mathbf{F}}{\partial t}\cdot\frac{\partial^2 \mathbf{w}^{\varepsilon}}{\partial t^2}dx.$$

In the same way one may prove.

Theorem 8.2 *For all $\varepsilon > 0$ and for an arbitrary time interval $[0, T]$ there exists a unique generalized solution $\{\mathbf{w}^{\varepsilon},\ p^{\varepsilon}\}$ of the problem (II) and*

$$\max_{0<t<T}\int_{\Omega}\left(|p^{\varepsilon}(\mathbf{x},t)|^2 + \left|\frac{\partial \mathbf{w}^{\varepsilon}}{\partial t}(\mathbf{x},t)\right|^2 + (1-\chi^{\varepsilon})\bar{\alpha}_{\lambda}\left|\mathbb{D}(x,\mathbf{w}^{\varepsilon})\right|^2\right)dx$$

$$+\max_{0<t<T}\int_{\Omega(f)}\left(|p^{\varepsilon}(\mathbf{x},t)|^2 + \left|\frac{\partial \mathbf{w}^{\varepsilon}}{\partial t}(\mathbf{x},t)\right|^2\right)dx + \int_0^T\int_{\Omega(f)}\chi^{\varepsilon}\bar{\alpha}_{\mu}\left|\mathbb{D}\left(x,\frac{\partial \mathbf{w}^{\varepsilon}}{\partial t}\right)\right|^2 dxdt$$

$$+\max_{0<t<T}\int_{\Omega}\left(\left|\frac{\partial p^{\varepsilon}}{\partial t}(\mathbf{x},t)\right|^2 + \left|\frac{\partial^2 \mathbf{w}^{\varepsilon}}{\partial t^2}(\mathbf{x},t)\right|^2 + (1-\chi^{\varepsilon})\bar{\alpha}_{\lambda}\left|\mathbb{D}\left(x,\frac{\partial \mathbf{w}^{\varepsilon}}{\partial t}\right)\right|^2\right)dx$$

$$+\max_{0<t<T}\int_{\Omega(f)}\left(\left|\frac{\partial p^{\varepsilon}}{\partial t}(\mathbf{x},t)\right|^2 + \left|\frac{\partial^2 \mathbf{w}^{\varepsilon}}{\partial t^2}(\mathbf{x},t)\right|^2\right)dx + \int_0^T\int_{\Omega(f)}\chi^{\varepsilon}\bar{\alpha}_{\mu}\left|\mathbb{D}\left(x,\frac{\partial^2 \mathbf{w}^{\varepsilon}}{\partial t^2}\right)\right|^2 dxdt$$

$$+\int_{\Omega_T} \chi^\varepsilon \bar{\alpha}_\mu \left(\left| \mathbb{D}\left(x, \frac{\partial \mathbf{w}^\varepsilon}{\partial t}\right) \right|^2 + \left| \mathbb{D}\left(x, \frac{\partial^2 \mathbf{w}^\varepsilon}{\partial t^2}\right) \right|^2 \right) dxdt \leqslant C_0 F^2, \tag{8.0.26}$$

where the constant C_0 is independent of the small parameter ε and the criteria $\bar{\alpha}_\lambda$, $\bar{\alpha}_\lambda^{(0)}$, $\bar{\alpha}_\mu$.

Theorem 8.3 *For all $\varepsilon > 0$ and for an arbitrary time interval $[0, T]$ there exists a unique generalized solution $\{\mathbf{w}^\varepsilon,\ p^\varepsilon\}$ of the problem (II) and*

$$\max_{0<t<T} \int_\Omega \left(|p^\varepsilon(\mathbf{x}, t)|^2 + \left| \frac{\partial \mathbf{w}^\varepsilon}{\partial t}(\mathbf{x}, t) \right|^2 + (1-\chi^\varepsilon)\bar{\alpha}_\lambda \left| \mathbb{D}(x, \mathbf{w}^\varepsilon) \right|^2 \right) dx$$

$$+ \max_{0<t<T} \int_G \left(|p^\varepsilon(\mathbf{x}, t)|^2 + \left| \frac{\partial \mathbf{w}^\varepsilon}{\partial t}(\mathbf{x}, t) \right|^2 + (1-\chi_0^\varepsilon)\bar{\alpha}_\lambda^{(0)} \left| \mathbb{D}(x, \mathbf{w}^\varepsilon) \right|^2 \right) dx$$

$$\max_{0<t<T} \int_\Omega \left(\left| \frac{\partial p^\varepsilon}{\partial t}(\mathbf{x}, t) \right|^2 + \left| \frac{\partial^2 \mathbf{w}^\varepsilon}{\partial t^2}(\mathbf{x}, t) \right|^2 + (1-\chi^\varepsilon)\bar{\alpha}_\lambda \left| \mathbb{D}\left(x, \frac{\partial \mathbf{w}^\varepsilon}{\partial t}\right) \right|^2 \right) dx$$

$$+ \max_{0<t<T} \int_G \left(\left| \frac{\partial p^\varepsilon}{\partial t}(\mathbf{x}, t) \right|^2 + \left| \frac{\partial^2 \mathbf{w}^\varepsilon}{\partial t^2}(\mathbf{x}, t) \right|^2 + (1-\chi_0^\varepsilon)\bar{\alpha}_\lambda^{(0)} \left| \mathbb{D}\left(x, \frac{\partial \mathbf{w}^\varepsilon}{\partial t}\right) \right|^2 \right) dx$$

$$+ \int_{\Omega_T} \chi^\varepsilon \bar{\alpha}_\mu \left(\left| \mathbb{D}\left(x, \frac{\partial \mathbf{w}^\varepsilon}{\partial t}\right) \right|^2 + \left| \mathbb{D}\left(x, \frac{\partial^2 \mathbf{w}^\varepsilon}{\partial t^2}\right) \right|^2 \right) dxdt$$

$$+ \int_0^T \int_G \chi_0^\varepsilon \bar{\alpha}_\mu^{(0)} \left(\left| \mathbb{D}\left(x, \frac{\partial \mathbf{w}^\varepsilon}{\partial t}\right) \right|^2 + \left| \mathbb{D}\left(x, \frac{\partial^2 \mathbf{w}^\varepsilon}{\partial t^2}\right) \right|^2 \right) dxdt \leqslant C_0 F^2, \tag{8.0.27}$$

where the constant C_0 is independent of the small parameter ε and the criteria $\bar{\alpha}_\lambda$, $\bar{\alpha}_\lambda^{(0)}$, $\bar{\alpha}_\mu$.

8.1 Acoustics in an "Elastic Body–Poroelastic Medium" Configuration

8.1.1 Main Results

Theorem 8.4 *Let $\{\mathbf{w}^\varepsilon,\ p^\varepsilon\}$ be the weak solution of the problem (I) and*

$$0 < \lambda_0^{(0)} < \infty, \quad \mu_1 = \lambda_1 = \infty.$$

Then the limits $\mathbf{w}$ and p of the sequences $\{\mathbf{w}^\varepsilon\}$ and $\{p^\varepsilon\}$ satisfy the dynamic equation in the form of the integral identity

$$\int_{Q_T} \Big((1-\zeta)\lambda_0^{(0)}\mathbb{D}(x,\mathbf{w}) - p\,\mathbb{I}\Big) : \mathbb{D}(x,\boldsymbol{\varphi})dxdt$$
$$= \int_{Q_T} \hat{\rho}_s \left(\mathbf{F} - \frac{\partial^2 \mathbf{w}}{\partial t^2}\right)\cdot \boldsymbol{\varphi} dxdt \tag{8.1.1}$$

for any function $\boldsymbol{\varphi} \in \overset{\circ}{\mathbf{W}}{}_2^{1,0}(Q_T)$, *and the continuity equation in the form of the integral identity*

$$\int_{Q_T} \left(\left((1-\zeta)\left(\frac{1}{\bar{c}_s^{(0)}}\right)^2 + \zeta\left(\frac{m}{\bar{c}_f^2} + \frac{1-m}{\bar{c}_s^2}\right)\right)\frac{\partial p}{\partial t}\,\psi - \nabla\,\psi \cdot \frac{\partial \mathbf{w}}{\partial t}\right) dxdt = 0 \tag{8.1.2}$$

for any smooth function $\psi \in W_2^{1,0}(Q_T)$.

Here

$$\hat{\rho}_s = (1-\zeta(\mathbf{x}))\,\rho_s^{(0)} + \zeta(\mathbf{x})\hat{\rho}, \quad \hat{\rho} = m\,\rho_f + (1-m)\,\rho_s.$$

Relations (8.1.1)–(8.1.2) *are completed with the homogeneous boundary condition*

$$\mathbf{w}(\mathbf{x},t) = 0, \tag{8.1.3}$$

on the boundary $S_T \backslash \partial\Omega_T$, *and the homogeneous initial conditions*

$$\mathbf{w}(\mathbf{x},0) = \frac{\partial \mathbf{w}}{\partial t}(\mathbf{x},0) = 0, \;\; \mathbf{x} \in Q. \tag{8.1.4}$$

We refer to the problem (8.1.1)–(8.1.4) as the homogenized **model** $(\mathbb{ACM})_1$.

Note, that the integral identities (8.1.1), (8.1.2) are equivalent to Lamé's system

$$\frac{1}{(\bar{c}_s^{(0)})^2}\,p + \nabla\cdot\mathbf{w} = 0, \tag{8.1.5}$$

$$\rho_s^{(0)}\frac{\partial^2 \mathbf{w}}{\partial t^2} = \nabla\cdot\Big(\lambda_0^{(0)}\mathbb{D}(x,\mathbf{w}) - p\,\mathbb{I}\Big) + \rho_s^{(0)}\mathbf{F} \tag{8.1.6}$$

in the domain $\Omega_T^{(s)}$, and the acoustic system

$$\hat{\rho}\,\frac{\partial^2 \mathbf{w}}{\partial t^2} = -\nabla\,p + \hat{\rho}\,\mathbf{F}, \;\; \left(\frac{m}{\bar{c}_f^2} + \frac{1-m}{\bar{c}_s^2}\right)\frac{\partial p}{\partial t} + \nabla\cdot\left(\frac{\partial \mathbf{w}}{\partial t}\right) = 0 \tag{8.1.7}$$

in the domain Ω_T.

These differential equations are completed with the continuity conditions

$$\lim_{\substack{\mathbf{x}\to\mathbf{x}^0\\ \mathbf{x}\in\Omega^{(s)}}} \mathbf{w}(\mathbf{x},t)\cdot\mathbf{n}(\mathbf{x}^0) = \left(-\frac{1}{\hat{\rho}}\lim_{\substack{\mathbf{x}\to\mathbf{x}^0\\ \mathbf{x}\in\Omega}} \mathbf{w}(\mathbf{x},t)\cdot\mathbf{n}(\mathbf{x}^0)\right), \tag{8.1.8}$$

$$\lim_{\substack{\mathbf{x}\to\mathbf{x}^0\\ \mathbf{x}\in\Omega^{(s)}}} \left(\lambda_0^{(0)}\mathbb{D}\left(x,\mathbf{w}(\mathbf{x},t)\right) - p\left(\mathbf{x},t\right)\mathbb{I}\right)\cdot\mathbf{n}(\mathbf{x}^0) = \lim_{\substack{\mathbf{x}\to\mathbf{x}^0\\ \mathbf{x}\in\Omega}} p\left(\mathbf{x},t\right)\mathbf{n}(\mathbf{x}^0) \tag{8.1.9}$$

on the common boundary $S_T^{(0)}$, the boundary and initial conditions (8.1.3), (8.1.4), the boundary condition

$$\mathbf{w}(\mathbf{x},t)\cdot\mathbf{n}(\mathbf{x}) = 0 \tag{8.1.10}$$

on the boundary $S_T \backslash \partial\Omega_T^{(s)}$ and the initial conditions

$$p(\mathbf{x},0) = 0,\ \ \mathbf{w}(\mathbf{x},0) = \frac{\partial\mathbf{w}}{\partial t}(\mathbf{x},0) = 0,\ \ \mathbf{x}\in\Omega. \tag{8.1.11}$$

Theorem 8.5 *Let* $\{\mathbf{w}^\varepsilon,\ p^\varepsilon\}$ *be the weak solution of the problem (I) and*

$$0 < \lambda_0^{(0)} < \infty,\quad 0 \leqslant \mu_1,\ \lambda_1 < \infty.$$

Then the limits $\mathbf{w}$ *and* p *of the sequences* $\{\mathbf{w}^\varepsilon\}$ *and* $\{p^\varepsilon\}$ *satisfy in the domain* $\Omega_T^{(s)}$ *Lamé's system (8.1.5), (8.1.6), the boundary and initial conditions (8.1.3), (8.1.4), and the homogenized equation*

$$\hat{\rho}\left(\frac{m}{\bar{c}_f^2} + \frac{(1-m)}{\bar{c}_s^2}\right)\frac{\partial p}{\partial t} + \nabla\cdot\mathbf{f} = -\nabla\cdot\int_0^t m\,\mathbb{B}^{(a)}(\mu_1,\lambda_1;t-\tau)\cdot\nabla\,p\left(\mathbf{x},\tau\right)d\tau \tag{8.1.12}$$

in the domain Ω_T.

These differential equations are completed with the continuity condition

$$\lim_{\substack{\mathbf{x}\to\mathbf{x}^0\\ \mathbf{x}\in\Omega^{(s)}}} \frac{\partial\mathbf{w}}{\partial t}(\mathbf{x},t)\cdot\mathbf{n}(\mathbf{x}^0) = \left(\lim_{\substack{\mathbf{x}\to\mathbf{x}^0\\ \mathbf{x}\in\Omega}} \int_0^t \mathbb{B}^{(a)}(\mu_1,\lambda_1;t-\tau)\cdot\nabla\,p\left(\mathbf{x},\tau\right)d\tau + \mathbf{f}(\mathbf{x},t)\right)\cdot\mathbf{n}(\mathbf{x}^0), \tag{8.1.13}$$

and the continuity condition (8.1.9) on the common boundary $S_T^{(0)}$, the boundary condition

$$\int_0^t \mathbb{B}^{(a)}(\mu_1, \lambda_1; t-\tau) \cdot \nabla p\,(\mathbf{x}, \tau) d\tau \cdot \mathbf{n}(\mathbf{x}) = -\mathbf{f}\,(\mathbf{x}, t) \cdot \mathbf{n}(\mathbf{x}) \qquad (8.1.14)$$

on the boundary $\partial\Omega_T \backslash S_T^{(0)}$, and the initial condition

$$p(\mathbf{x}, 0) = 0, \;\; \mathbf{x} \in \Omega. \qquad (8.1.15)$$

In Eqs. (8.1.12)–(8.1.14)

$$\hat{\rho} = m\,\rho_f + (1-m)\,\rho_s,$$

the matrix $\mathbb{B}^{(a)}(\mu_1, \lambda_1; t)$ and the function $\mathbf{f}(\mathbf{x}, t)$ are given in Chap. 6 by the formulae (6.2.40) and (6.2.42).

We refer to the problem (8.1.3)–(8.1.6), (8.1.9), (8.1.11)–(8.1.13) as the homogenized **model** $(\mathbb{ACM})_2$.

Theorem 8.6 *Let $\{\mathbf{w}^\varepsilon,\; p^\varepsilon\}$ be the weak solution of the problem (I),*

$$0 < \lambda_0^{(0)} < \infty, \quad \mu_1 = \infty, \;\; 0 \leqslant \lambda_1 < \infty,$$

and $\mathbf{w}_f^\varepsilon = \mathbb{E}_{\Omega_f^\varepsilon}(\mathbf{w}^\varepsilon)$ (for definition of this extension see Chap. 6).

Then the limits $\mathbf{w}$ and p of the sequences $\{\mathbf{w}^\varepsilon\}$ and $\{p^\varepsilon\}$ satisfy in the domain $\Omega_T^{(s)}$ Lamé's system (8.1.5), (8.1.6) and the boundary and initial conditions (8.1.3), (8.1.4). In the domain Ω_T the pressure of the mixture p, and the limiting functions $\mathbf{w}^{(s)}$, and $\mathbf{w}_f$ of the sequences $\{(1-\chi^\varepsilon)\mathbf{w}^\varepsilon\}$ and $\{\mathbf{w}_f^\varepsilon\}$ satisfy the system of homogenized equations, consisting of the continuity equation

$$\left(\frac{m}{\bar{c}_f^2} + \frac{(1-m)}{\bar{c}_s^2}\right)\frac{\partial p}{\partial t} + \nabla \cdot \left(m\frac{\partial \mathbf{w}_f}{\partial t} + \frac{\partial \mathbf{w}^{(s)}}{\partial t}\right) = 0, \qquad (8.1.16)$$

the momentum balance equation

$$m\rho_f \frac{\partial \mathbf{w}_f}{\partial t} + \rho_s \frac{\partial \mathbf{w}^{(s)}}{\partial t} + \int_0^t \left(-\hat{\rho}\mathbf{F} + \nabla\, p\right)(\mathbf{x}, \tau) d\tau = 0, \qquad (8.1.17)$$

for the liquid component and the momentum balance equation

$$\frac{\partial \mathbf{w}^{(s)}}{\partial t} - (1-m)\frac{\partial \mathbf{w}_f}{\partial t} = -\int_0^t \mathbb{B}^{(s)}(\infty, \lambda_1; t-\tau) \cdot \left(\nabla p + \rho_s \left(\frac{\partial^2 \mathbf{w}_f}{\partial \tau^2} - \mathbf{F}\right)\right)(\mathbf{x}, \tau) d\tau \tag{8.1.18}$$

for the solid component.

The problem is completed with the boundary condition (8.1.9) *and the boundary condition*

$$\lim_{\substack{\mathbf{x} \to \mathbf{x}^0 \\ \mathbf{x} \in \Omega^{(s)}}} \frac{\partial \mathbf{w}}{\partial t}(\mathbf{x}, t) \cdot \mathbf{n}(\mathbf{x}^0) = \left(\lim_{\substack{\mathbf{x} \to \mathbf{x}^0 \\ \mathbf{x} \in \Omega}} \left(m\frac{\partial \mathbf{w}_f}{\partial t} + \frac{\partial \mathbf{w}^{(s)}}{\partial t}\right)(\mathbf{x}, t)\right) \cdot \mathbf{n}(\mathbf{x}^0) \tag{8.1.19}$$

on the common boundary $S_T^{(0)}$, the boundary condition

$$\left(m\frac{\partial \mathbf{w}_f}{\partial t} + \frac{\partial \mathbf{w}^{(s)}}{\partial t}\right) \cdot \mathbf{n}(\mathbf{x}) = 0 \tag{8.1.20}$$

on the boundary $S_T \backslash \partial\Omega_T^{(s)}$, and the initial conditions

$$\mathbf{w}_f(\mathbf{x}, 0) = \mathbf{w}^{(s)}(\mathbf{x}, 0) = 0, \;\; \mathbf{x} \in \Omega. \tag{8.1.21}$$

In (8.1.17)–(8.1.18)

$$\hat{\rho} = m\,\rho_f + (1-m)\,\rho_s,$$

and the matrix $\mathbb{B}^{(s)}(\infty, \lambda_1; t)$ has been defined in Chap. 3 by formulae (3.2.47) and (3.2.54).

We refer to the problem (8.1.3)–(8.1.6), (8.1.9), (8.1.16)–(8.1.21) as the homogenized **model** $(\mathbb{ACM})_3$.

Theorem 8.7 *Let $\{\mathbf{w}^\varepsilon,\ p^\varepsilon\}$ be the weak solution of the problem (I),*

$$0 < \lambda_0^{(0)} < \infty, \quad \lambda_1 = \infty, \;\; 0 \leqslant \mu_1 < \infty,$$

and $\mathbf{w}_s^\varepsilon = \mathbb{E}_{\Omega_s^\varepsilon}(\mathbf{w}^\varepsilon)$ (for definition of this extension see Chap. 6).

Then the limits $\mathbf{w}$ and p of the sequences $\{\mathbf{w}^\varepsilon\}$ and $\{p^\varepsilon\}$ satisfy in the domain $\Omega_T^{(s)}$ Lamé's system (8.1.5), (8.1.6) and the boundary and initial conditions (8.1.3), (8.1.4).

In the domain Ω_T the pressure of the mixture p, and the limiting functions $\mathbf{w}^{(f)}$, and w_s of the sequences $\{\chi^\varepsilon \mathbf{w}^\varepsilon\}$ and $\{\mathbf{w}_s^\varepsilon\}$ satisfy the system of homogenized *equations, consisting of the continuity equation*

$$\left(\frac{m}{\bar{c}_f^2}+\frac{(1-m)}{\bar{c}_s^2}\right)\frac{\partial p}{\partial t}+\nabla\cdot\left(\frac{\partial \mathbf{w}^{(f)}}{\partial t}+(1-m)\frac{\partial \mathbf{w}_s}{\partial t}\right)=0, \tag{8.1.22}$$

the momentum balance equation

$$\rho_f\frac{\partial \mathbf{w}^{(f)}}{\partial t}+(1-m)\rho_s\frac{\partial \mathbf{w}_s}{\partial t}=\int_0^t\left(\hat{\rho}\mathbf{F}-\nabla p\right)(\mathbf{x},\tau)d\tau, \tag{8.1.23}$$

for the solid component and the momentum balance equation

$$\frac{\partial \mathbf{w}^{(f)}}{\partial t}-m\,\frac{\partial \mathbf{w}_s}{\partial t}$$
$$=-\int_0^t\mathbb{B}^{(f)}(\mu_1,\infty;t-\tau)\cdot\left(\nabla p+\rho_f\left(\frac{\partial^2\mathbf{w}_s}{\partial\tau^2}-\mathbf{F}\right)\right)(\mathbf{x},\tau)d\tau \tag{8.1.24}$$

for the liquid component.

The problem is completed with the boundary condition (8.1.9) and the boundary condition

$$\lim_{\substack{\mathbf{x}\to\mathbf{x}^0\\ \mathbf{x}\in\Omega^{(s)}}}\frac{\partial\,\mathbf{w}}{\partial t}(\mathbf{x},t)\cdot\mathbf{n}(\mathbf{x}^0)=\left(\lim_{\substack{\mathbf{x}\to\mathbf{x}^0\\ \mathbf{x}\in\Omega}}\left(\frac{\partial \mathbf{w}^{(f)}}{\partial t}+(1-m)\frac{\partial \mathbf{w}_s}{\partial t}\right)(\mathbf{x},t)\right)\cdot\mathbf{n}(\mathbf{x}^0) \tag{8.1.25}$$

on the common boundary $S_T^{(0)}$, the boundary condition

$$\left(\frac{\partial \mathbf{w}^{(f)}}{\partial t}+(1-m)\frac{\partial \mathbf{w}_s}{\partial t}\right)\cdot\mathbf{n}(\mathbf{x})=0 \tag{8.1.26}$$

on the boundary $S_T\backslash\partial\Omega_T^{(s)}$, and the initial conditions

$$\mathbf{w}_s(\mathbf{x},0)=\mathbf{w}^{(f)}(\mathbf{x},0)=0,\ \ \mathbf{x}\in\Omega. \tag{8.1.27}$$

In (8.1.23)–(8.1.24)

$$\hat{\rho}=m\,\rho_f+(1-m)\,\rho_s,$$

and the matrix $\mathbb{B}^{(f)}(\mu_1,\infty;t)$ has been defined in Chap. 3 by formulae (3.2.70) and (3.2.76).

We refer to the problem (8.1.3)–(8.1.6), (8.1.9), (8.1.22)–(8.1.27) as the homogenized **model** $(\mathbb{ACM})_4$.

Theorem 8.8 *Let $\{\mathbf{w}^\varepsilon,\ p^\varepsilon\}$ be the weak solution of the problem (I),*

$$0 < \lambda_0^{(0)} < \infty, \quad \mu_1 = \infty, \ 0 < \lambda_0 < \infty,$$

and $\mathbf{w}_s^\varepsilon = \mathbb{E}_{\Omega_s^\varepsilon}(\mathbf{w}^\varepsilon)$.

Then the limits $\mathbf{w}$ and p of the sequences $\{\mathbf{w}^\varepsilon\}$ and $\{p^\varepsilon\}$ satisfy in the domain $\Omega_T^{(s)}$ Lamé's system (8.1.5), (8.1.6) and the boundary and initial conditions (8.1.3), (8.1.4).

In the domain Ω_T the limit w_s (the solid displacement) of the sequence $\{\mathbf{w}_s^\varepsilon\}$ satisfies the homogenized equation

$$\hat{\rho}\frac{\partial^2 \mathbf{w}_s}{\partial t^2} = \nabla \cdot \big(\lambda_0\, \mathfrak{N}_3^s : \mathbb{D}(x, \mathbf{w}_s)\big) + \hat{\rho}\mathbf{F}, \tag{8.1.28}$$

completed with the initial and boundary conditions

$$\mathbf{w}_s(\mathbf{x}, 0) = \frac{\partial \mathbf{w}_s}{\partial t}(\mathbf{x}, 0) = 0, \ \mathbf{x} \in \Omega, \tag{8.1.29}$$

$$\mathbf{w}_s(\mathbf{x}, t) = 0, \ \mathbf{x} \in \partial\Omega \backslash S^{(0)}, \ t \in (0, T). \tag{8.1.30}$$

On the common boundary $S_T^{(0)}$ the continuity conditions

$$\lim_{\substack{\mathbf{x} \to \mathbf{x}^0 \\ \mathbf{x} \in \Omega^{(s)}}} \mathbf{w}(\mathbf{x}, t) = \lim_{\substack{\mathbf{x} \to \mathbf{x}^0 \\ \mathbf{x} \in \Omega}} \mathbf{w}_s(\mathbf{x}, t) \tag{8.1.31}$$

and

$$\lim_{\substack{\mathbf{x} \to \mathbf{x}^0 \\ \mathbf{x} \in \Omega^{(s)}}} \Big(\lambda_0^{(0)}\mathbb{D}\big(x, \mathbf{w}(\mathbf{x}, t)\big) - p\,(\mathbf{x}, t)\,\mathbb{I}\Big) \cdot \mathbf{n}(\mathbf{x}^0)$$

$$= \lim_{\substack{\mathbf{x} \to \mathbf{x}^0 \\ \mathbf{x} \in \Omega}} \Big(\lambda_0\, \mathfrak{N}_3^s : \mathbb{D}\big(x, \mathbf{w}_s(\mathbf{x}, t)\big)\Big) \cdot \mathbf{n}(\mathbf{x}^0) \tag{8.1.32}$$

hold true.

In (8.1.28)

$$\hat{\rho} = m\,\rho_f + (1 - m)\,\rho_s,$$

and the symmetric strictly positive definite constant fourth-rank tensor $\mathfrak{N}_3^s$ is given in Chap. 1 by the formula (1.3.39).

We refer to the problem (8.1.3)–(8.1.6), (8.1.9), (8.1.28)–(8.1.32) as the homogenized **model** $(\mathbb{ACM})_5$.

Theorem 8.9 *Let* $\{\mathbf{w}^\varepsilon,\ p^\varepsilon\}$ *be the weak solution of the problem (I),*

$$0<\lambda_0^{(0)}<\infty,\quad 0<\mu_1<\infty,\ 0<\lambda_0<\infty,$$

and $\mathbf{w}_s^\varepsilon=\mathbb{E}_{\Omega_s^\varepsilon}\big(\mathbf{w}^\varepsilon\big)$.

Then the limits $\mathbf{w}$ *and* p *of the sequences* $\{\mathbf{w}^\varepsilon\}$ *and* $\{p^\varepsilon\}$ *satisfy in the domain* G_T *Lamé's system (8.1.5), (8.1.6) and the boundary and initial conditions (8.1.3), (8.1.4).*

In the domain Ω_T *the limits* p_f *(the liquid pressure),* $\mathbf{w}^f$ *(the liquid displacement), and* $\mathbf{w}_s$ *(the solid displacement) of the sequences* $\{\chi^\varepsilon\, p^\varepsilon\}$, $\{\chi^\varepsilon\,\mathbf{w}^\varepsilon\}$, *and* $\{\mathbf{w}_s^\varepsilon\}$ *satisfy the system of the homogenized equations, consisting of the continuity equation*

$$\frac{m}{c_f^2}p_f+\nabla\cdot\mathbf{w}^{(f)}=\mathbb{C}_0^s:\mathbb{D}(x,\mathbf{w}_s)+\frac{c_0^s}{\lambda_0}p_f, \tag{8.1.33}$$

the momentum balance equation

$$\rho_f\frac{\partial^2\mathbf{w}^{(f)}}{\partial t^2}+\rho_s\frac{\partial^2\mathbf{w}_s}{\partial t^2}=\nabla\cdot\big(\lambda_0\,\mathfrak{N}_2^s:\mathbb{D}(x,\mathbf{w}_s)-p_f\,\mathbb{C}_1^s\big)+\hat{\rho}\mathbf{F}, \tag{8.1.34}$$

for the solid component, the momentum balance equation

$$\begin{aligned}&-\int_0^t\mathbb{B}^{(f)}(\mu_1,\infty;t-\tau)\cdot\Big(\nabla p_f+\rho_f\Big(\frac{\partial^2\mathbf{w}_s}{\partial\tau^2}-\mathbf{F}\Big)\Big)(\mathbf{x},\tau)d\tau\\&=\frac{\partial\mathbf{w}^{(f)}}{\partial t}-m\,\frac{\partial\mathbf{w}_s}{\partial t}\end{aligned} \tag{8.1.35}$$

for the liquid component, the continuity condition (8.1.31) and the continuity conditions

$$\lim_{\substack{\mathbf{x}\to\mathbf{x}^0\\ \mathbf{x}\in\Omega^{(s)}}}\mathbf{w}(\mathbf{x},t)\cdot\mathbf{n}(\mathbf{x}^0)=\lim_{\substack{\mathbf{x}\to\mathbf{x}^0\\ \mathbf{x}\in\Omega}}\big(\mathbf{w}^{(f)}(\mathbf{x},t)+(1-m)\mathbf{w}_s(\mathbf{x},t)\big)\cdot\mathbf{n}(\mathbf{x}^0), \tag{8.1.36}$$

and

$$\begin{aligned}&\lim_{\substack{\mathbf{x}\to\mathbf{x}^0\\ \mathbf{x}\in\Omega^{(s)}}}\Big(\lambda_0^{(0)}\mathbb{D}\big(x,\mathbf{w}(\mathbf{x},t)\big)-p\,(\mathbf{x},t)\,\mathbb{I}\Big)\cdot\mathbf{n}(\mathbf{x}^0)\\&\qquad=\lim_{\substack{\mathbf{x}\to\mathbf{x}^0\\ \mathbf{x}\in\Omega}}\Big(\lambda_0\,\mathfrak{N}_2^s:\mathbb{D}\big(x,\mathbf{w}_s(\mathbf{x},t)\big)-p_f\,\mathbb{C}_1^s\big)\Big)\cdot\mathbf{n}(\mathbf{x}^0)\end{aligned} \tag{8.1.37}$$

on the common boundary $S_T^{(0)}$, *the homogeneous boundary and initial conditions (8.1.29) and (8.1.30) for the solid displacements, and the homogeneous boundary and initial conditions*

$$\mathbf{w}^{(f)}(\mathbf{x},t)\cdot\mathbf{n}(\mathbf{x})=0,\ \ \mathbf{x}\in\partial\Omega\backslash S^{(0)},\ \ t\in(0,T), \tag{8.1.38}$$

$$\mathbf{w}^{(f)}(\mathbf{x},0)=0,\ \ \mathbf{x}\in\Omega. \tag{8.1.39}$$

for the liquid displacements.

In (8.1.33)–(8.1.35)

$$\hat{\rho}=m\,\rho_f+(1-m)\,\rho_s,$$

the symmetric strictly positively definite constant fourth-rank tensor $\mathfrak{N}_2^s$, *the matrices* $\mathbb{C}_0^s$ *and* $\mathbb{C}_1^s$, *and the constant* c_0^s *are given in Chapter I by formulae (1.3.26), (1.3.27) and, (1.3.31) and the matrix* $\mathbb{B}^{(f)}(\mu_1,\infty;t)$ *has been defined in Chap. 3 by the formula (3.2.70).*

We refer to the problem (8.1.3)–(8.1.6), (8.1.9), (8.1.29)–(8.1.31), (8.1.33)–(8.1.39) as the homogenized **model** $(\mathbb{ACM})_6$.

8.1.2 Proofs of Theorems 8.4–8.7

The proofs of these theorems are standard and repeat the proofs of the corresponding theorems in the previous chapters, because we can prove the statements separately in each of the domains G and Ω. Thus, the main problem here is the boundary conditions on the common boundary $S^{(0)}$. These boundary conditions follow from the limiting integral identity (8.1.2), and the integral identity

$$\int_{Q_T}\big((1-\zeta)\lambda_0^{(0)}\mathbb{D}(x,\mathbf{w})-p\,\mathbb{I}\big):\mathbb{D}(x,\boldsymbol{\varphi})dxdt$$
$$=\int_{Q_T}\int_Y\rho_{(s)}(\mathbf{x},\mathbf{y})\left(\mathbf{F}-\frac{\partial^2\mathbf{w}}{\partial t^2}(\mathbf{x},t,\mathbf{y})\right)\cdot\boldsymbol{\varphi}(\mathbf{x},t)dydxdt, \tag{8.1.40}$$

where $\mathbf{W}(\mathbf{x},t,\mathbf{y})$ is the two-scale limit of the sequence $\{\mathbf{w}^{\varepsilon}\}$, and

$$\rho_{(s)}(\mathbf{x},\mathbf{y})=\big(1-\zeta(\mathbf{x})\big)\rho_s^{(0)}+\zeta(\mathbf{x})\big(\rho_f\,\chi(\mathbf{y})+\big(1-\chi(\mathbf{y})\big)\,\rho_s\big).$$

For all cases (8.1.40) implies the dynamic Lamé's equation (8.1.5) in the domain G_T and the boundary condition (8.1.9) on the common boundary $S^{(0)}$. The integral identity (8.1.2) implies the continuity equation (8.1.6) in the domain G_T, the continuity equation

$$\left(\frac{m}{\bar{c}_f^2} + \frac{1-m}{\bar{c}_s^2}\right)\frac{\partial^2 p}{\partial t^2} + \nabla \cdot \frac{\partial^2 \mathbf{w}}{\partial t^2} = 0, \tag{8.1.41}$$

in the domain Ω_T, and the boundary condition

$$\lim_{\substack{\mathbf{x}\to\mathbf{x}^0\\ \mathbf{x}\in\Omega^{(s)}}} \frac{\partial^2 \mathbf{w}}{\partial t^2}(\mathbf{x},t)\cdot\mathbf{n}(\mathbf{x}^0) = \lim_{\substack{\mathbf{x}\to\mathbf{x}^0\\ \mathbf{x}\in\Omega}} \frac{\partial^2 \mathbf{w}}{\partial t^2}(\mathbf{x},t)\cdot\mathbf{n}(\mathbf{x}^0) \tag{8.1.42}$$

on the common boundary $S^{(0)}$.

All the differences are concentrated in the dynamic equation in the domain Ω_T and in the representation of the velocity of the mixture $\dfrac{\partial \mathbf{w}}{\partial t}$.

8.1.2.1 Proof of Theorem 8.4

For this case $\mathbf{W}(\mathbf{x},t,\mathbf{y}) = \mathbf{w}(\mathbf{x},t)$ and the integral identity (8.1.40) implies the dynamic equation

$$\hat{\rho}\,\frac{\partial^2 \mathbf{w}}{\partial t^2} = -\nabla\, p + \hat{\rho}\,\mathbf{F} \tag{8.1.43}$$

in the domain Ω_T.

Relations (8.1.41)–(8.1.43) evidently imply the acoustic equation (8.1.7) in the domain Ω_T and the boundary condition (8.1.8) on the boundary $S^{(0)}$.

8.1.2.2 Proof of Theorem 8.5

To obtain the dynamic equation here we simply repeat the proof of Theorem 6.6. That is, we firstly derive the microscopic dynamic equation (6.2.32) with $\tilde{P} = p$ and then the microscopic continuity equation (6.2.33). These relations result in the representation

$$\frac{\partial\, \mathbf{w}}{\partial t}(\mathbf{x},t) = \int_0^t \mathbb{B}^{(a)}(\mu_1,\lambda_1;t-\tau)\cdot\nabla\, p\,(\mathbf{x},\tau)d\tau + \mathbf{f}(\mathbf{x},t) \tag{8.1.44}$$

of the velocity in the mixture.

This representation, the continuity equation (8.1.41), and the boundary condition (8.1.42) imply the equation (8.1.12) and the boundary condition (8.1.13).

8.1.2.3 Proof of Theorem 8.6

For this case the velocity of the mixture is given by the formula

$$\frac{\partial\,\mathbf{W}}{\partial t}(\mathbf{x},t,\mathbf{y}) = \chi(\mathbf{y})\frac{\partial \mathbf{w}_f}{\partial t}(\mathbf{x},t) + \big(1-\chi(\mathbf{y})\big)\frac{\partial \mathbf{W}}{\partial t}(\mathbf{x},t,\mathbf{y}),$$

$$\frac{\partial\,\mathbf{w}}{\partial t} = m\frac{\partial \mathbf{w}_f}{\partial t} + \frac{\partial \mathbf{w}^{(s)}}{\partial t}, \quad \mathbf{w}^{(s)}(\mathbf{x},t) = \int_Y \big(1-\chi(\mathbf{y})\big)\mathbf{W}(\mathbf{x},t,\mathbf{y})dy,$$

which together with (8.1.41) result in (8.1.16).

The integral identity (8.1.40) implies the dynamic equation (8.1.17) for the liquid component.

To find the representation (8.1.18) we use the microscopic dynamic equation

$$\rho_s\frac{\partial^2 \mathbf{W}^{(s)}}{\partial t^2} = \frac{\lambda_1}{2}\Delta_y \mathbf{W}^{(s)} - \nabla_y \Pi^{(s)} - \nabla\,p, \quad \mathbf{W}^{(s)} = \big(1-\chi(\mathbf{y})\big)\mathbf{W},$$

for the solid component, the microscopic continuity equation

$$\nabla\cdot\mathbf{W}^{(s)} = 0,$$

in the domain Y_s, and the corresponding boundary and initial conditions. This problem has been already solved in Chap. 3 (see the proof of Theorem 3.4). The rest of the proof is the same as the proofs of previous theorems.

8.1.2.4 Proof of Theorem 8.7

Here the velocity of the mixture is given by the formula

$$\frac{\partial\,\mathbf{W}}{\partial t}(\mathbf{x},t,\mathbf{y}) = \chi(\mathbf{y})\frac{\partial \mathbf{W}}{\partial t}(\mathbf{x},t,\mathbf{y}) + \big(1-\chi(\mathbf{y})\big)\frac{\partial \mathbf{w}_s}{\partial t}(\mathbf{x},t),$$

$$\frac{\partial\,\mathbf{w}}{\partial t} = \frac{\partial \mathbf{w}^{(f)}}{\partial t} + (1-m)\frac{\partial \mathbf{w}_s}{\partial t}, \quad \mathbf{w}^{(f)}(\mathbf{x},t) = \int_Y \chi(\mathbf{y})\mathbf{W}(\mathbf{x},t,\mathbf{y})dy,$$

which together with (8.1.41) result in (8.1.22).

The integral identity (8.1.40) implies the dynamic equation (8.1.23) for the solid component. The representation (8.1.24) has been already obtained in Chap. 3 (see the proof of Theorem 3.5). The rest of the proof is the same as for the previous theorems.

8.1.3 Proofs of Theorems 8.8 and 8.9

For these cases the two-scale limit $P(\mathbf{x},t,\mathbf{y})$ of the sequence $\{p^\varepsilon\}$ is given by

$$(1-\zeta)\,p + \zeta\,\chi(\mathbf{y})\,p_f(\mathbf{x},t) + \zeta\,\big(1-\chi(\mathbf{y})\big)\,P_s(\mathbf{x},t,\mathbf{y}).$$

For $\mu_1 = \infty$ the two-scale limit $\mathbf{w}$ of $\{\mathbf{w}^\varepsilon\}$ is given by

$$\mathbf{W}(\mathbf{x}, t, \mathbf{y}) = \mathbf{w}_s(\mathbf{x}, t),$$

and for $\mu_1 < \infty$

$$\mathbf{W}(\mathbf{x}, t, \mathbf{y}) = \chi(\mathbf{y})\mathbf{W}(\mathbf{x}, t, \mathbf{y}) + \big(1 - \chi(\mathbf{y})\big)\mathbf{w}_s(\mathbf{x}, t).$$

The two-scale limit of the sequence $\{\mathbf{w}_s^\varepsilon\}$ is equal to $\mathbf{w}_s(\mathbf{x}, t)$, and the two-scale limit of the sequence $\{\mathbb{D}(x, \mathbf{w}_s^\varepsilon)\}$ is given by

$$\mathbb{D}\big(x, \mathbf{w}_s(\mathbf{x}, t)\big) + \mathbb{D}\big(y, \mathbf{U}(\mathbf{x}, t, \mathbf{y})\big)$$

(see proofs of Theorem 1.6 and Theorem 1.7).

The integral identity (8.1.40) is replaced by

$$\begin{aligned}\int_{Q_T} \Big(& (1 - \zeta)\lambda_0^{(0)}\mathbb{D}(x, \mathbf{w}) + \zeta\, \lambda_0\big((1 - m)\mathbb{D}(x, \mathbf{w}_s) \\ & + \langle \mathbb{D}(y, \mathbf{U})\rangle_{Y_s}\big)\Big) : \mathbb{D}(x, \boldsymbol{\varphi})dxdt - \int_{Q_T} p\, \nabla \cdot \boldsymbol{\varphi} dxdt \\ = \int_{Q_T} \int_Y & \rho_{(s)}(\mathbf{x}, \mathbf{y}) \left(\mathbf{F} - \frac{\partial^2 \mathbf{W}}{\partial t^2}(\mathbf{x}, t, \mathbf{y})\right) dy \cdot \boldsymbol{\varphi}(\mathbf{x}, t)dxdt \end{aligned} \tag{8.1.45}$$

with smooth functions $\boldsymbol{\varphi}$, vanishing on the boundary ∂Q, and the integral identity (8.1.2) is replaced by

$$\int_{Q_T} \left(\eta \int_Y \left(\frac{1 - \zeta}{\left(\bar{c}_s^{(0)}\right)^2} + \left(\frac{\chi}{\bar{c}_f^2} + \frac{1 - \chi}{\bar{c}_s^2}\right) \zeta \right) \frac{\partial P}{\partial t} dy - \nabla \eta \cdot \frac{\partial \mathbf{w}}{\partial t} \right) dxdt = 0 \tag{8.1.46}$$

with smooth functions η.

In (8.1.45)

$$\rho_{(s)}(\mathbf{x}, \mathbf{y}) = \big(1 - \zeta(\mathbf{x})\big)\rho_s^{(0)} + \zeta(\mathbf{x})\big(\rho_f \chi(\mathbf{y}) + \big(1 - \chi(\mathbf{y})\big)\rho_s\big).$$

As before, relations (8.1.45) and (8.1.46) result in Lamé's system (8.1.5) and (8.1.6) in $\Omega_T^{(s)}$ and the boundary conditions (8.1.32), (8.1.36) and (8.1.37) on the common boundary $S_T^{(0)}$.

Equations (8.1.28), (8.1.32), (8.1.33) and (8.1.34) and the boundary and initial conditions (8.1.29), (8.1.30), (8.1.38) and (8.1.39) are already derived in Chap. 1 and Chap. 6 (see proofs of Theorems 1.6–1.8 and 6.9–6.11).

Finally, the boundary condition (8.1.31) is a consequence of the smoothness of $\mathbf{w}_s$ and $\mathbf{w}$ and is derived as well as the boundary condition (8.1.30).

8.2 Acoustics in a "Liquid–Poroelastic Medium" Configuration

8.2.1 Main Results

Theorem 8.10 *Let $\{\mathbf{w}^{\varepsilon},\ p^{\varepsilon}\}$ be the weak solution of the problem (II) and*

$$\mu_1 = \lambda_1 = \infty.$$

Then the limits $\mathbf{v} = \dfrac{\partial \mathbf{w}}{\partial t}$ (the liquid velocity) and p (the pressure) of the sequences $\left\{\dfrac{\partial \mathbf{w}^{\varepsilon}}{\partial t}\right\}$ and $\{p^{\varepsilon}\}$ satisfy the system of acoustic equations

$$\rho_f \frac{\partial \mathbf{v}}{\partial t} + \nabla\, p = \rho_f \mathbf{F}, \tag{8.2.1}$$

$$\frac{1}{\bar{c}_f^2}\frac{\partial\, p}{\partial t} + \nabla \cdot \mathbf{v} = 0 \tag{8.2.2}$$

in the domain $\Omega^{(f)}$ for $t > 0$, and the system of acoustic equations

$$\hat{\rho}\frac{\partial \mathbf{v}}{\partial t} + \nabla\, p = \hat{\rho}\mathbf{F}, \tag{8.2.3}$$

$$\left(\frac{m}{\bar{c}_f^2} + \frac{1-m}{\bar{c}_s^2}\right)\frac{\partial\, p}{\partial t} + \nabla \cdot \mathbf{v} = 0 \tag{8.2.4}$$

in the domain Ω for $t > 0$.

Relations (8.2.1)–(8.2.4) are completed with the homogeneous boundary condition

$$\mathbf{v}(\mathbf{x}, t) \cdot \mathbf{n}(\mathbf{x}) = 0, \tag{8.2.5}$$

on the boundary S_T, the homogeneous initial conditions

$$p(\mathbf{x}, 0) = 0,\ \ \mathbf{v}(\mathbf{x}, 0) = 0,\ \ \mathbf{x} \in Q, \tag{8.2.6}$$

and the continuity conditions

$$\lim_{\substack{\mathbf{x}\to\mathbf{x}^0\\ \mathbf{x}\in\Omega^{(f)}}} \mathbf{v}(\mathbf{x}, t)\cdot\mathbf{n}(\mathbf{x}^0) = \lim_{\substack{\mathbf{x}\to\mathbf{x}^0\\ \mathbf{x}\in\Omega}} \mathbf{v}(\mathbf{x}, t)\cdot\mathbf{n}(\mathbf{x}^0), \tag{8.2.7}$$

$$\lim_{\substack{\mathbf{x}\to\mathbf{x}^0\\ \mathbf{x}\in\Omega^{(f)}}} p(\mathbf{x}, t) = \lim_{\substack{\mathbf{x}\to\mathbf{x}^0\\ \mathbf{x}\in\Omega}} p(\mathbf{x}, t) \tag{8.2.8}$$

on the common boundary $S^{(0)}_T$.

Here

$$\hat{\rho} = m\,\rho_f + (1-m)\,\rho_s,$$

$\mathbf{n}(\mathbf{x})$ *is the normal to* S *at* $\mathbf{x} \in S$*, and* $\mathbf{n}(\mathbf{x}^0)$ *is the normal to* $S^{(0)}$ *at* $\mathbf{x}^0 \in S^{(0)}$.

We refer to the problem (8.2.1)–(8.2.8) as the homogenized **model** $(\mathbb{ACM})_7$.

Theorem 8.11 *Let* $\{\mathbf{w}^\varepsilon,\ p^\varepsilon\}$ *be the weak solution of the problem (II) and*

$$0 \leqslant \mu_1,\ \lambda_1 < \infty.$$

Then the limits $\mathbf{v} = \dfrac{\partial \mathbf{w}}{\partial t}$ *(the liquid velocity) and* p *(the pressure) of the sequences* $\left\{\dfrac{\partial \mathbf{w}^\varepsilon}{\partial t}\right\}$ *and* $\{p^\varepsilon\}$ *satisfy in the domain* $\Omega^{(f)}_T$ *the system of acoustic equations (8.2.1), (8.2.2), and the system of acoustic equations in the domain* Ω_T*, consisting of the momentum balance equation in the form*

$$\mathbf{v}(\mathbf{x},t) = \int_0^t \mathbb{B}^{(a)}(\mu_1,\lambda_1;t-\tau)\cdot\nabla\,p\,(\mathbf{x},\tau)d\tau + \mathbf{f}(\mathbf{x},t), \tag{8.2.9}$$

and continuity equation (8.2.4).

The differential equations are completed with the boundary and initial conditions (8.2.5), (8.2.6), and the continuity conditions (8.2.7) and (8.2.8).

The matrix $\mathbb{B}^{(a)}(\mu_1,\lambda_1;t)$ *and the function* $\mathbf{f}(\mathbf{x},t)$ *are given in Chap. 6 by the formulae (6.2.40), (6.2.42).*

We refer to the problem (8.2.1), (8.2.2), (8.2.4)–(8.2.9) as the homogenized **model** $(\mathbb{ACM})_8$.

Theorem 8.12 *Let* $\{\mathbf{w}^\varepsilon,\ p^\varepsilon\}$ *be the weak solution of the problem (II),*

$$\mu_1 = \infty,\ 0 \leqslant \lambda_1 < \infty,$$

and $\mathbf{w}^\varepsilon_f = \mathbb{E}_{\Omega^\varepsilon_f}\left(\mathbf{w}^\varepsilon\right)$ *(for definition of this extension see Chap. 6).*

Then the limits $\mathbf{v} = \dfrac{\partial \mathbf{w}}{\partial t}$ *(the liquid velocity) and* p *(the pressure) of the sequences* $\left\{\dfrac{\partial \mathbf{w}^\varepsilon}{\partial t}\right\}$ *and* $\{p^\varepsilon\}$*, where*

$$\mathbf{v} = (1-\zeta)\mathbf{v} + \zeta\,m\frac{\partial \mathbf{w}_f}{\partial t} + \zeta\,\frac{\partial \mathbf{w}^{(s)}}{\partial t} = (1-\zeta)\mathbf{v} + \zeta\,m\,\mathbf{v}_f + \zeta\,\mathbf{v}^{(s)}, \tag{8.2.10}$$

and $\mathbf{w}^{(s)}$ *and* $\mathbf{w}_f$ *are the limits of the sequences* $\{(1-\chi^\varepsilon)\mathbf{w}^\varepsilon\}$ *and* $\{\mathbf{w}^\varepsilon_f\}$*, satisfy in the domain* $\Omega^{(f)}_T$ *the system of acoustic equations (8.2.1), (8.2.2), and the system of* *acoustic equations in the domain Ω_T, consisting of the momentum balance equation*

$$m\rho_f\,\mathbf{v}_f+\rho_s\,\mathbf{v}^{(s)}+\int_0^t\big(-\hat{\rho}\mathbf{F}+\nabla p\big)(\mathbf{x},\tau)d\tau=0, \tag{8.2.11}$$

for the liquid component, the momentum balance equation

$$\mathbf{v}^{(s)}-(1-m)\mathbf{v}_f$$
$$=-\int_0^t\mathbb{B}^{(s)}(\infty,\lambda_1;t-\tau)\cdot\Big(\nabla p+\rho_s\big(\frac{\partial\mathbf{v}_f}{\partial\tau}-\mathbf{F}\big)\Big)(\mathbf{x},\tau)d\tau \tag{8.2.12}$$

for the solid component, and the continuity equation

$$\left(\frac{m}{\bar{c}_f^2}+\frac{(1-m)}{\bar{c}_s^2}\right)\frac{\partial p}{\partial t}+\nabla\cdot\big(m\,\mathbf{v}_f+\mathbf{v}^{(s)}\big)=0. \tag{8.2.13}$$

The problem is completed with the boundary and initial conditions (8.2.5), (8.2.6), and the continuity conditions (8.2.7) and (8.2.8).

In (8.2.11)–(8.2.12)

$$\hat{\rho}=m\,\rho_f+(1-m)\,\rho_s,$$

and the matrix $\mathbb{B}^{(s)}(\infty,\lambda_1;t)$ has been defined in Chap. 3 by the formulae (3.2.47) and (3.2.54).

We refer to the problem (8.2.1), (8.2.2), (8.2.5)–(8.2.8), (8.2.10)–(8.2.13) as the homogenized **model** $(\mathbb{ACM})_9$.

Theorem 8.13 *Let $\{\mathbf{w}^\varepsilon,\ p^\varepsilon\}$ be the weak solution of the problem (II),*

$$\lambda_1=\infty,\ 0\leqslant\mu_1<\infty,$$

and $\mathbf{w}_s^\varepsilon=\mathbb{E}_{\Omega_s^\varepsilon}\big(\mathbf{w}^\varepsilon\big)$ (for definition of this extension see Chap. 6*).*

Then the limits $\mathbf{v}=\dfrac{\partial\mathbf{w}}{\partial t}$ (the liquid velocity) and p (the pressure) of the sequences $\left\{\dfrac{\partial\mathbf{w}^\varepsilon}{\partial t}\right\}$ and $\{p^\varepsilon\}$, where

$$\begin{aligned}\mathbf{v}&=(1-\zeta)\mathbf{v}+\zeta\,\frac{\partial\mathbf{w}^{(f)}}{\partial t}+\zeta\,(1-m)\frac{\partial\mathbf{w}_s}{\partial t}\\&=(1-\zeta)\mathbf{v}+\zeta\,\mathbf{v}^{(f)}+\zeta\,\mathbf{v}_s,\end{aligned} \tag{8.2.14}$$

and $\mathbf{w}^{(f)}$ and $\mathbf{w}_s$ are the limits of the sequences $\{\chi^\varepsilon\mathbf{w}^\varepsilon\}$ and $\{\mathbf{w}_s^\varepsilon\}$, satisfy in the domain $\Omega_T^{(f)}$ the system of acoustic equations (8.2.1), (8.2.2), and the system of acoustic equations in the domain Ω_T, consisting of the continuity equation

$$\left(\frac{m}{\bar{c}_f^2}+\frac{(1-m)}{\bar{c}_s^2}\right)\frac{\partial p}{\partial t}+\nabla\cdot\left(\mathbf{v}^{(f)}+\mathbf{v}_s\right)=0, \tag{8.2.15}$$

the momentum balance equation

$$\rho_f\,\mathbf{v}^{(f)}+(1-m)\rho_s\,\mathbf{v}_s=\int_0^t\left(\hat{\rho}\mathbf{F}-\nabla\,p\right)(\mathbf{x},\tau)d\tau, \tag{8.2.16}$$

for the solid component and the momentum balance equation

$$\mathbf{v}^{(f)}-m\,\mathbf{v}_s$$
$$=-\int_0^t\mathbb{B}^{(f)}(\mu_1,\infty;t-\tau)\cdot\left(\nabla\,p+\rho_f\left(\frac{\partial\mathbf{v}_s}{\partial\tau}-\mathbf{F}\right)\right)(\mathbf{x},\tau)d\tau \tag{8.2.17}$$

for the liquid component.

The problem is completed with the boundary and initial conditions (8.2.5), (8.2.6), and the continuity conditions (8.2.7) and (8.2.8).

In (8.2.16)–(8.2.17)

$$\hat{\rho}=m\,\rho_f+(1-m)\,\rho_s,$$

and the matrix $\mathbb{B}^{(f)}(\mu_1,\infty;t)$ has been defined in Chap. 3 by the formulae (3.2.70) and (3.2.76).

We refer to the problem ((8.2.1), (8.2.2), (8.2.5)–(8.2.8), (8.2.14)–(8.2.17) as the homogenized **model** $(\mathbb{ACM})_{10}$.

Theorem 8.14 *Let $\{\mathbf{w}^\varepsilon,\ p^\varepsilon\}$ be the weak solution of the problem (II),*

$$\mu_1=\infty,\ \ 0<\lambda_0<\infty,$$

and $\mathbf{w}_s^\varepsilon=\mathbb{E}_{\Omega_s^\varepsilon}\left(\mathbf{w}^\varepsilon\right)$.

Then the limits $\mathbf{v}=\dfrac{\partial\mathbf{w}}{\partial t}$ (the liquid velocity), p (the pressure), and $\mathbf{w}_s$ (the solid displacement) of the sequences $\left\{\dfrac{\partial\mathbf{w}^\varepsilon}{\partial t}\right\}$, $\{p^\varepsilon\}$, and $\{\mathbf{w}_s^\varepsilon\}$, where

$$\mathbf{v}=(1-\zeta)\mathbf{v}+\zeta\,\frac{\partial\mathbf{w}_s}{\partial t}=(1-\zeta)\mathbf{v}+\zeta\,\mathbf{v}_s, \tag{8.2.18}$$

satisfy in the domain $\Omega_T^{(f)}$ the system of acoustic equations (8.2.1), (8.2.2), and Lamé's equation

$$\hat{\rho}\frac{\partial^2\mathbf{w}_s}{\partial t^2}=\nabla\cdot\left(\lambda_0\,\mathfrak{N}_3^s:\mathbb{D}(x,\mathbf{w}_s)\right)+\hat{\rho}\mathbf{F} \tag{8.2.19}$$

in the domain Ω_T, completed with the homogeneous boundary condition

$$\mathbf{v}(\mathbf{x},t)\cdot\mathbf{n}(\mathbf{x}) = 0, \tag{8.2.20}$$

on the boundary $\partial\Omega^{(f)}\backslash S^{(0)}$ *for* $t>0$, *the homogeneous initial conditions*

$$p(\mathbf{x},0) = 0,\;\; \mathbf{v}(\mathbf{x},0) = 0 \tag{8.2.21}$$

for the liquid velocity and the pressure in the domain $\Omega^{(f)}$, *the homogeneous boundary condition*

$$\mathbf{w}_s(\mathbf{x},t) = 0, \tag{8.2.22}$$

on the boundary $\partial\Omega\backslash S^{(0)}$ *for* $t>0$, *and the homogeneous initial conditions*

$$\mathbf{w}_s(\mathbf{x},0) = \frac{\partial \mathbf{w}_s}{\partial t}(\mathbf{x},0) = 0 \tag{8.2.23}$$

for the solid displacement in Ω.

On the common boundary $S_T^{(0)}$ *the continuity conditions*

$$\lim_{\substack{\mathbf{x}\to\mathbf{x}^0\\ \mathbf{x}\in\Omega^{(f)}}} \mathbf{v}(\mathbf{x},t)\cdot\mathbf{n}(\mathbf{x}^0) = \lim_{\substack{\mathbf{x}\to\mathbf{x}^0\\ \mathbf{x}\in\Omega}} \frac{\partial \mathbf{w}_s}{\partial t}(\mathbf{x},t)\cdot\mathbf{n}(\mathbf{x}^0) \tag{8.2.24}$$

and

$$-\lim_{\substack{\mathbf{x}\to\mathbf{x}^0\\ \mathbf{x}\in\Omega^{(f)}}} p(\mathbf{x},t)\,\mathbf{n}(\mathbf{x}^0) = \lim_{\substack{\mathbf{x}\to\mathbf{x}^0\\ \mathbf{x}\in\Omega}} \Big(\lambda_0\,\mathfrak{N}_3^s : \mathbb{D}\big(x,\mathbf{w}_s(\mathbf{x},t)\big)\Big)\cdot\mathbf{n}(\mathbf{x}^0), \tag{8.2.25}$$

hold true.

Here $\mathbf{n}(\mathbf{x}^0)$ *is the normal vector to* $S^{(0)}$ *at* $\mathbf{x}^0\in S^{(0)}$.

In (8.2.19)

$$\hat{\rho} = m\,\rho_f + (1-m)\,\rho_s,$$

and the symmetric strictly positive definite constant fourth-rank tensor $\mathfrak{N}_3^s$ *is given in Chap. 1 by the formula (1.3.39).*

We refer to the problem (8.2.1), (8.2.2), (8.2.18)–(8.2.25) as the homogenized **model** $(\mathbb{ACM})_{11}$.

Theorem 8.15 *Let* $\{\mathbf{w}^\varepsilon,\; p^\varepsilon\}$ *be the weak solution of the problem (II),*

$$0<\mu_1<\infty,\;\; 0<\lambda_0<\infty,$$

and $\mathbf{w}_s^\varepsilon = \mathbb{E}_{\Omega_s^\varepsilon}\big(\mathbf{w}^\varepsilon\big)$.

Then the limits $\mathbf{v} = \dfrac{\partial \mathbf{w}}{\partial t}$ *(the liquid velocity) and* p *(the pressure) of the sequences* $\left\{\dfrac{\partial \mathbf{w}^{\varepsilon}}{\partial t}\right\}$ *and* $\{p^{\varepsilon}\}$*, where*

$$\mathbf{v} = (1-\zeta)\mathbf{v} + \zeta\,\frac{\partial \mathbf{w}^{(f)}}{\partial t} + \zeta\,(1-m)\frac{\partial \mathbf{w}_s}{\partial t} = (1-\zeta)\mathbf{v} + \zeta\,\mathbf{v}^{(f)} + \zeta\,\mathbf{v}_s, \quad (8.2.26)$$

satisfy the system of acoustic equations (8.2.1), (8.2.2) in the domain $\Omega_T^{(f)}$*, and the boundary and initial conditions (8.2.20)–(8.2.21).*

In the domain Ω_T *the limiting functions* p_f *(liquid pressure),* $\mathbf{w}^{(f)}$ *(liquid displacement), and* $\mathbf{w}_s$ *(solid displacement) of the sequences* $\{\zeta\,\chi^{\varepsilon}\,p^{\varepsilon}\}$*,* $\{\zeta\,\chi^{\varepsilon}\,\mathbf{w}^{\varepsilon}\}$*, and* $\{\zeta\,\mathbf{w}_s^{\varepsilon}\}$ *satisfy the system of homogenized equations, consisting of the continuity equation*

$$\frac{1}{c_f^2}p_f + \nabla\cdot\mathbf{w}^{(f)} = \mathbb{C}_0^s : \mathbb{D}(x,\mathbf{w}_s) + \frac{c_0^s}{\lambda_0}p_f, \quad (8.2.27)$$

the momentum balance equation

$$\rho_f\frac{\partial^2\mathbf{w}^{(f)}}{\partial t^2} + \rho_s\frac{\partial^2\mathbf{w}_s}{\partial t^2} = \nabla\cdot\left(\lambda_0\,\mathfrak{N}_2^s : \mathbb{D}(x,\mathbf{w}_s) - p_f\,\mathbb{C}_1^s\right) + \hat{\rho}\mathbf{F}, \quad (8.2.28)$$

for the solid component, and the momentum balance equation

$$-\int_0^t \mathbb{B}^{(f)}(\mu_1,\infty;t-\tau)\cdot\left(\nabla p_f + \rho_f\left(\frac{\partial^2\mathbf{w}_s}{\partial\tau^2} - \mathbf{F}\right)\right)(\mathbf{x},\tau)d\tau = \frac{\partial\mathbf{w}^{(f)}}{\partial t} - m\,\frac{\partial\mathbf{w}_s}{\partial t} \quad (8.2.29)$$

for the liquid component.

These differential equations are completed with the continuity conditions

$$\lim_{\substack{\mathbf{x}\to\mathbf{x}^0\\ \mathbf{x}\in\Omega^{(f)}}} \mathbf{v}(\mathbf{x},t)\cdot\mathbf{n}(\mathbf{x}^0) = \lim_{\substack{\mathbf{x}\to\mathbf{x}^0\\ \mathbf{x}\in\Omega}}\left(\mathbf{v}^{(f)}(\mathbf{x},t) + (1-m)\frac{\partial\mathbf{w}_s}{\partial t}(\mathbf{x},t)\right)\cdot\mathbf{n}(\mathbf{x}^0), \quad (8.2.30)$$

and

$$-\lim_{\substack{\mathbf{x}\to\mathbf{x}^0\\ \mathbf{x}\in\Omega^{(f)}}} p(\mathbf{x},t)\,\mathbf{n}(\mathbf{x}^0) = \lim_{\substack{\mathbf{x}\to\mathbf{x}^0\\ \mathbf{x}\in\Omega}}\left(\lambda_0\,\mathfrak{N}_2^s : \mathbb{D}\big(x,\mathbf{w}_s(\mathbf{x},t) - p_f\,\mathbb{C}_1^s\big)\right)\cdot\mathbf{n}(\mathbf{x}^0) \quad (8.2.31)$$

on the common boundary $S_T^{(0)}$*, the homogeneous boundary and initial conditions (8.2.22) and (8.2.23) for the solid displacement, and the homogeneous boundary and initial conditions*

$$\mathbf{w}^{(f)}(\mathbf{x},t)\cdot\mathbf{n}(\mathbf{x})=0,\ \ \mathbf{x}\in\partial\Omega\backslash S^{(0)},\ \ t\in(0,T), \tag{8.2.32}$$

$$\mathbf{w}^{(f)}(\mathbf{x},0)=0,\ \ \mathbf{x}\in\Omega \tag{8.2.33}$$

for the liquid displacement.

In (8.2.27)–(8.2.32) $\mathbf{n}(\mathbf{x}^0)$ *is the normal vector to* $S^{(0)}$ *at* $\mathbf{x}^0\in S^{(0)}$, $\mathbf{n}(\mathbf{x})$ *is the normal vector to* $\partial\Omega$ *at* $\mathbf{x}\in\partial\Omega$,

$$\hat{\rho}=m\,\rho_f+(1-m)\,\rho_s,$$

the symmetric strictly positively definite constant fourth-rank tensor $\mathfrak{N}_2^s$, *matrices* $\mathbb{C}_0^s$ *and* $\mathbb{C}_1^s$, *and the constant* c_0^s *are given in Chap. 1 by the formulae (1.3.26), (1.3.27) and (1.3.31), and the matrix* $\mathbb{B}^{(f)}(\mu_1,\infty;t)$ *is defined in Chap. 3 by the formulae (3.2.70) and (3.2.76).*

We refer to the problem (8.2.1), (8.2.2), (8.2.20), (8.2.21), (8.2.26)–(8.2.33) as the homogenized **model** $(\mathbb{ACM})_{12}$.

8.2.2 Proofs of Theorems 8.10–8.13

The proofs of these theorems are standard and repeat the proofs of the corresponding theorems in previous chapters, because we can prove all the statements, except the validity of the continuity conditions on the common boundary, separately in each of domains $\Omega^{(f)}$ and Ω. Thus, the main problem here is the continuity conditions on the common boundary $S^{(0)}$. These continuity conditions follow from the limiting integral identity

$$-\int_{Q_T}p\,(\nabla\cdot\boldsymbol{\varphi})\,dxdt=\int_{Q_T}\int_Y\rho_{(f)}(\mathbf{x},\mathbf{y})\big(\mathbf{F}-\frac{\partial^2\mathbf{w}}{\partial t^2}(\mathbf{x},t,\mathbf{y})\big)\cdot\boldsymbol{\varphi}(\mathbf{x},t)dydxdt, \tag{8.2.34}$$

for any smooth function $\boldsymbol{\varphi}\in\overset{\circ}{\mathbf{W}}{}_2^{1,0}(Q_T)$, and the integral identity

$$\int_{Q_T}\Bigg(\Bigg((1-\zeta)\frac{1}{\bar{c}_f^2}+\zeta\left(\frac{m}{\bar{c}_f^2}+\frac{1-m}{\bar{c}_s^2}\right)\Bigg)\frac{\partial p}{\partial t}\,\psi-\nabla\psi\cdot\frac{\partial\mathbf{w}}{\partial t}\Bigg)dxdt=0 \tag{8.2.35}$$

for any smooth function $\psi\in W_2^{1,0}(Q_T)$.

Here $\mathbf{W}(\mathbf{x},t,\mathbf{y})$ is a two-scale limit of the sequence $\{\mathbf{w}^{\varepsilon}\}$, and

$$\rho_{(f)}(\mathbf{x},\mathbf{y})=\big(1-\zeta(\mathbf{x})\big)\rho_f+\zeta(\mathbf{x})\big(\rho_f\,\chi(\mathbf{y})+\big(1-\chi(\mathbf{y})\big)\,\rho_s\big).$$

For all cases (8.2.34) and (8.2.35) imply the system of acoustic equations (8.2.1) and (8.2.2) in the domain $\Omega_T^{(f)}$, the continuity conditions (8.2.7) and (8.2.8) on the common boundary $S^{(0)}$, and the continuity equation

$$\Big(\frac{m}{\bar{c}_f^2}+\frac{1-m}{\bar{c}_s^2}\Big)\frac{\partial p}{\partial t}+\nabla\cdot\frac{\partial \mathbf{w}}{\partial t}=0 \tag{8.2.36}$$

in the domain Ω_T.

All the differences are concentrated in the dynamic equation in the domain Ω_T and in the representation of the velocity of the mixture $\dfrac{\partial \mathbf{w}}{\partial t}$.

8.2.2.1 Proof of Theorem 8.10

For this case $\mathbf{W}(\mathbf{x},t,\mathbf{y}) = \mathbf{w}(\mathbf{x},t)$ and the integral identity (8.2.34) implies the dynamic equation

$$\hat{\rho}\,\frac{\partial^2 \mathbf{w}}{\partial t^2}=-\nabla\, p+\hat{\rho}\,\mathbf{F} \tag{8.2.37}$$

in the domain Ω_T.

8.2.2.2 Proof of Theorem 8.11

Here to obtain the dynamic equation we simply repeat the proof of Theorem 6.6. That is we derive the microscopic dynamic equation (6.2.32) with $\tilde{P} = p$ and the microscopic continuity equation (6.2.33), which result in the representation

$$\frac{\partial \mathbf{w}}{\partial t}(\mathbf{x},t)=\int_0^t \mathbb{B}^{(a)}(\mu_1,\lambda_1;t-\tau)\cdot\nabla\, p\,(\mathbf{x},\tau)d\tau+\mathbf{f}(\mathbf{x},t) \tag{8.2.38}$$

of the velocity of the mixture.

8.2.2.3 Proof of Theorem 8.12

For this case the velocity of the mixture is given by the formula

$$\frac{\partial \mathbf{W}}{\partial t}(\mathbf{x},t,\mathbf{y})=\chi(\mathbf{y})\frac{\partial \mathbf{w}_f}{\partial t}(\mathbf{x},t)+\big(1-\chi(\mathbf{y})\big)\frac{\partial \mathbf{W}}{\partial t}(\mathbf{x},t,\mathbf{y}),$$

$$\frac{\partial \mathbf{w}}{\partial t}=m\frac{\partial \mathbf{w}_f}{\partial t}+\frac{\partial \mathbf{w}^{(s)}}{\partial t},\quad \mathbf{w}^{(s)}(\mathbf{x},t)=\int_Y\big(1-\chi(\mathbf{y})\big)\mathbf{W}(\mathbf{x},t,\mathbf{y})dy.$$

The integral identity (8.2.34) implies the dynamic equation (8.2.11) for the liquid component.

To find the representation (8.2.12) we use the microscopic dynamic equation

$$\rho_s \frac{\partial^2 \mathbf{W}^{(s)}}{\partial t^2} = \lambda_1 \triangle_y \mathbf{W}^{(s)} - \nabla_y \Pi^{(s)} - \nabla\, p, \quad \mathbf{W}^{(s)} = \big(1 - \chi(\mathbf{y})\big)\mathbf{W},$$

for the solid component, the microscopic continuity equation

$$\nabla \cdot \mathbf{W}^{(s)} = 0,$$

in the domain Y_s, and the corresponding boundary and initial conditions. This problem has been already solved in Chap. 3 (see the proof of Theorem 3.4). The rest of the proof is the same as for the previous theorems.

8.2.2.4 Proof of Theorem 8.13

Here the velocity of the mixture is given by the formula

$$\frac{\partial \mathbf{W}}{\partial t}(\mathbf{x}, t, \mathbf{y}) = \chi(\mathbf{y}) \frac{\partial \mathbf{W}}{\partial t}(\mathbf{x}, t, \mathbf{y}) + \big(1 - \chi(\mathbf{y})\big) \frac{\partial \mathbf{w}_s}{\partial t}(\mathbf{x}, t),$$

$$\frac{\partial\, \mathbf{w}}{\partial t} = \frac{\partial \mathbf{w}^{(f)}}{\partial t} + (1-m)\frac{\partial \mathbf{w}_s}{\partial t}, \quad \mathbf{w}^{(f)}(\mathbf{x}, t) = \int_Y \chi(\mathbf{y}) \mathbf{W}(\mathbf{x}, t, \mathbf{y}) dy.$$

The integral identity (8.2.34) implies the dynamic equation (8.2.16) for the solid component. The representation (8.2.17) has been already obtained in Chap. 3 (see the proof of Theorem 3.5). The rest of the proof is the same as for the previous theorems.

8.2.3 Proofs of Theorems 8.14 and 8.15

For these cases the two-scale limit $P(\mathbf{x}, t, \mathbf{y})$ of the sequence $\{p^\varepsilon\}$ is given by

$$(1-\zeta)\, p + \frac{\zeta}{m} \chi(\mathbf{y})\, p_f(\mathbf{x}, t) + \zeta\, \big(1 - \chi(\mathbf{y})\big)\, P_s(\mathbf{x}, t, \mathbf{y}).$$

For $\mu_1 = \infty$ the two-scale limit $\mathbf{W}$ of the sequence $\{\mathbf{w}^\varepsilon\}$ is given by

$$\mathbf{W}(\mathbf{x}, t, \mathbf{y}) = \mathbf{w}_s(\mathbf{x}, t),$$

and for $\mu_1 < \infty$ it is given by

$$\mathbf{W}(\mathbf{x}, t, \mathbf{y}) = \chi(\mathbf{y}) \mathbf{W}(\mathbf{x}, t, \mathbf{y}) + \big(1 - \chi(\mathbf{y})\big) \mathbf{w}_s(\mathbf{x}, t).$$

The two-scale limit of $\{\mathbf{w}_s^\varepsilon\}$ is equal to $\{\mathbf{w}_s(\mathbf{x},t)\}$, and the two-scale limit of the sequence $\{\mathbb{D}(x,\mathbf{w}_s^\varepsilon)\}$ is given by

$$\mathbb{D}\big(x,\mathbf{w}_s(\mathbf{x},t)\big)+\mathbb{D}\big(y,\mathbf{U}(\mathbf{x},t,\mathbf{y})\big)$$

(see proofs of Theorems 1.6 and 1.7).

The integral identity (8.2.34) is replaced by

$$\int_{Q_T}\Big(\zeta\,\lambda_0\big(m\mathbb{D}(x,\mathbf{w})+\langle\mathbb{D}\big(y,\mathbf{U}\big)\rangle_{Y_s}\big)-p\,\mathbb{I}\Big):\mathbb{D}(x,\boldsymbol{\varphi})dxdt$$
$$=\int_{Q_T}\int_Y\rho_{(f)}(\mathbf{x},\mathbf{y})\left(\mathbf{F}-\frac{\partial^2\mathbf{W}}{\partial t^2}(\mathbf{x},t,\mathbf{y})\right)\cdot\boldsymbol{\varphi}(\mathbf{x},t)dydxdt \qquad (8.2.39)$$

with the smooth functions $\boldsymbol{\varphi}$, vanishing on the boundary ∂Q, and the integral identity (8.2.35) is replaced by

$$\int_{Q_T}\left(\eta\int_Y\left(\frac{1-\zeta}{\bar{c}_f^2}+\left(\frac{\chi}{\bar{c}_f^2}+\frac{1-\chi}{\bar{c}_s^2}\right)\zeta\right)\frac{\partial P}{\partial t}dy-\nabla\eta\cdot\frac{\partial\mathbf{w}}{\partial t}\right)dxdt=0 \quad (8.2.40)$$

with the smooth functions η.

In (8.2.39)

$$\rho_{(f)}(\mathbf{x},\mathbf{y})=\big(1-\zeta(\mathbf{x})\big)\rho_f+\zeta(\mathbf{x})\big(\rho_f\chi(\mathbf{y})+\big(1-\chi(\mathbf{y})\big)\rho_s\big).$$

For $\mu_1=\infty$

$$\mathbf{W}(\mathbf{x},t,\mathbf{y})=\mathbf{w}_s(\mathbf{x},t),$$

and for $\mu_1<\infty$

$$\mathbf{W}(\mathbf{x},t,\mathbf{y})=\chi(\mathbf{y})\mathbf{W}(\mathbf{x},t,\mathbf{y})+\big(1-\chi(\mathbf{y})\big)\mathbf{w}_s(\mathbf{x},t).$$

As before, the relations (8.2.39) and (8.2.40) result in the system of acoustic equations (8.2.1), (8.2.2) in $\Omega_T^{(f)}$, the homogenized momentum balance equations (8.2.19) and (8.2.27), the continuity equation (8.2.28) in Ω_T, the continuity conditions (8.2.24), (8.2.25), (8.2.30) and (8.2.31) on the common boundary $S_T^{(0)}$, and the boundary conditions (8.2.20) and (8.2.32).

The boundary and initial conditions (8.2.21), (8.2.22), (8.2.23) and (8.2.33) and the equation (8.2.29) have been already derived in the previous sections (see also Chaps. 1, 4 and 6).

8.3 Acoustics in a "Poroelastic Medium–Poroelastic Medium" Configuration

8.3.1 Main Results

In this section we restrict ourselves to the only two cases
(1) $\lambda_0^{(0)}=0,\ \lambda_1^{(0)}<\infty,\ 0<\lambda_0<\infty,\ \mu_0=0,\ \mu_1<\infty,$
and
(2) $\lambda_0^{(0)}=0,\ \lambda_1^{(0)}=\infty,\ 0<\lambda_0<\infty,\ \mu_0=0,\ \mu_1<\infty.$
The remaining cases can be treated similarly.

Theorem 8.16 *Let* $\{\mathbf{w}^\varepsilon,\ p^\varepsilon\}$ *be the weak solution of the problem (III),*

$$\lambda_0^{(0)}=0,\ \lambda_1^{(0)}<\infty,\ 0<\lambda_0<\infty,\ \mu_0=0,\ \mu_1<\infty,$$

and $\mathbf{w}_s^\varepsilon=\mathbb{E}_{\Omega_s^\varepsilon}\big(\mathbf{w}^\varepsilon\big)$.

Then the limits $\mathbf{v}=\dfrac{\partial\mathbf{w}}{\partial t}$ *(the liquid velocity) and* p *(the pressure) of the sequences* $\left\{\dfrac{\partial\mathbf{w}^\varepsilon}{\partial t}\right\}$ *and* $\{p^\varepsilon\}$ *satisfy in the domain* G_T *the system of acoustic equations, consisting of the momentum balance equation in the form*

$$\mathbf{v}(\mathbf{x},t)=\int_0^t\mathbb{B}_0^{(a)}(\mu_1,\lambda_1^{(0)};t-\tau)\cdot\nabla p(\mathbf{x},\tau)d\tau+\mathbf{f}(\mathbf{x},t),\tag{8.3.1}$$

and continuity equation

$$\left(\frac{m}{\bar{c}_f^2}+\frac{1-m}{\big(\bar{c}_s^{(0)}\big)^2}\right)\frac{\partial p}{\partial t}+\nabla\cdot\mathbf{v}=0.\tag{8.3.2}$$

In the domain Ω_T *the limiting functions* $m\,p_f$ *(liquid pressure),* $\mathbf{w}^{(f)}$ *(liquid displacements), and* $\mathbf{w}_s$ *(solid displacements) of the sequences* $\{\zeta\,\chi^\varepsilon\,p^\varepsilon\}$, $\{\zeta\,\chi^\varepsilon\,\mathbf{w}^\varepsilon\}$, *and* $\{\zeta\,\mathbf{w}_s^\varepsilon\}$ *satisfy the system of homogenized equations, consisting of the continuity equation*

$$\frac{m}{c_f^2}p_f+\nabla\cdot\mathbf{w}^{(f)}=\mathbb{C}_0^s:\mathbb{D}(x,\mathbf{w}_s)+\frac{c_0^s}{\lambda_0}p_f,\tag{8.3.3}$$

the momentum balance equation

$$\rho_f\frac{\partial^2\mathbf{w}^{(f)}}{\partial t^2}+\rho_s\frac{\partial^2\mathbf{w}_s}{\partial t^2}=\nabla\cdot\big(\lambda_0\,\mathfrak{N}_2^s:\mathbb{D}(x,\mathbf{w}_s)-p_f\,\mathbb{C}_1^s\big)+\hat{\rho}\mathbf{F},\tag{8.3.4}$$

for the solid component, and the momentum balance equation

$$-\int_0^t \mathbb{B}^{(f)}(\mu_1,\infty;t-\tau)\cdot\left(\nabla p_f+\rho_f\left(\frac{\partial^2\mathbf{w}_s}{\partial\tau^2}-\mathbf{F}\right)\right)(\mathbf{x},\tau)d\tau=\frac{\partial\mathbf{w}^{(f)}}{\partial t}-m\frac{\partial\mathbf{w}_s}{\partial t} \tag{8.3.5}$$

for the liquid component.

The differential equations are completed with the homogeneous boundary conditions

$$\mathbf{v}(\mathbf{x},t)\cdot\mathbf{n}(\mathbf{x})=0,\quad \mathbf{x}\in\partial G\backslash S^{(0)},\ t>0, \tag{8.3.6}$$

$$\mathbf{w}^{(f)}(\mathbf{x},t)\cdot\mathbf{n}(\mathbf{x})=0,\quad \mathbf{x}\in\partial\Omega\backslash S^{(0)},\ t>0, \tag{8.3.7}$$

$$\mathbf{w}_s(\mathbf{x},t)=0,\quad \mathbf{x}\in\partial\Omega\backslash S^{(0)},\ t>0, \tag{8.3.8}$$

the homogeneous initial conditions

$$p(\mathbf{x},0)=0,\ \mathbf{x}\in G, \tag{8.3.9}$$

$$\mathbf{w}^{(f)}(\mathbf{x},0)=\mathbf{w}_s(\mathbf{x},0)=\frac{\partial\mathbf{w}_s}{\partial t}(\mathbf{x},0)=0,\ \mathbf{x}\in\Omega, \tag{8.3.10}$$

and the continuity conditions

$$\lim_{\substack{\mathbf{x}\to\mathbf{x}^0\\ \mathbf{x}\in G}}\mathbf{v}(\mathbf{x},t)\cdot\mathbf{n}(\mathbf{x}^0)=\lim_{\substack{\mathbf{x}\to\mathbf{x}^0\\ \mathbf{x}\in\Omega}}(\mathbf{v}^{(f)}(\mathbf{x},t)+(1-m)\frac{\partial\mathbf{w}_s}{\partial t}(\mathbf{x},t))\cdot\mathbf{n}(\mathbf{x}^0), \tag{8.3.11}$$

and

$$-\lim_{\substack{\mathbf{x}\to\mathbf{x}^0\\ \mathbf{x}\in G}}p(\mathbf{x},t)\,\mathbf{n}(\mathbf{x}^0)=\lim_{\substack{\mathbf{x}\to\mathbf{x}^0\\ \mathbf{x}\in\Omega}}\Big(\lambda_0\,\mathfrak{N}_2^s:\mathbb{D}\big(x,\mathbf{w}_s(\mathbf{x},t)\big)-p_f\,\mathbb{C}_1^s\Big)\Big)\cdot\mathbf{n}(\mathbf{x}^0) \tag{8.3.12}$$

on the common boundary $S_T^{(0)}$.

In (8.3.1)–(8.3.12) $\mathbf{n}(\mathbf{x}^0)$ is the normal vector to $S^{(0)}$ at $\mathbf{x}^0\in S^{(0)}$, $\mathbf{n}(\mathbf{x})$ is the normal vector to ∂Q at $\mathbf{x}\in\partial Q$, and

$$\hat{\rho}=m\,\rho_f+(1-m)\,\rho_s.$$

The matrix $\mathbb{B}_0^{(a)}(\mu_1,\lambda_1^{(0)};t)$ and the function $\mathbf{f}(\mathbf{x},t)$ are given in Chap. 6 by formulae (6.2.40), (6.2.42), where instead of χ, ρ_s, and λ_1 one must consider $\chi^{(0)}$, $\rho_s^{(0)}$, and $\lambda_1^{(0)}$.

The symmetric strictly positively definite constant fourth-rank tensor $\mathfrak{N}_2^s$, matrices $\mathbb{C}_0^s$ and $\mathbb{C}_1^s$, and the constant c_0^s are given in Chapter I by formulae (1.3.26), (1.3.27) and (1.3.31), and the matrix $\mathbb{B}^{(f)}(\mu_1,\infty;t)$ is defined in Chap. 3 by the formulae (3.2.70) and (3.2.76).

We refer to the problem (8.3.1)–(8.3.12) as the homogenized **model** $(\mathbb{ACM})_{13}$.

Theorem 8.17 *Let $\{\mathbf{w}^{\varepsilon},\ p^{\varepsilon}\}$ be the weak solution of the problem (III),*

$$\lambda_0^{(0)} = 0,\ \lambda_1^{(0)} = \infty,\ 0 < \lambda_0 < \infty,\ \mu_0 = 0,\ \mu_1 < \infty,$$

and $\mathbf{w}_s^{\varepsilon} = \mathbb{E}_{\Omega_s^{\varepsilon}}\big(\mathbf{w}^{\varepsilon}\big)$, $\mathbf{u}_s^{\varepsilon} = \mathbb{E}_{G_s^{\varepsilon}}\big(\mathbf{w}^{\varepsilon}\big)$.

Then the pressure p of the mixture in G_T, the pressure p_f of the liquid in Ω_T, and the velocity $\left\{\dfrac{\partial \mathbf{w}}{\partial t}\right\}$ of the medium

$$\mathbf{w} = (1-\zeta)(\mathbf{u}^{(f)} + \mathbf{u}_s) + \zeta\,(\mathbf{w}^{(f)} + \mathbf{w}_s),$$

where p, $\mathbf{u}^{(f)}$ (the liquid displacement in G_T), $\mathbf{u}_s$ (the solid displacement in G_T), $m\,p_f$, $\mathbf{w}^{(f)}$ (the liquid displacement in Ω_T), and $\mathbf{w}_s$ (the solid displacement in Ω_T) are limits of the sequences $\{(1-\zeta)\,p^{\varepsilon}\}$, $\{(1-\zeta)\,\chi_0^{\varepsilon}\,\mathbf{w}^{\varepsilon}\}$, $\{(1-\zeta)\,\mathbf{u}_s^{\varepsilon}\}$ $\{\zeta\,\chi^{\varepsilon}\,p^{\varepsilon}\}$, $\{\zeta\,\chi^{\varepsilon}\,\mathbf{w}^{\varepsilon}\}$, and $\{\zeta\,\mathbf{w}_s^{\varepsilon}\}$, satisfy in the domain G_T the system of acoustic equations, consisting of the continuity equation

$$\left(\frac{m_0}{\bar{c}_f^{\,2}} + \frac{(1-m_0)}{\left(\bar{c}_s^{(0)}\right)^2}\right)\frac{\partial p}{\partial t} + \nabla\cdot\left(\frac{\partial \mathbf{u}^{(f)}}{\partial t} + (1-m_0)\frac{\partial \mathbf{u}_s}{\partial t}\right) = 0, \tag{8.3.13}$$

the momentum balance equation

$$\rho_f\frac{\partial \mathbf{u}^{(f)}}{\partial t} + (1-m_0)\rho_s^{(0)}\frac{\partial \mathbf{u}_s}{\partial t} = \int_0^t\left(\hat{\rho}^{(0)}\mathbf{F} - \nabla\,p\right)(\mathbf{x},\tau)d\tau, \tag{8.3.14}$$

for the solid component, and the momentum balance equation

$$\frac{\partial \mathbf{u}^{(f)}}{\partial t} - m_0\frac{\partial \mathbf{u}_s}{\partial t} = -\int_0^t \mathbb{B}_0^{(f)}(\mu_1,\infty;t-\tau)\cdot\left(\nabla\,p + \rho_f\left(\frac{\partial^2 \mathbf{u}_s}{\partial \tau^2} - \mathbf{F}\right)\right)(\mathbf{x},\tau)\,d\tau \tag{8.3.15}$$

for the liquid component.

In the domain Ω_T the limiting functions satisfy the system of homogenized equations, consisting of the continuity equation (8.3.3), the momentum balance equation 8.3.4) for the solid component, and the momentum balance equation (8.3.5) for the liquid component.

The differential equations are completed with the homogeneous boundary and initial conditions (8.3.6)–(8.3.10), the homogeneous condition (8.3.10), the homogeneous initial conditions

$$\mathbf{u}^{(f)}(\mathbf{x},0) = \mathbf{u}_s(\mathbf{x},0) = 0,\ \mathbf{x}\in G, \tag{8.3.16}$$

and the continuity conditions (8.3.11) and (8.3.12) on the common boundary $S_T^{(0)}$.

In (8.3.13)–(8.3.16)

$$\hat{\rho}^{(0)} = m_0\,\rho_f + (1 - m_0)\rho_s^{(0)}, \quad m_0 = \int_Y \chi^{(0)}(\mathbf{y})dy,$$

the matrix $\mathbb{B}_0^{(f)}(\mu_1, \infty; t)$ *has been defined in Chap. 3 by formulae (3.2.70) and (3.2.76), where instead of* χ *one must consider* $\chi^{(0)}$.

We refer to the problem (8.1.3)–(8.1.6), (8.1.9), (8.1.11)–(8.1.13) as the homogenized **model** $(\mathbb{ACM})_{14}$.

The proofs of these theorems repeat the proofs of the similar statements in this chapter.

Chapter 9
Double Porosity Models for Acoustics

In this chapter we consider acoustics in porous media with the double porosity geometry, where the liquid domain is composed by a periodic system of pores with the dimensionless size ε and a periodic system of cracks with the dimensionless size δ (see Chap. 4). The liquid domain Ω^{δ}_{f}, which is a subdomain of a bounded domain Ω with the Lipschitz continuous boundary $S = \partial\Omega$, is defined in the following way. Let K be the unit cube, $K = Z_f \cup Z_s \cup \gamma_c$, where Z_f and Z_s are open sets, the common boundary $\gamma_c = \partial Z_f \cap \partial Z_s$ is the Lipschitz continuous surface, and the periodic repetition in $\mathbb{R}^3$ of the domain Z_s is a connected domain with the Lipschitz continuous boundary. The elementary cell Z_f models the crack space Ω^{δ}_{c} : the domain Ω^{δ}_{c} is the intersection of the cube Ω with the periodic repetition in $\mathbb{R}^3$ of the elementary cell δZ_f. In the same way we define the pore space Ω^{ε}_{p}: $K = Y_f \cup Y_s \cup \gamma_p$, γ_p is the Lipschitz continuous surface, the periodic repetition in $\mathbb{R}^3$ of the domain Y_s is the connected domain with the Lipschitz continuous boundary, and Ω^{ε}_{p} is the intersection of $\Omega\backslash\Omega^{\delta}_{c}$ with the periodic repetition in $\mathbb{R}^3$ of the elementary cell εY_f. Finally, we put $\Omega^{\delta}_{f} = \Omega^{\varepsilon}_{p} \cup \Omega^{\delta}_{c}$, $\Omega^{\delta}_{s} = \Omega\backslash\overline{\Omega^{\delta}_{f}}$ is a solid skeleton, and $\Gamma^{\delta} = \partial\Omega^{\delta}_{s} \cap \partial\Omega^{\delta}_{f}$ is the "solid skeleton–liquid domain" interface.

We also may characterize the liquid and solid domains using the indicator functions in Ω. Let $\varsigma(\mathbf{x})$ be the indicator function of the domain Ω in $\mathbb{R}^3$. That is $\varsigma(\mathbf{x}) = 1$ if $\mathbf{x} \in \Omega$ and $\varsigma(\mathbf{x}) = 0$ if $\mathbf{x} \in \mathbb{R}^3\backslash\Omega$. Let also $\chi_p(\mathbf{y})$ be the 1-periodic extension of the indicator function of the domain Y_f in K and $\chi_c(\mathbf{z})$ be the 1-periodic extension of the indicator function of the domain Z_f in K. Then $\chi^{\delta}_{c}(\mathbf{x}) = \varsigma(\mathbf{x})\chi_c\left(\frac{\mathbf{x}}{\delta}\right)$ stands for the indicator function of the domain Ω^{δ}_{c}, $\chi^{\delta,\varepsilon}_{p}(\mathbf{x}) = \varsigma(\mathbf{x})\left(1 - \chi_c\left(\frac{\mathbf{x}}{\delta}\right)\right)\chi_p\left(\frac{\mathbf{x}}{\varepsilon}\right)$ stands for the indicator function of the domain Ω^{ε}_{p} and

$$\tilde{\chi}^{\delta,\varepsilon}(\mathbf{x}) = \varsigma(\mathbf{x})\tilde{\chi}\left(\frac{\mathbf{x}}{\varepsilon}, \frac{\mathbf{x}}{\delta}\right), \quad \tilde{\chi}(\mathbf{y}, \mathbf{z}) = \chi_c(\mathbf{z}) + \big(1 - \chi_c(\mathbf{z})\big)\chi_p(\mathbf{y})$$

stands for the indicator function of the liquid domain Ω^{δ}_{f} (Figs. 9.1, 9.2 and 9.3).

A. Meirmanov, *Mathematical Models for Poroelastic Flows*, Atlantis Studies in Differential Equations 1, DOI: 10.2991/978-94-6239-015-7_9,

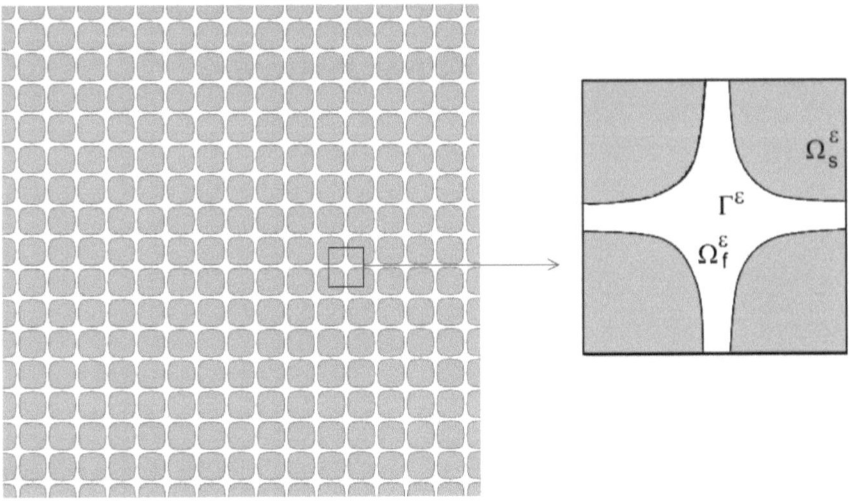

Fig. 9.1 Single porosity geometry

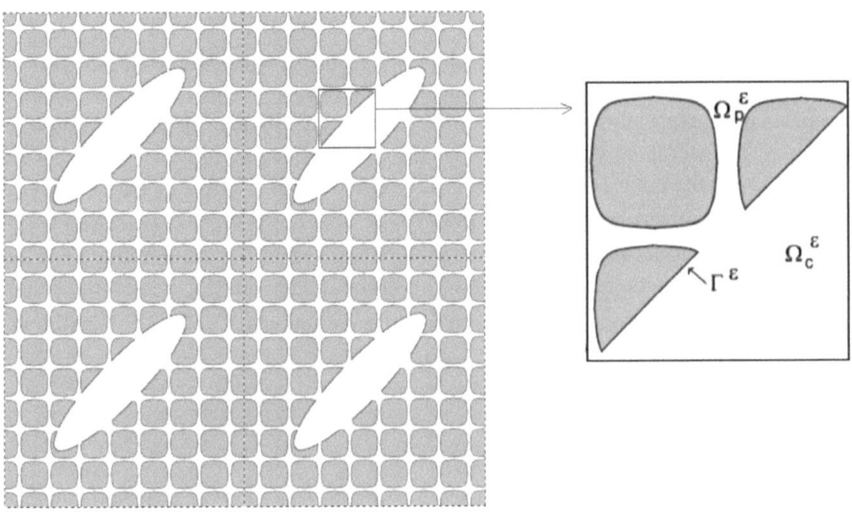

Fig. 9.2 Double porosity geometry: isolated fractures

9.1 Acoustics in a Slightly Compressible Liquid and an Elastic Solid Skeleton

We consider the model $\mathbb{M}_{20}$ of isothermal acoustics as the basic mathematical model at the microscopic level:

$$\frac{1}{\bar{\alpha}_q^{\delta,\varepsilon}} q + \nabla \cdot \mathbf{w} = 0, \tag{9.1.1}$$

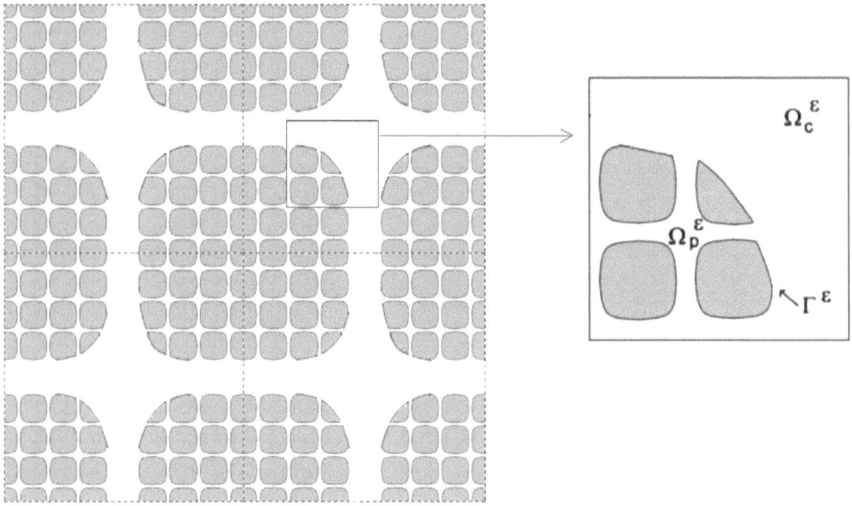

Fig. 9.3 Double porosity geometry: connected fracture space

$$\rho^{\delta,\varepsilon}\frac{\partial^2 \mathbf{w}}{\partial t^2} = \nabla\cdot\mathbb{P} + \rho^{\delta,\varepsilon}\mathbf{F}, \tag{9.1.2}$$

$$\mathbb{P} = \tilde{\chi}^{\delta}\bar{\alpha}_\mu\mathbb{D}\Big(x, \frac{\partial \mathbf{w}}{\partial t}\Big) + (1-\tilde{\chi}^{\delta,\varepsilon})\lambda_0\,\mathbb{D}(x,\mathbf{w}) - q\,\mathbb{I}, \tag{9.1.3}$$

$$\mathbf{w}(\mathbf{x},t) = 0,\ \ (\mathbf{x},t)\in S_T = S\times(0,T), \tag{9.1.4}$$

$$\mathbf{w}(\mathbf{x},0) = \frac{\partial \mathbf{w}}{\partial t}(\mathbf{x},0) = 0,\ \ \mathbf{x}\in\Omega, \tag{9.1.5}$$

where

$$\rho^{\delta,\varepsilon} = \rho_f\tilde{\chi}^{\delta,\varepsilon} + \rho_s(1-\tilde{\chi}^{\delta,\varepsilon}),\quad \bar{\alpha}_q^{\delta,\varepsilon} = c_f^2\tilde{\chi}^{\delta,\varepsilon} + c_s^2(1-\tilde{\chi}^{\delta,\varepsilon}).$$

In Chap. 4 under the condition $\varepsilon = \delta^r, r > 1$, we have used the three-scale convergent method to derive the homogenized models. In this chapter we apply the method of reiterated homogenization, suggested in [11, 30, 34]. Firstly for fixed $\delta > 0$ we consider the joint motion of the solid skeleton and the liquid in pores, and approximate the system (9.1.1)–(9.1.5), describing this motion, by some homogenized system letting $\varepsilon \to 0$.

Let us consider the case, when

$$\bar{\alpha}_\mu = \mu_2\delta^2,\quad 0 < \mu_2 < \infty. \tag{9.1.6}$$

By supposition

$$\delta = \frac{l_c}{L}, \quad \varepsilon = \frac{l_p}{L},$$

where l_c is the crack characteristic size, l_p is the pore characteristic size, and L is the characteristic size of the entire porous body.

For the liquid in pores we represent the criterion $\bar{\alpha}_\mu$ as

$$\bar{\alpha}_\mu = \mu_2 \frac{l_c^2}{L^2} \frac{L^r}{l_p^r} \varepsilon^r = \tilde{\mu}\, \varepsilon^r,$$

with some $1 < r < 2$, fix μ_2, $\tilde{\mu}$ and δ, and letting ε be variable.

Under this supposition we apply the first homogenization procedure as $\varepsilon \to 0$ and as the result will get the homogenized model $(\mathbb{IA})_{15}$ in the domain $\Omega^\delta_{s,p} = \Omega \backslash \overline{\Omega^\delta_c}$, describing the acoustics in a mixture of the solid and the liquid in pores (see Theorem 6.11):

$$\hat{\rho} \frac{\partial^2 \mathbf{w}_s}{\partial t^2} = \nabla \cdot \big(\lambda_0\, \mathfrak{N}^s_3 : \mathbb{D}(x, \mathbf{w}_s) \big) + \hat{\rho} \mathbf{F}. \tag{9.1.7}$$

Adding the Stokes system in the crack space Ω^δ_c we arrive at the system

$$\tilde{\rho}^\delta \left(\frac{\partial^2 \mathbf{w}}{\partial t^2} - \mathbf{F} \right) \\ = \nabla \cdot \left(\chi^\delta_c \left(\mu_2\, \delta^2\, \mathbb{D}\left(x, \frac{\partial \mathbf{w}}{\partial t} \right) - q_c\, \mathbb{I} \right) + (1 - \chi^\delta_c) \lambda_0\, \mathfrak{N}^s_3 : \mathbb{D}(x, \mathbf{w}) \right), \tag{9.1.8}$$

$$\chi^\delta_c (q_c + c_f^2\, \nabla \cdot \mathbf{w}_c) = 0 \tag{9.1.9}$$

in the domain $\Omega_T = \Omega \times (0, T)$, describing at the microscopic level acoustics in the liquid in cracks and in the mixture of the solid skeleton and the liquid in pores.

The differential equations are completed with boundary and initial conditions (9.1.4) and (9.1.5).

In (9.1.8) and (9.1.9)

$$\mathbf{w} = \chi^\delta_c\, \mathbf{w}_c + (1 - \chi^\delta_c) \mathbf{w}_s, \quad \tilde{\rho}^\delta = \rho_f \chi^\delta_c + \hat{\rho}(1 - \chi^\delta_c),$$

$\mathbf{w}_c$ are displacements of the liquid in crack space, $\mathbf{w}_s$ are displacements of the solid skeleton, which coincide with displacements of the mixture of the solid skeleton and the liquid in pores, q_c is the pressure of the liquid in cracks, and the symmetric strictly positive definite constant fourth-rank tensor $\mathfrak{N}^s_3$ is given in Chap. 1 by formula (1.3.39).

The second homogenization as $\delta \to 0$ leads to the desired double porosity model for acoustics in a mixture of a slightly compressible liquid and an elastic solid skeleton, and completes the method of reiterated homogenization.

9.1.1 Statement of the Problem and Main Results

As usual, we define a weak solution to the problem (9.1.4), (9.1.5), (9.1.8) and (9.1.9) as a pair of functions $\{\mathbf{w}^\delta,\ q_c^\delta\}$, which satisfy the regularity conditions

$$\frac{\partial \mathbf{w}^\delta}{\partial t},\ \mathbf{w}^\delta \in \overset{\circ}{\mathbf{W}}{}_2^{1,0}(\Omega_T),\ q_c^\delta \in L_2(\Omega_T),$$

the continuity equation (9.1.9) for the liquid in cracks in the usual sense a.e. in Ω_T, and the integral identity

$$\int_{\Omega_T}\Big(\chi_c^\delta\Big(\mu_2\,\delta^2\,\mathbb{D}\Big(x,\frac{\partial \mathbf{w}^\delta}{\partial t}\Big) - q_c^\delta\,\mathbb{I}\Big) \tag{9.1.10}$$

$$+ (1-\chi_c^\delta)\lambda_0\,\mathfrak{N}_3^s : \mathbb{D}(x,\mathbf{w}^\delta)\Big) : \mathbb{D}(x,\varphi)dxdt$$

$$= \int_{\Omega_T}\tilde{\rho}^\delta\Big(\frac{\partial\varphi}{\partial t}\cdot\frac{\partial \mathbf{w}^\delta}{\partial t} + \mathbf{F}\cdot\varphi\Big)dxdt$$

for all functions φ, such that $\varphi \in \mathbf{w}_2^{1,0}(\Omega_T)$, $\dfrac{\partial\varphi}{\partial t} \in \mathbf{L}_2(\Omega_T)$, $\varphi(\mathbf{x},t) = 0$ on the boundary S_T, and $\varphi(\mathbf{x},T) = 0$ for $\mathbf{x} \in \Omega$.

Theorem 9.1 *Let*

$$\int_{Q_T} |\mathbf{F}(\mathbf{x},\mathbf{t})|^2 dxdt = F^2 < \infty.$$

Then for all $\delta > 0$ and for the arbitrary time interval $[0,T]$ there exists a unique weak solution of the problem (9.1.4), (9.1.5), (9.1.8), (9.1.9) and

$$\max_{0<t<T}\int_\Omega\Big(|q_c^\delta(\mathbf{x},t)|^2 + \Big|\frac{\partial \mathbf{w}^\delta}{\partial t}(\mathbf{x},t)\Big|^2 + (1-\chi_c^\delta)\lambda_0\big(\mathfrak{N}_3^s : \mathbb{D}(x,\mathbf{w}^\delta)\big) : \mathbb{D}(x,\mathbf{w}^\delta)\Big)dx \tag{9.1.11}$$

$$+ \int_{\Omega_T}\mu_2\,\delta^2\,\chi_c^\delta\,\Big|\mathbb{D}\Big(x,\frac{\partial \mathbf{w}^\delta}{\partial t}\Big)\Big|^2 dxdt \leqslant C_0 F^2,$$

where C_0 is independent of δ.

The proof of this theorem repeats the proofs of the similar theorems in previous chapters and is based on the energy equality

$$\frac{1}{2}\frac{d}{dt}\int_{\Omega}\Big(\tilde{\rho}^{\delta}\left|\frac{\partial \mathbf{w}^{\delta}}{\partial t}\right|^{2}+(1-\chi_{c}^{\delta})\lambda_{0}|\mathbb{D}(x,\mathbf{w}^{\delta})|^{2}+\chi_{c}^{\delta}\frac{1}{c_{f}^{2}}|q_{c}^{\delta}|^{2}\Big)dx$$
$$+\int_{\Omega}\mu_{2}\,\delta^{2}\,\chi_{c}^{\delta}\big(\mathfrak{N}_{3}^{s}:\mathbb{D}(x,\mathbf{w}^{\delta})\big):\mathbb{D}(x,\mathbf{w}^{\delta})dx=\int_{\Omega}\tilde{\rho}^{\delta}\,\mathbf{F}\cdot\frac{\partial \mathbf{w}^{\delta}}{\partial t}dx.$$

Theorem 9.2 *Let $\{\mathbf{w}^{\delta},\, q_{c}^{\delta}\}$ be a weak solution of the problem and $\bar{\mathbf{w}}_{s}^{\delta}=\mathbb{E}_{\Omega_{s}^{\delta}}(\mathbf{w}^{\delta})$ be an extension (1.2.9).*

Then up to some subsequences the sequence $\{\chi_{c}^{\delta}\mathbf{w}^{\delta}\}$ converges weakly in $\mathbf{L}_{2}(\Omega_{T})$ to the displacements $\mathbf{w}_{c}$ of the liquid in cracks, the sequence $\{\chi_{c}^{\delta}q_{c}^{\delta}\}$ converges weakly in $L_{2}(\Omega_{T})$ to the pressure q_{c} of the liquid in cracks and two-scale in $L_{2}(\Omega_{T})$ to $q_{c}(\mathbf{x},t)\chi_{c}(\mathbf{z})$, and the sequence $\{\bar{\mathbf{w}}_{s}^{\delta}\}$ converges weakly in $\mathbf{w}_{2}^{1,0}(\Omega_{T})$ and two-scale in $\mathbf{L}_{2}(\Omega_{T})$ to the displacements $\bar{\mathbf{w}}_{s}(\mathbf{x},t)$ of the solid skeleton.

The limiting functions solve the system of homogenized equations in the domain Ω_{T}, consisting of the homogenized continuity equation

$$\frac{1}{\bar{c}_{f}^{2}}\,q_{c}+\nabla\cdot\mathbf{w}_{c}=\mathbb{C}_{0}^{c}:\mathbb{D}(x,\bar{\mathbf{w}}_{s})+\frac{c_{0}^{c}}{\lambda_{0}}\,q_{c},\tag{9.1.12}$$

the homogenized momentum balance equation

$$\rho_{f}\frac{\partial^{2}\mathbf{w}_{c}}{\partial t^{2}}+\rho_{s}\frac{\partial^{2}\bar{\mathbf{w}}_{s}}{\partial t^{2}}=\nabla\cdot\big(\lambda_{0}\,\mathfrak{B}^{(c)}:\mathbb{D}(x,\bar{\mathbf{w}}_{s})-q_{c}\,\mathbb{C}_{1}^{c}\big)+\hat{\rho}\mathbf{F},\tag{9.1.13}$$

for the solid component and the homogenized momentum balance equation

$$-\int_{0}^{t}\mathbb{B}^{(f)}(\mu_{1},\infty;t-\tau)\cdot\left(\frac{1}{m_{c}}\nabla q_{c}+\rho_{f}\left(\frac{\partial^{2}\bar{\mathbf{w}}_{s}}{\partial\tau^{2}}-\mathbf{F}\right)\right)(\mathbf{x},\tau)d\tau$$
$$=\frac{\partial\mathbf{w}_{c}}{\partial t}-m_{c}\frac{\partial\bar{\mathbf{w}}_{s}}{\partial t}\tag{9.1.14}$$

for the liquid component.

The differential equations are completed with the homogeneous initial and boundary conditions

$$\bar{\mathbf{w}}_{s}(\mathbf{x},t)=0,\ \ \mathbf{x}\in S,\ \ t\in(0,T),\tag{9.1.15}$$

$$\bar{\mathbf{w}}_{s}(\mathbf{x},0)=\frac{\partial\bar{\mathbf{w}}_{s}}{\partial t}(\mathbf{x},0)=0,\ \ \mathbf{x}\in\Omega\tag{9.1.16}$$

for the displacement of the mixture, and the homogeneous boundary condition

$$\mathbf{w}_{c}(\mathbf{x},t)\cdot\mathbf{n}(\mathbf{x})=0,\ \ \mathbf{x}\in S,\ \ t\in(0,T)\tag{9.1.17}$$

for the displacement $\mathbf{w}_c$ of the liquid in cracks. In (9.1.12)–(9.1.14) the matrix $\mathbb{B}^{(f)}(\mu_1,\infty;t)$ has been defined in Chap. 3 by formulae (3.2.70) and (3.2.76), and matrices $\mathbb{C}_0^c$ and $\mathbb{C}_1^c$, the constant c_0^c, and the constant symmetric and strictly positively definite fourth-rank tensor $\mathfrak{B}^{(c)}$ are given below by formulae (9.1.28)–(9.1.30).

9.1.2 Proof of Theorem 9.2

The estimates (9.1.11) guarantee the weak and two-scale convergence in $L_2(\Omega_T)$ and $\mathbf{L}_2(\Omega_T)$ of the sequences $\{\chi_c^\delta q_c^\delta\}$, $\left\{\chi_c^\delta \dfrac{\partial \mathbf{w}^\delta}{\partial t}\right\}$, $\{\bar{\mathbf{w}}_s^\delta\}$, and $\{\mathbb{D}(x,\bar{\mathbf{w}}_s^\delta)\}$:

$$\chi_c^\delta \frac{\partial \mathbf{w}^\delta}{\partial t} \rightharpoonup \mathbf{V}_c(\mathbf{x},t),\quad \chi_c^\delta \frac{\partial \mathbf{w}^\delta}{\partial t} \stackrel{t.-s.}{\longrightarrow} \mathbf{V}_c(\mathbf{x},\mathbf{z},t),\quad \chi_c^\delta q_c^\delta \stackrel{t.-s.}{\longrightarrow} \chi_c(\mathbf{z})\, q_c(\mathbf{x},t),$$

$$\bar{\mathbf{w}}_s^\delta \stackrel{t.-s.}{\longrightarrow} \bar{\mathbf{w}}_s(\mathbf{x},t),\quad \mathbb{D}(x,\bar{\mathbf{w}}_s^\delta) \stackrel{t.-s.}{\longrightarrow} \mathbb{D}\big(x,\bar{\mathbf{w}}_s(\mathbf{x},t)\big) + \mathbb{D}\big(z,\bar{\mathbf{U}}(\mathbf{x},\mathbf{z},t)\big).$$

These limiting functions satisfy the macroscopic momentum balance equation

$$\begin{aligned}\tilde{\rho}\frac{\partial^2 \bar{\mathbf{w}}_s}{\partial t^2} &- \tilde{\rho}\mathbf{F} + \nabla\, q_c \\ &= \nabla\cdot\Big(\lambda_0\, \mathfrak{N}_s^3 : \big((1-m_c)\mathbb{D}(x,\bar{\mathbf{w}}_s) + \langle \mathbb{D}(z,\bar{\mathbf{U}})\rangle_{Z_s}\big)\Big)\end{aligned} \tag{9.1.18}$$

in the domain Ω_T, and the microscopic momentum balance equation

$$\nabla_z\cdot\Big((1-\chi_c(\mathbf{z}))\mathfrak{N}_s^3 : \big(\mathbb{D}(x,\bar{\mathbf{w}}_s) + \mathbb{D}(z,\bar{\mathbf{U}})\big) + \frac{1}{\lambda_0\, m_c}\, q_c\, \mathbb{I}\Big) = 0 \tag{9.1.19}$$

in the unit cube K for the displacement vector of the mixture in the solid skeleton and in the liquid in pores.

For the velocity of the liquid in cracks one has the macroscopic continuity equation

$$\frac{1}{c_f^2}\frac{\partial q_c}{\partial t} + \nabla\cdot\mathbf{V}_c = \left\langle \nabla_z\cdot\frac{\partial \bar{\mathbf{U}}}{\partial t}\right\rangle_{Z_s} \tag{9.1.20}$$

in the domain Ω_T with the boundary condition

$$\mathbf{V}_c\cdot\mathbf{n} = 0,\ \ \mathbf{x}\in S, \tag{9.1.21}$$

the microscopic continuity equation

$$\nabla_z\cdot\mathbf{V}_c = 0 \tag{9.1.22}$$

and the microscopic momentum balance equation

$$\rho_f \frac{\partial \mathbf{V}_c}{\partial t} = \mu_2 \, \triangle_z \mathbf{V}_c - \nabla_z \Pi - \frac{1}{m_c} \nabla \, q_c + \rho_f \mathbf{F} \tag{9.1.23}$$

in the liquid domain Z_f, completed with the boundary condition

$$\mathbf{V}_c(\mathbf{x}, \mathbf{z}, t) = \frac{\partial \bar{\mathbf{w}}_s}{\partial t}(\mathbf{x}, t), \;\; \mathbf{z} \in \gamma_c. \tag{9.1.24}$$

The problem (9.1.22)–(9.1.24) has been already studied in previous Chaps. 3 and 6 and

$$\begin{aligned} \mathbf{V}_c = \langle \mathbf{V}_c \rangle_{Z_s} &= m \, \frac{\partial \bar{\mathbf{w}}_s}{\partial t} \\ &- \int_0^t \mathbb{B}^{(f)}(\mu_2, \infty; t-\tau) \cdot \left(\frac{1}{m_c} \nabla \, q_c + \rho_f \left(\frac{\partial^2 \bar{\mathbf{w}}_s}{\partial \tau^2} - \mathbf{F} \right) \right) (\mathbf{x}, \tau) d\tau, \end{aligned} \tag{9.1.25}$$

where the matrix $\mathbb{B}^{(f)}(\mu_1, \infty; t)$ is defined by (3.2.70) and (3.2.76) in Chap. 3.

To solve (9.1.19) we look for the solution in the form

$$\bar{\mathbf{U}} = \sum_{i,j=1}^{3} \mathbf{U}_a^{(ij)} \, (\mathbf{z}) D_{ij}(\mathbf{x}, t) + \frac{1}{\lambda_0 \, m_c} q_c \, \mathbf{U}_a^{(0)}(\mathbf{z}) \, q_c(\mathbf{x}, \mathbf{t}),$$

where

$$D_{ij}(\mathbf{x}, t) = \frac{1}{2} \Big(\frac{\partial u_i}{\partial x_j}(\mathbf{x}, t) + \frac{\partial u_j}{\partial x_i}(\mathbf{x}, t) \Big), \quad \bar{\mathbf{w}}_s = (u_1, \, u_2, \, u_3),$$

and the 1-periodic in $\mathbf{z}$ functions $\mathbf{U}_a^{(ij)} i, \, j = 1, 2, 3$ and $\mathbf{U}_a^{(0)}$ satisfy the following periodic boundary-value problem

$$\left. \begin{aligned} &\nabla_z \cdot \big(\mathfrak{N}_s^3 : \mathbb{D}(z, \mathbf{U}_a^{(ij)}) \big) = 0, \;\; \mathbf{z} \in Z_s, \\ &\Big(\mathfrak{N}_s^3 : \big(\mathbb{D}(z, \mathbf{U}_a^{(ij)}) + \mathbb{J}^{ij} \big) \Big) \cdot \nu = 0, \;\; \mathbf{z} \in \gamma_c, \end{aligned} \right\} \tag{9.1.26}$$

$$\left. \begin{aligned} &\nabla_z \cdot \big(\mathfrak{N}_s^3 : \mathbb{D}(z, \mathbf{U}_a^{(0)}) \big) = 0, \;\; \mathbf{z} \in Z_s, \\ &\big(\mathfrak{N}_s^3 : \mathbb{D}(z, \mathbf{U}_a^{(ij)}) + \mathbb{I} \big) \cdot \nu = 0, \;\; \mathbf{z} \in \gamma_c. \end{aligned} \right\} \tag{9.1.27}$$

Then

$$\langle \mathbb{D}(z, \bar{\mathbf{U}}) \rangle_{Z_s} = \left(\sum_{i,j=1}^{3} \langle \mathbb{D}(z, \mathbf{U}_a^{(ij)}) \rangle_{Z_s} \mathbb{J}^{ij} \right) : \mathbb{D}(x, \bar{\mathbf{w}}_s) + \frac{1}{\lambda_0 \, m_c} \langle \mathbb{D}(z, \mathbf{U}_a^{(0)}) \rangle_{Z_s} \, q_c,$$

$$\langle \nabla_z \cdot \bar{\mathbf{U}} \rangle_{Z_s} = \left(\sum_{i,j=1}^{3} \langle \nabla_z \cdot \mathbf{U}_a^{(ij)} \rangle_{Z_s} \mathbb{J}^{ij} \right) : \mathbb{D}(x, \bar{\mathbf{w}}_s) + \frac{1}{\lambda_0 m_c} \langle \nabla_z \cdot \mathbf{U}_a^{(0)} \rangle_{Z_s} q_c,$$

and

$$\mathfrak{B}^{(c)} = \mathfrak{N}_s^3 : \left((1 - m_c)\mathbb{J} + \sum_{i,j=1}^{3} \langle \mathbb{D}(z, \mathbf{U}_a^{(ij)}) \rangle_{Z_s} \mathbb{J}^{ij} \right), \tag{9.1.28}$$

$$\mathbb{C}_1^c = \frac{1}{m_c} \langle \mathbb{D}(z, \mathbf{U}_a^{(0)}) \rangle_{Z_s}, \quad \mathbb{C}_0^c = \sum_{i,j=1}^{3} \langle \nabla_z \cdot \mathbf{U}_a^{(ij)} \rangle_{Z_s} \mathbb{J}^{ij}, \tag{9.1.29}$$

$$c_0^c = \frac{1}{m_c} \langle \nabla_z \cdot \mathbf{U}_a^{(0)} \rangle_{Z_s}. \tag{9.1.30}$$

The symmetry of the tensor $\mathfrak{B}^{(c)}$ is proved in the same way as in Chap. 4.

Chapter 10
Diffusion and Convection in Porous Media

The model $\mathbb{M}_{27}$ consists of the stationary Stokes equations

$$\nabla \cdot \big(\varepsilon^2 \mu_1(c)\mathbb{D}(x, \mathbf{v}) + (\nu_0 \nabla \cdot \mathbf{v} - p)\mathbb{I}\big) + (\rho_f + \delta\, c)\mathbf{F} = 0, \tag{10.0.1}$$

$$\frac{\partial p}{\partial t} + c_f^2\, \nabla \cdot \mathbf{v} = 0 \tag{10.0.2}$$

for a weakly compressible liquid, and the diffusion-convection equation

$$\frac{\partial c}{\partial t} + \nabla \cdot (c\, \mathbf{v} - D_0\, \nabla\, c) = 0 \tag{10.0.3}$$

in the liquid domain Ω_f^ε for $t > 0$.

The differential equations are completed with the boundary conditions

$$\mathbf{v}(\mathbf{x}, t) = 0, \tag{10.0.4}$$

$$\nabla\, c(\mathbf{x}, t) \cdot \mathbf{n}(\mathbf{x}) = 0, \tag{10.0.5}$$

on the boundary $S^\varepsilon = \partial \Omega_f^\varepsilon$ for $t > 0$, and the initial conditions

$$c(\mathbf{x}, 0) = c_0(\mathbf{x}), \;\; \mathbf{x} \in \Omega_f^\varepsilon, \tag{10.0.6}$$

$$p(\mathbf{x}, 0) = 0, \;\; \mathbf{x} \in \Omega_f^\varepsilon. \tag{10.0.7}$$

In Eq. (10.0.5) $\mathbf{n}$ is the unit normal vector to the boundary S^ε.

The model $\mathbb{M}_{25}$ takes into account the movement of a solid skeleton, namely, the concentration of the admixture c, the displacement of the continuous medium $\mathbf{w}$, and the pressure p satisfy the diffusion-convection equation

A. Meirmanov, *Mathematical Models for Poroelastic Flows*,
Atlantis Studies in Differential Equations 1, DOI: 10.2991/978-94-6239-015-7_10,

$$\frac{\partial c}{\partial t}+\frac{\partial \mathbf{w}}{\partial t}\cdot\nabla c=D_0\,\triangle c \tag{10.0.8}$$

in the pore space Ω_f^{ε} for $t>0$, the momentum balance equation

$$\nabla\cdot\left(\chi^{\varepsilon}\mu_0\mathbb{D}\left(x,\frac{\partial \mathbf{w}}{\partial t}\right)+(1-\chi^{\varepsilon})\lambda_0\mathbb{D}(x,\mathbf{w})-p\mathbb{I}\right)+\rho^{\varepsilon}\mathbf{F}=0, \tag{10.0.9}$$

and the continuity equation

$$\nabla\cdot\mathbf{w}=0 \tag{10.0.10}$$

in the domain Ω for $t>0$, the normalization condition

$$\int_{\Omega}p\,(\mathbf{x},t)dx=0, \tag{10.0.11}$$

the boundary and initial conditions (10.0.5), (10.0.6) for the concentration c, and the boundary and initial conditions

$$\mathbf{w}=0,\ \ \mathbf{x}\in S=\partial\Omega,\ \ t>0, \tag{10.0.12}$$

$$\mathbf{w}(\mathbf{x},0)=0,\ \ \mathbf{x}\in\Omega_f^{\varepsilon} \tag{10.0.13}$$

for the displacement $\mathbf{w}$.

In Eq. (10.0.9)

$$\rho^{\varepsilon}=\chi^{\varepsilon}(\rho_f+\delta\,c)+\rho_s(1-\chi^{\varepsilon}).$$

Throughout this chapter we impose Assumptions 0.1 and 3.1.

10.1 Diffusion-Convection in an Absolutely Rigid Skeleton

In this section we suppose that $c_0(\mathbf{x})$ and $\mathbf{F}(\mathbf{x},t)$ are the measurable functions,

$$0\leqslant c_0(\mathbf{x})\leqslant 1,\ \int_{\Omega_T}|\mathbf{F}(\mathbf{x},t)|^2dxdt\leqslant F^2<\infty, \tag{10.1.1}$$

$$0<\mu_*\leqslant\mu_1(c)\leqslant\mu_*^{-1},\ \ \mu_0\in C^2[0,\infty), \tag{10.1.2}$$

and $\nu_0,\ c_f^2,\ D_0,\ \mu_*$ are positive constants.

10.1.1 Statement of the Problem and Main Results

Definition 10.1 We say that the triple of functions $\{\mathbf{v}^\varepsilon,\ p^\varepsilon,\ c^\varepsilon\}$ is *a weak solution of the problem* (10.0.1)–(10.0.7) if

$$\mathbf{v}^{\varepsilon} \in \mathbf{L}_2\left((0,T);\overset{\circ}{\mathbf{W}}{}_2^1\left(\Omega_f^{\varepsilon}\right)\right),\quad c^{\varepsilon} \in L_2\left((0,T);W_2^1\left(\Omega_f^{\varepsilon}\right)\right),$$
$$\frac{\partial p^{\varepsilon}}{\partial t} \in L_2\left((0,T);L_2\left(\Omega_f^{\varepsilon}\right)\right),$$

and the integral identities

$$\int_0^T\int_{\Omega_f^\varepsilon}\left(c^\varepsilon\frac{\partial\xi}{\partial t}+\left(c^\varepsilon\mathbf{v}^\varepsilon-D_0\nabla c^\varepsilon\right)\cdot\nabla\xi+\xi\,c^\varepsilon\nabla\cdot\mathbf{v}^\varepsilon\right)dxdt$$
$$=-\int_{\Omega_f^\varepsilon}c_0(\mathbf{x})\,\xi(\mathbf{x},0)\,dx, \tag{10.1.3}$$

$$\int_0^T\int_{\Omega_f^\varepsilon}\left(\varepsilon^2\mu_1(c^\varepsilon)\,\mathbb{D}(x,\mathbf{v}^\varepsilon)-\left(p^\varepsilon-\nu_0\nabla\cdot\mathbf{v}^\varepsilon\right)\mathbb{I}\right):\mathbb{D}(x,\varphi)dxdt$$
$$=\int_0^T\int_{\Omega_f^\varepsilon}\left(\rho_f+\delta\,c^\varepsilon\right)\mathbf{F}\cdot\varphi\,dxdt, \tag{10.1.4}$$

and

$$\int_0^T\int_{\Omega_f^\varepsilon}\left(\frac{\partial\psi}{\partial t}p^\varepsilon+c_f^2\,\mathbf{v}^\varepsilon\cdot\nabla\psi\right)dxdt=0 \tag{10.1.5}$$

hold true for any smooth functions ξ, ψ and φ, such that $\xi(\mathbf{x},T)=\psi(\mathbf{x},T)=0$ and $\varphi(\mathbf{x},t)=0$ for $\mathbf{x}\in S^\varepsilon$.

Note that the integral identity (10.1.3) contains the differential equation (10.0.3) in the pore space, the boundary condition (10.0.5) on the boundary $S^{(\varepsilon)}$, and the initial condition (10.0.6). The boundary condition (10.0.4) is already included into the corresponding functional space for $\mathbf{v}$, and the initial condition (10.0.7) is already included into the integral identity (10.1.5).

Theorem 10.1 *The problem (10.0.1)–(10.0.7) has at least one weak solution* $\{\mathbf{v}^\varepsilon,\ p^\varepsilon,\ c^\varepsilon\}$, *such that*

$$0\leqslant c^\varepsilon(\mathbf{x},t)\leqslant 1,\ \ \mathbf{x}\in\Omega_f^\varepsilon,\ t>0, \tag{10.1.6}$$

$$\int_0^T\int_{\Omega_f^\varepsilon}|\nabla c^\varepsilon|^2dxdt\leqslant C,\ F^2 \tag{10.1.7}$$

$$\int_0^T \int_{\Omega_f^\varepsilon} \left(\varepsilon^2 |\nabla \mathbf{v}^\varepsilon|^2 + |\mathbf{v}^\varepsilon|^2 + \left| \frac{\partial p^\varepsilon}{\partial t} \right|^2 + (\nabla \cdot \mathbf{v}^\varepsilon)^2 \right) dxdt \leqslant C F^2, \tag{10.1.8}$$

where C is independent of ε.

By *Homogenization* we mean the limiting procedure as $\varepsilon \searrow 0$. But in our method that is possible only for functions, defined in the whole domain Ω for $t > 0$. So, we first extend the functions $\mathbf{v}^\varepsilon$, p^ε, and c^ε onto Ω for $t > 0$, and only after that apply the homogenization theory.

The functions $\mathbf{v}^\varepsilon$, $\nabla \cdot \mathbf{v}^\varepsilon$, and p^ε are extended in a trivial way by setting $\tilde{\mathbf{v}}^\varepsilon = \mathbf{v}^\varepsilon$, $\tilde{p}^\varepsilon = p^\varepsilon$ in Ω_f^ε for $t > 0$, and $\tilde{\mathbf{v}}^\varepsilon = 0$, $\nabla \cdot \tilde{\mathbf{v}}^\varepsilon = 0$, and $\tilde{p}^\varepsilon = 0$ outside Ω_f^ε.

For the functions c^ε the extension result [1] (see also Appendix B) states that there exists an extension

$$\tilde{c}^\varepsilon = \tilde{\mathbb{E}}_{\Omega_f^\varepsilon}(c^\varepsilon), \tag{10.1.9}$$

such that

$$c^\varepsilon = \tilde{c}^\varepsilon, \quad \mathbf{x} \in \Omega_f^\varepsilon, \ t \in (0, T), \tag{10.1.10}$$

and

$$\int_\Omega |\tilde{c}^\varepsilon(\mathbf{x}, t)|^2 dx \leqslant C \int_{\Omega_f^\varepsilon} |c^\varepsilon(\mathbf{x}, t)|^2 dx,$$
$$\int_\Omega |\nabla \tilde{c}^\varepsilon(\mathbf{x}, t)|^2 dx \leqslant C \int_{\Omega_f^\varepsilon} |\nabla c^\varepsilon(\mathbf{x}, t)|^2 dx, \quad t \in (0, T), \tag{10.1.11}$$

where C is independent of ε and $t \in (0, T)$.

Theorem 10.2 *Let $\{\mathbf{v}^\varepsilon, \ p^\varepsilon, \ c^\varepsilon\}$ be the weak solution to the problem (10.0.1)–(10.0.7). Then*

(I) there exists a subsequence of small parameters $\{\varepsilon > 0\}$ as $\varepsilon \searrow 0$, such that

(1) the sequence $\{\tilde{\mathbf{v}}^\varepsilon\}$ converges weakly in $\mathbf{L}_2(\Omega_T)$ to the function $\mathbf{v}$,
(2) the sequence $\{\nabla \cdot \tilde{\mathbf{v}}^\varepsilon\}$ converges weakly in $\mathbf{L}_2(\Omega_T)$ to the function $\nabla \cdot \mathbf{v}$,
(3) the sequence $\{\tilde{p}^\varepsilon\}$ converges weakly in $L_2(\Omega_T)$ to the function p,
(4) the sequence $\{\tilde{c}^\varepsilon\}$ converges weakly in $W_2^{1,0}(\Omega_T)$ and strongly in $L_2(\Omega_T)$ to the function c.

(II) The triple of limiting functions $\{\mathbf{v}, \ p, \ c\}$ is the weak solution of the diffusion-convection problem for a compressible liquid in an absolutely rigid solid skeleton, which consists of the dynamic equations

$$\mathbf{v} = \frac{1}{\mu_1(c)} \mathbb{B} \left(-\frac{1}{m} \nabla q + (\rho_f + \delta\, c) \mathbf{F} \right), \tag{10.1.12}$$

$$q = p + \frac{\nu_0}{c_f^2}\frac{\partial p}{\partial t}, \tag{10.1.13}$$

$$\frac{\partial p}{\partial t} + c_f^2\,\nabla\cdot\mathbf{v} = 0 \tag{10.1.14}$$

for the velocity $\mathbf{v}$ *and pressure* p *of a slightly compressible liquid, and the diffusion-convection equation*

$$m\,\frac{\partial c}{\partial t} + \mathbf{v}\cdot\nabla c = D_0\,\nabla\cdot\left(\mathbb{B}^{(c)}\nabla c\right) \tag{10.1.15}$$

for the concentration c *of an admixture in the domain* Ω *for* $t > 0$.

The problem is endowed with the homogeneous boundary conditions

$$\mathbf{v}(\mathbf{x},t)\cdot\mathbf{n}(\mathbf{x}) = 0, \tag{10.1.16}$$

$$\nabla\,c(\mathbf{x},t)\cdot\mathbf{n}(\mathbf{x}) = 0 \tag{10.1.17}$$

on the boundary S *for* $t > 0$, *and the initial conditions*

$$p(\mathbf{x},0) = 0, \quad c(\mathbf{x},0) = c_0(\mathbf{x}) \quad \mathbf{x}\in\Omega. \tag{10.1.18}$$

In Eqs. (10.1.12)–(10.1.18)

$$m = \langle\chi\rangle_Y = \int_Y \chi(\mathbf{y})dy$$

is the porosity, the symmetric and strictly positively definite constant matrix $\mathbb{B}$ *is given below by formula (10.1.58) (see also (1.1.27) in Theorem 1.1, Chap. 1), the symmetric and strictly positively definite constant matrix* $\mathbb{B}^{(c)}$ *is defined below by formula (10.1.61), and* $\mathbf{n}$ *is the unit outward normal vector to the boundary* S.

We refer to the problem (10.1.12)–(10.1.18) as the homogenized **model** $(\mathbb{DCARS})_1$.

Theorem 10.3 *Let* $\{\mathbf{v}^{(k)},\ p^{(k)},\ c^{(k)}\}$ *be the weak solution of the problem (10.0.1)–(10.0.7) with* $c_f^2 = k$. *Then*

(I) there exists a subsequence $k_n \to \infty$ *such that*

(1) the sequence $\{\mathbf{v}^{(k_n)}\}$ *converges weakly in* $\mathbf{L}_2(\Omega_T)$ *to the function* $\mathbf{v}^{(\infty)}$,
(2) the sequence $\{p^{(k_n)}\}$ *converges weakly in* $L_2(\Omega_T)$ *to the function* $p^{(\infty)}$,
(3) the sequence $\{c^{(k_n)}\}$ *converges weakly in* $W_2^{1,0}(\Omega_T)$ *and strongly in* $L_2(\Omega_T)$ *to the function* $c^{(\infty)}$;

(II) the triple of limiting functions $\{\mathbf{v}^{(\infty)},\ p^{(\infty)},\ c^{(\infty)}\}$ *is the weak solution of the diffusion-convection problem for an incompressible liquid in an absolutely rigid*

solid skeleton, which consists of the Darcy system of filtration with the variable viscosity

$$\mathbf{v}^{(\infty)} = \frac{1}{\mu_1(c^{(\infty)})}\mathbb{B}\left(-\frac{1}{m}\nabla p^{(\infty)} + \left(\rho_f + \delta\, c^{(\infty)}\right)\mathbf{F}\right), \tag{10.1.19}$$

$$\nabla \cdot \mathbf{v}^{(\infty)} = 0 \tag{10.1.20}$$

for the velocity $\mathbf{v}^{(\infty)}$ *and pressure* $p^{(\infty)}$ *of an incompressible liquid in the domain* Ω *for* $t > 0$*, the diffusion-convection equation (10.1.15) with the velocity field* $\{\mathbf{v}^{(\infty)}\}$ *for the concentration* $c^{(\infty)}$*, the boundary conditions (10.1.16) and (10.1.17), and the initial condition (10.1.18) for the concentration.*

We refer to the problem (10.1.15)–(10.1.20) as the homogenized **model** $(\mathbb{DCARS})_2$.

Theorem 10.4 *Let* $\{\mathbf{v}^{(\lambda)},\ p^{(\lambda)},\ c^{(\lambda)}\}$ *be the weak solution to the problem (10.0.1)–(10.0.7) with* $\nu_0 = \lambda$*. Then*

(I) there exists a subsequence $\lambda_n \to 0$*, such that*

(1) the sequence $\{\mathbf{v}^{(\lambda_n)}\}$ *converges weakly in* $\mathbf{L}_2(\Omega_T)$ *to the function* $\mathbf{v}^{(0)}$*,*
(2) the sequence $\{p^{(\lambda_n)}\}$ *converges weakly in* $L_2(\Omega_T)$ *to the function* $p^{(0)}$*,*
(3) the sequence $\{c^{(\lambda_n)}\}$ *converges weakly in* $W_2^{1,0}(\Omega_T)$ *and strongly in* $L_2(\Omega_T)$ *to the function* $c^{(0)}$*;*

(II) the triple of limiting functions $\{\mathbf{v}^{(0)},\ p^{(0)},\ c^{(0)}\}$ *is the weak solution of the diffusion-convection problem for a slightly compressible liquid in absolutely rigid solid skeleton, which consists of the Darcy system of filtration with the variable viscosity*

$$\mathbf{v}^{(0)} = \frac{1}{\mu_1(c^{(0)})}\mathbb{B}\left(-\frac{1}{m}\nabla p^{(0)} + \left(\rho_f + \delta\, c^{(0)}\right)\mathbf{F}\right), \tag{10.1.21}$$

$$\frac{\partial p^{(0)}}{\partial t} + c_f^2\, \nabla \cdot \mathbf{v}^{(0)} = 0 \tag{10.1.22}$$

for the velocity $\mathbf{v}^{(0)}$ *and pressure* $p^{(0)}$ *of a slightly compressible liquid in the domain* Ω *for* $t > 0$*, the diffusion-convection equation (10.1.15) with the velocity field* $\{\mathbf{v}^{(0)}\}$ *for the concentration* $c^{(0)}$*, the boundary conditions (10.1.16) and (10.1.17), and the initial condition (10.1.18) for the concentration.*

We refer to the problem (10.1.15)–(10.1.18), (10.1.21), (10.1.22) as the homogenized **model** $(\mathbb{DCARS})_3$.

10.1.2 Proof of Theorem 10.1

Let us divide the proof into several stages. As the first step we consider an approximate problem, where the velocity $\mathbf{v}$ (for the moment we omit the index ε) in the convection-diffusion equation is replaced by its approximation

$$\begin{aligned}\mathbf{v}_{(h)} &= \mathbb{M}^h(\mathbf{v})\\ &= \frac{1}{h^4}\int_{-\infty}^{\infty} J\left(\frac{t-\tau}{h}\right)\left(\int_{\mathbb{R}^3} J\left(\frac{|\mathbf{z}-\mathbf{x}|}{h}\right)\bar{\mathbf{v}}(\mathbf{z},\tau)dz\right)d\tau. \quad (10.1.23)\end{aligned}$$

In Eq. (10.1.23)

$$\bar{\mathbf{v}}(\mathbf{x},t) = \begin{cases} \mathbf{v}(\mathbf{x},t) & \textit{if } x\in\Omega_f^\varepsilon,\ 0<t<T,\\ 0 & \textit{if } x\in\mathbb{R}^3\backslash\Omega_f^\varepsilon,\ t>0,\\ 0 & \textit{if } x\in\Omega_f^\varepsilon,\ \textit{and } t\geqslant T,\ \textit{or } t\leqslant 0,\end{cases}$$

and $J(s)$ is an infinitely smooth function, such that

$$J(s)=0,\ \text{ if } |s|>1,\ \text{ and } \int_{\infty}^{\infty} J(s)ds\int_{\mathbb{R}^3} J(|\mathbf{x}|)dx = 1.$$

By the well-known properties of the mollifiers $\mathbb{M}^h$ [3]

(1) $\mathbf{v}_{(h)}\in\mathbf{C}^\infty\big(\mathbb{R}^3\times(-\infty,\infty)\big)$;
(2) if $\mathbf{v}\in\mathbf{L}_2(\Omega_T)$, then $\mathbf{v}_{(h)}\to\mathbf{v}$ strongly in $\mathbf{L}_2(\Omega_T)$ as $h\to 0$;
(3) if $\mathbf{v}\in\overset{\circ}{\mathbf{W}}{}_2^{1,0}(\Omega_T^\varepsilon)$, then $\nabla\cdot\mathbf{v}_{(h)}\to\nabla\cdot\mathbf{v}$ strongly in $\overset{\circ}{\mathbf{W}}{}_2^{1,0}(\Omega_T^\varepsilon)$ as $h\to 0$.

More precisely, we look for the solution $\{\mathbf{v}^{\varepsilon,h},\ p^{\varepsilon,h},\ c^{\varepsilon,h}\}$ of the system of differential equations

$$\begin{aligned}&\nabla\cdot\big(\varepsilon^2\mu_1(c^{\varepsilon,h})\mathbb{D}(x,\mathbf{v}^{\varepsilon,h}) + (\nu_0\nabla\cdot\mathbf{v}^{\varepsilon,h} - p^{\varepsilon,h})\mathbb{I}\big)\\ &\quad + (\rho_f+\delta c^{\varepsilon,h})\mathbf{F} = 0, \qquad (10.1.24)\end{aligned}$$

$$\frac{\partial p^{\varepsilon,h}}{\partial t} + c_f^2\,\nabla\cdot\mathbf{v}^{\varepsilon,h} = 0, \qquad (10.1.25)$$

$$\frac{\partial c^{\varepsilon,h}}{\partial t} + \mathbf{v}_{(h)}^{\varepsilon,h}\cdot\nabla c^{\varepsilon,h} = D_0\,\triangle\, c^{\varepsilon,h} \qquad (10.1.26)$$

in the domain Ω_f^ε for $t>0$, satisfying the following boundary and initial conditions

$$\mathbf{v}^{\varepsilon,h}(\mathbf{x},t)=0,\ \ \mathbf{x}\in S^\varepsilon,\ \ t>0, \qquad (10.1.27)$$

$$\nabla c^{\varepsilon,h}(\mathbf{x},t)\cdot\mathbf{n}(\mathbf{x})=0,\ \ \mathbf{x}\in S^{\varepsilon},\ \ t>0, \tag{10.1.28}$$

$$c^{\varepsilon,h}(\mathbf{x},0)=c_0^h(\mathbf{x}),\ \ \mathbf{x}\in\Omega_f^{\varepsilon}, \tag{10.1.29}$$

$$p^{\varepsilon,h}(\mathbf{x},0)=0,\ \ \mathbf{x}\in\Omega_f^{\varepsilon}. \tag{10.1.30}$$

In Eq. (10.1.26) $\mathbf{v}_{(h)}^{\varepsilon,h}=\mathbb{M}^h(\mathbf{v}^{\varepsilon,h})$ and

$$c_0^h\in\overset{\circ}{C}{}^{\infty}(\Omega_f^{\varepsilon}),\ \ 0\leqslant c_0^h(\mathbf{x})\leqslant 1,\ \ c_0^h(\mathbf{x})\to c_0(\mathbf{x})\ \text{as}\ h\to 0\ \text{a.e. in}\ \Omega^{\varepsilon}.$$

To solve (10.1.24)–(10.1.30) we fix the set

$$\mathfrak{M}=\{\bar{c}\in C\big([0,T];C(\overline{\Omega}_f^{\varepsilon})\big):0\leqslant\bar{c}(\mathbf{x},t)\leqslant 1\}$$

and consider the first auxiliary problem

$$\nabla\cdot\big(\varepsilon^2\mu_1(\bar{c})\mathbb{D}(x,\mathbf{u})+(\nu_0\nabla\cdot\mathbf{u}-q)\mathbb{I}\big)+(\rho_f+\delta\,\bar{c})\mathbf{F}=0, \tag{10.1.31}$$

$$\frac{\partial q}{\partial t}+c_f^2\,\nabla\cdot\mathbf{u}=0 \tag{10.1.32}$$

for $\mathbf{x}\in\Omega_f^{\varepsilon}$ and $t>0$, and

$$\mathbf{u}(\mathbf{x},t)=0,\ \ \mathbf{x}\in S^{\varepsilon},\ \ t>0;\ \ q(\mathbf{x},0)=0,\ \ \mathbf{x}\in\Omega_f^{\varepsilon}. \tag{10.1.33}$$

For all $\bar{c}\in\mathfrak{M}$ this problem defines the nonlinear operator

$$\mathbf{u}=\mathbb{A}_1(\bar{c}),\ \ \mathbb{A}_1:\mathfrak{M}\to\mathbf{L}_2\big((0,T);\overset{\circ}{\mathbf{W}}{}_2^1(\Omega_f^{\varepsilon})\big).$$

Next we consider the second auxiliary problem

$$\frac{\partial c}{\partial t}+\mathbf{u}_{(h)}\cdot\nabla c=D_0\,\triangle\,c \tag{10.1.34}$$

for $\mathbf{x}\in\Omega_f^{\varepsilon}$ and $t>0$, and

$$\nabla c(\mathbf{x},t)\cdot\mathbf{n}(\mathbf{x})=0,\ \ \mathbf{x}\in S^{\varepsilon},\ \ t>0, \tag{10.1.35}$$

$$c(\mathbf{x},0)=c_0^h(\mathbf{x}),\ \ \mathbf{x}\in\Omega^{\varepsilon}, \tag{10.1.36}$$

where

$$\mathbf{u}_{(h)}=\mathbb{M}^h(\mathbf{u}),\ \ \mathbf{u}=\mathbb{A}_1(\tilde{c}).$$

The problem (10.1.34)–(10.1.36) defines the nonlinear operator $\mathbb{A}_2$ which, due to the maximum principle, transforms $\overset{\circ}{\mathbf{W}}{}_2^{1,0}(\Omega_T^\varepsilon)$ into the set $\mathfrak{M}$:

$$c = \mathbb{A}_2(\mathbf{u}), \;\; \mathbb{A}_2 : \overset{\circ}{\mathbf{W}}{}_2^{1,0}(\Omega_T^\varepsilon) \to \mathfrak{M}.$$

Thus, the nonlinear operator $\mathbb{A} = \mathbb{A}_2 \cdot \mathbb{A}_1$ transforms the set $\mathfrak{M}$ into itself. It is clear that all of the fixed points $c^{\varepsilon,h}$ of the operator $\mathbb{A}$ define solutions $\{\mathbf{v}^{\varepsilon,h},\; p^{\varepsilon,h},\; c^{\varepsilon,h}\}$ to the problem (10.1.24)–(10.1.30). To prove the existence of at least one fixed point of $\mathbb{A}$ we have to show that $\mathbb{A}$ is a completely continuous operator.

The weak solutions to the problems (10.1.24)–(10.1.30) and (10.1.31)–(10.1.33) are defined in the same way as the weak solution to the problem (10.0.1)–(10.0.7).

Lemma 10.1 *For any $\bar{c} \in \mathfrak{M}$ the problem (10.1.31)–(10.1.33) has a unique weak solution $\{\boldsymbol{u},\, q\}$, such that*

$$\int_0^T \int_{\Omega_f^\varepsilon} \left(\varepsilon^2 |\nabla \mathbf{u}|^2 + |\mathbf{u}|^2 + \left| \frac{\partial q}{\partial t} \right|^2 + (\nabla \cdot \mathbf{u})^2 \right) dx dt \leqslant C F^2, \tag{10.1.37}$$

and for any $\bar{c}_1,\, \bar{c}_2 \in \mathfrak{M}$

$$\int_0^T \int_{\Omega_f^\varepsilon} \left(\varepsilon^2 |\nabla (\mathbf{u}_1 - \mathbf{u}_2)|^2 + |\mathbf{u}_1 - \mathbf{u}_2|^2 \right) dx dt$$

$$\leqslant C F^2 \left(\max_{\Omega_f^\varepsilon \times (0,T)} |\bar{c}_1(\mathbf{x}, t) - \bar{c}_2(\mathbf{x}, t)| \right)^2, \tag{10.1.38}$$

where C is independent of ε and h, and $\mathbf{u}_i = \mathbb{A}_1(\bar{c}_i),\;\; i = 1, 2$.

Proof The proof of the first part of this lemma is standard. It can be based on the Galerkin method coupled with the energy estimate

$$\int_0^T \int_{\Omega_f^\varepsilon} (\varepsilon^2 |\nabla \mathbf{u}|^2 + \left| \frac{\partial q}{\partial t} \right|^2 + (\nabla \cdot \mathbf{u})^2) d \leqslant C F^2. \tag{10.1.39}$$

The latter is the result of the multiplication of (10.1.31) by $\mathbf{u}$, and the use of (10.1.32), Hölder's, Korn's, and Friedrichs-Poincaré's inequalities. Note that we may extend all functions outside Ω_f^ε onto some cube $Q \supset \Omega_f^\varepsilon$ as zero, and apply Korn's inequality for Q. Thus, the constant C in (10.1.39) is independent of ε.

The estimate

$$\int_0^T \int_{\Omega_f^\varepsilon} |\mathbf{u}|^2 dx dt \leqslant C \varepsilon^2 \int_0^T \int_{\Omega_f^\varepsilon} |\nabla \mathbf{u}|^2 dx dt$$

has been already proved in Chap. 1, Theorem 1.1.

The proof of the second part of the lemma is also standard. We consider the initial—boundary value problem for the difference $\hat{\mathbf{u}} = \mathbf{u}_1 - \mathbf{u}_2$, multiply the differential equation for $\hat{\mathbf{u}}$ by $\hat{\mathbf{u}}$, integrate the result by parts over domain Ω^{ε}, and use Hölder's inequality.

Lemma 10.2 *The problem (10.1.34)–(10.1.36) has a unique solution* $c \in C^1\left((0,T); C^{2,1}\left(\overline{\Omega}^{\varepsilon}_f\right)\right)$, *such that*

$$0 \leqslant c(\mathbf{x},t) \leqslant 1, \;\; \mathbf{x} \in \Omega^{\varepsilon}_f, \;\; t > 0, \tag{10.1.40}$$

$$\int_0^T \int_{\Omega^{\varepsilon}_f} |\nabla c|^2 dxdt \leqslant C F^2 \tag{10.1.41}$$

where C is independent of ε and h, and

$$\max_{\Omega^{\varepsilon}_f \times (0,T)} \left(\left| \frac{\partial c}{\partial t}(\mathbf{x},t) \right| + |\nabla c\,(\mathbf{x},t)| \right) \leqslant N(h). \tag{10.1.42}$$

If $c_i = \mathbb{A}_2(\mathbf{u}_i)$, $\mathbf{u}_i = \mathbb{A}_1(\bar{c}_i), \;\; i = 1,2,$ *for* $\bar{c}_1, \; \bar{c}_2 \in \mathfrak{M}$, *then*

$$\max_{0<t<T} \int_{\Omega^{\varepsilon}_f} |c_1(\mathbf{x},t) - c_2(\mathbf{x},t)|^2 dx + \int_0^T \int_{\Omega^{\varepsilon}_f} |\nabla(c_1 - c_2)|^2 dxdt$$
$$\leqslant N(h) \int_0^T \int_{\Omega^{\varepsilon}_f} |(\mathbf{u}_1 - \mathbf{u}_2)|^2 dxdt. \tag{10.1.43}$$

Proof The existence of a unique solution of (10.1.34)–(10.1.36) follows from ([61], Sect. 5, Chap. III). Moreover, this solution satisfies the maximum principle (10.1.40). In fact, let

$$c^+(\mathbf{x},t) = \max\{c\,(\mathbf{x},t) - 1; 0\}.$$

Then, using mollifiers in the same way as in the proof of Theorem 7.1 (Chap. III, [61]), we arrive at the equality

$$\frac{1}{2}\int_{\Omega^{\varepsilon}_f} |c^+(\mathbf{x},t_0)|^2 dx + D_0 \int_0^{t_0} \int_{\Omega^{\varepsilon}_f} |\nabla c^+|^2 dxdt$$
$$= -\int_0^{t_0} \int_{\Omega^{\varepsilon}_f} \left(\mathbf{u}_{(h)} \cdot \nabla c^+\right) c^+ \, dxdt = I, \tag{10.1.44}$$

where we have used the initial condition (10.1.36) ($c^+(\mathbf{x},0) = 0$), and the evident relations

$$\nabla c \cdot \nabla c^+ = \nabla c^+ \cdot \nabla c^+, \quad c^+ \nabla c = c^+ \nabla c^+.$$

Applying to the right-hand side of (10.1.44) Hölder's and Cauchy's inequalities, and the boundedness of $\mathbf{u}_{(h)}$ we arrive at

$$I \leqslant \frac{D_0}{2}\int_0^{t_0}\int_{\Omega_f^\varepsilon} |\nabla c^+|^2 dxdt + N^2(h)\frac{1}{2\,D_0}\int_0^{t_0}\int_{\Omega_f^\varepsilon} |c^+|^2 dxdt,$$

and

$$\frac{1}{2}\int_{\Omega_f^\varepsilon} |c^+(\mathbf{x}, t_0)|^2 dx \leqslant N^2(h)\frac{1}{2\,D_0}\int_0^{t_0}\int_{\Omega_f^\varepsilon} |c^+(\mathbf{x}, t)|^2 dxdt,$$

which implies the equality $c^+(\mathbf{x}, t) \equiv 0$, and the right-hand side inequality in (10.1.40).

The left-hand side inequality in (10.1.40) is proved in the same way, if we put

$$c^-(\mathbf{x}, t) = \min\{c\,(\mathbf{x}, t); 0\}.$$

Estimate (10.1.41) follows from the energy equality

$$\frac{1}{2}\int_{\Omega_f^\varepsilon} |c\,(\mathbf{x}, t_0)|^2 dx + D_0\int_0^{t_0}\int_{\Omega_f^\varepsilon} |\nabla c|^2\, dxdt = -\int_0^{t_0}\int_{\Omega_f^\varepsilon} \left(\mathbf{u}_{(h)}\cdot\nabla c\right) c\, dxdt$$

for c, similarly to (10.1.44), after applying (10.1.40).

The boundedness of c, the infinite smoothness of $\mathbf{u}_{(h)}$ and c_0^h, and the local estimates for linear parabolic equations (Sect. 10, Chap. IV, [61]) imply the infinite smoothness of c inside Ω_f^ε for $0 \leqslant t \leqslant T$. The boundary condition (10.1.35) permits us to extend the solution of (10.1.34)–(10.1.36) onto the small neighborhood outside of Ω_f^ε for $0 \leqslant t \leqslant T$ as an even function. Applying again local estimates for linear parabolic equations, we conclude that c is infinitely smooth in the closure of $\Omega_f^\varepsilon \times (0, T)$, which, in particular, implies (10.1.42).

The proof of (10.1.43) is straightforward. The integral identity for the difference $\tilde{c} = c_1 - c_2$ has the form

$$\int_0^T\int_{\Omega_f^\varepsilon}\left(-\tilde{c}\,\frac{\partial\xi}{\partial t} + D_0\,\nabla\tilde{c}\cdot\nabla\xi\right) dxdt = \int_0^T\int_{\Omega_f^\varepsilon} \xi\,\left(\tilde{\mathbf{u}}\cdot\nabla c_2 - (\mathbf{u}_1)_{(h)}\cdot\nabla\tilde{c}\right) dx,$$

where $\tilde{\mathbf{u}} = (\mathbf{u}_2)_{(h)}(\mathbf{x}, t) - (\mathbf{u}_1)_{(h)}(\mathbf{x}, t)$.

As before, this identity results:

$$\frac{1}{2}\int_{\Omega_f^\varepsilon} |\tilde{c}(\mathbf{x}, t)|^2 dx + D_0\int_0^t\int_{\Omega_f^\varepsilon} |\nabla\tilde{c}\,(\mathbf{x}, \tau)|^2 dxd\tau$$
$$= \int_0^t\int_{\Omega_f^\varepsilon} \tilde{c}(\mathbf{x}, \tau)\,\left(\tilde{\mathbf{u}}(\mathbf{x}, \tau)\cdot\nabla c_2(\mathbf{x}, \tau) - (\mathbf{u}_1)_{(h)}(\mathbf{x}, \tau)\cdot\nabla\tilde{c}\,(\mathbf{x}, \tau)\right) dxd\tau \equiv I.$$

The estimate (10.1.43) follows from the last equality and Gronwall's inequality, if we estimate the right-hand side I using Hölder's and Cauchy's inequalities. We also will use the estimates

$$|\nabla c_2(\mathbf{x},t)| \leqslant N(h), \quad |(\mathbf{u}_1)_{(h)}(\mathbf{x},t)| \leqslant N(h), \quad \mathbf{x} \in \Omega^{\varepsilon}_f, \quad 0 < t < T,$$
$$\int_{\Omega^{\varepsilon}_f} |\tilde{\mathbf{u}}(\mathbf{x},t)|^2 dx \leqslant C_0 \int_{\Omega^{\varepsilon}_f} |\mathbf{u}_1(\mathbf{x},t) - \mathbf{u}_2(\mathbf{x},t)|^2 dx,$$

which are based on the properties of the mollifiers.

In fact,

$$I \leqslant \frac{1}{2D_0}\Big(\max_{\mathbf{x}\in\Omega^{\varepsilon}_f, 0<\tau<T} |(\mathbf{u}_1)_{(h)}(\mathbf{x},t)|^2\Big)\int_0^t \int_{\Omega^{\varepsilon}_f} |\tilde{c}(\mathbf{x},\tau)|^2 dx d\tau$$
$$+ N(h)\Big(\int_0^t \int_{\Omega^{\varepsilon}_f} |\tilde{c}(\mathbf{x},\tau)|^2 dx d\tau\Big)^{\frac{1}{2}}\Big(\int_0^t \int_{\Omega^{\varepsilon}_f} |\tilde{\mathbf{u}}(\mathbf{x},\tau)|^2 dx d\tau\Big)^{\frac{1}{2}}$$
$$+ \frac{D_0}{2}\int_0^t \int_{\Omega^{\varepsilon}_f} |\nabla \tilde{c}(\mathbf{x},\tau)|^2 dx d\tau,$$

and

$$\frac{1}{2}\int_{\Omega^{\varepsilon}_f} |\tilde{c}(\mathbf{x},t)|^2 dx + \frac{D_0}{2}\int_0^t \int_{\Omega^{\varepsilon}_f} |\nabla \tilde{c}\,(\mathbf{x},\tau)|^2 dx d\tau$$
$$\leqslant N(h)\Big(\int_0^t \int_{\Omega^{\varepsilon}_f} |\tilde{c}(\mathbf{x},\tau)|^2 dx d\tau + \int_0^t \int_{\Omega^{\varepsilon}_f} |\tilde{\mathbf{u}}(\mathbf{x},\tau)|^2 dx d\tau\Big).$$

Now, to prove the solvability of (10.1.24)–(10.1.30) we just apply Schauder's fixed point theorem [55].

Indeed, the estimates (10.1.38), (10.1.42), (10.1.43) and the interpolation inequality

$$\max_{\Omega^{\varepsilon}_f \times (0,T)} |v(\mathbf{x},t)|^2 \leqslant 2\Big(\int_0^T \int_{\Omega^{\varepsilon}_f} |v|^2 dx dt\Big)^{\frac{1}{2}}\Big(\int_0^T \int_{\Omega^{\varepsilon}_f} \Big|\frac{\partial v}{\partial t}\Big|^2 dx dt\Big)^{\frac{1}{2}}$$
$$\leqslant C\Big(\int_0^T \int_{\Omega^{\varepsilon}_f} |v|^2 dx dt\Big)^{\frac{1}{2}} \max_{\Omega^{\varepsilon}_f \times (0,T)} \Big|\frac{\partial v}{\partial t}(\mathbf{x},t)\Big|$$

for any smooth function v, such that $v(\mathbf{x},0) = 0$, prove the continuity of $\mathbb{A}$ if we take into account the equality

$$(c_1 - c_2)(\mathbf{x},0) = 0 \text{ for } c_1 = \mathbb{A}_2(\mathbf{u}_1), \ c_2 = \mathbb{A}_2(\mathbf{u}_2).$$

The estimate (10.1.42) shows that $\mathbb{A}$ is a compact operator. Therefore $\mathbb{A}$ is a completely continuous operator on the set $\mathfrak{M}$. Next, the estimate (10.1.40) shows that $\mathbb{A}$ transforms the set $\mathfrak{M}$ into itself. Finally, $\mathfrak{M}$ is a closed convex set, which is enough for existence at least one fixed point of $\mathbb{A}$ in $\mathfrak{M}$.

It is clear that all of the fixed points of $\mathbb{A}$ preserve estimates (10.1.37), (10.1.40), and (10.1.41). Thus, the following lemma holds true.

Lemma 10.3 *There exists at least one weak solution* $\{\mathbf{v}^{\varepsilon,h},\ p^{\varepsilon,h},\ c^{\varepsilon,h}\}$ *of the problem (10.1.24)–(10.1.30), such that*

$$\int_0^T \int_{\Omega_f^\varepsilon} \left(\varepsilon^2 |\nabla \mathbf{v}^{\varepsilon,h}|^2 + |\mathbf{v}^{\varepsilon,h}|^2 + \left| \frac{\partial p^\varepsilon}{\partial t} \right|^2 + (\nabla \cdot \mathbf{v}^{\varepsilon,h})^2 \right) dx dt \leqslant C F^2, \tag{10.1.45}$$

$$0 \leqslant c^{\varepsilon,h}(\mathbf{x}, t) \leqslant 1, \ \mathbf{x} \in \Omega_f^\varepsilon, \ t > 0, \tag{10.1.46}$$

$$\int_0^T \int_{\Omega_f^\varepsilon} |\nabla c^{\varepsilon,h}|^2 dx dt \leqslant C F^2 \tag{10.1.47}$$

where C is independent of ε *and h.*

As the last step in the proof of Theorem 10.1 we pass to the limit as $h \to 0$ in the corresponding integral identity, namely the following lemma holds true.

Lemma 10.4 *There exists at least one weak solution* $\{\mathbf{v}^\varepsilon,\ p^\varepsilon,\ c^\varepsilon\}$ *of the problem (10.0.1)–(10.0.7), such that*

$$\int_0^T \int_{\Omega_f^\varepsilon} \left(\varepsilon^2 |\nabla \mathbf{v}^\varepsilon|^2 + |\mathbf{v}^\varepsilon|^2 + \left| \frac{\partial p^\varepsilon}{\partial t} \right|^2 + (\nabla \cdot \mathbf{v}^\varepsilon)^2 \right) dx dt \leqslant C F^2, \tag{10.1.48}$$

$$0 \leqslant c^\varepsilon(\mathbf{x}, t) \leqslant 1, \ \mathbf{x} \in \Omega_f^\varepsilon, \ t > 0, \tag{10.1.49}$$

$$\int_0^T \int_{\Omega_f^\varepsilon} |\nabla c^\varepsilon|^2 dx dt \leqslant C F^2 \tag{10.1.50}$$

where C is independent of ε.

Proof To prove the lemma we just have to find the convergent subsequences and pass to the limit as $h \searrow 0$ in the integral identities (10.1.3) and (10.1.4) corresponding to the dynamic equations for $\mathbf{v}^{\varepsilon,h}$ and $p^{\varepsilon,h}$, and in the integral identity, corresponding to the diffusion-convection equation, which we rewrite as

$$\int_0^T \int_{\Omega_f^\varepsilon} \left(c^{\varepsilon,h} \frac{\partial \xi}{\partial t} + (c^{\varepsilon,h} \nabla \xi)_{(h)} \cdot \mathbf{v}^{\varepsilon,h} - D_0 \nabla c^{\varepsilon,h} \cdot \nabla \xi + (\xi c^{\varepsilon,h})_{(h)} \nabla \cdot \mathbf{v}^{\varepsilon,h} \right) dx dt$$
$$+ \int_\Omega \chi^\varepsilon c_0^h(\mathbf{x}) \xi(\mathbf{x}, 0) dx = 0, \qquad (10.1.51)$$

for any smooth functions ξ, such that $\xi(\mathbf{x}, T) = 0$.

Here we have used the well-known property of the mollifiers that

$$\int_Q u(\mathbf{x}) \, (v)_{(h)}(\mathbf{x}) dx = \int_Q (u)_{(h)}(\mathbf{x}) \, v(\mathbf{x}) dx.$$

The weak compactness of $\{p^{\varepsilon,h}\}$ and $\{\nabla \cdot \mathbf{v}^{\varepsilon,h}\}$ in $L_2\big((0,T); L_2(\Omega_f^\varepsilon)\big)$ and the weak compactness of $\{c^{\varepsilon,h}\}$ and $\{\mathbf{v}^{\varepsilon,h}\}$ in $L_2\big((0,T); W_2^1(\Omega_f^\varepsilon)\big)$ and $L_2\big((0,T); \overset{\circ}{W}{}_2^1(\Omega_f^\varepsilon)\big)$ correspondingly follow from the estimates (10.1.45)–(10.1.47). The strong compactness of $\{c^{\varepsilon,h}\}$ in $L_2\big((0,T); L_2(\Omega_f^\varepsilon)\big)$ follows from the same estimates and from Aubin's compactness lemma [12, 68].

The limit in (10.1.3), (10.1.4) and (10.1.51) does not cause problems. We just note that the product $(c^{\varepsilon,h} \nabla \xi)_{(h)} \cdot \mathbf{v}^{\varepsilon,h}$ converges to $c^\varepsilon \nabla \xi \cdot \mathbf{v}^\varepsilon$ and the product $(\xi \, c^{\varepsilon,h})_{(h)} \nabla \cdot \mathbf{v}^{\varepsilon,h}$ converges to $\xi \, c^\varepsilon \, \nabla \cdot \mathbf{v}^\varepsilon$ due to the strong convergence of $\{c^{\varepsilon,h}\}$ in $L_2\big((0,T); L_2(\Omega_f^\varepsilon)\big)$.

10.1.3 Proof of Theorem 10.2

The main problem here is the strong compactness of $\{\tilde{c}^\varepsilon\}$ in $L_2(\Omega_T)$. This follows from the estimates (10.1.6), (10.1.7), the diffusion-convection equation (10.1.3), the compactness lemma (see [9, 84]), and the properties of the corresponding extensions (see also Appendix B).

The boundedness and the weak compactness in $\mathbf{L}_2(\Omega_T)$ of $\{\tilde{\mathbf{v}}^\varepsilon\}$ follow from the estimates (10.1.8).

Let

$$q^\varepsilon = p^\varepsilon - \nu_0 \nabla \cdot \mathbf{v}^\varepsilon = p^\varepsilon + \frac{\nu_0}{c_f^2} \frac{\partial p^\varepsilon}{\partial t}, \qquad (10.1.52)$$

and $\tilde{q}^\varepsilon$ be an extension of q^ε from Ω_f^ε onto Ω: $\tilde{q}^\varepsilon = 0$ outside of Ω_f^ε for $t > 0$.

The weak compactness of $\{\tilde{p}^\varepsilon\}$, $\{\tilde{q}^\varepsilon\}$, and $\{\nabla \cdot \tilde{\mathbf{v}}^\varepsilon\}$ in $L_2(\Omega_T)$ follow from the estimates (10.1.8) and the properties of the corresponding extensions.

Using (10.1.52) and the extensions $\tilde{p}^\varepsilon$, $\tilde{q}^\varepsilon$, and $\tilde{\mathbf{v}}^\varepsilon$ of the functions p^ε, q^ε, and $\mathbf{v}^\varepsilon$, we rewrite the integral identities (10.1.3) and (10.1.4) as

$$\int_0^T \int_\Omega \chi^\varepsilon \Big(\tilde{c}^\varepsilon \frac{\partial \xi}{\partial t} + (\tilde{c}^\varepsilon \tilde{\mathbf{v}}^\varepsilon - D_0 \nabla \tilde{c}^\varepsilon) \cdot \nabla \xi + \xi \, \tilde{c}^\varepsilon \nabla \cdot \tilde{\mathbf{v}}^\varepsilon \Big) dx dt$$
$$= - \int_\Omega \tilde{c}_0(\mathbf{x}) \, \xi(\mathbf{x}, 0) \, dx, \tag{10.1.53}$$

$$\int_0^T \int_\Omega \chi^\varepsilon \big(\varepsilon^2 \mu_1(\tilde{c}^\varepsilon) \, \mathbb{D}(x, \tilde{\mathbf{v}}^\varepsilon) - \tilde{q}^\varepsilon \mathbb{I} \big) : \mathbb{D}(x, \varphi) dx dt$$
$$= \int_0^T \int_\Omega (\rho_f + \delta \, \tilde{c}^\varepsilon) \mathbf{F} \cdot \varphi \big) \, dx dt. \tag{10.1.54}$$

Here

$$\tilde{c}_0 = c_0 \text{ in } \Omega_f^\varepsilon, \; \tilde{c}_0 = 0 \text{ in } \Omega_s^\varepsilon, \; \chi^\varepsilon \tilde{\mathbf{v}}^\varepsilon = \tilde{\mathbf{v}}^\varepsilon, \; \chi^\varepsilon \nabla \cdot \tilde{\mathbf{v}}^\varepsilon = \nabla \cdot \tilde{\mathbf{v}}^\varepsilon.$$

The homogenization of the dynamic equations repeats the similar result in Chap. 1. In fact, the weak limit in the relation (10.1.52) and in the continuity equation (10.1.5) result in Eqs. (10.1.13), (10.1.14), the boundary condition (10.1.16), and the first initial condition in (10.1.18).

If $P(\mathbf{x}, t, \mathbf{y})$ and $Q(\mathbf{x}, t, \mathbf{y})$ are two-scale limits of $\{\tilde{p}^\varepsilon\}$ and $\{\tilde{q}^\varepsilon\}$ respectively, then

$$Q = P + \frac{\nu_0}{c_f^2} \frac{\partial P}{\partial t}, \quad P(\mathbf{x}, t, \mathbf{y}) = \frac{1}{m} p(\mathbf{x}, t) \chi(\mathbf{y}).$$

Finally, let $\mathbf{v}(\mathbf{x}, t, \mathbf{y})$ be the two-scale limit of $\{\tilde{\mathbf{v}}^\varepsilon\}$. Then

$$\frac{1}{2} \mu_1(c) \triangle_y \mathbf{v} - \nabla_y Q - \frac{1}{m} \nabla q + (\rho_f + \delta \, c) \mathbf{F} = 0, \quad \mathbf{y} \in Y_f, \tag{10.1.55}$$

$$\nabla \cdot \mathbf{v} = 0, \quad \mathbf{y} \in Y_f. \tag{10.1.56}$$

$$\mathbf{v} = 0, \quad \mathbf{y} \in \gamma. \tag{10.1.57}$$

We look for the solution of the problem (10.1.55)–(10.1.57) in the form

$$\mathbf{v} = \frac{2}{\mu_1(c)} \Big(\sum_{i=1}^{3} \mathbf{v}^{(i)} \otimes \mathbf{e}_i \Big) \cdot \Big(-\frac{1}{m} \nabla q + (\rho_f + \delta \, c) \mathbf{F} \Big),$$

where $\mathbf{e}_1, \; \mathbf{e}_2, \; \mathbf{e}_3$ is a standard Cartesian basis.

Then

$$\mathbb{B} = \sum_{i=1}^{3}\left(\int_{Y_f} \mathbf{v}^{(i)}(\mathbf{y})dy\right)\otimes \mathbf{e}_i = \sum_{i=1}^{3}\langle \mathbf{v}^{(i)}\rangle_{Y_f}\otimes \mathbf{e}_i, \qquad (10.1.58)$$

where $\mathbf{v}^{(i)}$ are solutions to the periodic boundary—value problem (1.1.26).

The homogenization of the diffusion-convection equation for c^{ε} is also standard. In fact, due to the smoothness of the solution

$$\tilde{c}^{\varepsilon}(\mathbf{x},t) \rightharpoonup c(\mathbf{x},t) \text{ weakly in } W_2^{1,0}(\Omega_T),$$

$$\tilde{c}^{\varepsilon}(\mathbf{x},t) \to c(\mathbf{x},t) \text{ two-scale in } L_2(\Omega_T),$$

$$\nabla \tilde{c}^{\varepsilon} \to \nabla c + \nabla_y C \text{ two-scale in } \mathbf{L}L_2(\Omega_T),$$

The two-scale limit in (10.1.53) with the test functions $\xi = \xi(\mathbf{x},t)$ and $\xi = \varepsilon\xi_0\left(\mathbf{x},\frac{\mathbf{x}}{\varepsilon},t\right)$, where $\xi_0(\mathbf{x},t,\mathbf{y})$ is an arbitrary 1-periodic in $\mathbf{y}$ function, results in

$$m\frac{\partial c}{\partial t} + \mathbf{v}\cdot\nabla c = D_0\nabla\cdot\left(m\nabla c + \langle\nabla_y C\rangle_{Y_f}\right), \qquad (10.1.59)$$

and

$$\nabla\cdot\left(\chi(\mathbf{y})(\nabla c + \nabla_y C)\right) = 0, \quad \mathbf{y}\in Y. \qquad (10.1.60)$$

As usual, we look for the solution of (10.1.60) in the form

$$C(\mathbf{x},t,\mathbf{y}) = \sum_{i=1}^{3} C^{(i)}(\mathbf{y})\,\frac{\partial c}{\partial x_i}(\mathbf{x},t).$$

Then,

$$\mathbb{B}^{(c)} = m\mathbb{I} + \left(\sum_{i=1}^{3}\langle\nabla_y C^{(i)}(\mathbf{y})\rangle_{Y_f}\otimes\mathbf{e}_i\right), \qquad (10.1.61)$$

where

$$\nabla\cdot\left(\chi(\mathbf{y})(\mathbf{e}_i + \nabla_y C^{(i)})\right) = 0, \quad \mathbf{y}\in Y.$$

10.1.4 Proof of Theorem 10.3

Let

$$\mathbf{w}(\mathbf{x},t) = \int_0^t \mathbf{v}(\mathbf{x},\tau)d\tau. \qquad (10.1.62)$$

Then the system (10.1.12)–(10.1.14) takes the form

$$m\mu_1(c)\big(\mathbb{B}^{(f)}\big)^{-1}\cdot\mathbf{v}=\nabla\big(c_f^2\,(\nabla\cdot\mathbf{w})+\nu_0(\nabla\cdot\mathbf{v})\big)+m(\rho_f+\delta\,c)\mathbf{F}. \quad (10.1.63)$$

Multiplication of (10.1.63) by $\mathbf{v}$ and integration by parts over Ω result in the energy equality

$$\int_\Omega\Big(m\,\mu_1(c)\mathbf{v}\cdot\big(\mathbb{B}^{(f)}\big)^{-1}\cdot\mathbf{v}+\nu_0(\nabla\cdot\mathbf{v})^2-m(\rho_f+\delta\,c)\mathbf{F}\cdot\mathbf{v}\Big)dx$$
$$+\frac{c_f^2}{2}\frac{d}{dt}\int_\Omega(\nabla\cdot\mathbf{w})^2dx=0, \quad (10.1.64)$$

and, consequently, the a priori estimate

$$\int_{\Omega_T}\big(|\mathbf{v}|^2+\nu_0(\nabla\cdot\mathbf{v})^2\big)dxdt+c_f^2\max_{0<t<T}\int_\Omega(\nabla\cdot\mathbf{w})^2dx\leqslant C\,F^2, \quad (10.1.65)$$

where C is independent of c_f^2 and ν_0.

Coming back to (10.1.12) and using (10.1.65) we get

$$\int_{\Omega_T}|\nabla\,q|^2dxdt\leqslant C\,F^2. \quad (10.1.66)$$

Equations (10.1.13), (10.1.14) and the boundary condition (10.1.16) provide the equality

$$\int_\Omega q(\mathbf{x},t)dx=0.$$

Therefore,

$$\int_{\Omega_T}|q|^2dxdt\leqslant C\,F^2 \quad (10.1.67)$$

(see [61]). The combination of (10.1.65) and (10.1.67) gives us

$$\int_{\Omega_T}|p|^2dxdt\leqslant C\,FC\,F^2. \quad (10.1.68)$$

Finally, for the concentration c the estimates (10.1.6) and (10.1.7) with the constant C independent of c_f^2 and ν_0 hold true.

Now we are ready to pass to the limit as $k=c_f^2\to\infty$. On the basis of the estimates (10.1.6), (10.1.7), (10.1.65)–(10.1.68) we may choose some subsequences $\{\mathbf{v}^{(k_n)}\}$, $\{q^{(k_n)}\}$, and $\{c^{(k_n)}\}$ such that the sequence $\{\mathbf{v}^{(k_n)}\}$ converges weakly in $\mathbf{L}_2(\Omega_T)$ to the function $\mathbf{v}^{(\infty)}$, the sequence $\{q^{(k_n)}\}$ converges weakly in $L_2(\Omega_T)$ to the function $p^{(\infty)}$, the sequence $\{c^{(k_n)}\}$ converges weakly in $W_2^{1,0}(\Omega_T)$ and strongly in $L_2(\Omega_T)$

to the function $c^{(\infty)}$, the sequence $\{\nabla \cdot \mathbf{v}^{(k_n)}\}$ converges weakly in $L_2(\Omega_T)$ to the function $\nabla \cdot \mathbf{v}^{(\infty)}$, and the sequence $\{\nabla \cdot \mathbf{w}^{(k_n)}\}$ converges strongly in $L_2(\Omega_T)$ to zero.

It is clear that the relation $\nabla \cdot \mathbf{w}^{(\infty)} = 0$ and the relation (10.1.62) for $\mathbf{w}^{(\infty)}$ and $\mathbf{v}^{(\infty)}$ imply the continuity equation $\nabla \cdot \mathbf{v}^{(\infty)} = 0$, and that the concentration $c^{(\infty)}$ satisfies the diffusion-convection equation (10.1.15) with the velocity field $\{\mathbf{v}^{(\infty)}\}$.

To prove Darcy's law (10.1.19) it is enough to fulfill the limiting procedure as $k_n \to \infty$ in the integral identity

$$\int_{\Omega_T} \Big(\mu_1(c^{(k_n)})\varphi \cdot \big(\mathbb{B}^{(f)}\big)^{-1} \cdot \mathbf{v}^{(k_n)} - \frac{1}{m} q^{(k_n)}(\nabla \cdot \varphi) - \rho_f \mathbf{F} \cdot \varphi\Big) dx dt = 0.$$

10.1.5 Proof of Theorem 10.4

The proof of Theorem 10.4 repeats the proof of Theorem 10.3 with evident changes. Note, that due to (10.1.65) $\nu_0 \nabla \cdot \mathbf{v}^{(\delta)} \to 0$ as $\delta \to 0$ strongly in $L_2\big((0,T); L_2(\Omega)\big)$.

10.2 Diffusion-Convection in Poroelastic Media

10.2.1 Statement of the Problem and Main Results

Throughout this section we suppose that $\dfrac{1}{\varepsilon}$ is an integer, Ω is a cube whose edge length is also an integer, and

$$c_0 \in L_2(\Omega), \quad 0 \leqslant c_0(\mathbf{x}) \leqslant 1, \tag{10.2.1}$$

$$\mathbf{F} = \mathbf{F}(\mathbf{x}), \quad \sup_{\mathbf{x} \in \Omega} \big(|\mathbf{F}(\mathbf{x})| + |\nabla \mathbf{F}(\mathbf{x})|\big) = F < \infty, \tag{10.2.2}$$

Definition 10.2 We say that the triple of functions $\{\mathbf{w}^{\varepsilon}, p^{\varepsilon}, c^{\varepsilon}\}$ is *a weak solution* of the problem (10.0.5), (10.0.6), (10.0.8)–(10.0.13), if

$$\mathbf{w}^{\varepsilon} \in \overset{\circ}{\mathbf{W}}{}^{1,0}_2(\Omega_T), \quad \frac{\partial \mathbf{w}^{\varepsilon}}{\partial t} \in \mathbf{L}_2\big((0,T); \mathbf{W}^1_2(\Omega^{\varepsilon}_f)\big),$$

$$c^{\varepsilon} \in L_2\big((0,T); W^1_2(\Omega^{\varepsilon}_f)\big), \quad p^{\varepsilon} \in L_2(\Omega_T),$$

and continuity equation (10.0.10) in Ω for $t > 0$, the normalization condition (10.0.11), the initial condition (10.0.13), and the integral identities

$$\int_0^T \int_{\Omega_f^\varepsilon} \left(-c^\varepsilon \frac{\partial \xi}{\partial t} + \xi \frac{\partial \mathbf{w}^\varepsilon}{\partial t} \cdot \nabla c^\varepsilon + D_0 \nabla c^\varepsilon \cdot \nabla \xi \right) dxdt$$

$$= \int_{\Omega_f^\varepsilon} c_0(\mathbf{x})\, \xi(\mathbf{x}, 0)\, dx, \tag{10.2.3}$$

$$\int_0^T \int_\Omega \left(\chi^\varepsilon(\mathbf{x})\mu_0\, \mathbb{D}\left(x, \frac{\partial \mathbf{w}^\varepsilon}{\partial t}\right) + \left(1 - \chi^\varepsilon(\mathbf{x})\right)\lambda_0\, \mathbb{D}(x, \mathbf{w}^\varepsilon) - p^\varepsilon \mathbb{I} \right) : \mathbb{D}(x, \varphi) \tag{10.2.4}$$

$$dxdt = \int_0^T \int_\Omega \rho^\varepsilon \mathbf{F} \cdot \varphi\, dxdt, \tag{10.2.4}$$

hold true the for any smooth functions ξ and φ, such that $\xi(\mathbf{x}, T) = 0$ and $\varphi(\mathbf{x}, t) = 0$ for $\mathbf{x} \in S$ and $t > 0$.

The integral identity (10.2.4) shows, that $\mathbf{w}^\varepsilon$ possesses different smoothness in domains Ω_f^ε and Ω_s^ε. To preserve the best properties, which the solution has in the liquid part, we will use extension results [1, 36, 89] (see also Appendix B): there exists a linear extension operator

$$\mathbb{E}_{\Omega_f^\varepsilon} : \mathbf{W}_2^1(\Omega_f^\varepsilon) \to \mathbf{W}_2^1(\Omega), \quad \mathbf{v}^\varepsilon = \mathbb{E}_{\Omega_f^\varepsilon}\left(\frac{\partial \mathbf{w}^\varepsilon}{\partial t}\right),$$

such that

$$\frac{\partial \mathbf{w}^\varepsilon}{\partial t}(\mathbf{x}, t) = \mathbf{v}^\varepsilon(\mathbf{x}, t), \quad \mathbf{x} \in \Omega_f^\varepsilon, \ t \in (0, T), \tag{10.2.5}$$

$$\int_\Omega |\mathbf{v}^\varepsilon(\mathbf{x}, t)|^2 dx \leqslant C_0 \int_{\Omega_f^\varepsilon} \left|\frac{\partial \mathbf{w}^\varepsilon}{\partial t}(\mathbf{x}, t)\right|^2 dx,$$

$$\int_\Omega \left|\mathbb{D}\left(x, \mathbf{v}^\varepsilon(\mathbf{x}, t)\right)\right|^2 dx \leqslant C_0 \int_{\Omega_f^\varepsilon} \left|\mathbb{D}\left(x, \frac{\partial \mathbf{w}^\varepsilon}{\partial t}(\mathbf{x}, t)\right)\right|^2 dx, t \in (0, T), \tag{10.2.6}$$

where C_0 is independent of ε and $t \in (0, T)$.

We additionally suppose that the geometry of the elementary cell Y_f permits us to choose an extension operator such that the function $\mathbf{v}^\varepsilon$ vanishes at the boundary $S = \partial\Omega$ and we may apply the embedding [3]

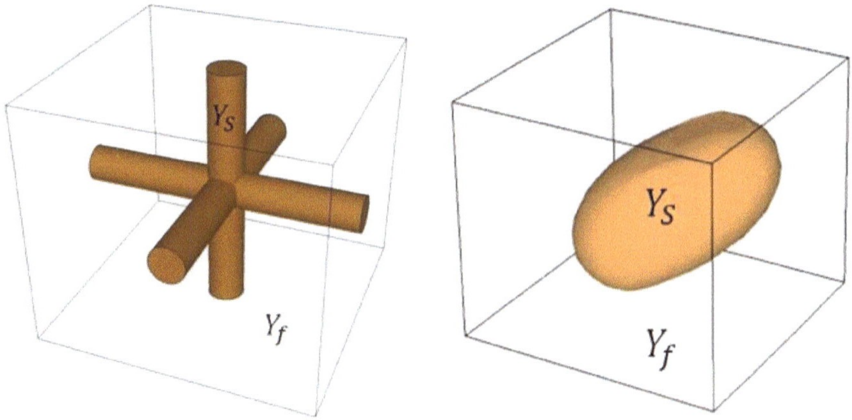

Fig. 10.1 Structures 1 and 2

$$\int_{\Omega} |\mathbf{v}^{\varepsilon}(\mathbf{x},t)|^2 dx \leqslant C_0 \int_{\Omega} |\nabla \mathbf{v}^{\varepsilon}(\mathbf{x},t)|^2 dx \\ \leqslant C_0^2 \int_{\Omega} |\mathbb{D}\big(x, \mathbf{v}^{\varepsilon}(\mathbf{x},t)\big)|^2 dx, \tag{10.2.7}$$

where we have used Korn's inequality for the domain Ω. For example, this is possible for the structures presented by Fig. 10.1 (see Lemma B.10).

Theorem 10.5 *Under the conditions (10.2.1) and (10.2.2) the problem (10.0.5), (10.0.6), (10.0.8)–(10.0.13) has at least one weak solution* $\{\mathbf{w}^{\varepsilon},\ p^{\varepsilon},\ c^{\varepsilon}\}$, *such that*

$$\max_{0<t<T} \int_{\Omega_f^{\varepsilon}} |c^{\varepsilon}(\mathbf{x},t)|^2 dx + \int_0^T \int_{\Omega_f^{\varepsilon}} |\nabla c^{\varepsilon}(\mathbf{x},t)|^2 dx dt \leqslant C_0 F^2, \tag{10.2.8}$$

$$0 \leqslant c^{\varepsilon}(\mathbf{x},t) \leqslant 1, \ \mathbf{x} \in \Omega_f^{\varepsilon},\ t>0, \tag{10.2.9}$$

$$\max_{0<t<T} \int_{\Omega} \big(|\mathbf{w}^{\varepsilon}(\mathbf{x},t)|^2 + |\nabla \mathbf{w}^{\varepsilon}(\mathbf{x},t)|^2\big) dx \\ + \int_0^T \int_{\Omega} \big(\mathbf{v}^{\varepsilon}(\mathbf{x},t)|^2 + |\nabla \mathbf{v}^{\varepsilon}(\mathbf{x},t)|^2 + |p^{\varepsilon}(\mathbf{x},t)|^2\big) dx dt \leqslant C_0 F^2, \tag{10.2.10}$$

$$\max_{0<t<T} \int_{\Omega} \big(|\mathbf{v}^{\varepsilon}(\mathbf{x},t)|^2 + |\nabla \mathbf{v}^{\varepsilon}(\mathbf{x},t)|^2\big) dx \leqslant C(C_0, F), \tag{10.2.11}$$

$$\int_{\Omega} |\nabla \mathbf{v}^{\varepsilon}(\mathbf{x},t_1) - \nabla \mathbf{v}^{\varepsilon}(\mathbf{x},t_2)|^2 dx \leqslant C(C_0, F)\,|t_1 - t_2|, \tag{10.2.12}$$

where $\mathbf{v}^{\varepsilon} = \mathbb{E}_{\Omega_f^{\varepsilon}}\left(\dfrac{\partial \mathbf{w}^{\varepsilon}}{\partial t}\right)$, *and the constant* C_0 *is independent of the small parameter* ε *and* $t \in (0, T)$.

Theorem 10.6 *Under the conditions of Theorem 10.5 the sequences* $\{\mathbf{w}^{\varepsilon}\}$, $\{\mathbf{v}^{\varepsilon}\}$, *and* $\{\tilde{c}^{\varepsilon}\}$ *converge weakly as* $\varepsilon \to 0$ *(up to some subsequences) in* $\overset{\circ}{\mathbf{W}}{}_2^{1,0}(\Omega_T)$ *and* $\overset{\circ}{W}{}_2^{1,0}(\Omega_T)$ *to the functions* $\mathbf{w}$, $\mathbf{v} = \dfrac{\partial \mathbf{w}}{\partial t}$, *and* c *respectively, and the sequence* $\{p^{\varepsilon}\}$ *converges weakly as* $\varepsilon \to 0$ *(up to some subsequences) in* $L_2(\Omega_T)$ *to the function* p.

These limiting functions satisfy in the domain Ω *for* $t > 0$ *the system of differential equations, consisting of the homogenized momentum balance equation*

$$\nabla \cdot \widehat{\mathbb{P}} + \hat{\rho}\,(c)\,\mathbf{F} = 0, \tag{10.2.13}$$

where

$$\widehat{\mathbb{P}} = -p\,\mathbb{I} + \mathfrak{N}_1 : \mathbb{D}(x, \mathbf{v}) + \mathfrak{N}_2 : \mathbb{D}(x, \mathbf{w}) + \int_0^t \mathfrak{N}_3(t-\tau) : \mathbb{D}(x, \mathbf{w}(\mathbf{x}, \tau))d\tau,$$

the continuity equation

$$\nabla \cdot \mathbf{v} = 0, \tag{10.2.14}$$

and the homogenized diffusion-convection equation

$$m\,\frac{\partial c}{\partial t} + \mathbf{v} \cdot \left(\mathbb{B}^{(c)} \nabla c\right) = D_0\, \nabla \cdot \left(\mathbb{B}^{(c)} \nabla c\right). \tag{10.2.15}$$

The differential equations (10.2.13)–(10.2.15) are completed with the boundary conditions

$$\mathbf{w} = 0, \tag{10.2.16}$$

$$\nabla c \cdot \mathbf{n} = 0, \tag{10.2.17}$$

on the outer boundary $S = \partial \Omega$ *for* $t > 0$, *the initial conditions*

$$\mathbf{w}(\mathbf{x}, 0) - 0, \tag{10.2.18}$$

$$c\,(\mathbf{x}, 0) = m\, c_0\,(\mathbf{x}) \tag{10.2.19}$$

in the domain Ω, *and the normalization condition*

$$\int_{\Omega} p\,(\mathbf{x}, t)dx = 0. \tag{10.2.20}$$

In Eqs. (10.2.13)–(10.2.19) $\mathbf{n}$ *is the unit outward normal vector to the boundary* S,

$$\hat{\rho}\,(c) = m\,(\rho_f + \delta\,c) + (1 - m)\,\rho_s, \quad m = \int_Y \chi(\mathbf{y})dy,$$

the fourth-rank constant tensors $\mathfrak{N}_1$, $\mathfrak{N}_2$, *and the fourth-rank tensor* $\mathfrak{N}_3(t)$ *have been defined in the Chap. 1 by formulae (1.4.30), where the tensor* $\mathfrak{N}_1$ *is symmetric and strictly positively definite, and the symmetric and strictly positively definite constant matrix* $\mathbb{B}^{(c)}$ *has been defined in Theorem 10.2 by formula (10.1.61).*

If the domain Y_f *is symmetric under rotations by the angle* $\dfrac{\pi}{2}$ *around the principal axes of a Cartesian coordinate system, then the matrix* $\mathbb{B}^{(c)}$ *will be diagonal.*

We refer to the problem (10.2.13)–(10.2.19) as the homogenized **model** ($\mathbb{DCPEM}$).

10.2.2 Proof of Theorem 10.5

The proof is based on Schauder's fixed point theorem [55]. Let us divide this proof into several steps.

First, we consider the auxiliary problem, consisting of the dynamic equations

$$\nabla\cdot\Bigg(\chi^\varepsilon\,\mu_0\,\mathbb{D}\Bigg(x,\frac{\partial\mathbf{w}^{(h)}}{\partial t}\Bigg) + (1-\chi^\varepsilon)\,\lambda_0\,\mathbb{D}\left(x,\mathbf{w}^{(h)}\right) - p^{(h)}\,\mathbb{I}\Bigg)$$
$$= -\rho_h(x,c^{(h)})\mathbf{F}, \tag{10.2.21}$$

$$\nabla\cdot\mathbf{w}^{(h)} = 0 \tag{10.2.22}$$

in the domain Ω for $t > 0$, the modified diffusion-convection equation

$$\frac{\partial c^{(h)}}{\partial t} + \widehat{\mathbf{v}^{(h)}}\cdot\nabla\,c^{(h)} = D_0\,\triangle\,c^{(h)} \tag{10.2.23}$$

in the domain Ω_f^ε for $t > 0$, the boundary condition

$$\mathbf{w}^{(h)}(\mathbf{x},t) = 0 \tag{10.2.24}$$

on the boundary $S = \partial\Omega$ for $t > 0$, the boundary condition

$$\nabla\,c^{(h)}(\mathbf{x},t)\cdot\mathbf{n}(\mathbf{x}) = 0 \tag{10.2.25}$$

on the boundary $\partial\Omega_f^\varepsilon$ for $t > 0$, the normalization condition

$$\int_\Omega p^{(h)}\,(\mathbf{x},t)dx = 0, \tag{10.2.26}$$

and the initial conditions

$$\chi^{\varepsilon}(\mathbf{x})\mathbf{w}^{(h)}(\mathbf{x},0)=0, \qquad (10.2.27)$$

$$c^{(h)}(\mathbf{x},0)=c_0(\mathbf{x}). \qquad (10.2.28)$$

In Eqs. (10.2.21)–(10.2.28) $\mathbf{n}$ is the unit outward normal vector to the boundary $\partial\Omega^{\varepsilon}$,

$$\rho_h(x,c^{(h)})=\chi^{\varepsilon}\left(\rho_f+\delta\,(c^{(h)})_h\right)+(1-\chi^{\varepsilon})\rho_s,$$

and

$$(c^{(h)})_h(\mathbf{x},t)=\frac{1}{h}\int_t^{t+h}c^{(h)}(\mathbf{x},\tau)\,d\tau,$$

$$\widehat{\mathbf{v}^{(h)}}(\mathbf{x},t)=\frac{1}{h^4}\int_t^{t+h}\int_{\mathbb{R}^3}\eta\left(\frac{|\mathbf{x}-\mathbf{z}|}{h}\right)\mathbf{v}^{\varepsilon}(\mathbf{z},\tau)\,dz$$

are mollifiers [3, 61] with the infinitely smooth, nonnegative, even, and finite in $(-1,\,1)$ function $\eta(x)$, such that

$$\int_{-1}^{1}\eta(x)dx=1,$$

and $\mathbf{v}^{\varepsilon}=\mathbb{E}_{\Omega_f^{\varepsilon}}\left(\dfrac{\partial\mathbf{w}^{\varepsilon}}{\partial t}\right)$.

To solve (10.2.21)–(10.2.28) we choose the set

$$\mathfrak{M}=\left\{\bar{c}\in L_2\big((0,T);L_2(\Omega_f^{\varepsilon})\big):\ \int_0^T\int_{\Omega_f^{\varepsilon}}|\bar{c}(\mathbf{x},t)|^2dxdt\leqslant T\,|\Omega_f^{\varepsilon}|\right\},$$

where $|\Omega_f^{\varepsilon}|$ ia the Lebesgue measure of the set Ω_f^{ε}, and for $\bar{c}\in\mathfrak{M}$ consider the second auxiliary problem, consisting of the dynamic equations

$$\nabla\cdot\left(\chi^{\varepsilon}\mu_0\,\mathbb{D}\left(x,\frac{\partial\mathbf{w}}{\partial t}\right)+(1-\chi^{\varepsilon})\lambda_0\,\mathbb{D}(x,\mathbf{w})-p\,\mathbb{I}\right)+\rho_h(x,\bar{c})\mathbf{F}=0, \qquad (10.2.29)$$

$$\nabla\cdot\mathbf{w}=0 \qquad (10.2.30)$$

in the domain Ω for $t>0$, where

$$\rho_h(x,\bar{c})=\chi^{\varepsilon}\left(\rho_f+\delta\,(\bar{c})_h\right)+\chi_s^{\varepsilon}\rho_s,$$

completed with the normalization condition (10.2.26), and the boundary and initial conditions (10.2.24) and (10.2.27).

For all $\bar{c} \in \mathfrak{M}$ this problem defines the linear operators

$$\frac{\partial \mathbf{w}}{\partial t} = \mathbb{A}_0(\bar{c}), \ \mathbf{v} = \mathbb{A}_1(\bar{c}) = \big(\mathbb{E}_{\Omega_f^\varepsilon} \circ \mathbb{A}_0\big)(\bar{c}),$$
$$\mathbb{A}_1 : \mathfrak{M} \to \mathbf{W}_2^{1,0}(\Omega_T).$$

Lemma 10.5 *Under the conditions of Theorem 10.5 for any $\tilde{c} \in \mathfrak{M}$ the problem (10.2.24), (10.2.26), (10.2.27), (10.2.29), and (10.2.30) has a unique weak solution $\{\mathbf{w}, \ p\}$, such that*

$$\max_{0<t<T} \int_\Omega \big(|\mathbf{w}(\boldsymbol{x}, t)|^2 + |\nabla \, \mathbf{w}(\boldsymbol{x}, t)|^2\big) dx + \int_0^T \int_\Omega \big(|\mathbf{v}(\boldsymbol{x}, t)|^2 + |\nabla \, \mathbf{v}(\boldsymbol{x}, t)|^2 + |p(\boldsymbol{x}, t)|^2\big) dx dt \leqslant C_0 \, F^2 \left(\int_0^T \int_{\Omega_f^\varepsilon} |\bar{c}|^2 dx dt + 1\right), \tag{10.2.31}$$

where C_0 is independent of ε and h.

Proof The solvability of the problem (10.2.24), (10.2.26), (10.2.27), (10.2.29) and (10.2.30) is quite standard and we just show, how to derive the basic a priori estimates.

We put $\varphi = \dfrac{\partial \mathbf{w}}{\partial t}$ in the integral identity

$$\int_0^t \int_\Omega \left(\chi^\varepsilon \mu_0 \, \mathbb{D}\left(x, \frac{\partial \mathbf{w}}{\partial t}\right) + (1 - \chi^\varepsilon)\lambda_0 \, \mathbb{D}(x, \mathbf{w}) - p \, \mathbb{I}\right) : \mathbb{D}(x, \varphi) dx d\tau = \int_0^t \int_\Omega \rho_h(x, \bar{c})\mathbf{F} \cdot \varphi dx d\tau, \tag{10.2.32}$$

corresponding to (10.2.29), and get

$$\mu_0 \int_0^t \int_{\Omega_f^\varepsilon} \left|\mathbb{D}\left(x, \frac{\partial \mathbf{w}}{\partial \tau}(\mathbf{x}, \tau)\right)\right|^2 dx d\tau + \frac{\lambda_0}{2} \int_{\Omega_s^\varepsilon} \big|\mathbb{D}\big(x, \mathbf{w}(\mathbf{x}, t)\big)\big|^2 dx \equiv I = \int_0^t \int_\Omega \left(\rho_h(x, \bar{c})\mathbf{F} \cdot \frac{\partial \mathbf{w}}{\partial \tau}\right) (\mathbf{x}, \tau) dx d\tau. \tag{10.2.33}$$

We estimate from below the left-hand side of (10.2.33) using (10.2.6), the evident inequality

$$\int_\Omega \chi^\varepsilon |\mathbb{D}\big(x, \mathbf{w}(\mathbf{x}, t)\big)|^2 dx \leqslant C_0 \int_0^t \int_\Omega \chi^\varepsilon \left|\mathbb{D}\left(x, \frac{\partial \mathbf{w}}{\partial \tau}(\mathbf{x}, \tau)\right)\right|^2 dx d\tau,$$

and Korn's inequality, as

$$
\frac{\mu_0}{2\,C_0^2}\int_0^t\int_\Omega |\nabla\,\mathbf{v}(\mathbf{x},\tau)|^2 dx d\tau + \min\left(\frac{\mu_0}{2\,C_0^2},\frac{\lambda_0}{2\,C_0}\right)\int_\Omega |\nabla\,\mathbf{w}(\mathbf{x},t)|^2 dx
$$

$$
\leqslant \frac{\mu_0}{2\,C_0}\int_0^t\int_\Omega |\mathbb{D}(x,\mathbf{v}(\mathbf{x},\tau))|^2 dx d\tau + \min\left(\frac{\mu_0}{2\,C_0},\frac{\lambda_0}{2}\right)\int_\Omega |\mathbb{D}(x,\mathbf{w}(\mathbf{x},t))|^2 dx \leqslant I,
$$

where $\mathbf{v} = \mathbb{E}_{\Omega_f^\varepsilon}\left(\dfrac{\partial \mathbf{w}}{\partial t}\right)$.

Now we estimate from above the right-hand side of (10.2.33) using (10.2.7), the integration by parts, Hölder's, and Cauchy's inequalities with parameter β, and the imbedding theorem for the function $\mathbf{w}$ in Ω [3]:

$$
I = \int_0^t\int_{\Omega_f^\varepsilon} \rho_h(x,\bar{c})\mathbf{F}\cdot\mathbf{v} dx d\tau + \int_\Omega (1-\chi^\varepsilon)\rho_s \mathbf{F}\cdot\mathbf{w} dx
$$

$$
\leqslant \beta\int_0^t\int_\Omega |\mathbf{v}(\mathbf{x},\tau)|^2 dx d\tau + \beta\int_\Omega |\mathbf{w}(\mathbf{x},t)|^2 dx + \frac{1}{\beta}\,C_0 F^2\left(\int_0^T\int_{\Omega_f^\varepsilon} |\bar{c}|^2 dx dt + 1\right)
$$

$$
\leqslant \beta\,C_0\int_0^t\int_\Omega |\nabla\,\mathbf{v}(\mathbf{x},\tau)|^2 dx d\tau + \beta\,C_0\int_\Omega |\nabla\,\mathbf{w}(\mathbf{x},t)|^2 dx
$$

$$
+ \frac{1}{\beta}\,C_0 F^2\left(\int_0^T\int_{\Omega_f^\varepsilon} |\bar{c}|^2 dx dt + 1\right).
$$

Gathering all together we have

$$
\frac{\mu_0}{2\,C_0^2}\int_0^t\int_\Omega |\nabla\,\boldsymbol{v}(\boldsymbol{x},\tau)|^2 dx d\tau + \min\left(\frac{\mu_0}{2\,C_0^2},\frac{\lambda_0}{2\,C_0}\right)\int_\Omega |\nabla\,\boldsymbol{w}(\boldsymbol{x},t)|^2 dx
$$

$$
\leqslant \beta\,C_0\int_0^t\int_\Omega |\nabla\,\mathbf{v}(\mathbf{x},\tau)|^2 dx d\tau + \beta\,C_0\int_\Omega |\nabla\,\mathbf{w}(\mathbf{x},t)|^2 dx
$$

$$
+ \int_0^t\int_\Omega |\mathbf{w}(\mathbf{x},\tau)|^2 dx d\tau + \frac{1}{\beta}\,C_0 F^2\left(\int_0^T\int_{\Omega_f^\varepsilon} |\bar{c}|^2 dx dt + 1\right).
$$

The desired estimate (10.2.31) for the functions $\mathbf{w}$ and $\mathbf{v}$ follows now from the last inequality, if we put there

$$
\beta = \min\left(\frac{\mu_0}{4\,C_0^2},\frac{\lambda_0}{4\,C_0}\right)
$$

and use Gronwall's inequality.

The pressure p is estimated from the Eq. (10.2.29) as a linear bounded functional

$$\int_{\Omega} p\,\nabla\cdot\varphi dx = \int_{\Omega}\big(\mathbb{P}:\mathbb{D}(x,\varphi)-\rho_h(x,\bar{c})\mathbf{F}\cdot\varphi\big)dx,$$

defined for $\varphi\in\overset{\circ}{\mathbf{W}}{}_2^{1,0}(\Omega_T)$, with

$$\mathbb{P}=\chi^{\varepsilon}\mu_0\,\mathbb{D}\left(x,\frac{\partial\mathbf{w}}{\partial t}\right)+(1-\chi^{\varepsilon})\,\lambda_0\,\mathbb{D}(x,\mathbf{w})\in L_2\big((0,T);L_2(\Omega),$$

and satisfying the normalization condition (10.2.26).

Lemma 10.6 *Under the conditions of Theorem 10.5 $\mathbb{A}_1(\bar{c})$ is a continuous operator. If $\mathbf{v}_1=\mathbb{A}_1(\bar{c}_1)$, $\mathbf{v}_2=\mathbb{A}_1(\bar{c}_2)$, and $\tilde{\mathbf{v}}=\mathbf{v}_1-\mathbf{v}_2$, then*

$$\int_0^T\int_{\Omega}\big(|\tilde{\mathbf{v}}(\mathbf{x},t)|^2+|\nabla\,\tilde{\mathbf{v}}(\mathbf{x},t)|^2\big)dxdt$$
$$\leqslant C_0\,F^2\int_0^T\int_{\Omega_f^{\varepsilon}}|\bar{c}_1-\bar{c}_2|^2dxdt. \tag{10.2.34}$$

The statement of the lemma follows from the linearity of $\mathbb{A}_1$ and the estimate (10.2.31).

As the next step we consider the solutions of the differential equation

$$\frac{\partial c}{\partial t}+\widehat{\mathbf{v}}\cdot\nabla c=D_0\,\triangle c \tag{10.2.35}$$

in the domain Ω_f^{ε} for $t>0$, satisfying the boundary condition

$$\nabla c\cdot\mathbf{n}=0 \tag{10.2.36}$$

on the boundary $\partial\Omega_f^{\varepsilon}$ for $t>0$, and the initial condition

$$c\,(\mathbf{x},0)=c_0\,(\mathbf{x}). \tag{10.2.37}$$

In Eqs. (10.2.35), (10.2.36) $\mathbf{v}=\mathbb{A}_1(\bar{c})$, $\bar{c}\in\mathfrak{M}$, and $\mathbf{n}$ is the unit outward normal vector to the boundary $\partial\Omega_f^{\varepsilon}$.

By the properties of the mollifiers, the function $\widehat{\mathbf{v}}$ is bounded and has bounded first derivatives. Thus, due to well-known results for the linear parabolic equations [61], the problem (10.2.35)–(10.2.37) has a unique weak solution $c=\mathbb{A}_2(\mathbf{v})$ such that

$$0\leqslant\,c(\mathbf{x},t)\leqslant\,1,\;\;\mathbf{x}\in\Omega_f^{\varepsilon},\;t>0, \tag{10.2.38}$$

and

$$\max_{0<t<T} \int_{\Omega_f^\varepsilon} |c(\mathbf{x},t)|^2 dx + \int_0^T \int_{\Omega_f^\varepsilon} |\nabla c(\mathbf{x},t)|^2 dxdt \leqslant C_0 F^2. \tag{10.2.39}$$

In fact, let

$$c^+(\mathbf{x},t) = \max\{c(\mathbf{x},t) - 1; 0\}.$$

Then, using the mollifiers in the same way as in the proof of Theorem 7.1 (Chap. 3, [61]), we arrive at the equality

$$\frac{1}{2}\int_{\Omega_f^\varepsilon} |c^+(\mathbf{x},t_0)|^2 dx + D_0 \int_0^{t_0}\int_{\Omega_f^\varepsilon} |\nabla c^+|^2 dxdt$$
$$= -\int_0^{t_0}\int_{\Omega_f^\varepsilon} \left(\widehat{\mathbf{v}} \cdot \nabla c^+\right) c^+ \, dxdt = I, \tag{10.2.40}$$

where we have used the initial condition (10.2.37) ($c^+(\mathbf{x},0) = 0$), and the evident relations

$$\nabla c \cdot \nabla c^+ = \nabla c^+ \cdot \nabla c^+, \quad c^+ \nabla c = c^+ \nabla c^+.$$

Applying to the right-hand side of (10.2.40) Hölder's and Cauchy's inequalities, and the boundedness of $\widehat{\mathbf{v}}$ we get

$$I \leqslant \frac{D_0}{2}\int_0^{t_0}\int_{\Omega_f^\varepsilon} |\nabla c^+|^2 dxdt + N^2(h)\frac{1}{2\,D_0}\int_0^{t_0}\int_{\Omega_f^\varepsilon} |c^+|^2 dxdt,$$

and

$$\frac{1}{2}\int_{\Omega_f^\varepsilon} |c^+(\mathbf{x},t_0)|^2 dx \leqslant N^2(h)\frac{1}{2\,D_0}\int_0^{t_0}\int_{\Omega_f^\varepsilon} |c^+(\mathbf{x},t)|^2 dxdt,$$

which implies the equality $c^+(\mathbf{x},t) \equiv 0$, and the validity of the right-hand side inequality in (10.2.38).

The left-hand side inequality in (10.2.38) is proved in the same way, if we consider

$$c^-(\mathbf{x},t) = \min\{c(\mathbf{x},t); 0\}.$$

The estimate (10.2.39) follows from the equality for c

$$\frac{1}{2}\int_{\Omega_f^\varepsilon} |c(\mathbf{x},t_0)|^2 dx + D_0 \int_0^{t_0}\int_{\Omega_f^\varepsilon} |\nabla c|^2 \, dxdt = -\int_0^{t_0}\int_{\Omega_f^\varepsilon} \left(\widehat{\mathbf{v}} \cdot \nabla c\right) c \, dxdt,$$

similarly to (10.2.40), if we use the estimates (10.2.38) and the well-known property of the mollifiers

$$\int_0^T \int_\Omega |\widehat{\mathbf{v}}|^2 \, dxdt \leqslant \int_0^T \int_\Omega |\mathbf{v}|^2 \, dxdt.$$

Lemma 10.7 *For any $h > 0$ $\mathbb{A}_2$ is a continuous operator. That is, if $c_i = \mathbb{A}_2(\mathbf{v}_i)$, $\mathbf{v}_i = \mathbb{A}_1(\bar{c}_i)$, $i = 1, 2$, for $\bar{c}_1, \bar{c}_2 \in \mathfrak{M}$, and $\tilde{c} = c_1 - c_2$, then*

$$\max_{0<t<T} \int_{\Omega_f^\varepsilon} |\tilde{c}\,(\boldsymbol{x}, t)|^2 dx + \int_0^T \int_{\Omega_f^\varepsilon} |\nabla \tilde{c}\,(\boldsymbol{x}, t)|^2 dx dt$$
$$\leqslant N(h) \left(\int_0^T \int_{\Omega} \chi^\varepsilon |\mathbf{v}_1(\boldsymbol{x}, t) - \mathbf{v}_2(\boldsymbol{x}, t)|^2 dx dt \right)^{\frac{1}{2}}, \quad (10.2.41)$$

where $N(h)$ depends on the parameter h.

Proof The proof of this lemma is straightforward. The integral identity for the difference $\tilde{c}$ has the form

$$\int_0^T \int_{\Omega_f^\varepsilon} \left(-\tilde{c}\,\frac{\partial \xi}{\partial t} + D_0 \nabla \tilde{c} \cdot \nabla \xi \right) dx dt = \int_0^T \int_{\Omega_f^\varepsilon} \xi \left(\tilde{\mathbf{v}} \cdot \nabla c_2 - \widehat{\mathbf{v}}_1 \cdot \nabla \tilde{c} \right) dx,$$

where $\tilde{\mathbf{v}} = \widehat{\mathbf{v}}_2(\mathbf{x}, t) - \widehat{\mathbf{v}}_1(\mathbf{x}, t)$.

As before, this identity results:

$$\frac{1}{2} \int_{\Omega_f^\varepsilon} |\tilde{c}(\mathbf{x}, t)|^2 dx + D_0 \int_0^t \int_{\Omega_f^\varepsilon} |\nabla \tilde{c}\,(\mathbf{x}, \tau)|^2 dx d\tau$$
$$= \int_0^t \int_{\Omega_f^\varepsilon} \tilde{c}(\mathbf{x}, \tau) \left(\tilde{\mathbf{v}}(\mathbf{x}, \tau) \cdot \nabla c_2(\mathbf{x}, \tau) - \widehat{\mathbf{v}}_1(\mathbf{x}, \tau) \cdot \nabla \tilde{c}\,(\mathbf{x}, \tau) \right) dx d\tau = I.$$

The estimate (10.2.41) follows from the last equality, if we estimate its right-hand side I using Hölder, Cauchy, and Gronwall's inequalities, and the estimates

$$|\tilde{c}(\mathbf{x}, t)| \leqslant 2, \quad |\widehat{\mathbf{v}}_1(\mathbf{x}, t)| \leqslant N(h), \quad \mathbf{x} \in \Omega_f^\varepsilon, \ 0 < t < T,$$
$$\int_{\Omega_f^\varepsilon} |\tilde{\mathbf{v}}(\mathbf{x}, t)|^2 dx \leqslant C_0 \int_{\Omega_f^\varepsilon} |\mathbf{v}_1(\mathbf{x}, t) - \mathbf{v}_2(\mathbf{x}, t)|^2 dx.$$

In fact,

$$I \leqslant \frac{1}{2D_0} \left(\max_{\mathbf{x} \in \Omega_f^\varepsilon, 0<\tau<T} |\widehat{\mathbf{v}}_1(\mathbf{x}, t)|^2 \right) \int_0^t \int_{\Omega_f^\varepsilon} |\tilde{c}(\mathbf{x}, \tau)|^2 dx d\tau$$
$$+ 2 \left(\int_0^t \int_{\Omega_f^\varepsilon} |\nabla c_2(\mathbf{x}, \tau)|^2 dx d\tau \right)^{\frac{1}{2}} \left(\int_0^t \int_{\Omega_f^\varepsilon} |\tilde{\mathbf{v}}(\mathbf{x}, \tau)|^2 dx d\tau \right)^{\frac{1}{2}}$$
$$+ \frac{D_0}{2} \int_0^t \int_{\Omega_f^\varepsilon} |\nabla \tilde{c}(\mathbf{x}, \tau)|^2 dx d\tau,$$

and

$$\frac{1}{2}\int_{\Omega_f^\varepsilon}|\tilde{c}(\mathbf{x},t)|^2dx+\frac{D_0}{2}\int_0^t\int_{\Omega_f^\varepsilon}|\nabla\,\tilde{c}\,(\mathbf{x},\tau)|^2dxd\tau$$
$$\leqslant N(h)\int_0^t\int_{\Omega_f^\varepsilon}|\tilde{c}(\mathbf{x},\tau)|^2dxd\tau+N(h)\left(\int_0^t\int_{\Omega_f^\varepsilon}|\tilde{\mathbf{v}}(\mathbf{x},\tau)|^2dxd\tau\right)^{\frac{1}{2}}.$$

Let $\mathbb{A}=\mathbb{A}_2\cdot\mathbb{A}_1$. The maximum principle (10.2.38) shows that $\mathbb{A}$ transforms the set $\mathfrak{M}$ into itself. It is clear that all the fixed points $c^{(h)}$ of the operator $\mathbb{A}$ define the solutions $\{\mathbf{w}^{(h)},\ p^{(h)},\ c^{(h)}\}$ of the auxiliary problem $(\mathbb{AP})$.

To prove the existence of at least one fixed point of $\mathbb{A}$ we have to show that $\mathbb{A}$ is a completely continuous operator. Lemmas 10.6 and 10.7 prove the continuity of $\mathbb{A}$. The compactness of $\mathbb{A}$ follows from the estimate

$$\int_0^T\int_\Omega\chi^\varepsilon\left|\mathbb{D}\left(x,\frac{\partial\mathbf{v}}{\partial t}\right)\right|^2dxdt$$
$$\leqslant\frac{\delta}{h}\left(\int_0^T\int_{\Omega_f^\varepsilon}|\bar{c}|^2dxdt\right)^{\frac{1}{2}}\left(\int_0^T\int_\Omega\chi^\varepsilon\left|\frac{\partial\mathbf{v}}{\partial t}\right|^2dxdt\right)^{\frac{1}{2}},$$

which is the result of differentiation of (10.2.29) with respect to time, multiplication by $\dfrac{\partial^2\mathbf{w}}{\partial t^2}$ and integration by parts over Ω (formal derivation). For the rigorous derivation we, as before, must use mollifiers.

In the usual way this last relation implies

$$\int_0^T\int_\Omega\chi^\varepsilon\left|\frac{\partial\mathbf{v}}{\partial t}\right|^2dxdt\leqslant\int_0^T\int_\Omega\left|\frac{\partial\mathbf{v}}{\partial t}\right|^2dxdt\leqslant C_0\int_0^T\int_\Omega\left|\nabla\left(\frac{\partial\mathbf{v}}{\partial t}\right)\right|^2dxdt$$
$$\leqslant C_0^2\int_0^T\int_\Omega\left|\mathbb{D}\left(x,\frac{\partial\mathbf{v}}{\partial t}\right)\right|^2dxdt\leqslant C_0^3\int_0^T\int_\Omega\chi^\varepsilon\left|\mathbb{D}\left(x,\frac{\partial\mathbf{v}}{\partial t}\right)\right|^2dxdt,$$

$$\int_0^T\int_\Omega\chi^\varepsilon\left|\mathbb{D}\left(x,\frac{\partial\mathbf{v}}{\partial t}\right)\right|^2dxdt\leqslant C_0^3\frac{\delta^2}{h^2}\int_0^T\int_\Omega\chi^c\,|\bar{c}|^2dxdt=N(h),$$
$$\max_{0<t<T}\int_\Omega|\nabla\,\mathbf{v}(\mathbf{x},t)|^2dx\leqslant C_0\int_0^T\int_\Omega\left|\nabla\left(\frac{\partial\mathbf{v}}{\partial t}\right)\right|^2dxdt\leqslant N(h),$$
$$\int_0^T\int_\Omega\left|\frac{\partial\mathbf{v}}{\partial t}\right|^2dxdt\leqslant C_0\int_0^T\int_\Omega\left|\nabla\left(\frac{\partial\mathbf{v}}{\partial t}\right)\right|^2dxdt\leqslant N(h).\tag{10.2.42}$$

Now, let the sequence $\{\bar{c}^k\}$ be weakly convergent in $L_2\big((0,T); L_2(\Omega_f^\varepsilon)\big)$ to $\bar{c}$. Then on the basis of (10.2.42) we may extract some subsequence $\{\mathbf{v}^{k_n}\}$, which converges strongly in $\mathbf{L}_2(\Omega_T)$ and weakly in $\mathbf{W}_2^{1,0}(\Omega_T)$ to $\mathbf{v} = \mathbb{A}_1(\bar{c})$. Lemma 10.7 guarantees the strong convergence of $\{\bar{c}^k\}$ in $L_2\big((0,T); L_2(\Omega_f^\varepsilon)\big)$ to $\bar{c}$, which proves the compactness of $\mathbb{A}$.

Finally, $\mathfrak{M}$ is a closed convex set, and that is enough for existence at least one fixed point of $\mathbb{A}$ in $\mathfrak{M}$ [55].

It is clear, that all the fixed points of $\mathbb{A}$ preserve the estimates (10.2.31), (10.2.38), and (10.2.39). Thus, the following lemma holds true

Lemma 10.8 *Under the conditions of Theorem 10.5 there exists at least one weak solution $\{\mathbf{w}^{(h)},\ p^{(h)},\ c^{(h)}\}$ of the problem (10.2.21)–(10.2.28), such that*

$$\int_0^T \int_\Omega \Big(|\mathbf{v}^{(h)}(\boldsymbol{x},t)|^2 + |\nabla\, \mathbf{v}^{(h)}(\boldsymbol{x},t)|^2 + |p^{(h)}(\boldsymbol{x},t)|^2\Big)\, dxdt$$
$$+ \max_{0<t<T} \int_\Omega \big(|\mathbf{w}^{(h)}(\boldsymbol{x},t)|^2 + |\nabla\, \mathbf{w}^{(h)}(\boldsymbol{x},t)|^2\big)dx \leqslant C_0\, F^2, \quad (10.2.43)$$

$$0 \leqslant c^{(h)}(\boldsymbol{x},t) \leqslant 1, \;\; \boldsymbol{x} \in \Omega_f^\varepsilon, \;\; t>0, \quad (10.2.44)$$

$$\int_0^T \int_{\Omega_f^\varepsilon} |\nabla\, c^{(h)}\,(\boldsymbol{x},t)|^2 dxdt \leqslant C_0\, F^2, \quad (10.2.45)$$

$$\max_{0<t<T} \int_\Omega \big(|\mathbf{v}^{(h)}(\boldsymbol{x},t)|^2 + |\nabla\, \mathbf{v}^{(h)}(\boldsymbol{x},t)|^2\big)dx \leqslant C(C_0, F), \quad (10.2.46)$$

$$\int_\Omega |\nabla \mathbf{v}^{(h)}(\boldsymbol{x},t_1) - \nabla \mathbf{v}^{(h)}(\boldsymbol{x},t_2)|^2 dx \leqslant C(C_0, F)\, |t_1 - t_2|, \quad (10.2.47)$$

where C_0 is independent of ε, h, and $t \in (0,T)$.

Proof Note, that the right-hand side of (10.2.21) possesses the bounded time derivative (the bounds of any norms obviously depend on the parameter h). Therefore the solution $\{\mathbf{w}^{(h)},\ p^{(h)}\}$ of (10.2.21) has the additional smoothness

$$\chi^\varepsilon\, \nabla \left(\frac{\partial^2 \mathbf{w}^{(h)}}{\partial t^2}\right),\; \chi_s^\varepsilon\, \nabla \left(\frac{\partial \mathbf{w}^{(h)}}{\partial t}\right) \in \mathbf{L}_2(\Omega_T), \;\; \frac{\partial p^{(h)}}{\partial t} \in L_2(\Omega_T),$$

and we may differentiate with respect to time the integral identity corresponding to (10.2.21):

$$\int_0^{t_0}\int_\Omega\left(\chi^\varepsilon \mu_0 \mathbb{D}\left(x, \frac{\partial^2 \mathbf{w}^{(h)}}{\partial t^2}\right) + \chi_s^\varepsilon \lambda_0 \mathbb{D}\left(x, \frac{\partial \mathbf{w}^{(h)}}{\partial t}\right)\right) : \mathbb{D}(x,\varphi)dxdt$$
$$- \int_0^{t_0}\int_\Omega \frac{\partial p^{(h)}}{\partial t}\nabla\cdot\varphi dxdt = \delta\int_0^{t_0}\int_\Omega \chi^\varepsilon (\mathbf{F}\cdot\varphi)\frac{\partial}{\partial t}\left((c^{(h)})_h\right)dxdt = I.$$

All statements of the lemma, except the estimates (10.2.46) and (10.2.47) are already proved.

To prove (10.2.46) we rewrite the right hand-side of the last identity as

$$I = \delta\int_0^{t_0}\int_{\Omega_f^\varepsilon}(\mathbf{F}\cdot\varphi)\left(\frac{\partial c^{(h)}}{\partial t}\right)_h dxdt = \delta\int_0^{t_0}\int_{\Omega_f^\varepsilon}\frac{\partial c^{(h)}}{\partial t}(\mathbf{F}\cdot\varphi)_{\bar{h}}\,dxdt,$$

where

$$(u)_{\bar{h}}(\mathbf{x},t) = \frac{1}{h}\int_{t-h}^{t} c^{(h)}(\mathbf{x},\tau)\,d\tau, \text{ and } \varphi(\mathbf{x},t) = 0 \text{ for } t<0.$$

Next we use the convection-diffusion equation (10.2.23) and express the time derivative of $c^{(h)}$:

$$I = -\delta\int_0^{t_0}\int_{\Omega_f^\varepsilon}\left((\widehat{\mathbf{v}^{(h)}}\cdot\nabla c^{(h)})(\mathbf{F}\cdot\varphi)_{\bar{h}} + D_0\nabla c^{(h)}\cdot\nabla(\mathbf{F}\cdot\varphi)_{\bar{h}}\right)dxdt.$$

Finally into the identity obtained we put $\varphi = \dfrac{\partial \mathbf{w}^{(h)}}{\partial t}$

$$\begin{aligned}\frac{\mu_0}{2}\int_\Omega \chi^\varepsilon\left|\mathbb{D}(x,\mathbf{v}^{(h)}(\mathbf{x},t_0))\right|^2 dx + \lambda_0\int_0^{t_0}\int_\Omega \chi_s^\varepsilon\left|\mathbb{D}\left(x,\frac{\partial \mathbf{w}^{(h)}}{\partial t}\right)\right|^2 dxdt \\ = I = -\delta\int_0^{t_0}\int_{\Omega_f^\varepsilon}(\widehat{\mathbf{v}^{(h)}}\cdot\nabla c^{(h)})(\mathbf{F}\cdot\mathbf{v}^{(h)})_{\bar{h}}\,dxdt \\ -\delta\int_0^{t_0}\int_{\Omega_f^\varepsilon} D_0\nabla c^{(h)}\cdot\nabla(\mathbf{F}\cdot\mathbf{v}^{(h)})_{\bar{h}}\,dxdt.\end{aligned} \tag{10.2.48}$$

We estimate the right hand-side I in the usual way (see the proof of the estimate (10.2.31)):

$$
\begin{aligned}
I \leqslant\, & C_0 \max|\mathbf{F}| \Big(\int_0^{t_0}\int_\Omega |\mathbf{v}^{(h)}|^4 dxdt\Big)^{\frac{1}{2}} \Big(\int_0^{t_0}\int_{\Omega_f^\varepsilon} |\nabla c^{(h)}|^2 dxdt\Big)^{\frac{1}{2}} \\
&+ C_0 \max|\mathbf{F}| \Big(\int_0^{t_0}\int_\Omega |\nabla \mathbf{v}^{(h)}|^2 dxdt\Big)^{\frac{1}{2}} \Big(\int_0^{t_0}\int_{\Omega_f^\varepsilon} |\nabla c^{(h)}|^2 dxdt\Big)^{\frac{1}{2}} \\
&+ C_0 \max|\nabla \mathbf{F}| \Big(\int_0^{t_0}\int_\Omega |\mathbf{v}^{(h)}|^2 dxdt\Big)^{\frac{1}{2}} \Big(\int_0^{t_0}\int_{\Omega_f^\varepsilon} |\nabla c^{(h)}|^2 dxdt\Big)^{\frac{1}{2}} \\
\leqslant\, & C_0^2 F \int_0^{t_0}\int_\Omega |\nabla \mathbf{v}^{(h)}|^2 dxdt \Big(\int_0^{t_0}\int_{\Omega_f^\varepsilon} |\nabla c^{(h)}|^2 dxdt + 1\Big) \\
&+ C_0 F \int_0^{t_0}\int_\Omega |\mathbf{v}^{(h)}|^2 dxdt + \int_0^{t_0}\int_{\Omega_f^\varepsilon} |\nabla c^{(h)}|^2 dxdt) \leqslant C(C_0, F), \quad (10.2.49)
\end{aligned}
$$

where we have used (10.2.43). Thus, (10.2.48) and (10.2.49) result in (10.2.46).

The estimate (10.2.47) is proved similarly to the previous one. In fact, we may rewrite the corresponding to (10.2.21) integral identity as

$$
\mu_0 \int_\Omega \chi^\varepsilon \Big(\mathbb{D}\big(x, \mathbf{v}^{(h)}(\mathbf{x}, t_2)\big) - \mathbb{D}\big(x, \mathbf{v}^{(h)}(\mathbf{x}, t_1)\big)\Big) : \mathbb{D}\big(x, \varphi(\mathbf{x})\big) dx = I_1 + I_2,
$$

where

$$
\begin{aligned}
I_1 &= -\lambda_0 \int_{t_1}^{t_2}\int_\Omega \chi_s^\varepsilon\, \mathbb{D}(x, \mathbf{w}^{(h)}) : \mathbb{D}(x, \varphi) dxdt, \\
I_2 &= \delta \int_{t_1}^{t_2}\int_\Omega \chi^\varepsilon (\mathbf{F}\cdot\varphi) \Big(\frac{\partial c^{(h)}}{\partial t}\Big)_h dxdt,
\end{aligned}
$$

and estimate I_2 as above, using the diffusion-convection equation (10.2.23):

$$
\begin{aligned}
I_2 &= \delta \int_{t_1}^{t_2} \Big(\int_{\Omega_f^\varepsilon} (\mathbf{F}\cdot\varphi) \frac{\partial c^{(h)}}{\partial t} dx\Big)_h dt \\
&= -\delta \int_{t_1}^{t_2} \Big(\int_{\Omega_f^\varepsilon} \Big((\widehat{\mathbf{v}^{(h)}}\cdot \nabla c^{(h)})(\mathbf{F}\cdot\varphi) + D_0 \nabla c^{(h)} \cdot \nabla(\mathbf{F}\cdot\varphi)\Big) dx\Big)_h dt \\
&= -\delta \int_{t_1}^{t_2}\int_{\Omega_f^\varepsilon} \Big((\widehat{\mathbf{v}^{(h)}}\cdot \nabla c^{(h)}))_h(\mathbf{F}\cdot\varphi) + D_0 \nabla (c^{(h)})_h \cdot \nabla(\mathbf{F}\cdot\varphi)\Big) dxdt \\
&\leqslant C_0 \max|\mathbf{F}| \Big(\int_\Omega |\varphi|^4 dx\Big)^{\frac{1}{4}} \int_{t_1}^{t_2} \Big(\int_\Omega |\mathbf{v}^{(h)}|^4 dx\Big)^{\frac{1}{4}} \Big(\int_{\Omega_f^\varepsilon} |\nabla c^{(h)}|^2 dx\Big)^{\frac{1}{2}} dt \\
&\quad + C_0 \max|\mathbf{F}| \Big(\int_\Omega |\nabla\varphi|^2 dx\Big)^{\frac{1}{2}} \int_{t_1}^{t_2} \Big(\int_{\Omega_f^\varepsilon} |\nabla c^{(h)}|^2 dx\Big)^{\frac{1}{2}} dt
\end{aligned}
$$

$$
\begin{aligned}
&+ C_0 \max |\nabla \mathbf{F}| \left(\int_\Omega |\varphi|^2 dx\right)^{\frac{1}{2}} \int_{t_1}^{t_2} \left(\int_{\Omega_f^\varepsilon} |\nabla c^{(h)}|^2 dx\right)^{\frac{1}{2}} dt \\
&\leqslant C_0^3 F \left(\int_\Omega |\nabla \varphi|^2 dx\right)^{\frac{1}{2}} \max_t \left(\int_\Omega |\nabla \mathbf{v}^{(h)}|^2 dx\right)^{\frac{1}{2}} \int_{t_1}^{t_2} \left(\int_{\Omega_f^\varepsilon} |\nabla c^{(h)}|^2 dx\right)^{\frac{1}{2}} dt \\
&+ C_0^2 F \left(\int_\Omega |\nabla \varphi|^2 dx\right)^{\frac{1}{2}} \int_{t_1}^{t_2} \left(\int_{\Omega_f^\varepsilon} |\nabla c^{(h)}|^2 dx\right)^{\frac{1}{2}} dt \\
&\leqslant C(C_0, F) \left(\int_\Omega |\nabla \varphi|^2 dx\right)^{\frac{1}{2}} \left(\int_0^T \int_{\Omega_f^\varepsilon} |\nabla c^{(h)}|^2 dx dt\right)^{\frac{1}{2}} |t_2 - t_1|^{\frac{1}{2}} \\
&\leqslant C(C_0, F) \left(\int_\Omega |\nabla \varphi|^2 dx\right)^{\frac{1}{2}} |t_2 - t_1|^{\frac{1}{2}}.
\end{aligned}
$$

For I_1 one has

$$
I_1 \leqslant \lambda_0 \left(\int_\Omega |\nabla \varphi|^2 dx\right)^{\frac{1}{2}} \left(\int_0^T \int_\Omega |\nabla \mathbf{w}^{(h)}|^2 dx dt\right)^{\frac{1}{2}} |t_2 - t_1|^{\frac{1}{2}}.
$$

Thus,

$$
\begin{aligned}
&\int_\Omega \chi^\varepsilon \left(\mathbb{D}\big(x, \mathbf{v}^{(h)}(\mathbf{x}, t_2)\big) - \mathbb{D}\big(x, \mathbf{v}^{(h)}(\mathbf{x}, t_1)\big)\right) : \mathbb{D}\big(x, \varphi(\mathbf{x})\big) dx \\
&\qquad \leqslant C(C_0, F) \left(\int_\Omega |\nabla \varphi|^2 dx\right)^{\frac{1}{2}} |t_2 - t_1|^{\frac{1}{2}} \\
&\qquad \leqslant C_0, C(C_0, F) \left(\int_\Omega |\mathbb{D}\big(x, \varphi)|^2 dx\right)^{\frac{1}{2}} |t_2 - t_1|^{\frac{1}{2}} \\
&\qquad \leqslant C_0, C(C_0, F) \left(\int_\Omega \chi^\varepsilon |\mathbb{D}\big(x, \varphi)|^2 dx\right)^{\frac{1}{2}} |t_2 - t_1|^{\frac{1}{2}}
\end{aligned}
$$

and for

$$
\varphi(\mathbf{x}) = \mathbf{v}^{(h)}(\mathbf{x}, t_2) - \mathbf{v}^{(h)}(\mathbf{x}, t_1)
$$

we arrive at

$$\int_{\Omega} \chi^{\varepsilon} \left|\mathbb{D}\big(x, \mathbf{v}^{(h)}(\mathbf{x}, t_2)\big) - \mathbb{D}\big(x, \mathbf{v}^{(h)}(\mathbf{x}, t_1)\big)\right|^2 dx \leqslant C(C_0, F)\,|t_2 - t_1|.$$

As the last step in the proof of Theorem 10.5 we pass to the limit as $h \to 0$ in the corresponding integral identities. The following lemma holds true

Lemma 10.9 *Under the conditions of Theorem 10.5 there exists at least one weak solution $\{\mathbf{w}^{\varepsilon},\ p^{\varepsilon},\ c^{\varepsilon}\}$ of the problem (10.0.5), (10.0.6), (10.0.8)–(10.0.13) and the estimates (10.2.8)–(10.2.12) hold true.*

Proof Lemma 10.8 provides the weak compactness of $\{\mathbf{w}^{(h)}\}$, $\{\mathbf{v}^{(h)}\}$ in $\mathbf{W}_2^{1,0}(\Omega_T)$, and the weak compactness of $\{p^{(h)}\}$ in $L_2(\Omega_T)$. That is, up to some subsequences

$$\mathbf{w}^{(h)} \rightharpoonup \mathbf{w}^{\varepsilon},\ \ \mathbf{v}^{(h)} \rightharpoonup \mathbf{v}^{\varepsilon},\ \text{ weakly in } \mathbf{W}_2^{1,0}(\Omega_T) \text{ as } h \to 0,$$

$$p^{(h)} \rightharpoonup p^{\varepsilon} \text{ weakly in } L_2(\Omega_T) \text{ as } h \to 0.$$

The same Lemma 10.8 implies the boundedness and the weak compactness of $c^{(h)}$ in $W_2^{1,0}(\Omega_T)$:

$$c^{(h)} \rightharpoonup c^{\varepsilon} \text{ weakly in } W_2^{1,0}(\Omega_T) \text{ as } h \to 0.$$

Passing to the limit as $h \to 0$ in the integral identity

$$\int_0^T \int_{\Omega} \Big(\chi^{\varepsilon}\,\mu_0\,\mathbb{D}(x, \mathbf{v}^{(h)}) + (1-\chi^{\varepsilon})\,\lambda_0\,\mathbb{D}(x, \mathbf{w}^{(h)}) - p^{(h)}\,\mathbb{I}\Big) : \mathbb{D}(x, \varphi)dxd\tau$$
$$= \int_0^T \int_{\Omega} \Big((\chi^{\varepsilon}\,\rho_0 + (1-\chi^{\varepsilon})\rho_s)\,(\mathbf{F}\cdot\varphi) + \delta\,\chi^{\varepsilon}\,c^{(h)}\big(\mathbf{F}\cdot(\varphi)_{\bar{h}}\big)\Big)\,dxd\tau$$

and taking into account the equality

$$\frac{\partial \mathbf{w}^{\varepsilon}}{\partial t}(\mathbf{x}, t) = \mathbf{v}^{\varepsilon}(\mathbf{x}, t), \quad \mathbf{x} \in \Omega^{\varepsilon}_f,\ t \in (0, T), \tag{10.2.50}$$

we get (10.2.4).

Note that in the product $c^{(h)}\big(\mathbf{F}\cdot(\varphi)_{\bar{h}}\big)$ the sequence $\{(\varphi)_{\bar{h}}\}$ converges strongly in $\mathbf{L}_2(\Omega_T)$ to φ.

The last relation (10.2.50) is the simple consequence of the integral identity

$$\int_0^T \int_{\Omega} \chi^{(h)}\left(\mathbf{w}^{(h)}\cdot\frac{\partial\varphi}{\partial t} + \mathbf{w}^{(h)}\cdot\varphi\right)dxdt = 0$$

for any smooth functions φ, such that $\varphi(\mathbf{x}, 0) = \varphi(\mathbf{x}, T) = 0$.

The limit as $h \to 0$ in

$$\int_0^T \int_\Omega \mathbf{w}^{(h)} \cdot \nabla \xi dxdt = 0$$

for any smooth ξ results in the continuity equation (10.0.10).

Finally, the main problem for the diffusion-convection equation (10.2.23) is the limit in the product $\widehat{\mathbf{v}^{(h)}} \cdot \nabla c^{(h)}$, where the sequence $\{\nabla c^{(h)}\}$ converges only weakly in $\mathbf{L}_2(\Omega_T)$. Thanks to estimates (10.2.46) and (10.2.47) there exists some subsequence of $\{\widehat{\mathbf{v}^{(h)}}\}$, which converges strongly in $\mathbf{L}_2(\Omega_T)$. This fact provides the weak convergence in $\mathbf{L}_2(\Omega_T)$ of the product $\widehat{\mathbf{v}^{(h)}} \cdot \nabla c^{(h)}$ to the corresponding product $\mathbf{v}^\varepsilon \cdot \nabla c^\varepsilon$.

So, these estimates and the diagonal procedure permit us to find some subsequences of $\{\mathbf{v}^{(h)}(,t)\}$ and $\{\nabla \mathbf{v}^{(h)}(,t)\}$, which converge weakly in $\mathbf{L}_2(\Omega)$ for almost all $t \in (0,T)$ to $\mathbf{v}^\varepsilon(\mathbf{x},t)$ and $\nabla \mathbf{v}^\varepsilon(\mathbf{x},t)$ respectively. Due to the embedding theorem [3], the weak convergence of $\{\nabla \mathbf{v}^{(h)}(,t)\}$ in $\mathbf{L}_2(\Omega)$ for almost all $t \in (0,T)$ results in the strong convergence of $\{\mathbf{v}^{(h)}(,t)\}$ in $\mathbf{L}_2(\Omega)$ for almost all $t \in (0,T)$. Thus, the sequence of the bounded functions

$$f^{(h)}(t) = \int_\Omega |\mathbf{v}^{(h)}(\mathbf{x},t) - \mathbf{v}^\varepsilon(\mathbf{x},t)|^2 dx, \quad |f^{(h)}(t)| \leqslant C(C_0, F),$$

converges almost everywhere in $(0,T)$ to zero. According to Lebesque's theorem (or the dominated convergence theorem, [3])

$$\int_0^T f^{(h)}(t)dt = \int_0^T \int_\Omega |\mathbf{v}^{(h)}(\mathbf{x},t) - \mathbf{v}^\varepsilon(\mathbf{x},t)|^2 dxdt \to 0 \text{ as } h \to 0.$$

The validity of the estimates (10.2.8)–(10.2.12) follow from the properties of the weakly and strongly convergent sequences.

In particular, by construction, the sequence $\{u^{(h)}(\mathbf{x})\}$, where

$$u^{(h)}(\mathbf{x}) = \nabla \mathbf{v}^{(h)}(\mathbf{x},t_1) - \nabla \mathbf{v}^{(h)}(\mathbf{x},t_2)$$

weakly converges to the function

$$u^\varepsilon(\mathbf{x}) = \nabla \mathbf{v}^\varepsilon(\mathbf{x},t_1) - \nabla \mathbf{v}^\varepsilon(\mathbf{x},t_2)$$

for almost all $t_1,\ t_2 \in (0,T)$.

Therefore,

$$\int_\Omega |u^\varepsilon(\mathbf{x})|^2 dx \leqslant \limsup_{h \to 0} \int_\Omega |u^{(h)}(\mathbf{x})|^2 dx \leqslant C(C_0, F)\,|t_1 - t_2|$$

for almost all $t_1,\ t_2 \in (0,T)$.

10.2.3 Proof of Theorem 10.6

The proof is based on the principles of compactness, and the two-scale convergent method [88].

First of all, using the extension operators, we rewrite the identities (10.2.3) and (10.2.4) as

$$\int_0^T \int_\Omega \chi^\varepsilon \left(-\tilde{c}^\varepsilon \frac{\partial \xi}{\partial t} + \xi \frac{\partial \mathbf{w}^\varepsilon}{\partial t} \cdot \nabla \tilde{c}^\varepsilon + D_0 \nabla \tilde{c}^\varepsilon \cdot \nabla \xi \right) dxdt = \int_\Omega \chi^\varepsilon c_0(\mathbf{x}) \xi(\mathbf{x}, 0)\, dx \tag{10.2.51}$$

$$\int_0^T \int_\Omega \left(\chi^\varepsilon \mu_0 \mathbb{D}(x, \mathbf{v}^\varepsilon) + (1-\chi^\varepsilon)\lambda_0 \mathbb{D}(x, \mathbf{w}^\varepsilon) - p^\varepsilon \mathbb{I}\right) : \mathbb{D}(x, \varphi) dxdt = \int_0^T \int_\Omega \left(\chi^\varepsilon(\rho_0 + \delta c^\varepsilon) + \rho_s(1-\chi^\varepsilon)\right)\mathbf{F} \cdot \varphi\, dxdt, \tag{10.2.52}$$

where $\mathbf{v}^\varepsilon = \mathbb{E}_{\Omega_f^\varepsilon}\left(\dfrac{\partial \mathbf{w}^\varepsilon}{\partial t}\right)$ and $\tilde{c}^\varepsilon = \tilde{\mathbb{E}}_{\Omega_f^\varepsilon}(c^\varepsilon)$.

We also rewrite the continuity equation (10.0.10) as the integral identity

$$\int_0^T \int_\Omega \mathbf{w}^\varepsilon \cdot \nabla \psi\, dxdt = 0, \tag{10.2.53}$$

which holds true for any smooth functions ψ.

Next on the base of the estimates (10.2.8)–(10.2.12), we state that there exist the functions

$$c,\ p \in L_2(\Omega_T),\ \nabla c,\ \mathbf{v} = \frac{\partial \mathbf{w}}{\partial t},\ \nabla \mathbf{v} \in \mathbf{L}_2(\Omega_T),$$
$$C,\ P \in L_2(\Omega_T \times Y),\ \nabla_y C,\ \mathbf{V},\ \nabla_y \mathbf{V} \in \mathbf{L}_2(\Omega_T \times Y),$$

and the convergent subsequences $\{\tilde{c}^\varepsilon\}$, $\{\nabla \tilde{c}^\varepsilon\}$, $\{p^\varepsilon\}$, $\{\mathbf{w}^\varepsilon\}$, $\{\mathbb{D}(x, \mathbf{w}^\varepsilon)\}$, $\{\mathbf{v}^\varepsilon\}$, and $\{\mathbb{D}(x, \mathbf{v}^\varepsilon)\}$ such that

$$\begin{aligned}
&\tilde{c}^\varepsilon(\mathbf{x}, t) \rightharpoonup c(\mathbf{x}, t) \text{ weakly in } W_2^{1,0}(\Omega_T),\\
&\tilde{c}^\varepsilon(\mathbf{x}, t) \to c(\mathbf{x}, t) \text{ two-scale in } L_2(\Omega_T),\\
&\nabla \tilde{c}^\varepsilon \to \nabla c + \nabla_y C \text{ two-scale in } \mathbf{L}_2(\Omega_T),\\
&p^\varepsilon(\mathbf{x}, t) \rightharpoonup p(\mathbf{x}, t) \text{ weakly in } L_2(\Omega_T),\\
&p^\varepsilon(\mathbf{x}, t) \to P(\mathbf{x}, t, \mathbf{y}) \text{ two-scale in } L_2(\Omega_T),
\end{aligned}$$

$$
\begin{gathered}
\mathbf{w}^{\varepsilon}(\mathbf{x},t) \rightharpoonup \mathbf{w}(\mathbf{x},t) \text{ weakly in } \overset{\circ}{\mathbf{W}}{}_2^{1,0}(\Omega_T),\\
\mathbb{D}\left(x,\mathbf{w}^{\varepsilon}\right) \to \mathbb{D}(x,\mathbf{w}) + \mathbb{D}(y,\mathbf{W}) \text{ two-scale in } \mathbf{L}_2(\Omega_T),\\
\mathbf{v}^{\varepsilon}(\mathbf{x},t) \rightharpoonup \mathbf{v}(\mathbf{x},t) \text{ weakly in } \overset{\circ}{\mathbf{W}}{}_2^{1,0}(\Omega_T),\\
\mathbb{D}\left(x,\mathbf{v}^{\varepsilon}\right) \to \mathbb{D}(x,\mathbf{v}) + \mathbb{D}(y,\mathbf{V}) \text{ two-scale in } \mathbf{L}_2(\Omega_T),\\
\mathbf{V}(\mathbf{x},t,\mathbf{y}) = \frac{\partial \mathbf{W}}{\partial t}(\mathbf{x},t,\mathbf{y}),\\
\mathbf{v}^{\varepsilon}(\mathbf{x},t) \to \mathbf{v}(\mathbf{x},t) \text{ and strongly in } \mathbf{L}_2(\Omega_T)
\end{gathered}
$$

as $\varepsilon \to 0$.

The last statement follows from the estimates (10.2.11) and (10.2.12). In fact, these estimates and the diagonal procedure permit us to find some subsequences of $\{\mathbf{v}^{\varepsilon}(,t)\}$ and $\{\nabla\,\mathbf{v}^{\varepsilon}(,t)\}$, which converge weakly in $\mathbf{L}_2(\Omega)$ for almost all $t \in (0,T)$ to $\mathbf{v}(\mathbf{x},t)$ and $\nabla\,\mathbf{v}(\mathbf{x},t)$ respectively. Due to the embedding theorem [3], the weak convergence of $\{\nabla\,\mathbf{v}^{\varepsilon}(,t)\}$ in $\mathbf{L}_2(\Omega)$ for almost all $t \in (0,T)$ results in the strong convergence of $\{\mathbf{v}^{\varepsilon}(,t)\}$ in $\mathbf{L}_2(\Omega)$ for almost all $t \in (0,T)$. Thus, the sequence of the bounded functions

$$
f^{\varepsilon}(t) = \int_{\Omega} |\mathbf{v}^{\varepsilon}(\mathbf{x},t) - \mathbf{v}(\mathbf{x},t)|^2 dx, \quad |f^{\varepsilon}(t)| \leqslant C(C_0,F),
$$

converges almost everywhere in $(0,T)$ to zero. According to Lebesque's theorem (or the dominated convergence theorem, [3])

$$
\int_0^T f^{\varepsilon}(t)dt = \int_0^T \int_{\Omega} |\mathbf{v}^{\varepsilon}(\mathbf{x},t) - \mathbf{v}(\mathbf{x},t)|^2 dxdt \to 0 \text{ as } h \to 0.
$$

The following statements, except the last one, are already known (see Chap. 1 and the first part of the present chapter).

Lemma 10.10 *The limiting functions satisfy in the domain Ω for $t > 0$ the system of macroscopic equations*

$$
\nabla \cdot \mathbf{v} = 0, \tag{10.2.54}
$$

$$
\nabla \cdot \widehat{\mathbb{P}} + \hat{\rho}(c)\,\mathbf{F} = 0, \tag{10.2.55}
$$

$$
\begin{aligned}
\widehat{\mathbb{P}} = \mu_0 &\left(m\,\mathbb{D}(x,\mathbf{v}) + \left\langle \mathbb{D}\left(y, \frac{\partial \mathbf{W}}{\partial t}\right)\right\rangle_{Y_f}\right)\\
&+ \lambda_0\big((1-m)\,\mathbb{D}(x,\mathbf{w}) + \langle \mathbb{D}(y,\mathbf{W})\rangle_{Y_s}\big) - p\,\mathbb{I},
\end{aligned}
$$

$$
m\,\frac{\partial c}{\partial t} + \mathbf{v}\cdot\big(m\,\nabla c + \langle \nabla_y C\rangle_{Y_f}\big) = D_0\,\nabla\cdot\big(m\,\nabla c + \langle \nabla_y C\rangle_{Y_f}\big). \tag{10.2.56}
$$

Lemma 10.11 *The limiting functions satisfy in the domain Y for $\boldsymbol{x} \in \Omega$ and $t > 0$ the system of microscopic equations*

$$\nabla_y \cdot \mathbf{W} = 0, \tag{10.2.57}$$

$$\nabla_y \cdot \tilde{\mathbb{P}} = 0, \tag{10.2.58}$$

$$\tilde{\mathbb{P}} = \mu_0 \chi(\mathbf{y}) \left(\mathbb{D}(x, \mathbf{v}) + \mathbb{D}\left(y, \frac{\partial \mathbf{W}}{\partial t}\right)\right) \\ + \lambda_0 \big(1 - \chi(\mathbf{y})\big)\big(\mathbb{D}(x, \mathbf{w}) + \mathbb{D}(y, \mathbf{W})\big) - P\,\mathbb{I},$$

$$\nabla_y \cdot \Big(\chi(\mathbf{y})\big(\nabla c + \nabla_y C\big)\Big) = 0. \tag{10.2.59}$$

Lemma 10.12 *The relations*

$$\widehat{\mathbb{P}} = -p\,\mathbb{I} + \mathfrak{N}_1 : \mathbb{D}(x, \mathbf{v}) + \mathfrak{N}_2 : \mathbb{D}(x, \mathbf{w}) \\ + \int_0^t \mathfrak{N}_3(t - \tau) : \mathbb{D}(x, \mathbf{w}(\boldsymbol{x}, \tau))d\tau, \tag{10.2.60}$$

and

$$m\,\nabla c + \langle \nabla_y C \rangle_{Y_f} = \mathbb{B}^{(c)} \nabla c, \tag{10.2.61}$$

hold true.

Here fourth-rank constant tensors $\mathfrak{N}_1$, $\mathfrak{N}_2$, and fourth-rank tensor $\mathfrak{N}_3(t)$ have been defined in Chap. 1 by formulae (1.4.30), and the constant matrix $\mathbb{B}^{(c)}$ has been defined in Theorem 10.2 by formula (10.1.61).

Lemma 10.13 *Let the domain Y_f be symmetric with respect to rotations of the angle $\pi/2$ around the principal axes of the Cartesian coordinate system. Then the matrix $\mathbb{B}^{(c)}$ is diagonal.*

Proof Let $\mathbb{T}_2$ be the rotation of the angle $\pi/2$ around the axes y_3 with the direction $\mathbf{e}_3$, such that $\mathbb{T}_2 \cdot \mathbf{e}_2 = \mathbf{e}_1$. Here $\{\mathbf{e}_1, \mathbf{e}_2, \mathbf{e}_3\}$ are the orthogonal basis of the Cartesian coordinate system. If $\mathbf{z} = \mathbb{T}_2 \cdot \mathbf{y}$, $\mathbf{y} \in Y$, then, by conditions of the lemma, $\mathbb{T}_2 : Y_f \to Y_f$ and $\chi(\mathbf{y}) = \chi(\mathbf{z})$.

Let us recall, that the 1-periodic solution C of the boundary-value problem (10.2.59) has the form

$$C = \sum_{i=1}^{3} C^{(i)}(\mathbf{y}) \frac{\partial c}{\partial x_i},$$

where $C^{(i)}(\mathbf{y})$, $i = 1, 2, 3$, are solutions of the periodic boundary-value problem

$$\nabla_y \cdot \big(\chi(\mathbf{y})(\mathbf{e}_i + \nabla_y C^{(i)})\big) = 0 \tag{10.2.62}$$

in the domain Y.

For the function $\widetilde{C}^{(2)}(\mathbf{z}) = C^{(2)}(\mathbf{y})$ we have the periodic boundary-value problem

$$\nabla_z \cdot \left(\chi(\mathbf{z})(\mathbb{T}_2 \cdot \mathbf{e}_2 + \mathbb{T}_2 \cdot \mathbb{T}_2^* \cdot \nabla_z \widetilde{C}^{(2)})\right) = \nabla_z \cdot \left(\chi(\mathbf{z})(\mathbf{e}_1 + \nabla_z \widetilde{C}^{(2)})\right) = 0 \quad (10.2.63)$$

in the unit cube $Z = Y$.

The problem (10.2.63) coincides with the problem (10.2.62) and due to the uniqueness of this problem

$$C^{(2)}(\mathbf{y}) = \widetilde{C}^{(2)}(\mathbf{z}) = C^{(1)}(\mathbf{z}) = C^{(1)}(\mathbb{T}_2 \cdot \mathbf{y}). \quad (10.2.64)$$

In the same way we obtain

$$C^{(3)}(\mathbf{y}) = \widetilde{C}^{(3)}(\mathbf{x}) = C^{(1)}(\mathbf{x}) = C^{(1)}(\mathbb{T}_3 \cdot \mathbf{y}) \quad (10.2.65)$$

for the rotation $\mathbf{x} = \mathbb{T}_3\mathbf{y}$ around the axes y_1, such that $\mathbb{T}_3 \cdot \mathbf{e}_3 = \mathbf{e}_1$.

To find $\mathbb{B}_0$ we put

$$\mathbf{a}_1 = \langle \nabla_y C^{(1)}(\mathbf{y}) \rangle_{Y_f},$$

and calculate

$$\mathbf{a}_2 = \langle \nabla_y C^{(2)}(\mathbf{y}) \rangle_{Y_f} = \mathbb{T}_2^* \langle \nabla_z C^{(1)}(\mathbf{z}) \rangle_{Y_f} = \mathbb{T}_2^* \cdot \mathbf{a}_1,$$

and

$$\mathbf{a}_3 = \langle \nabla_y C^{(3)}(\mathbf{y}) \rangle_{Y_f} = \mathbb{T}_3^* \langle \nabla\, C^{(1)}(\mathbf{x}) \rangle_{Y_f} = \mathbb{T}_3^* \cdot \mathbf{a}_1.$$

Let $\mathbf{a}_1 = \alpha_1\mathbf{e}_1 + \alpha_2\mathbf{e}_2 + \alpha_3\mathbf{e}_3$. Then,

$$\begin{aligned}
\mathbf{a}_2 &= \alpha_1 \mathbb{T}_2^* \cdot \mathbf{e}_1 + \alpha_2 \mathbb{T}_2^* \cdot \mathbf{e}_2 + \alpha_3 \mathbb{T}_2^* \cdot \mathbf{e}_3 = \alpha_1\mathbf{e}_2 - \alpha_2\mathbf{e}_1 + \alpha_3\mathbf{e}_3,\\
\mathbf{a}_3 &= \alpha_1 \mathbb{T}_3^* \cdot \mathbf{e}_1 + \alpha_2 \mathbb{T}_3^* \cdot \mathbf{e}_2 + \alpha_3 \mathbb{T}_3^* \cdot \mathbf{e}_3 = \alpha_1\mathbf{e}_3 + \alpha_2\mathbf{e}_2 - \alpha_3\mathbf{e}_1,
\end{aligned}$$

where we have used the evident properties of the transformations $\mathbb{T}_2$ and $\mathbb{T}_3$:

$$\mathbb{T}_2^* \cdot \mathbf{e}_1 = \mathbf{e}_2,\ \mathbb{T}_2^* \cdot \mathbf{e}_2 = -\mathbf{e}_1,\ \mathbb{T}_2^* \cdot \mathbf{e}_3 = \mathbf{e}_3,$$

and

$$\mathbb{T}_3^* \cdot \mathbf{e}_1 = \mathbf{e}_3,\ \mathbb{T}_3^* \cdot \mathbf{e}_2 = \mathbf{e}_2,\ \mathbb{T}_3^* \cdot \mathbf{e}_3 = -\mathbf{e}_1.$$

Thus,

$$\begin{aligned}
\mathbb{B}_0 &= \mathbf{a}_1 \otimes \mathbf{e}_1 + \mathbf{a}_2 \otimes \mathbf{e}_2 + \mathbf{a}_3 \otimes \mathbf{e}_3\\
&= \alpha_1\mathbf{e}_1 \otimes \mathbf{e}_1 + \alpha_1\mathbf{e}_2 \otimes \mathbf{e}_2 + \alpha_1\mathbf{e}_3 \otimes \mathbf{e}_3 + \alpha_2\mathbf{e}_2 \otimes \mathbf{e}_1 - \alpha_2\mathbf{e}_1 \otimes \mathbf{e}_2\\
&\quad + \alpha_3\mathbf{e}_3 \otimes \mathbf{e}_1 - \alpha_3\mathbf{e}_1 \otimes \mathbf{e}_3 + \alpha_3\mathbf{e}_3 \otimes \mathbf{e}_2 + \alpha_2\mathbf{e}_2 \otimes \mathbf{e}_3\\
&= \alpha_1(\mathbf{e}_1 \otimes \mathbf{e}_1 + \mathbf{e}_2 \otimes \mathbf{e}_2 + \mathbf{e}_3 \otimes \mathbf{e}_3),
\end{aligned}$$

due to the symmetry of $\mathbb{B}_0$, which implies $\alpha_2 = \alpha_3 = 0$.

Chapter 11
The Muskat Problem

Here as the basic mathematical model at the microscopic level in the domain Ω for $t > 0$ we consider the mathematical model $\mathbb{M}_{28}$ of the joint motion of two incompressible immiscible liquids, consisting of the dynamic equations

$$\nabla \cdot \left(\chi^{\varepsilon}\, \mu_0\, \mathbb{D}\left(x, \frac{\partial \mathbf{w}^{\varepsilon}}{\partial t}\right) + (1-\chi^{\varepsilon})\lambda_0\, \mathbb{D}(x, \mathbf{w}^{\varepsilon}) - p^{\varepsilon}\, \mathbb{I}\right) + \rho^{\varepsilon}\, \mathbf{F} = 0, \tag{11.0.1}$$

$$\nabla \cdot \left(\frac{\partial \mathbf{w}^{\varepsilon}}{\partial t}\right) = 0 \tag{11.0.2}$$

for the displacement vector $\mathbf{w}^{\varepsilon}$ and the pressure p^{ε} of the medium, completed with the Cauchy problem

$$\frac{\partial \rho^{\varepsilon}}{\partial t} + \frac{\partial \mathbf{w}^{\varepsilon}}{\partial t} \cdot \nabla\, \rho \equiv \frac{\partial \rho^{\varepsilon}}{\partial t} + \nabla \cdot \left(\rho^{\varepsilon}\, \frac{\partial \mathbf{w}^{\varepsilon}}{\partial t}\right) = 0, \tag{11.0.3}$$

$$\rho^{\varepsilon}(\mathbf{x}, 0) = \rho_0^{\varepsilon}(\mathbf{x}) \equiv \rho_f^{(0)}(\mathbf{x})\, \chi^{\varepsilon} + \rho_s\,(1-\chi^{\varepsilon}) \tag{11.0.4}$$

for the density ρ of the medium.

The differential equations are endowed with the boundary condition

$$\mathbf{v}^{\varepsilon}(\mathbf{x}, t) = 0, \quad \mathbf{x} \in \Omega,\ t > 0 \tag{11.0.5}$$

on the boundary $S = \partial\Omega$ of the domain Ω, the initial condition

$$\chi^{\varepsilon}(\mathbf{x})\mathbf{w}^{\varepsilon}(\mathbf{x}, 0) = 0, \quad \mathbf{x} \in \Omega, \tag{11.0.6}$$

and the normalization condition

$$\int_{\Omega} p^{\varepsilon}(\mathbf{x}, t)dx = 0. \tag{11.0.7}$$

A. Meirmanov, *Mathematical Models for Poroelastic Flows*,
Atlantis Studies in Differential Equations 1, DOI: 10.2991/978-94-6239-015-7_11,

Throughout this chapter we impose Assumption 0.1 on the structure of the pore space and additionally suppose that the solid skeleton is disconnected: $Y_s \subset Y_f$.

We also suppose that conditions

$$0 < \lambda_0,\ \mu_0,\ \rho_s < \infty, \tag{11.0.8}$$

$$0 < \rho_f^- = const \leqslant \rho_f^{(0)}(\mathbf{x}) \leqslant \rho_f^+ = const < \infty, \tag{11.0.9}$$

and

$$\max_{(\mathbf{x},t)\in\Omega_T} \left(|\mathbf{F}(\mathbf{x},t)| + |\nabla \mathbf{F}(\mathbf{x},t)| + \left| \frac{\partial \mathbf{F}}{\partial t}(\mathbf{x},t) \right| \right) = F < \infty \tag{11.0.10}$$

hold true.

Recall that Eq. (11.0.1) is understood in the sense of distributions. It contains Stokes equations in the liquid part, Lamé's equations in the solid part, and the continuity condition

$$\begin{aligned} &\lim_{\substack{\mathbf{x}\to\mathbf{x}^0 \\ \mathbf{x}\in\Omega_s^\varepsilon}} \Big(\lambda_0 \mathbb{D}\big(x, \mathbf{w}^\varepsilon(\mathbf{x},t)\big) - p^\varepsilon(\mathbf{x},t)\,\mathbb{I}\Big)\cdot \mathbf{n}(\mathbf{x}^0) \\ &\qquad = \lim_{\substack{\mathbf{x}\to\mathbf{x}^0 \\ \mathbf{x}\in\Omega_f^\varepsilon}} \Big(\mu_0 \mathbb{D}\Big(x, \frac{\partial \mathbf{w}^\varepsilon}{\partial t}(\mathbf{x},t)\Big) - p^\varepsilon(\mathbf{x},t)\,\mathbb{I}\Big)\cdot \mathbf{n}(\mathbf{x}^0) \end{aligned} \tag{11.0.11}$$

on the common boundary Γ^ε "solid skeleton-pore space".

11.1 Statement of the Problem and Main Results

First of all let us look at the dynamic equation (11.0.1). The smoothness of the solution with respect to time in the solid part depends on the smoothness of the term $\rho^\varepsilon\,\mathbf{F}$ with respect to time. But the density ρ^ε might be some step function, which is sufficient for the existence of the derivative $\dfrac{\partial \mathbf{w}^\varepsilon}{\partial t}$ in the liquid part, but does not guarantee the existence of this derivative in the solid part.

Our aim is to get the homogenized model, which is asymptotically closed to (11.0.1)–(11.0.7). Therefore we may make a small change to the model and in the transport equation (11.0.3) instead of $\dfrac{\partial \mathbf{w}^\varepsilon}{\partial t}$ consider some function $\mathbf{v}^\varepsilon$, which coincides with $\dfrac{\partial \mathbf{w}^\varepsilon}{\partial t}$ in the liquid part and is asymptotically close to $\dfrac{\partial \mathbf{w}^\varepsilon}{\partial t}$ as $\varepsilon \to 0$ in the solid part.

To do that, we extend the liquid velocity $\dfrac{\partial \mathbf{w}^\varepsilon}{\partial t}$ from the liquid part Ω_f^ε to the solid part Ω_s^ε using the extension Lemma B.10: for any $\varepsilon > 0$ there exists an extension

$\mathbf{v}^\varepsilon \in \overset{\circ}{\mathbf{W}}{}^1_2(\Omega)$ of the function $\dfrac{\partial \mathbf{w}^\varepsilon}{\partial t}$ from the domain Ω^ε_f onto domain Ω such that

$$\chi^\varepsilon(\mathbf{x})\left(\mathbf{v}^\varepsilon(\mathbf{x},t) - \frac{\partial \mathbf{w}^\varepsilon}{\partial t}(\mathbf{x},t)\right) = 0, \quad \mathbf{x} \in \Omega,\ t > 0,$$

and

$$\int_\Omega |\mathbf{v}^\varepsilon|^2 dx \leqslant C \int_{\Omega^\varepsilon_f} \left|\frac{\partial \mathbf{w}^\varepsilon}{\partial t}\right|^2 dx, \tag{11.1.1}$$

$$\int_\Omega |\mathbb{D}(x,\mathbf{v}^\varepsilon)|^2 dx \leqslant C \int_{\Omega^\varepsilon_f} \left|\mathbb{D}\left(x, \frac{\partial \mathbf{w}^\varepsilon}{\partial t}\right)\right|^2 dx, \tag{11.1.1}$$

where C is independent of ε and t.

We fix this linear continuous operator from $\mathbf{W}^1_2(\Omega^\varepsilon_f)$ to $\mathbf{W}^1_2(\Omega)$ and denote it as

$$\mathbf{v}^\varepsilon = \mathbb{E}_{\Omega^\varepsilon_f}\left(\frac{\partial \mathbf{w}^\varepsilon}{\partial t}\right), \tag{11.1.2}$$

and, in what follows, call this function $\mathbf{v}^\varepsilon$ the *liquid velocity*.

Note, that due to the continuity Eq. (11.0.2) in the domain Ω_T and the structure of the pore space we may choose the extension operator such that the function $\mathbf{v}^\varepsilon$ will be solenoidal:

$$\nabla \cdot \mathbf{v}^\varepsilon = 0, \ \ (\mathbf{x},t) \in \Omega_T. \tag{11.1.3}$$

The results of Sect. 1.4 show that $\dfrac{\partial \mathbf{w}^\varepsilon}{\partial t}$ and $\mathbf{v}^\varepsilon$ are asymptotically closed as $\varepsilon \to 0$.

Definition 11. 1 We say that the triple of functions $\{\mathbf{w}^\varepsilon, p^\varepsilon, \rho^\varepsilon\}$ is a weak solution to the problem (11.0.1)–(11.0.7), if

$$\mathbf{w}^\varepsilon \in \overset{\circ}{\mathbf{W}}{}^{1,0}_2(\Omega_T), \ \chi^\varepsilon \nabla\left(\frac{\partial \mathbf{w}^\varepsilon}{\partial t}\right) \in \mathbf{L}_2(\Omega_T), \ \rho^\varepsilon, p^\varepsilon \in L_2(\Omega_T),$$

$\mathbf{w}^\varepsilon$ is solenoidal in $\Omega_T = \Omega \times (0,T)$, the pressure p satisfies the normalization condition (11.0.7), and the integral identities

$$\int_{\Omega_T} \rho^\varepsilon \left(\frac{\partial \xi}{\partial t} + \frac{\partial \mathbf{w}^\varepsilon}{\partial t} \cdot \nabla \xi\right) dxdt + \int_\Omega \rho^\varepsilon_0(\mathbf{x})\, \xi(\mathbf{x},0)\, dx = 0, \tag{11.1.4}$$

and

$$\int_{\Omega_T}\Big(-\chi^\varepsilon\alpha_\mu\,\mathbb{D}(x,\mathbf{w}^\varepsilon):\mathbb{D}\Big(x,\frac{\partial\varphi}{\partial t}\Big)+(1-\chi^\varepsilon)\alpha_\lambda\,\mathbb{D}(x,\mathbf{w}^\varepsilon):\mathbb{D}(x,\varphi)\Big)\,dxdt$$
$$=\int_{\Omega_T}\big(p^\varepsilon\,\nabla\cdot\varphi+\rho^\varepsilon\mathbf{F}\cdot\varphi\big)dxdt, \qquad (11.1.5)$$

hold true for any smooth functions ξ and φ, such that $\xi(\mathbf{x},T)=0$ and $\varphi(\mathbf{x},t)=0$ for $\mathbf{x}\in S$.

In (11.1.4) $\mathbf{v}^\varepsilon$ is the extension of $\dfrac{\partial\mathbf{w}^\varepsilon}{\partial t}$ from the liquid part Ω_f^ε onto the solid part Ω_s^ε given by (11.1.2) and

$$\mathbf{v}^\varepsilon\in\overset{\circ}{\mathbf{W}}{}_2^{1,0}(\Omega_T).$$

Note, that integral identity (11.1.5) contains differential equations in the pore space and in the solid skeleton, the boundary condition (11.0.11) on the common boundary Γ^ε and the initial condition (11.0.6).

Theorem 11.1 *Let conditions (11.0.8)–(11.0.11) hold true.*

Then the problem (11.0.1)–(11.0.7) has at least one weak solution $\{\mathbf{w}^\varepsilon,\,p^\varepsilon,\,\rho^\varepsilon\}$, such that

$$\max_{0<t<T}\int_\Omega\Big(|\mathbf{w}^\varepsilon|^2+|\mathbf{v}^\varepsilon|^2+|\nabla\mathbf{w}^\varepsilon(\mathbf{x},t)|^2\Big)dx+\int_{\Omega_T}|p^\varepsilon|^2dxdt\leqslant C, \qquad (11.1.6)$$

$$\max_{0<t<T}\int_\Omega\Big(|\nabla\mathbf{v}^\varepsilon|^2+\chi^\varepsilon\Big|\mathbb{D}\Big(x,\frac{\partial\mathbf{w}^\varepsilon}{\partial t}(\mathbf{x},t)\Big)\Big|^2\Big)dx+\int_{\Omega_T}\Big|\nabla\frac{\partial\mathbf{w}^\varepsilon}{\partial t}\Big|^2dxdt\leqslant C, \qquad (11.1.7)$$

$$\int_\Omega|\nabla\mathbf{v}^\varepsilon(\mathbf{x},t_1)-\nabla\mathbf{v}^\varepsilon(\mathbf{x},t_2)|^2dx\leqslant C\,|t_1-t_2|^{\frac{1}{2}},\ t_1,\,t_2\in(0,T), \qquad (11.1.8)$$

$$0\leqslant\rho^\varepsilon(\mathbf{x},t)\leqslant\max(\rho_f^+,\rho_s)=\rho_0^+, \qquad (11.1.9)$$

where $C=C(\rho_0^+,F,T)$ is independent of ε, and

$$\mathbf{v}^\varepsilon=\mathbb{E}_{\Omega_f^\varepsilon}\Big(\frac{\partial\mathbf{w}^\varepsilon}{\partial t}\Big).$$

Theorem 11.2 *Under the conditions of Theorem 2.1 let $\{\mathbf{w}^\varepsilon,\,p^\varepsilon,\,\rho^\varepsilon\}$ be the weak solution to the problem (11.0.1)–(11.0.7). Then there exists a subsequence of small parameters $\{\varepsilon>0\}$ as $\varepsilon\searrow0$, such that the sequence $\{\mathbf{w}^\varepsilon\}$ converges weakly in $\overset{\circ}{\mathbf{W}}{}_2^{1,0}(\Omega_T)$ and strongly in $\mathbf{L}_2(\Omega_T)$ to the function $\mathbf{w}$, the sequence $\{\rho^\varepsilon\}$ converges weakly in $L_2(\Omega_T)$ to the function ρ, the sequence $\{p^\varepsilon\}$ converges weakly in $L_2(\Omega_T)$ to* *the function p, the sequence* $\left\{\dfrac{\partial \mathbf{w}^{\varepsilon}}{\partial t}\right\}$, *converges weakly in* $\overset{\circ}{\mathbf{W}}{}_2^{1,0}(\Omega_T)$ *and strongly in* $\mathbf{L}_2(\Omega_T)$ *to the function* $\dfrac{\partial \mathbf{w}}{\partial t}$, *the sequence* $\{\mathbf{v}^{\varepsilon}\}$, *where* $\mathbf{v}^{\varepsilon} = \mathbb{E}_{\Omega_f^{\varepsilon}}\left(\dfrac{\partial \mathbf{w}^{\varepsilon}}{\partial t}\right)$, *converges strongly in* $\mathbf{L}_2(\Omega_T)$ *to the function* $\dfrac{\partial \mathbf{w}}{\partial t}$.

The triple of limiting functions $\{\mathbf{w},\ p,\ \rho\}$ *is a weak solution to the Muskat problem for viscoelastic filtration, which consists of the dynamic equations*

$$\nabla \cdot \left(\widehat{\mathbb{P}}(\mathbf{w})\right) - \nabla p + \rho \mathbf{F} = 0, \tag{11.1.10}$$

$$\widehat{\mathbb{P}}(\mathbf{w}) = \mathfrak{M}_0 : \mathbb{D}\left(x, \frac{\partial \mathbf{w}}{\partial t}\right) + \mathfrak{M}_1 : \mathbb{D}(x, \mathbf{w}) + \int_0^t \mathfrak{M}_2(t-\tau) : \mathbb{D}(x, \mathbf{w}(\mathbf{x}, \tau) d\tau,$$

$$\nabla \cdot \mathbf{w} = 0, \tag{11.1.11}$$

for the displacement $\mathbf{w}$ *and the pressure p of the mixture of the solid skeleton and the liquid in pores, and the transport equation*

$$\frac{\partial \rho}{\partial t} + \nabla \cdot \left(\frac{\partial \mathbf{w}}{\partial t} \rho\right) = 0, \tag{11.1.12}$$

for the density ρ *of the liquid mixture in the domain* Ω_T.

The problem is endowed with the homogeneous boundary condition

$$\mathbf{w}(\mathbf{x}, t) = 0, \quad \mathbf{x} \in S,\ t > 0, \tag{11.1.13}$$

and the initial conditions

$$\mathbf{w}(\mathbf{x}, 0) = 0, \quad \rho(\mathbf{x}, 0) = \rho_0(\mathbf{x}) \equiv \rho_f^{(0)}(\mathbf{x})\, m + \rho_s\,(1-m) \quad \mathbf{x} \in \Omega. \tag{11.1.14}$$

In (11.1.10) the fourth-rank constant tensors $\mathfrak{M}_0$, $\mathfrak{M}_1$, *and the fourth-rank tensor* $\mathfrak{M}_2(t)$ *are defined in Chap. 1 by the formulae (1.4.30), and the tensor* $\mathfrak{M}_0$ *is symmetric and strictly positively definite.*

11.2 Proof of Theorem 11.1

We prove the existence of the solution to the problem (11.0.1)–(11.0.7) using the Schauder Fixed Point Theorem, mollifiers, and viscosity solutions method. For the correct limiting procedure we have to derive a priori estimates, independent of the parameters of the approximation. First of all we approximate the density in the dynamic equation using a mollifier with respect to time. That gives us the additional smoothness of the solutions to the dynamic equations with respect to time. Next we approximate the transport equation for the density by the diffusion–convection equation with a small diffusion (viscosity). It gives us the additional smoothness of the density with respect to time and spatial variables. Then we prove the existence of the solution $\{\rho_{(\delta,h)}, \mathbf{u}_{(\delta,h)}, p_{(\delta,h)}\}$ to the double approximate problem using the Schauder Fixed Point Theorem. To pass to the limit as $\delta \to 0$ we need estimates independent of δ (Lemma 11.2). Here we essentially use the uniform boundedness of $\frac{\partial \rho_{(\delta,h)}}{\partial t}$ in a dual space $L_2\big((0,T); W_2^{-1}(\Omega)\big)$ and we also prove that these estimates are independent of h and ε. As we have mentioned above, the smoothness of the solution must provide the convergence (at least weak) of the product $\rho_{(\delta,h)} \cdot \mathbf{v}_{(\delta,h)}$, where $\mathbf{v}_{(\delta,h)} = \mathbb{E}_{\Omega_f^\varepsilon}(\frac{\partial \mathbf{u}_{(\delta,h)}}{\partial t})$. For the fixed $h > 0$ the sequence $\{\rho_{(\delta,h)}\}$ is compact in $L_2(\Omega_T)$ and the sequence $\{\mathbf{v}_{(\delta,h)}\}$ is weakly compact in $\mathbf{L}_2(\Omega_T)$. Thus, we may pass to the limit as $\delta \to 0$ and get the solution $\{\rho_{(h)}, \mathbf{u}_{(h)}, p_{(h)}\}$ of the approximate problem. As a last step we have to pass to the limit as $h \to 0$. But the bounded sequence $\{\rho_{(h)}\}$ is no longer a compact set in $L_2(\Omega_T)$. Therefore, we must prove the strong compactness in $\mathbf{L}_2(\Omega_T)$ of the sequence $\{\mathbf{v}_{(h)}\}$. Functions $\mathbf{v}_{(h)}$ have spatial derivatives uniformly bounded in $\mathbf{L}_2(\Omega_T)$. So, the sequence $\{\mathbf{v}_{(h)}\}$ is a compact set in $\mathbf{L}_2(\Omega)$ for any fixed $t \in (0,T)$. To state that $\{\mathbf{v}_{(h)}\}$ is a compact set in $\mathbf{L}_2(\Omega_T)$ one has to have some smoothness of $\mathbf{v}_{(h)}$ with respect to time [12, 68]. But the proper dynamic equations do not directly provide this smoothness and we must find another way, which has been realized in Lemma 11.2.

We divide the proof of the theorem into several independent steps.

First we solve the double approximation problem

$$\nabla \cdot \big(\mathbb{P}^{(\varepsilon)}(\mathbf{u})\big) - \nabla p + (\rho)_{(\delta)}\mathbf{F} = 0, \quad \nabla \cdot \mathbf{u} = 0, \tag{11.2.1}$$

$$\frac{\partial \rho}{\partial t} + \mathbf{v} \cdot \nabla \rho = h\,\Delta \rho, \tag{11.2.2}$$

$$\mathbf{u}|_S = 0, \;\; \chi^\varepsilon \mathbf{u}|_{t=0} = 0, \;\; \rho|_S = 0, \;\; \rho|_{t=0} = \rho_h^{(0)}(\mathbf{x}), \tag{11.2.3}$$

$$\rho_h^{(0)} \in \overset{\circ}{C}{}^{\infty}(\Omega), \;\; 0 \leqslant \rho_h^{(0)} \leqslant \rho_0^+, \;\; \rho_h^{(0)} \to \rho^{(0)} \text{ a.e. in } \Omega,$$

for $h > 0$ and $\delta > 0$ in Ω_T, and then pass to the limit as $\delta \to 0$ and $h \to 0$.

In (11.2.1), (11.2.2)

$$\mathbb{P}^{(\varepsilon)}(\mathbf{u}) = \chi^\varepsilon \alpha_\mu \mathbb{D}\left(x, \frac{\partial \mathbf{u}}{\partial t}\right) + (1-\chi^\varepsilon)\alpha_\lambda \mathbb{D}(x, \mathbf{u}),$$

$$(\rho)_{(\delta)} = \frac{1}{\delta}\int_{t-\delta}^{t} \hat{\rho}(\mathbf{x},\tau)d\tau, \;\; \mathbf{v} = \mathbb{E}_{\Omega_f^\varepsilon}\left(\frac{\partial \mathbf{u}}{\partial t}\right),$$

$$\hat{\rho}(\mathbf{x},t)=\rho(\mathbf{x},t),\ \text{ for } t>0,\ \text{ and } \hat{\rho}(\mathbf{x},t)=\rho_h^{(0)}(\mathbf{x}) \text{ for } t<0.$$

By the properties of mollifiers the function $(\rho)_{(\delta)}$ possesses time derivative $\frac{\partial}{\partial t}\big((\rho)_{(\delta)}\big)\in L_2(\Omega_T)$. This fact guarantees the additional smoothness of the solution $\mathbf{u}$ with respect to time.

In the standard way we may define a weak solution to the problems (11.2.1)–(11.2.3) as functions $\mathbf{u}$, p and ρ satisfying the corresponding integral identities.

Lemma 11.1 *For given $h>0$ and $\delta>0$ the problems (11.2.1)–(11.2.3) has at least one weak solution.*

Proof To solve the problems (11.2.1)–(11.2.3) we fix the set

$$\mathfrak{M}=\{\sigma\in L_2(\Omega_T):0\leqslant\sigma(\mathbf{x},t)\leqslant\rho_0^+ \text{ a.e. in } \Omega_T\},$$

and consider in Ω_T the initial boundary-value problem for the linear system

$$\nabla\cdot\big(\mathbb{P}^{(\varepsilon)}(\mathbf{u})\big)-\nabla p+(\sigma)_{(\delta)}\mathbf{F}=0,\quad \nabla\cdot\mathbf{u}=0, \tag{11.2.4}$$

with the homogeneous boundary and initial conditions (11.2.3) for the function $\mathbf{u}$.

The solvability of this problem is standard and follows from the estimates

$$\max_{0<t<T}\Big(\|\chi^{\varepsilon}\mathbb{D}(x,\mathbf{u})\|_{2,\Omega}(t)\Big)+\|(1-\chi^{\varepsilon})\mathbb{D}(x,\mathbf{u})\|_{2,\Omega_T}+\|p\|_{2,\Omega_T}\leqslant C, \tag{11.2.5}$$

$$\max_{0<t<T}\left(\left\|\chi^{\varepsilon}\mathbb{D}\left(x,\frac{\partial\mathbf{u}}{\partial t}\right)\right\|_{2,\Omega}(t)\right)+\left\|(1-\chi^{\varepsilon})\mathbb{D}\left(x,\frac{\partial\mathbf{u}}{\partial t}\right)\right\|_{2,\Omega_T}\leqslant\frac{C}{\delta}, \tag{11.2.6}$$

where $C=C(\rho_0^+,F,T)$.

To obtain the first estimate we multiply Eq. (11.2.4) by $\mathbf{u}$ and integrate by parts over domain Ω using Hölder's, Korn's, Friedrichs-Poincaré and Gronwall's inequalities. To get the second estimate we differentiate the Eq. (11.2.4) with respect to time, multiply the result by $\frac{\partial\mathbf{u}}{\partial t}$ and integrate by parts over domain Ω, repeating the same procedure as above.

Therefore, $\mathbf{u}$ and $\mathbf{v}=\mathbb{E}_{\Omega_f^{\varepsilon}}\left(\frac{\partial\mathbf{u}}{\partial t}\right)$ are the linear continuous operators on σ: $\mathbf{u}=\Phi_0(\sigma)$, $\mathbf{v}=\Phi(\sigma)$. In particular,

$$\max_{0<t<T}\Big(\|\mathbf{v}\|_{2,\Omega}(t)+\|\nabla\mathbf{v}\|_{2,\Omega}(t)\Big)\leqslant\frac{C}{\delta}F\,\|\sigma\|_{2,\Omega_T}. \tag{11.2.7}$$

In (11.2.7) we have used Korn's inequality and estimates (11.2.6) and (11.1.1).

Next we consider in Ω_T the linear problem for the parabolic equation (11.2.2) with the conditions (11.2.3) for the function ρ, where $\mathbf{v} = \mathbb{E}_{\Omega^\varepsilon_f}\left(\dfrac{\partial \mathbf{u}}{\partial t}\right)$ and $\mathbf{u} = \Phi(\sigma)$. As before, the solvability of this problem is a simple consequence of the maximum principle

$$0 \leqslant \rho(\mathbf{x}, t) \leqslant \rho_0^+, \text{ a.e. in } \Omega_T, \tag{11.2.8}$$

the energy equality

$$\int_\Omega |\rho(\mathbf{x}, t)|^2 dx + 2h \int_0^t \int_\Omega |\nabla \rho(\mathbf{x}, \tau)|^2 dx d\tau = \int_\Omega |\rho_h^{(0)}(\mathbf{x})|^2 dx,$$

and the estimate

$$\max_{0<t<T} \int_\Omega |\rho(\mathbf{x}, t)|^2 dx + 2h \int_0^T \int_\Omega |\nabla \rho(\mathbf{x}, t)|^2 dx dt \leqslant |\rho_0^+|^2 \tag{11.2.9}$$

(for more details see [61]).

For fixed h the problems (11.2.2), (11.2.3) defines some continuous operator $\rho = \Psi(\mathbf{v})$. In fact, if $\rho_1 = \Psi(\mathbf{v}_1)$ and $\rho_2 = \Psi(\mathbf{v}_2)$, then for $\tilde{\rho} = \rho_1 - \rho_2$ and $\tilde{\mathbf{v}} = \mathbf{v}_1 - \mathbf{v}_2$ one has

$$\frac{1}{2}\frac{d}{dt} \int_\Omega |\tilde{\rho}(\mathbf{x}, t)|^2 dx + h \int_\Omega |\nabla \tilde{\rho}|^2 dx = J,$$

$$J = -\int_\Omega \tilde{\rho}\, \nabla \rho_2 \cdot \tilde{\mathbf{v}} dx = \int_\Omega \rho_2\, \nabla \tilde{\rho} \cdot \tilde{\mathbf{v}}\, dx.$$

$$|J| \leqslant \frac{h}{2} \int_\Omega |\nabla \tilde{\rho}(\mathbf{x}, t)|^2 dx + \frac{1}{2h} |\rho_0^+|^2 \int_\Omega |\tilde{\mathbf{v}}|^2 dx.$$

Thus,

$$\|\Psi(\mathbf{v}_1) - \Psi(\mathbf{v}_2)\|_{2,\Omega_T} \leqslant \frac{\rho_0^+}{\sqrt{h}} \|\mathbf{v}_1 - \mathbf{v}_2\|_{2,\Omega_T}. \tag{11.2.10}$$

Let now $\rho = \Lambda(\sigma) \equiv \Psi\big(\Phi(\sigma)\big)$. Operator Λ is a continuous in $\mathfrak{M}$ due to the estimates (11.2.7) and (11.2.10), and transforms $\mathfrak{M}$ into itself due to the estimates (11.2.8). The set $\mathfrak{M}$ is obviously convex and closed in $L_2(\Omega_T)$.

Moreover, by the well-known properties of the solutions to the linear parabolic equation (11.2.2), the embedding theorems [61], and the estimate (11.2.7)

$$\mathbf{v} \in L_6(\Omega_T), \;\; \rho \in W_6^{2,1}(\Omega_T) \subset H^{\beta, \frac{\beta}{2}}(\overline{\Omega_T}).$$

This last smoothness property of the function ρ means that the operator Λ is completely continuous. Applying Schauder's Fixed Point Theorem [55] we get at least one fixed point $\rho_{(\delta,h)}$ of the operator Λ, which defines the solution $\{\rho_{(\delta,h)},\ \mathbf{u}_{(\delta,h)},\ p_{(\delta,h)}\}$ to the problem (11.2.1)–(11.2.3), satisfying the estimates (11.2.5)–(11.2.9).

Now we derive the basic a priori estimates, which permits us to pass to the limit in (11.2.1)–(11.2.3) as $\delta \to 0$ and $h \to 0$.

Lemma 11.2 *The solution $\{\rho_{(\delta,h)},\ \mathbf{u}_{(\delta,h)},\ p_{(\delta,h)}\}$ of the problem (11.2.1)–(11.2.3) satisfies the estimates (11.2.8), (11.2.9) for the density $\rho_{(\delta,h)}$, the estimate (11.2.5), and the estimates*

$$\max_{0<t<T}\int_{\Omega}|\nabla\mathbf{v}_{(\delta,h)}(\mathbf{x},t)|^2dx+\int_0^T\int_{\Omega}\Big(|p_{(\delta,h)}|^2+\Big|\nabla\frac{\partial\mathbf{u}_{(\delta,h)}}{\partial t}\Big|^2\Big)dxdt\leqslant C, \tag{11.2.11}$$

for the pressure $p_{(\delta,h)}$, the liquid velocity $\mathbf{v}_{(\delta,h)}$ and the displacement $\mathbf{u}_{(\delta,h)}$, where the constant $C=C(\rho_0^+,F,T)$ is independent of ε, δ and h.

Proof For the moment we omit indexes h and δ. As we have mentioned above, the approximation of ρ by $(\rho)_{(\delta)}$ results in the additional smoothness of the solution $\mathbf{u}$ of the problem (11.2.1)–(11.2.3) with respect to time. Now we prove that this additional smoothness does not depend on the small parameters ε, δ and h.

To do that we first multiply the Eq. (11.2.1) by $\dfrac{\partial\mathbf{u}}{\partial t}$, integrate by parts over domain Ω, pass the time derivative in the term

$$\int_{\Omega}(\rho)_{(\delta)}\mathbf{F}\cdot\frac{\partial\mathbf{u}}{\partial t}dx$$

from $\mathbf{u}$ onto $(\rho)_{(\delta)}\mathbf{F}$ and express the time derivative

$$\frac{\partial(\rho)_{(\delta)}}{\partial t}=\frac{1}{\delta}\int_{t-\delta}^{t}\frac{\partial\rho}{\partial\tau}(\mathbf{x},\tau)d\tau,\ \text{ for } t>\delta,$$

and

$$\frac{\partial(\rho)_{(\delta)}}{\partial t}=\frac{1}{\delta}\int_{0}^{t}\frac{\partial\rho}{\partial\tau}(\mathbf{x},\tau)d\tau,\ \text{ for } t<\delta,$$

using (11.2.2) in the form

$$\frac{\partial\rho}{\partial t}=\nabla\cdot\big(h\,\nabla\rho-\rho\,\mathbf{v}\big).$$

We have

$$\alpha_\mu\int_{\Omega}\chi^{\varepsilon}\Big|\mathbb{D}\Big(x,\frac{\partial\mathbf{u}}{\partial t}\Big)\Big|^2dx+\frac{\alpha_\lambda}{2}\frac{d}{dt}\int_{\Omega}(1-\chi^{\varepsilon})|\mathbb{D}(x,\mathbf{u})|^2dx=I_0(t), \tag{11.2.12}$$

where

$$-I_0(t) = I_{0,0}(t) + \int_\Omega (\rho)_{(\delta)} \left(\frac{\partial \mathbf{F}}{\partial t} \cdot \mathbf{u} \right) dx,$$

$$I_{0,0}(t) = \int_\Omega \left(\frac{1}{\delta} \int_{t-\delta}^{t} \big(h\nabla\rho - \rho\,\mathbf{v}\big)(\mathbf{x},\tau)d\tau \right) \cdot \nabla\big(\mathbf{F}\cdot\mathbf{u}\big)(\mathbf{x},t)dx, \quad \text{for } t > \delta,$$

and

$$I_{0,0}(t) = \int_\Omega \left(\frac{1}{\delta} \int_{0}^{t} \big(h\nabla\rho - \rho\,\mathbf{v}\big)(\mathbf{x},\tau)d\tau \right) \cdot \nabla\big(\mathbf{F}\cdot\mathbf{u}\big)(\mathbf{x},t)dx, \quad \text{for } t < \delta.$$

Everything that we have done to get (11.2.12) is just a formal procedure, but it can be done rigorously using the corresponding integral identities.

It is easy to see that for any positive γ

$$\begin{aligned}\int_0^t |I_0(\tau)|d\tau \leqslant \gamma \int_0^t \int_\Omega |\mathbf{v}|^2 dx d\tau &\\ + \left(\frac{1}{4\gamma} + \frac{1}{2} \right) (\rho_0^+ F)^2 \int_0^t \int_\Omega \big(|\nabla \mathbf{u}|^2 + |\mathbf{u}|^2\big) dx d\tau &\\ + \frac{h}{2} \int_0^t \int_\Omega h|\nabla\rho|^2 dx d\tau + 2\rho_0^+ F \left(\int_0^t \int_\Omega |\mathbf{u}|^2 dx d\tau \right)^{\frac{1}{2}}. &\end{aligned}$$

After integrating (11.2.12) with respect to time and taking into account the estimates (11.2.5), (11.2.8) and (11.0.8) one has

$$\begin{aligned}\int_0^t \int_\Omega \chi^\varepsilon \left| \mathbb{D}\left(x, \frac{\partial \mathbf{u}}{\partial t}(\mathbf{x},\tau)\right) \right|^2 dx d\tau + \int_\Omega |\mathbb{D}(x, \mathbf{u}(\mathbf{x},t))|^2 dx &\\ \leqslant C\gamma \int_0^t \int_\Omega |\mathbf{v}(\mathbf{x},\tau)|^2 dx d\tau + C\left(\frac{1}{\gamma} + 1\right) (\rho_0^+ F + 1)^4 + C\,(\rho_0^+)^2. &\end{aligned}$$

Choosing $C\gamma \leqslant \dfrac{1}{2}$ we get

$$\int_0^T \int_\Omega \chi^\varepsilon \left| \mathbb{D}\left(x, \frac{\partial \mathbf{u}}{\partial t}(\mathbf{x},t)\right) \right|^2 dx dt \leqslant C(\rho_0^+ F + 1)^4, \tag{11.2.13}$$

where C is independent of ε, δ and h.

This last estimate, estimate (11.1.1), and Korn's inequality imply

$$\int_0^T \int_\Omega |\nabla \mathbf{v}(\mathbf{x},t)|^2 dx dt \leqslant C\,(\rho_0^+ F + 1)^4. \tag{11.2.14}$$

Due to the inclusion $\mathbf{v} \in L_2\big((0,T); \overset{\circ}{W}{}_2^1(\Omega)\big)$ we may use the Friedrichs-Poincaré inequality

$$\int_0^T \int_\Omega |\mathbf{v}(\mathbf{x},t)|^2 dxdt \leqslant C \int_0^T \int_\Omega |\nabla \mathbf{v}(\mathbf{x},t)|^2 dxdt, \tag{11.2.15}$$

with constant C independent of ε.

Estimates (11.2.14) and (11.2.15) result in

$$\int_0^T \int_\Omega |\mathbf{v}(\mathbf{x},t)|^2 dxdt \leqslant C\,(\rho_0^+ F + 1)^4. \tag{11.2.16}$$

Now, we repeat the same procedure for the time derivatives, namely, we differentiate the Eq. (11.2.1) with respect to time, multiply the result by $\dfrac{\partial \mathbf{u}}{\partial t}$, and integrate over domain Ω:

$$\frac{\alpha_\mu}{2}\frac{d}{dt}\int_\Omega \chi^\varepsilon \left|\mathbb{D}\left(x, \frac{\partial \mathbf{u}}{\partial t}\right)\right|^2 dx + \alpha_\lambda \int_\Omega (1-\chi^\varepsilon)\left|\mathbb{D}\left(x, \frac{\partial \mathbf{u}}{\partial t}\right)\right|^2 dx = I_1(t),$$

$$-I_1(t) = I_{1,0}(t) + \int_\Omega (\rho)_{(\delta)} \left(\frac{\partial \mathbf{F}}{\partial t} \cdot \frac{\partial \mathbf{u}}{\partial t}\right) dx, \tag{11.2.17}$$

$$I_{1,0}(t) = \int_\Omega \left(\frac{1}{\delta}\int_{t-\delta}^t (h\nabla\rho - \rho\,\mathbf{v})\,(\mathbf{x},\tau)d\tau\right) \cdot \nabla\left(\mathbf{F}\cdot\frac{\partial \mathbf{u}}{\partial t}\right)(\mathbf{x},t)dx, \ \text{ for } t > \delta,$$

$$I_{1,0}(t) = \int_\Omega \left(\frac{1}{\delta}\int_0^t (h\nabla\rho - \rho\,\mathbf{v})\,(\mathbf{x},\tau)d\tau\right) \cdot \nabla\left(\mathbf{F}\cdot\frac{\partial \mathbf{u}}{\partial t}\right)(\mathbf{x},t)dx, \ \text{ for } t < \delta,$$

$$\int_0^t |I_1(\tau)|d\tau \leqslant (\gamma + h)\int_0^t \int_\Omega \left(\left|\nabla\left(\frac{\partial \mathbf{u}}{\partial t}\right)\right|^2 + \left|\frac{\partial \mathbf{u}}{\partial t}\right|^2\right) dxd\tau$$
$$+ (\rho_0^+ \cdot F)^2 \left(\int_0^t \int_\Omega \left(\frac{1}{4\gamma}|\mathbf{v}|^2 + h|\nabla\rho|^2\right) dxd\tau + 1\right).$$

After integrating (11.2.17) with respect to time and using the Friedrichs-Poincaré inequality

$$\int_\Omega \left|\frac{\partial \mathbf{u}}{\partial t}\right|^2 dx \leqslant C \int_\Omega \left|\nabla \frac{\partial \mathbf{u}}{\partial t}\right|^2 dx,$$

Korn's inequality

$$\int_{\Omega}\left|\nabla\frac{\partial\mathbf{u}}{\partial t}\right|^2 dx \leqslant C\int_{\Omega}\left|\mathbb{D}\left(x,\frac{\partial\mathbf{u}}{\partial t}\right)\right|^2 dx,$$

the evident decomposition

$$\int_{\Omega}\left|\mathbb{D}\left(x,\frac{\partial\mathbf{u}}{\partial t}\right)\right|^2 dx = \int_{\Omega}\chi^{\varepsilon}\left|\mathbb{D}\left(x,\frac{\partial\mathbf{u}}{\partial t}\right)\right|^2 dx + \int_{\Omega}(1-\chi^{\varepsilon})\left|\mathbb{D}\left(x,\frac{\partial\mathbf{u}}{\partial t}\right)\right|^2 dx,$$

and the estimates (11.2.9) and (11.2.16) we have

$$\frac{\alpha_{\mu}}{2}\int_{\Omega}\chi^{\varepsilon}\left|\mathbb{D}\left(x,\frac{\partial\mathbf{u}}{\partial t}(\mathbf{x},t)\right)\right|^2 dx + \alpha_{\lambda}\int_0^t\int_{\Omega}(1-\chi^{\varepsilon})\left|\mathbb{D}\left(x,\frac{\partial\mathbf{u}}{\partial \tau}(\mathbf{x},\tau)\right)\right|^2 dxd\tau$$
$$\leqslant (\gamma+h)C\left(\int_{\Omega}\chi^{\varepsilon}\left|\mathbb{D}\left(x,\frac{\partial\mathbf{u}}{\partial t}\right)\right|^2 dx + \int_{\Omega}(1-\chi^{\varepsilon})\left|\mathbb{D}\left(x,\frac{\partial\mathbf{u}}{\partial t}\right)\right|^2 dx\right) + C\,(\rho_0^+\cdot F)^6.$$

Therefore

$$\max_{0<t<T}\int_{\Omega}\chi^{\varepsilon}\left|\mathbb{D}\left(x,\frac{\partial\mathbf{u}}{\partial t}(\mathbf{x},t)\right)\right|^2 dx + \int_0^T\int_{\Omega}\left|\nabla\frac{\partial\mathbf{u}}{\partial t}\right|^2 dxdt \leqslant C\,(\rho_0^+F+1)^6. \tag{11.2.18}$$

The last estimate, the estimate (11.1.1), and Korn's inequality imply

$$\max_{0<t<T}\int_{\Omega}|\nabla\mathbf{v}(\mathbf{x},t)|^2 dx \leqslant C\,(\rho_0^+F+1)^6. \tag{11.2.19}$$

The estimate (11.2.11) for the pressure p follows from Eq. (11.2.1) as an estimate for the bounded linear functional acting in the space $L_2\Big((0,T);\overset{\circ}{W}{}_2^1(\Omega)\Big)$ in the form

$$\int_{\Omega_T} p\,\nabla\cdot\varphi\, dxdt = \int_{\Omega_T}\big(\mathbb{P}^{(\varepsilon)}(\mathbf{u}):\mathbb{D}(x,\varphi) + (\rho)_{(\delta)}\mathbf{F}\cdot\varphi\big)dxdt, \tag{11.2.20}$$

and the estimates (11.2.8) and (11.2.18).

As the last step we pass to the limit as $\delta\to 0$. We do that in the integral identity (11.2.20) and in the integral identity

$$\int_{\Omega_T}\left(\rho\left(\frac{\partial\psi}{\partial t}+\mathbf{v}\cdot\nabla\psi\right) - h\nabla\rho\cdot\nabla\psi\right)dxdt = -\int_{\Omega}\rho_h^{(0)}(\mathbf{x})\psi(\mathbf{x},0)dx, \tag{11.2.21}$$

with arbitrary smooth functions φ and ψ. The functions φ and ψ vanish at S and the function ψ vanishes at $t=T$.

Estimates (11.2.8), (11.2.9), and (11.2.11) guarantee the inclusion

$$\frac{\partial \rho_{(\delta,h)}}{\partial t} \in L_2\Big((0,T);\, W_2^{-1}(\Omega)\Big),$$

and the uniform boundedness in $L_2\Big((0,T);\, W_2^{-1}(\Omega)\Big)$ with respect to δ, h, and ε.

On the basis of these estimates, the above mentioned inclusion, and the well-known compactness results [68], we may choose some subsequence from $\{\delta > 0\}$, such that the sequences

$$\{p_{(\delta,h)}\}, \{\nabla \mathbf{u}_{(\delta,h)}\}, \left\{\nabla \frac{\partial \mathbf{u}_{(\delta,h)}}{\partial t}\right\}, \ \{\mathbf{v}_{(\delta,h)}\} \text{ and } \{\nabla \rho_{(\delta,h)}\}$$

converge weakly in $L_2(\Omega_T)$ and $\mathbf{L}_2(\Omega_T)$ as $\delta \to 0$ to the functions

$$p_{(h)}, \ \ \nabla \mathbf{u}_{(h)}, \ \ \nabla \frac{\partial \mathbf{u}_{(h)}}{\partial t}, \ \ \mathbf{v}_{(h)} = \mathbb{E}_{\Omega_f^\varepsilon}\left(\frac{\partial \mathbf{u}_{(h)}}{\partial t}\right) \ \text{ and } \nabla \rho_{(h)}$$

correspondingly, and the sequence $\{\rho_{(\delta,h)}\}$ converges strongly in $L_2(\Omega_T)$ as $\delta \to 0$ to the function $\rho_{(h)}$.

Passing to the limit as $\delta \to 0$ in the integral identities (11.2.20) and (11.2.21) we conclude that the limiting functions $\{\mathbf{u}_{(h)}, p_{(h)}, \rho_{(h)}\}$ are the weak solution to the approximation problem

$$\nabla \cdot \big(\mathbb{P}^{(\varepsilon)}(\mathbf{u}_{(h)})\big) - \nabla p_{(h)} + \rho_{(h)}\,\mathbf{F} = 0, \quad \nabla \cdot \mathbf{u}_{(h)} = 0, \tag{11.2.22}$$

$$\frac{\partial \rho_{(h)}}{\partial t} + \mathbf{v}_{(h)} \cdot \nabla \rho_{(h)} = h\,\Delta \rho_{(h)}, \tag{11.2.23}$$

completed with the initial and boundary conditions (11.2.3).

Finally, to prove Theorem 11.1 we have to pass to the limit as $h \to 0$ in (11.2.3), (11.2.22), and (11.2.23).

To do that we derive the main a priori estimate.

Lemma 11.3 *The solutions $\{\rho_{(h)}, \mathbf{u}_{(h)}, p_{(h)}\}$ of the problems (11.2.22), (11.2.23), and (11.2.3) satisfy the estimates (11.2.8), (11.2.9), (11.2.11) and the estimate*

$$\int_\Omega |\nabla \mathbf{v}_{(h)}(\mathbf{x}, t_1) - \nabla \mathbf{v}_{(h)}(\mathbf{x}, t_2)|^2 dx \leqslant C|t_1 - t_2|^{\frac{1}{2}}, \tag{11.2.24}$$

with constant C independent of h and ε.

Proof As before, we omit for the moment the index h. It is clear, that we have to prove only the estimate (11.2.24). In the same way as in Lemma 11.2 we get the following integral identity

$$\alpha_\mu \frac{d}{dt} \int_\Omega \chi^\varepsilon \mathbb{D}\left(x, \frac{\partial \mathbf{u}}{\partial t}(\mathbf{x}, t)\right) : \mathbb{D}\left(x, \varphi(\mathbf{x})\right) dx = I_2(t), \tag{11.2.25}$$

where

$$I_2(t) = -\alpha_\lambda \int_\Omega (1-\chi^\varepsilon)\mathbb{D}\left(x, \frac{\partial \mathbf{u}}{\partial t}(\mathbf{x}, t)\right) : \mathbb{D}\big(x, \varphi(\mathbf{x})\big)dx$$
$$+ \int_\Omega \big(h\nabla\rho - \rho\,\mathbf{v}\big)(\mathbf{x}, t)\cdot\nabla\big(\mathbf{F}(\mathbf{x}, t)\cdot\varphi(\mathbf{x})\big)dx + \int_\Omega \rho(\mathbf{x}, t)\left(\frac{\partial \mathbf{F}}{\partial t}(\mathbf{x}, t)\cdot\varphi(\mathbf{x})\right)dx.$$

This last integral identity holds true for any solenoidal function $\varphi \in \overset{\circ}{\mathbf{W}}{}_2^1(\Omega)$.

After integrating the right-hand side of (11.2.25) over the interval (t_1, t_2) and using the estimates (11.2.8), (11.2.9), and (11.2.11) we arrive at

$$\int_{t_1}^{t_2} |I_2(t)|dt \leqslant C\int_{t_1}^{t_2} \|h|\nabla\rho| + |\mathbf{v}| + \left|\nabla\frac{\partial \mathbf{u}}{\partial t}\right| + 1\|_{2,\Omega}(t)\,dt\,\|\nabla\varphi\|_{2,\Omega}$$
$$\leqslant C\,|t_1 - t_2|^{\frac{1}{2}}\,\|\nabla\varphi\|_{2,\Omega}.$$

Therefore,

$$\int_\Omega \chi^\varepsilon \mathbb{D}\left(x, \frac{\partial \tilde{\mathbf{u}}}{\partial t}\right) : \mathbb{D}\big(x, \varphi\big)dx \leqslant C\,|t_1 - t_2|^{\frac{1}{2}}\,\|\nabla\varphi\|_{2,\Omega},$$

where

$$\frac{\partial \tilde{\mathbf{u}}}{\partial t} = \frac{\partial \mathbf{u}}{\partial t}(\mathbf{x}, t_2) - \frac{\partial \mathbf{u}}{\partial t}(\mathbf{x}, t_1).$$

In particular, for

$$\varphi = \tilde{\mathbf{v}} = \mathbf{v}(\mathbf{x}, t_2) - \mathbf{v}(\mathbf{x}, t_1), \quad \|\nabla\varphi\|_{2,\Omega} \leqslant C, \quad \forall\, t_1,\, t_2 \in (0, T),$$

and

$$\int_\Omega \chi^\varepsilon \mathbb{D}\left(x, \frac{\partial \tilde{\mathbf{u}}}{\partial t}\right) : \mathbb{D}\big(x, \tilde{\mathbf{v}}\big)dx \leqslant C\,|t_1 - t_2|^{\frac{1}{2}}.$$

But, by the definition of the extension $\mathbf{v}$

$$\chi^\varepsilon \mathbb{D}\big(x, \tilde{\mathbf{v}}\big) = \chi^\varepsilon \mathbb{D}\left(x, \frac{\partial \tilde{\mathbf{u}}}{\partial t}\right).$$

Thus,

$$\int_\Omega \chi^\varepsilon \mathbb{D}\left(x, \frac{\partial \tilde{\mathbf{u}}}{\partial t}\right) : \mathbb{D}\left(x, \frac{\partial \tilde{\mathbf{u}}}{\partial t}\right)dx = \int_\Omega \chi^\varepsilon \mathbb{D}\left(x, \frac{\partial \tilde{\mathbf{u}}}{\partial t}\right) : \mathbb{D}\big(x, \tilde{\mathbf{v}}\big)dx \leqslant C\,|t_1 - t_2|^{\frac{1}{2}}.$$

The statement of the lemma follows now from the last estimate, estimate (11.1.1) and Korn's inequality.

Lemmas 11.2 and 11.3 permit us to find some subsequence from $\{h > 0\}$, such that the sequences

$$\{\mathbb{D}(x, \mathbf{u}_{(h)})\}, \ \left\{\chi^{\varepsilon}\mathbb{D}\left(x, \frac{\partial \mathbf{u}_{(h)}}{\partial t}\right)\right\}, \ \{\mathbf{v}_{(h)}\}, \ \{\mathbb{D}(x, \mathbf{v}_{(h)})\}, \ \{p_{(h)}\} \text{ and } \{\rho_{(h)}\}$$

converge weakly in $\mathbf{L}_2(\Omega_T)$ and $L_2(\Omega_T)$ as $h \to 0$ to the functions

$$\mathbb{D}(x, \mathbf{w}^{\varepsilon}), \ \chi^{\varepsilon}\mathbb{D}\left(x, \frac{\partial \mathbf{w}^{\varepsilon}}{\partial t}\right), \ \mathbf{v}^{\varepsilon}, \ \mathbb{D}(x, \mathbf{w}^{\varepsilon}), \ p^{\varepsilon} \text{ and } \rho^{\varepsilon}$$

correspondingly [68], and the sequence $\{\mathbf{v}_{(h)}\}$ converges strongly in $\mathbf{L}_2(\Omega_T)$ as $h \to 0$ to the function $\mathbf{v}^{\varepsilon} = \mathbb{E}_{\Omega_f^{\varepsilon}}\left(\dfrac{\partial \mathbf{w}^{\varepsilon}}{\partial t}\right)$.

In fact, to prove the last statement we fix the countable set $(t_{(k)})_{k=1}^{\infty}$, which is dense in $(0, T)$ and choose the subsequence from $\{h > 0\}$, such that the sequences $\{\nabla \mathbf{v}_{(h)}(\mathbf{x}, t_{(k)})\}$ converge weakly in $\mathbf{L}_2(\Omega)$ as $h \to 0$ for all $k = 1, 2, 3, \ldots$. That is possible due to the estimate (11.2.11) and the standard diagonal procedure. The last fact and the estimate (11.2.24) guarantee the weak convergence in $\mathbf{L}_2(\Omega)$ of the sequences $\{\nabla \mathbf{v}_{(h)}(\mathbf{x}, t)\}$ for all $t \in (0, T)$. Now we apply the completely continuous imbedding of $\overset{\circ}{\mathbf{W}}{}^1_2(\Omega)$ into $\mathbf{L}_2(\Omega)$ (see [68]) and conclude that the sequence' $\{\mathbf{v}_{(h)}(\mathbf{x}, t)\}$ converges strongly in $\mathbf{L}_2(\Omega)$ for all $t \in (0, T)$. The limiting procedure in the integral identities (11.2.20) and (11.2.21) as $h \to 0$ proves the statement of Theorem 11.1.

11.3 Proof of Theorem 11.2

To prove this theorem we only have to pass to the limit as $\varepsilon \to 0$ in the corresponding integral identities (11.1.4) and (11.1.5). It is already known, how to pass to the limit in the convection term $\rho^{\varepsilon} \cdot \mathbf{v}^{\varepsilon}$ in (11.1.4). The limit in (11.1.5) and in the continuity equation repeats the same procedure as in Chap. 1.4 (proof of Theorem 1.11) if we take into account the estimates (11.1.6)–(11.1.9).

Appendix A
Elements of Continuum Mechanics

A.1 Subject and Method of Continuum Mechanics

1. The subjects of study in continuum mechanics are physical bodies (physical continua), having the properties of *continuous media* and *internal mobility*. A physical continuum is a medium field with a continuous matter such that every part of the medium, however small, is itself a continuum and entirely filled with the matter. The property of internal mobility (or deformation) consists in translating separated parts of the physical continuum with respect to each other, but keeping the external form invariant.

Strictly speaking, by virtue of the atomic-molecular construction of any matter, no such physical bodies do not exist. When we talk about a physical continuum we suppose that the property of continuous matter is approximately true. That means that we regard the scale of molecular process to be less than minimum scale of interactions being studied. These scales are distinguished for different conditions. For example, the average distance between particles (molecules) of air near of the Earth is $l \sim 10^{-6}$ cm, but in the atmosphere at the height of 60 km it is $l \sim 10^{-3}$ cm and in outer space it is $l \sim 1$ cm. If one considers that the lower bound of length L on which processes are studied in these media is equal to 10^{-1}, 10^{2} and 10^{5} cm, then for all those cases we have $\dfrac{l}{L} \sim 10^{-5}$. Therefore media in outer space can be regarded as physical continua in the same meaning that we assume for air near the Earth.

So the continuum hypothesis implies that a very small volume will contain a large number of molecules. For example, $V = 1\,\text{cm}^3$ of air contains $N = 2.687 \times 10^{19}$ molecules under normal conditions (from Avogadro's hypothesis). Thus, in a cube with 0.001 cm sides there are 2.687×10^{10} molecules—which is a very large number. We are not interested in the properties of each molecule at any point $\mathbf{x}$ but rather in the average over a large number of molecules in the neighborhood of the point $\mathbf{x}$. Mathematically, the association of averaged values of properties at a point $\mathbf{x}$ gives rise to a continuum of points and numbers. In summary, the continuum hypothesis

A. Meirmanov, *Mathematical Models for Poroelastic Flows*,
Atlantis Studies in Differential Equations 1, DOI: 10.2991/978-94-6239-015-7,

implies the postulate: **Matter is continuously distributed throughout the region under consideration because there are a large number of molecules even in macroscopically small volumes.**

Conceptually continuum media are separated into gases, liquids and solids. This is a conditional distinction, depending on the statistical aspects of the molecular motions in the media. For example, in gases, the molecules are far apart-having an average separation between molecules of the order of 3.5×10^{-7} cm. The cohesive forces between the molecules are weak. The molecules randomly collide and exchange their momentum, heat, and other properties giving rise to viscosity, thermal conductivity, etc. These effects, though molecular in origin, are considered to be physical properties of the medium itself. In liquids, the separation between molecules is much smaller and the cohesive forces between a molecule and its neighbors are quite strong. Again, the averaged molecular properties resulting from these cohesive forces are taken as the properties of the medium. While air and water are treated by the same continuum hypothesis, the effects of their motions are different due to the differences in their molecular properties, e.g. viscosity, thermal conductivity, etc.

2. Continuum mechanics describes the global behavior of gases, liquids or solids under the influence of external disturbances.

The concept of a physical continuum makes the powerful methods of calculus available for the study of nonuniform distributions of physical variables and provides an easily visualized physical model that closely approximates observations of bulk matter. The problems of continuum mechanics are multiform.

Continuum mechanics is a foundation for the understanding of many aspects of the applied sciences and engineering. It is a subject of enormous interest in numerous fields such as biology, biomedicine, geophysics, meteorology, physical chemistry, plasma physics and almost all branches of engineering.

Continuum mechanics is separated into experimental physical and theoretical parts. We will consider only theoretical continuum mechanics.

The **method of theoretical continuum mechanics** consists of constructing a mathematical model of the behavior of continuous media. A mathematical model is a system of relationships (equations and inequalities) between values, which characterize the different properties of media. Usually they are differential (finite) equations. The initial and boundary data are added to these equations. The mathematical model has to have the property of correctness. That means that the solutions of its component equations have to exist, to be unique, and to be stable. For some models there is no strict proof of correctness: in these cases one has to use the criteria of real-life experience. Physical experiments serve as tests for the validity of a theoretical model.

After constructing a mathematical model we produce purely mathematical methods to study it. For achieve this we use analytical and numerical methods. Because of the difficulty of solving equations in continuum mechanics there are various methods of simplifications.

3. To understand better the physical foundations of the construction of mathematical models of continuum mechanics we firstly consider a molecular (microscopic) description.

Let some volume V of a continuum medium contain N molecules μ_i $(i = 1, 2, \ldots, N)$ with coordinates of position $\mathbf{x}$ and mass m_i. The motion $\mathbf{x}_i(t)$ of molecule μ_i obeys Newton's Second Law

$$m_i \frac{d^2\mathbf{x}_i}{dt^2} = \mathbf{f}_i, \quad \mathbf{x}_i(t_0) = \mathbf{x}_i^0, \quad \frac{d\mathbf{x}_i}{dt}(t_0) = \mathbf{v}_i^0, \quad (i = 1, 2, \ldots, N),$$

where $\mathbf{f}_i$ is a force, which acts on the molecule μ_i. A solution of these equations defines the position and velocity of molecule μ_i at any moment of time t. If we were able to solve these equations we could answer any question about the behavior of media in the volume V.

However this method is impracticable, because the number N is very large and we do not know exactly the forces $\mathbf{f}_i$. Therefore in continuum mechanics we adopt a macroscopic viewpoint: we ignore all the fine details of the molecular or atomic structure and, for the purpose of study, we replace the microscopic medium with a hypothetical continuum in which the basic values are replaced by average values.

To distinguish the continuum or macroscopic model from a microscopic one the concept of the mean free path plays a fundamental role. This concept can be defined as the average distance that a molecule travels between successive collisions with other molecules. The ratio of the mean free path λ to the characteristic length L of the physical boundaries of interest, called the Knudsen number $K_n = \frac{\lambda}{L}$, may be used to distinguish between macroscopic and microscopic models. Based on the Knudsen number the motion regimes are grouped as:

(a) continuum ($K_n < 0.1$);
(b) rarefied gas ($0.1 < K_n < 5$);
(c) free molccular flow ($K_n > 5$).

Regimes (a) and (b) are macroscopic models. All these regimes are encountered in real life.

Two macroscopic theories are the most prevalent: the molecular-kinetic theory and the phenomenological theory.

In the molecular-kinetic theory all average values are described with the help of a theoretical probability approach. The mathematical model takes the form of the Boltzmann equation. We study the phenomenological theory.

4. The basis of the phenomenological theory is that each point of a body V is represented by its density, velocity and other mechanical values. These values are defined as the limits of some average values, which are determined in the following way. Let molecules μ_i $(i = 1, 2, \ldots, N)$ from volume V have mass m_i, velocity $\mathbf{v}_i$ and internal energy U_i. With the help of these values one calculates the macro-characteristics of the volume V:

$$M = \sum_{i=1}^{N} m_i \quad \text{is the mass,}$$

$$\mathbf{K} = \sum_{i=1}^{N} m_i \mathbf{v}_i \quad \text{is the impulse,}$$

$$E = \sum_{i=1}^{N} \left(U_i + \frac{1}{2} m_i |\mathbf{v}_i|^2 \right) \quad \text{is the total energy.}$$

Then

$$\rho_* = \frac{m}{|V|} \quad \text{is the average density,}$$

$$\mathbf{v}_* = \frac{\mathbf{K}}{|V|} \quad \text{is the average velocity,}$$

$$U_* = \frac{U}{|V|} \quad \text{is the average energy.}$$

Here

$$U = \sum_{i=1}^{N} \left(U_i + \frac{1}{2} m_i |\mathbf{v}_i - \mathbf{v}_*|^2 \right) \text{ and } |V| \text{ is the volume of } V.$$

The macroscopic characteristics of the volume V can be expressed by means of the average values:

$$M = |V|\, \rho_*, \quad \mathbf{K} = |V|\, \rho_*\, \mathbf{v}_*, \quad E = |V| \left(U_* + \frac{1}{2} \rho_*\, |\mathbf{v}_*|^2 \right).$$

The hypothesis of the physical continuum allows us to give the point $\mathbf{x}$ the "limit" values of its averages, for example,

$$\rho = \lim \rho_*, \quad \mathbf{v} = \lim \mathbf{v}_*,$$

where the volume V vanishes such that $\mathbf{x} \in V$. A mathematical model takes the form of conservation laws that describe the changes of macroscopic characteristics with respect to time.

We will construct the phenomenological theory of continuum mechanics as the theory of a mathematical structure. This mathematical structure is based on the following system of axioms.

A.2 Basic Definitions and Axioms

1. A continuous medium is a part of physical space, changing with time. That means that a continuous medium is a part of Euclidean three dimensional space $\mathbb{R}^3$, and that time is independent of events. We use the non-relativistic Newtonian approach, i.e. time is absolute.

Axiom 1 (Axiom of space-time)
Continuous medium is a subset of three dimensional Euclidean affine space. Time is absolute. A Euclidean-affine space is a curvature-free space in which a set of rectangular Cartesian coordinates can always be introduced on a global scale. It is a linear three dimensional space over a field of real numbers $\mathbb{R}$. In this space the origin point O is fixed. Open connected sets $\Omega \subset \mathbb{R}^3$ are regarded as a positions (configurations) of the continuous medium.

Axiom 2 (Axiom of mass and internal energy)
Set $\Omega \subset \mathbb{R}^3$ is called a material domain (or medium) if an additive positive function of sets $M(\omega)$ is defined on it, which is called mass. It is supposed that for any (nonempty) volume $\omega \subset \Omega$ its mass is $M(\omega) > 0$. The additiveness of mass means that if $\omega_1 \subset \Omega$, $\omega_2 \subset \Omega$ and $\omega_1 \cap \omega_2 = \emptyset$, then $M(\omega_1 \cup \omega_2) = M(\omega_1) + M(\omega_2)$. Besides mass we determine another additive function of a set, which we call the internal energy and we denote it by E_i.

A medium Ω is called a *material continuum*, if functions M and E_i are differentiable on Ω and their densities (volume densities) are bounded.

The volume density of mass is denoted by ρ and it is called a *density of media* (or simply *density*). The volume density of energy is denoted by ρU and U is called the *specific internal energy* (internal energy per unit mass). The following formula

$$M(\omega) = \int_{\omega} \rho \, dx, \quad E_i(\omega) = \int_{\omega} \rho \, U \, dx$$

determines the connection between the additive functions of set ω and its volume density.

Axiom 3 (Axiom of material continuum)
A continuous medium is material continuum. The transition of continuous medium from position Ω_1 into position Ω_2 is called its motion. The motion of a continuous medium depends on time t, which varies in some interval $(0, t_0) \subset \mathbb{R}$. The position of medium at the moment of time t is denoted by Ω_t. For all $t \in (0, t_0)$ we consider one-parametrical family of movements γ_t from position Ω_0 to Ω_t. That means that we have a mapping: $\gamma : \Omega_0 \times (0, t_0) \longrightarrow \Omega_t$. We denote $\gamma_t(\xi) = \gamma(\xi, t)$ for all $\xi \in \Omega_0$ and $\gamma_\xi(t) = \gamma(\xi, t)$ for all $t \in (0, t_0)$ or we will write

$$\gamma_t : \Omega_0 \longrightarrow \Omega_t, \quad \gamma_\xi : (0, t_0) \longrightarrow \Omega_t.$$

Set

$$\{\mathbf{x} \in \mathbb{R}^3 : \mathbf{x} = \gamma_\xi(t),\ t \in (0, t_0)\}$$

is called a trajectory of point $\xi \in \Omega_0$.

Axiom 4 (Axiom of movement)

For every $t \in (0, t_0)$ *there exists the movement* γ_t *of a continuous medium from position* Ω_0 *to position* Ω_t *and a mapping* $\gamma_t : \Omega_0 \longrightarrow \Omega_t$ *is a homeomorphism; for all points* $\xi \in \Omega_0$ *the mapping* $\gamma_\xi : (0, t_0) \longrightarrow \Omega_t$ *is a continuous and piecewise continuously differentiable function on* $(0, t_0)$.

This axiom allows one to postulate a point of a continuous medium.

A material point (or particle) of a continuous medium is called a point $\mathbf{x} = \gamma(\xi, t) \in \Omega_t$, which is obtained as a result of movement of the fixed point $\xi \in \Omega_0$. Every particle describes in $\mathbb{R}^3$ the trajectory of this point.

A set of points which consists of the same particles for all $t \in (0, t_0)$ is called *a material volume* ω_t. By virtue of axiom 3 for all $\xi \in \Omega_0$ and all (except maybe for a finite number) values $t \in (0, t_0)$ there exists a derivative $\dfrac{\partial \gamma}{\partial t}(\xi, t)$.

A derivative $\dfrac{\partial \gamma}{\partial t}(\xi, t)$ is called the *velocity* of point $\xi \in \Omega_0$ and it is denoted by

$$\mathbf{v} = \frac{\partial}{\partial t}\gamma(\xi, t).$$

2. Let F be either a scalar or vector or tensor function of position $\mathbf{x}$ and time t, representing some physical property of the movement. There are two ways to describe a field F in the moving continuous medium. The first one is called *Eulerian*. It consists of giving a value of field F at the position t as a function of $\mathbf{x} \in \mathbb{R}^3$ and time $t \in (0, t_0)$, i.e. it has a value $F(\mathbf{x}, t)$.

The second way is called *Lagrangian*. In this case the given field is considered as a function of each particle $\xi \in \Omega_0$ at the moment of time $t \in (0, t_0)$. Let it be $\overset{0}{F}(\xi, t)$. The functions $F(\mathbf{x}, t)$ and $\overset{0}{F}(\xi, t)$ are connected by identity

$$F(\mathbf{x}, t) = \overset{0}{F}(\xi, t). \tag{A.2.1}$$

There are two possible time derivatives:

$$\frac{\partial F}{\partial t}(\mathbf{x}, t) \text{ and } \frac{\partial \overset{0}{F}}{\partial t}(\xi, t).$$

A value $\dfrac{\partial F}{\partial t}(\mathbf{x}, t)$ is the rate of change of field F measured by an observer stationed at the fixed point $\mathbf{x} \in \Omega_t$ and it is a local time variation of F.

On the other hand, $\frac{\partial \overset{0}{F}}{\partial t}(\xi, t)$ is a rate of change of $\overset{0}{F}(\xi, t)$ measured by an observer moving with the particle. The differentiation (A.2.1) with respect to time gives

$$\frac{\partial \overset{0}{F}}{\partial t} = \frac{\partial F}{\partial t} + \nabla F \cdot \mathbf{v} \overset{df}{=} \frac{dF}{dt}.$$

A value $\frac{dF}{dt}$ is called a total derivative (material or substantial derivative, or the derivative following the motion).

In particular, if $F = \mathbf{x} = \gamma(\xi, t)$ we obtain the formula for the definition of velocity

$$\mathbf{v} = \frac{\partial \gamma}{\partial t}(\xi, t) = \frac{d\mathbf{x}}{dt}.$$

Coordinates (ξ, t) are called *material* or *Lagrangian coordinates* and $(\mathbf{x}, t)$ are called *spatial* or *Eulerian coordinates.*

The difference between these descriptions is crucial. For example, if the field of a vector of velocity is known in a Lagrange description, i.e. we have a vectorial function $\overset{\mathbf{0}}{\mathbf{v}}(\xi, t)$, then we can find trajectories of particles (and that means we can find the movement of the continuous medium)

$$\mathbf{x} = \xi + \int_0^t \overset{\mathbf{0}}{\mathbf{v}}(\xi, \tau) d\tau.$$

And if we know a field $\mathbf{v}$ in a Eulerian description (meaning that we have $\mathbf{v} = \mathbf{v}(\mathbf{x}, t)$), then the same problem of determination of trajectories gives us the Cauchy problem for the system of ordinary differential equations

$$\frac{d\mathbf{x}}{dt} = \mathbf{v}(\mathbf{x}, t), \quad \mathbf{x}(\xi, 0) = \xi. \tag{A.2.2}$$

In spite of the simplicity of the first problem a Lagrangian description is not always convenient. In particular, the main differential equations of continuum mechanics are simpler as Eulerian description.

In the Eulerian description a map $\gamma : \Omega_0 \times (0, t_0) \longrightarrow \Omega_t$ is obtained as a solution of the Cauchy problem (A.2.2). If the vectorial function $\mathbf{v}(\mathbf{x}, t)$ is continuously differentiable, then for such solution there exists the Jacobian $J = det(\frac{\partial \mathbf{x}}{\partial \xi})$. For the Jacobian we have a kinematic formula, known as *Euler's formula*:

$$\frac{dJ}{dt} = J \nabla \cdot \mathbf{v}.$$

3. In addition to the main numerical characteristics of material media (mass and energy) there are the following additive functions of the set $\omega \subset \Omega$:

(i) linear momentum:

$$\mathbf{K}(\omega) = \int_\omega \rho\, \mathbf{v}\, dx,$$

(ii) angular momentum:

$$\mathbf{H}(\omega) = \int_\omega \rho\, (\mathbf{x} \times \mathbf{v})\, dx,$$

(iii) kinetic energy:

$$E_k(\omega) = \int_\omega \frac{1}{2}\rho\, |\mathbf{v}|^2\, dx,$$

(iv) total energy:

$$E(\omega) = E_k(\omega) + E_i(\omega).$$

The changes of these magnitudes under movement are the result of force and energetic changes in the volume ω. These actions are realized with the help of new magnitudes: resultant force $\mathbf{F}(\omega)$, resultant moment $\mathbf{G}(\omega)$ and power N.

If we take these magnitudes for a fixed moving material volume ω_t, then they will be only functions of time t. A following axiom determine the relations between them.

Axiom 5 (Balance and poise)
For arbitrary moving material volume ω_t and in any time $t \in (0, t_0)$ we have

$$\frac{d}{dt}M(\omega_t) = 0, \quad \frac{d}{dt}\mathbf{K}(\omega_t) = \mathbf{F}(\omega_t),$$

$$\frac{d}{dt}\mathbf{H}(\omega_t) = \mathbf{G}(\omega_t), \quad \frac{d}{dt}E(\omega_t) = N(\omega_t).$$

Sometimes this axiom is called the *hardening principle*, because these equalities are fulfilled for the movement of rigid bodies.

3. Further, we have to specify the right-hand sides in the formulae of Axiom 4. At first we define the concept of resultant forces. We will consider two types of forces which act on a material volume ω:

(a) body forces,
(b) surface forces.

The body forces are forces of an extensive character acting on the bulk portions of the continuous medium and arise from some external cause. Examples of external causes are the force of gravity, forces of electric and magnetic origin acting on a continuous medium carrying charged particles, etc. The body force is proportional to the volume of a continuous medium and therefore it is expressed as a force per unit volume.

An additive vectorial function $\mathbf{F}_e$ having a density (body force per unit mass) is called an *external body force*. If we denote the body force per unit mass by the symbol $\mathbf{f}(\mathbf{x}, t)$, then the body force per unit volume will be $\rho\,\mathbf{f}$. Therefore, the external body force acting on the volume ω is given by formula

$$\mathbf{F}_e(\omega) = \int_\omega \rho\,\mathbf{f}\,dx.$$

And the moment of external body force acting on any material volume ω is defined by the formula

$$\mathbf{G}_e(\omega) = \int_\omega \rho\,(\mathbf{x} \times \mathbf{f})\,dx.$$

Surface forces are forces of an intensive or local nature. They arise from mechanical interaction between contiguous portions of continuous media. To explain the phenomena from a continuum point of view, we consider two adjacent portions of continuous media separated by an imaginary surface drawn between the media.

At the separating surface there exists a direct mechanical contact between the particles of the media on the two sides of the surface, giving rise to forces of action and reaction. If the continuous medium on one side is imagined to have been replaced by the force system which it has produced, then at each point of the imaginary surface there will be a force vector.

An internal surface force acts on a volume ω only through its surface $\partial\omega$. In order to define it we consider cross section Σ of Ω by some plane dividing Ω into two parts Ω_1 and Ω_2.

The additive vector-function $\mathbf{F}_i$ of sets $\sigma \subset \Sigma$ is called *an internal surface force* acting through a cross section Σ from the side Ω_2 on the Ω_1.

Axiom 6 (Of internal surface forces)

An internal surface force is defined for any cross section Σ of Ω and it has a density (surface) on Σ.

Remark This axiom is named the Cauchy Stress Principle and it asserts the existence and differentiability of this force.

Let $\mathbf{n}$ be a local outward drawn unit normal vector of Σ directed on the side of Ω_2 (positive side of Ω_1). We denote the density of the internal surface force by $\mathbf{p}_n$.

A vector $\mathbf{p}_n$ is called the *stress vector* of surface forces acting on Ω_1 through the area with the normal $\mathbf{n}$.

And for $\sigma \subset \Sigma$ the force, which acts on part Ω_1 from the side of part Ω_2 through an area σ is equal to

$$\mathbf{F}_i(\sigma) = \int_\sigma \mathbf{p}_n\,d\sigma.$$

The value

$$\mathbf{F}_i(\omega) = \int_{\partial\omega} \mathbf{p}_n \, d\sigma$$

is called the *internal surface force* acting on volume $\omega \subset \Omega$ from the side of Ω. Here $\mathbf{n}$ is positive outward drawn unit normal vector to the surface volume ω.

The value

$$\mathbf{G}_i(\omega) = \int_{\partial\omega} (\mathbf{x} \times \mathbf{p}_n) \, d\sigma$$

is called *a moment of internal surface force*, acting on the volume ω.

Axiom 7 (Of forces and moments)

The (main) resultant force and resultant moment, acting on any material volume $\omega \subset \Omega$ is given by the formulae:

$$\mathbf{F}(\omega) = \mathbf{F}_i(\omega) + \mathbf{F}_e(\omega) = \int_{\partial\omega} \mathbf{p}_n \, d\sigma + \int_{\omega} \rho \, \mathbf{f} dx,$$

$$\mathbf{G}(\omega) = \mathbf{G}_i(\omega) + \mathbf{G}_e(\omega) = \int_{\partial\omega} (\mathbf{x} \times \mathbf{p}_n) \, d\sigma + \int_{\omega} \rho \, (\mathbf{x} \times \mathbf{f}) \, dx.$$

4. In contrast to forces and moments, acting on the volume ω, the power, brought into the volume ω, depends not only on the forces acting, but also on the heat output and on external heat sources. So, an additive scalar function Q of sets $\sigma \subset \Sigma$ is called a *heat output* through area Σ from the part Ω_2 into Ω_1.

Axiom 8 (Of heat output)

A heat output is defined for any cross section Σ of Ω and it has a density (surface density) on Σ. The surface density of heat output is denoted by q_n and the value

$$Q(\sigma) = \int_{\sigma} q_n d\sigma$$

gives the heat output from the Ω_2 into Ω_1 through the area $\sigma \subset \Sigma$.

The value

$$Q(\omega) = \int_{\partial\omega} q_n d\sigma$$

is called the *heat output* into volume $\omega \subset \Omega$ from the domain $\Omega \setminus \bar{\omega}$. Here $\mathbf{n}$ is a positive outwardly-directed unit vector normal to the volume surface $\partial\omega$.

Axiom 9 (Of energy transfer)

The power $N(\omega)$ passing into any volume ω is equal to

$$N(\omega) = N_e(\omega) + N_i(\omega) + Q(\omega) = \int_{\omega} \rho \, (\mathbf{v} \cdot \mathbf{f}) dx + \int_{\partial\omega} \mathbf{v} \cdot \mathbf{p}_n d\sigma + \int_{\partial\omega} q_n d\sigma.$$

We can summarize the previous axioms and definitions as the following classical mathematical model of moving continuous media.

Mathematical model (M_1) (Integral conservation laws)

In a moving continuous medium for any moving volume $\omega_t \subset \Omega_t$ and at any moment of time $t \in (0, t_0)$ the following equalities hold true:

$$\frac{d}{dt}\int_{\omega_t} \rho\, dx = 0,$$

$$\frac{d}{dt}\int_{\omega_t} \rho\, \mathbf{v}\, dx = \int_{\partial\omega_t} \mathbf{p}_n\, d\sigma + \int_{\omega_t} \rho \mathbf{f}\, dx,$$

$$\frac{d}{dt}\int_{\omega_t} \rho\, (\mathbf{x} \times \mathbf{v})\, dx = \int_{\partial\omega_t} (\mathbf{x} \times \mathbf{p}_n)\, d\sigma + \int_{\omega_t} \rho\, (\mathbf{x} \times \mathbf{f})\, dx,$$

$$\frac{d}{dt}\int_{\omega_t} \rho \left(\frac{1}{2}|\mathbf{v}|^2 + U\right) dx = \int_{\partial\omega_t} (\mathbf{v} \cdot \mathbf{p}_n)\, d\sigma + \int_{\omega_t} \rho\, (\mathbf{v} \cdot \mathbf{f})\, dx + \int_{\partial\omega_t} q_n\, d\sigma.$$

Each of these equalities is called the *conservation law* of the corresponding mechanical value: conservation law of mass, conservation law of linear momentum, conservation law of angular momentum, conservation law of energy.

Finally, we may formulate the following definition:

A moving continuous medium is an object satisfying the Axioms **A1–A9**. *The mathematical model consists of four conservation laws.*

A.3 Continuous Motion

1. The main functions (magnitudes) related to a moving continuous medium: density ρ, specific internal energy U, velocity $\mathbf{v}$, stress $\mathbf{p}_n$, with a normal vector $\mathbf{n}$, a density of heat output q_n, and a density of external body forces $\mathbf{f}$, will be further considered using a Eulerian description. This means that these functions are functions of $(\mathbf{x}, t)$ in a domain $W \in \mathbb{R}^4(\mathbf{x}, t)$. The magnitudes $\mathbf{p}_n$ and q_n depend on the unit vector $\mathbf{n} \in \mathbb{R}^3$ (point of a unit sphere S_1) and therefore they are given on the product $W \times S_1$.

At first we study a class of movements of continuous media where the main magnitudes are sufficiently smooth functions.

2. A movement of a continuous medium is called *continuous* in a domain W if the functions ρ, U, $\mathbf{v}$, $\mathbf{p}_n$, q_n are continuously differentiable functions in W, the functions $\mathbf{p}_n$ and q_n are continuous in $W \times S_1$, and the function $\mathbf{f}$ is continuous in W.

3. Let us consider the derivative

$$\frac{d}{dt}I = \frac{d}{dt}\int_{\omega_t} \rho F dx,$$

where the function $F(\mathbf{x}, t)$ is continuously differentiable and the movement of the continuous medium is continuous. In order to calculate this value we perform a transition to the Lagrange system of coordinates $\mathbf{x} = \gamma(\xi, t)$. The integral has the form

$$I = \int_{\omega_0} \overset{0}{\rho}(\xi, t) \overset{0}{F}(\xi, t) \overset{0}{J}(\xi, t) d\xi.$$

By virtue of the Theorem of Real Analysis we can replace the integral and calculate the derivative

$$\frac{d}{dt} I = \int_{\omega_0} \frac{\partial}{\partial t}\Big(\overset{0}{\rho}(\xi, t) \overset{0}{F}(\xi, t) \overset{0}{J}(\xi, t)\Big) d\xi.$$

On the strength of Euler's formula

$$\frac{\partial}{\partial t}\Big(\overset{0}{\rho}(\xi, t) \overset{0}{F}(\xi, t) \overset{0}{J}(\xi, t)\Big) = \frac{d}{dt}\Big(\rho F J\Big) = J\Big(\frac{d}{dt}(\rho F) + \rho\, F\, \nabla \cdot \mathbf{v}\Big),$$

we obtain

$$\int_{\omega_0} J\Big(\frac{d}{dt}(\rho F) + \rho\, F\, \nabla \cdot \mathbf{v}\Big)\, d\xi = \int_{\omega_t} \Big(\frac{d}{dt}(\rho F) + \rho\, F\, \nabla \cdot \mathbf{v}\Big)\, dx$$

and

$$\frac{d}{dt} \int_{\omega_t} \rho F dx = \int_{\omega_t} \Big(\frac{d}{dt}(\rho F) + \rho\, F\, \nabla \cdot \mathbf{v}\Big)\, dx. \tag{A.3.1}$$

4. For $F = 1$ one has

$$\frac{d}{dt} \int_{\omega_t} \rho\, dx = \int_{\omega_t} \Big(\frac{d\rho}{dt} + \rho\, \nabla \cdot \mathbf{v}\Big)\, dx = 0.$$

Because ω_t is an arbitrary volume then by virtue of the Theorem of Real Analysis

$$\frac{d\rho}{dt} + \rho\, \nabla \cdot \mathbf{v} = 0. \tag{A.3.2}$$

This equation is called a *continuity equation*. It is equivalent to the mass conservation law in the class of continuous motions.

The continuity equation permits the simplifying

$$\frac{d}{dt} \int_{\omega_t} \rho\, F\, dx = \int_{\omega_t} \rho\, \frac{dF}{dt}\, dx. \tag{A.3.3}$$

5. By virtue of (A.3.3) the equation of linear momentum takes the form

$$\int_{\omega_t} \rho\, \frac{d\mathbf{v}}{dt}\, dx = \int_{\partial\omega_t} \mathbf{p}_n\, d\sigma + \int_{\omega_t} \rho\, \mathbf{f}\, dx.$$

The last formula implies

Theorem 1 (The first fundamental theorem of continuum mechanics)
There exists a tensor field of second order $\mathbb{P}$ *in* W *such that for all* $(\mathbf{x}, t) \in W$

$$\mathbf{p}_n = \mathbb{P}\langle \mathbf{n} \rangle.$$

The tensor $\mathbb{P}$ is called a *stress tensor.*

Using the Gauss-Ostrogradsky Theorem, we have

$$\int_{\partial \omega_t} \mathbf{p}_n \, d\sigma = \int_{\omega_t} \nabla \cdot \mathbb{P} \, dx.$$

Hence, the equation of linear momentum is reduced to

$$\int_{\omega_t} \left(\rho \frac{d\mathbf{v}}{dt} - \nabla \cdot \mathbb{P} - \rho \, \mathbf{f} \right) dx = 0.$$

For a continuous motion the integrand function is continuous. Since this equation is valid for any volume ω_t, we get the differential form of the conservation law of linear momentum

$$\rho \frac{d\mathbf{v}}{dt} = \nabla \cdot \mathbb{P} + \rho \, \mathbf{f}. \tag{A.3.4}$$

In the same way we may consider the angular momentum equation and prove

Theorem 2 *For a continuous motion the conservation law of angular momentum equation is fulfilled if and only if the stress tensor* $\mathbb{P}$ *is a symmetric tensor, i.e.,* $\mathbb{P} = \mathbb{P}^*$.

6. For a continuous motion the conservation law of energy is reduced to the equation

$$\int_{\omega_t} \left(\rho \frac{d}{dt} \left(\frac{1}{2} |\mathbf{v}|^2 + U \right) - \nabla \cdot (\mathbb{P}\langle \mathbf{v} \rangle) - \rho \, (\mathbf{v} \cdot \mathbf{f}) \right) dx = \int_{\partial \omega_t} q_n \, d\sigma.$$

This representation for q_n gives

Theorem 3 *For a continuous motion in* W *there exists a vector field* $\mathbf{q}$ *in* W*, such that for all* $(\mathbf{x}, t) \subset W$

$$q_n = -\mathbf{q} \cdot \mathbf{n}. \tag{A.3.5}$$

Vector $\mathbf{q}$ is called *a heat output rate vector* (or *heat flux).*

Introducing the heat output rate vector allows the transformation of the surface integral into the volume integral:

$$\int_{\partial \omega_t} q_n \, d\sigma = - \int_{\partial \omega_t} \mathbf{q} \cdot \mathbf{n} \, d\sigma = - \int_{\partial \omega_t} \nabla \cdot \mathbf{q} dx,$$

and the conservation law of energy becomes

$$\int_{\omega_t} \left(\rho \frac{d}{dt}\Big(\frac{1}{2}|\mathbf{v}|^2 + U\Big) - \nabla \cdot \big(\mathbb{P}\langle \mathbf{v}\rangle\big) - \rho(\mathbf{v}\cdot\mathbf{F}) + \nabla \cdot \mathbf{q}\right) dx = 0.$$

For a continuous motion the last equation is equivalent to

$$\rho \frac{d}{dt}\Big(\frac{1}{2}|\mathbf{v}|^2 + U\Big) = \nabla \cdot \big(\mathbb{P}\langle \mathbf{v}\rangle\big) + \rho(\mathbf{v}\cdot\mathbf{F}) - \nabla \cdot \mathbf{q}.$$

We simplify this equation using the relationships

$$\frac{d}{dt}|\mathbf{v}|^2 = 2\mathbf{v}\cdot\frac{d\mathbf{v}}{dt},$$

$$\nabla \cdot \big(\mathbb{P}\langle \mathbf{v}\rangle\big) = \mathbf{v}\cdot\big(\nabla\cdot\mathbb{P}\big) + \mathbb{P} : \big(\mathbb{D}(x,\mathbf{v})\big),$$

where $\mathbb{D}$ is a rate-of-strain tensor

$$2\,\mathbb{D}(x,\mathbf{v}) = \nabla\mathbf{v} + \big(\nabla\mathbf{v}\big)^*,$$

and differential equation (A.3.4):

$$\rho\frac{dU}{dt} = \mathbb{P} : \mathbb{D}(x,\mathbf{v}) - \nabla\cdot\mathbf{q}. \tag{A.3.6}$$

This equation is called an *energy equation* (or *heat flux equation*).

Thus, for an arbitrary continuous motion of a continuous medium described by model (M_1), there exist continuously differentiable fields of a symmetric stress tensor $\mathbb{P}$ and a vector of heat output rate in which the integral conservation laws are equivalent to the system of differential equations

$$\frac{d\rho}{dt} + \rho\,\nabla\cdot\mathbf{v} = 0,$$

$$\rho\frac{d\mathbf{v}}{dt} = \nabla\cdot\mathbb{P} + \rho\,\mathbf{f},$$

$$\rho\frac{dU}{dt} = \mathbb{P} : \mathbb{D}(x,\mathbf{v}) - \nabla\cdot\mathbf{q}.$$

This system of partial differential equations is called the model (M_2) of continuous motion of continuum mechanics. If we assume that the body force is prescribed, then the model (M_2) consists of five independent (scalar) equations involving fourteen unknown variables, namely, ρ, $\mathbf{v}$, $\mathbb{P}$, U, $\mathbf{q}$.

A model is called *closed* if the number of unknown variables is equal to the number of equations in the model. And so, we have the problem of closing the model (M_2). This problem has to be solved on the basis of an additional information about continuous media.

A.4 Elements of Thermodynamics

1. Thermodynamics studies the relations between the heat energy and other kinds of energies and gives (reciprocal) formulae to convert of one kind of energy into another. For example, if a body is heated, then strains and stresses are developed. Conversely, if a body is strained rapidly, then heat is generated inside the body.

The main concept of thermodynamics is the physical body state. A phenomenological description of the state is given with the help of various functions called state variables. For example, as mentioned before, the density ρ (or specific volume $V = \frac{1}{\rho}$), the internal energy U are parameters of the state of a continuous medium. Also the absolute temperature $\theta > 0$, specific entropy S and pressure p are basic state variables.

Let $\mathbf{z} = (z_1, z_2, \dots, z_k)$ be a set of the main state variables of continuous medium: other state variables are functions of these variables. Such a medium is called a *k-parameter medium.*

Each point $\mathbf{z}$ characterizes the state of the given continuous medium and we briefly call this point the *state* (of the medium). Usually, the set of all such points (a space of states) form some manifold Z. Suppose that in the space of state variables Z we may choose a path (oriented curve) $l(\mathbf{z}^{(1)}, \mathbf{z}^{(2)})$, which provides a change of variables from one state $\mathbf{z}^{(1)}$ to another state $\mathbf{z}^{(2)}$. These paths (changes of states) are called *processes*. If for a process from $\mathbf{z}^{(1)}$ into $\mathbf{z}^{(2)}$ there exists a process from $\mathbf{z}^{(2)}$ into $\mathbf{z}^{(1)}$, then such a process is called *reversible*, otherwise it is called *irreversible*.

2. Generally speaking, the heat Q, which is the energy of the chaotic motion of molecules, is not a state variable. This heat Q, obtained by the medium after a change of state from $\mathbf{z}^{(1)}$ to $\mathbf{z}^{(2)}$ via process $l(\mathbf{z}^{(1)}, \mathbf{z}^{(2)})$ depends on this process.

Let us consider all states $\mathbf{z}'$ in some small neighborhood of the state $\mathbf{z}$. Then for smooth processes $l(\mathbf{z}, \mathbf{z}')$ one has

$$dQ = \sum_{m=1}^{k} B_m(\mathbf{z})dz_m,$$

where $dz_m = z'_m - z_m$. In this representation the quantity Q depends on the process, if the right-hand side of the formula is not a total differential. But one may prove that there exists a state parameter θ (temperature), such that for reversible processes the quotient $\frac{dQ}{\theta}$ becomes the total differential of another state parameter S. This function is called *entropy*. Entropy is regarded as a measure of change of energy dissipation with respect to temperature and defined as

$$S' - S = \int_{l(\mathbf{z},\mathbf{z}')} \frac{dQ}{\theta}$$

for any reversible process $l(\mathbf{z}, \mathbf{z}')$.

3. The first law of thermodynamics states that the equality

$$dQ = dA + dU$$

always holds true, expressing the energy conservation law: if some physical body receives heat dQ, then this body will do mechanical work dA, and its internal energy increases by dU.

The second law of thermodynamics is based on the concept of entropy associated with irreversible thermodynamic processes and states that

$$\theta dS \geqslant dQ$$

for irreversible processes and

$$\theta dS = dQ$$

for reversible processes.

Axiom 10 (Thermodynamics axiom)

For a continuous medium the first and the second laws of thermodynamics apply. Thus, for reversible thermodynamic processes *the basic thermodynamic identity*

$$\theta dS = dA + dU$$

holds true.

4. Heat fluxes from one part of a continuous medium to another part are described empirically by Fourier's law, expressing the very simple fact that these fluxes are the results of differences of the temperature in different parts of the medium.

Axiom 11 (Fourier's law)

The heat flux is proportional to the temperature gradient:

$$\mathbf{q} = -\varkappa \nabla \theta. \tag{A.4.1}$$

The coefficient of heat conductivity $\varkappa$ is always positive. In models of continuum mechanics it is considered as a known function of other state variables. Therefore, the energy equation has the form

$$\rho \frac{dU}{dt} = \mathbb{P} : \mathbb{D} + \nabla \cdot (\varkappa \nabla \theta). \tag{A.4.2}$$

A.5 Some Classical Models of Continuum Mechanics

1. Differential equations of mathematical model (M_2) are universal, that is, they are valid for all continuous media. On the other hand, all additional relations (axioms), that we have to formulate to close the mathematical model (M_2), depend on the given continuous medium. For example, for highly mobile continuous media, such as liquids or gases, the stress tensor depends on the rate of strain tensor and independent thermodynamic state variables. We do not formalize the derivation of a closed mathematical model of liquid or gas, and assume that this relation is linear:

$$\mathbb{P} = 2\mu\mathbb{D}(x, \mathbf{v}) + \big(-p + \nu\nabla\cdot\mathbf{v} - \gamma_f(\theta - \theta_0)\big)\mathbb{I}. \tag{A.5.1}$$

The scalar invariants μ and ν depend on the thermodynamic state variables and they are called *the first (dynamic) and second (volume or bulk) coefficients of viscosity*, respectively, and are supposed to be given. The quantity θ_0 is a given average temperature.

Axiom (A.5.1) allows one to calculate

$$\mathbb{P} : \mathbb{D}(x, \mathbf{v}) = \big(-p - \gamma_f\vartheta\big)\nabla\cdot\mathbf{v} + \Phi,$$

where the function

$$\Phi_f = 2\mu\mathbb{D}(x, \mathbf{v}) : \mathbb{D}(x, \mathbf{v}) + \nu(\nabla\cdot\mathbf{v})^2$$

is called the *dissipation function*, and for the constant viscosity μ

$$\nabla\cdot\mathbb{P} = \mu\Delta\mathbf{v} - \nabla p + (\nu + \mu)\nabla(\nabla\cdot\mathbf{v}) - \gamma_f\nabla\vartheta.$$

To complete the model we postulate the basic thermodynamic identity

$$\vartheta dS = dU + pdV. \tag{A.5.2}$$

The last identity is more conveniently written in the form

$$d\Psi = Vdp - Sd\vartheta \tag{A.5.3}$$

for the thermodynamic potential

$$\Psi = U + pV - \vartheta S,$$

which is supposed to be a known function of variables p and ϑ.

Using the thermodynamic identity (A.5.2) and the continuity equation, we make another simplifying transformation. We express the material derivative of the internal energy as a similar derivative of the entropy:

$$\rho\frac{dU}{dt} = \rho\vartheta\frac{dS}{dt} - p\rho\frac{dV}{dt} = \rho\vartheta\frac{dS}{dt} + \frac{p}{\rho}\frac{d\rho}{dt} = \rho\vartheta\frac{dS}{dt} - p\nabla\cdot\mathbf{v}.$$

Thus, we finally obtain *the classical model* $\mathbb{M}_3$ of fluids and gases

$$\left.\begin{aligned}
&\frac{d\rho}{dt} + \rho\,\nabla\cdot\mathbf{v} = 0,\\
&\rho\,\frac{d\mathbf{v}}{dt} = \nabla\cdot\mathbb{P} + \rho\mathbf{F},\\
&\mathbb{P} = 2\mu\mathbb{D}(x,\mathbf{v}) + \big(-p + \nu(\nabla\cdot\mathbf{v}) - \gamma_f(\theta-\theta_0)\big)\mathbb{I},\\
&\rho\,\vartheta\frac{dS}{dt} = \nabla\cdot(\varkappa_f\nabla\theta) - \gamma_f\vartheta(\nabla\cdot\mathbf{v}) + \Phi_f,\\
&\Phi_f = 2\mu\big(\mathbb{D}(x,\mathbf{v})\big):\big(\mathbb{D}(x,\mathbf{v})\big) + \nu(\nabla\cdot\mathbf{v})^2,\\
&\Psi = \Psi(\vartheta,p),\quad V = \frac{1}{\rho} = \frac{\partial\Psi}{\partial p},\quad S = -\frac{\partial\Psi}{\partial\vartheta},
\end{aligned}\right\}\tag{A.5.4}$$

where the function $\Psi(p,\vartheta)$ is supposed to be known.

Of course, the exact mathematical model $\mathbb{M}_3$ is too complicated, so in practical applications one usually uses a different simplified submodel.

First, we consider the *isothermal model*, where the stress tensor does not depend on the temperature ($\gamma_f = 0$). In this case the system is decoupled, that is, the equations of motion are solved independently of the heat equation.

The next simplification is the assumption that the medium is *incompressible*. That is $\rho = \rho_0 = const$. Otherwise we say that the medium is *compressible*.

The incompressibility assumption, and the assumption $\mu = const$ essentially simplify the original model $\mathbb{M}_3$. The mathematical model obtained is $\mathbb{M}_4$

$$\left.\begin{aligned}
&\nabla\cdot\mathbf{v} = 0,\\
&\rho_0\frac{d\mathbf{v}}{dt} = \mu\,\Delta\mathbf{v} - \nabla p + \rho_0\mathbf{F},
\end{aligned}\right\}\tag{A.5.5}$$

called the *Navier-Stokes equations*.

Finally, the most simple model is a linearization of the Navier-Stokes equations in a state of rest, when the material derivative of the velocity is approximated by the partial derivative in time of the velocity:

$$\left.\begin{aligned}
&\nabla\cdot\mathbf{v} = 0,\\
&\rho_0\frac{\partial\mathbf{v}}{\partial t} = \mu\,\Delta\mathbf{v} - \nabla p + \rho_0\mathbf{F}.
\end{aligned}\right\}\tag{A.5.6}$$

This mathematical model $\mathbb{M}_5$ is called the *Stokes equations*.

Coming back to the full non-isothermal model $\mathbb{M}_3$ we consider its linear approximation, which is called the mathematical model $\mathbb{M}_6$ of a *weakly compressible thermofluid*. In this model the density of the liquid is approximated by a linear function of pressure, and the left hand-side of the heat equation, containing the material derivative of the entropy, approximated by the partial derivative of the temperature with respect to time, and in the nonlinear term on the right hand-side of the heat equation the temperature ϑ is replaced by its mean value ϑ_0:

$$\rho \sim \rho_f^0 + \frac{1}{c_f^2}(p - p_0), \quad \rho\vartheta\frac{dS}{dt} \sim c_{p,f}\frac{\partial\vartheta}{\partial t}, \quad \vartheta\nabla\cdot\mathbf{v} \sim \vartheta_0\nabla\cdot\mathbf{v}, \quad \Phi_f \sim 0,$$

where p_0 is the atmospheric pressure, c_f and $c_{p,f}$ are the speed of sound and the specific heat capacity correspondingly in the liquid in consideration and ρ_f^0 is the mean dimensionless density of the liquid.

With these assumptions the model $\mathbb{M}_6$ takes the form

$$\left.\begin{aligned} &\frac{\partial p}{\partial t} + \rho_f^0\, c_f^2\, \nabla\cdot\mathbf{v} = 0,\\ &\rho_f^0\frac{\partial\mathbf{v}}{\partial t} = \nabla\cdot\mathbb{P} + \rho_f^0\mathbf{F},\\ &\mathbb{P} = 2\mu\,\mathbb{D}(x,\mathbf{v}) + \big(-p + \nu(\nabla\cdot\mathbf{v}) - \gamma_f(\theta-\theta_0)\big)\mathbb{I},\\ &c_{p,f}\,\frac{\partial\vartheta}{\partial t} = \nabla\cdot\big(\varkappa_f\nabla\vartheta\big) - \gamma_f\,\vartheta_0\,(\nabla\cdot\mathbf{v}). \end{aligned}\right\} \tag{A.5.7}$$

2. As a rule, a mathematical model describing the behavior of an elastic solid medium is considered in Lagrangian coordinates. The modeling of such a medium is based on the following postulate: the stress state of the medium is determined by the strain tensor and, possibly, the temperature. Of course, a similar construction is also possible in Eulerian coordinates. These non-linear mathematical models are very complicated and the most natural way to simplify the models is the method of linearization. This method is based on the assumption of the existence of an equilibrium state, when the movement of the medium is equal to zero and the assumption that small perturbations of the medium lead to small displacements. Mathematically, these assumptions are equivalent to assumptions of the existence of stationary solutions and their stability with respect to small perturbations of the incoming data. Under these assumptions, the original mathematical model is reliably approximated by its linear version. As might be expected, linear versions of models in the Lagrange variables match those in Eulerian version. Since the main purpose of this book is to simulate the joint motion of solid and liquid media near the equilibrium, then as the basis for the description of deformable solids we take the Euler description and modification of the axioms of thermodynamics (A.5.2), as formulated in the preceding paragraph for liquids and gases:

$$d\Psi = Vdp - Sd\vartheta, \qquad \Psi = U + pV - \vartheta S.$$

A continuous medium is called a *deformable or elastic solid body*, if the stress tensor is determined by the axiom

$$\frac{d\mathbb{P}}{dt} = 2\lambda\mathbb{D}(x,\mathbf{v}) - \Big(\frac{dp}{dt} + \gamma_s\frac{d\vartheta}{dt}\Big)\mathbb{I}. \tag{A.5.8}$$

Thus, the mathematical model $\mathbb{M}_7$ of a deformable elastic body is described by a system of differential equations

$$\left.\begin{aligned}
&\frac{d\rho}{dt} + \rho\nabla\cdot\mathbf{v} = 0,\\
&\rho\frac{d\mathbf{v}}{dt} = \nabla\cdot\mathbb{P} + \rho\mathbf{F},\\
&\rho\frac{dU}{dt} = \mathbb{P} : \big(\mathbb{D}(x,\mathbf{v})\big) + \nabla\cdot(\varkappa\nabla\vartheta),
\end{aligned}\right\} \tag{A.5.9}$$

completed by Eq. (A.5.8), defining the stress tensor, and equations

$$U = \Psi - p\frac{\partial\Psi}{\partial p} - \vartheta\frac{\partial\Psi}{\partial\vartheta}, \quad \frac{1}{\rho} = \frac{\partial\Psi}{\partial p}, \tag{A.5.10}$$

where the function $\Psi(p,\vartheta)$ is a known function of its arguments.

As in the case of models of liquids and gases, we consider an approximate model of a deformable solid body, based on linearization of the original nonlinear model. First, we linearize the constitutive equation (A.5.8), writing it in the form

$$\frac{\partial\mathbb{P}}{\partial t} = 2\lambda\mathbb{D}(x,\mathbf{v}) - \Big(\frac{\partial p}{\partial t} + \gamma_s\frac{\partial\vartheta}{\partial t}\Big)\mathbb{I}.$$

Next we define the **displacement vector w** by the formula

$$\mathbf{v} = \frac{\partial\mathbf{w}}{\partial t}.$$

Then the linearization of (A.5.8) takes the final form

$$\mathbb{P} = 2\lambda\mathbb{D}(x,\mathbf{w}) - \big(p + \gamma_s\vartheta\big)\mathbb{I}.$$

In the same way as before, we arrive at the following form of linearization of the model $\mathbb{M}_7$ in a state of rest:

$$\left.\begin{aligned}
&\frac{\partial p}{\partial t} + \rho_s^0 c_s^2\nabla\cdot\mathbf{v} = 0, \quad \mathbf{v} = \frac{\partial\mathbf{w}}{\partial t},\\
&\rho_s^0\frac{\partial\mathbf{v}}{\partial t} = \lambda\Delta\mathbf{w} - \nabla p - \gamma_s\nabla\vartheta + \rho_s^0\mathbf{F},\\
&c_{p,s}\frac{\partial\vartheta}{\partial t} = \nabla\cdot\big(\varkappa_s\nabla\vartheta\big) - \gamma_s\,\vartheta_0\,(\nabla\cdot\mathbf{v}),
\end{aligned}\right\} \tag{A.5.11}$$

where c_s and $c_{p,s}$ are the speed of sound and the specific heat capacity in the solid under consideration and ρ_s^0 is the mean dimensionless density of the solid.

We call this system a model $\mathbb{M}_8$ of the **linear thermoelasticity**.

For $\gamma_s = 0$ the dynamic equations

$$\left.\begin{aligned}&\frac{\partial p}{\partial t}+\rho_s^0 c_s^2\nabla\cdot\mathbf{v}=0,\quad \mathbf{v}=\frac{\partial \mathbf{w}}{\partial t},\\ &\rho_s^0\frac{\partial \mathbf{v}}{\partial t}=\lambda\Delta\mathbf{w}-\nabla p+\rho_s^0\mathbf{F}\end{aligned}\right\}\tag{A.5.12}$$

becomes independent of the heat equation

$$c_{p,s}\frac{\partial\vartheta}{\partial t}=\nabla\cdot\big(\varkappa_s\nabla\vartheta\big).\tag{A.5.13}$$

We call the system (A.5.12) the *Lamé equations of linear elasticity*.

A.6 Shock Relations

1. It is easy to see the integral laws of conservation of the mathematical model $\mathbb{M}_1$ of moving continuous media can be written in the form of an abstract scalar conservation law

$$\frac{d}{dt}\int_{\omega}\rho u dx+\int_{\partial\omega}\mathbf{X}\cdot\mathbf{n}d\sigma=\int_{\omega}Y\,dx\tag{A.6.1}$$

for any individual volume ω.

For example, for the vector law of conservation of momentum, where $\mathbf{v}=(v_1,v_2,v_3)$, $\mathbb{P}=(P_{ij})$, $i,j=1,2,3$ and $\mathbf{F}=(F_1,F_2,F_3)$, the equation for the first component of the velocity vector has the form (A.6.1) with $u=v_1$, $\mathbf{X}=(-P_{11},-P_{12},-P_{13})$ and $Y=\rho F_1$.

A characteristic feature of such a description of the motion of a continuous medium is a minimum requirement of the smoothness of the basic variables describing the motion of a continuous medium, namely, the model requires only the *summability* of all terms of the model variables. There may be other equivalent formulations of the model of moving continuous media. But, before formulating an equivalent mathematical model, let us for the moment come back to the continuous movement of a continuum. As above, each of the differential equations of mathematical model $\mathbb{M}_2$ can be written in the form of an abstract scalar equation

$$\rho\frac{du}{dt}+\nabla\cdot\mathbf{X}=Y.\tag{A.6.2}$$

If we use an equivalent form of the continuity equation

$$\frac{\partial\rho}{\partial t}+\nabla\cdot(\rho\mathbf{v})=0,$$

then the Eq. (A.6.2) can be rewritten as

$$\frac{\partial}{\partial t}(\rho u) + \nabla \cdot \big(\mathbf{X} + \rho u \mathbf{v}\big) = Y. \tag{A.6.3}$$

Coming back to the model $\mathbb{M}_2$ we see that one of its equivalent forms is the following

$$\left.\begin{aligned}
&\frac{\partial \rho}{\partial t} + \nabla \cdot (\rho \mathbf{v}) = 0,\\
&\frac{\partial}{\partial t}(\rho \mathbf{v}) + \nabla \cdot \big(\rho \mathbf{v} \otimes \mathbf{v} - \mathbb{P}\big) = \rho \mathbf{F},\\
&\frac{\partial}{\partial t}\Big(\rho(U + \frac{1}{2}|\mathbf{v}|^2)\Big) - \rho \mathbf{F} \cdot \mathbf{v}\\
&\quad = -\nabla \cdot \Big(\rho \mathbf{v}(U + \frac{1}{2}|\mathbf{v}|^2) - \varkappa \nabla \vartheta - \mathbb{P} \cdot \mathbf{v}\Big),
\end{aligned}\right\} \tag{A.6.4}$$

where $\mathbf{a} \otimes \mathbf{b}$ denotes the tensor, which is defined for any two vectors $\mathbf{a}$ and $\mathbf{b}$ so that for any vector $\mathbf{c}$ the action of the tensor $\mathbf{a} \otimes \mathbf{b}$ to the vector $\mathbf{c}$ is given by the formula

$$\big(\mathbf{a} \otimes \mathbf{b}\big)\langle \mathbf{c} \rangle = \mathbf{a}(\mathbf{b} \cdot \mathbf{c}).$$

Now we go back to the model $\mathbb{M}_1$ of moving continuous media. To do this, we multiply Eq. (A.6.3) by an arbitrary smooth function φ, compactly supported in W (i..e, equal to zero outside some compact set lying strictly inside the domain W) and integrate over the domain W. Then use the Gauss-Ostrogradsky Theorem of and pass the differentiation from the functions u and $(\mathbf{F} + u\mathbf{v})$ onto the function φ:

$$\int_W \Big(\rho u \frac{\partial \varphi}{\partial t} + \nabla \varphi \cdot \big(\mathbf{X} + \rho u \mathbf{v}\big) + Y\varphi\Big) r dx dt = 0. \tag{A.6.5}$$

The integral identity (A.6.5) holds for all continuously differentiable functions φ, finite in the domain W, and it is obviously equivalent to the integral identity (A.6.1).

Thus, an equivalent form of the mathematical model $\mathbb{M}_1$ of moving continuous media has the form:

$$\left.\begin{aligned}
&\int_W \Big(\rho \frac{\partial \varphi}{\partial t} + \nabla \varphi \cdot \big(\rho \mathbf{v}\big)\Big) dx dt = 0,\\
&\int_W \Big(\rho \mathbf{v} \cdot \frac{\partial \psi}{\partial t} + \mathbb{D}(x, \psi) : \big(\rho \mathbf{v} \otimes \mathbf{v} - \mathbb{P}\big) + \rho \mathbf{F} \cdot \psi\Big) dx dt = 0,\\
&\int_W \Big(\rho\big(U + \frac{1}{2}|\mathbf{v}|^2\big) \frac{\partial \xi}{\partial t} + \rho \mathbf{F} \cdot \mathbf{v} \xi\Big) dx dt\\
&\qquad + \int_W \nabla \xi \cdot \Big(\rho \mathbf{v}\big(U + \frac{1}{2}|\mathbf{v}|^2\big) - \varkappa \nabla \vartheta - \mathbb{P} \cdot \mathbf{v}\Big) dx dt = 0,
\end{aligned}\right\} \tag{A.6.6}$$

where arbitrary scalar functions φ and ξ and an arbitrary vector-function ψ are continuously differentiable and compactly supported in the domain W.

2. There exists another equivalent form of the mathematical model $\mathbb{M}_1$, namely, let us integrate the Eq. (A.6.3) over an arbitrary domain $G \subset W$ with the Lipschitz boundary Γ:

$$\int_G \Big(\frac{\partial}{\partial t}(\rho u) + \nabla \cdot \big(\mathbf{X} + \rho u \mathbf{v}\big) - Y\Big) dxdt = 0.$$

Next we use the Gauss-Ostrogradsky Theorem and get

$$\int_\Gamma \Big(\rho u\big(\mathbf{l} + \mathbf{v}\big) + \mathbf{X}\Big) \cdot \nu d\Gamma = \int_G Y dxdt, \tag{A.6.7}$$

where $\mathbf{l}$ is an unit vector of the time axis, and ν is the outward unit normal vector to the surface Γ.

Equation (A.6.7) for the arbitrary domain $G \subset W$ is equivalent to identity (A.6.5) for the arbitrary function φ and both of these equations are equivalent to the Eq. (A.6.1) for any individual volume ω.

Recall that in the identity (A.6.7) Γ is the three-dimensional hypersurface of the four-dimensional space-time and the unit outward normal ν to the surface Γ is a four-dimensional vector:

$$\nu = (\nu_1, \nu_2, \nu_3, \nu_4),$$

and

$$\nu \cdot \mathbf{l} = \nu_4, \quad \mathbf{v} = (\nu_1, \nu_2, \nu_3, 0).$$

Now, using Eq. (A.6.7) let us derive one more equivalent form of the mathematical model $\mathbb{M}_1$:

$$\left.\begin{aligned}
&\int_\Gamma \rho\big(\mathbf{l} + \mathbf{v}\big) \cdot \nu d\Gamma = 0, \\
&\int_\Gamma \Big(\rho \mathbf{v} \otimes \big(\mathbf{l} + \mathbf{v}\big) - \mathbb{P}\Big) \cdot \nu d\Gamma = \int_G \rho \mathbf{F} dxdt, \\
&\int_\Gamma \Big(\rho\big(U + \frac{1}{2}|\mathbf{v}|^2\big)\big(\mathbf{l} + \mathbf{v}\big) - \mathbb{P} \cdot \mathbf{v} - \varkappa \nabla \vartheta\Big) \cdot \nu d\Gamma \\
&\qquad = \int_G \rho \mathbf{F} \cdot \mathbf{v} dxdt,
\end{aligned}\right\} \tag{A.6.8}$$

where identities (A.6.8) hold true for any four dimensional domain $G \subset \mathbb{R}^3 \times (0, t_0)$ with the Lipschitz boundary Γ.

Definition A.1 The motion of a continuous medium is called *generalized motion*, if the functions ρ, U, $\mathbb{P}$, $\mathbf{v}$, q are bounded measurable functions of the independent variables (x, t) and for them integral relations (A.6.8) are satisfied for any four dimensional domain $G \subset \mathbb{R}^3 \times (0, t_0)$ with the Lipschitz boundary Γ.

3. The mathematical model $\mathbb{M}_1$ of moving continuous media is too complicated for analysis and so far there is no strong result for the generalized motion of continuous media. Therefore, the study of more simple sub-models of the model $\mathbb{M}_1$ is very important. One of these sub-models of the model $\mathbb{M}_1$ is the mathematical model $\mathbb{M}_9$ of generalized motion with a strong discontinuity.

Let motion be considered in the domain $W \subset \mathbb{R}^3 \times (0, t_0)$ where this domain is divided by some smooth surface $\Pi \subset \mathbb{R}^3 \times (0, t_0)$ into two domains W_1 and W_2.

Definition A.2 The generalized motion of a continuous medium is called a *motion with strong discontinuity* if in each domain W_1 and W_2 the motion is a continuous one and the functions ρ, U, $\mathbb{P}$, $\mathbf{v}$, q have continuous limit values on the surface Π, which are, generally speaking, different for W_1 and W_2.

In this case the hypersurface Π is called a *surface of strong discontinuity*.

By definition, in each of the domains W_1 and W_2 the differential equations (A.6.4) hold true. It turns out that the main characteristics of the motion satisfy additional relations on the surface of strong discontinuity Π.

To derive these relations, we consider an arbitrary domain $G \subset W$ with smooth boundary Γ. Let $G_1 = G \cap W_1 \neq \emptyset$, $G_2 = G \cap W_2 \neq \emptyset$, $\Gamma_1 = \Gamma \cap \bar{G}_1$, $\Gamma_2 = \Gamma \cap \bar{G}_2$ and $\gamma = G \cap \Pi$. We apply the identity (A.6.7) for each of the domains G, G_1 and G_2:

$$\int_{\Gamma} \Big(\rho u(\mathbf{l} + \mathbf{v}) + \mathbf{X}\Big) \cdot \nu d\Gamma = \int_G Y dx dt,$$

$$\int_{\Gamma_1} \Big(\rho u(\mathbf{l} + \mathbf{v}) + \mathbf{X}\Big) \cdot \nu d\Gamma + \int_{\gamma} \Big(\rho_1 u_1(\mathbf{l} + \mathbf{v}_1) + \mathbf{X}_1\Big) \cdot \nu d\gamma = \int_{G_1} Y dx dt,$$

$$\int_{\Gamma_2} \Big(\rho u(\mathbf{l} + \mathbf{v}) + \mathbf{X}\Big) \cdot \nu d\Gamma - \int_{\gamma} \Big(\rho_2 u_2(\mathbf{l} + \mathbf{v}_2) + \mathbf{X}_2\Big) \cdot \nu d\gamma = \int_{G_2} Y dx dt,$$

where ν is the unit normal to the surface γ, $\rho_1, u_1, \mathbf{v}_1, \mathbf{X}_1$ $\rho_2, u_2, \mathbf{v}_2, \mathbf{X}_2$ are the limit values on the surface γ from the domain G_1 and from the domain G_2 respectively. Since

$$\int_{\Gamma} \Big(\rho u(\mathbf{l} + \mathbf{v}) + \mathbf{X}\Big) \cdot \nu d\Gamma$$
$$= \int_{\Gamma_1} \Big(\rho u(\mathbf{l} + \mathbf{v}) + \mathbf{X}\Big) \cdot \nu d\Gamma + \int_{\Gamma_2} \Big(\rho u(\mathbf{l} + \mathbf{v}) + \mathbf{X}\Big) \cdot \nu d\Gamma,$$

then subtracting from the first identity the last two identities we obtain

$$\int_{\gamma} \Big(\rho_1 u_1(\mathbf{l} + \mathbf{v}_1) + \mathbf{X}_1\Big) \cdot \nu d\gamma = \int_{\gamma} \Big(\rho_2 u_2(\mathbf{l} + \mathbf{v}_2) + \mathbf{X}_2\Big) \cdot \nu d\gamma$$

for an arbitrary subset $\gamma \in \Pi$. Hence

$$\Big(\rho_1 u_1 (\mathbf{l} + \mathbf{v}_1) + \mathbf{X}_1\Big) \cdot \nu = \Big(\rho_2 u_2 (\mathbf{l} + \mathbf{v}_2) + \mathbf{X}_2\Big) \cdot \nu$$

for all $(\mathbf{x}, t) \in \Pi$.

The last equation can be written as

$$\big[\rho u (\mathbf{l} + \mathbf{v}) + \mathbf{X}\big] \cdot \nu = 0, \quad (\mathbf{x}, t) \in \Pi. \tag{A.6.9}$$

Here [ ] is a symbol of jump of the function φ at Π:

$$[\varphi] = \varphi_1 - \varphi_2,$$

where φ_1 and φ_2 are limit values of the function φ from different sides of the surface Π.

The condition (A.6.9) gives us the missing relations on the surface of strong discontinuity, completing the mathematical model $\mathbb{M}_9$ of generalized motion with a strong discontinuity:

$$\left.\begin{aligned} &\big[\rho(\mathbf{l} + \mathbf{v})\big] \cdot \nu = 0, \quad (\mathbf{x}, t) \in \Pi, \\ &\big[\rho \mathbf{v} \otimes (\mathbf{l} + \mathbf{v}) - \mathbb{P}\big] \cdot \nu = 0, \quad (\mathbf{x}, t) \in \Pi, \\ &\big[\rho\big(U + \frac{1}{2}|\mathbf{v}|^2\big)(\mathbf{l} + \mathbf{v}) - \mathbb{P} \cdot \mathbf{v} - \varkappa \nabla \vartheta\big] \cdot \nu = 0, \quad (\mathbf{x}, t) \in \Pi. \end{aligned}\right\} \tag{A.6.10}$$

Conditions (A.6.10) are called **shock relations**.

Note that we may get the same differential equation (A.6.4) and shock relations (A.6.10) if, as the basic equations of the generalized motion of the continuum, we use Eq. (A.6.6).

4. Shock relations can be rewritten in terms of the space $\mathbb{R}^3$. More precisely, in terms of the two-dimensional surface

$$\Pi(t) = \{(\mathbf{x}, t) \in \Pi \mid t = const\} \subset \mathbb{R}^3,$$

which is a cross section of the surface of strong discontinuity Π by the plane $\{t = const\}$, namely, let $\mathbf{n}$ be the unit vector normal to the surface $\Pi(t)$ at a given point $\mathbf{x} \in \Pi(t)$. It is clear that this analysis has a local nature. Therefore we can assume that in the neighborhood of $(\mathbf{x}, t) \in \Pi$ the surface Π can be represented as

$$\Pi : \quad h(\mathbf{x}, t) = 0.$$

Then

$$\nu = \left(\frac{\nabla h}{\sqrt{|\nabla h|^2 + h_t^2}}, \frac{h_t}{\sqrt{|\nabla h|^2 + h_t^2}}\right), \quad \mathbf{n} = \left(\frac{\nabla h}{|\nabla h|}, 0\right),$$

where

$$\nabla h = \left(\frac{\partial h}{\partial x_1}, \frac{\partial h}{\partial x_2}, \frac{\partial h}{\partial x_3}\right) \in \mathbb{R}^3, \quad h_t = \frac{\partial h}{\partial t}.$$

Simple calculations show that

$$\nu = (-V_n \mathbf{l} + \mathbf{n}) \sin\alpha, \tag{A.6.11}$$

where

$$V_n = -\frac{h_t}{|\nabla h|}, \quad \sin\alpha = (\nu \cdot \mathbf{n}) = \frac{|\nabla h|}{\sqrt{|\nabla h|^2 + h_t^2}}.$$

The value V_n is called the *velocity of replacement of the surface $\Pi(t)$ in the direction of normal* $\mathbf{n}$.

This value can be defined geometrically. Indeed, we consider a line L from the point $\mathbf{x} \in \Pi(t)$ in the direction of the normal $\mathbf{n}$ to the surface of $\Pi(t)$. Next, we pass to the surface $\Pi(t + \delta t)$. Let the point $\mathbf{x} + \delta\mathbf{x}$ be the intersection of this surface with the line L. Then

$$V_n = \lim_{\delta t \to 0} \frac{\delta\mathbf{x} \cdot \mathbf{n}}{\delta t}.$$

Substituting the expression (A.6.11) into the equality (A.6.9) we get

$$\left[\rho u\left(v_n - V_n\right) + X_n\right] = 0, \quad (\mathbf{x}, t) \in \Pi, \tag{A.6.12}$$

where

$$v_n = \mathbf{v} \cdot \mathbf{n}, \quad X_n = \mathbf{X} \cdot \mathbf{n}.$$

As above, the condition (A.6.12) allows to express the shock relations in terms of the space $\mathbb{R}^3$:

$$\left.\begin{aligned}
&\left[\rho(v_n - V_n)\right] = 0,\\
&\left[\rho\mathbf{v}(v_n - V_n) - \mathbb{P} \cdot \mathbf{n}\right] = 0,\\
&\left[\rho\left(U + \frac{1}{2}|\mathbf{v}|^2\right)(v_n - V_n) - \mathbf{v} \cdot (\mathbb{P} \cdot \mathbf{n}) - \varkappa\nabla\vartheta \cdot \mathbf{n}\right] = 0,
\end{aligned}\right\} \tag{A.6.13}$$

for all $(\mathbf{x}, t) \in \Pi$.

5. A typical type of strong discontinuity is one in which $v_n = V_n$. That is, the velocity of the particles of the medium in the direction of the normal $\mathbf{n}$ coincides with the velocity of the surface $\Pi(t)$. Therefore, there is no exchange of particles between domains W_1 and W_2 and the surface $\Pi(t)$ is a material one. So, $\Pi(t)$ is the surface of contact between two different states of a continuum. For example, water and air or water and solid. Such a surface of strong discontinuity is called *contact discontinuity*. The equations of contact discontinuity are:

$$\left.\begin{array}{l} v_n = V_n, \\ [\mathbb{P} \cdot \mathbf{n}] = 0, \\ [\mathbf{v}] \cdot (\mathbb{P} \cdot \mathbf{n}) + [\varkappa \nabla \vartheta \cdot \mathbf{n}] = 0, \end{array}\right\} \tag{A.6.14}$$

for all $(\mathbf{x}, t) \in \Pi$.

A.7 Joint Motion of an Elastic Sold and a Viscous Liquid

1. *Heterogeneous continuous media* are those continuous media which consist of two or more different components (phases) and in any sufficiently small amount of a continuum there might be different phases. The minimum size of this volume is different in various heterogeneous media, but usually it is in the range from several microns to several (first) tens of microns. Examples of such continuous media are the motion of solid micro-particles in a liquid or gas or the movement of fluid in the micro-pores of a deformable elastic body. There are two different approaches to the description of heterogeneous media.

The first approach is based upon the notion of a continuous medium as a kind of conglomerate, where at each point all phases of such a medium are present. In this approach, the main difficulty is physical modeling, namely, the choice of axioms that define the dependence of the stress tensor on the basic characteristics of the motion and thermodynamic relations.

The second approach is based on precise physical modeling with further simplification by mathematical models using the methods of mathematical analysis. As a rule, the differential equations of the exact mathematical model contain some small parameter. Therefore, the main methods of simplifying the exact mathematical models are the methods of linearization and homogenization. Roughly speaking, these are methods of constructing approximate mathematical models from the original one, when the small parameter tends to zero. From a physical point of view a heterogeneous medium is an example of generalized motion with a strong discontinuity, considered in the preceding paragraph. In this approach we must keep in mind the limits of the application of physical models and the limits of applicability of methods of mathematical analysis. For example, physical experiments show that the basic phenomenological models of continuous media are still applicable at scales of a few microns. It is clear, that the mathematical part of the second approach depends on the method chosen. The more precise and more rigorous method provides a more trustable mathematical model.

Our aim is to obtain the mathematical models of the joint motion of an elastic porous body and a liquid that fills the pores and cracks, that is those voids that appear in the solid body during the time of its formation. The proper elastic body is called a *solid skeleton* (or an *elastic skeleton*, or simply a *skeleton*). For the modeling of such a motion we use the second approach, when a two-component continuum (solid-liquid) is described by the mathematical model $\mathbb{M}_9$ of a generalized motion with a strong discontinuity.

Let $\Omega \subset \mathbb{R}^3$ be the domain occupied by a continuous medium, $G_s(t)$—the domain occupied by a solid skeleton, $G_f(t)$—the pore space occupied by the fluid and $\Pi(t)$—the boundary between the solid skeleton and pore space:

$$\overline{\Omega} = G_s(t) \cup G_f(t) \cup \Pi(t).$$

Let additionally $\mathbf{w}_s$, $\mathbf{v}_s$, ρ_s, $\mathbb{P}_s$, p_s, U_s and ϑ_s be the displacements, the velocity, the density, the stress tensor, the pressure, the specific internal energy and the temperature in the solid skeleton, and $\mathbf{w}_f$, $\mathbf{v}_f$, ρ_f, $\mathbb{P}_f$, p_f, U_f and ϑ_f be the displacements, the velocity, the density, the stress tensor, the pressure, the specific internal energy and the temperature in the liquid.

We suppose that the stress tensor $\tilde{P}$, the displacement $\mathbf{w}$, velocity $\mathbf{v}$, density ρ, pressure p, specific internal energy U, and temperature ϑ of the continuum are given by formulae

$$\tilde{P} = \tilde{\chi}\tilde{P}_f + (1 - \tilde{\chi})\tilde{P}_s,$$

$$\mathbb{P}_f = 2\mu\,\mathbb{D}(x, \mathbf{v}_f) + \big(-p_f + \nu(\nabla \cdot \mathbf{v}_f) - \gamma_f(\vartheta_f - \vartheta_0)\big)\mathbb{I},$$

$$\mathbb{P}_s = 2\lambda\mathbb{D}(x, \mathbf{w}_s) - \big(p_s + \gamma_s(\vartheta_s - \vartheta_0)\big)\mathbb{I},$$

$$\mathbf{w} = \tilde{\chi}\mathbf{w}_f + (1 - \tilde{\chi})\mathbf{w}_s,$$

$$\mathbf{v} = \tilde{\chi}\mathbf{v}_f + (1 - \tilde{\chi})\mathbf{v}_s,$$

$$\rho = \tilde{\chi}\rho_f + (1 - \tilde{\chi})\rho_s, \;\; p = \tilde{\chi}p_f + (1 - \tilde{\chi})p_s,$$

$$U = \tilde{\chi}U_f + (1 - \tilde{\chi})U_s, \;\; \vartheta = \tilde{\chi}\vartheta_f + (1 - \tilde{\chi})\vartheta_s,$$

where the characteristic function $\tilde{\chi}$ of the domain $G_f(t)$ is defined as

$$\tilde{\chi}(\mathbf{x}, t) = 1, \;\text{if}\,(\mathbf{x}, t) \in G_f(t) \;\text{and}\; \tilde{\chi}(\mathbf{x}, t) = 0, \;\text{if}\,(\mathbf{x}, t) \in G_s(t).$$

Then Eq. (A.6.6) for the joint motion of the solid skeleton and the liquid take the form

$$\left.\begin{array}{l}
\int_0^{t_0}\int_\Omega\Big(\rho\frac{\partial\varphi}{\partial t}+\nabla\varphi\cdot(\rho\mathbf{v})\Big)dxdt=0,\\
\int_0^{t_0}\int_\Omega\Big(\rho\mathbf{v}\cdot\frac{\partial\psi}{\partial t}+\mathbb{D}(x,\psi):\big(\rho\mathbf{v}\otimes\mathbf{v}-\mathbb{P}\big)+\rho\mathbf{F}\cdot\psi\Big)dxdt=0,\\
\int_0^{t_0}\int_\Omega\big(\rho(U+\frac{1}{2}|\mathbf{v}|^2)\frac{\partial\xi}{\partial t}+\rho\mathbf{F}\cdot\mathbf{v}\xi\big)dxdt\\
\qquad+\int_0^{t_0}\int_\Omega\nabla\xi\cdot\big(\rho\mathbf{v}(U+\frac{1}{2}|\mathbf{v}|^2)-\varkappa\nabla\vartheta-\mathbb{P}\cdot\mathbf{v}\big)dxdt=0,
\end{array}\right\}\tag{A.7.1}$$

for arbitrary smooth functions φ, ψ and ξ with a compact support in $\Omega\times(0,t_0)$.

By definition, in the domain $G_f(t)$ the equations of the continuous motion for the liquid component

$$\left.\begin{array}{l}
\frac{d\rho_f}{dt}+\rho_f\nabla\cdot\mathbf{v}_f=0,\\
\rho_f\frac{d\mathbf{v}_f}{dt}=\nabla\cdot\mathbb{P}_f+\rho_f\mathbf{F},\\
\mathbb{P}_f=2\mu\mathbb{D}(x,\mathbf{v}_f)+\big(-p_f+\nu\nabla\cdot\mathbf{v}_f-\gamma_f\vartheta_f\big)\mathbb{I},\\
\rho_f\vartheta_f\frac{dS_f}{dt}=\nabla\cdot(\varkappa_f\nabla\vartheta)-\gamma_f\vartheta_f\nabla\cdot\mathbf{v}_f+\Phi_f,\\
\Phi_f=2\mu\big(\mathbb{D}(x,\mathbf{v}_f)\big):\big(\mathbb{D}(x,\mathbf{v}_f)\big),\\
\Psi_f=\Psi_f(\vartheta_f,p_f),\quad\frac{1}{\rho_f}=\frac{\partial\Psi_f}{\partial p_f},\quad S_f=-\frac{\partial\Psi_f}{\partial\vartheta_f},
\end{array}\right\}\tag{A.7.2}$$

hold true and in the domain $G_s(t)$ the equations of the continuous motion for the solid component

$$\left.\begin{array}{l}
\frac{d\rho_s}{dt}+\rho_s\nabla\cdot\mathbf{v}_s=0,\\
\rho_s\frac{d\mathbf{v}_s}{dt}=\nabla\cdot\mathbb{P}_s+\rho_s\mathbf{F},\\
\frac{d\mathbb{P}_s}{dt}=2\lambda\mathbb{D}(x,\mathbf{v}_s)-\big(\frac{dp_s}{dt}+\gamma_s\frac{d\vartheta_s}{dt}\big)\mathbb{I},\\
\rho_s\vartheta_s\frac{dS_s}{dt}=\nabla\cdot(\varkappa_s\nabla\vartheta)+p_s\nabla\cdot\mathbf{v}_s+\mathbb{P}_s:\mathbb{D}(x,\mathbf{v}_s),\\
\Psi_s=\Psi_s(\vartheta_s,p_s),\quad\frac{1}{\rho_s}=\frac{\partial\Psi_s}{\partial p_s},\quad S_s=-\frac{\partial\Psi_s}{\partial\vartheta_s},
\end{array}\right\}\tag{A.7.3}$$

hold true.

On the common boundary $\Pi(t)$ "solid–liquid" one has equations of contact discontinuity

$$\left.\begin{array}{l}\mathbf{v}_s \cdot \mathbf{n} = \mathbf{v}_f \cdot \mathbf{n} = V_n,\\ \mathbb{P}_s \cdot \mathbf{n} = \mathbb{P}_f \cdot \mathbf{n},\\ \mathbf{v}_s \cdot (\mathbb{P}_s \cdot \mathbf{n}) + \varkappa_s \nabla \vartheta_s \cdot \mathbf{n} = \mathbf{v}_f \cdot (\mathbb{P}_f \cdot \mathbf{n}) + \varkappa_f \nabla \vartheta_f \cdot \mathbf{n}.\end{array}\right\} \tag{A.7.4}$$

The condition that the surface $\Pi(t)$ is a surface of contact discontinuity means that $\Pi(t)$ is a material surface. That is

$$\frac{d\tilde{\chi}}{dt} \equiv \frac{\partial \tilde{\chi}}{\partial t} + \mathbf{v} \cdot \nabla \tilde{\chi} = 0. \tag{A.7.5}$$

Equations (A.7.1)–(A.7.5) form the mathematical model $\mathbb{M}_{10}$ of **joint motion with a contact discontinuity of an elastic solid and a viscous liquid**.

This model is an example of the mathematical problem with **a free (unknown) boundary** $\Pi(t)$.

2. Equations (A.7.1)–(A.7.5) are very difficult to analyse mathematically and we need additional physical assumptions to simplify the mathematical model $\mathbb{M}_{10}$.

In classical continuum mechanics such a simplification is a linearization of the original nonlinear model in the rest state. Here the basic physical assumption is the postulate of the smallness of the deviations of the basic characteristics of the motion from these characteristics in a state of rest. For a heterogeneous continuum elastic porous body filled with liquid, such a physical postulate is justified because, for example, the velocity of fluid in pores is about 3–6 m/year.

Let δ be a small characteristic size, defining the deviations of the basic characteristics of the motion from those characteristics in a state of rest. Then, as usual, we neglect all terms of order δ^2.

All values with a dash stand for the linear part (with respect to the small parameter δ) of deviations of the main characteristics of the medium from the main characteristics of the medium in the rest state. In addition, let $\chi_0(\mathbf{x})$ be the characteristic function of the liquid domain at the initial time and ϑ_0 be the mean (constant) temperature, and ρ_s^0 and ρ_f^0 be the average density of the solid skeleton and the liquid, respectively. Then

$$\tilde{\chi}(\mathbf{x}, t) = \chi_0(\mathbf{x}) + o(\delta), \quad p = \bar{p} + o(\delta), \quad \bar{p} = \chi_0 \bar{p}_f + (1 - \chi_0)\bar{p}_f,$$

$$\rho = \rho^0(\mathbf{x}) + \frac{1}{c^2}\bar{p} + o(\delta), \quad \rho^0 = \chi_0 \rho_f^0 + (1 - \chi_0)\rho_s^0, \quad \frac{1}{c^2} = \frac{\chi_0}{c_f^2} + \frac{(1 - \chi_0)}{c_s^2},$$

$$\mathbf{v} = \bar{\mathbf{v}} + o(\delta), \quad \bar{\mathbf{v}} = \chi_0 \bar{\mathbf{v}}_f + (1 - \chi_0)\bar{\mathbf{v}}_s,$$

$$\vartheta = \vartheta_0 + \bar{\vartheta} + o(\delta), \quad \bar{\vartheta} = \chi_0 \bar{\vartheta}_f + (1 - \chi_0)\bar{\vartheta}_s,$$

$$\rho\vartheta \frac{dS}{dt} = \eta \frac{\partial \bar{\vartheta}}{\partial t} + o(\delta), \quad \eta = \chi_0 c_{p,f} + (1 - \chi_0)c_{p,s},$$

$$\vartheta_i \nabla \cdot \mathbf{v}_i + \Phi_i = \vartheta_0 \nabla \cdot \mathbf{v}_i + o(\delta), \quad i = f, s,$$

$$\mathbb{P} = \bar{\mathbb{P}} + o(\delta), \quad \bar{\mathbb{P}} = \chi_0 \bar{\mathbb{P}}^f + (1 - \chi_0)\bar{\mathbb{P}}^s,$$

$$\bar{\mathbb{P}}^f = 2\mu \mathbb{D}(x, \bar{\mathbf{v}}_f) - \big(\bar{p}_f + \gamma_f \bar{\vartheta}_f - \nu \nabla \cdot \bar{\mathbf{v}}_f\big)\mathbb{I}, \tag{A.7.6}$$

$$\bar{\mathbb{P}}^s = 2\lambda \mathbb{D}(x, \bar{\mathbf{w}}_s) - \big(\bar{p}_s + \gamma_s \bar{\vartheta}_s\big)\mathbb{I}, \quad \bar{\mathbf{v}} = \frac{\partial \bar{\mathbf{w}}}{\partial t}, \tag{A.7.7}$$

where

$$\lim_{\delta \to 0} \frac{o(\delta)}{\delta} = 0.$$

First of all, we note that the unknown domains $G_s(t)$ and $G_f(t)$ in the linear model are replaced by their initial positions $G_s(0) = \Omega_s$ and $G_f(0) = \Omega_f$. Similarly, the free boundary $\Pi(t)$ in the linear model is replaced by its initial position $\Pi(0) = \Gamma$. Thus, in the domain Ω_s the main characteristics describing the behavior of the elastic skeleton satisfy the differential equations of the model $\mathbb{M}_8$ of linear thermoelasticity

$$\left.\begin{aligned}
&\frac{1}{c_s^2}\frac{\partial \bar{p}_s}{\partial t} + \rho_s^0 \nabla \cdot \bar{\mathbf{v}}_s = 0,\\
&\rho_s^0 \frac{\partial \bar{\mathbf{v}}_s}{\partial t} = \nabla \cdot \mathbb{P}^s + \rho_s^0 \mathbf{F},\\
&c_{p,s}\frac{\partial \bar{\vartheta}_s}{\partial t} = \nabla \cdot \big(\varkappa_s \nabla \bar{\vartheta}_s\big) - \gamma_s \vartheta_0 \nabla \cdot \bar{\mathbf{v}}_s,
\end{aligned}\right\} \tag{A.7.8}$$

and in the domain Ω_f the main characteristics describing the behavior of the liquid, satisfy the differential equations of the model $\mathbb{M}_6$ of a weakly compressible thermofluid

$$\left.\begin{aligned}
&\frac{1}{c_f^2}\frac{\partial \bar{p}_f}{\partial t} + \rho_f^0 \nabla \cdot \bar{\mathbf{v}}_f = 0,\\
&\rho_f^0 \frac{\partial \bar{\mathbf{v}}_f}{\partial t} = \nabla \cdot \bar{\mathbb{P}}^f + \rho_f^0 \mathbf{F},\\
&c_{p,f}\frac{\partial \bar{\vartheta}_f}{\partial t} = \nabla \cdot \big(\varkappa_f \nabla \bar{\vartheta}_f\big) - \gamma_f \vartheta_0 \nabla \cdot \bar{\mathbf{v}}_f.
\end{aligned}\right\} \tag{A.7.9}$$

Relations (A.7.6)–(A.7.9) are completed with linearized equations of contact discontinuity.

Since the free boundary is substituted by its initial position, we may use only one condition of the law of conservation of mass on the free boundary. Namely, the condition

$$\bar{\mathbf{v}}_s \cdot \mathbf{n} = \bar{\mathbf{v}}_f \cdot \mathbf{n}, \quad \mathbf{x} \in \Gamma.$$

The obtained mathematical model is still incomplete. Therefore we postulate additional the conditions

$$\bar{\mathbf{v}}_s = \bar{\mathbf{v}}_f, \quad \mathbf{x} \in \Gamma, \tag{A.7.10}$$

$$\bar{\vartheta}_s = \bar{\vartheta}_f, \quad \mathbf{x} \in \Gamma \tag{A.7.11}$$

on the contact surface Γ.

The last two conditions allow one to consider generalized motion of continuous media with strong discontinuity, for continuous velocity and temperature. In this case we may define all first order differential operators and omit indices s and f:

$$\bar{\mathbf{v}} = \chi_0 \bar{\mathbf{v}} + (1 - \chi_0)\bar{\mathbf{v}}, \quad \bar{\vartheta} = \chi_0 \bar{\vartheta} + (1 - \chi_0)\bar{\vartheta}.$$

With this assumptions the linearized equations of the contact discontinuity take the final form

$$\bar{\mathbb{P}}^s \cdot \mathbf{n} = \bar{\mathbb{P}}^f \cdot \mathbf{n}, \quad \mathbf{x} \in \Gamma, \tag{A.7.12}$$

$$\varkappa_s \nabla \bar{\vartheta}_s \cdot \mathbf{n} = \varkappa_f \nabla \bar{\vartheta}_f \cdot \mathbf{n}, \quad \mathbf{x} \in \Gamma. \tag{A.7.13}$$

Equations (A.7.6)–(A.7.13), completed with the corresponding boundary conditions on the outer boundary $S = \partial \Omega$ and the initial conditions at $t = 0$, form the closed linear mathematical model $\mathbb{M}_{11}$ of a joint nonisothermal motion of a solid elastic body and a weakly compressible viscous fluid.

The model $\mathbb{M}_{11}$ can be written in an equivalent form as a system of integral identities. Namely, the functions $\bar{p}$, $\bar{\vartheta}$, $\bar{\mathbf{w}}$ and $\bar{\mathbf{v}} = \dfrac{\partial \bar{\mathbf{w}}}{\partial t}$ satisfy in the domain $\Omega = \Omega_f \cup \Gamma \cup \Omega_s$ for $t \in (0, t_0)$ the continuity equation

$$\bar{p} + c^2(\mathbf{x})\rho^0(\mathbf{x}) \nabla \cdot \bar{\mathbf{w}} = 0, \tag{A.7.14}$$

almost everywhere in $\Omega \times (0, t_0)$ and the integral identities

$$\left.\begin{aligned} &\int_0^{t_0} \int_\Omega \Big(\rho^0(\mathbf{x})\bar{\mathbf{v}} \cdot \frac{\partial \psi}{\partial t} - \mathbb{D}(x, \psi) : \bar{\mathbb{P}} + \rho^0(\mathbf{x})\mathbf{F} \cdot \psi\Big) dxdt = 0, \\ &\int_0^{t_0} \int_\Omega \Big(\eta(\mathbf{x})\bar{\vartheta} \frac{\partial \xi}{\partial t} - \nabla \xi \cdot \varkappa \nabla \bar{\vartheta} - \gamma(\mathbf{x})\vartheta_0 \xi \nabla \cdot \bar{\mathbf{v}}\Big) dxdt = 0 \end{aligned}\right\} \tag{A.7.15}$$

for any smooth and finite in $\Omega \times (0, t_0)$ functions φ ψ.

In (A.7.15)

$$\gamma(\mathbf{x}) = \gamma_f \chi_0(\mathbf{x}) + \gamma_s\big(1 - \chi_0(\mathbf{x})\big), \ \rho^0(\mathbf{x}) = \rho_f^0 \chi_0(\mathbf{x}) + \rho_s^0\big(1 - \chi_0(\mathbf{x})\big),$$

$$\eta(\mathbf{x}) = c_{p,f} \chi_0(\mathbf{x}) + c_{p,s}\big(1 - \chi_0(\mathbf{x})\big).$$

We also will use the differential form of equations for the model $\mathbb{M}_{11}$, when some equations are understood in the sense of distributions, namely, we use the form

$$\rho^0(\mathbf{x})\frac{\partial \bar{\mathbf{v}}}{\partial t} = \nabla \cdot \bar{\mathbb{P}} + \rho^0(\mathbf{x})\mathbf{F},$$

$$\eta(\mathbf{x})\frac{\partial \bar{\vartheta}}{\partial t} = \nabla \cdot (\varkappa \nabla \bar{\vartheta}) - \gamma(\mathbf{x})\vartheta_0 \nabla \cdot \bar{\mathbf{v}},$$

which is a formal representation of the integral identities (A.7.15).

3. The equations of the model $\mathbb{M}_{11}$ are written in dimensional form, while for further mathematical analysis the dimensionless form of equations is more convenient. To rewrite the equations in a dimensionless form we make the following change of variables

$$\mathbf{x} \to \frac{\mathbf{x}}{L}, \quad t \to \frac{t}{\tau}, \quad \bar{\mathbf{w}} \to \frac{\mathbf{w}}{L}, \quad \bar{\vartheta} \to \frac{\vartheta}{\vartheta_0}, \quad \mathbf{F} \to \frac{\mathbf{F}}{g}, \quad \rho \to \frac{\rho}{\rho^0},$$

where L is the characteristic size of the domain under consideration, g is the value of acceleration due to gravity, ρ^0 is the mean density of water, and τ is the characteristic time of the process.

To avoid new symbols we keep the same notations for domains occupied by solid skeleton, liquid and the common "solid skeleton–liquid" boundary Ω_s, Ω_f and Γ.

In dimensionless variables the differential equation of the model $\mathbb{M}_{11}$ in the domain Ω for $t > 0$ takes the form

$$p + \tilde{\alpha}_p \nabla \cdot \mathbf{w} = 0, \tag{A.7.16}$$

$$\alpha_\tau \tilde{\rho}\frac{\partial^2 \mathbf{w}}{\partial t^2} = \nabla \cdot \mathbb{P} + \tilde{\rho}\mathbf{F}, \tag{A.7.17}$$

$$\tilde{\eta}_0 \frac{\partial \vartheta}{\partial t} = \nabla \cdot (\tilde{\alpha}_\varkappa \nabla \vartheta) - \gamma_0\, \tilde{\alpha}_\theta\, \nabla \cdot \frac{\partial \mathbf{w}}{\partial t}, \tag{A.7.18}$$

where the dimensionless stress tensor of the medium

$$\mathbb{P} = \chi_0 \mathbb{P}^f + (1 - \chi_0)\mathbb{P}^s \tag{A.7.19}$$

coincides with the dimensionless viscous stress tensor

$$\mathbb{P}^f = \alpha_\mu \mathbb{D}\Big(x, \frac{\partial \mathbf{w}}{\partial t}\Big) - \Big(p + \alpha_{\vartheta f}\vartheta - \alpha_\nu \nabla \cdot \frac{\partial \mathbf{w}}{\partial t}\Big)\mathbb{I} \tag{A.7.20}$$

in the liquid and with the dimensionless elastic stress tensor

$$\mathbb{P}^s = \alpha_\lambda \mathbb{D}(x, \mathbf{w}) - \big(p + \alpha_{\vartheta,s}\vartheta\big)\mathbb{I} \tag{A.7.21}$$

in the solid skeleton.

In Eqs. (A.7.16)–(A.7.21)

$$\tilde{\rho} = \chi_0 \rho_f + (1 - \chi_0)\rho_s, \quad \tilde{\eta}_0 = \chi_0\, \eta_{p,f} + (1 - \chi_0)\, \eta_{p,s},$$

$$\tilde{\alpha}_p = \chi_0 \alpha_{p,f} + (1 - \chi_0)\alpha_{p,s},$$

$$\tilde{\alpha}_{\varkappa} = \chi_0 \alpha_{\varkappa,f} + (1 - \chi_0)\alpha_{\varkappa,s}, \quad \tilde{\alpha}_{\vartheta} = \chi_0 \alpha_{\vartheta,f} + (1 - \chi_0)\alpha_{\vartheta,s}.$$

Dimensionless criteria α_i $(i = \tau, \mu, \ldots)$ are defined by formulae:

$$\alpha_\tau = \frac{L}{g\tau^2}, \quad \alpha_\mu = \frac{2\mu}{\tau L g \rho_0}, \quad \alpha_\lambda = \frac{2\lambda}{L g \rho_0}, \quad \alpha_\nu = \frac{\nu}{\tau L g \rho_0},$$

$$\alpha_{\vartheta,j} = \frac{\vartheta_0 \gamma_i}{L g \rho_0}, \quad \alpha_{\varkappa,j} = \frac{\varkappa_j \tau}{L^2 \gamma_f}, \quad \alpha_{p,j} = \rho_j \frac{c_j^2}{Lg},$$

$$\gamma_0 = \frac{L g \rho_0}{\vartheta_0 \gamma_f}, \quad \rho_j = \frac{\rho_j^0}{\rho_0}, \quad \eta_{p,i} = \frac{c_{p,i}}{\gamma_f}, \quad j = f, s,$$

where c_f and c_s are the speed of compressive sound waves in the liquid and in the solid respectively, and ρ_f and ρ_s are the respective mean dimensionless densities of the liquid in pores and the solid skeleton correlated with the mean density of water ρ^0.

Equations of contact discontinuity (A.7.12) and (A.7.13) transform to

$$[\mathbb{P}] \cdot \mathbf{n} = \left(\lim_{\substack{\mathbf{x} \to \mathbf{x}_0 \\ \mathbf{x} \in \Omega_s}} \mathbb{P}(\mathbf{x}, t) - \lim_{\substack{\mathbf{x} \to \mathbf{x}_0 \\ \mathbf{x} \in \Omega_f}} \mathbb{P}(\mathbf{x}, t) \right) \cdot \mathbf{n}(\mathbf{x}_0) = 0, \tag{A.7.22}$$

$$[\alpha_{\varkappa} \nabla \vartheta] \cdot \mathbf{n}(\mathbf{x}_0) = 0 \tag{A.7.23}$$

for $\mathbf{x}_0 \in \Gamma$.

As before, Eqs. (A.7.17), (A.7.18), (A.7.22), and (A.7.23) are completed with conditions

$$[\vartheta(\mathbf{x}_0, t)] = 0, \;\; [\mathbf{w}(\mathbf{x}_0, t)] = 0, \;\; \mathbf{x}_0 \in \Gamma,$$

which are understood as the corresponding integral identities

$$\int_0^{t_0} \int_\Omega \left(\tilde{\rho}(\mathbf{x}) \frac{\partial \mathbf{w}}{\partial t} \cdot \frac{\partial \psi}{\partial t} - \mathbb{D}(x, \psi) : \mathbb{P} + \tilde{\rho}(\mathbf{x}) \mathbf{F} \cdot \psi \right) dx dt = 0, \tag{A.7.24}$$

$$\int_0^{t_0} \int_\Omega \left(\tilde{\eta}_0(\mathbf{x}) \vartheta \frac{\partial \xi}{\partial t} - \tilde{\alpha}_{\varkappa} \nabla \xi \cdot \nabla \vartheta - \gamma_0\, \tilde{\alpha}_\theta \xi \nabla \cdot \frac{\partial \mathbf{w}}{\partial t} \right) dx dt = 0, \tag{A.7.25}$$

for any smooth and finite in $\Omega \times (0, t_0)$ functions φ and ψ.

Differential equations (A.7.16)–(A.7.18) form a mathematical model $\mathbb{M}_{12}$ **of small perturbations in the joint motion of a non-isothermal viscous fluid and a non-isothermal elastic body**, which is just the dimensionless form of the model $\mathbb{M}_{11}$.

4. The model $\mathbb{M}_{12}$ describes all possible motions of a mixture of a solid and a liquid. The given physical process is characterized by parameters: the characteristic time of the process, the characteristic size of the domain under consideration, viscosities, the speed of sound and so on, which enter into dimensionless criteria α_i $(i = \tau,\ \mu,\ \lambda, \ldots)$.

Therefore, a given physical process corresponds to the given set of dimensionless criteria. We also may characterize physical processes using these dimensionless criteria. Thus, the characteristic time of filtration processes of underground liquids is some month and the characteristic size of the physical domains there is about one thousands meters. Therefore,

$$\alpha_\tau \sim 0,$$

and we may postulate that for liquid filtration in an elastic solid skeleton

$$\alpha_\tau = 0. \tag{A.7.26}$$

The corresponding mathematical model $\mathbb{M}_{13}$ **of a filtration of a compressible thermo-fluid in a thermo-elastic solid skeleton** consists of the following differential equations

$$\frac{1}{\tilde{\alpha}_p} p + \nabla \cdot \mathbf{w} = 0, \tag{A.7.27}$$

$$\nabla \cdot \mathbb{P} + \tilde{\rho}\mathbf{F} = 0, \tag{A.7.28}$$

$$\tilde{\eta}_0 \frac{\partial \vartheta}{\partial t} = \nabla \cdot (\tilde{\alpha}_\varkappa \nabla \vartheta) - \gamma_0\, \tilde{\alpha}_\vartheta\, \nabla \cdot \frac{\partial \mathbf{w}}{\partial t}, \tag{A.7.29}$$

$$\mathbb{P} = \chi_0 \alpha_\mu \mathbb{D}\Big(x, \frac{\partial \mathbf{w}}{\partial t}\Big) + (1-\chi_0)\alpha_\lambda \mathbb{D}(x, \mathbf{w}) - \big(p + \tilde{\alpha}_\vartheta\, \vartheta - \chi_0 \alpha_\nu \nabla \cdot \frac{\partial \mathbf{w}}{\partial t}\big)\mathbb{I}. \tag{A.7.30}$$

Next we consider different sub-models of the model $\mathbb{M}_{13}$ with simplifying physical assumptions. The first sub-model is the model $\mathbb{M}_{14}$ of isothermal filtration (or, simply, **liquid filtration**), which corresponds to the assumption

$$\gamma_0 \alpha_{\vartheta,f} = 0, \quad \gamma_0 \alpha_{\vartheta,s} = 0. \tag{A.7.31}$$

In this model the temperature is defined independently by the heat equation

$$\tilde{\eta}_0 \frac{\partial \vartheta}{\partial t} = \nabla \cdot (\tilde{\alpha}_\varkappa \nabla \vartheta), \tag{A.7.32}$$

and the motion of the medium is defined by the system

$$\frac{1}{\tilde{\alpha}_p} p + \nabla \cdot \mathbf{w} = 0, \tag{A.7.33}$$

$$\nabla \cdot \mathbb{P} + \tilde{\rho}\mathbf{F} = 0, \tag{A.7.34}$$

$$\mathbb{P} = \chi_0 \alpha_\mu \mathbb{D}\Big(x, \frac{\partial \mathbf{w}}{\partial t}\Big) + (1 - \chi_0)\alpha_\lambda \mathbb{D}(x, \mathbf{w}) - \big(p - \chi_0 \alpha_\nu \nabla \cdot \frac{\partial \mathbf{w}}{\partial t}\big)\mathbb{I}. \tag{A.7.35}$$

The second model is a sub-model of the model $\mathbb{M}_{14}$, which describes a filtration of an incompressible liquid.

As is well-known, **the measure of incompressibility of a given medium is the speed of sound in the medium**. Incompressible media have an infinite speed of sound. Therefore, for long-term physical processes the behavior of acoustic waves is not so important, and for many real liquids we may accept the assumption that the given liquid is incompressible. On the other hand, as a rule, the speed of sound in the solid skeleton two or three times more than the speed of sound in the liquid. Thus we also may accept the assumption that the given solid skeleton is incompressible and together these give

$$\frac{1}{\alpha_{p,f}} \sim 0, \quad \frac{1}{\alpha_{p,s}} \sim 0.$$

The corresponding axiom

$$\alpha_{p,f} = \alpha_{p,s} = \infty \tag{A.7.36}$$

picks outs the class of physical media (**incompressible media**), which is described by the mathematical model $\mathbb{M}_{15}$ of **the filtration of an incompressible liquid**:

$$\nabla \cdot \mathbf{w} = 0, \tag{A.7.37}$$

$$\nabla \cdot \mathbb{P} + \tilde{\rho}\mathbf{F} = 0, \tag{A.7.38}$$

$$\mathbb{P} = \chi_0 \alpha_\mu \mathbb{D}\Big(x, \frac{\partial \mathbf{w}}{\partial t}\Big) + (1 - \chi_0)\alpha_\lambda \mathbb{D}(x, \mathbf{w}) - p\,\mathbb{I}. \tag{A.7.39}$$

Sometimes we do not use the properties of the solid skeleton and may simplify the model $\mathbb{M}_{12}$ using postulate

$$\alpha_\lambda = \infty, \tag{A.7.40}$$

which means that the solid skeleton is an absolutely rigid solid body. That axiom transforms the initial model into the mathematical model $\mathbb{M}_{16}$ of **the filtration of a compressible thermo-fluid in an non-isothermal absolutely rigid solid skeleton**:

$$\chi_0\Big(\frac{1}{\alpha_{p,f}} p + \nabla \cdot \mathbf{w}\Big) = 0, \tag{A.7.41}$$

$$\chi_0\big(\nabla \cdot \mathbb{P} + \rho_f \mathbf{F}\big) = 0, \tag{A.7.42}$$

$$\tilde{\eta}_0 \frac{\partial \vartheta}{\partial t} = \nabla \cdot (\tilde{\alpha}_{\varkappa} \nabla \vartheta) - \gamma_0\, \chi_0 \alpha_{\vartheta,f}\, \nabla \cdot \frac{\partial \mathbf{w}}{\partial t}, \tag{A.7.43}$$

$$\mathbb{P} = \alpha_\mu \mathbb{D}\Big(x, \frac{\partial \mathbf{w}}{\partial t}\Big) - \Big(p + \tilde{\alpha}_\vartheta\, \vartheta - \alpha_\nu \nabla \cdot \frac{\partial \mathbf{w}}{\partial t}\Big)\mathbb{I}, \tag{A.7.44}$$

where Eq. (A.7.42) is equivalent to the integral identity

$$\int_0^{t_0} \int_{\Omega_f} \Big(\mathbb{D}(x, \psi) : \bar{\mathbb{P}} - \rho_f \mathbf{F} \cdot \psi\Big) dx dt = 0 \tag{A.7.45}$$

for any smooth function ψ, finite in Ω_f. This model is completed with the additional boundary condition

$$\mathbf{w}(\mathbf{x}, t) = 0, \quad \mathbf{x} \in \Gamma. \tag{A.7.46}$$

The sub-model $\mathbb{M}_{17}$ of the model $\mathbb{M}_{16}$, describing isothermal motion consists of equations

$$\chi_0\Big(\frac{1}{\alpha_{p,f}}\, p + \nabla \cdot \mathbf{w}\Big) = 0, \tag{A.7.47}$$

$$\chi_0\big(\nabla \cdot \mathbb{P} + \rho_f \mathbf{F}\big) = 0, \tag{A.7.48}$$

$$\mathbb{P} = \alpha_\mu \mathbb{D}\Big(x, \frac{\partial \mathbf{w}}{\partial t}\Big) - \Big(p - \alpha_\nu \nabla \cdot \frac{\partial \mathbf{w}}{\partial t}\Big)\mathbb{I}, \tag{A.7.49}$$

$$\mathbf{w}(\mathbf{x}, t) = 0, \quad \mathbf{x} \in \Gamma, \tag{A.7.50}$$

and we call this motion as the **filtration of a compressible liquid in an absolutely rigid solid skeleton**.

Finally, the **filtration of incompressible liquid in an absolutely rigid solid skeleton** is described by the model $\mathbb{M}_{18}$:

$$\chi_0 \nabla \cdot \mathbf{w} = 0, \tag{A.7.51}$$

$$\chi_0\Big(\nabla \cdot \mathbb{P} + \rho_f \mathbf{F}\Big) = 0, \tag{A.7.52}$$

$$\mathbb{P} = \alpha_\mu \mathbb{D}\Big(x, \frac{\partial \mathbf{w}}{\partial t}\Big) - p\,\mathbb{I}, \tag{A.7.53}$$

$$\mathbf{w}(\mathbf{x}, t) = 0, \quad \mathbf{x} \in \Gamma. \tag{A.7.54}$$

4. For short-term processes like acoustic processes or hydraulic shock in porous media

$$\alpha_\tau \sim \infty,$$

and we use one more renormalization by setting

$$\alpha_\tau \mathbf{w} \longrightarrow \mathbf{w},$$

which transforms the model $\mathbb{M}_{12}$ to the model $\mathbb{M}_{19}$ of **non-isothermal short-term processes**:

$$p + \tilde{\bar{\alpha}}_p \nabla \cdot \mathbf{w} = 0, \tag{A.7.55}$$

$$\tilde{\rho} \frac{\partial^2 \mathbf{w}}{\partial t^2} = \nabla \cdot \mathbb{P} + \tilde{\rho} \mathbf{F}, \tag{A.7.56}$$

$$\tilde{\eta}_0 \frac{\partial \vartheta}{\partial t} = \nabla \cdot (\tilde{\alpha}_\varkappa \nabla \vartheta) - \bar{\gamma}_0 \tilde{\alpha}_\vartheta \nabla \cdot \frac{\partial \mathbf{w}}{\partial t}, \tag{A.7.57}$$

$$\mathbb{P} = \chi_0 \mathbb{P}^f + (1 - \chi_0) \mathbb{P}^s, \tag{A.7.58}$$

$$\mathbb{P}^f = \bar{\alpha}_\mu \mathbb{D}\Big(x, \frac{\partial \mathbf{w}}{\partial t}\Big) - \big(p + \tilde{\alpha}_{\vartheta f} \vartheta - \bar{\alpha}_\nu \nabla \cdot \frac{\partial \mathbf{w}}{\partial t}\big)\mathbb{I}, \tag{A.7.59}$$

$$\mathbb{P}^s = \bar{\alpha}_\lambda \mathbb{D}(x, \mathbf{w}) - \big(p + \tilde{\alpha}_{\vartheta,s} \vartheta\big)\mathbb{I}. \tag{A.7.60}$$

In Eqs. (A.7.55)–(A.7.60)

$$\tilde{\bar{\alpha}}_p = \chi_0 \bar{\alpha}_{p,f} + (1 - \chi_0) \bar{\alpha}_{p,s}, \quad \tilde{\bar{\alpha}}_\vartheta = \chi_0 \bar{\alpha}_{\vartheta,f} + (1 - \chi_0) \bar{\alpha}_{\vartheta,s}.$$

Dimensionless criteria $\bar{\alpha}_i$ ($i = \mu, \nu, \lambda, \ldots$) are defined by formulae:

$$\bar{\alpha}_\mu = \frac{2\mu\tau}{L^2 \rho_0}, \quad \bar{\alpha}_\nu = \frac{\nu\tau}{L^2 \rho_0}, \quad \bar{\alpha}_\lambda = \frac{2\lambda\tau^2}{L^2 \rho_0},$$

$$\bar{\gamma}_0 = \rho_0 \frac{g^2 \tau^2}{\vartheta_0 \gamma_f}, \quad \bar{\alpha}_{p,j} = \rho_j c_j^2 \frac{\tau^2}{L^2}, \quad j = f, s.$$

As before, we may consider a sub-model $\mathbb{M}_{20}$ of **isothermal short-term processes** under the assumption (A.7.31), which gives

$$p + \tilde{\bar{\alpha}}_p \nabla \cdot \mathbf{w} = 0, \tag{A.7.61}$$

$$\tilde{\rho} \frac{\partial^2 \mathbf{w}}{\partial t^2} = \nabla \cdot \mathbb{P} + \tilde{\rho} \mathbf{F}, \tag{A.7.62}$$

$$\mathbb{P} = \chi_0 \mathbb{P}^f + (1 - \chi_0) \mathbb{P}^s, \tag{A.7.63}$$

$$\mathbb{P}^f = \bar{\alpha}_\mu \mathbb{D}\Big(x, \frac{\partial \mathbf{w}}{\partial t}\Big) - \Big(p - \bar{\alpha}_\nu \nabla \cdot \frac{\partial \mathbf{w}}{\partial t}\Big)\mathbb{I}, \tag{A.7.64}$$

$$\mathbb{P}^s = \bar{\alpha}_\lambda \mathbb{D}(x, \mathbf{w}) - p\,\mathbb{I}. \tag{A.7.65}$$

Finally, we may simplify the last model and consider the model $\mathbb{M}_{21}$ of **short-term processes in incompressible media**:

$$\nabla \cdot \mathbf{w} = 0, \tag{A.7.66}$$

$$\tilde{\rho}\frac{\partial^2 \mathbf{w}}{\partial t^2} = \nabla \cdot \mathbb{P} + \tilde{\rho}\mathbf{F}, \tag{A.7.67}$$

$$\mathbb{P} = \chi_0 \bar{\alpha}_\mu \mathbb{D}\Big(x, \frac{\partial \mathbf{w}}{\partial t}\Big) + (1 - \chi_0)\bar{\alpha}_\lambda \mathbb{D}(x, \mathbf{w}) - p\,\mathbb{I}. \tag{A.7.68}$$

and its sub-model $\mathbb{M}_{22}$ of **short-term processes in an absolutely rigid solid skeleton**:

$$\nabla \cdot \mathbf{w} = 0, \tag{A.7.69}$$

$$\chi_0\Big(\tilde{\rho}\frac{\partial^2 \mathbf{w}}{\partial t^2} - \nabla \cdot \mathbb{P} - \tilde{\rho}\mathbf{F}\Big) = 0, \tag{A.7.70}$$

$$\mathbb{P} = \bar{\alpha}_\mu \mathbb{D}\Big(x, \frac{\partial \mathbf{w}}{\partial t}\Big) - p\,\mathbb{I}. \tag{A.7.71}$$

5. It is easy to prove that the model $\mathbb{M}_{13}$ is an asymptotic limit of the model $\mathbb{M}_{12}$ as α_τ goes to zero, the model $\mathbb{M}_{15}$ is an asymptotic limit of the model $\mathbb{M}_{14}$ as $\alpha_{p,f}$ and $\alpha_{p,s}$ go to infinity, the model $\mathbb{M}_{16}$ is an asymptotic limit of the model $\mathbb{M}_{12}$ as α_λ goes to infinity, and the model $\mathbb{M}_{18}$ is an asymptotic limit of the model $\mathbb{M}_{17}$ as $\alpha_{p,f}$ goes to infinity.

Under assumptions (A.7.31) the model $\mathbb{M}_{12}$ transforms to the model $\mathbb{M}_{23}$ **of small perturbations in joint motion of isothermal viscous fluid and isothermal elastic body**, consisting of the following equations

$$p + \tilde{\alpha}_p \nabla \cdot \mathbf{w} = 0, \tag{A.7.72}$$

$$\alpha_\tau \tilde{\rho}\frac{\partial^2 \mathbf{w}}{\partial t^2} = \nabla \cdot \mathbb{P} + \tilde{\rho}\mathbf{F}, \tag{A.7.73}$$

$$\mathbb{P} = \chi_0 \mathbb{P}^f + (1 - \chi_0)\mathbb{P}^s, \tag{A.7.74}$$

$$\mathbb{P}^f = \alpha_\mu \mathbb{D}\Big(x, \frac{\partial \mathbf{w}}{\partial t}\Big) - \Big(p - \alpha_\nu \nabla \cdot \frac{\partial \mathbf{w}}{\partial t}\Big)\mathbb{I}, \tag{A.7.75}$$

$$\mathbb{P}^s = \alpha_\lambda \mathbb{D}(x, \mathbf{w}) - p\,\mathbb{I}. \tag{A.7.76}$$

The limit as $\tilde{\alpha}_p \to \infty$ in $\mathbb{M}_{23}$ results in the model

$$\nabla \cdot \mathbf{w} = 0, \tag{A.7.77}$$

$$\alpha_\tau \tilde{\rho} \frac{\partial^2 \mathbf{w}}{\partial t^2} = \nabla \cdot \mathbb{P} + \tilde{\rho} \mathbf{F}, \tag{A.7.78}$$

$$\mathbb{P} = \chi_0 \alpha_\mu \mathbb{D}\Big(x, \frac{\partial \mathbf{w}}{\partial t}\Big) + (1 - \chi_0) \alpha_\lambda \mathbb{D}(x, \mathbf{w}) - p\,\mathbb{I}, \tag{A.7.79}$$

which we call the model $\mathbb{M}_{24}$ **of small perturbations in joint motion of isothermal incompressible viscous fluid and isothermal incompressible elastic body**.

The limit as $\alpha_\tau \to 0$ in $\mathbb{M}_{24}$ results in the model $\mathbb{M}_{15}$.

6. Diffusion-convection processes in porous media $\Omega \subset \mathbb{R}^3$ are described by the diffusion-convection equation

$$\frac{\partial c}{\partial t} + \mathbf{v} \cdot \nabla c = D \triangle c, \tag{A.7.80}$$

for the concentration c of an admixture in the liquid domain Ω_f (pore space).

Here D is the given diffusion coefficient, and $\mathbf{v}$ is the velocity of the liquid.

If we consider the most general case of the motion of continuous media, which is a generalized motion with strong discontinuity, then the boundary condition on the surface of strong discontinuity $\Gamma_t = \partial \Omega_s \cap \partial \Omega_f$ (the common boundary "pore space–solid skeleton") at the time $t > 0$ has the form

$$c\,\big((\mathbf{v} \cdot \mathbf{n}) - V_n\big) = D\,(\nabla c \cdot \mathbf{n}). \tag{A.7.81}$$

In (A.7.81) $\mathbf{n}$ is a unit normal vector to Γ_t, and V_n is a velocity of replacement of surface Γ_t in the direction of normal $\mathbf{n}$.

In the general case the velocity field is defined by the mathematical model $\mathbb{M}_{10}$, which is a free boundary problem. In particular, one of the boundary condition at the free surface of a contact discontinuity has the form

$$\mathbf{v} \cdot \mathbf{n} = V_n, \tag{A.7.82}$$

and for this case (A.7.81) transforms to

$$\nabla c \cdot \mathbf{n} = 0. \tag{A.7.83}$$

It is clear, that even if one knows how to solve the free boundary problem that arises, this mathematical model obviously would not be suitable for practical use, since the function χ_0 changes its value from 0 to 1 on the scale of a few microns. Thus, the most suitable way to get a practically significant mathematical model is a homogenization. But in this case the mathematical problem becomes absolutely unsolvable. To get something more simple we use the mathematical model $\mathbb{M}_{15}$, where the characteristic function $\tilde{\chi}$ of the liquid domain Ω_f is approximated by its value at initial time:

$$\tilde{\chi} \simeq \chi_0(\mathbf{x}),$$

and

$$\mathbf{v} \simeq \frac{\partial \mathbf{w}}{\partial t},$$

In dimensionless variables this model, coupled with a convection-diffusion equation has the form

$$\nabla \cdot \mathbf{w} = 0, \tag{A.7.84}$$

$$\nabla \cdot \mathbb{P} + \tilde{\rho}\mathbf{F} = 0, \tag{A.7.85}$$

$$\mathbb{P} = \chi_0 \alpha_\mu \mathbb{D}\Big(x, \frac{\partial \mathbf{w}}{\partial t}\Big) + (1 - \chi_0)\alpha_\lambda \mathbb{D}(x, \mathbf{w}) - p\,\mathbb{I}, \tag{A.7.86}$$

$$\frac{\partial c}{\partial t} + \frac{\partial \mathbf{w}}{\partial t} \cdot \nabla c = \alpha_D \, \triangle c, \tag{A.7.87}$$

where

$$\alpha_D = \frac{D\tau}{L^2}, \quad \alpha_\mu = \alpha_\mu(c), \quad \tilde{\rho} = \chi_0(\rho_f + \delta c) + (1 - \chi_0)\rho_s.$$

We must complete the model with the boundary conditions on the common (and fixed) boundary Γ. The boundary conditions for dynamic equations have already been discussed.

For the convection-diffusion equation one has a choice. By supposition, $V_n = 0$, and this postulate and the boundary condition (A.7.81) imply

$$\Big(\alpha_D \nabla c - c \frac{\partial \mathbf{w}}{\partial t}\Big) \cdot \mathbf{n} = 0. \tag{A.7.88}$$

That is the first option. The second option is the condition (A.7.83). Thus, for the same process we have two different models: the mathematical model $\mathbb{M}_{25}$, consisting of (A.7.83)–(A.7.87), and the mathematical model $\mathbb{M}_{26}$, consisting of (A.7.84)–(A.7.88). The difference because both of the boundary conditions for concentration are not quite exact in our approximation. But there is a case when both models coincide.

In fact, for an absolutely rigid solid skeleton $\mathbf{w} = \mathbf{v} = 0$ in the solid part. Hence (A.7.83) coincides with (A.7.88). We consider the model $\mathbb{M}_{17}$

$$\chi_0\Big(\frac{1}{\alpha_{p,f}} p + \nabla \cdot \mathbf{w}\Big) = 0, \tag{A.7.89}$$

$$\chi_0\Big(\nabla\cdot\mathbb{P}+(\rho_f+\delta\, c)\mathbf{F}\Big)=0, \tag{A.7.90}$$

$$\mathbb{P}=\alpha_\mu(c)\mathbb{D}\Big(x,\frac{\partial\mathbf{w}}{\partial t}\Big)-\Big(p-\alpha_\nu\nabla\cdot\frac{\partial\mathbf{w}}{\partial t}\Big)\mathbb{I}, \tag{A.7.91}$$

coupled with (A.7.83) and (A.7.87), and refer to the model obtained as the mathematical model $\mathbb{M}_{27}$.

7. The joint motion of two incompressible immiscible liquids with the same viscosity and different constant densities is described at the microscopic level by the dynamic equations

$$\nabla\cdot\big(\tilde{\chi}\,\mathbb{P}_f+(1-\tilde{\chi})\mathbb{P}_s\big)+\big(\rho_f\,\tilde{\chi}+\rho_s\,(1-\tilde{\chi})\big)\mathbf{F}=0, \tag{A.7.92}$$

$$\nabla\cdot\mathbf{v}=0, \tag{A.7.93}$$

$$\frac{\partial\tilde{\chi}}{\partial t}+\mathbf{v}\cdot\nabla\,\tilde{\chi}=0,\quad \tilde{\chi}(\mathbf{x},0)=\chi_0(\mathbf{x}), \tag{A.7.94}$$

completed with the Cauchy problem

$$\frac{\partial\rho_f}{\partial t}+\mathbf{v}\cdot\nabla\,\rho_f=0,\quad \rho_f(\mathbf{x},0)=\rho_f^{(0)}(\mathbf{x}) \tag{A.7.95}$$

for the density ρ_f of the nonhomogeneous liquid in the liquid domain Ω_f for $t>0$.

The last problem is equivalent to the Cauchy problem

$$\frac{\partial\rho}{\partial t}+\mathbf{v}\cdot\nabla\,\rho=0,\quad \rho(\mathbf{x},0)=\rho_f^{(0)}(\mathbf{x})\,\chi_0(\mathbf{x})+\rho_s\,\big(1-\chi_0(\mathbf{x})\big) \tag{A.7.96}$$

for the density ρ of the medium.

Let Π_0 be a smooth surface dividing Ω into two subdomains Ω^+ and Ω^- and

$$\rho_f^{(0)}(\mathbf{x})=\rho_f^+=const\ \text{for}\ \mathbf{x}\in\Omega^+,\quad \rho_f^{(0)}(\mathbf{x})=\rho_f^-=const\ \text{for}\ \mathbf{x}\in\Omega^-.$$

Then for the smooth velocity field $\mathbf{v}(\mathbf{x},t)$ there exists a smooth surface of the strong discontinuity $\Pi(t)$, $\Pi(0)=\Pi_0$, dividing Ω_f into two subdomains $\Omega_f^+(t)$ and $\Omega_f^-(t)$, such that

$$\rho(\mathbf{x},t)=\rho_f^+\ \text{for}\ \mathbf{x}\in\Omega_f^+(t),\ \text{and}\ \rho(\mathbf{x},t)=\rho_f^-\ \text{for}\ \mathbf{x}\in\Omega^-.$$

That is, the problem (A.7.92)–(A.7.94), (A.7.96) really describes the joint motion of two immiscible incompressible liquids with the different constant densities separated by the free boundary $\Pi(t)$.

It is obvious that the resulting problem is too complicated. To simplify the model and get simpler, but still reasonable one, we replace (A.7.92)–(A.7.94) by the system

$$\nabla \cdot \left(\chi_0 \, \alpha_\mu \, \mathbb{D}\Big(x, \frac{\partial \mathbf{w}}{\partial t}\Big) + (1 - \chi_0)\alpha_\lambda \, \mathbb{D}(x, \mathbf{w}) - p\,\mathbb{I} \right) + \rho \, \mathbf{F} = 0, \qquad \text{(A.7.97)}$$

$$\nabla \cdot \mathbf{v} = 0, \quad \mathbf{v} = \frac{\partial \mathbf{w}}{\partial t}, \qquad \text{(A.7.98)}$$

completed with the Cauchy problem

$$\frac{\partial \rho}{\partial t} + \mathbf{v} \cdot \nabla \rho = 0, \quad \rho(\mathbf{x}, 0) = \rho_f^{(0)}(\mathbf{x}) \, \chi_0 + \rho_f \, (1 - \chi_0) \qquad \text{(A.7.99)}$$

for the density ρ of the medium.

We refer to the model (A.7.97)–(A.7.99) as the mathematical model $\mathbb{M}_{28}$.

Appendix B
Auxiliary Mathematical Topics

B.1 Hilbert Spaces

A *Hilbert space* is a complete inner product space. That is to say, firstly the space H is a real linear space provided with an inner product, denoted (u, v), for u and v in H, satisfying the following defining conditions:

$$(u, u) \geqslant 0, \quad (u, u) = 0 \Leftrightarrow u = 0, \tag{B.1.1}$$

$$(u, v) = (v, u), \quad (u + v, w) = (u, v) + (u, w), \tag{B.1.2}$$

$$(\alpha u, v) = \alpha(u, v), \quad \forall \alpha \in \mathbb{R}. \tag{B.1.3}$$

To such an inner product is assigned a norm, by

$$\|u\| = \sqrt{(u, u)}. \tag{B.1.4}$$

Then it is easy to verify that

$$\|u\| \geqslant 0, \quad \|u\| = 0 \Leftrightarrow u = 0,$$

$$\|\alpha u\| = |\alpha| \|u\|.$$

Finally, the triangle inequality

$$\|u + v\| \leqslant \|u\| + \|v\|$$

is a consequence of Cauchy's inequality

$$|(u, v)| \leqslant \|u\| \cdot \|v\|. \tag{B.1.5}$$

A. Meirmanov, *Mathematical Models for Poroelastic Flows*,
Atlantis Studies in Differential Equations 1, DOI: 10.2991/978-94-6239-015-7,

A sequence $\{u_n\}$, $u_n \in H$ converges to $u \in H$:

$$u_n \to u, \quad \text{as } n \to \infty,$$

if

$$\|u - u_n\| \to 0 \quad \text{as } n \to \infty. \tag{B.1.6}$$

We also say that convergence (B.1.6) is a *strong convergence*.

A sequence $\{u_n\}$ is a Cauchy sequence provided that

$$\|u_n - u_m\| \to 0, \quad \text{as } n,\, m \to \infty.$$

Completeness is the property that any Cauchy sequence converges.

The element $u \in H$ is an *accumulation point* of a set $M \subset H$, if there exists a sequence $\{u_n\}$, $u_n \in M$, such that

$$u_n \to u, \quad \text{as } n \to \infty.$$

A *closure* $\overline{M}$ of a set M is a set of all the accumulation points of M.

M is *closed* if $\overline{M} = M$.

We say that M is *compact* if any infinite subset of M contains a convergent sequence.

A set M is *dense* in H, if $\overline{M} = H$.

A Hilbert space H is *separable*, if there exists a countable set dense in H.

A set $L(M)$ of all finite linear combinations

$$\sum_{i=1}^{m} u_{\alpha_i},$$

of a set $M = \{u_\alpha\}$ is said a *linear span* of M.

We say u and v are *orthogonal* if$(u, v) = 0$.

A set M is *orthonormal*, if

$$\|u\| = 1 \text{ for all } u \in M \quad \text{and} \quad (u, v) = 0 \text{ for all } u,\, v \in M,\ u \neq v.$$

We call an orthonormal set M a *basis* if it is not contained in any larger orthonormal set.

Lemma B.1 *Any separable Hilbert space contains a countable basis.*

Let $M = \{e_n\}_{n=1}^{\infty}$ be a basis. We denote

$$u = \sum_{n=1}^{\infty} c_n e_n,$$

if

$$\sum_{n=1}^{m} c_n e_n \to u, \quad \text{as } m \to \infty.$$

Lemma B.2 *Let H be a separable Hilbert space. Then for any basis $M = \{e_n\}_{n=1}^{\infty}$ and for any $u \in H$*

$$u = \sum_{n=1}^{\infty} c_n e_n, \quad \textit{where} \quad c_n = (u, e_n). \tag{B.1.7}$$

A sequence $\{u_n\}$ is said to be *weakly convergent* to u:

$$u_n \rightharpoonup u, \ \text{ as } n \to \infty,$$

if

$$(u_n - u, v) \to 0, \ \text{ as } n \to \infty, \ \forall v \in H.$$

Finally, we say that K is a *weakly compact set* if any sequence $\{u_n\}$, $u_n \in K$, contains a weakly convergent subsequence:

$$u_{n_m} \rightharpoonup u \in K, \ \text{ as } \ m \to \infty.$$

Theorem B.1 *In a separable Hilbert space any bounded closed ball*

$$B = \{u \in H : \|u\| \leqslant C\},$$

is a weakly compact set.

B.2 Sobolev Spaces for Scalar Functions

Let $\Omega \in \mathbb{R}^n$ be a bounded set with a Lipschitz continuous boundary $S = \partial\Omega$. Then for any $u \in C^1(\overline{\Omega})$ and for any $v \in \overset{\circ}{C}{}^1(\Omega)$

$$\int_{\Omega} \Big(\frac{\partial u}{\partial x_i} v + u \frac{\partial v}{\partial x_i}\Big) dx = 0, \ i = 1, \dots, n. \tag{B.2.1}$$

We say that a function $u(\mathbf{x})$, $\ \mathbf{x} \in \Omega$, is an element of the Hilbert space $L_2(\Omega)$ with an inner product

$$(u, v) = \int_{\Omega} u(\mathbf{x})\, v(\mathbf{x}) dx,$$

if

$$\|u\|_{2,\Omega} = \Big(\int_\Omega u^2(\mathbf{x})dx\Big)^{\frac{1}{2}} < \infty.$$

The Cauchy inequality (B.1.5) for $L_2(\Omega)$ has the form

$$\Big|\int_\Omega u(\mathbf{x})\, v(\mathbf{x})dx\Big| \leqslant \Big(\int_\Omega u^2(\mathbf{x})dx\Big)^{\frac{1}{2}}\Big(\int_\Omega v^2(\mathbf{x})dx\Big)^{\frac{1}{2}}. \tag{B.2.2}$$

We also call this inequality the *Hölder inequality*.

A function $v_i \in L_2(\Omega)$ is a weak derivative of a function $u \in L_2(\Omega)$ with respect to a variable x_i, if

$$\int_\Omega \Big(\frac{\partial \varphi}{\partial x_i} u + v_i \frac{\partial \varphi}{\partial x_i}\Big)dx = 0, \ \ \forall \varphi \in \overset{\circ}{C}{}^1 (\Omega). \tag{B.2.3}$$

As usual we denote

$$v_i = \frac{\partial u}{\partial x_i}.$$

Now we define a Sobolev space $W_2^1(\Omega)$ as a linear inner product space of all functions $u \in L_2(\Omega)$ with weak derivatives

$$\frac{\partial u}{\partial x_i} \in L_2(\Omega), \ \ i = 1, \dots, n,$$

with the inner product

$$\langle u, v\rangle = \int_\Omega \Big(u \cdot v + \nabla u \cdot \nabla v\Big)dx \tag{B.2.4}$$

and the norm

$$\|u\|^{(1)}_{2,\Omega} = \Big(\int_\Omega \Big(u^2 + |\nabla u|^2\Big)dx\Big)^{\frac{1}{2}}, \tag{B.2.5}$$

where

$$\nabla u = \Big(\frac{\partial u}{\partial x_1}, \dots, \frac{\partial u}{\partial x_n}\Big),$$

and

$$\nabla u \cdot \nabla v = \sum_{i=1}^n \frac{\partial u}{\partial x_i} \cdot \frac{\partial v}{\partial x_i}, \ \ |\nabla u|^2 = \nabla u \cdot \nabla u.$$

There is an equivalent definition of the Sobolev space $W_2^1(\Omega)$ as a closure of the inner product space of all functions $u \in C^1(\Omega)$ with inner product (B.2.4).

That is, for any $u \in W_2^1(\Omega)$ there exists a sequence $\{u_n\}$, $u_n \in C^1(\Omega)$, such that

$$\|u - u_n\|_{2,\Omega}^{(1)} \to 0 \quad \text{as } n \to \infty.$$

Therefore a Sobolev space $W_2^1(\Omega)$ is complete inner product space, or a complete Hilbert space.

Lemma B.3 *For all $u \in W_2^1(\Omega)$ the trace*

$$u_S(\mathbf{x}) = u(\mathbf{x}), \quad \mathbf{x} \in S = \partial\Omega$$

is well defined and

$$\Big(\int_S |u|^2 ds\Big)^{\frac{1}{2}} \leqslant C\|u\|_{2,\Omega}^{(1)}, \tag{B.2.6}$$

where C depends only on the geometry of Ω and does not depend on u.

Moreover, if $\mathbf{n}(\mathbf{x})$ is an outward unit normal to S at $\mathbf{x} \in S$, then

$$\int_S \big|u_S(\mathbf{x}) - u\big(\mathbf{x} - \varepsilon\mathbf{n}(\mathbf{x})\big)\big|^2 ds \to 0, \ \textit{as } \varepsilon \to 0.$$

Lemma B.4 *The imbedding operator $W_2^1(\Omega) \to L_2(\Omega)$ is completely continuous. That is, any weakly convergent sequence in $W_2^1(\Omega)$ converges strongly in $L_2(\Omega)$.*

In the same way we define a space $\overset{\circ}{W}{}_2^1(\Omega)$ as a closure of the inner product space of all functions $u \in \overset{\circ}{C}{}^1(\Omega)$, vanishing at $\partial\Omega$, with inner product (B.2.4).

It is easy to see, that for any function $u \in W_2^1(\Omega)$ and for any function $v \in \overset{\circ}{W}{}_2^1(\Omega)$ formula (B.2.1) of integration by parts is still valid.

Lemma B.5 ***(Friedrichs–Poincaré's inequality for the scalar functions)***

For all $u \in \overset{\circ}{W}{}_2^1(\Omega)$

$$\int_\Omega |u|^2 dx \leqslant C\int_\Omega |\nabla u|^2 dx, \tag{B.2.7}$$

where C depends only on the geometry of Ω and does not depend on u and

$$\int_S \big|u\big(\mathbf{x} - \varepsilon\mathbf{n}(\mathbf{x})\big)\big|^2 ds \to 0, \ \textit{as } \varepsilon \to 0,$$

where $\mathbf{n}(\mathbf{x})$ is an outward unit normal to S at $\mathbf{x} \in S$.

For functions $u(\mathbf{x}, t)$, where $(\mathbf{x}, t) \in \Omega_T = \Omega \times (0, T)$, we define the space $L_2\big((0, T); L_2(\Omega)\big) = L_2(\Omega_T)$ as a Hilbert space with the inner product

$$(u, v) = \left(\int_0^T \left(\int_\Omega u(\mathbf{x}, t)\, v(\mathbf{x}, t) dx\right) dt\right)^{\frac{1}{2}}$$
$$= \left(\int_{\Omega_T} u(\mathbf{x}, t)\, v(\mathbf{x}, t) dx dt\right)^{\frac{1}{2}} \tag{B.2.8}$$

and the norm

$$\|u\|_{2,\Omega_T} = \left(\int_{\Omega_T} u^2(\mathbf{x}, t) dx dt\right)^{\frac{1}{2}}. \tag{B.2.9}$$

The Hölder inequality for $L_2(\Omega_T)$ has the form

$$\left|\int_{\Omega_T} u(\mathbf{x}, t)\, v(\mathbf{x}, t) dx dt\right|$$
$$\leqslant \left(\int_{\Omega_T} u^2(\mathbf{x}, t) dx dt\right)^{\frac{1}{2}} \left(\int_{\Omega_T} v^2(\mathbf{x}, t) dx dt\right)^{\frac{1}{2}}. \tag{B.2.10}$$

We also define spaces

$$L^2\big((0, T); W_2^1(\Omega)\big) = W_2^{1,0}(\Omega_T) \text{ and } L^2\big((0, T); \overset{\circ}{W}{}_2^{1}(\Omega)\big) = \overset{\circ}{W}{}_2^{1,0}(\Omega_T)$$

as the Hilbert spaces with the inner product

$$\langle u, v\rangle = \left(\int_{\Omega_T} (u(\mathbf{x}, t)\, v(\mathbf{x}, t) + \nabla u(\mathbf{x}, t) \cdot \nabla v(\mathbf{x}, t)) dx dt\right)^{\frac{1}{2}} \tag{B.2.11}$$

and the norm

$$\|u\|_{2,\Omega_T}^{(1)} = \left(\int_{\Omega_T} (u^2(\mathbf{x}, t) + |\nabla u(\mathbf{x}, t)|^2) dx dt\right)^{\frac{1}{2}}. \tag{B.2.12}$$

For any two functions $u,\ v \in W_2^1(\Omega)$ the Stokes (Gauss-Ostrogradsky) Theorem has the form

$$\int_\Omega \left(\frac{\partial u}{\partial x_i} v + u \frac{\partial v}{\partial x_i}\right) dx = \int_S u\, v\, \nu_i ds,\ i = 1, \dots, n, \tag{B.2.13}$$

where $\nu = (\nu_1, \dots, \nu_n)$ is the unit outward normal to the boundary $S = \partial\Omega$.

B.3 Sobolev Spaces for Vector Functions

We say that a vector function $\mathbf{u}(\mathbf{x}) = (u_1(\mathbf{x}), \ldots, u_n(\mathbf{x})),\ \mathbf{x} \in \Omega$, is an element of the Hilbert space $\mathbf{L}_2(\Omega)$ with an inner product

$$(\mathbf{u}, \mathbf{v}) = \int_\Omega \mathbf{u}(\mathbf{x}) \cdot \mathbf{v}(\mathbf{x}) dx,$$

if

$$\|\mathbf{u}\|_{2,\Omega} = \Big(\int_\Omega |\mathbf{u}(\mathbf{x})|^2 dx\Big)^{\frac{1}{2}} < \infty.$$

Here

$$\mathbf{u} \cdot \mathbf{v} = \sum_{i=1}^{n} u_i v_i, \quad |\mathbf{u}|^2 = \mathbf{u} \cdot \mathbf{u}.$$

The Hölder inequality for $\mathbf{L}_2(\Omega)$ has the form

$$\Big|\int_\Omega \mathbf{u}(\mathbf{x}) \cdot \mathbf{v}(\mathbf{x}) dx\Big| \leqslant \Big(\int_\Omega |\mathbf{u}(\mathbf{x})|^2 dx\Big)^{\frac{1}{2}} \Big(\int_\Omega |\mathbf{v}(\mathbf{x})|^2 dx\Big)^{\frac{1}{2}}. \tag{B.3.1}$$

Now we introduce a Sobolev space $\mathbf{W}_2^1(\Omega)$ ($\overset{\circ}{\mathbf{W}}{}_2^1(\Omega)$) for vector functions $\mathbf{u}$ as a closure of the space $\mathbf{C}^1(\overline{\Omega})$ ($\overset{\circ}{\mathbf{C}}{}^1(\Omega)$) of all vector functions, continuously differentiable in Ω, with the inner product

$$\langle \mathbf{u}, \mathbf{v} \rangle = \int_\Omega \Big(\mathbf{u}(\mathbf{x}) \cdot \mathbf{v}(\mathbf{x}) + \nabla \mathbf{u}(\mathbf{x}) : \nabla \mathbf{v}(\mathbf{x})\Big) dx \tag{B.3.2}$$

and the norm

$$\|\mathbf{u}\|_{2,\Omega}^{(1)} = \big(\langle \mathbf{u}, \mathbf{u} \rangle\big)^{\frac{1}{2}}. \tag{B.3.3}$$

The second-rank tensor (matrix) $\nabla \mathbf{u}$ is defined dy

$$\nabla \mathbf{u} = \frac{\partial \mathbf{u}}{\partial \mathbf{x}} = \Big(\frac{\partial u_i}{\partial x_j}\Big), \quad i, j = 1, \ldots, n$$

and

$$\nabla \mathbf{u} : \nabla \mathbf{v} = \sum_{i,j=1}^{n} \frac{\partial u_i}{\partial x_j} \cdot \frac{\partial v_i}{\partial x_j}, \quad |\nabla \mathbf{u}|^2 = \nabla \mathbf{u} : \nabla \mathbf{u}.$$

Lemma B.6 ***(Korn's inequality)***
Let $\mathbf{u} \in \mathbf{W}_2^1(\Omega)$ and $\mathbf{u} = 0$ on the part $S_0 \subset S = \partial\Omega$ with a strictly positive measure. Then

$$\int_{\Omega} |\nabla \mathbf{u}|^2 dx \leqslant C \int_{\Omega} \mathbb{D}(x, \mathbf{u}) : \mathbb{D}(x, \mathbf{u}) dx = C \int_{\Omega} |\mathbb{D}(x, \mathbf{u})|^2 dx, \tag{B.3.4}$$

where C depends only on the geometry of Ω and does not depend on $\mathbf{u}$, and the symmetric second-rank tensor $\mathbb{D}(x, \mathbf{u})$ is defined by

$$\mathbb{D}(x, \mathbf{u}) = \frac{1}{2}\Big(\frac{\partial \mathbf{u}}{\partial \mathbf{x}} + \frac{\partial \mathbf{u}}{\partial \mathbf{x}}^*\Big).$$

More precisely, Korn's inequality is valid for any subset $V \subset \mathbf{W}_2^1(\Omega)$, such that the equality

$$\mathbb{D}(x, \mathbf{u}) = 0 \quad \text{for } \mathbf{u} \in V,$$

implies $\mathbf{u} = 0$.

Lemma B.7 ***(Friedrichs-Poincaré's inequality for the vector functions)***

For all $\mathbf{u} \in \overset{\circ}{\mathbf{W}}{}_2^1(\Omega)$

$$\int_{\Omega} |\mathbf{u}|^2 dx \leqslant C \int_{\Omega} |\mathbb{D}(x, \mathbf{u})|^2 dx, \tag{B.3.5}$$

where C depends only on the geometry of Ω and does not depend on $\mathbf{u}$ and

$$\int_S \big|\mathbf{u}\big(\mathbf{x} - \varepsilon \mathbf{n}(\mathbf{x})\big)\big|^2 ds \to 0, \ \ as\ \varepsilon \to 0,$$

where $\mathbf{n}(\mathbf{x})$ is an outward unit normal to S at $\mathbf{x} \in S$.

For functions $\mathbf{u}(\mathbf{x}, t)$, where $(\mathbf{x}, t) \in \Omega_T$, we define the space $\mathbf{L}_2((0, T); L_2(\Omega)) = \mathbf{L}_2(\Omega_T)$ as a Hilbert space with an inner product

$$(\mathbf{u}, \mathbf{v}) = \int_{\Omega_T} \mathbf{u}(\mathbf{x}, t) \cdot \mathbf{v}(\mathbf{x}, t) dx dt$$

and the norm

$$\|\mathbf{u}\|_{2,\Omega_T} = \Big(\int_{\Omega_T} |\mathbf{u}(\mathbf{x}, t)|^2 dx dt\Big)^{\frac{1}{2}}.$$

The Hölder inequality for $\mathbf{L}_2(\Omega_T)$ has the form

$$\Big|\int_{\Omega_T} \mathbf{u}(\mathbf{x}, t) \cdot \mathbf{v}(\mathbf{x}, t) dx dt\Big| \leqslant \Big(\int_{\Omega_T} |\mathbf{u}(\mathbf{x}, \mathbf{t})|^2 dx dt\Big)^{\frac{1}{2}} \Big(\int_{\Omega_T} |\mathbf{v}(\mathbf{x}, t)|^2 dx dt\Big)^{\frac{1}{2}}. \tag{B.3.6}$$

Let $\mathbf{C}^{1,0}(\overline{\Omega}_T)$ be the space of all vector functions, continuously differentiable with respect to spatial variables, and $\overset{\circ}{\mathbf{C}}{}^{1,0}(\Omega_T)$ be a subspace of $\mathbf{C}^{1,0}(\overline{\Omega}_T)$, consisting of all functions vanishing at the boundary $S = \partial\Omega$.

The spaces

$$\mathbf{L}^2\big((0,T);\mathbf{W}_2^1(\Omega)\big) = \mathbf{W}_2^{1,0}(\Omega_T) \text{ and } \mathbf{L}^2\big((0,T);\overset{\circ}{\mathbf{W}}{}_2^1(\Omega)\big) = \overset{\circ}{\mathbf{W}}{}_2^{1,0}(\Omega_T)$$

are the closure of the inner product spaces of all functions from $\mathbf{C}^{1,0}(\overline{\Omega}_T)$ and $\overset{\circ}{\mathbf{C}}{}^{1,0}(\Omega_T)$ respectively with the inner product

$$\langle\mathbf{u},\mathbf{v}\rangle = \int_{\Omega_T}\Big(\mathbf{u}(\mathbf{x},t)\cdot\mathbf{v}(\mathbf{x},t) + \nabla\mathbf{u}(\mathbf{x},t) : \nabla\mathbf{v}(\mathbf{x},t)\Big)dx \tag{B.3.7}$$

and the norm

$$\|\mathbf{u}\|_{2,\Omega_T}^{(1)} = \big(\langle\mathbf{u},\mathbf{u}\rangle\big)^{\frac{1}{2}}. \tag{B.3.8}$$

For any function $\mathbf{u} \in \mathbf{W}_2^1(\Omega)$ the Stokes (Gauss-Ostrogradsky) Theorem takes the form

$$\int_\Omega \nabla\cdot\mathbf{u}dx = \int_S \mathbf{u}\cdot\nu ds, \tag{B.3.9}$$

where ν is the unit outward normal to the boundary $S = \partial\Omega$.

More generally, for any second-rank tensor $\mathbb{P} \in \mathbf{W}_2^1(\Omega)$ the Stokes (Gauss-Ostrogradsky) Theorem has the form

$$\int_\Omega \nabla\cdot\mathbb{P}dx = \int_S \mathbb{P}\cdot\nu ds, \tag{B.3.10}$$

where ν is the unit outward normal to the boundary $S = \partial\Omega$, and $\mathbf{a}\cdot(\nabla\cdot\mathbb{P}) = \nabla\cdot(\mathbb{P}^*\cdot\mathbf{a})$ for any constant vector $\mathbf{a}$.

B.4 Periodic Structures

In this section we discuss the properties of Sobolev spaces in periodic domains Ω_f^ε or Ω_s^ε, which have been defined in Chap. 1.

B.4.1 Extension Results

The following statements are valid due to the well-known results from [1, 36, 53, 89]. We formulate them in the forms that are appropriate for us.

Lemma B.8 ***(Extension lemma for the scalar functions*** *[1, 53])*
Suppose that Assumptions 0.1 and 1.1 regarding the geometry of a periodic structure hold true (the domain Ω_s is a connected set) and $w \in W_2^1(\Omega)$.
Then there exists an extension

$$w_s = E_{\Omega_s^\varepsilon}(w), \quad \mathbb{E}_{\Omega_s^\varepsilon} : W_2^1(\Omega_s^\varepsilon) \to W_2^1(\Omega), \tag{B.4.1}$$

from the domain Ω_s^ε onto the whole domain Ω such that

$$\big(1 - \chi^\varepsilon(\mathbf{x})\big)\big(w(\mathbf{x}, t) - w_s(\mathbf{x}, t)\big) = 0, \quad \mathbf{x} \in \Omega, \ t \in (0, T), \tag{B.4.2}$$

and

$$\int_\Omega |w_s(\mathbf{x}, t)|^2 dx \leqslant C_0 \int_{\Omega_s^\varepsilon} |w(\mathbf{x}, t)|^2 dx,$$
$$\int_\Omega |\nabla w_s(\mathbf{x}, t)|^2 dx \leqslant C_0 \int_{\Omega_s^\varepsilon} |\nabla w(\mathbf{x}, t)|^2 dx, \quad t \in (0, T), \tag{B.4.3}$$

where C_0 is independent of ε and $t \in (0, T)$.

Lemma B.9 ***(Extension lemma for the vector functions*** *[36, 89])*
Suppose that Assumptions 0.1 and 1.1 on the geometry of periodic structure hold true (the domain Ω_s is a connected set) and $\mathbf{w} \in \mathbf{W}_2^1(\Omega)$.
Then there exists an extension

$$\mathbf{w}_s = \mathbb{E}_{\Omega_s^\varepsilon}(\mathbf{w}), \quad \mathbb{E}_{\Omega_s^\varepsilon} : \mathbf{W}_2^1(\Omega_s^\varepsilon) \to \mathbf{W}_2^1(\Omega), \tag{B.4.4}$$

from the domain Ω_s^ε onto the whole domain Ω such that

$$\big(1 - \chi^\varepsilon(\mathbf{x})\big)\big(\mathbf{w}(\mathbf{x}, t) - \mathbf{w}_s(\mathbf{x}, t)\big) = 0, \quad \mathbf{x} \in \Omega, \ t \in (0, T), \tag{B.4.5}$$

and

$$\int_\Omega |\mathbf{w}_s(\mathbf{x}, t)|^2 dx \leqslant C_0 \int_{\Omega_s^\varepsilon} |\mathbf{w}(\mathbf{x}, t)|^2 dx,$$
$$\int_\Omega |\mathbb{D}(x, \mathbf{w}_s)|^2 dx \leqslant C_0 \int_{\Omega_s^\varepsilon} |\mathbb{D}(x, \mathbf{w})|^2 dx, \quad t \in (0, T), \tag{B.4.6}$$

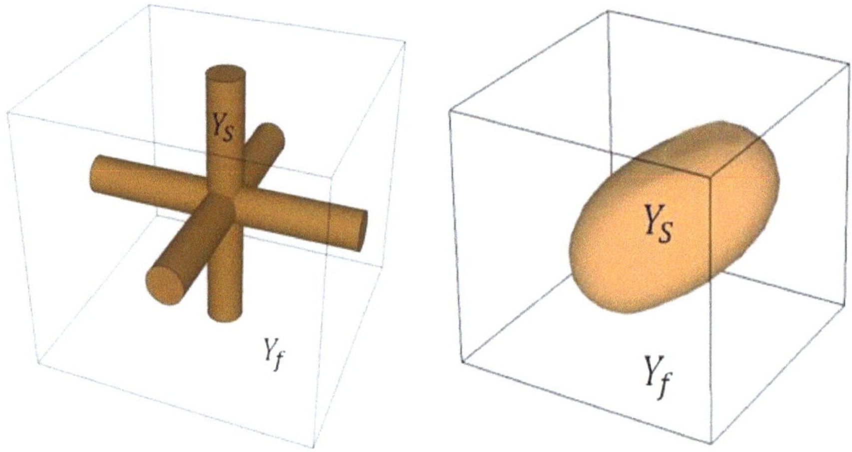

Fig. B.1 *Left* Connected solid and liquid parts. *Right* Disconnected solid part

where C_0 is independent of ε and $t \in (0, T)$.

For $w \in \overset{\circ}{W}{}_2^{1}(\Omega)$ ($\mathbf{w} \in \overset{\circ}{\mathbf{W}}{}_2^{1}(\Omega)$) these statements do not guarantee the inclusion $w_s \in \overset{\circ}{W}{}_2^{1}(\Omega)$ ($\mathbf{w}_s \in \overset{\circ}{\mathbf{W}}{}_2^{1}(\Omega)$). But for the special geometry of the pore space the extension permits this inclusion, namely, the following lemma holds true.

Let the first geometry of the pore space be represented by Fig. B.1(left), and the second geometry be represented by Fig. B.1(right).

Lemma B.10 ***(Extension lemma for the special geometry** [58, 89])*

Let Assumption 0.1 hold and $\mathbf{w} \in \overset{\circ}{\mathbf{W}}{}_2^{1}(\Omega)$.

Then for the first geometry of the pore space there exist extensions $\mathbf{w}_s = \mathbb{E}_{\Omega_s^\varepsilon}(\mathbf{w})$ from Ω_s^ε onto Ω and $\mathbf{w}_f = \mathbb{E}_{\Omega_f^\varepsilon}(\mathbf{w})$ from Ω_f^ε onto Ω such that $\mathbf{w}_s,\ \mathbf{w}_f \in \overset{\circ}{\mathbf{W}}{}_2^{1}(\Omega)$ and the estimates B.12 for $\mathbf{w}_s$ and $\mathbf{w}_f$ hold true.

For the second geometry there exists the extension $\mathbf{w}_f = \mathbb{E}_{\Omega_f^\varepsilon}(\mathbf{w})$ from Ω_f^ε onto Ω such that $\mathbf{w}_f \in \overset{\circ}{\mathbf{W}}{}_2^{1}(\Omega)$ and the estimates B.12 for $\mathbf{w}_f$ holds true.

If for the second geometry additionally $\nabla \cdot \mathbf{w} = 0$ in Ω, then $\nabla \cdot \mathbf{w}_f = 0$ in Ω.

Sometimes we do not need the homogeneous boundary condition $\mathbf{w}_s = 0$ on $\partial\Omega$, but we do need the estimate

$$\int_\Omega |\mathbf{w}_s|^2 dx \leqslant C \int_\Omega |D(x, \mathbf{w}_s)|^2 dx \tag{B.4.7}$$

with the constant C independent of ε.

For this case we prove the following statement.

Lemma B.11 *Under the conditions of Lemma B.9 let* $\mathbf{w} \in \overset{\circ}{\mathbf{W}}{}_2^1(\Omega)$.
Then the estimate (B.4.7) holds true.

Proof Let Q be a cube, $\Omega \subset Q$, and $\mathbf{u}$ be an extension of $\mathbf{w}$ such that $\mathbf{u} = 0$ for $\mathbf{u} \in Q\backslash\Omega$. The inclusion $\mathbf{w} \in \overset{\circ}{\mathbf{W}}{}_2^1(\Omega)$ implies $\mathbf{u} \in \overset{\circ}{\mathbf{W}}{}_2^1(Q)$. For the domain Q we may define the solid part Q_s^ε and the extension $\mathbf{u}_s$ in the same way as for the domain Ω we have defined the solid part Ω_s^ε and the extension $\mathbf{w}_s$. It is clear that $\mathbf{u}_s \in \overset{\circ}{\mathbf{W}}{}_2^1(Q)$ for sufficiently large Q_s.

Thus, we may apply the Friedrichs–Poincaré inequality

$$\int_Q |\mathbf{u}_s|^2 dx \leqslant C \int_Q |D(x, \mathbf{u}_s)|^2 dx.$$

It is also clear that

$$\mathbf{u}_s = \mathbf{w}_s \text{ in } \Omega \text{ and } \int_\Omega |\mathbf{w}_s|^2 dx \leqslant \int_Q |\mathbf{u}_s|^2 dx.$$

Therefore

$$\int_\Omega |\mathbf{w}_s|^2 dx \leqslant C \int_Q |D(x, \mathbf{u}_s)|^2 dx.$$

Lemma B.9 states that

$$\int_Q |D(x, \mathbf{u}_s)|^2 dx \leqslant C \int_{Q_s^\varepsilon} |D(x, \mathbf{u})|^2 dx.$$

But

$$\begin{aligned}\int_{Q_s^\varepsilon} |D(x, \mathbf{u})|^2 dx &= \int_{\Omega_s^\varepsilon} |D(x, \mathbf{w})|^2 dx \\ &= \int_{\Omega_s^\varepsilon} |D(x, \mathbf{w}_s)|^2 dx \leqslant \int_\Omega |D(x, \mathbf{w}_s)|^2 dx.\end{aligned}$$

Gathering all together we arrive at the desired estimate (B.4.7).

Lemma B.12 ***(Korn's inequality for periodic structures)***
Under the conditions of Lemma B.9 let $\mathbf{w} \in \mathbf{W}_2^1(\Omega)$.
Then

$$\int_{\Omega_f^\varepsilon} |\nabla \mathbf{w}|^2 dx \leqslant C \int_{\Omega_f^\varepsilon} |D(x, (\mathbf{w})|^2 dx \tag{B.4.8}$$

for the connected set Ω_f^ε, *and*

$$\int_{\Omega_s^\varepsilon} |\nabla \mathbf{w}|^2 dx \leqslant C \int_{\Omega_s^\varepsilon} |D(x, (\mathbf{w})|^2 dx, \tag{B.4.9}$$

for the connected set Ω_s^ε.

In (B.4.8), (B.4.9) the constant C *is independent of* ε.

Proof To prove the lemma we use Lemma B.9 and conclude that there exist functions $\mathbf{w}_f \in \mathbf{W}_2^1(\Omega)$ for the connected set Ω_f^ε and $\mathbf{w}_s \in \mathbf{W}_2^1(\Omega)$ for the connected set Ω_s^ε such that

$$\mathbf{w}_f = \mathbf{w} \text{ in } \Omega_f^\varepsilon, \quad \mathbf{w}_s = \mathbf{w} \text{ in } \Omega_s^\varepsilon$$

and

$$\int_\Omega |D(x, \mathbf{w}_f)|^2 dx \leqslant C \int_{\Omega_f^\varepsilon} |D(x, \mathbf{w})|^2 dx,$$

$$\int_\Omega |D(x, \mathbf{w}_s)|^2 dx \leqslant C \int_{\Omega_s^\varepsilon} |D(x, \mathbf{w})|^2 dx,$$

where C is independent of ε.

Next we use the standard Korn inequality (Lemma B.6):

$$\int_\Omega |\nabla \mathbf{w}_f|^2 dx \leqslant C \int_\Omega |D(x, \mathbf{w}_f)|^2 dx,$$

$$\int_\Omega |\nabla \mathbf{w}_s|^2 dx \leqslant C \int_\Omega |D(x, \mathbf{w}_s)|^2 dx.$$

Finally, we apply the evident relations

$$\int_{\Omega_f^\varepsilon} |\nabla \mathbf{w}|^2 dx \leqslant \int_{\Omega_f^\varepsilon} |\nabla \mathbf{w}|^2 dx + \int_{\Omega_s^\varepsilon} |\nabla \mathbf{w}_f|^2 dx = \int_\Omega |\nabla \mathbf{w}_f|^2 dx$$

and

$$\int_{\Omega_s^\varepsilon} |\nabla \mathbf{w}|^2 dx \leqslant \int_{\Omega_s^\varepsilon} |\nabla \mathbf{w}|^2 dx + \int_{\Omega_f^\varepsilon} |\nabla \mathbf{w}_s|^2 dx = \int_\Omega |\nabla \mathbf{w}_s|^2 dx,$$

which result in the statements of the lemma.

B.5 Multi-Scale Convergence

B.5.1 Two-Scale Convergence

The method of two-scale convergence was proposed by G. Nguetseng [89] and has been applied to a wide range of homogenization problems (see, for example, the survey [70]).

A sequence $\{w^{\varepsilon}\} \subset L_2(\Omega_T)$ is said to be *two-scale convergent* to a function $\widetilde{W}(\mathbf{x}, t, \mathbf{y}, \tau) \in L_2(\Omega_T \times Y)$, 1-periodic in the variables $(\mathbf{y}, \tau) \in Y \times (0, 1)$, if and only if for any function $\sigma = \sigma(\mathbf{x}, t, \mathbf{y}, \tau)$, 1-periodic in $(\mathbf{y}, \tau)$

$$\int_{\Omega_T} w^{\varepsilon}(\mathbf{x}, t)\sigma\left(\mathbf{x}, t, \frac{\mathbf{x}}{\varepsilon}, \frac{t}{\varepsilon}\right)dxdt \\ \to \int_{\Omega_T}\left(\int_0^1\int_Y \widetilde{W}(\mathbf{x}, t, \mathbf{y}, \tau)\sigma(\mathbf{x}, t, \mathbf{y}, \tau)dyd\tau\right)dxdt \tag{B.5.1}$$

as $\varepsilon \to 0$.

In what follows we restrict ourself to the test functions $\sigma = \sigma(\mathbf{x}, t, \mathbf{y})$. Then the relation B.5.1 takes the form

$$\int_{\Omega_T} w^{\varepsilon}(\mathbf{x}, t)\sigma\left(\mathbf{x}, t, \frac{\mathbf{x}}{\varepsilon}\right)dxdt \\ \to \int_{\Omega_T}\left(\int_Y W(\mathbf{x}, t, \mathbf{y})\sigma(\mathbf{x}, t, \mathbf{y})dy\right)dxdt, \tag{B.5.2}$$

where

$$W(\mathbf{x}, t, \mathbf{y}) = \int_0^1 \widetilde{W}(\mathbf{x}, t, \mathbf{y}, \tau)d\tau.$$

The existence and main properties of weakly convergent sequences are established by the following fundamental theorem [70, 89]:

Theorem B.2 ***(Nguetseng's theorem** for scalar functions)*

1.*Any sequence $\{w^{\varepsilon}\}$ bounded in $L_2(\Omega_T)$ contains a subsequence, two-scale convergent to some function $W \in L_2(\Omega_T \times Y)$, 1-periodic in $\mathbf{y}$.*

2.*Let sequences $\{w^{\varepsilon}\}$ and $\{\varepsilon\nabla w^{\varepsilon}\}$ be uniformly bounded in $L_2(\Omega_T)$.*

Then there exist a function $W = W(\mathbf{x}, t, \mathbf{y})$ 1-periodic in $\mathbf{y}$, and a subsequence $\{w^{\varepsilon}\}$ such that $W, \nabla_y W \in L_2(\Omega_T \times Y)$, and the subsequences $\{w^{\varepsilon}\}$ and $\{\varepsilon\nabla w^{\varepsilon}\}$ two-scale converge to W and $\nabla_y W$ respectively.

3. *Let sequences $\{w^{\varepsilon}\}$ and $\{\nabla w^{\varepsilon}\}$ be bounded in $L_2(\Omega_T)$.*

Then there exist the functions $w \in L_2(\Omega_T)$ and $W \in L_2(\Omega_T \times Y)$ and a subsequence from $\{\nabla w^{\varepsilon}\}$ such that the function W is 1-periodic in $\mathbf{y}$, $\nabla w \in L_2(\Omega_T)$, $\nabla_y W \in L_2(\Omega_T \times Y)$, and the subsequence $\{\nabla w^{\varepsilon}\}$ two-scale converges to the function $\nabla w(\mathbf{x}, t) + \nabla_y W(\mathbf{x}, t, \mathbf{y})$.

Theorem B.3 *(**Nguetseng's theorem** for vector functions)*
1. Any sequence $\{\mathbf{w}^\varepsilon\}$ bounded in $\mathbf{L}_2(\Omega_T)$ contains a subsequence, two-scale convergent to some function $\mathbf{W} \in \mathbf{L}_2(\Omega_T \times Y)$, 1-periodic in $\mathbf{y}$.
2. Let sequences $\{\mathbf{w}^\varepsilon\}$ and $\{\varepsilon\mathbb{D}(x, \mathbf{w}^\varepsilon)\}$ be uniformly bounded in $\mathbf{L}_2(\Omega_T)$. Then there exists a function $\mathbf{W} = \mathbf{W}(\mathbf{x}, t, \mathbf{y})$, 1-periodic in $\mathbf{y}$, and a subsequence $\{\mathbf{w}^\varepsilon\}$ such that $\mathbf{W}$, $\nabla_y \mathbf{W} \in \mathbf{L}_2(\Omega_T \times Y)$, and the subsequences $\{\mathbf{w}^\varepsilon\}$ and $\{\varepsilon\mathbb{D}(x, \mathbf{w}^\varepsilon)\}$ two-scale converge in $\mathbf{L}_2(\Omega_T)$ to $\mathbf{W}$ and $\mathbb{D}(y, \mathbf{W})$ respectively.
3. Let sequences $\{\mathbf{w}^\varepsilon\}$ and $\{D(x, \mathbf{w}^\varepsilon)\}$ be bounded in $\mathbf{L}_2(\Omega_T)$. Then there exist the functions $\mathbf{w} \in \mathbf{L}_2(\Omega_T)$ and $\mathbf{W} \in \mathbf{L}_2(\Omega_T \times Y)$ and a subsequence from $\{\mathbb{D}(x, \mathbf{w}^\varepsilon)\}$ such that the function $\mathbf{W}$ is 1-periodic in $\mathbf{y}$, $\{\mathbb{D}(x, \mathbf{w})\} \in \mathbf{L}_2(\Omega_T)$, $D(y, \mathbf{W}) \in \mathbf{L}_2(\Omega_T \times Y)$, and the subsequence $\{\mathbb{D}(x, \mathbf{w}^\varepsilon)\}$ two-scale converges to the function $\mathbb{D}(x, \mathbf{w}) + D(y, \mathbf{W})$.
4. Let $\sigma \in L_2(Y)$ and $\sigma^\varepsilon(\mathbf{x}) = \sigma\left(\frac{\mathbf{x}}{\varepsilon}\right)$. Assume that a sequence $\{w^\varepsilon\} \subset L_2(\Omega_T)$ two-scale converges to $W \in L_2(\Omega_T \times Y)$. Then the sequence $\{\sigma^\varepsilon w^\varepsilon\}$ two-scale converges to the function σW.

Lemma B.13 *Let a sequence $\{w^\varepsilon(\mathbf{x}, t)\}$ weakly converge in $L_2(\Omega_T)$ to $w(\mathbf{x}, t)$ and two-scale converges to $W(\mathbf{x}, t, \mathbf{y})$,*

$$\alpha(\varepsilon)\|\nabla w^\varepsilon\|_{2,\Omega_T} \leqslant C,$$

where C is independent of the small parameter ε, and

$$\lim_{\varepsilon\to 0}\frac{\alpha(\varepsilon)}{\varepsilon} = \infty,$$

Then $W(\mathbf{x}, t, \mathbf{y}) = w(\mathbf{x}, t)$.

Proof Let $\Psi(\mathbf{x}, t, \mathbf{y})$ be an arbitrary smooth scalar function periodic in $\mathbf{y}$. The sequence $\{\sigma_j^\varepsilon\}$, where

$$\sigma_j^\varepsilon = \int_{\Omega_T} \alpha(\varepsilon)\frac{\partial w^\varepsilon}{\partial x_j}(\mathbf{x}, t)\Psi\left(\mathbf{x}, t, \frac{\mathbf{x}}{\varepsilon}\right)dxdt$$

is uniformly bounded in ε.

Therefore,

$$\int_{\Omega_T} \varepsilon\frac{\partial w^\varepsilon}{\partial x_j}(\mathbf{x}, t)\Psi\left(\mathbf{x}, t, \frac{\mathbf{x}}{\varepsilon}\right)dxdt = \frac{\varepsilon}{\alpha(\varepsilon)}\sigma_j^\varepsilon \to 0$$

as $\varepsilon \searrow 0$, which is equivalent to

$$\int_{\Omega_T}\int_Y W(\mathbf{x}, t, \mathbf{y})\frac{\partial \Psi}{\partial y_j}(\mathbf{x}, t, \mathbf{y})dydxdt = 0, \quad j = 1, \ldots, n,$$

or

$$W(\mathbf{x}, t, \mathbf{y}) = w(\mathbf{x}, t).$$

The following lemma shows the limiting property of the weakly convergent sequences $\{w_s^\varepsilon\}$ for $w^\varepsilon \in \overset{\circ}{W}{}_2^1(\Omega)$. The lemma is valid both for scalar and vector functions. So we will consider only scalar functions.

Lemma B.14 *Under the conditions of Lemma B.8 let* $w^\varepsilon \in \overset{\circ}{W}{}_2^{1,0}(\Omega_T)$ *and the sequence* $\{w_s^\varepsilon\}$ *converge weakly in* $W_2^{1,0}(\Omega_T)$ *and two-scale in* $L_2(\Omega_T)$ *to* $w_s(\mathbf{x}, t)$.

Then $w_s \in \overset{\circ}{W}{}_2^1(\Omega_T)$.

Proof By construction the function w_s^ε vanishes at the part $S_s^\varepsilon = \partial S \cap \overline{\Omega_s^\varepsilon}$ of the boundary S.

Let us choose the function $\mathbf{u}(\mathbf{y})$, 1-periodic and solenoidal in the unit cube Y, such that *supp u* $\subset Y_f$. Then the function $h(\mathbf{x}, t)\mathbf{u}\big(\frac{\mathbf{x}}{\varepsilon}\big)$ vanishes at the part $S_f^\varepsilon = \partial S \cap \overline{\Omega_f^\varepsilon}$ of the boundary S for any smooth function $h(\mathbf{x}, t)$ and the product $h(\mathbf{x}, t)w_s^\varepsilon(\mathbf{x}, t)\mathbf{u}\big(\frac{\mathbf{x}}{\varepsilon}\big)$ vanishes at the boundary S.

Therefore we may apply the formula (B.2.1) for integration by parts, which in this case takes the form

$$\int_{\Omega_T} \Big(h(\mathbf{x}, t)\, \nabla w_s^\varepsilon(\mathbf{x}, t)\Big) \cdot \mathbf{u}\big(\frac{\mathbf{x}}{\varepsilon}\big) + w_s^\varepsilon(\mathbf{x}, t)\Bigg(\mathbf{u}\big(\frac{\mathbf{x}}{\varepsilon}\big) \cdot \nabla h(\mathbf{x}, t)\Bigg) dxdt = 0. \quad \text{(B.5.3)}$$

Let $\nabla w_s(\mathbf{x}, t) + \nabla_y W_s(\mathbf{x}, \mathbf{y}, t)$ be the two-scale limit of the sequence $\{\nabla w_s^\varepsilon\}$. Then the limit as $\varepsilon \to 0$ in (B.15) results in

$$\int_{\Omega_T} \big(h\, (\nabla w_s \cdot \langle \mathbf{u}\rangle_{Y_f} + \langle \nabla_y W_s \cdot \mathbf{u}\rangle_{Y_f}) + w_s(\langle \mathbf{u}\rangle_{Y_f} \cdot \nabla h)\big) dxdt = 0.$$

Here

$$\langle \mathbf{u}\rangle_{Y_f} = \int_{Y_f} \mathbf{u} dy \text{ and } \langle \nabla_y W_s \cdot \mathbf{u}\rangle_{Y_f} = \int_{Y_f} \nabla_y W_s \cdot \mathbf{u} dy = 0$$

due to condition $\nabla_y \cdot \mathbf{u} = 0$.

Let

$$\langle \mathbf{u}\rangle_{Y_f} = \mathbf{e}. \quad \text{(B.5.4)}$$

Then the last identity takes the form

$$\int_{\Omega_T}\big(h\,(\nabla w_s\cdot\mathbf{e})+w_s(\mathbf{e}\cdot\nabla h)\big)dxdt=\int_{\Omega_T}\nabla\cdot(h\,w_s\,\mathbf{e})dxdt=0.$$

Applying the Stokes formula we obtain

$$\int_{S_T}h(\mathbf{x},t)w_s(\mathbf{x},t)(\mathbf{e}\cdot\mathbf{n})d\sigma=0, \tag{B.5.5}$$

where $S_T=S\times(0,T)$ and $\mathbf{n}$ is the unit normal to the to the boundary S.

In the following lemma we prove that for any unit vector $\mathbf{e}$ there exists a solenoidal function $\mathbf{u}(\mathbf{y})$ with *supp* $u\subset Y_f$ satisfying (B.5.4). Then due to the arbitrary choice of the functions h and vectors $\mathbf{e}$ the identity (B.5.5) implies $w_s(\mathbf{x},t)=0$ for $\mathbf{x}\in S$.

Lemma B.15 *For any unit vector* $\mathbf{e}$ *there exists a solenoidal vector function* $\mathbf{u}(\mathbf{y})$ *satisfying (B.5.4) and the condition* $\operatorname{supp}u\subset Y_f$.

Proof Let $B\subset Y_f$ be a ball and $\mathbf{u}(\mathbf{y})$ be a nontrivial solution of the problem

$$\triangle\mathbf{u}-\nabla p=\mathbf{f},\ \ \mathbf{y}\in B, \tag{B.5.6}$$

$$\nabla\cdot\mathbf{u}=0,\ \ \mathbf{y}\in B, \tag{B.5.7}$$

$$\mathbf{u}=0,\ \ \mathbf{y}\in\partial B \tag{B.5.8}$$

with some fixed function $\mathbf{f}(\mathbf{y})$.

We may always assume that

$$\int_B\mathbf{u}dy=\mathbf{e}_0,\ \ \text{with } |\mathbf{e}_0|=1.$$

Let $\mathbb{T}$ be the orthogonal matrix and $\mathbb{T}\cdot\mathbf{e}_0=\mathbf{e}$.

Then in the new variables $\mathbf{z}=\mathbb{T}\cdot\mathbf{y}$ the function $\mathbf{v}(\mathbf{z})=\mathbb{T}\cdot\mathbf{u}(\mathbf{y})$ satisfies the problem

$$\triangle_z\mathbf{v}-\nabla_z q=\mathbf{F},\ \ \mathbf{z}\in B, \tag{B.5.9}$$

$$\nabla_z\cdot\mathbf{v}=0,\ \ \mathbf{v}\in B, \tag{B.5.10}$$

$$\mathbf{v}=0,\ \ \mathbf{z}\in\partial B \tag{B.5.11}$$

where $\mathbf{F}(\mathbf{z})=\mathbb{T}\cdot\mathbf{f}$ and $q(\mathbf{z})=p(\mathbf{y})$. By the construction

$$\mathbf{e}=\mathbb{T}\cdot\mathbf{e}_0=\int_B\mathbb{T}\cdot\mathbf{u}dy=\int_B\mathbf{v}dz.$$

B.5.2 Three-Scale Convergence

The method of three-scale convergence was proposed by G. Allaire and M. Briane [6] and G. Nguetseng [70].

Let $\varepsilon = \delta^r$, $r > 1$.

We say that the sequence $\{\mathbf{w}^\delta\}$ three-scale converges in $\mathbf{L}_2(\Omega_T)$ to the function $\mathbf{W}(\mathbf{x}, t, \mathbf{y}, \mathbf{z})$, 1-periodic in the variables $\mathbf{y}$ and $\mathbf{z}$, if

$$\int_{\Omega_T} \mathbf{w}^\delta(\mathbf{x}, t)\, \varphi\Big(\mathbf{x}, t, \frac{\mathbf{x}}{\varepsilon}, \frac{\mathbf{x}}{\delta}\Big) dxdt$$
$$\to \int_{\Omega_T} \Big(\int_Y \int_Z \mathbf{W}(\mathbf{x}, t, \mathbf{y}, \mathbf{z})\, \varphi(\mathbf{x}, t, \mathbf{y}, \mathbf{z}) dzdy\Big) dxdt, \quad \text{(B.5.12)}$$

for any smooth function $\varphi(\mathbf{x}, t, \mathbf{y}, \mathbf{z})$, 1-periodic in $\mathbf{y}$ and $\mathbf{z}$.

Theorem B.4 *(Three-scale convergence)*
1. Any sequence $\{\mathbf{w}^\delta\}$ bounded in $\mathbf{L}_2(\Omega_T)$ contains a subsequence, three-scale convergent to some function $\mathbf{W} \in \mathbf{L}_2(\Omega_T \times Y \times Y)$, 1-periodic in $\mathbf{y}$ and $\mathbf{z}$.
2. Let sequences $\{\mathbf{w}^\delta\}$ and $\{D(x, \mathbf{w}^\delta)\}$ be bounded in $\mathbf{L}_2(\Omega_T)$.

Then there exist a subsequence from $\{\mathbb{D}(x, \mathbf{w}^\delta)\}$ and the functions $\mathbf{w} \in \mathbf{L}_2(\Omega_T)$ and $\mathbf{W}_p$, $\mathbf{W}_c$, $\in \mathbf{L}_2(\Omega_T \times Y \times Y)$, $\mathbf{W}_p$ and $\mathbf{W}_c$ are 1-periodic in $\mathbf{y}$ and $\mathbf{z}$, $\mathbb{D}(x, \mathbf{w}) \in \mathbf{L}_2(\Omega_T)$, $\mathbb{D}(y, \mathbf{W}_p)$, $\mathbb{D}(y, \mathbf{W}_c) \in \mathbf{L}_2(\Omega_T \times Y \times Y)$, such that the subsequence $\{\mathbb{D}(x, \mathbf{w}^\delta)\}$ three-scale converges to the function $\mathbb{D}(x, \mathbf{w}) + \mathbb{D}(y, \mathbf{W}_p) + \mathbb{D}(z, \mathbf{W}_c)$.

B.6 Some Compactness Results

In this section we formulate some compactness results for the bounded sequences $\{u^k(\mathbf{x}, t)\}$ in $L_2(\Omega_T)$, which are compact in $L_2(\Omega)$ for any fixed $t \in (0, T)$.

We say that the function $u \in L_2(\Omega_T)$ possesses a time derivative $\dfrac{\partial u}{\partial t}$, bounded in $L_2\big((0, T);\ W_2^{-1}(\Omega)\big)$, if

$$\Big|\int_{\Omega_T} u \frac{\partial \varphi}{\partial t} dxdt\Big| \leqslant C\, \|\varphi\|^{(1)}_{2,\Omega_T}$$

for any smooth function φ.

Lemma B.16 ***(Aubin's compactness lemma** [9])*

Let the sequence $\{u^k(\mathbf{x},t)\}$ *be bounded in* $L_\infty\big((0,T);L_2(\Omega)\big)\cap W_2^{1,0}(\Omega_T)$ *and weakly convergent in the space* $L_2(\Omega_T)\cap W_2^{1,0}(\Omega_T)$ *to the function* $u(\mathbf{x},t)$*. Let also the sequence* $\left\{\dfrac{\partial u^k}{\partial t}(\mathbf{x},t)\right\}$ *be bounded in the space* $L_2\big((0,T);\ W_2^{-1}(\Omega)\big)$*.*

Then the sequence $\{u^k(\mathbf{x},t)\}$ *converges strongly in* $L_2(\Omega_T)$ *to its weak limit* $u(\mathbf{x},t)$*.*

Lemma B.17 ***(Compactness in periodic structures** [84])*

Let the sequence $\{u^k(\mathbf{x},t)\}$ *be bounded in* $L_\infty\big((0,T);L_2(\Omega)\big)\cap W_2^{1,0}(\Omega_T)$ *and weakly convergent in the space* $L_2(\Omega_T)\cap W_2^{1,0}(\Omega_T)$ *to the function* $u(\mathbf{x},t)$*. Let also the sequence* $\left\{\chi^\varepsilon(\mathbf{x})\dfrac{\partial u^k}{\partial t}(\mathbf{x},t)\right\}$ *be bounded in the space* $L_2\big((0,T);\ W_2^{-1}(\Omega)\big)$*, where* $\chi^k(\mathbf{x})=\chi(k\mathbf{x})$*,* $\chi(\mathbf{y})$ *is a measurable bounded function, 1-periodic in the variable* $\mathbf{y}$*, such that*

$$\langle\,\chi\,\rangle_Y=\int_Y \chi(\mathbf{y})dy=m\neq 0,$$

and Y *is the unit cube in* R^n*.*

Then the sequence $\{u^k(\mathbf{x},t)\}$ *converges strongly in* $L_2(\Omega_T)$ *to its weak limit* $u(\mathbf{x},t)$*.*

References

1. Acerbi, E., Chiado Piat, V., Dal Maso, G., & Percivale, D. (1992). An extension theorem from connected sets and homogenization in general periodic domains. *Nonlinear Analysis, 18*, 481–496.
2. Adachi, J.I., Detournay, E., & Peirce, A. P. (2010). Analysis of the classical pseudo-3D model for hydraulic fracture with equilibrium height growth across stress barriers. *International Journal of Rock Mechanics and Mining Sciences, 47*, 625–639.
3. Adams, R. E. (1975). *Sobolev spaces*. New York: Academic Press.
4. Allaire, G. (1989). Homogenization of the Stokes flow in a connected porous medium. *Asymptotic Analysis, 2*, 203–222.
5. Allaire, G. (1992). Homogenization and two-scale convergence. *SIAM J. Math. Anal., 23*, 1482–1518.
6. Allaire, G., & Briane, M. (1996). Multisale convergence and reiterated homogenization. *Proceedings of the Royal Society of Edinburgh, 126A*, 297–342.
7. Allaire, G., & Capdeboscq, Y. (2000). Homogenization of a spectral problem in neutronic multigroup diffusion. *Computer Methods in Applied Mechanics and Engineering, 187*, 91–117.
8. Amaziane, B., Goncharenko, M., & Pankratov, L. (2007). Homogenization of a convection–diffusion equation in perforated domains with a weak adsorption. *Zeitschrift für Angewandte Mathematik und Physik, 58*, 592–611.
9. Amaziane, B., Antontsev, S., Pankratov, L., & Piatnitski, A. (2010). Homogenization of immiscible compressible two-phase flow in porous media: Application to gas migration in a nuclear waste repository. *SIAM Journal on Multiscale Modeling and Simulation, 8*(5), 2023–2047.
10. Antontsev, S., Meirmanov, A., & Yurinsky, V. (2000). A free boundary problem for Stokes equations: Classical solutions. *Interfaces and Free Boundaries, 2*, 413–424.

11. Arbogast, T., Douglas, J., & Hornung, U. (1990). Derivation of the double porosity model of single phase flow via homogenization theory. *SIAM Journal on Mathematical Analysis, 21*(4), 823–836.
12. Aubin, J. P. (1963). Un théoréme de compacité. *Comptes Rendus de l'Académie des Sciences, 256*, 5042–5044.
13. Auriault, J.-L. (1997). Poroelastic media. In U. Hornung (Ed.), *Homogenization and porous media*. Interdisciplinary applied mathematics (pp. 163–182). Berlin: Springer.
14. Bakhvalov, N. S., & Panasenko, G. (1989). *Homogenization: Averaging processes in periodic media*. New York: Springer.
15. Barenblatt, G. I., Zheltov, Iu P, & Kochina, I. N. (1960). Basic concepts in the theory of seepage of homogeneous liquids in fissures rocks. *Journal of Applied Mathematics and Mechanics, 24*, 1286–1303.
16. Barenblatt, G. I., Yentov, V. M., & Ryzhik, V. M. (1972). *Theory of nonstationary liquid and gas filtration*. Moscow: Nedra. (Russian).
17. Bensoussan, A., Lions, J.-L., & Papanicolau, G. (1978). *Asymptotic analysis for periodic structure*. Amsterdam: North Holland.
18. Bhiladvala, R. B., & Wang, Z. J. (2004). Effect of fluids on the Q factor and resonance frequency of oscillating micrometer and nanometer scale beams. *Physical Review E, 69*, 036307.
19. Biot, M. (1941). General theory of three dimensional consolidation. *Journal of Applied Physics, 12*, 155–164.
20. Biot, M. (1955). Theory of elasticity and consolidation for a porous anisotropic solid. *Journal of Applied Physics, 26*, 182–185.
21. Biot, M. (1955). Theory of propagation of elastic waves in a fluid-saturated porous solid. I. Low-frequency range. *Journal of the Acoustical Society of America, 28*, 168–178.
22. Biot, M. (1955). Theory of propagation of elastic waves in a fluid-saturated porous solid. II. Higher frequency range. *Journal of the Acoustical Society of America, 28*, 179–191.
23. Biot, M. (1962). Generalized theory of acoustic propagation in porous dissipative media. *Journal of the Acoustical Society of America, 34*, 1256–1264.
24. Biot, M., & Willis, D. (1957). The elastic coefficients of the theory of consolidation. *Journal of Applied Mechanics, 24*, 594–601.
25. Booth, R. J. S. (2011). Asymptotics for the Muskat problem. *Journal of Engineering Mathematics*,. doi:10.1007/s1066501093703.
26. Bourgeat, A., Gipouloux, O., & Marušić-Paloka, E. (2001). Mathematical modelling and numerical simulation of a non-Newtonian viscous flow through a thin filter. *SIAM Journal on Applied Mathematics, 62*(2), 597–626.
27. Bourgeat, A., Gipouloux, O., & Marušić-Paloka, E. (2003). Filtration law for polymer flow through porous media. *Multiscale Modeling and Simulation, 1*(3), 132–157.
28. Bourgeat, A., Jurak, M., & Piatnitski, A. L. (2003). Averaging a transport equation with small diffusion and oscillating velocity. *Mathematical Methods in the Applied Sciences, 26*, 95–117.
29. Bourgeat, A., & Marušić-Paloka, E. (2005). A homogenized model of an underground waste repository including a disturbed zone. *Multiscale Modeling and Simulation, 3*(4), 918–939.
30. Bourgeat, A., Pankratov, L., & Panfilov, M. (2004). Study of the double porosity model versus the fissures thickness. *Asymptotic Analysis, 38*, 129–141.
31. Buchanan, J. L., & Gilbert, R. P. (1997). Transition loss in the farfield for an ocean with a Biot sediment over an elastic substrate. *Zeitschrift für Angewandte Mathematik und Mechanik, 77*, 121–135.
32. Buckingham, M. J. (1998). Seismic wave propagation in rocks and marine sediments: A new theoretical approach. In A. Alippi & G. B. Cannelli (Eds.), *Underwater acoustics* (Vol. I). Rome: CNR-IDAC.
33. Burridge, R., & Keller, J. B. (1981). Poroelasticity equations derived from microstructure. *Journal of the Acoustical Society of America, 70*(4), 1140–1146.
34. Chen, Z. (2007). Homogenization and simulation for compositional flow in naturally fractured reservoirs. *Journal of Mathematical Analysis and Applications, 326*, 31–75.

35. Clopeau, Th, Ferrin, J. L., Gilbert, R. P., & Mikelić, A. (2001). Homogenizing the acoustic properties of the seabed: Part II. *Mathematical and Computer Modelling*, *33*, 821–841.
36. Conca, C. (1985). On the application of the homogenization theory to a class of problems arising in fluid mechanics. *Journal de Mathématiques Pures et Appliquées*, *64*, 31–75.
37. Coussy, O. (2004). *Poromechanics*. Chichester: Wiley.
38. Darcy, H. (1856). *Les fontaines publiques de la ville de Dijon*. Paris: Dalmont.
39. de Swaan, A. (1976). Analytic solutions for determining naturally fractured reservoir properties by well testing. *Society of Petroleum Engineers Journal*, *16*, 117–122.
40. Escher, J., Matioc, A., & Matioc, B. (2011). Modelling and analysis of the Muskat problem for thin fluid layers. *Journal of Mathematical Fluid Mechanics* (Springer, Basel AG). doi:10.1007/s0002101100532.
41. Fahuai, Yi. (2003). Global classical solution of Muskat free boundary problem. *Journal of Mathematical Analysis and Applications*, *288*, 442–461.
42. Fasano, A, Mikelić, A., & Primicerio, M. (1998). Homogenization of flows through porous media with grains. *Advances in Mathematical Sciences and Applications*, *8*, 1–31.
43. Ferrin, J. L., & Mikelić, A. (2003). Homogenizing the acoustic properties of a porous matrix containing an incompressible inviscid fluids. *Mathematical Models and Methods in Applied Sciences*, *26*, 831–859.
44. Galtsev, O., & Meirmanov, A. (2011). Numerical homogenization in the Rayleigh-Taylor problem of filtration two immiscible incompressible liquids. *Mathematical Modelling*, *23*(10), 33–43. (Russian).
45. Gilbert, R. P., & Lin, J. Z. (1997). Acoustic waves in shallow inhomogeneous oceans with a poro-elastic seabed. *Zeitschrift für Angewandte Mathematik und Mechanik*, *79*(4), 1–12.
46. Gilbert, R. P., & Mikelić, A. (2000). Homogenizing the acoustic properties of the seabed: Part I. *Nonlinear Analysis*, *40*, 185–212.
47. Horgan, C. O. (1995). Korn's inequalities and their applications in continuum mechanics. *SIAM Review*, *37*(4), 491–511.
48. Hornung, U. (1997). *Homogenization and porous media*. New York: Springer.
49. Hornung, U., & Jäger, W. (1991). Diffusion, convection, absorbtion and reaction of chemicals in porous media. *Journal of Differential Equations*, 92(2), 199–225.
50. Jäger, W., & Mikelić, A. (1994). On the flow conditions at the boundary between a porous medium and an impervious solid. In *Progress in PDE: The Metz surveys 3*. Pitman research notes in mathematics (Vol. 314). London: Longman Scientific and Technical.
51. Jäger, W., & Mikelić, A. (1996). On the boundary conditions at the contact interface between a porous medium and a free fluid. *Annali della Scuola Normale Superiore di Pisa. Classe di Scienze. Serie IV*, *XXIII*(Fasc. 3), 403–465.
52. Jäger, W., & Mikelić A. (1999). On the boundary conditions at the contact interface between two porous media. In W. Jäger, et al. (Eds.), *PDE, theory and numerical solution*. Chapman and Hall/CRC research notes in mathematics (Vol. 406). London: CRC Press.
53. Jikov, V. V., Kozlov, S. M., & Oleinik, O. A. (1994). *Homogenization of differential operators and integral functionals*. New York: Springer.
54. Kazemi, H. (1969). Pressure transient analysis of naturally fractured reservoirs with uniform fracture distribution. *Society of Petroleum Engineers Journal*, *9*, 451–462.
55. Kirk, W. A., & Sims, B. (2001). *Handbook of metric fixed point theory*. London: Kluwer Academic.
56. Kolmogorov, A. N., & Fomin S. V. (1999). *Elements of the theory of functions and functional analysis*. Mineola: Dover Publications.
57. Kovalyshen, Y., & Detournay, E. (2010). A reexamination of the classical PKN model of hydraulic fracture. *Transport in Porous Media*, *81*, 317–339.
58. Ladyzhenskaya, O. A. (1969). *The mathematical theory of viscous incompressible flow*. New York: Gordon and Breach.
59. Ladyzhenskaya, O. A. (1985). *The boundary-value problems of mathematical physics*. New York: Springer.

60. Ladyzhenskaya, O. A., & Uraltseva N. N. (1968). *Linear and quasilinear elliptic equations*. New York: Academic Press.
61. Ladyzhenskaya, O. A., Solonnikov, V. A., & Uraltseva, N. N. (1968). *Linear and quasilinear equations of parabolic type*. Translations of mathematical monographs (Vol. 23). Providence, Rhode Island: American Mathematical Society.
62. Landau, L. D., & Lifschitz, E. M. (1995). *Fluid mechanics: Course of theoretical physics* (Vol. 6). Oxford: Butterworth-Heinemann. Landau, L. D., & Lifschitz, E. M. (1995). *Theory of elasticity: Course of theoretical physics* (Vol. 7). Oxford: Butterworth-Heinemann.
63. Levy, T. (1977). Acoustic phenomena in elastic porous media. *Mechanics Research Communications*, *4*(4), 253–257.
64. Levy, T. (1987). Fluids in porous media and suspensions. In E. Sanchez-Palencia & A. Zaoui (Eds.), *Homogenization techniques for composite media*. Lecture notes in physics (Vol. 272). Berlin: Springer.
65. Levy, T. (1979). Propagation waves in a fluid-saturated porous elastic solid. *International Journal of Engineering Science*, *17*, 1005–1014.
66. Levy, T., & Sanchez-Palencia, E. (1977). Equations and interface conditions for acoustic phenomena in porous media. *Journal of Mathematical Analysis and Applications*, *61*, 813–834.
67. Lewis, D. G. (1950). The instability of liquid surface when accelerated in a direction perpendicular to their plans. II. *Proceedings of the Royal Society of London. Series A*, *202*(1068), 81–96.
68. Lions, J. L. (1969). *Quelques méthodes de résolution des problémes aux limites non linéaire*. Dunod, Paris: Gauthier-Villars.
69. Lissandrello, C., Yakhot, V., & Ekinci, K. L. (2012). Crossover from hydrodynamics to the kinetic regime in confined nanoflows. *Physical Review Letters*, *108*, 084501.
70. Lukkassen, D., Nguetseng, G., & Wall, P. (2002). Two-scale convergence. *International Journal of Pure and Applied Mathematical Sciences*, *2*(1), 35–86.
71. Marušić-Paloka, E., & Piatnitski, A. (2005). Homogenization of a nonlinear convection-diffusion equation with rapidly oscillating coefficients and strong convection. *Journal of the London Mathematical Society*, *72*(2), 371–409.
72. Massoud, M. (2005). *Engineering thermofluids, thermodynamics, fluid mechanics, and heat transfer*. Berlin: Springer.
73. Meirmanov, A. (2007). Nguetseng's two-scale convergence method for filtration and seismic acoustic problems in elastic porous media. *Siberian Mathematical Journal*, *48*, 519–538.
74. Meirmanov, A. (2008). Acoustic and filtration properties of a thermoelastic porous medium: Biot's equations of thermo-poroelasticity. *Sbornik Mathematics*, *199*(3), 1–24.
75. Meirmanov, A. (2008). Homogenized models for filtration and for acoustic wave propagation in thermo-elastic porous media. *European Journal of Applied Mathematics*, *19*, 259–284.
76. Meirmanov, A. (2008). A description of seismic acoustic wave propagation in porous media via homogenization. *SIAM Journal on Mathematical Analysis*, *40*(3), 1272–1289.
77. Meirmanov, A. (2008). Darcy's law for a compressible thermofluid. *Asymptotic Analysis*, *58*(4), 191–209.
78. Meirmanov, A. (2010). Double porosity models in incompressible poroelastic media. *Mathematical Models and Methods in Applied Sciences*, *20*(4), 635–659.
79. Meirmanov, A. (2009). Derivation of equations of seismic and acoustic wave propagation and equations of filtration via homogenization of periodic structures. *Journal of Mathematical Sciences*, *163*(2), 111–172.
80. Meirmanov, A. (2011). The Muskat problem for a viscoelastic filtration. *Interfaces and Free Boundaries*, *13*(4), 463–484.
81. Meirmanov, A. (2013). A diffusion-convection in poroelastic media. Part I: The microscopic description. *Advances in Mathematical Sciences and Applications*, *22*(1), 259–280.
82. Meirmanov, A. (2013). A diffusion-convection in poroelastic media. Part II: The macroscopic description. *Advances in Mathematical Sciences and Applications*, *22*(1), 281–294.

83. Meirmanov, A. M., & Sazhenkov, S. A. (2007). Generalized solutions to the linearized equations of thermoelastic solid and viscous thermofluid. *Electronic Journal of Differential Equations*, *2007*(41), 1–29.
84. Meirmanov, A. M., & Zimin, R. (2011). Some compactness result for periodic structures and its application to the homogenization of a diffusion-convection equation. *Electronic Journal of Differential Equations*, *2011*(115), 1–11.
85. Meirmanov, A. M., & Zimin, R. (2012). Mathematical models of a diffusion-convection in porous media. *Electronic Journal of Differential Equations*, *2012*(105), 1–16.
86. Mikelić, A., & Paoli, L. (1999). Homogenization of the inviscid incompressible fluid flow trough a 2D porous medium. *Proceedings of the AMS*, *17*, 2019–2028.
87. Muskat, M. (1934). Two-fluid system in porous media. The encroachment of water into an oil sand. *Physics*, *5*, 250–264.
88. Nguetseng, G. (1989). A general convergence result for a functional related to the theory of homogenization. *SIAM Journal on Mathematical Analysis*, *20*, 608–623.
89. Nguetseng, G. (1990). Asymptotic analysis for a stiff variational problem arising in mechanics. *SIAM Journal on Mathematical Analysis*, *21*, 1394–1414.
90. Ovsiannikov, L. V. (1977). *Introduction to mechanics of continuous media, Parts 1, 2*. Novosibirsk: Novosibirsk State University. (Russian).
91. Polubarinova-Kochina, P. Ya. (1962). *Theory of ground water movement*. Princeton, NJ: Princeton University Press.
92. Pride, S.R., & Berryman, J.G. (2003). Linear dynamics of double-porosity dual-permeability materials. I. Governing equations and acoustic attenuation. *Physical Review E*, *68*, 036603.
93. Pride, S. R., & Berryman, J. G. (2003). Linear dynamics of double-porosity dual-permeability materials. II. Fluid transport equations. *Physical Review E*, *68*, 036604.
94. Radkevich, E. (1995). On the spectrum of the pencil in the Verigin-Muskat problem. *Sbornik: Mathematics*, *80*(1), 33–74.
95. Strutt, J. W. (Lord Rayleigh) (1984). *Theory of sound* (Vol. 2). New York, NY: Dover Publications.
96. Sanchez-Hubert, J. (1980). Asymptotic study of the macroscopic behavior of a solid-liquid mixture. *Mathematical Methods in the Applied Sciences*, *2*, 1–18.
97. Sanchez-Palencia, E. (1980). *Non-homogeneous media and vibration theory*. Lecture notes in physics (Vol. 129). Berlin: Springer.
98. Siegel, M., Caflish, R. E., & Howison, S. (2004). Global existence, singular solutions, and ill-posedness for the Muskat problem. *Communications on Pure and Applied Mathematics*, *LVII*, 1–38.
99. Tartar, L. (1977). *Cours Peccot*. Paris: Collége de France.
100. Taylor, G. (1950). Instability of liquid surface when accelerated in a direction perpendicular to their plans. I. *Proceedings of the Royal Society of London. Series A*, *201*(1065), 192–196.
101. Terzaghi, K. (1923). Die Berechnung der Durchlassigkeitsziffer des Tones aus dem Verlauf der hydrodynamischen Spannungsercheinungen, Sitzung berichte. Akademie der Wissenschaften, Wien Mathematisch-Naturwissenschaftliche Klasse (Vol. 132, no IIa, pp. 104–124).
102. Terzaghi, K., Braselton, R., & Mesri, G. (1996). *Soil mechanics in engineering practice*. New York: Wiley-Interscience.
103. Tolstoy, I. (Ed.). (1992). *Acoustics, elasticity, and thermodynamics of porous media. Twenty-one papers by M.A. Biot*. New York: Acoustical Society of America.
104. Warren, J. E., & Root, P. J. (1963). The behaviour of naturally fractured reservoirs. *Society of Petroleum Engineers Journal*, *3*, 235–255.
105. Weiguoab, L., & Yangshenga, Z. (2005). A mathematical model for solid liquid and mass transfer coupling and numerical simulation for hydraulic fracture in rock salt. *Progress in Natural Science*, *15*(8), 742–748.

MIX
Papier aus verantwortungsvollen Quellen
Paper from responsible sources
FSC® C105338

If you have any concerns about our products,
you can contact us on
ProductSafety@springernature.com

In case Publisher is established outside the EU,
the EU authorized representative is:
Springer Nature Customer Service Center GmbH
Europaplatz 3, 69115 Heidelberg, Germany

Printed by Libri Plureos GmbH
in Hamburg, Germany